Contents

Volcanism 70

Igneous Rocks and Intrusive Igneous Activity 46

The Changing Earth
Exploring Geology and Evolution

James S. Monroe
Reed Wicander
CENTRAL MICHIGAN UNIVERSITY

West Publishing Company
MINNEAPOLIS/SAINT PAUL · NEW YORK · LOS ANGELES · SAN FRANCISCO

PRODUCTION CREDITS

COPYEDITING	Mary Russell
COVER DESIGN	Diane Beasley
INTERIOR DESIGN	Diane Beasley
COMPOSITION	Carlisle Communications
ARTWORK	Carlyn Iverson, Precision Graphics, Publication Services, Rolin Graphics, and Alice B. Thiede and William A. Thiede, Carto-Graphics. Individual credits follow the index.
INDEX	Terry Casey
COVER IMAGE	©GEOPIC

The cover image is a satellite view of the Turfan Depression and Heavenly Mountains in north-central China. The numerous streams draining the southern slopes of the Heavenly Mountains flow into a gravel plain and recharge the groundwater system of the area. The areas in red in the southern part of the image are cultivated land. A large sand desert composed of complex dunes is visible in the lower right hand corner of the cover.

WEST'S COMMITMENT TO THE ENVIRONMENT

In 1906, West Publishing Company began recycling materials left over from the production of books. This began a tradition of efficient and responsible use of resources. Today, up to 95 percent of our legal books and 70 percent of our college and school texts are printed on recycled, acid-free stock. West also recycles nearly 22 million pounds of scrap paper annually—the equivalent of 181,717 trees. Since the 1960s, West has devised ways to capture and recycle waste inks, solvents, oils, and vapors created in the printing process. We also recycle plastics of all kinds, wood, glass, corrugated cardboard, and batteries, and have eliminated the use of styrofoam book packaging. We at West are proud of the longevity and the scope of our commitment to our environment.

Production, prepress, printing, and binding by West Publishing Company.

Printed in the United States of America

01 00 99 98 97 96 95   8 7 6 5 4 3 2 1

Library of Congress Cataloging–in–Publication Data

Monroe, James S. (James Stewart), 1938–
     The changing earth : exploring geology and evolution / James S. Monroe, Reed Wicander.
        p.   cm.
     ISBN 0–314–02833–1 (soft)
     1. Geology.   I. Wicander, Reed, 1946–   . II. Title.
QE28.M686   1993
550––dc20   93–43307
                                                          CIP

Brief Contents

CHAPTER
5

Weathering, Soil, Sediment, and Sedimentary Rocks 94

CHAPTER
6

Metamorphism and Metamorphic Rocks
   128

CHAPTER
11

· ·

Mass Wasting 266

CHAPTER
12

· ·

Running Water 298

CHAPTER 17

CHAPTER 18

CHAPTER
19

A History of the Universe, Solar System, and Planets   492

CHAPTER
20

Precambrian Earth and Life History   520

CHAPTER
23

Mesozoic Earth and Life History 636

CHAPTER
24

Cenozoic Earth and Life History 676

CHAPTER
25
....

Geology, the Environment, and Natural Resources 712

Preface

The Earth is a dynamic planet that has changed continuously during its 4.6 billion years of existence. The size, shape, and geographic distribution of the continents and ocean basins have changed through time, as have the atmosphere and biota. Over the past 20 years, bold new theories and discoveries concerning the Earth's origin and how it works have sparked a renewed interest in geology. We have become increasingly aware of how fragile our planet is and, more importantly, how interdependent all of its various systems are. We have learned that we cannot continually pollute our environment and that our natural resources are limited and, in most cases, nonrenewable. Furthermore, we are coming to realize how central geology is to our everyday lives. For these and other reasons, geology is one of the most important college or university courses a student can take.

The Changing Earth: Exploring Geology and Evolution is designed for an introductory course in geology and is written with the student in mind. One of the problems with any introductory science course is that the students are overwhelmed by the amount of material that must be learned. Furthermore, most of the material does not seem to be linked by any unifying theme and does not always appear to be relevant to their lives.

One of the goals of this book is to provide students with an understanding of the principles and processes of geology and how geology relates to the human experience; that is, how geology affects not only individuals, but society in general. It is also our intention to provide students with an overview of the geologic and biologic history of the Earth, not as a set of encyclopedic facts to memorize, but rather as a continuum of interrelated events that reflect the underlying geologic and biologic principles and processes that have shaped

our planet and life upon it. With these goals in mind, we introduce the major themes of the book in the first chapter to provide students with an overview of the subject and enable them to see how the various systems of the Earth are interrelated. We also discuss the economic and environmental aspects of geology throughout the book rather than treating these topics in separate chapters. In this way students can see, through relevant and interesting examples, how geology impacts our lives.

▼ TEXT ORGANIZATION

Plate tectonic theory is the unifying theme of geology and this book. This theory has revolutionized geology because it provides a global perspective of the Earth and allows geologists to treat many seemingly unrelated geologic phenomena as part of a total planetary system. Because plate tectonic theory is so important, it is introduced in Chapter 1, and is discussed in most subsequent chapters in terms of the subject matter of that chapter.

We have organized *The Changing Earth: Exploring Geology and Evolution* into several informal categories. Chapter 1 is an introduction to geology, its relevance to the human experience, plate tectonic theory, the rock cycle, organic evolution, and geologic time and uniformitarianism. Chapters 2–6 examine the Earth's materials (minerals and igneous, sedimentary, and metamorphic rocks) and the geologic processes associated with them including the role of plate tectonics in their origin and distribution. Chapters 7–10 deal with the related topics of earthquakes and the Earth's interior, the sea floor, plate tectonics and deformation, mountain building, and the continents. Chapters 11–16 cover the Earth's surface processes. Chapter 17 discusses geologic time, introduces several dating methods, and explains how geologists correlate rocks. Chapter 18 explores fossils and evolution, and Chapter 19 discusses the origin of the universe, the solar system and planets, and the Earth's place in the evolution of this larger system. Chapters 20 through 24 constitute our chronological treatment of the geologic and biologic history of the Earth. These chapters are arranged so that the geologic history is followed by a discussion of the biologic history during that time interval. We believe that this format allows easier integration of life history with geologic history. In the final chapter, we bring everything together to show the importance of geology in environmental issues, the search for natural resources, and how the study of the Earth's history can provide a perspective for such questions as global warming.

We have found that presenting the material in this order works well for most students. We know, however, that many instructors prefer an entirely different order of topics depending on the emphasis in their course. We have therefore written this book so that instructors can present the chapters in any order that suits the needs of their course.

CHAPTER ORGANIZATION

All chapters have the same organizational format. Each chapter opens with a photograph relating to the chapter content, a detailed outline, and a Prologue, which is intended to stimulate interest in the chapter by discussing some aspect of the material.

The text is written in a clear informal style, making it easy for students to comprehend. Numerous color diagrams, many of them block diagrams, and photographs complement the text, providing a visual presentation of the concepts and information presented. Each chapter contains at least one Perspective that presents a brief discussion of an interesting aspect of geology or geological research. Each of the chapters on geologic history in the second half of the text contains a final section on mineral resources characteristic of that time period. These sections provide applied economic material of interest to students.

The end-of-chapter materials begin with a concise review of important concepts and ideas in the Chapter Summary. The Important Terms, which are printed in boldface type in the chapter text, are listed at the end of each chapter for easy review, and a full glossary of important terms appears at the end of the text. The Review Questions are another important feature of this

book; they include multiple-choice questions as well as short answer and essay questions. Each chapter concludes with an up-to-date list of Additional Readings, most of which are written at a level appropriate for beginning students interested in pursuing a particular topic.

SPECIAL FEATURES

This book contains a number of special features that set it apart from other geology textbooks. Among them are a critical thinking and study skills section, the chapter prologues, the integration of economic and environmental geologic issues throughout the book, a capstone chapter, and a set of multiple-choice questions with answers for each chapter.

Study Skills

Immediately following the Preface is a section devoted to developing critical thinking and study skills. This section contains hints to help students improve their study habits, prepare for exams, and generally get the most out of every course they take. While these tips can be helpful in any course, many of them are particularly relevant to geology. Whether you are just beginning college or about to graduate, take a few minutes to read over this section as these suggestions can help you in your studies and later in life.

Prologues

The introductory prologues are designed to provide an introduction to the chapter material by focusing on a particularly interesting aspect of the chapter. Prologue topics include the Loma Prieta earthquake (Chapter 7), the story of Floyd Collins (Chapter 13), the Burgess Shale fauna (Chapter 22), and the story of Dinosaur National Monument (Chapter 23).

Economic and Environmental Geology

The topics of environmental and economic geology are discussed throughout the text. Integrating economic and environmental geology with the chapter material helps students see the importance and relevance of geology to their lives. In addition, many chapters close with a section on resources, further emphasizing the importance of geology in today's world.

Figures

Many of the illustrations depicting geologic processes or events are block diagrams rather than cross sections so that students can more easily visualize the salient features of these processes and events. Our color paleogeographic maps are designed to illustrate clearly and accurately the geography during the various geologic periods. Full color scenes showing associations of plants and animals are based on the most current interpretations. Great care has been taken to ensure that the art and captions provide an attractive, informative, and accurate illustration program.

Figure and Table Reference System

A color cue (◆) will be found in the text next to the first reference for each figure, and a (●) next to the reference for each table. This system is designed to help students quickly return to their place in the text when they interrupt their reading to examine an illustration or table.

Perspectives

The chapter perspectives focus on aspects of environmental, economic, or planetary geology, as well as current geological research. They were chosen to provide students with an overview of the many fascinating aspects of geology. Examples include asbestos (Chapter 6), the supercontinent cycle (Chapter 9), cladistics and cladograms (Chapter 18), and warm-blooded dinosaurs (Chapter 23). The perspectives can be assigned as part of the chapter reading, used as the basis for lecture or discussion topics, or even used as the starting point for student papers.

Capstone Chapter

A capstone chapter brings together the various economic and environmental aspects of geology discussed throughout the book and shows how geology plays an important role in society today. In addition, such topics as acid rain and global warming are covered and we discuss how geology applies to these problems.

Review Questions

Most geology books have a set of review questions at the end of each chapter. This book, however, includes not only the usual essay and thought-provoking questions, but also a set of multiple-choice questions, a feature not found in other geology textbooks. The answers to the multiple-choice questions are at the end of the book so that students can check their answers and increase their confidence before taking an examination.

Summary Tables

Of particular assistance to students are the end-of-chapter summary tables found in chapters 21–23. These tables are designed to give an overall perspective of the geologic and biologic events that occurred during a particular time interval and to show how these events are interrelated. The emphasis in these tables is on the geologic evolution of North America. Global tectonic events and sea-level changes are also incorporated into these tables to provide global insights.

Appendixes

With few exceptions, units of length, mass, and volume are given in metric units in the text. Appendix A is an *English-Metric Conversion Chart*. The terminology of biological classification is complex, and though we have minimized it in the text, some is necessary. A *Classification of Organisms* appears in Appendix B.

Glossary

It is important for students to comprehend geologic terminology if they are to understand geologic principles and concepts. Accordingly, an alphabetical and comprehensive glossary of all important terms appears at the back of the book.

⋟ INSTRUCTOR ANCILLARY MATERIALS

To assist you in teaching this course and supplying your students with the best in teaching aids, West Publishing Company has prepared a complete supplemental package available to all adopters.

The Comprehensive Instructor's Manual will include teaching ideas, lecture outlines (including notes on figures and photographs available as slides), teaching tips, Consider This lecture questions, Enrichment Topics, global examples, lists for slides, transparency masters and acetates as well as a computerized test bank with 3,000 test items.

West's Geology Videodisc can be used in lectures. The videodisc includes, among other things, a wealth of images from around the country organized by region, images from the textbook, animated sequences, and clips from appropriate topical films. Also, videotapes are available to qualified adopters.

Three slide sets will be provided to qualified adopters. The first set will include 150 of the most important and attractive figures and photographs of rocks and minerals, as well as photographs from the book, and the second set will contain over 450 slides illustrating important geologic features. The majority of these photographs will be from North America, but examples from around the world and the solar system will also be provided. The third set contains photos and art for coverage of historical geology.

A set of 200 full-color transparency acetates provide clear and effective illustrations of important artwork and maps from the text.

Great Ideas for Teaching Geology has been prepared to provide analogies, demonstrations, and the like to enhance lectures and discussion sections.

A Newsletter will be provided to adopters twice each year to update the book with recent and relevant research disclosures. This will ensure that your students have the most current information available.

We have incorporated much of the material usually found in study guides into the book itself. This saves students time and money and also makes the book a more valuable learning tool.

Lastly, tutorial software is available for students' review and for lecture presentations. Contact your local West sales representative for details.

ACKNOWLEDGMENTS

As the authors, we are, of course, responsible for the organization, style, and accuracy of the text, and any mistakes, omissions, or errors are our responsibility. We wish to express our sincere appreciation to the numerous geologists who reviewed our physical and historical geology books. In addition, we would like to particularly thank Dr. Mary Lou Bevier, Department of Geological Sciences, The University of British Columbia, Vancouver, Canada, and Dr. Gerald Osborn, Department of Geology and Geophysics, The University of Calgary, Alberta, Canada for reviewing the final chapter and making numerous useful and helpful suggestions.

We are grateful to the members of the Geology Department of Central Michigan University for providing us with photographs. They are Richard V. Dietrich (Professor Emeritus), Eric Johnson, David J. Matty, Jane M. Matty, Wayne E. Moore (Professor Emeritus), and Stephen D. Stahl. Bruce M. C. Pape of the Geography Department also provided photographs. We also thank Dawn Anderson and Kathleen Butzier for typing many of the copyright permission forms and Mrs. Martha Brian of the Geology Department, whose general efficiency was invaluable during the preparation of this book.

We are also grateful for the generosity of the various agencies and individuals from many countries who provided photographs.

Special thanks must go to Jerry Westby, college editorial manager for West Publishing Company, a friend and colleague who made many valuable suggestions and patiently guided us through the entire project. His continued encouragement helped us produce the best possible book. We are also equally indebted to our production editor, Laura Evans, whose attention to detail and consistency made our job so much easier. Diane Beasley created another successful design for our book.

We would also like to thank Mary Russell for her copyediting skills. Because geology is such a visual science, we extend special thanks to Carlyn Iverson who rendered the reflective art and to the artists at Precision Graphics and Publication Services who were responsible for much of the rest of the art program as well as Alice B. Thiede who rendered many of the paleogeographic maps throughout the text. They did an excellent job, and we enjoyed working with them. We would also like to acknowledge our promotion manager, Seán P. Berres for his help in the development of the promotional material associated with this book, and Maureen Rosener, marketing manager, who developed the excellent videodisc that accompanies this book.

As always, our families were patient and encouraging when most of our spare time and energy were devoted to this book. We thank them for their continued support and understanding.

Developing Critical Thinking and Study Skills

⩣ INTRODUCTION

College is a demanding and important time, a time when your values will be challenged, and you will try out new ideas and philosophies. You will make personal and career decisions that will affect your entire life. One of the most important lessons you can learn in college is how to balance your time among work, study, and recreation. If you develop good time management and study skills early in your college career, you will find that your college years will be successful and rewarding.

This section offers some suggestions to help you maximize your study time and develop critical thinking and study skills that will benefit you, not only in college, but throughout your life. While mastering the content of a course is obviously important, learning how to study and to think critically is, in many ways, far more important. Like most things in life, learning to think critically and study efficiently will initially require additional time and effort, but once mastered, these skills will save you time in the long run.

You may already be familiar with many of the suggestions and may find that others do not directly apply to you. Nevertheless, if you take the time to read this section and apply the appropriate suggestions to your own situation, we are confident that you will become a better and more efficient student, find your classes more rewarding, have more time for yourself, and get better grades. We have found that the better students are usually also the busiest. Because these students are busy with work or extracurricular activities, they have had to learn to study efficiently and manage their time effectively.

One of the keys to success in college is avoiding procrastination. While procrastination provides temporary satisfaction because you have avoided doing something you did not want to do, in the long run it leads to stress. While a small amount of stress can be beneficial, waiting until the last minute usually leads to mistakes and a subpar performance. By setting clear, specific goals and working toward them on a regular basis, you can greatly reduce the temptation to procrastinate. It is better to work efficiently for short periods of time than to put in long, unproductive hours on a task, which is usually what happens when you procrastinate.

Another key to success in college is staying physically fit. It is easy to fall into the habit of eating junk food and never exercising. To be mentally alert, you must be physically fit. Try to develop a program of regular exercise. You will find that you have more energy, feel better, and study more efficiently.

⩣ GENERAL STUDY SKILLS

Most courses, and geology in particular, build upon previous material, so it is extremely important to keep up with the coursework and set aside regular time for study in each of your courses. Try to follow these hints, and you will find you do better in school and have more time for yourself:

- Develop the habit of studying on a daily basis.

- Set aside a specific time each day to study. Some people are day people, and others are night people. Determine when you are most alert and use that time for study.

- Have an area dedicated for study. It should include a well-lighted space with a desk and the study materials you need, such as a dictionary, thesaurus, paper, pens and pencils, and a computer if you have one.

- Study for short periods and take frequent breaks, usually after an hour of study. Get up and move around and do something completely different. This will help you stay alert, and you'll return to your studies with renewed vigor.

- Try to review each subject every day or at least the day of the class. Develop the habit of reviewing lecture material from a class the same day.

- Become familiar with the vocabulary of the course. Look up any unfamiliar words in the glossary of your textbook or in a dictionary. Learning the language of the discipline will help you learn the material.

⧨ GETTING THE MOST FROM YOUR NOTES

If you are to get the most out of a course and do well on exams, you must learn to take good notes. This does not mean you should try to take down every word your professor says. Part of being a good note taker is knowing what is important and what you can safely leave out.

Early in the semester, try to determine whether the lecture will follow the textbook or be predominantly new material. If much of the material is covered in the textbook, your notes do not have to be as extensive or detailed as when the material is new. In any case, the following suggestions should make you a better note taker and enable you to derive the maximum amount of information from a lecture:

- Regardless of whether the lecture discusses the same material as the textbook or supplements the reading assignment, read or scan the chapter the lecture will cover before class. This way you will be somewhat familiar with the concepts and can listen critically to what is being said rather than trying to write down everything. Later a few key words or phrases will jog your memory as to what was said.

- Before each lecture, briefly review your notes from the previous lecture. Doing this will refresh your memory and provide a context for the new material.

- Develop your own style of note taking. Do not try to write down every word. These are notes you're taking, not a transcript. Learn to abbreviate and develop your own set of abbreviations and symbols for common words and phrases: for example, w/o (without), w (with), = (equals), ∧ (above or increases), ∨ (below or decreases), < (less than), > (greater than), & (and), u (you).

- Geology lends itself to many abbreviations that can increase your note-taking capability: for example, pt (plate tectonics), iggy (igneous), meta (metamorphic),

sed (sedimentary), rx (rock or rocks), ss (sandstone), my (million years), gts (geologic time scale), Є (Cambrian), and dino (dinosaur).

- Rewrite your notes soon after the lecture. Rewriting your notes helps reinforce what you heard and gives you an opportunity to determine whether you understand the material.

- By learning the vocabulary of the discipline before the lecture, you can cut down on the amount you have to write—you won't have to write down a definition if you already know the word.

- Learn the mannerisms of the professor. If he or she says something is important or repeats a point, be sure to write it down and highlight it in some way. Students have told me (RW) that when I stated something twice during a lecture, they knew it was important and probably would appear on a test. (They were usually right!)

- Check any unclear points in your notes with a classmate or look them up in your textbook. Pay particular attention to the professor's examples. These usually elucidate and clarify an important point and are easier to remember than an abstract concept.

- Go to class regularly, and sit near the front of the class if possible. It is easier to hear and see what is written on the board or projected onto the screen, and there are fewer distractions.

- If the professor allows it, tape record the lecture, but don't use the recording as a substitute for notes. Listen carefully to the lecture and write down the important points; then fill in any gaps when you replay the tape.

- If your school allows it, and they are available, buy class lecture notes. These are usually taken by a graduate student who is familiar with the material; typically they are quite comprehensive. Again use these notes to supplement your own.

- Ask questions. If you don't understand something, ask the professor. Many students are reluctant to do this, especially in a large lecture hall, but if you don't understand a point, other people are probably confused as well. If you can't ask questions during a lecture, talk to the professor after the lecture or during office hours.

⧨ GETTING THE MOST OUT OF WHAT YOU READ

The old adage that "you get out of something what you put into it" is very true when it comes to reading textbooks. By carefully reading your text and following these suggestions, you can greatly increase your understanding of the subject:

- Look over the chapter outline to see what the material is about and how it flows from topic to topic. If you have time, skim through the chapter before you start to read in depth.

- Pay particular attention to the tables, charts, and figures. They contain a wealth of information in abbreviated form and illustrate important concepts and ideas. Geology, in particular, is a visual science, and the figures and photographs will help you visualize what is being discussed in the text and provide actual examples of features such as faults or unconformities.

- As you read your textbook, highlight or underline key concepts or sentences, but make sure you don't highlight everything. Make notes in the margins. If you don't understand a term or concept, look it up in the glossary.

- Read the chapter summary carefully. Be sure you understand all of the key terms, especially those in bold face or italic type. Because geology builds on previous material, it is imperative that you understand the terminology.

- Go over the end-of-chapter questions. Write out your answers as if you were taking a test. Only when you see your answer in writing will you know if you really understood the material.

▼ DEVELOPING CRITICAL THINKING SKILLS

Few things in life are black and white, and it is important to be able to examine an issue from all sides and come to a logical conclusion. One of the most important things you will learn in college is to think critically and not accept everything you read and hear at face value. Thinking critically is particularly important in learning new material and relating it to what you already know. Although you can't know everything, you can learn to question effectively and arrive at conclusions consistent with the facts. Thus, these suggestions for critical thinking can help you in all your courses:

- Whenever you encounter new facts, ideas, or concepts, be sure you understand and can define all of the terms used in the discussion.

- Determine how the facts or information was derived. If the facts were derived from experiments, were the

experiments well executed and free of bias? Can they be repeated? The controversy over cold fusion is an excellent example. Two scientists claim to have produced cold fusion reactions using simple experimental laboratory apparatus, yet other scientists have as yet been unable to achieve the same reaction by repeating the experiments.

- Do not accept any statement at face value. What is the source of the information? How reliable is the source?

- Consider whether the conclusions follow from the facts. If the facts do not appear to support the conclusions, ask questions and try to determine why they don't. Is the argument logical or is it somehow flawed?

- Be open to new ideas. After all, the underlying principles of plate tectonic theory were known early in this century, yet were not accepted until the 1970s in spite of overwhelming evidence.

- Look at the big picture to determine how various elements are related. For example, how will constructing a dam across a river that flows to the sea affect the stream's profile? What will be the consequences to the beaches that will be deprived of sediment from the river? One of the most important lessons you can learn from your geology course is how interrelated the various systems of the Earth are. When you alter one feature, you affect numerous other features as well.

▼ IMPROVING YOUR MEMORY

Why do you remember some things and not others? The reason is that the brain stores information in different ways and forms, making it easy to remember some things and difficult to remember others. Because college requires that you learn a vast amount of information, any suggestions that can help you retain more material will help you in your studies:

- Pay attention to what you read or hear. Focus on the task at hand, and avoid daydreaming. Repetition of any sort will help you remember material. Review the previous lecture before going to class, or look over the last chapter before beginning the next. Ask yourself questions as you read.

- Use mnemonic devices to help you learn unfamiliar material. For example, the order of the Paleozoic periods (Cambrian, Ordovician, Silurian, Devonian, Mississippian, Pennsylvanian, and Permian) of the geologic time scale can be remembered by the phrase, *Campbell's Onion Soup Does Make Peter Pale*, or the order of the Cenozoic epochs (Paleocene, Eocene, Oligocene, Miocene, Pliocene, and Pleistocene) can be remembered by the phrase, *Put Eggs On My Plate Please*. Using rhymes can also be helpful.

- Look up the roots of important terms. If you understand where a word comes from, its meaning will be easier to remember. For example, *pyroclastic* comes from *pyro* meaning fire and *clastic* meaning broken pieces. Hence a pyroclastic rock is one formed by volcanism and composed of pieces of other rocks. We have provided the roots of many important terms throughout this text to help you remember their definitions.

- Outline the material you are studying. This will help you see how the various components are interrelated. Learning a body of related material is much easier than learning unconnected and discrete facts. Looking for relationships is particularly helpful in geology because so many things are interrelated. For example, plate tectonics explains how mountain building, volcanism, and earthquakes are all related (Chapter 9). The rock cycle relates the three major groups of rocks to each other and to subsurface and surface processes (Chapter 1).

- Use deductive reasoning to tie concepts together. Remember that geology builds on what you learned previously. Use that material as your foundation and see how the new material relates to it.

- Draw a picture. If you can draw a picture and label its parts, you probably understand the material. Geology lends itself very well to this type of memory device because so much is visual. For example, instead of memorizing a long list of glacial terms, draw a picture of a glacier and label its parts and the type of topography it forms.

- Focus on what is important. You can't remember everything, so focus on the important points of the lecture or the chapter. Try to visualize the big picture, and use the facts to fill in the details.

▼ PREPARING FOR EXAMS

For most students, tests are the critical part of a course. To do well on an exam, you must be prepared. These suggestions will help you focus on preparing for the examination:

- The most important advice is to study regularly rather than try to cram everything into one massive study session. Get plenty of rest the night before an exam, and stay physically fit to avoid becoming susceptible to minor illnesses that sap your strength and lessen your ability to concentrate on the subject at hand.

- Set up a schedule so that you cover small parts of the material on a regular basis. Learning some concrete examples will help you understand and remember the material.

- Review the chapter summaries. Construct an outline to make sure you understand how everything fits together. Drawing diagrams will help you remember key points. Make up flash cards to help you remember terms and concepts.

- Form a study group, but make sure your group focuses on the task at hand, not on socializing. Quiz each other and compare notes to be sure you have covered all the material. We have found that students dramatically improved their grades after forming or joining a study group.

- Write out answers to all of the end-of-chapter questions. Review the key terms. Go over all of the key points the professor emphasized in class.

- If you have any questions, visit the professor or teaching assistant. If review sessions are offered, be sure to attend. If you are having problems with the material, ask for help as soon as you have difficulty. Don't wait until the end of the semester.

- If old exams are available, look at them to see what is emphasized and what type of questions are asked. Find out whether the exam will be all objective or all essay or a combination. If you have trouble with a particular type of question (such as multiple choice or essay), practice answering questions of that type— your study group or a classmate may be able to help.

▼ TAKING EXAMS

It is now time to take the exam. The most important thing to remember is not to panic. This, of course, is easier said than done. Almost everyone suffers from test anxiety to some degree. Usually, it passes as soon as the exam begins, but in some cases, it is so debilitating that the individual does not perform as well as he or she could. If you are one of those people, get help as soon as possible. Most colleges and universities have a program to help students overcome test anxiety or at least keep it in check. Don't be afraid to seek help if you suffer test anxiety. Your success in college depends to a large extent on how well you perform on exams, so by not seeking help, you are only hurting yourself. In addition, the following suggestions may be helpful:

- First of all, relax. Then look over the exam briefly to see its format and determine which questions are worth the most points. If it helps, quickly jot down any information you are afraid you might forget or particularly want to remember for a question.
- Answer the questions that you know the best first. Make sure, however, that you don't spend too much time on any one question or on one that is worth only a few points.
- If the exam is a combination of multiple choice and essay, answer the multiple-choice questions first. If you are not sure of an answer, go on to the next one. Sometimes the answer to one question can be found in another question. Furthermore, the multiple-choice questions may contain many of the facts needed to answer some of the essay questions.
- Read the question carefully and answer only what it asks. Save time by not repeating the question as your opening sentence to the answer. Get right to the point. Jot down a quick outline for longer essay questions to make sure you cover everything.
- If you don't understand a question, ask the examiner. Don't assume anything. After all, it is your grade that will suffer if you misinterpret the question.
- If you have time, review your exam to make sure you covered all the important points and answered all the questions.
- If you have followed our suggestions, by the time you finish the exam, you should feel confident that you did well and will have cause for celebration.

⩬ CONCLUDING COMMENTS

We hope that the suggestions we have offered will be of benefit to you not only in this course, but throughout your college career. While it is difficult to break old habits and change a familiar routine, we are confident that following these suggestions will make you a better student. Furthermore, many of the suggestions will help you work more efficiently, not only in college, but also throughout your career. Learning is a lifelong process that does not end when you graduate. The critical thinking skills that you learn now will be invaluable throughout your life, both in your career and as an informed citizen.

The Changing Earth
Exploring Geology and Evolution

CHAPTER

1

Apollo 17 view of the Earth. Almost the entire coastline of Africa is clearly shown in this view with Madagascar visible off its eastern coast. The Arabian Peninsula can be seen at the northeastern edge of Africa, while the Asian mainland is on the horizon toward the northeast. The present location of continents and ocean basins is the result of plate movement. The interaction of plates through time has affected the physical and biological history of the Earth.

Understanding the Earth: An Introduction

Prologue

If we could film the history of the Earth from its beginning 4.6 billion years ago to the present, and speed up that film so it could be shown as a feature movie, we would see a planet undergoing remarkable change. Its geography would change as continents moved about its surface, and as a result of these movements, ocean basins would open and close. Mountain ranges would form along continental margins or where continents collided with each other. Oceanic and atmospheric circulation patterns would shift in response to the movement of continents. Massive ice sheets would form, grow, and then melt away. At other times, our film would show extensive swamps or vast interior deserts.

If we focused our imaginary camera closer to the Earth's surface, we would capture a breathtaking panorama of different organisms. We would witness the first living cells evolving from a primordial organic soup sometime between 4.6 and 3.5 billion years ago. About 2 billion years later, cells with a nucleus evolved. Next (approximately 700 million years ago), the first multicelled soft-bodied organisms evolved in the oceans, then animals with hard parts, and then animals with backbones.

As we scanned the various continents, we would see an essentially bare landscape until about 450 million years ago. At that time, the landscape would seem to come to life as plants and animals moved out of the water and onto the land. From then on, the landscape would be dominated by insects, amphibians, reptiles, birds, and mammals. In our film, humans and their history would occupy only the last second or so of the movie. However, the impact of human activity on the global ecosystem far exceeds the effects caused by all of the other life forms that came before.

Three interrelated themes run through the preceding discussion. The first is that the Earth is composed of a series of moving plates whose interactions have affected its physical and biological history. The second is that the Earth's biota has evolved. The third is that the physical and biological changes that occurred took place over long periods of time.

The first of these themes, *plate tectonics,* provides geologists with a unifying theory that explains the Earth's internal workings and accounts for a variety of apparently unrelated geologic features and events. The second theme, the theory of *organic evolution,* explains how life has changed through time, based on the idea that all living organisms are the evolutionary descendants of life-forms that existed in the past. Finally, the third theme, the concept of *geologic time,* allows geologists to show how small, almost imperceptible changes over vast lengths of time have resulted in significant changes.

These three interrelated themes, plate tectonics, organic evolution, and geologic time, are central to our understanding of the workings of our planet and an appreciation of its history. As you read this book, keep in mind that the different topics you are studying are parts of dynamic interrelated systems, and not isolated pieces of information. For example, volcanic eruptions are the result of complex interactions involving the Earth's interior and surface. These eruptions not only have an immediate effect on the surrounding area, but also contribute to climatic changes that affect the entire planet.

☰ INTRODUCTION

The Earth is unique among the planets of our solar system in that it supports life and has oceans of water, a hospitable atmosphere, and a variety of climates. It is ideally suited for life as we know it because of a combination of factors, including its distance from the Sun and the evolution of its interior, crust, oceans, and atmosphere. Over time, changes in the Earth's atmosphere, oceans, and, to some extent, its crust have been influenced by life processes. In turn, these physical changes have affected the evolution of life.

The Earth is not a simple, unchanging planet. Rather, it is a complex dynamic body in which innumerable interactions are occurring among its many components. The continuous evolution of the Earth and its life makes geology an exciting and ever-changing science in which new discoveries are continually being made.

☰ WHAT IS GEOLOGY?

Just what is geology and what is it that geologists do? **Geology,** from the Greek *geo* and *logos,* is defined as the study of the Earth. It is generally divided into two broad areas—physical geology and historical geology. *Physical geology* is the study of Earth materials, such as minerals and rocks, as well as the processes operating within the Earth and upon its surface. *Historical geology* examines the origin and evolution of the Earth, its continents, oceans, atmosphere, and life.

The discipline of geology is so broad that it is subdivided into many different fields or specialties. • Table 1.1 shows many of the diverse fields of geology and their relationship to the sciences of astronomy, physics, chemistry, and biology.

Nearly every aspect of geology has some economic or environmental relevance. Many geologists are involved in exploration for mineral and energy resources. Geologists use their specialized knowledge to locate the natural resources on which our industrialized society is based. As the world demand for these nonrenewable resources increases, geologists are applying the basic principles of geology in increasingly sophisticated ways (◆ Fig. 1.1).

Although locating mineral and energy resources is extremely important, geologists are also being asked to use their expertise to help solve many of our environmental problems. Geologists are involved in finding groundwater for the ever-burgeoning needs of communities and industries and in monitoring surface and underground water pollution and suggesting ways to clean it up. Geological engineers help find safe loca-

• **TABLE 1.1** Specialties of Geology and Their Broad Relationship to the Other Sciences

Specialty	Area of Study	Related Science
Geochronology	Time and history of the Earth	
Planetary geology	Geology of the planets	Astronomy
Paleontology	Fossils	Biology ·
Economic geology	Mineral and energy resources	
Environmental geology	Environment	
Geochemistry	Geology of chemical change	
Hydrogeology	Water resources	Chemistry
Mineralogy	Minerals	
Petrology	Rocks	
Geophysics	Earth's interior	
Structural geology	Rock deformation	Physics
Seismology	Earthquakes	
Geomorphology	Landforms	
Oceanography	Oceans	
Paleogeography	Ancient geographic features and locations	
Stratigraphy/Sedimentology	Layered rocks and sediments	

(a)

(b)

◆ FIGURE 1.1 (*a*) Geologists measuring the amount of erosion on a glacier in Alaska. (*b*) Geologists increasingly use computers in their search for petroleum and other natural resources.

tions for dams, waste disposal sites, and power plants, and design earthquake-resistant buildings.

Geologists are also involved in making short- and long-range predictions about earthquakes and volcanic eruptions and the potential destruction that may result. They are working with civil defense planners to help draw up contingency plans should such natural disasters occur.

As this brief survey illustrates, geologists are employed in a wide variety of pursuits. As the world's population increases and greater demands are made on the Earth's limited resources, the need for geologists and their expertise will become even greater.

⩔ GEOLOGY AND THE HUMAN EXPERIENCE

Many people are surprised at the extent to which we depend on geology in our everyday lives and also at the numerous references to geology in the arts, music, and literature. Rocks and landscapes are realistically represented in many sketches and paintings. Examples by famous artists include Leonardo da Vinci's *Virgin of the Rocks* and *Virgin and Child with Saint Anne,* Giovanni Bellini's *Saint Francis in Ecstasy* and *Saint Jerome,* and Asher Brown Durand's *Kindred Spirits* (◆ Fig. 1.2).

Ferde Grofé's *Grand Canyon Suite* was, no doubt, inspired by the grandeur and timelessness of Arizona's Grand Canyon and its vast rock exposures. The rocks on the Island of Staffa in the Inner Hebrides provided the inspiration for Felix Mendelssohn's famous *Hebrides* Overture.

References to geology abound in *The German Legends of the Brothers Grimm* and Jules Verne's *Journey to the Center of the Earth* describes an expedition into the Earth's interior. On one level, the poem "Ozymandias" by Percy B. Shelley deals with the fact that nothing lasts forever and even solid rock eventually disintegrates under the ravages of time and weathering. References to geology can even be found in comics, two of the best known being *B.C.* by Johnny Hart and *The Far Side* by Gary Larson.

Geology has also played an important role in history. Wars have been fought for the control of such natural resources as oil, gas, gold, silver, diamonds, and other valuable minerals. Empires throughout history have risen and fallen on the distribution and exploitation of natural resources. The configuration of the Earth's surface (topography), which is shaped by geologic agents, plays a critical role in military tactics. Natural barriers such as mountain ranges and rivers have frequently served as political boundaries.

◆ FIGURE 1.2 *Kindred Spirits* by Asher Brown Durand (1849) realistically depicts the layered rocks occurring along gorges in the Catskill Mountains of New York State. Asher Brown Durand was one of numerous artists of the nineteenth-century Hudson River School, who were known for their realistic landscapes.

◆ FIGURE 1.3 As these headlines from various newspapers indicate, geology affects our everyday lives.

⤳ HOW GEOLOGY AFFECTS OUR EVERYDAY LIVES

Destructive volcanic eruptions, devastating earthquakes, disastrous landslides, large sea waves, floods, and droughts are headline-making events that affect many people (◆ Fig. 1.3). Although we are unable to prevent most of these natural disasters, the more we know about them, the better we are able to predict, and possibly control, the severity of their impact. The environmental movement has forced everyone to take a closer look at our planet and the delicate balance between its various systems.

Most readers of this book will not become professional geologists. However, everyone should have a basic understanding of the geological processes that ultimately affect all of us. Such an understanding of geology is important so that one can avoid, for example, building in an area prone to landslides or flooding. Just ask anyone who purchased a home in the Portuguese Bend area of southern California during the 1950s or who built along a lakeshore and later saw the lake level rise and the beach and sometimes even their house disappear.

As society becomes increasingly complex and technologically oriented, we, as citizens, need an understanding of science so that we can make informed choices about those things that affect our lives. We are already aware of some of the negative aspects of an industrialized society, such as problems relating to solid waste disposal, contaminated groundwater, and acid rain. We are also learning the impact that humans, in increasing numbers, have on the environment and that we can no longer ignore the role that we play in the dynamics of the global ecosystem.

Most people do not realize how much geology affects their lives. For many people, the connection between geology and such well-publicized problems as nonrenewable energy and mineral resources, let alone waste disposal and pollution, is simply too far removed or too complex to be fully appreciated. But consider for a moment just how dependent we are on geology in our daily routines.

Much of the electricity for our appliances comes from the burning of coal, oil, or natural gas or from uranium consumed in nuclear-generating plants. It is geologists who locate the coal, petroleum, and uranium. The copper or other metal wires through which electricity travels are manufactured from materials found as the result of mineral exploration. The buildings that we live and work in owe their very existence to geological resources. A few examples are the concrete foundation (concrete is a mixture of clay, sand, or gravel, and limestone), the drywall (made largely from the mineral gypsum), the windows (the mineral quartz is the principal ingredient in the manufacture of glass), and the metal or plastic plumbing fixtures inside the building (the metals are from ore deposits and the plastics are most likely manufactured from petroleum distillates of crude oil).

Furthermore, when we go to work, the car or public transportation we use is powered and lubricated by some type of petroleum by-product and is constructed of metal alloys and plastics. And the roads or rails we ride over come from geologic materials, such as gravel, asphalt, concrete, or steel. All of these items are the result of processing geologic resources.

It is quite apparent that as individuals and societies, the standard of living we enjoy is directly dependent on the consumption of geologic materials. Therefore, we need to be aware of geology and of how our use and misuse of geologic resources may affect the delicate balance of nature and irrevocably alter our culture as well as our environment.

▼ THE EARTH AS A DYNAMIC PLANET

The Earth is a dynamic planet that has continuously changed during its 4.6-billion-year existence. The size, shape, and geographic distribution of continents and ocean basins have changed through time, the composition of the atmosphere has evolved, and life-forms existing today differ from those that lived during the past. We can easily visualize how mountains and hills are worn down by erosion and how landscapes are changed by the forces of wind, water, and ice. Volcanic eruptions and earthquakes reveal an active interior, and folded and fractured rocks indicate the tremendous power of the Earth's internal forces.

The Earth consists of three concentric layers: the core, the mantle, and the crust (◆ Fig. 1.4). This orderly division results from density differences between the layers as a function of variations in composition, temperature, and pressure.

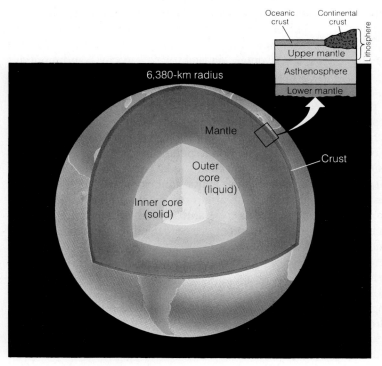

◆ FIGURE 1.4 A cross section of the Earth illustrating the core, mantle, and crust. The enlarged portion shows the relationship between the lithosphere, composed of the continental crust, oceanic crust, and upper mantle, and the underlying asthenosphere and lower mantle.

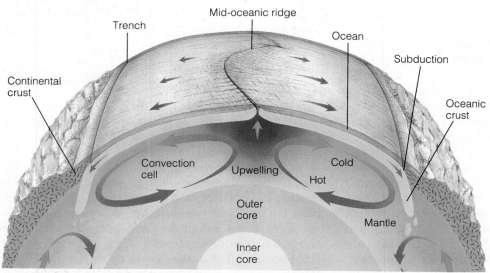

◆ FIGURE 1.5 The Earth's plates are thought to move as a result of underlying mantle convection cells in which warm material from deep within the Earth rises toward the surface, cools, and then, upon losing heat, descends back downward into the interior. The movement of these convection cells is believed to be the mechanism responsible for the movement of the Earth's plates, as shown in this diagrammatic cross section.

The **core** has a calculated density of 10 to 13 grams per cubic centimeter (g/cm^3) and occupies about 16% of the Earth's total volume. Seismic (earthquake) data indicate that the core consists of a small, solid inner core and a larger, apparently liquid, outer core. Both are thought to consist largely of iron and a small amount of nickel.

The **mantle** surrounds the core and comprises about 83% of the Earth's volume. It is less dense than the core ($3.3–5.7$ g/cm^3) and is thought to be composed largely of *peridotite,* a dark, dense igneous rock containing abundant iron and magnesium. The mantle can be divided into three distinct zones based on physical characteristics. The lower mantle is solid and forms most of the volume of the Earth's interior. The **asthenosphere** surrounds the lower mantle. It has the same composition as the lower mantle but behaves plastically and slowly flows. Partial melting within the asthenosphere generates *magma* (molten material), some of which rises to the Earth's surface because it is less dense than the rock from which it was derived. The solid upper mantle surrounds the asthenosphere. The upper mantle and the overlying crust constitute the **lithosphere,** which is broken into numerous individual pieces called **plates** that move over the asthenosphere as a result of underlying *convection cells* (◆ Fig. 1.5).

Interactions of these plates are responsible for such phenomena as earthquakes, volcanic eruptions, and the formation of mountain ranges and ocean basins.

The **crust,** the outermost layer of the Earth, consists of two types. The *continental crust* is thick (20–90 km), has an average density of 2.7 g/cm^3, and contains considerable silicon and aluminum. The *oceanic crust* is thin (5–10 km), denser than continental crust (3.0 g/cm^3), and is composed of the igneous rock *basalt.*

Since the widespread acceptance of plate tectonic theory about 25 years ago, geologists have viewed the Earth from a global perspective in which all of its systems are interconnected. Thus, the distribution of mountain chains, major fault systems, volcanoes, and earthquakes, the origin of new ocean basins, the movement of continents, and several other geological processes and features are perceived to be interrelated.

▼ GEOLOGY AND THE FORMULATION OF THEORIES

The term **theory** has various meanings. In colloquial usage, it means a speculative or conjectural view of something—hence the widespread belief that scientific theories are little more than unsubstantiated wild guesses. In scientific usage, however, a theory is a

THE FORMULATION
OF PLATE TECTONIC THEORY

· ·

The idea that continents moved during the past goes back to the time when people first noticed that the margins of eastern South America and western Africa looked as if they fit together. Geologists also noticed that similar or identical fossils occur on widely separated continents, that the same types of rocks from the same time period are found on different continents, and that ancient rocks and features indicating former glacial conditions occur in today's tropical areas. As more and more facts were gathered, hypotheses were proposed to explain them. In 1912, Alfred Wegener, a German meteorologist, amassed a tremendous amount of geological, paleontological, and climatological data that indicated continents moved through time; he proposed the hypothesis of *continental drift* to explain and synthesize this myriad of facts.

Wegener stated that at one time all of the continents were united into one single supercontinent that he named *Pangaea*. Pangaea later broke apart, and the individual continents drifted to their current locations. The continental drift hypothesis explained why the shorelines of different continents fit together, how different mountain ranges were once part of a larger continuous mountain range, why the same fossil animals and plants are found on different continents, and why rocks indicating glacial conditions are now found on continents located in the tropics.

Wegener's hypothesis and its predictability could be tested by asking what type of rocks or fossils would one expect to find at a given location on a continent if that continent was in the tropics 180 million years ago. To test the hypothesis of continental drift, all researchers had to do was to go into the field and examine the rocks and fossils for a particular time period on any continent to see if they indicated what the hypothesis predicted for the proposed location of that continent. In almost all cases, the data fit the hypothesis. However, there was one problem with Wegener's hypothesis: it did not explain how continents moved over oceanic crust and what the mechanism of continental movement was.

During the late 1950s and early 1960s, new data about the sea floor emerged that enabled geologists to propose the hypothesis of *sea-floor spreading*. This hypothesis suggested that the continents and segments of oceanic crust move together as single units, and that some type of thermal convection cell system operating within the Earth was the mechanism responsible for plate movements.

Sea-floor spreading and continental drift were then combined into a single hypothesis in which moving rigid plates are composed of continental and/or oceanic crust and the underlying upper mantle. These plates are bounded by mid-oceanic ridges, oceanic trenches, faults, and mountain belts. In this hypothesis, plates move away from mid-oceanic ridges and toward oceanic trenches. Furthermore, new crust is added along the mid-oceanic ridges and consumed or destroyed along oceanic trenches, and mountain chains are formed adjacent to the oceanic trenches.

According to this later hypothesis, Europe and North America should be steadily moving away from each other at a rate of up to several centimeters per year. Precise measurements of continental positions by satellites have verified this, thus confirming the validity of the plate movement hypothesis.

Furthermore, if plates are moving away from mid-oceanic ridges as predicted by the plate tectonic hypothesis, then rocks of the oceanic crust should become progressively older with increasing distance from the mid-oceanic ridges. To test this prediction, deep-sea sediment and oceanic crust were drilled as part of a massive scientific study of the ocean basins called the *Deep Sea Drilling Project*. Analysis of the oceanic crust and the layer of sediment immediately above it showed that the age of the oceanic crust does indeed increase with distance from the mid-oceanic ridges, and that the oldest oceanic crust is adjacent to the continental margins.

With the confirmation of these and other predictions of the plate tectonic hypothesis, most geologists accept that the hypothesis is correct and therefore call it the *plate tectonic theory*. Its acceptance has been so widespread not only because of the overwhelming evidence supporting it but also because it appears to explain the relationships between many seemingly unrelated geologic features and events.

coherent explanation for one or several related natural phenomena that is supported by a large body of objective evidence. From a theory are derived predictive statements that can be tested by observation and/or experiment so that their validity can be assessed. The law of universal gravitation is an example of a theory describing the attraction between masses (an apple and the Earth in the popularized account of Newton and his discovery).

Theories are formulated through the process known as the **scientific method.** This method is an orderly, logical approach that involves gathering and analyzing the facts or data about the problem under consideration. Tentative explanations or **hypotheses** are then formulated to explain the observed phenomena. Next, the hypotheses are tested to see if what they predicted actually occurs in a given situation. Finally, if one of the hypotheses is found, after repeated tests, to explain the phenomena, then that hypothesis is proposed as a theory. One should remember, however, that in science, even a theory is still subject to further testing and refinement as new data become available.

The fact that a scientific theory can be tested and is subject to such testing separates science from other forms of human inquiry. Because scientific theories can be tested, they have the potential of being supported or even proved wrong. Accordingly, science must proceed without any appeal to beliefs or supernatural explanations, not because such beliefs or explanations are necessarily untrue, but because we have no way to investigate them. For this reason, science makes no claim about the existence or nonexistence of a supernatural or spiritual realm.

Each scientific discipline has certain theories that are of particular importance for that discipline. In geology, the formulation of plate tectonic theory has changed the way geologists view the Earth. Geologists now view Earth history in terms of interrelated events that are part of a global pattern of change.

Before plate tectonic theory was generally accepted by geologists, however, numerous interrelated hypotheses were proposed and tested. Thus, the evolution of this theory illustrates the scientific method at work (see Perspective 1.1).

☙ PLATE TECTONIC THEORY

The acceptance of **plate tectonic theory** is recognized as a major milestone in the geological sciences. It is comparable to the revolution caused by Darwin's the-ory of evolution in biology. Plate tectonics has provided a framework for interpreting the composition, structure, and internal processes of the Earth on a global scale. It has led to the realization that the continents and ocean basins are part of a lithosphere-atmosphere-hydrosphere (water portion of the planet) system that evolved together with the Earth's interior.

According to plate tectonic theory, the lithosphere is divided into plates that move over the asthenosphere (◆ Fig. 1.6). Zones of volcanic activity, earthquake activity, or both mark most plate boundaries. Along these boundaries, plates diverge, converge, or slide sideways past each other.

At **divergent plate boundaries,** plates move apart as magma rises to the surface from the asthenosphere (◆ Fig. 1.7). The magma solidifies to form rock, which attaches to the moving plates. The margins of divergent plate boundaries are marked by mid-oceanic ridges in oceanic crust and are recognized by linear rift valleys where newly forming divergent boundaries occur beneath continental crust. The separation of South America from Africa and the formation of the South Atlantic Ocean occurred along a divergent plate boundary, the Mid-Atlantic Ridge (Fig. 1.6).

Plates move toward one another along **convergent plate boundaries** (Fig. 1.7). When an oceanic plate meets a continental plate, for example, the denser oceanic plate sinks beneath the continental plate along what is known as a **subduction zone.** As it descends into the Earth, it becomes hotter until it melts, or partially melts, thus generating a magma. As this magma rises, it may erupt at the Earth's surface, forming a chain of volcanoes. The Andes Mountains on the west coast of South America are a good example of a volcanic mountain range formed as a result of subduction of the Nazca plate beneath the South American plate along a convergent plate boundary (Fig. 1.6).

Crustal production and consumption occur at divergent and convergent plate boundaries, respectively. In contrast, **transform plate boundaries** are sites where plates slide sideways past each other and crust is neither produced nor destroyed (Fig. 1.7). The San Andreas fault in California is a transform plate boundary separating the Pacific plate from the North American plate (Fig. 1.6). The earthquake activity along the San Andreas fault results from the Pacific plate moving northward relative to the North American plate.

A revolutionary concept when it was proposed in the 1960s, plate tectonic theory has had significant and far-reaching consequences in all fields of geology be-

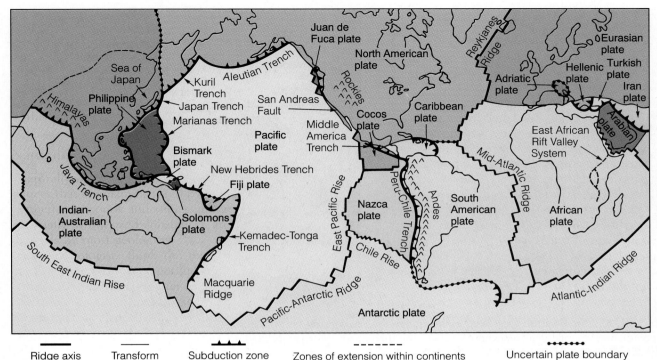

▬▬▬	▬ ▬	▬▲▬▲▬	- - - - - - -	••••••
Ridge axis	Transform	Subduction zone	Zones of extension within continents	Uncertain plate boundary

◆ FIGURE 1.6 The Earth's lithosphere is divided into rigid plates of various sizes that move over the asthenosphere.

cause it provides the basis for relating many seemingly unrelated geologic phenomena. Its impact has been particularly notable in the interpretation of Earth history. For example, the Appalachian Mountains in eastern North America and the mountain ranges of Greenland, Scotland, Norway, and Sweden are not the result of unrelated mountain-building episodes, but rather are part of a larger mountain-building event that involved the closing of an ancient "Atlantic Ocean" and the formation of the supercontinent Pangaea about 245 million years ago.

⩔ THE ROCK CYCLE

A **rock** is an aggregate of minerals. **Minerals** are naturally occurring, inorganic, crystalline solids that have definite physical and chemical properties. Minerals are composed of elements such as oxygen, silicon, and aluminum, and elements are made up of atoms, the smallest particles of matter that still retain the characteristics of an element. More than 3,500 minerals have been identified and described, but only about a dozen

make up the bulk of the rocks in the Earth's crust (see Table 2.6).

Geologists recognize three major groups of rocks—*igneous, sedimentary,* and *metamorphic*—each of which is characterized by its mode of formation. Each group contains a variety of individual rock types that differ from one another on the basis of composition or texture, that is, the size, shape, and arrangement of mineral grains.

The **rock cycle** is a way of viewing the interrelationships between the Earth's internal and external processes (◆ Fig. 1.8). It relates the three rock groups to each other; to surficial processes such as weathering, transportation, and deposition; and to internal processes such as magma generation and metamorphism. Plate movement is the mechanism responsible for recycling rock materials and therefore drives the rock cycle.

Igneous rocks result from the crystallization of magma, or the accumulation and consolidation of volcanic ejecta such as ash. As magma cools, minerals crystallize, and the resulting rock is characterized by interlocking mineral grains. Magma that cools slowly

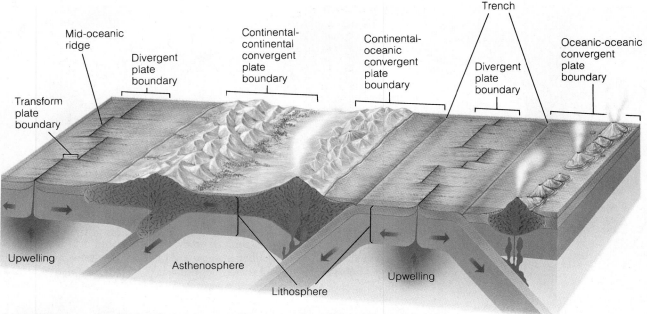

Transform plate boundary

Mid-oceanic ridge

Divergent plate boundary

Continental-continental convergent plate boundary

Continental-oceanic convergent plate boundary

Trench

Divergent plate boundary

Oceanic-oceanic convergent plate boundary

Upwelling

Asthenosphere

Lithosphere

Upwelling

◆ FIGURE 1.7 An idealized cross section illustrating the relationship between the lithosphere and the underlying asthenosphere and the three principal types of plate boundaries: divergent, convergent, and transform.

beneath the Earth's surface produces *intrusive igneous rocks* (◆ Fig. 1.9a), while magma that cools at the Earth's surface produces *extrusive igneous rocks* (◆ Fig. 1.9b).

Rocks exposed at the Earth's surface are broken into particles and dissolved by various weathering processes. The particles and dissolved material may be transported by wind, water, or ice and eventually deposited as *sediment.* This sediment may then be compacted or cemented into sedimentary rock.

Sedimentary rocks originate by consolidation of rock fragments, precipitation of mineral matter from solution, or compaction of plant or animal remains (◆ Fig. 1.9c and d). Because sedimentary rocks form at or near the Earth's surface, geologists can make inferences about the environment in which they were deposited, the type of transporting agent, and perhaps even something about the source from which the sediments were derived. Accordingly, sedimentary rocks are very useful for interpreting Earth history.

Metamorphic rocks result from the alteration of other rocks, usually beneath the Earth's surface, by heat, pressure, and the chemical activity of fluids. For example, marble, a rock preferred by many sculptors and builders, is a metamorphic rock produced when

the agents of metamorphism are applied to the sedimentary rocks limestone or dolostone. Metamorphic rocks are either *foliated* (◆ Fig. 1.9e) or *nonfoliated* (◆ Fig. 1.9f). Foliation is the parallel alignment of minerals due to pressure. It gives the rock a layered or banded appearance.

The Rock Cycle and Plate Tectonics

Interactions among plates determine, to a certain extent, which of the three rock groups will form (◆ Fig. 1.10). For example, weathering and erosion produce sediments that are transported by agents such as running water from the continents to the oceans, where they are deposited and accumulate. These sediments, some of which may be lithified and become sedimentary rock, become part of a moving plate along with the underlying oceanic crust. When plates converge, heat and pressure generated along the plate boundary may lead to igneous activity and metamorphism within the descending oceanic plate, thus producing various igneous and metamorphic rocks.

Some of the sediment and sedimentary rock is subducted and melts, while other sediments and sedimentary rocks along the boundary of the nonsubducted

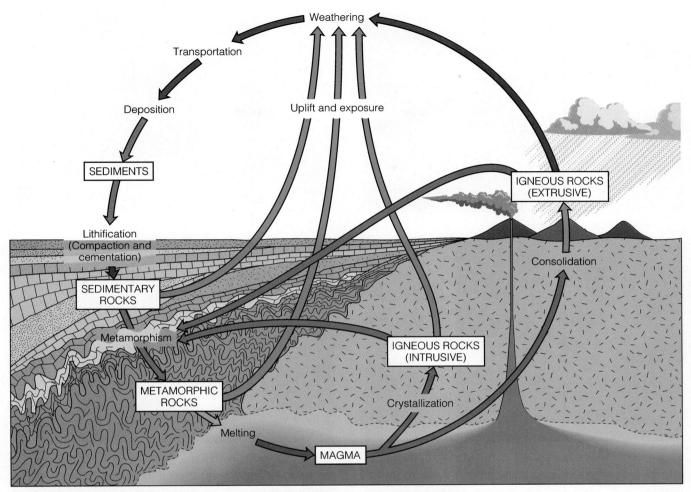

◆ FIGURE 1.8 The rock cycle showing the interrelationships between the Earth's internal and external processes and how each of the three major rock groups is related to the others.

plate are metamorphosed by the heat and pressure generated along the converging plate boundary. Later, the mountain range or chain of volcanic islands formed along the convergent plate boundary will once again be weathered and eroded, and the new sediments will be transported to the ocean to begin yet another cycle.

⬇ ORGANIC EVOLUTION

Plate tectonic theory provides us with a model for understanding the internal workings of the Earth and their effect on the Earth's surface. The theory of **organic evolution** provides the conceptual framework for understanding the history of life.

The publication in 1859 of Darwin's *On the Origin of Species by Means of Natural Selection* revolutionized biology and marked the beginning of modern evolutionary biology. With its publication, most naturalists recognized that evolution provided a unifying theory that explained an otherwise encyclopedic collection of biologic facts.

The central thesis of organic evolution is that all living organisms are related, and that they have descended with modifications from organisms that lived during the past. When Darwin proposed his theory of organic evolution he cited a wealth of **supporting** evidence, including the way organisms are classified, embryology, comparative anatomy, the geographic distribution of

◆ FIGURE 1.9 Hand specimens of common igneous (*a, b*), sedimentary (*c, d*), and metamorphic (*e, f*) rocks. (*a*) Granite, an intrusive igneous rock. (*b*) Basalt, an extrusive igneous rock. (*c*) Conglomerate, a sedimentary rock formed by the consolidation of rock fragments. (*d*) Marine limestone, a sedimentary rock formed by the extraction of mineral matter from seawater by organisms or by the inorganic precipitation of the mineral calcite from seawater. (*e*) Gneiss, a foliated metamorphic rock. (*f*) Marble, a nonfoliated metamorphic rock. (Photos courtesy of Sue Monroe.)

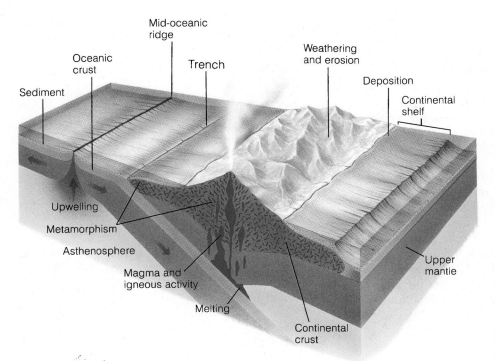

◆ FIGURE 1.10 Plate tectonics and the rock cycle. The cross section shows how the three major rock groups, igneous, metamorphic, and sedimentary, are recycled through both the continental and oceanic regions.

organisms, and, to a limited extent, the fossil record. Furthermore, Darwin proposed that *natural selection,* which results in the survival to reproductive age of those organisms best adapted to their environment, is the mechanism that accounts for evolution.

Perhaps the most compelling evidence in favor of evolution can be found in the fossil record. Just as the rock record allows geologists to interpret physical events and conditions in the geologic past, fossils provide us with a record of life in the past and evidence that organisms have evolved through time.

Fossils, which are the remains or traces of once-living organisms, not only provide evidence that evolution has occurred, but also demonstrate that the Earth has a history extending beyond that recorded by humans. The succession of fossils in the rock record provides geologists with a means for dating rocks and allowed a relative geologic time scale to be constructed in the 1800s.

▼ GEOLOGIC TIME AND UNIFORMITARIANISM

An appreciation of the immensity of geologic time is central to understanding the evolution of the Earth and its biota. Indeed, time is one of the main aspects that sets geology apart from the other sciences. Most people have difficulty comprehending geologic time because they tend to think in terms of the human perspective—seconds, hours, days, and years. Ancient history is what occurred hundreds or even thousands of years ago. When geologists talk of ancient geologic history, however, they are referring to events that happened hundreds of millions or even billions of years ago. To a geologist, recent geologic events are those that occurred within the last million years or so.

One popular analogy geologists use to convey the immensity of geologic time is to compare the history of the Earth to a calendar year (● Table 1.2). The time when the Earth formed 4.6 billion years ago corresponds to 12:00 A.M., January 1. On this calendar, we see that the oldest fossils, simple, microscopic bacteria, which first appeared about 3.6 billion years ago, are in mid-March; dinosaurs, which existed between 230 million and 66 million years ago, are between December 11 and December 26; and all of recorded human history occurs during the last few seconds of December 31. Furthermore, all of the scientific and technological discoveries that have brought us to our present level of knowledge take place in the final second of the year!

The **geologic time scale** resulted from the work of many nineteenth-century geologists who pieced together information from numerous rock exposures and constructed a sequential chronology based on changes in the Earth's biota through time. However, with the discovery of radioactivity in 1895, and the development of various radiometric dating techniques, geologists have since been able to assign absolute age dates in years to the subdivisions of the geologic time scale (◆ Fig. 1.11).

One of the cornerstones of geology is the **principle of uniformitarianism.** It is based on the premise that present-day processes have operated throughout geologic time. Therefore, in order to understand and interpret the rock record, we must first understand present-day processes and their results.

Uniformitarianism is a powerful principle that allows us to use present-day processes as the basis for interpreting the past and for predicting potential future events. We should keep in mind that uniformitarianism does not exclude such sudden or catastrophic events as meteorite impacts, volcanic eruptions, earthquakes, landslides, or flooding. These are processes that shape our modern world, and, in fact, some geologists view the history of the Earth as a series of such short-term or punctuated events. Such a view is certainly in keeping with the modern principle of uniformitarianism.

Furthermore, uniformitarianism does not require that the rates and intensities of geological processes be constant through time. We know that volcanic activity was more intense in North America 5 to 10 million years ago than it is today, and that glaciation has been more prevalent during the last several million years than in the previous 300 million years.

What uniformitarianism means is that even though the rates and intensities of geological processes have varied during the past, the physical and chemical laws of nature have remained the same and cannot be violated. Although the Earth is in a dynamic state of change and has been ever since it was formed, the processes that shaped it during the past are the same ones that are in operation today.

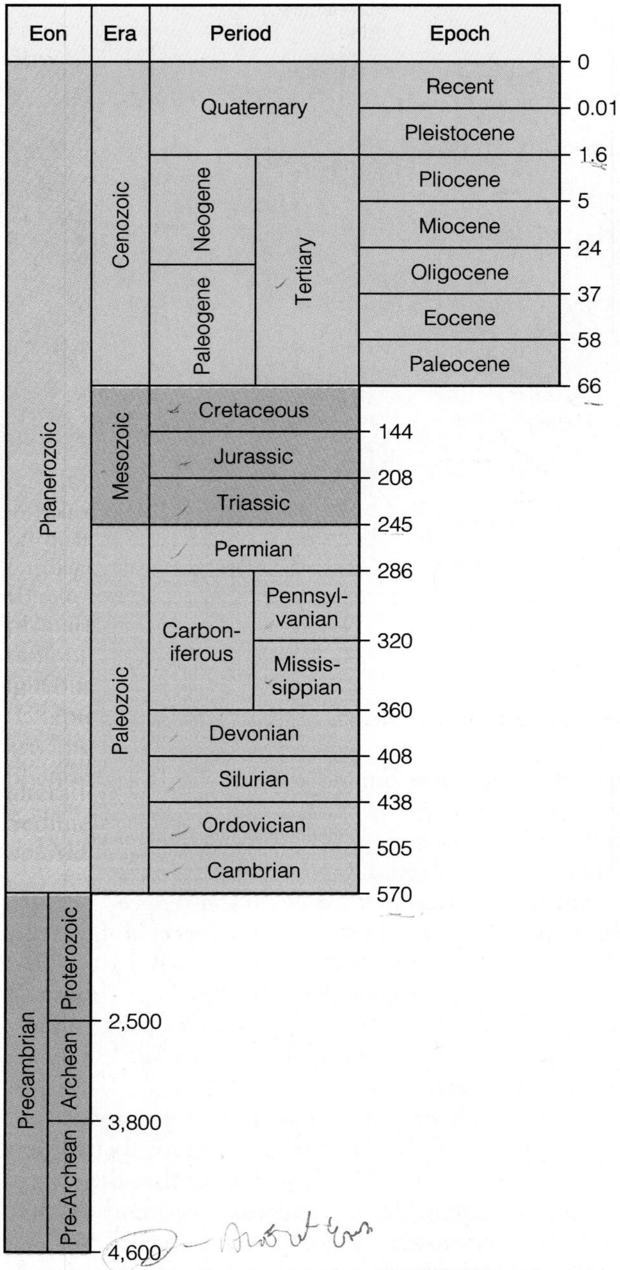

Eon	Era	Period		Epoch	(millions of years)
Phanerozoic	Cenozoic	Quaternary		Recent	0
				Pleistocene	0.01
		Neogene	Tertiary	Pliocene	1.6
				Miocene	5
		Paleogene		Oligocene	24
				Eocene	37
				Paleocene	58
					66
	Mesozoic	Cretaceous			
		Jurassic			144
		Triassic			208
	Paleozoic	Permian			245
		Carboniferous	Pennsylvanian		286
			Mississippian		320
		Devonian			360
		Silurian			408
		Ordovician			438
		Cambrian			505
					570
Precambrian	Proterozoic				2,500
	Archean				3,800
	Pre-Archean				4,600

◆ FIGURE 1.11 The geologic time scale. Numbers to the right of the columns are ages in millons of years before the present.

● **TABLE 1.2** Geologic Time and Significant Events in Earth History Condensed into One Calendar Year

Years before Present	Event	Days since January 1	Date and Time
10,000	Ice Age ends	364.9	December 31, 11:58:51 P.M.
1.6 million	Ice Age begins	364.9	December 31, 8:57:11 P.M.
4 million	First humans	364.6	December 31, 4:11:29 P.M.
53 million	First horses	360.8	December 27, 7:04:10 P.M.
66 million	Dinosaurs become extinct	359.8	December 26, 6:18:47 P.M.
115 million	First flowering plants	355.9	December 22, 9:00:00 A.M.
145 million	First birds	353.5	December 20, 11:52:10 A.M.
222 million	First mammals	347.4	December 14, 9:14:05 A.M.
230 million	First dinosaurs	344.8	December 11, 7:08:52 P.M.
360 million	First amphibians	336.4	December 3, 10:26:02 A.M.
430 million	First land plants	330.9	November 27, 9:07:50 P.M.
510 million	First fish	324.5	November 21, 12:46:57 P.M.
700 million	First multicellular animals	309.5	November 4, 10:57:23 A.M.
3.5 billion	Oldest fossils	71.0	March 13, 1:08:52 A.M.
4.6 billion	Earth formed	0	January 1, 12:00 A.M.

Chapter Summary

1. Geology is the study of the Earth. It is divided into two broad areas: physical geology is the study of the composition of Earth materials as well as the processes that operate within the Earth and upon its surface; historical geology examines the origin and evolution of the Earth, its continents, oceans, atmosphere, and life.

2. Geology is part of the human experience. We can find examples of it in the arts, music, and literature. A basic understanding of geology is also important for dealing with the many environmental problems and issues facing society.

3. Geologists engage in a variety of occupations, the main one being exploration for mineral and energy resources. They are also becoming increasingly involved in environmental issues and making short- and long-range predictions of the potential dangers from such natural disasters as volcanic eruptions and earthquakes.

4. The Earth is differentiated into layers. The outermost layer is the crust, which is divided into continental and oceanic crust. Below the crust is the upper mantle. The crust and upper mantle, or lithosphere, overlie the asthenosphere, a zone that behaves plastically. The asthenosphere is underlain by the solid lower mantle. The Earth's core consists of an outer liquid portion and an inner solid portion.

5. The Earth's crust and upper mantle form the lithosphere, which is broken into a series of plates that diverge, converge, and slide sideways past one another.

6. The scientific method is an orderly, logical approach that involves gathering and analyzing facts about a particular phenomenon, formulating hypotheses to explain the phenomenon, testing the hypotheses, and finally proposing a theory. A theory is a testable explanation for some natural phenomenon that has a large body of supporting evidence.

7. Plate tectonic theory provides a unifying explanation for many geological features and events. Plates can move away from each other, toward each other, or slide past each other. The interaction between plates is responsible for volcanic eruptions, earthquakes, the formation of mountain ranges and ocean basins, and the recycling of rock materials.

8. Igneous, sedimentary, and metamorphic rocks comprise the three major groups of rocks. Igneous rocks result from the crystallization of magma or consolidation of volcanic ejecta. Sedimentary rocks are formed by the consolidation of rock fragments, precipitation of mineral matter from solution, or compaction of plant or animal remains. Metamorphic rocks are produced from other rocks by heat, pressure, and chemically active fluids, generally beneath the Earth's surface.

9. The rock cycle illustrates the interactions between internal and external Earth processes and shows how the three rock groups are interrelated.
10. The central thesis of the theory of organic evolution is that all living organisms evolved (descended with modifications) from organisms that existed in the past.
11. Time sets geology apart from the other sciences, except astronomy. The geologic time scale is the calendar geologists use to date past events and is divided into eras, periods, and epochs.
12. The principle of uniformitarianism is basic to the interpretation of Earth history. This principle holds that the laws of nature have been constant through time and that the same processes operating today have operated in the past, albeit at different rates.

Important Terms

asthenosphere
convergent plate boundary
core
crust
divergent plate boundary
fossil
geologic time scale
geology
hypothesis

igneous rock
lithosphere
mantle
metamorphic rock
mineral
organic evolution
plate tectonic theory
plate

principle of uniformitarianism
rock
rock cycle
scientific method
sedimentary rock
subduction zone
theory
transform plate boundary

Review Questions

1. The study of Earth materials is:
   a. ___ paleontology; b. ___ stratigraphy; c. ___ physical geology; d. ___ historical geology; e. ___ environmental geology.
2. Which of the following statements about a mineral is not true?
   a. ___ it is organic; b. ___ it has definite physical and chemical properties; c. ___ it is naturally occurring; d. ___ it is a crystalline solid; e. ___ none of these.
3. Which of the following is not a subdivision of geology?
   a. ___ paleontology; b. ___ astronomy; c. ___ mineralogy; d. ___ petrology; e. ___ stratigraphy.
4. Into how many major concentric layers is the Earth divided?
   a. ___ 1; b. ___ 2; c. ___ 3; d. ___ 4; e. ___ 5.
5. The Earth's core is inferred to be:
   a. ___ hollow; b. ___ composed of rock with a high silica content; c. ___ completely molten; d. ___ composed mostly of iron and nickel; e. ___ completely solid.
6. The asthenosphere:
   a. ___ lies beneath the lithosphere; b. ___ is composed primarily of peridotite; c. ___ behaves plastically and flows slowly; d. ___ is the zone over which plates move; e. ___ all of these.
7. The layer between the core and the crust is the:
   a. ___ mantle; b. ___ lithosphere; c. ___ hydrosphere; d. ___ biosphere; e. ___ asthenosphere.
8. What fundamental process is believed to be responsible for plate motion?
   a. ___ hot spot activity; b. ___ subduction; c. ___ spreading ridges; d. ___ convection cells; e. ___ density differences.
9. Which of the following statements about a scientific theory is not true?
   a. ___ it is an explanation for some natural phenomenon; b. ___ it has a large body of supporting evidence; c. ___ it is a conjecture or guess; d. ___ it is testable; e. ___ none of these.
10. The man who proposed the hypothesis of continental drift was:
    a. ___ Hutton; b. ___ Wegener; c. ___ Hess; d. ___ Lyell; e. ___ Lovelock.
11. Mid-oceanic ridges are examples of what type of boundary?
    a. ___ divergent; b. ___ convergent; c. ___ transform; d. ___ subduction; e. ___ answers (b) and (d).
12. The San Andreas fault separating the Pacific plate from the North American plate is an example of what type of boundary?
    a. ___ divergent; b. ___ convergent; c. ___ transform; d. ___ subduction; e. ___ answers (b) and (d).
13. A plate is composed of the:
    a. ___ core and lower mantle; b. ___ lower mantle and asthenosphere; c. ___ asthenosphere and upper mantle;

d. ___ upper mantle and crust; e. ___ continental and oceanic crust.

14. Which of the following is not a major rock group?
a. ___ volcanic; b. ___ igneous; c. ___ metamorphic; d. ___ sedimentary; e. ___ none of these.

15. Which rocks form from the cooling of a magma?
a. ___ igneous; b. ___ metamorphic; c. ___ sedimentary; d. ___ all of these; e. ___ none of these.

16. The premise that present-day processes have operated throughout geologic time is known as the principle of:
a. ___ plate tectonics; b. ___ sea-floor spreading; c. ___ continental drift; d. ___ organic evolution; e. ___ uniformitarianism.

17. The rock cycle implies that:
a. ___ metamorphic rocks are derived from magma; b. ___ any rock type can be derived from any other rock type; c. ___ igneous rocks only form beneath the Earth's surface; d. ___ sedimentary rocks only form from the weathering of igneous rocks; e. ___ all of these.

18. That all living organisms are the descendants of different life-forms that existed in the past is the central claim of:

a. ___ organic evolution; b. ___ the principle of fossil succession; c. ___ the principle of uniformitarianism; d. ___ plate tectonics; e. ___ none of these.

19. Why is it important for people to have a basic understanding of geology?

20. Describe some of the ways in which geology affects our everyday lives.

21. Explain both the difference between physical and historical geology and how they are related.

22. Name the major layers of the Earth, and describe their general composition.

23. Describe the scientific method, and explain how it may lead to a scientific theory. Define scientific theory.

24. Briefly discuss the theory of organic evolution.

25. Briefly describe the plate tectonic theory, and explain why it is a unifying theory of geology.

26. What are the three types of plate boundaries?

27. What are the three major groups of rocks?

28. Describe the rock cycle, and explain how it may be related to plate tectonics.

29. What is the principle of uniformitarianism? Does it allow for catastrophic events? Explain.

Additional Readings

Albritton, C. C., Jr. 1980. *The abyss of time.* San Francisco: Freeman, Cooper.

Dietrich, R. V. 1989. Rock music. *Earth Science* 42, no. 2: 24–25.

_____. 1990. Rocks depicted in painting and sculpture. *Rocks & Minerals* 65, no. 3: 224–36.

_____. 1992. How can I get others interested in rocks? *Rocks & Minerals* 67, no. 2:124–28.

Dietrich, R. V., and B. J. Skinner. 1990. *Gems, granites, and gravels.* New York: Cambridge Univ. Press.

Ernst, W. G. 1990. *The dynamic planet.* Irvington, N.Y.: Columbia Univ. Press.

Mirsky, A. 1989. Geology in our everyday lives. *Journal of Geological Education* 37, no. 1: 9–12.

Rhodes, F. H. T., and R. O. Stone. 1981. *Language of the Earth.* Elmsford, N.Y.: Pergamon.

Siever, R. 1983. The dynamic Earth. *Scientific American* 249, no. 3: 46–55.

Wicander, R., and J. S. Monroe. 1993. *Historical geology: Evolution of the Earth and life through time.* 2d ed. St. Paul, Minn.: West Publishing.

CHAPTER

2

Chapter Outline

"Steamboat"—red and green tourmaline and colorless quartz crystals. From the Tourmaline
King mine, near Pala, San Diego County, California. The specimen is about 28 cm high.
(National Museum of Natural History, Smithsonian Institution.)

Minerals

Prologue

Among the hundreds of minerals used by humans none is so highly prized and eagerly sought as gold (◆ Fig. 2.1). This deep yellow mineral has been the cause of feuds and wars and was one of the incentives for the exploration of the Americas. Gold has been mined for at least 6,000 years, and archeological evidence indicates that people in Spain possessed small quantities of gold 40,000 years ago. Probably no other substance has caused so much misery, but at the same time provided so many benefits for those who possessed it.

Why is gold so highly prized? Certainly not for use in tools or weapons, for it is too soft and pliable to hold a cutting edge. Furthermore, it is too heavy to be practical for most utilitarian purposes (it weighs about twice as much as lead). During most of historic time, gold has been used for jewelry, ornaments, and ritual objects and has served as a symbol of wealth and as a monetary standard. Gold is so desired for several reasons: (1) its pleasing appearance, (2) the ease with which it can be worked, (3) its durability, and (4) its scarcity (it is much rarer than silver).

Central and South American natives used gold extensively long before the arrival of Europeans. In fact, the Europeans' lust for gold was responsible for the ruthless conquest of the natives in those areas. In the United States, gold was first profitably mined in North Carolina in 1801 and in Georgia in 1829, but the truly spectacular finds occurred in California in 1848. This latter discovery culminated in the great gold rush of 1849 when tens of thousands of people flocked to California to find riches. Unfortunately, only a few found what they sought. Nevertheless, during the five years from 1848 to 1853, which constituted the gold rush proper, more than $200 million in gold was recovered.

Another gold rush occurred in 1876 following the report by Lieutenant Colonel George Armstrong Custer that "gold in satisfactory quantities can be obtained in the Black Hills [South Dakota]." The flood of miners into the Black Hills, the Holy Wilderness of the Sioux Indians, resulted in the Indian War during which Custer and some 260 of his men were annihilated at the Battle of the Little Bighorn in Montana in June 1876. Despite this stunning victory, the Sioux could not sustain a war against the U.S. Army, and in September 1876, they were forced to relinquish the Black Hills.

◆ FIGURE 2.1 Specimen of gold from Grass Valley, California. (National Museum of Natural History, Smithsonian Institution.)

Canada, too, has had its gold rushes. The earliest gold discovery was in 1850 in the Queen Charlotte Islands on the Pacific coast, and by 1858 about 10,000 men and women were panning for gold there. The greatest Canadian gold rush occurred from 1897 to 1899 in the Klondike region of the Yukon Territory. As many as 35,000 men and women poured in to this remote, hostile region. In fact, Dawson City grew so rapidly that food shortages during the autumn and winter of 1897 resulted in the evacuation of hundreds of people. Just as in other gold rushes, the local merchants made out better than the miners, most of whom barely eked out a living.

For 50 years following the California gold rush, the United States led the world in gold production, and it still produces a considerable amount, mostly from mines in Nevada and South Dakota. Currently, however, the leading producer is South Africa with the former Soviet Union a distant second, followed by Canada, where gold valued at $2.38 billion was mined in 1990. Much gold still is used for jewelry, but in contrast to its earlier uses, gold now is used in the chemical industry, and for gold plating, electrical circuitry, and glass making. Consequently, the quest for gold has not ceased or even abated. In many industrialized nations, including the United States, domestic production cannot meet the demand, and much of the gold used must be imported.

⩔ INTRODUCTION

The term "mineral" commonly brings to mind dietary substances that are essential for good nutrition such as calcium, iron, potassium, and magnesium. These substances are actually chemical elements, not minerals in the geologic sense. Mineral is also sometimes used to refer to any substance that is neither animal nor vegetable. Such usage implies that minerals are inorganic substances, which is correct, but not all inorganic substances are minerals. Water, for example, is not a mineral even though it is inorganic and is composed of the same chemical elements as ice, which is a mineral. Ice is, of course, a solid whereas water is a liquid; minerals are solids rather than liquids or gases. In fact, geologists have a very specific definition of the term **mineral**: a naturally occurring, inorganic crystalline solid. Crystalline means it has a regular internal structure. Furthermore, a mineral has a narrowly defined chemical composition and characteristic physical properties such as density, color, and hardness. Most rocks are solid aggregates of one or more minerals, and thus minerals are the building blocks of rocks.

Obviously, minerals are important to geologists as the constituents of rocks, but they are important for other reasons as well. Gemstones such as diamond and topaz are minerals, and rubies are simply red-colored varieties of the mineral corundum. The sand used in the manufacture of glass is composed of the mineral quartz, and ore deposits are natural concentrations of economically valuable minerals. Indeed, industrialized societies depend directly upon finding and using mineral resources such as iron, copper, gold, and many others.

⩔ MATTER AND ITS COMPOSITION

Anything that has mass and occupies space is *matter*. The atmosphere, water, plants and animals, and minerals and rocks are all composed of matter. Matter occurs in one of three states or phases, all of which are important in geology: *solids, liquids,* and *gases* (⬤ Table 2.1). Atmospheric gases and liquids such as surface water and groundwater will be discussed later in this book, but here we are concerned chiefly with solids because all minerals are solids.

⬤ **TABLE 2.1** Phases or States of Matter

Phase	Characteristics	Examples
Solid	Rigid substance that retains its shape unless distorted by a force	Minerals, rocks, iron, wood
Liquid	Flows easily and conforms to the shape of the containing vessel; has a well-defined upper surface and greater density than a gas	Water, lava, wine, blood, gasoline
Gas	Flows easily and expands to fill all parts of a containing vessel; lacks a well-defined upper surface; is compressible	Helium, nitrogen, air, water vapor

Elements and Atoms

All matter is made up of chemical **elements**, each of which is composed of incredibly small particles called **atoms**. Atoms are the smallest units of matter that retain the characteristics of an element. Ninety-one naturally occurring elements have been discovered, some of which are listed in ● Table 2.2, and more than a dozen additional elements have been made in laboratories. Each naturally occurring element and most artificially produced ones have a name and a chemical symbol (Table 2.2).

Atoms consist of a compact **nucleus** composed of one or more **protons**, which are particles with a positive electrical charge, and **neutrons**, which are electrically neutral (◆ Fig. 2.2). The nucleus of an atom makes up most of its mass. Encircling the nucleus are negatively charged particles called **electrons**. Electrons orbit rapidly around the nucleus at specific distances in one or more **electron shells** (Fig. 2.2). The number of protons in the nucleus of an atom determines what the element is and determines the **atomic number** for that element.

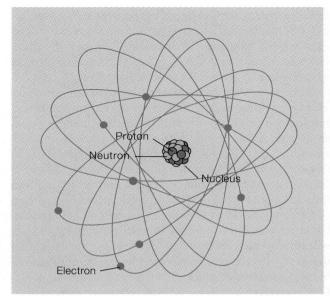

◆ FIGURE 2.2 The structure of an atom. The dense nucleus consisting of protons and neutrons is surrounded by a cloud of orbiting electrons.

● **TABLE 2.2** Symbols, Atomic Numbers, and Electron Configurations for Some of the Naturally Occurring Elements

Element	Symbol	Atomic Number	Number of Electrons in Each Shell			
			1	2	3	4
Hydrogen	H	1	1			
Helium	He	2	2			
Lithium	Li	3	2	1		
Beryllium	Be	4	2	2		
Boron	B	5	2	3		
Carbon	C	6	2	4		
Nitrogen	N	7	2	5		
Oxygen	O	8	2	6		
Fluorine	F	9	2	7		
Neon	Ne	10	2	8		
Sodium	Na	11	2	8	1	
Magnesium	Mg	12	2	8	2	
Aluminum	Al	13	2	8	3	
Silicon	Si	14	2	8	4	
Phosphorus	P	15	2	8	5	
Sulfur	S	16	2	8	6	
Chlorine	Cl	17	2	8	7	
Argon	Ar	18	2	8	8	
Potassium	K	19	2	8	8	1
Calcium	Ca	20	2	8	8	2

For example, each atom of the element hydrogen (H) has one proton in its nucleus and thus has an atomic number of 1. Helium (He) possesses 2 protons, carbon (C) has 6, and uranium (U) has 92, so their atomic numbers are 2, 6, and 92, respectively (Table 2.2).

Atoms are also characterized by their **atomic mass number**, which is determined by adding together the number of protons and neutrons in the nucleus (electrons contribute negligible mass to an atom) (Fig. 2.2). However, not all atoms of the same element have the same number of neutrons in their nuclei. In other words, atoms of the same element may have different atomic mass numbers. For example, different carbon (C) atoms have atomic mass numbers of 12, 13, and 14. All of these atoms possess 6 protons, otherwise they would not be carbon, but the number of neutrons varies. Forms of the same element with different atomic mass numbers are **isotopes** (◆ Fig. 2.3).

A number of elements have a single isotope but many, such as uranium and carbon, have several (Fig. 2.3). Some isotopes are unstable and spontaneously change to a stable form. This process, called *radioactive decay*, occurs because the forces that bind the nucleus together are not strong enough. Such decay occurs at known rates and is the basis for several techniques for determining age that will be discussed in Chapter 17. All isotopes of an element behave the same chemically. For example, both carbon 12 and carbon 14 are present in carbon dioxide (CO_2).

Bonding and Compounds

The process whereby atoms are joined to other atoms is called **bonding**. When atoms of two or more different elements are bonded, the resulting substance is a **compound**. Thus, a chemical substance such as gaseous oxygen, which consists entirely of oxygen atoms, is an element, whereas ice, which consists of hydrogen and oxygen, is a compound. Most minerals are compounds although there are several important exceptions, such as gold and silver.

To understand bonding, it is necessary to delve deeper into the structure of atoms. Recall that negatively charged electrons in electron shells orbit the nuclei of atoms. With the exception of hydrogen, which has only one proton and one electron, the innermost electron shell of an atom contains no more than two electrons. The other shells contain various numbers of electrons, but the outermost shell never contains more than eight (Table 2.2). The electrons in the outermost shell are those that are usually involved in chemical bonding.

Two types of chemical bonds are particularly important in minerals, *ionic* and *covalent*, and many minerals contain both types of bonds. Two other types of chemical bonds, *metallic* and *van der Waals*, are much less common, but are extremely important in determining the properties of some very useful minerals.

Ionic Bonding

Notice in Table 2.2 that most atoms have fewer than eight electrons in their outermost electron shell. Some elements, however, including neon and argon, have complete outer shells containing eight electrons; they are known as the *noble gases*. The noble gases do not react readily with other elements to form compounds because of this electron configuration. Interactions among atoms tend to produce electron configurations similar to those of the noble gases. That is, atoms interact such that their outermost electron shell is filled with eight electrons, unless the first shell (with two electrons) is also the outermost electron shell as in helium.

One way in which the noble gas configuration can be attained is by the transfer of one or more electrons from one atom to another. Common salt, for example,

◆ FIGURE 2.3 Schematic representation of isotopes of carbon. A carbon atom has an atomic number of 6 and an atomic mass number of 12, 13, or 14 depending on the number of neutrons in its nucleus.

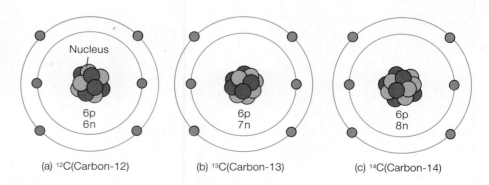

Nucleus

6p
6n

(a) ^{12}C(Carbon-12)

6p
7n

(b) ^{13}C(Carbon-13)

6p
8n

(c) ^{14}C(Carbon-14)

is composed of the elements sodium (Na) and chlorine (Cl), each of which is poisonous, but when combined chemically, they form the compound sodium chloride (NaCl), the mineral halite or common salt. Notice in ◆ Figure 2.4a that sodium has 11 protons and 11 electrons; thus, the positive electrical charges of the protons are exactly balanced by the negative charges of the electrons, and the atom is electrically neutral. Likewise, chlorine with 17 protons and 17 electrons is electrically neutral (Fig. 2.4a). However, neither sodium nor chlorine has eight electrons in its outermost electron shell; sodium has only one whereas chlorine has seven. In order to attain a stable configuration, sodium loses the electron in its outermost electron shell, leaving its next shell with eight electrons as the outermost one (Fig. 2.4a). However, sodium now has one fewer electron (negative charge) than it has protons (positive charge) so it is an electrically charged particle. Such a particle is an **ion** and, in the case of sodium, is symbolized Na^{+1}.

The electron lost by sodium is transferred to the outermost electron shell of chlorine, which had seven electrons to begin with. Thus, the addition of one more electron gives chlorine an outermost electron shell of eight electrons, the configuration of a noble gas. Its total number of electrons, however, is now 18, which exceeds by one the number of protons. Accordingly, chlorine also becomes an ion, but it is negatively charged (Cl^{-1}). An **ionic bond** forms between sodium and chlorine because of the attractive force between the positively charged sodium ion and the negatively charged chlorine ion (Fig. 2.4a).

In ionic compounds, such as sodium chloride (the mineral halite), the ions are arranged in a three-dimensional framework that results in overall electrical neutrality. In halite, sodium ions are bonded to chlorine ions on all sides, and chlorine ions are surrounded by sodium ions (◆ Fig. 2.4b).

Covalent Bonding

Covalent bonds form between atoms when their electron shells overlap and electrons are shared. For example,

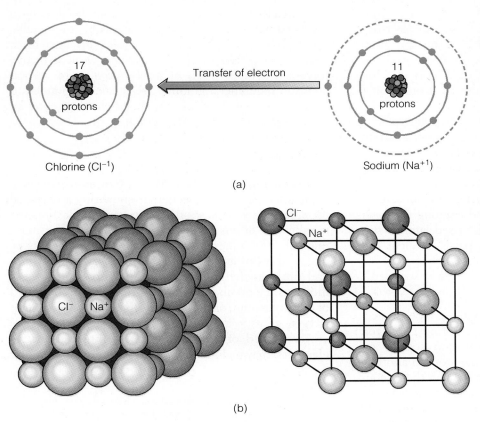

(a)

Chlorine (Cl^{-1})   Transfer of electron   Sodium (Na^{+1})

(b)

Cl⁻   Na⁺

◆ FIGURE 2.4 (*a*) Ionic bonding. The electron in the outermost shell of sodium is transferred to the outermost electron shell of chlorine. Once the transfer has occurred, sodium and chlorine are positively and negatively charged ions, respectively. (*b*) The crystal structure of sodium chloride, the mineral halite. The diagram on the left shows the relative sizes of sodium and chlorine ions, and the diagram on the right shows the locations of the ions in the crystal structure.

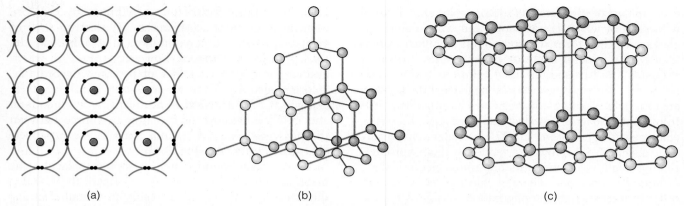

(a)	(b)	(c)

◆ FIGURE 2.5 (*a*) Covalent bonds formed by adjacent atoms sharing electrons in diamond. (*b*) The three-dimensional framework of carbon atoms in diamond. (*c*) Covalent bonding also occurs in graphite, but here the carbon atoms are bonded together to form sheets that are held to one another by van der Waals bonds. The sheets themselves are strong, but the bonds between sheets are weak.

atoms of the same element, such as oxygen in oxygen gas, cannot bond by transferring electrons from one atom to another. Carbon (C), which forms the minerals graphite and diamond, has four electrons in its outermost electron shell (◆ Fig. 2.5a). If these four electrons were transferred to another carbon atom, the atom receiving the electrons would have the noble gas configuration of eight electrons in its outermost electron shell, but the atom contributing the electrons would not.

In such situations, adjacent atoms share electrons by overlapping their electron shells. For example, a carbon atom in diamond shares all four of its outermost electrons with a neighbor to produce a stable noble gas configuration (Fig. 2.5a).

Covalent bonds are not restricted to substances composed of atoms of a single kind. Among the most common minerals, the silicates (discussed later in this chapter), the element silicon forms partly covalent and partly ionic bonds with oxygen.

Metallic and van der Waals Bonds

Metallic bonding results from an extreme type of electron sharing. The electrons of the outermost electron shell of such metals as gold, silver, and copper are readily lost and move about from one atom to another. This electron mobility accounts for the fact that metals have a metallic luster (their appearance in reflected light), provide good electrical and thermal conductivity, and can be easily reshaped. Only a few minerals possess metallic bonds, but those that do are very useful; copper, for example, is used for electrical wiring because of its high electrical conductivity.

Some electrically neutral atoms and molecules* have no electrons available for ionic, covalent, or metallic bonding. They nevertheless have a weak attractive force between them when in proximity. This weak attractive force is a *van der Waals* or *residual bond.* The carbon atoms in the mineral graphite are covalently bonded to form sheets, but the sheets are weakly held together by van der Waals bonds (◆ Fig. 2.5c).

▼ MINERALS

Most minerals are compounds of two or more elements. Mineral composition is generally shown by a chemical formula, which is a shorthand way of indicating the numbers of atoms of different elements composing a mineral. The mineral quartz, for example, consists of one silicon (Si) atom for every two oxygen (O) atoms, and thus has the formula SiO_2; the subscript number indicates the number of atoms. Orthoclase is composed of one potassium, one aluminum, three silicon, and eight oxygen atoms so its formula is $KAlSi_3O_8$. Some minerals are composed of a single element. Known as **native elements**, they include such minerals as gold (Au), silver

· · · · · · · · · ·

*A molecule is the smallest unit of a substance having the properties of that substance. A water molecule (H_2O), for example, possesses two hydrogen atoms and one oxygen atom.

(Ag), platinum (Pt), and graphite and diamond, both of which are composed of carbon (C).

Before we discuss minerals in more detail, let us recall our formal definition: a mineral is a naturally occurring, inorganic, crystalline solid, with a narrowly defined chemical composition and characteristic physical properties. The next sections will examine each part of this definition.

Naturally Occurring, Inorganic Substances

"Naturally occurring" excludes from minerals all substances that are manufactured by humans. Accordingly, synthetic diamonds and rubies and a number of other artificially synthesized substances are not regarded as minerals by most geologists.

Some geologists think the term "inorganic" in the mineral definition is superfluous. It does, however, remind us that animal matter and vegetable matter are not minerals. Nevertheless, some organisms such as corals and clams construct their shells of the compound calcium carbonate ($CaCO_3$), which is either aragonite or calcite, both of which are minerals.

The Nature of Crystals

By definition minerals are crystalline solids. A **crystalline solid** is a solid in which the constituent atoms are arranged in a regular, three-dimensional framework, (Fig. 2.4b). Under ideal conditions, such as in a cavity, mineral crystals can grow and form perfect crystals that possess planar surfaces (crystal faces), sharp corners, and straight edges (◆ Fig. 2.6). In other words, the regular geometric shape of a well-formed mineral crystal is the exterior manifestation of an ordered internal atomic arrangement. Not all rigid substances are crystalline solids, however; natural and manufactured glass, for example, lack the ordered arrangement of atoms and are said to be *amorphous,* meaning without form.

As early as 1669, a well-known Danish scientist, Nicholas Steno, determined that the angles of intersection of equivalent crystal faces on different specimens of quartz are identical. Since then the *constancy of interfacial angles* has been demonstrated for many other minerals, regardless of their size, shape, or geographic occurrence (◆ Fig. 2.7). Steno postulated that mineral crystals are composed of very small, identical building blocks and that the arrangement of these blocks determines the external form of the crystals. Such regularity of the external form of minerals must surely mean that external crystal form is controlled by internal structure.

Crystalline structure can be demonstrated even in minerals lacking obvious crystals. For example, many minerals possess a property called *cleavage,* meaning that they break or split along closely spaced, smooth planes. The fact that these minerals can be split along such smooth planar surfaces indicates that the mineral's internal structure controls such breakage. The behavior of light and X-ray beams transmitted through minerals also provides compelling evidence for an orderly arrangement of atoms within minerals.

Chemical Composition

The definition of a mineral contains the phrase "a narrowly defined chemical composition," because some minerals actually have a range of compositions. For many minerals the chemical composition is constant:

◆ FIGURE 2.6 Mineral crystals occur in a variety of shapes, several of which are shown here. (*a*) Cubic crystals typically develop in the minerals halite, galena, and pyrite. (*b*) Dodecahedron crystals such as those of garnet have 12 sides. (*c*) Diamond has octahedral or 8-sided crystals. (*d*) A prism terminated by pyramids is found in quartz.

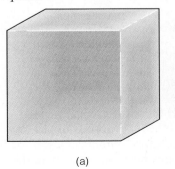
(a)

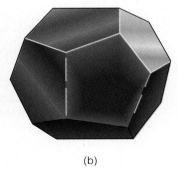
(b)

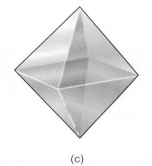
(c)

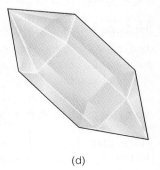
(d)

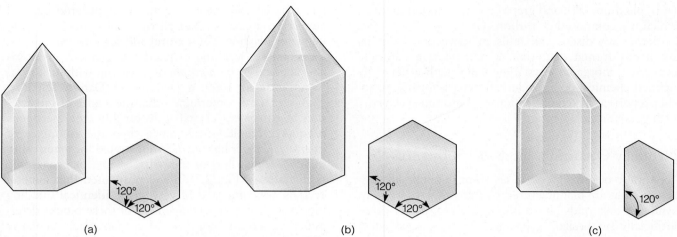

(a)   (b)   (c)

◆ FIGURE 2.7 Side views and cross sections of three quartz crystals showing the constancy of interfacial angles: (*a*) a well-shaped crystal; (*b*) a larger crystal; and (*c*) a poorly shaped crystal. The angles formed between equivalent crystal faces on different specimens of the same mineral are the same regardless of the size or shape of the specimens.

quartz is always composed of silicon and oxygen (SiO_2), and halite contains only sodium and chlorine ($NaCl$). Other minerals have a range of compositions because one element may substitute for another if the atoms of two or more elements are nearly the same size and the same charge. Notice in ◆ Figure 2.8 that iron and magnesium atoms are about the same size, and therefore they can substitute for one another. The chemical formula for the mineral olivine is $(Mg,Fe)_2SiO_4$, meaning that, in addition to silicon and oxygen, it may contain only magnesium, only iron, or a combination of both. A number of other minerals also have ranges of compositions, so these are actually mineral groups with several members.

Physical Properties

The last criterion in our definition of a mineral, "characteristic physical properties," refers to such properties as hardness, color, and crystal form. Such properties are controlled by composition and structure. We shall have more to say about physical properties of minerals later in this chapter.

⬇ MINERAL DIVERSITY

More than 3,500 minerals have been identified and described, but only a very few—perhaps two dozen—are particularly common. As we previously mentioned, 91 naturally occurring elements have been discovered. One might think that an extremely large number of minerals could be formed from so many elements, but several factors limit the number of possible minerals. For one thing, many combinations of elements are chemically impossible; there are no compounds composed of only potassium and sodium or of silicon and iron, for example. Another important factor restricting the number of common minerals is that only eight chemical elements make up the bulk of the Earth's crust (● Table 2.3). As Table 2.3 shows, oxygen and silicon constitute more than 74% (by weight) of the Earth's crust and nearly 84% of the atoms available to form compounds. By far the most common minerals in the Earth's crust consist of silicon and oxygen, combined with one or more of the other elements listed in Table 2.3.

⬇ MINERAL GROUPS

Objects such as organisms, rocks, and minerals can be classified in many ways, such as by color, size, shape, density, composition, structure, or a combination of such features. In any case, all classification schemes share two characteristics: (1) they systematically categorize similar objects, and (2) their purpose is to convey information. Geologists recognize mineral classes or groups, each of which contains members that share the

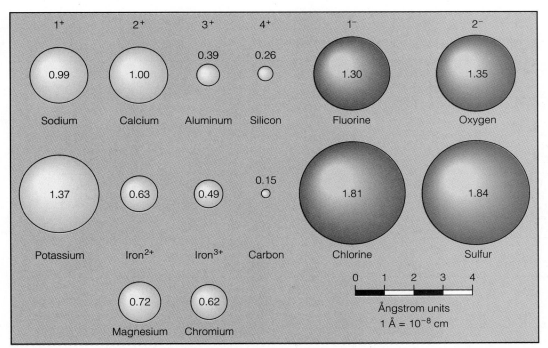

◆ FIGURE 2.8 Electrical charges and relative sizes of ions common in minerals. The numbers within the ions are the radii shown in Ångstrom units (1 Å = 10^{-10} m).

● **TABLE 2.3** Common Elements in the Earth's Crust

Element	Symbol	Percentage of Crust (by Weight)	Percentage of Crust (by Atoms)
Oxygen	O	46.6%	62.6%
Silicon	Si	27.7	21.2
Aluminum	Al	8.1	6.5
Iron	Fe	5.0	1.9
Calcium	Ca	3.6	1.9
Sodium	Na	2.8	2.6
Potassium	K	2.6	1.4
Magnesium	Mg	2.1	1.8
All others		1.5	0.1

same negatively charged ion or ion group; several of these mineral groups are listed in ● Table 2.4.

Silicate Minerals

Because silicon and oxygen are the two most abundant elements in the Earth's crust (Table 2.3), it is not surprising that many minerals contain these elements. A combination of silicon and oxygen is called **silica**, and the minerals containing silica are **silicates**. Quartz (SiO_2) is composed entirely of silicon and oxygen so it is pure silica. Most silicates, however, have one or more additional elements, as in orthoclase ($KAlSi_3O_8$) and olivine [$(Mg,Fe)_2SiO_4$]. Silicate minerals include about one-third of all known minerals, but their abundance is even more impressive when one considers that they

● **TABLE 2.4** Some of the Mineral Groups Recognized by Geologists

Mineral Group	Negatively Charged Ion or Ion Group	Examples	Composition
Carbonate	$(CO_3)^{-2}$	Calcite	$CaCO_3$
		Dolomite	$CaMg(CO_3)_2$
Halide	Cl^{-1}, F^{-1}	Halite	$NaCl$
		Fluorite	CaF_2
Native element	—	Gold	Au
		Silver	Ag
		Diamond	C
		Graphite	C
Oxide	O^{-2}	Hematite	Fe_2O_3
		Magnetite	Fe_3O_4
Silicate	$(SiO_4)^{-4}$	Quartz	SiO_2
		Potassium feldspar	$KAlSi_3O_8$
		Olivine	$(Mg,Fe)_2SiO_4$
Sulfate	$(SO_4)^{-2}$	Anhydrite	$CaSO_4$
		Gypsum	$CaSO_4 \cdot 2H_2O$
Sulfide	S^{-2}	Galena	PbS
		Pyrite	FeS_2

make up perhaps as much as 95% of the Earth's crust.

The basic building block of all silicate minerals is the **silica tetrahedron**, which consists of one silicon atom and four oxygen atoms (◆ Fig. 2.9). These atoms are arranged such that the four oxygen atoms surround a silicon atom, which occupies the space between the oxygen atoms; thus, a four-faced pyramidal structure is formed. The silicon atom has a positive charge of 4, while each of the four oxygen atoms has a negative charge of 2, resulting in an ion group with a total negative charge of 4 $(SiO_4)^{-4}$.

Because the silica tetrahedron has a negative charge, it does not exist in nature as an isolated ion group; rather it combines with positively charged ions or shares its oxygen atoms with other silica tetrahedra. In the simplest silicate minerals, the silica tetrahedra exist as single units bonded to positively charged ions. In minerals containing isolated tetrahedra, the silicon to oxygen ratio is 1:4, and the negative charge of the silica ion is balanced by positive ions (◆ Fig. 2.10a). Olivine $[(Mg,Fe)_2SiO_4]$, for example, has either two magne-

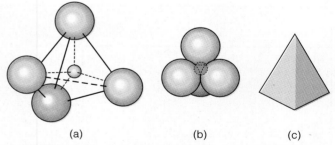

◆ FIGURE 2.9 The silica tetrahedron. (*a*) Expanded view showing oxygen atoms at the corners of a tetrahedron and a small silicon atom at the center. (*b*) View of the silica tetrahedron as it really exists with the oxygen atoms touching. (*c*) The silica tetrahedron represented diagramatically; the oxygen atoms are at the four points of the tetrahedron.

sium (Mg^{+2}) ions, two iron (Fe^{+2}) ions, or one of each to offset the −4 charge of the silica ion.

Silica tetrahedra may also be arranged so that they join together to form chains of indefinite length (◆ Fig.

			Formula of negatively charged ion group	Silicon to oxygen ratio	Example
(a)	Isolated tetrahedra		$(SiO_4)^{-4}$	1:4	Olivine
(b)	Continuous chains of tetrahedra	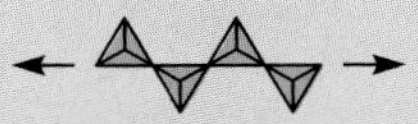 Single chain	$(SiO_3)^{-2}$	1:3	Pyroxene group
		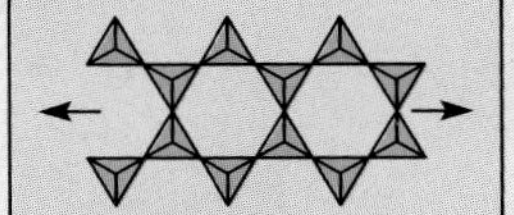 Double chain	$(Si_4O_{11})^{-6}$	4:11	Amphibole group
(c)	Continuous sheets	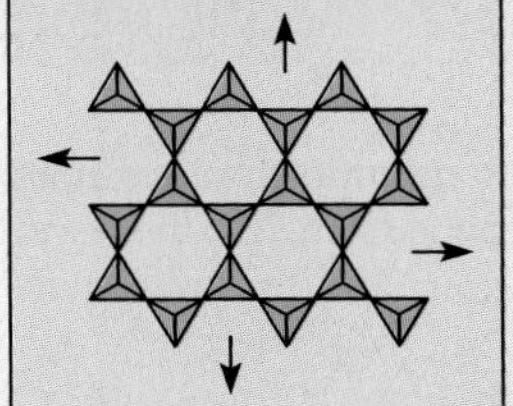	$(Si_4O_{10})^{-4}$	2:5	Micas
(d)	Three-dimensional networks	Too complex to be shown by a simple two-dimensional drawing	$(SiO_2)^0$	1:2	Quartz

◆ FIGURE 2.10 Structures of some of the common silicate minerals shown by various arrangements of silica tetrahedra: (*a*) isolated tetrahedra; (*b*) continuous chains; (*c*) continuous sheets; and (*d*) networks. The arrows adjacent to single-chain, double-chain, and sheet silicates indicate that these structures continue indefinitely in the directions shown.

2.10b). Single chains, as in the pyroxene minerals, form when each tetrahedron shares two of its oxygens with adjacent tetrahedra; the result is a silicon to oxygen ratio of 1:3. Enstatite, a pyroxene group mineral, reflects this ratio in its chemical formula, $MgSiO_3$. Individual chains, however, possess a net −2 electrical charge, so they are balanced by positive ions, such as Mg^{+2}, that link parallel chains together (Fig. 2.10b).

The amphibole group of minerals is characterized by a double-chain structure in which alternate tetrahedra in two parallel rows are cross-linked (Fig. 2.10b). The formation of double chains results in a silicon to oxygen ratio of 4:11, so that each double chain possesses a −6 electrical charge. Mg^{+2}, Fe^{+2}, and Al^{+2} are usually involved in linking the double chains together.

In sheet structure silicates, three oxygens of each tetrahedron are shared by adjacent tetrahedra (◆ Fig. 2.10c). Such structures result in continuous sheets of silica tetrahedra with silicon to oxygen ratios of 2:5.

Continuous sheets also possess a negative electrical charge that is satisfied by positive ions located between the sheets. This particular structure is what accounts for the characteristic sheet structure of the *micas*, such as biotite and muscovite, and the *clay minerals.*

Three-dimensional frameworks of silica tetrahedra form when all four oxygens of the silica tetrahedron are shared by adjacent tetrahedra (◆ Fig. 2.10d). Such sharing of oxygen atoms results in a silicon to oxygen ratio of 1:2, which is electrically neutral. Quartz is a common framework silicate.

Two subgroups of silicates are recognized, ferromagnesian and nonferromagnesian silicates. The **ferromagnesian silicates** are those containing iron (Fe), magnesium (Mg), or both. Such minerals are commonly dark colored and more dense than nonferromagnesian silicates. Some of the common ferromagnesian silicate minerals are olivine, the pyroxenes, the amphiboles, and biotite (◆ Fig. 2.11).

(a)   (b)

(c)   (d)

◆ FIGURE 2.11 Common ferromagnesian silicates: (a) olivine; (b) augite, a pyroxene group mineral; (c) hornblende, an amphibole group mineral; and (d) biotite mica. (Photo courtesy of Sue Monroe.)

(a)   (b)

(c)   (d)

◆ FIGURE 2.12 Common nonferromagnesian silicates: (a) quartz; (b) the potassium feldspar orthoclase; (c) plagioclase feldspar; and (d) muscovite mica. (Photo courtesy of Sue Monroe.)

The **nonferromagnesian silicates** lack iron and magnesium, are generally light colored, and are less dense than ferromagnesian silicates (◆ Fig. 2.12). The most common minerals in the Earth's crust are nonferromagnesian silicates known as *feldspars*. Feldspar is a general name, however, and two distinct groups are recognized, each of which includes several species. The *potassium feldspars* are represented by microcline and orthoclase ($KAlSi_3O_8$). The second group of feldspars, the *plagioclase feldspars,* range from calcium-rich ($CaAl_2Si_2O_8$) to sodium-rich ($NaAlSi_3O_8$) varieties.

Quartz (SiO_2) is another common nonferromagnesian silicate (see Perspective 2.1). It is a framework silicate that can usually be recognized by its glassy appearance and hardness (Fig. 2.12a). Another fairly common nonferromagnesian silicate is muscovite, which is a mica (Fig. 2.12d).

Carbonate Minerals

Carbonate minerals are those that contain the negatively charged carbonate ion $(CO_3)^{-2}$. An example is calcium carbonate ($CaCO_3$), the mineral *calcite* (Table 2.4). Calcite is the main constituent of the sedimentary rock *limestone*. A number of other carbonate minerals are known, but only one of these need concern us:

dolomite [$CaMg(CO_3)_2$] is formed by the chemical alteration of calcite by the addition of magnesium. Sedimentary rock composed of the mineral dolomite is *dolostone* (see Chapter 5).

Other Mineral Groups

In addition to silicates and carbonates, several other mineral groups are recognized (Table 2.4). The oxides consist of an element combined with oxygen as in *hematite* (Fe_2O_3). Hematite and another iron oxide called *magnetite* (Fe_3O_4) are both commonly present in small quantities in a variety of rocks. Rocks containing high concentrations of hematite and magnetite, such as those in the Lake Superior region of Canada and the United States, are important sources of iron ores for the manufacture of steel.

The sulfides have a positively charged ion combined with sulfur (S^{-2}), such as in the mineral galena (PbS), which contains lead (Pb) and sulfur (◆ Fig. 2.13a). Sulfates contain an element combined with the complex sulfate ion $(SO_4)^{-2}$; gypsum ($CaSO_4·2H_2O$) is a good example (◆ Fig. 2.13b). The halides contain halogen elements such as chlorine (Cl^{-1}) and fluorine (F^{-1}); examples include the minerals halite (NaCl) and fluorite (CaF_2) (◆ Fig. 2.13c).

(a)

(b)

(c)

◆ FIGURE 2.13 Representative examples of minerals from (a) the sulfides (galena—PbS); (b) the sulfates (gypsum— $CaSO_4 \cdot 2H_2O$); and (c) the halides (halite—NaCl).

All minerals possess characteristic physical properties that are determined by their internal structure and chemical composition. Many physical properties are remarkably constant for a given mineral species, but some, especially color, may vary. Though a professional geologist may use sophisticated techniques in studying and identifying minerals, most common minerals can be identified by using the following physical properties.

Color and Luster

For some minerals, especially those that have the appearance of metals, color is rather consistent, but for many others it varies because of minute amounts of impurities. Although the color of many minerals varies, some generalizations can be made. Ferromagnesian silicates are typically black, brown, or dark green, although olivine is olive green (Fig. 2.11). Nonferromagnesian silicates, on the other hand, can vary considerably in color, but are only rarely dark (Fig. 2.12).

Luster (not to be confused with color) is the appearance of a mineral in reflected light. Two major types of luster are recognized: *metallic* and *nonmetallic* (◆ Fig. 2.14). They are distinguished by observing the quality of light reflected from a mineral and determining if it has the appearance of a metal or a nonmetal. Several types of nonmetallic luster are recognized, including glassy or vitreous, greasy, waxy, brilliant (as in diamond), and dull or earthy.

◆ FIGURE 2.14 Luster is the appearance of a mineral in reflected light. Galena (left), the ore of lead, has the appearance of a metal and is said to have a metallic luster, whereas orthoclase has a nonmetallic luster.

QUARTZ—A COMMON USEFUL MINERAL

During the Middle Ages, quartz crystals were thought to be ice frozen so solidly that they would not melt (◆ Fig. 1). In fact, the term "crystal" is derived from a Greek word meaning ice. Even today, crystal refers not only to transparent quartz, but also to clear, colorless glass of high quality, such as crystal ware, crystal chandeliers, or the transparent glass or plastic cover of a watch or clock dial.

Quartz is a common mineral in the Earth's crust. Most of the sand on beaches, in sand dunes, and in stream channels is quartz. Sand deposits composed mostly of quartz are called silica sands and are used in the manufacture of glass. Quartz is also used in optical equipment, for abrasives such as sandpaper, and in the manufacture of steel alloys.

Quartz occurs in several color varieties. Milky white quartz is a common variety and frequently occurs as well-formed crystals. A milky white quartz crystal weighing 11.8 metric tons and measuring 3.5 m long and 1.7 m in diameter was discovered in Siberia. Color varieties of quartz include amethyst (purple), smoky (smoky brown to black), citrine (yellow to yellowish brown), and rose (pale pink to deep rose) (Fig. 1). Agate is a very finely crystalline variety of quartz commonly used as a decorative stone (Fig. 1d).

Colorless quartz in particular has been used as a semiprecious stone for jewelry. For example, the term "rhinestone" originally referred to transparent quartz crystals used for jewelry made in Germany. Herkimer "diamonds" are simply colorless quartz crystals from Herkimer County, New York. During the past, large, transparent quartz crystals were shaped into spheres for the fortune teller's crystal ball.

The property of piezoelectricity (which literally means "pressure" electricity) is what enables quartz to be such an accurate time-keeper. When pressure is applied to a quartz crystal, an electric current is generated. If an electric current is applied to a quartz crystal, the crystal expands and compresses extremely rapidly and regularly (about 100,000 times per second). In a quartz movement watch, a thin wafer of a quartz crystal vibrates because of the electrical current supplied by the watch's battery.

The first clock driven by a quartz crystal was developed in 1928. Today quartz clocks and watches are commonplace, and even inexpensive quartz timepieces are extremely accurate. Precision-manufactured quartz clocks used in astronomical observatories do not gain or lose more than one second every 10 years.

An interesting historical note regarding quartz is that during World War II the United States had difficulty obtaining quartz crystals from Brazil needed for making radios. This shortage prompted the development of artificially synthesized quartz, and now most of the quartz used in watches and clocks is synthetic.

Crystal Form

As previously noted, mineral crystals are rare. Thus, many mineral specimens will not show the perfect crystal form typical of that mineral species. Some minerals do typically occur as crystals. For example, 12-sided crystals of garnet are common, as are 6- and 12-sided crystals of pyrite (◆ Fig. 2.15). Minerals that grow in cavities or are precipitated from circulating hot water (hydrothermal solutions) in cracks and crevices in rocks also commonly occur as crystals.

Crystal form can be a very useful characteristic for mineral identification, but a number of minerals have the same crystal form. For example, pyrite (FeS_2), galena (PbS), and halite ($NaCl$) all occur as cubic crystals. However, such minerals can usually be easily identified by other properties such as color, luster, hardness, and density.

Cleavage and Fracture

Cleavage is a property of individual mineral crystals. Not all minerals possess cleavage, but those that do tend to break, or split, along a smooth plane or planes of weakness determined by the strength of the bonds

within the mineral crystal. Cleavage can be characterized in terms of quality (perfect, good, poor), direction, and angles of intersection of cleavage planes. Biotite, a common ferromagnesian silicate, has perfect cleavage in one direction (◆ Fig. 2.16a). The fact that biotite preferentially cleaves along a number of closely spaced, parallel planes is related to its structure; it is a sheet silicate with the sheets of silica tetrahedra weakly bonded to one another by iron and magnesium ions (Fig. 2.10c).

Feldspars possess two directions of cleavage that intersect at right angles (◆ Fig. 2.16b), and the mineral halite has three directions of cleavage, all of which intersect at right angles (◆ Fig. 2.16c). Calcite also possesses three directions of cleavage, but none of the intersection angles is a right angle, so cleavage fragments of calcite are rhombohedrons (◆ Fig. 2.16d). Minerals with four directions of cleavage include fluorite and diamond (◆ Fig. 2.16e). Ironically, diamond, the hardest mineral, can be easily cleaved (see Perspective 2.2). A few minerals such as sphalerite, an ore of zinc, have six directions of cleavage (◆ Fig. 2.16f).

(a)

(b)

◆ FIGURE 2.15 (*a*) Crystals of pyrite from Spain. (National Museum of Natural History, Smithsonian Institution.) (*b*) Garnet crystals from Alaska.

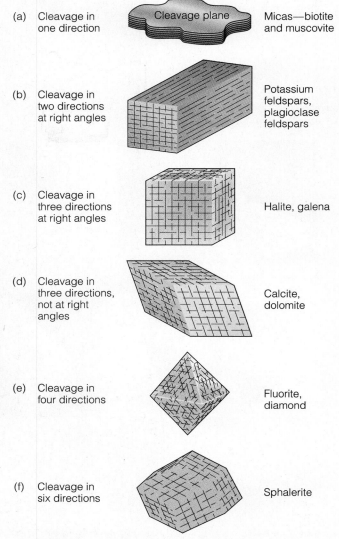

(a) Cleavage in one direction — Micas—biotite and muscovite

Cleavage plane

(b) Cleavage in two directions at right angles — Potassium feldspars, plagioclase feldspars

(c) Cleavage in three directions at right angles — Halite, galena

(d) Cleavage in three directions, not at right angles — Calcite, dolomite

(e) Cleavage in four directions — Fluorite, diamond

(f) Cleavage in six directions — Sphalerite

◆ FIGURE 2.16 Several types of mineral cleavage: (*a*) one direction; (*b*) two directions at right angles; (*c*) three directions at right angles; (*d*) three directions, not at right angles; (*e*) four directions; and (*f*) six directions.

Cleavage is a very important diagnostic property of minerals, and its recognition is essential in distinguishing between some minerals. For example, the pyroxene mineral augite and the amphibole mineral hornblende look much alike: both are generally dark green to black, have the same hardness, and possess two directions of cleavage. However, the cleavage planes of augite intersect at about 90°, whereas the cleavage planes of hornblende intersect at angles of 56° and 124° (◆ Fig. 2.17).

In contrast to cleavage, *fracture* is mineral breakage along irregular surfaces. Any mineral can be fractured if

enough force is applied. The fracture surfaces will not be smooth, however; they are commonly uneven or conchoidal (smoothly curved).

Hardness

Hardness is the resistance of a mineral to abrasion. An Austrian geologist, Friedrich Mohs, devised a relative hardness scale for 10 minerals. He arbitrarily assigned a

DIAMONDS AND PENCIL LEADS

You may be surprised to learn that diamonds and pencil "lead" (graphite) are composed of the same substance, carbon. Both diamonds and graphite are crystalline solids and are therefore minerals; because they each contain only a single element, they are also native elements. Other than composition, however, diamond and graphite have little in common: diamond is the hardest mineral, whereas graphite is so soft that it can be scratched by a fingernail; diamond may be colorless, red, yellow, blue, gray, or black, while graphite is invariably steel gray (◆ Fig. 1). Obviously, the same chemical substance occurs in vastly different forms, so what could possibly control such differences?

Diamond and graphite differ mostly because of their internal structure—both are crystalline but the atoms within crystals of diamond and graphite are arranged quite differently. Such minerals sharing the same composition but differing in structure are called *polymorphs* (poly = many; morph = shape or form). Notice in Figure 2.5 that in a diamond crystal the carbon atoms are arranged such that all of them are bonded to one another. In graphite the carbon atoms are bonded together to form sheets, but the sheets are weakly held together by van der Waals bonds

(Fig. 2.5c). Graphite can be used for pencil leads because it has good cleavage in one direction. When a pencil lead is moved across a piece of paper, small pieces of graphite flake off along the planes held together by van der Waals bonds and adhere to the paper.

Most of the diamonds mined are not of gem quality and are used in such industrial applications as diamond drill bits, diamond-tipped cutting blades, or abrasives. Most gem-quality diamonds are mined in South Africa, although in terms of total diamond production South Africa is in fifth place, with Australia being the largest producer.

How does one "cut" a diamond, the hardest substance known? Diamond cutting is actually done by several processes, one of which is cleaving. Diamond possesses four directions of cleavage, and if a diamond is cleaved such that all four cleavage planes are perfectly developed, the resulting "stone" will be shaped like two pyramids placed base to base. Diamonds are cleaved by placing a knife parallel with a cleavage plane and then tapping the knife with a mallet. Large diamonds are commonly preshaped by cleaving them into smaller pieces that are then further shaped by sawing and grinding with diamond dust.

(a)

(b)

◆ FIGURE 1 Two minerals composed of carbon. (*a*) Graphite. (Photo courtesy of Sue Monroe.) (*b*) The Oppenheimer diamond. (National Museum of Natural History, Smithsonian Institution.)

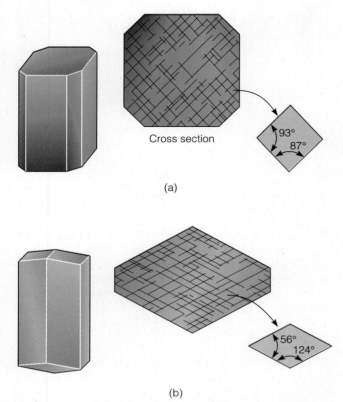

Cross section

(a)

(b)

◆ FIGURE 2.17 Cleavage in augite and hornblende. (*a*) Augite crystal and cross section of crystal showing cleavage. (*b*) Hornblende crystal and cross section of crystal showing cleavage.

hardness value of 10 to diamond, the hardest mineral known, and lesser values to the other minerals. Relative hardness can be determined easily by the use of Mohs hardness scale (● Table 2.5). For example, quartz will scratch fluorite but cannot be scratched by fluorite, gypsum can be scratched by a fingernail, and so on. Hardness is controlled mostly by internal structure. For example, both graphite and diamond are composed of carbon, but the former has a hardness of 1 to 2 whereas the latter has a hardness of 10.

Specific Gravity

The *specific gravity* of a mineral is the ratio of its weight to the weight of an equal volume of water. A mineral with a specific gravity of 3.0 is three times as heavy as water. Like all ratios, specific gravity is not expressed in units such as grams per cubic centimeter—it is a dimensionless number.

● **TABLE 2.5** Mohs Hardness Scale

Hardness	Mineral	Hardness of Some Common Objects
10	Diamond	
9	Corundum	
8	Topaz	
7	Quartz	
		Steel file (6½)
6	Orthoclase	
		Glass (5½–6)
5	Apatite	
4	Fluorite	
3	Calcite	Copper penny (3)
		Fingernail (2½)
2	Gypsum	
1	Talc	

Specific gravity varies in minerals depending upon their composition and structure. Among the common silicates, for example, the ferromagnesian silicates have specific gravities ranging from 2.7 to 4.3, whereas the nonferromagnesian silicates vary from 2.6 to 2.9. Obviously, the ranges of values overlap somewhat, but for the most part ferromagnesian silicates have greater specific gravities than nonferromagnesian silicates. In general, the metallic minerals, such as galena (7.58) and hematite (5.26), are heavier than nonmetals. Structure as a control of specific gravity is illustrated by the native element carbon (C): the specific gravity of graphite varies from 2.09 to 2.33; that of diamond is 3.5.

Other Properties

A number of other physical properties characterize some minerals. For example, talc has a distinctive soapy feel, graphite writes on paper, halite tastes salty, and magnetite is magnetic. Calcite possesses the property of *double refraction,* meaning that an object when viewed through a transparent piece of calcite will have a double image (◆ Fig. 2.18). Some minerals are plastic and, when bent into a new shape, will retain that shape, whereas others are flexible and, if bent, will return to their original position when the forces that bent them are removed.

A simple chemical test to identify the minerals calcite and dolomite involves applying a drop of dilute hydrochloric acid to the mineral specimen. If the mineral is calcite, it will react vigorously with the acid and release carbon dioxide, which causes the acid to bubble or effervesce. Dolomite, on the other hand, will not react with hydrochloric acid unless it is powdered.

⩔ IMPORTANT ROCK-FORMING MINERALS

Rocks are generally defined as solid aggregates of grains of one or more minerals. Two important exceptions to this definition are natural glass such as obsidian (see Chapter 3) and the sedimentary rock coal (see Chapter 5). Although it is true that many minerals occur in various kinds of rocks, only a few varieties are common enough to be designated as **rock-forming minerals**. Most of the others occur in such small amounts that they can be disregarded in the identification and classification of rocks; these are generally called *accessory minerals.* Granite, an igneous rock consisting largely of potassium feldspar and quartz, commonly contains such accessory minerals as sodium plagioclase, biotite, hornblende, muscovite, and, rarely, pyroxene (◆ Fig. 2.19).

Most rocks are composed of silicate minerals. However, only a few of the hundreds of known silicates are common in rocks, although many occur as accessories. The common rock-forming silicates are summarized in ● Table 2.6.

The most common nonsilicate rock-forming minerals are the two carbonates, calcite ($CaCO_3$) and dolomite [$CaMg(CO_3)_2$], the primary constituents of the sedimentary rocks limestone and dolostone, respectively. Among the sulfates and halides, gypsum ($CaSO_4 \cdot 2H_2O$) and halite (NaCl), respectively, are the only common rock-forming minerals.

⩔ MINERAL RESOURCES AND RESERVES

Geologists of the U.S. Geological Survey and the U.S. Bureau of Mines define a **resource** as follows:

A concentration of naturally occurring solid, liquid, or gaseous material in or on the Earth's crust in such form and amount that economic extraction of a commodity from the concentration is currently or potentially feasible.

◆ FIGURE 2.18 Calcite showing double refraction.

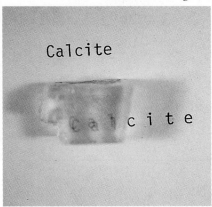

◆ FIGURE 2.19 The igneous rock granite is composed largely of potassium feldspar and quartz, lesser amounts of plagioclase feldspar, and accessory minerals such as biotite mica. (*a*) Hand specimen of granite. (*b*) Photomicrograph showing the various minerals.

(a)

(b)

● **TABLE 2.6** Rock-Forming Minerals

Mineral	Composition	Primary Occurrence
FERROMAGNESIAN SILICATES		
Olivine	$(Mg,Fe)_2SiO_4$	Igneous, metamorphic rocks
Pyroxene group		
Augite most common	Ca, Mg, Fe, Al silicate	Igneous, metamorphic rocks
Amphibole group		
Hornblende most common	Hydrous* Na, Ca, Mg, Fe, Al silicate	Igneous, metamorphic rocks
Biotite	Hydrous K, Mg, Fe silicate	All rock types
NONFERROMAGNESIAN SILICATES		
Quartz	SiO_2	All rock types
Potassium feldspar group		
Orthoclase, microcline	$KAlSi_3O_8$	All rock types
Plagioclase feldspar group	Varies from $CaAl_2Si_2O_8$ to $NaAlSi_3O_3$	All rock types
Muscovite	Hydrous K, Al silicate	All rock types
Clay mineral group	Varies	Soils and sedimentary rocks
CARBONATES		
Calcite	$CaCO_3$	Sedimentary rocks
Dolomite	$CaMg(CO_3)_2$	Sedimentary rocks
SULFATES		
Anhydrite	$CaSO_4$	Sedimentary rocks
Gypsum	$CaSO_4 \cdot 2H_2O$	Sedimentary rocks
HALIDES		
Halite	$NaCl$	Sedimentary rocks

*Contains elements of water in some kind of union.

Accordingly, resources include such substances as metals (*metallic resources*), sand, gravel, crushed stone, and sulfur (*nonmetallic resources*), and uranium, coal, oil, and natural gas (*energy resources*). However, an important distinction must be made between a resource, which is the total amount of a commodity whether discovered or undiscovered, and a **reserve**, which is that part of the resource base that can be extracted economically.

The amount of mineral resources used has steadily increased since Europeans settled this continent, and during 1988, each resident of North America used about 14 metric tons, a large part of which was bulk items such as sand, gravel, and crushed stone (◆ Fig. 2.20). It is no exaggeration to say that our industrialized societies are totally dependent on mineral resources. Unfortunately, they are being used at rates far faster than they form. Thus, mineral resources are nonrenewable, meaning that once the resources from a deposit have been exhausted, new supplies or suitable substitutes must be found.

For some mineral resources, adequate supplies are available for indefinite periods (sand and gravel, for example), whereas for others, supplies are limited or must be imported from other parts of the world (◆ Fig. 2.21). For example, the United States is almost totally dependent on imports of manganese, an essential element in the manufacture of steel. Even though the United States is a leading producer of gold, it still depends on imports for more than half of its gold needs. More than half of the crude oil used in the United States is imported, much of it from the Middle East where more than 50% of the Earth's proven reserves exist. A poignant reminder of our dependence on the availability of resources was the United States'

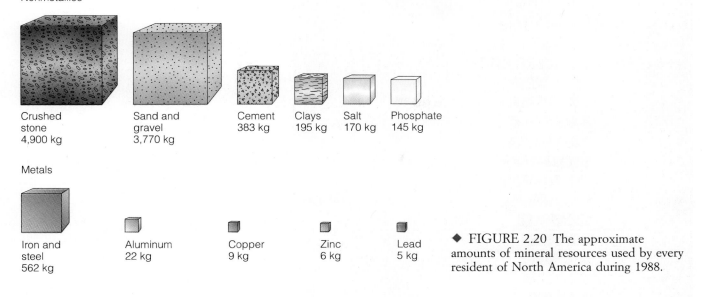

Nonmetallics

Crushed stone 4,900 kg

Sand and gravel 3,770 kg

Cement 383 kg

Clays 195 kg

Salt 170 kg

Phosphate 145 kg

Metals

Iron and steel 562 kg

Aluminum 22 kg

Copper 9 kg

Zinc 6 kg

Lead 5 kg

◆ FIGURE 2.20 The approximate amounts of mineral resources used by every resident of North America during 1988.

response to the takeover of Kuwait by Iraq during August 1990.

In terms of mineral and energy resources, Canada is more self-reliant than the United States. It meets most of its domestic needs for mineral resources, although it must import phosphate, chromium, manganese, and bauxite, the ore of aluminum. Canada also produces more crude oil and natural gas than it uses, and it is the world leader in the production and export of uranium.

What constitutes a resource as opposed to a reserve depends on several factors. For example, iron-bearing minerals occur in many rocks, but in quantities or ways that make their recovery uneconomical. As a matter of fact, most minerals that are concentrated in economic quantities are mined in only a few areas; 75% of all the metals mined in the world come from about 150 locations. Geographic location is also an important consideration. A mineral resource in a remote region may not be mined because transportation costs are too high, and what may be considered a resource in the United States or Canada may be mined in a third-world country where labor costs are low. The market price of a commodity is, of course, important in evaluating a potential resource. From 1935 to 1968, the United States government maintained the price of gold at $35 per troy ounce (= 31.1 g). When this restriction was removed and the price of gold became subject to supply and demand, the price rose (it reached an all-time high of $843 per troy ounce during January 1980). As a consequence, many marginal deposits became reserves and many abandoned mines were reopened.

Technological developments can also change the status of a resource. For example, the rich iron ores of the Great Lakes region of the United States and Canada had been depleted by World War II. However, the development of a method of separating the iron from previously unusable rocks and shaping it into pellets that are ideal for use in blast furnaces made it feasible to mine poorer grade ores.

Most of the largest and richest mineral deposits have probably already been discovered and, in some cases, depleted. In order to ensure continued supplies of essential minerals, geologists are using increasingly sophisticated geophysical and geochemical mineral exploration techniques. The U.S. Geological Survey and the U.S. Bureau of Mines continually assess the status of resources in view of changing economic and political conditions and developments in science and technology. In the following chapters, we will discuss the origin and distribution of various mineral resources and reserves.

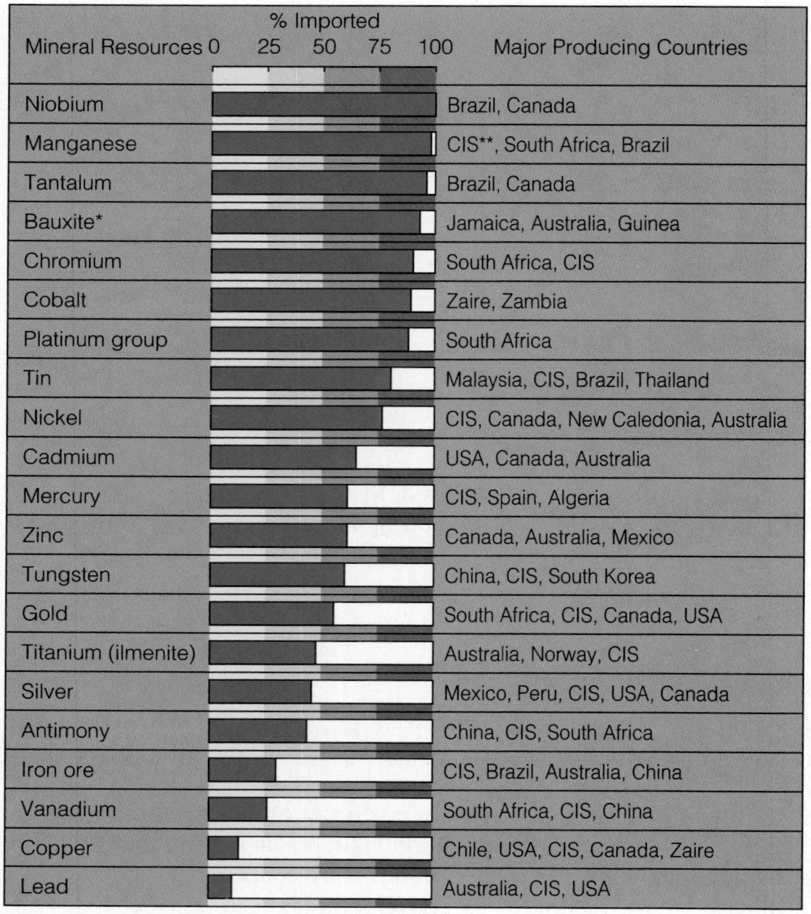

Mineral Resources	% Imported 0 25 50 75 100	Major Producing Countries
Niobium		Brazil, Canada
Manganese		CIS**, South Africa, Brazil
Tantalum		Brazil, Canada
Bauxite*		Jamaica, Australia, Guinea
Chromium		South Africa, CIS
Cobalt		Zaire, Zambia
Platinum group		South Africa
Tin		Malaysia, CIS, Brazil, Thailand
Nickel		CIS, Canada, New Caledonia, Australia
Cadmium		USA, Canada, Australia
Mercury		CIS, Spain, Algeria
Zinc		Canada, Australia, Mexico
Tungsten		China, CIS, South Korea
Gold		South Africa, CIS, Canada, USA
Titanium (ilmenite)		Australia, Norway, CIS
Silver		Mexico, Peru, CIS, USA, Canada
Antimony		China, CIS, South Africa
Iron ore		CIS, Brazil, Australia, China
Vanadium		South Africa, CIS, China
Copper		Chile, USA, CIS, Canada, Zaire
Lead		Australia, CIS, USA

* Ore of aluminum.
** Commonwealth of Independent States (includes much of the former Soviet Union).

◆ FIGURE 2.21 The percentages of some mineral resources imported by the United States. The lengths of the blue bars correspond to the percent of resources imported.

Chapter Summary

1. All matter is composed of chemical elements, each of which consists of atoms. Individual atoms consist of a nucleus, containing protons and neutrons, and electrons that circle the nucleus in electron shells.

2. Atoms are characterized by their atomic number (the number of protons in the nucleus) and their atomic mass number (the number of protons plus the number of neutrons in the nucleus).

3. Bonding is the process whereby atoms are joined to other atoms. If atoms of different elements are bonded, they form a compound. Ionic and covalent bonds are most common in minerals, but metallic and van der Waals bonds also occur in a few.

4. Most minerals are compounds, but a few, including gold and silver, are composed of a single element and are called native elements.

5. All minerals are crystalline solids, meaning that they possess an orderly internal arrangement of atoms.

6. Some minerals vary in chemical composition because atoms of different elements can substitute for one another provided that the electrical charge is balanced and the atoms are of about the same size.

7. Of the more than 3,500 known minerals, most are silicates. Ferromagnesian silicates contain iron (Fe) and magnesium (Mg), and nonferromagnesian silicates lack these elements.

8. In addition to silicates, several other mineral groups are recognized, including carbonates, oxides, sulfides, sulfates, and halides.
9. The physical properties of minerals such as color, hardness, cleavage, and crystal form are controlled by composition and structure.
10. A few minerals are common enough constituents of rocks to be designated rock-forming minerals.
11. Many resources are concentrations of minerals of economic importance.
12. Reserves are that part of the resource base that can be extracted economically.

Important Terms

atom
atomic mass number
atomic number
bonding
carbonate mineral
cleavage
compound
covalent bond
crystalline solid
electron

electron shell
element
ferromagnesian silicate
ion
ionic bond
isotope
mineral
native element
neutron
nonferromagnesian silicate

nucleus
proton
reserve
resource
rock
rock-forming mineral
silica
silica tetrahedron
silicate

Review Questions

1. The atomic number of an element is determined by the:
   a. ___ number of electrons in its outermost shell;
   b. ___ number of protons in its nucleus; c. ___ diameter of its most common isotope; d. ___ number of neutrons plus electrons in its nucleus; e. ___ total number of neutrons orbiting the nucleus.
2. To which of the following groups do most minerals in the Earth's crust belong?
   a. ___ oxides; b. ___ carbonates; c. ___ sulfates;
   d. ___ halides; e. ___ silicates.
3. When an atom loses or gains electrons, it becomes a(n):
   a. ___ isotope; b. ___ proton; c. ___ ion; d. ___ neutron;
   e. ___ native element.
4. The two most abundant elements in the Earth's crust are:
   a. ___ iron and magnesium; b. ___ carbon and potassium; c. ___ sodium and nitrogen; d. ___ silicon and oxygen; e. ___ sand and clay.
5. The sharing of electrons by adjacent atoms is a type of bonding called:
   a. ___ van der Waals; b. ___ covalent; c. ___ silicate;
   d. ___ tetrahedral; e. ___ ionic.
6. A chemical element is a substance made up of atoms, all of which have the same:
   a. ___ atomic mass number; b. ___ number of neutrons;
   c. ___ number of protons; d. ___ size; e. ___ weight.
7. Many minerals break along closely spaced planes and are said to possess:
   a. ___ specific gravity; b. ___ cleavage; c. ___ covalent bonds; d. ___ fracture; e. ___ double refraction.

8. The chemical formula for olivine is $(Mg,Fe)_2SiO_4$, which means that:
   a. ___ magnesium and iron can substitute for one another; b. ___ magnesium is more common than iron;
   c. ___ magnesium is heavier than iron; d. ___ all olivine contains both magnesium and iron; e. ___ more magnesium than iron occurs in the Earth's crust.
9. The basic building block of all silicate minerals is the:
   a. ___ silicon sheet; b. ___ oxygen silicon cube;
   c. ___ silica tetrahedron; d. ___ silicate double chain;
   e. ___ silica framework.
10. An example of a common nonferromagnesian silicate mineral is:
    a. ___ calcite; b. ___ quartz; c. ___ biotite; d. ___ hematite; e. ___ halite.
11. The ratio of a mineral's weight to the weight of an equal volume of water is its:
    a. ___ specific gravity; b. ___ luster; c. ___ hardness;
    d. ___ atomic mass number; e. ___ cleavage.
12. Those chemical elements having eight electrons in their outermost electron shell are the:
    a. ___ noble gases; b. ___ native elements; c. ___ carbonates; d. ___ halides; e. ___ isotopes.
13. Minerals are solids possessing an orderly internal arrangement of atoms, meaning that they are:
    a. ___ amorphous substances; b. ___ crystalline;
    c. ___ composed of at least three different elements;
    d. ___ composed of a single element; e. ___ ionic compounds.

14. The silicon ion has a positive charge of 4, and oxygen has a negative charge of 2. Accordingly, the ion group (SiO_4) has a:
    a. ___ positive charge of 2; b. ___ negative charge of 2;
    c. ___ negative charge of 1; d. ___ positive charge of 4;
    e. ___ negative charge of 4.
15. Calcite and dolomite are:
    a. ___ oxide minerals of great value; b. ___ ferromagnesian silicates possessing a distinctive sheet structure;
    c. ___ common rock-forming carbonate minerals;
    d. ___ minerals used in the manufacture of pencil leads;
    e. ___ important energy resources.
16. How does a crystalline solid differ from a liquid and a gas?
17. An atom of the element magnesium is shown below. If the two electrons in its outer electron shell are lost, what is the electrical charge of the magnesium ion?

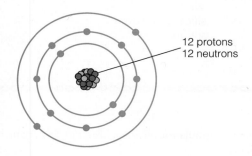

12 protons
12 neutrons

18. What is the atomic mass number of the magnesium atom shown above?
19. Compare ionic and covalent bonding.
20. Define compound and native element.
21. What accounts for the fact that some minerals have a range of chemical compositions?
22. Why are the angles between the same crystal faces on all specimens of a mineral species always the same?
23. What is a silicate mineral? How do the two subgroups of silicate minerals differ from one another?
24. In sheet silicates, individual sheets composed of silica tetrahedra possess a negative electrical charge. How is this negative charge satisfied?
25. What do all carbonate minerals have in common?
26. Describe the mineral property of cleavage, and explain what controls cleavage.
27. What are rock-forming minerals?

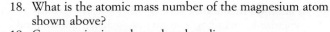

Additional Readings

Berry, L. G., B. Mason, and R. V. Dietrich. 1983. *Mineralogy.* 2d ed. San Francisco: W. H. Freeman.

Blackburn, W. H., and W. H. Dennen. 1988. *Principles of mineralogy.* Dubuque, Iowa: William C. Brown.

Dietrich, R. V., and B. J. Skinner. 1979. *Rocks and rock minerals.* New York: Wiley.

_____. 1990. *Gems, granites, and gravels: Knowing and using rocks and minerals.* New York: Cambridge Univ. Press.

Klein, C., and C. S. Hurlbut Jr. 1985. *Manual of mineralogy* (after James D. Dana). 20th ed. New York: Wiley.

Pough, F. H. 1987. *A field guide to rocks and minerals.* 4th ed. Boston: Houghton Mifflin.

Vanders, I., and P. F. Kerr. 1967. *Mineral recognition.* New York: Wiley.

CHAPTER
3

Intrusive igneous rock exposed in Yosemite National Park, California.

Igneous Rocks and Intrusive Igneous Activity

Prologue

About 45 to 50 million years ago, several small masses of molten rock were intruded into the Earth's crust in what is now northeastern Wyoming. These cooled and solidified, forming igneous rock bodies. The best known of these, Devil's Tower, was established as our first national monument by President Theodore Roosevelt in 1906. Devil's Tower is a remarkable landform. It rises nearly 260 m above its base (◆ Fig. 3.1) and is visible from 48 km away.

Devil's Tower and other similar nearby bodies are important in the legends of the Cheyenne and Lakota Sioux Indians. These native Americans call Devil's Tower *Mateo Tepee,* which means "Grizzly Bear Lodge." It was also called the "Bad God's Tower," and reportedly, "Devil's Tower" is a translation of this phrase. According to one Indian legend, the tower formed when the Great Spirit caused it to rise up from the ground, carrying with it several children who were trying to escape from a gigantic grizzly bear. Another legend tells of six brothers and a woman who were also being pursued by a grizzly bear. The youngest brother carried a small rock, and when he sang a song, the rock grew to the present size of Devil's Tower. In both legends, the bear's attempts to reach the Indians left deep scratch marks in the tower's rocks (◆ Fig. 3.2).

Geologists have a less dramatic explanation for the tower's origin. The near vertical striations (the bear's scratch marks) are simply the lines formed by the intersections of columnar joints. Columnar joints form in response to cooling and contraction in some igneous bodies and in some lava flows. Many of the columns are six sided, but columns with four, five, and seven sides occur as well. The larger columns measure about 2.5 m across. The pile of rubble at the tower's base is an accumulation of columns that have fallen from the tower.

Geologists agree that Devil's Tower originated as a small body intruded into the Earth's crust, and that subsequent erosion exposed it in its present form. The type of igneous body and the extent of its modification by erosion are debatable, however. Some geologists believe that Devil's Tower is the eroded remnant of a more extensive body of intrusive rock, whereas others think it is simply the remnant of the magma that solidified in a pipelike conduit of a volcano and that it has been little modified by erosion.

◆ FIGURE 3.1 Devil's Tower in Wyoming exhibits well-developed columnar jointing. (Photo courtesy of R. V. Dietrich.)

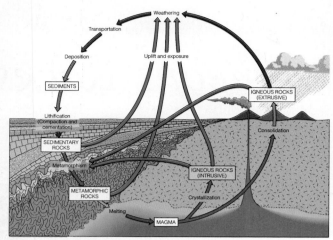

◆ FIGURE 3.3 The rock cycle, with emphasis on intrusive igneous rocks.

◆ FIGURE 3.2 An artist's rendition of a Cheyenne legend about the origin of Devil's Tower.

⬇ INTRODUCTION

Rocks resulting from volcanic eruptions are widespread, but they probably represent only a small portion of the total rocks formed by the cooling and crystallization of molten rock material called magma. Most magma cools below the Earth's surface and forms bodies of rock known as *plutons*. The same types of magmas are involved in both volcanism and plutonism, although some magmas, because of their greater mobility, more commonly reach the surface. Plutons typically underlie areas of extensive volcanism and were the sources of the overlying lavas and fragmental materials ejected from volcanoes during explosive eruptions. Furthermore, like volcanism, most plutonism occurs at or near plate margins. In this chapter we are concerned primarily with the textures, composition, and classification of igneous rocks and with plutonic or intrusive igneous activity (◆ Fig. 3.3). Although volcanism (Chapter 4) and intrusive igneous activity (Chapter 3) are discussed in separate chapters, they are related phenomena.

⬇ MAGMA AND LAVA

Magma is molten rock material below the Earth's surface, and **lava** is magma at the Earth's surface. Magma is less dense than the solid rock from which it was derived, thus it tends to move upward toward the surface. Some magma is erupted onto the surface as **lava flows,** and some is forcefully ejected into the atmosphere as particles called **pyroclastic materials** (from the Greek *pyro,* "fire," and *klastos, "broken").*

Igneous rocks (from the Latin *ignis,* meaning fire) form when magma cools and crystallizes, or when pyroclastic materials such as volcanic ash become consolidated. Magma extruded onto the Earth's surface as lava and pyroclastic materials forms **volcanic** or **extrusive igneous** rocks, whereas magma that crystallizes within the Earth's crust forms **plutonic** or **intrusive igneous** rocks. (Fig. 3.3).

Composition

Recall from Chapter 2 that the most abundant minerals in the Earth's crust are silicates, composed of silicon, oxygen and the other elements listed in Table 2.3. Accordingly, when crustal rocks melt and form magma, the magma is typically silica rich and also contains considerable aluminum, calcium, sodium, iron, magnesium and potassium as well as many other elements in lesser quantities. Not all magmas originate by melting of crustal rocks, however; some are derived from upper mantle rocks that are composed largely of ferromagnesian silicates. A magma from this source contains comparatively less silica and more iron and magnesium.

Although silica is the primary constituent of nearly all magmas, silica content varies and serves to distinguish **felsic, intermediate,** and **mafic** magmas (● Table 3.1). A felsic magma, for example, contains more than 65% silica and considerable sodium, potassium, and aluminum, but little calcium, iron, and magnesium. Cooling of felsic magma yields igneous rocks, such as rhyolite and granite, which are composed largely of the nonferromagnesian silicates potassium feldspar, sodium-rich plagioclase, and quartz (Table 3.1).

In contrast to felsic magmas, mafic magmas are silica poor, and contain proportionately more calcium, iron, and magnesium. When mafic magma cools and crystallizes, it yields igneous rocks such as basalt and gabbro, which contain high percentages of ferromagnesian silicates and calcium plagioclase (Table 3.1). As one would expect, igneous rocks that crystallize from intermediate magmas have mineral compositions that are intermediate between those of mafic and felsic rocks (Table 3.1).

Temperature

No direct measurements of temperatures of magma below the Earth's surface have been made. Erupting lavas generally have temperatures in the range of 1,000° to 1,200°C, although temperatures of 1,350°C have been recorded above Hawaiian lava lakes where volcanic gases reacted with the atmosphere.

Most direct temperature measurements have been taken at volcanoes characterized by little or no explosive activity where geologists can safely approach the lava. Therefore, little is known of the temperatures of felsic lavas, because eruptions of such lavas are rare, and when they do occur, they tend to be explosive. The temperatures of some lava domes, most of which are bulbous masses of felsic magma, have been measured at a distance by using an instrument called an optical pyrometer. The surfaces of these domes have temperatures up to 900°C, but the exterior of a dome is probably much cooler than its interior.

When Mount St. Helens erupted in 1980, it ejected felsic magma as particulate matter in pyroclastic flows. Two weeks later, these flows still had temperatures between 300° and 420°C.

Viscosity

Magma is also characterized by its **viscosity,** or resistance to flow. The viscosity of some liquids, such as water, is very low; thus, they are highly fluid and flow readily. The viscosity of some other liquids is so high, however, that they flow much more slowly. Motor oil and syrup flow readily when they are hot, but become stiff and flow very slowly when they are cold. Thus, one might expect that temperature controls the viscosity of magma, and such an inference is partly correct. We can generalize and say that hot lava flows more readily than cooler lava. However, temperature is not the only control of viscosity.

Magma viscosity is strongly controlled by silica content. In a felsic lava, numerous networks of silica tetrahedra retard flow, because the strong bonds of the networks must be ruptured for flow to occur. Mafic lavas, on the other hand, contain fewer silica tetrahedra networks and consequently flow more readily. Felsic lavas form thick, slow-moving flows, whereas mafic lavas tend to form thinner flows that move rather rapidly over great distances. One such flow in Iceland in 1783 flowed about 80 km, and some ancient flows in the state of Washington can be traced for more than 500 km.

⍗ IGNEOUS ROCKS

All intrusive and many extrusive igneous rocks form when minerals crystallize from magma. The process of crystallization involves the formation of crystal nuclei

● **TABLE 3.1** The Most Common Types of Magmas and Their Characteristics

Type of Magma	Silica Content (%)	Crystallizes to Form	
		Volcanic Rock	Plutonic Rock
Mafic	45—52%	Basalt	Gabbro
Intermediate	53—65	Andesite	Diorite
Felsic	> 65	Rhyolite	Granite

and subsequent growth of these nuclei. The atoms in a magma are in constant motion, but when cooling begins, some atoms bond to form small groups, or nuclei, whose arrangement of atoms corresponds to the arrangement in mineral crystals. As other atoms in the liquid chemically bond to these nuclei, they do so in an ordered geometric arrangement, and the nuclei grow into crystalline *mineral grains,* the individual particles that comprise a rock. During rapid cooling, the rate of nuclei formation exceeds the rate of growth, and an aggregate of many small grains results (◆ Fig. 3.4a). With slow cooling, the rate of growth exceeds the rate of nucleation, so relatively large grains form (◆ Fig. 3.4b).

Textures

Several textures of igneous rocks are related to the cooling history of a magma or lava. For example, rapid cooling, as occurs in lava flows or some near-surface intrusions, results in a fine-grained texture termed **aphanitic.** In an aphanitic texture, individual mineral grains are too small to be observed without magnification (◆ Fig. 3.5a). In contrast, igneous rocks with a

◆ FIGURE 3.4 The effect of the cooling rate of a magma on nucleation and growth of crystals: (*a*) Rapid cooling results in many small grains and a fine-grained texture. (*b*) Slow cooling results in a coarse-grained texture.

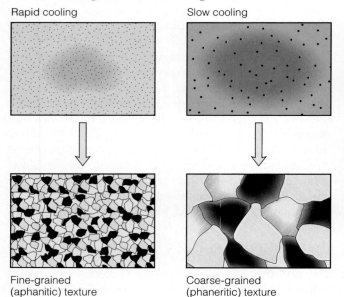

Rapid cooling   Slow cooling

Fine-grained (aphanitic) texture

(a)

Coarse-grained (phaneritic) texture

(b)

(a)

(b)

(c)

◆ FIGURE 3.5 Textures of igneous rocks. (*a*) Aphanitic or fine-grained texture in which individual minerals are too small to be seen without magnification. (*b*) Phaneritic or coarse-grained texture in which minerals are easily discerned without magnification. (*c*) Porphyritic texture consisting of minerals of markedly different sizes. (Photos courtesy of Sue Monroe.)

coarse-grained or **phaneritic** texture have mineral grains that are easily visible without magnification (◆ Fig. 3.5b). Such large mineral grains indicate slow cooling and generally an intrusive origin; a phaneritic texture can develop in the interiors of some thick lava flows as well.

Rocks with **porphyritic** textures have a somewhat more complex cooling history. Such rocks have a combination of mineral grains of markedly different sizes. The larger grains are **phenocrysts,** and the smaller ones are referred to as *groundmass* (◆ Fig. 3.5c). Suppose that a magma begins cooling slowly as an intrusive body, and that some mineral-crystal nuclei form and begin to grow. Suppose further that before the magma has completely crystallized, the remaining liquid phase and solid mineral grains within it are extruded onto the Earth's surface where it cools rapidly, forming an aphanitic texture. The resulting igneous rock would have large mineral grains (phenocrysts) suspended in a finely crystalline groundmass, and the rock would be characterized as a *porphyry.*

A lava may cool so rapidly that its constituent atoms do not have time to become arranged in the ordered, three-dimensional frameworks typical of minerals. As a consequence of such rapid cooling, a *natural glass* such as *obsidian* forms (◆ Fig. 3.6a). Even though obsidian is not composed of minerals, it is still considered to be igneous rock.

Some magmas contain large amounts of water vapor and other gases. These gases may be trapped in cooling lava where they form numerous small holes or cavities called **vesicles**; rocks possessing numerous vesicles are termed *vesicular,* as in vesicular basalt (◆ Fig. 3.6b). The igneous rock known as *scoria* contains more cavities than solid rock (◆ Fig. 3.6c).

A **pyroclastic** or **fragmental texture** characterizes igneous rocks formed by explosive volcanic activity. For example, ash may be discharged high into the atmosphere and eventually settle to the surface where it accumulates; if it is turned into solid rock, it is considered to be a pyroclastic igneous rock.

Composition

Magmas are characterized as mafic (45–52% silica), intermediate (53–65% silica), or felsic (> 65% silica) (see Table 3.1). The parent magma plays a significant role in determining the mineral composition of igneous rocks. However, it is possible for the same magma to yield different igneous rocks because its composition can change as a consequence of crystal settling, assimi-

(a)

(b)

(c)

◆ FIGURE 3.6 (*a*) The natural glass obsidian forms when lava cools too quickly for mineral crystals to form. (*b*) Vesicular texture. (*c*) Scoria contains more vesicles than solid rock. (Photos courtesy of Sue Monroe.)

lation, magma mixing, and the sequence in which minerals crystallize.

Bowen's Reaction Series

During the early part of this century, N. L. Bowen hypothesized that mafic, intermediate, and felsic magmas could all derive from a parent mafic magma. He knew that minerals do not all crystallize simultaneously from a magma. Based on his observations and laboratory experiments, Bowen proposed a mechanism, now called **Bowen's reaction series,** to account for the derivation of intermediate and felsic magmas from a basaltic (mafic) magma (◆ Fig. 3.7). Bowen's reaction series consists of two branches: a *discontinuous branch* and a *continuous branch* (Fig. 3.7). Crystallization of minerals occurs along both branches simultaneously, but for convenience we will discuss them separately.

In the discontinuous branch, which contains only ferromagnesian minerals, one mineral changes to another over specific temperature ranges (Fig. 3.7). As the temperature decreases, a temperature range is reached in which a given mineral begins to crystallize. However, a previously formed mineral reacts with the remaining liquid magma (the melt) such that it forms the next mineral in the sequence. For example, olivine [(Mg, Fe)$_2$SiO$_3$] is the first ferromagnesian mineral to crystallize. As the magma continues to cool, it reaches the temperature range at which pyroxene is stable; a reaction occurs between the olivine and the remaining melt, and pyroxene forms.

A similar reaction occurs between pyroxene and the melt as further cooling occurs, and the pyroxene structure is rearranged to form amphibole. Further cooling causes a reaction between the amphibole and the melt, and its structure is rearranged such that the sheet

◆ FIGURE 3.7 Bowen's reaction series. Note that it consists of a discontinuous branch and a continuous branch.

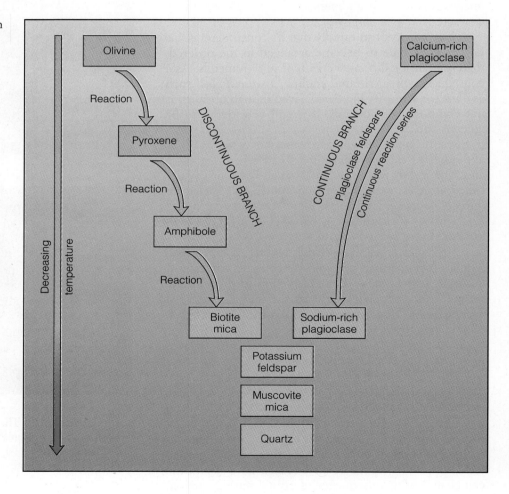

structure typical of biotite mica forms. Although the reactions just described tend to convert one mineral to the next in the series, the reactions are not always complete. For example, olivine might have a rim of pyroxene, indicating an incomplete reaction. In any case, by the time biotite has crystallized, essentially all magnesium and iron present in the original magma have been used up.

Plagioclase feldspars are the only minerals in the continuous branch of Bowen's reaction series (Fig. 3.7). Calcium-rich plagioclase crystallizes first. As cooling of the magma proceeds, calcium-rich plagioclase reacts with the melt, and plagioclase containing proportionately more sodium crystallizes until all of the calcium and sodium are used up. In many cases, however, cooling is too rapid for a complete transformation from calcium-rich to sodium-rich plagioclase to occur. Plagioclase forming under these conditions is *zoned,* meaning that it has a calcium-rich core surrounded by zones progressively richer in sodium.

Magnesium and iron on the one hand and calcium and sodium on the other are used up as crystallization occurs along the two branches in Bowen's reaction series. Accordingly, any magma left over is enriched in potassium, aluminum, and silicon. These elements combine to form potassium feldspar ($KAlSi_3O_8$), and if the water pressure is high, the sheet silicate muscovite mica will form. Any remaining magma is predominantly silicon and oxygen (silica) and forms the mineral quartz (SiO_2). The crystallization of potassium feldspar and quartz is not a true reaction series, however, because they form independently rather than from a reaction of the orthoclase with the remaining melt.

Crystal Settling

Crystal settling involves the physical separation of minerals by crystallization and gravitational settling (◆ Fig. 3.8). For example, olivine, the first ferromagnesian mineral to form in the discontinuous branch of Bowen's reaction series, has a specific gravity greater than that of the remaining magma and thus tends to sink downward in the melt. Accordingly, the remaining melt becomes relatively rich in silica, sodium, and potassium, because much of the iron and magnesium were removed when minerals containing these elements crystallized.

Although crystal settling does occur in magmas, it does not do so on the scale envisioned by Bowen. In some thick, tabular, intrusive igneous bodies called sills,

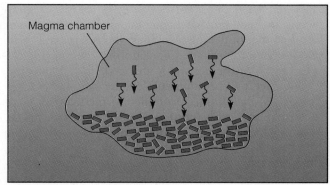

◆ FIGURE 3.8 Differentiation by crystal settling. Early-formed ferromagnesian minerals have a specific gravity greater than that of the magma so they settle and accumulate in the lower part of the magma chamber.

the first formed minerals in the reaction series are indeed concentrated. The lower parts of these bodies contain more olivine and pyroxene than the upper parts, which are less mafic. However, even in these bodies, crystal settling has yielded very little felsic magma from an original mafic magma.

If felsic magma could be derived on a large scale from mafic magma as Bowen believed, there should be far more mafic magma than felsic magma. In order to yield a particular volume of granite (a felsic igneous rock), about 10 times as much mafic magma would have to be present initially for crystal settling to yield the volume of granite in question. If this were so, then mafic intrusive igneous rocks should be much more common than felsic ones. However, just the opposite is the case. Thus, it appears that mechanisms other than crystal settling must account for the large volume of felsic magma. Partial melting of mafic oceanic crust and silica-rich sediments of continental margins during subduction yields magma richer in silica than the source rock. Furthermore, magma rising through the continental crust can absorb some felsic materials by *assimilation* and thus become more enriched in silica.

Assimilation

The composition of a magma can be changed by **assimilation,** a process whereby a magma reacts with preexisting rock, called *country rock,* with which it comes in contact (◆ Fig. 3.9). The walls of a volcanic conduit or magma chamber are, of course, heated by the adjacent magma, which may reach temperatures of

1,300°C. Some of these rocks can be partly or completely melted, provided their melting temperature is less than that of the magma. Since the assimilated rocks seldom have the same composition as the magma, the compositon of the magma is changed.

The fact that assimilation occurs can be demonstrated by *inclusions,* incompletely melted pieces of rock that are fairly common within igneous rocks. Many inclusions were simply wedged loose from the country rock as the magma forced its way into preexisting fractures (Figs. 3.9 and ◆ 3.10).

There is no doubt that assimilation occurs, but its effect on the bulk composition of most magmas must be slight. The reason is that the heat for melting must come from the magma itself, and this would have the effect of cooling the magma. Thus, only a limited amount of rock can be assimilated by a magma, and that amount is usually insufficient to bring about a major compositional change.

Neither crystal settling nor assimilation can produce a significant amount of felsic magma from a mafic one. However, both processes, if operating concurrently, can change the compositon of a mafic magma much more than either process acting alone. Some geologists believe that this is one way that many intermediate magmas form where oceanic lithosphere is subducted beneath continental lithosphere.

Magma Mixing

The fact that a single volcano can erupt lavas of different composition indicates that magmas of differing composition must be present. Thus, it seems likely that some of these magmas would come into contact and mix with one another. If this is the case, we would expect that the composition of the magma resulting from **magma mixing** would be a modified version of the parent magmas. For example, suppose that a rising mafic magma mixes with a felsic magma of about the same volume (◆ Fig. 3.11). The resulting "new" magma would have a more intermediate composition.

Classification of Igneous Rocks

Most igneous rocks are classified on the basis of textural features and composition (◆ Fig. 3.12). Notice in Figure 3.12 that all of the rocks, except peridotite, constitute pairs; the members of a pair have the same

◆ FIGURE 3.9 As magma moves upward, fragments of country rock are dislodged and settle into the magma. If they have a lower melting temperature than the magma, they may be incorporated into the magma by assimilation. Incompletely assimilated pieces of country rock are inclusions.

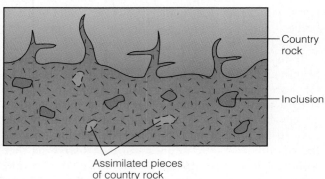

◆ FIGURE 3.10 Dark-colored inclusions in granitic rock in California. (Photo courtesy of David J. Matty.)

◆ FIGURE 3.11 Magma mixing. Two rising magmas mix and produce a magma with a composition different from either of the parent magmas.

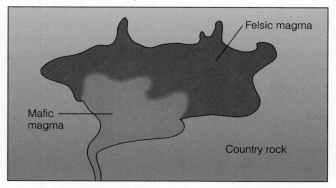

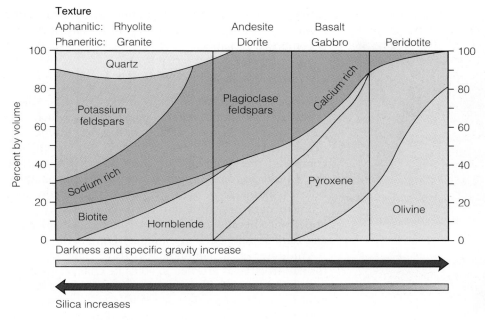

Texture

Aphanitic: Rhyolite   Andesite   Basalt

Phaneritic: Granite   Diorite   Gabbro   Peridotite

◆ FIGURE 3.12 Classification of igneous rocks. Diagram illustrates relative proportions of the chief mineral components of common igneous rocks.

composition but different textures. Thus, basalt and gabbro, andesite and diorite, and rhyolite and granite are compositional (mineralogical) equivalents, but basalt, andesite, and rhyolite are aphanitic and most commonly extrusive, whereas gabbro, diorite, and granite have phaneritic textures that generally indicate an intrusive origin.

The igneous rocks shown in Figure 3.12 are also differentiated by composition. Reading across the chart from rhyolite to andesite, to basalt, for example, the relative proportions of nonferromagnesian and ferromagnesian minerals differ. However, the differences in composition are gradual so that a compositional continuum exists. In other words, there are rocks whose compositions are intermediate between rhyolite and andesite, and so on.

Ultramafic Rocks

Ultramafic rocks are composed largely of ferromagnesian silicate minerals (◆ Fig. 3.13). For example, the ultramafic rock *peridotite* contains mostly olivine, lesser amounts of pyroxene, and generally a little plagioclase feldspar (Fig. 3.12). Another ultramafic rock (pyroxenite) is composed predominantly of pyroxene. Because these minerals are dark colored, the rocks are generally black or dark green. Peridotite is thought to be the rock type composing the upper mantle (see Chapter 7), but ultramafic rocks are rare at the Earth's surface. In

fact, ultramafic lava flows are rare in rocks younger than 2.5 billion years (see Perspective 3.1). Ultramafic rocks are generally thought to have originated by concentration of the early-formed ferromagnesian minerals that separated from mafic magmas.

Basalt-Gabbro

Basalt and *gabbro* (45-52% silica) are the fine-grained and coarse-grained rocks, respectively, that crystallize from mafic magmas (Fig. 3.13b and c). Thus, both have the same composition—mostly calcium-rich plagioclase and pyroxene, with smaller amounts of olivine and amphibole (Fig. 3.12). Because they contain a large proportion of ferromagnesian minerals, basalt and gabbro are dark colored; those that are porphyritic typically contain calcium plagioclase or olivine phenocrysts.

Basalt is generally considered to be the most common extrusive igneous rock. Extensive basalt lava flows were erupted in vast areas in Washington, Oregon, Idaho, and northern California (see Chapter 4). Oceanic islands such as Iceland, the Galapagos, the Azores, and the Hawaiian Islands are composed mostly of basalt. Furthermore, the upper part of the oceanic crust is composed almost entirely of basalt.

Gabbro is much less common than basalt, at least in the continental crust or where it can be easily observed. Small intrusive bodies of gabbro do occur in the continental crust, but intermediate to felsic intrusive

(a)

(b)

(c)

◆ FIGURE 3.13 (*a*) The ultramafic rock peridotite. Mafic igneous rocks: (*b*) basalt; and (*c*) gabbro. (Photos courtesy of Sue Monroe.)

rocks such as diorite and granite are much more common. However, the lower part of the oceanic crust is composed of gabbro.

Andesite-Diorite

Magmas intermediate in composition (53-65% silica) crystallize to form *andesite* and *diorite,* which are compositionally equivalent fine- and coarse-grained igneous rocks (◆ Fig. 3.14). Andesite and diorite are composed predominantly of plagioclase feldspar, with the typical ferromagnesian component being amphibole or biotite (Fig. 3.12). Andesite is generally medium to dark gray, but diorite has a salt and pepper appearance because of its white to light gray plagioclase and dark ferromagnesian minerals (Fig. 3.14).

Andesite is a common extrusive igneous rock formed from lavas erupted in volcanic chains at convergent plate margins. The volcanoes of the Andes Mountains of South America and the Cascade Range in western North America are composed in part of andesite. Intrusive bodies composed of diorite are fairly common in the continental crust. However, diorite is not nearly as abundant as granitic rocks and is uncommon outside the areas where andesites occur.

Rhyolite-Granite

Rhyolite and *granite* (> 65% silica) crystallize from felsic magmas and are therefore silica-rich rocks (◆ Fig. 3.15). Rhyolite and granite consist largely of potassium feldspar, sodium-rich plagioclase, and quartz, with perhaps some biotite and rarely amphibole (Fig. 3.12). Because nonferromagnesian minerals predominate, rhyolite and granite are generally light colored. Rhyolite is fine grained, although most often it contains phenocrysts of potassium feldspar or quartz, and granite is coarse grained. Granite porphyry is also fairly common.

Rhyolite lava flows are much less common than andesite and basalt flows. Recall that the greatest control of viscosity in a magma is the silica content. Thus, if a felsic magma rises to the surface, it begins to cool, the pressure on it decreases, and gases are released explosively, usually yielding rhyolitic pyroclastic particles. The rhyolitic lava flows that do occur are thick and highly viscous and thus move only short distances.

Among geologists, granite has come to mean any coarsely crystalline igneous rock with a composition corresponding to that of the field shown in Figure 3.12. Strictly speaking, not all rocks in this field are granites. For example, a rock with a composition close to the

Perspective 3.1

ULTRAMAFIC LAVA FLOWS

Geologists refer to the interval of geologic time from 3.8 to 2.5 billion years ago as the Archean Eon. Some of the most interesting rocks that formed during the Archean Eon are ultramafic lava flows because such flows are rare in younger rocks and none are forming at present. Archean ultramafic lava flows are generally parts of large, complex associations of rocks known as greenstone belts. An idealized greenstone belt consists of three major rock units: the lower and middle units are dominated by volcanic rocks, and the upper unit is composed mostly of sedimentary rocks. The lower volcanic units of some Archean greenstone belts contain ultramafic lava flows.

Why did ultramafic lava flows occur during early Earth history, but only rarely later? The answer is related to the heat produced within the Earth. When it first formed, the Earth possessed a considerable amount of residual heat inherited from the formative processes. Rock is a poor conductor of heat, so this primordial heat was slowly lost. Another source of heat within the Earth is related to the phenomenon of radioactive decay. Recall from Chapter 2 that as the isotopes of some elements decay to a more stable state, they generate heat. Thus, the Earth has a mechanism whereby heat can be generated internally. As the radioactive isotopes decay to a stable state, however, they become less abundant (except carbon 14, which is produced in the atmosphere). Accordingly, the heat generated within the Earth has decreased (◆Fig. 1).

In order to erupt, an ultramafic magma requires near-surface temperatures of more than 1,600°C; the surface temperatures of present-day basalt lava flows are generally between 1,000° and 1,200°C. During early Earth history, there was considerably more residual heat and more heat generated by radioactive decay; thus, the mantle was hotter than it is today—perhaps 300°C hotter (Fig. 1)—and ultramafic magmas could be erupted onto the surface. Since the amount of heat has decreased through time, the Earth has cooled, and the eruption of ultramafic lava flows has largely ceased.

◆ FIGURE 1 The ratio of heat produced by radioactive decay during the past and at the present. The shaded band encloses the ratios according to different models.

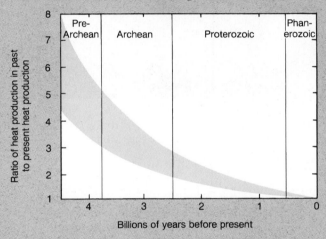

Billions of years before present

---

line separating granite and diorite is usually called *granodiorite*. To avoid the confusion that might result from introducing more rock names, we will follow the practice of referring to rocks to the left of the granite-diorite line in Figure 3.12 as *granitic*.

Granitic rocks are by far the most common intrusive igneous rocks, although they are restricted to the continents. Most granitic rocks were intruded at or near convergent plate margins during episodes of mountain building. When these mountainous regions are uplifted and eroded, the vast bodies of granitic rocks forming their cores are exposed. For example, the granitic rocks of the Sierra Nevada of California form a composite body measuring about 640 km long and 110 km wide, and the granitic rocks of the Coast Ranges of British Columbia, Canada, are even more voluminous.

(a)

(b)

◆ FIGURE 3.14 Intermediate igneous rocks: (*a*) andesite and (*b*) diorite. (Photos courtesy of Sue Monroe.)

(a)

(b)

◆ FIGURE 3.15 Felsic igneous rocks: (*a*) granite and (*b*) rhyolite. (Photos courtesy of Sue Monroe.)

Other Igneous Rocks

Some igneous rocks, including tuff, volcanic breccia, obsidian, and pumice, are identified solely by their textures. Much of the fragmental material erupted by volcanoes is *ash,* a designation for pyroclastic materials less than 2.0 mm in diameter, much of which is broken pieces or shards of volcanic glass. The consolidation of ash forms the pyroclastic rock *tuff* (◆ Fig. 3.16). Most tuff is silica rich and light colored and is appropriately called *rhyolite tuff.* Some ash flows are so hot that as they come to rest, the ash particles fuse together and form a *welded tuff.* Consolidated deposits of larger pyroclasts, such as cinders, blocks, and bombs, are *volcanic breccia.*

Both *obsidian* and *pumice* are varieties of volcanic glass (◆ Fig. 3.17). Obsidian may be black, dark gray, red, or brown, with the color depending on the presence of tiny particles of iron minerals. Obsidian breaks with the conchoidal fracture that is typical of glass (Fig. 3.17a). Analyses of numerous samples indicate that most obsidian is compositionally similar to rhyolite and thus has a high silica content.

Pumice is a variety of volcanic glass containing numerous bubble-shaped vesicles that develop when gas escapes through lava and forms a froth (Fig. 3.17b). Some pumice forms as crusts on lava flows, and some forms as particles erupted from explosive volcanoes. If pumice falls into water, it can be carried great distances because it is so porous and light that it floats.

◆ FIGURE 3.16 Exposure of tuff in Colorado. (Photo courtesy of David J. Matty.)

▼ INTRUSIVE IGNEOUS BODIES: PLUTONS

Unlike volcanism and the origin of extrusive or volcanic igneous rocks, which can be observed, intrusive igneous activity can be studied only indirectly. Intrusive igneous bodies called **plutons** form when magma cools and crystallizes within the Earth's crust (◆ Fig. 3.18). Although plutons can be observed after erosion has exposed them at the surface, we cannot duplicate the conditions that exist deep in the crust, except in small-scale laboratory experiments. Thus, geologists face a greater challenge in interpreting the mechanisms whereby plutons originate than in studying the origins of extrusive igneous rocks.

Several types of plutons are recognized, all of which are defined by their geometry (three-dimensional shape) and their relationship to the country rock (Fig. 3.18). Geometrically, plutons may be characterized as massive or irregular, tabular, cylindrical, or mushroom shaped. Plutons are also described as concordant or discordant. A **concordant** pluton, such as a sill, has boundaries that are parallel to the layering in the intruded rock or what is commonly called the *country rock.* A **discordant** pluton, such as a dike, has boundaries that cut across the layering of the country rock (Fig. 3.18).

Dikes and Sills

Both **dikes** and **sills** are tabular or sheetlike plutons, but dikes are discordant whereas sills are concordant (Fig. 3.18). Dikes are common intrusive features (◆ Fig.

(a)

(b)

◆ FIGURE 3.17 (*a*) Obsidian and (*b*) pumice. (Photos courtesy of Sue Monroe.)

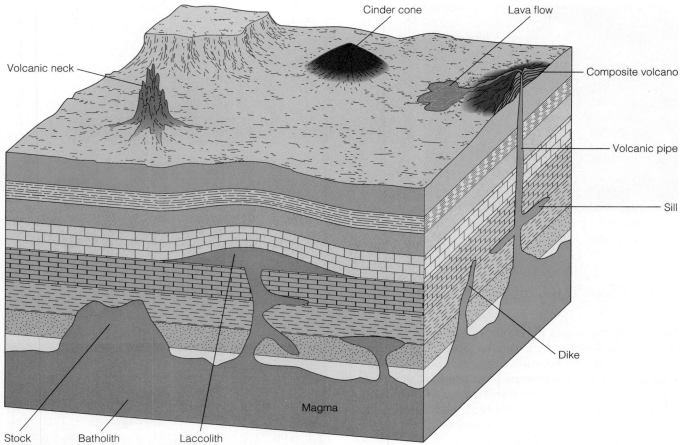

Volcanic neck

Cinder cone

Lava flow

Composite volcano

Volcanic pipe

Sill

Dike

Magma

Stock   Batholith   Laccolith

◆ FIGURE 3.18 Block diagram showing the various types of plutons. Notice that some of these plutons cut across the layering in the country rock and are thus discordant, whereas others parallel the layering and are concordant.

3.19). Many are small bodies measuring 1 or 2 m across, but they range from a few centimeters to more than 100 m thick. Dikes are emplaced within preexisting zones of weakness where fractures already exist or where the fluid pressure is great enough for them to form their own fractures during emplacement.

Erosion of the Hawaiian volcanoes exposes dikes in rift zones, the large fractures that cut across these volcanoes. The Columbia River basalts in Washington issued from long fissures, and the magma that cooled in the fissures formed dikes. Some of the large historic fissure eruptions are underlain by dikes; for example, dikes underlie both the Laki fissure eruption of 1783 in Iceland and the Eldgja fissure, also in Iceland, where eruptions occurred in A.D. 950 from a fissure 300 km long.

Sills are concordant plutons, many of which are a meter or less thick, although some are much thicker

◆ FIGURE 3.19 The dark layer cutting diagonally across the rock layers is a dike. The other dark layer is a sill because it parallels the layering in the country rock.

(Fig. 3.19). A well-known sill in the United States is the Palisades sill that forms the Palisades along the west side of the Hudson River in New York and New Jersey. It is exposed for 60 km along the river and is up to 300 m thick.

Most sills have been intruded into sedimentary rocks, but eroded volcanoes also reveal that sills are commonly injected into piles of volcanic rocks. In fact, some of the inflation of volcanoes preceding eruptions may be caused by the injection of sills (see Perspective 4.2).

In contrast to dikes, which follow zones of weakness, sills are emplaced when the fluid pressure is so great that the intruding magma actually lifts the overlying rocks. Because emplacement requires fluid pressure exceeding the force exerted by the weight of the overlying rocks, sills are typically shallow intrusive bodies.

Laccoliths

Laccoliths are similar to sills in that they are concordant, but instead of being tabular, they have a mushroomlike geometry (Fig. 3.18). They tend to have a flat floor and are domed up in their central part. Like sills, laccoliths are rather shallow intrusive bodies that actually lift up the overlying strata. In this case, however, the strata are arched upward over the pluton (Fig. 3.18). Most laccoliths are rather small bodies. The best-known laccoliths in North America are in the Henry Mountains of southeastern Utah.

Volcanic Pipes and Necks

The conduit connecting the crater of a volcano with an underlying magma chamber is a **volcanic pipe** (Fig. 3.18). In other words, it is the structure through which magma rises to the surface. When a volcano ceases to erupt, it is eroded as it is attacked by water, gases, and acids. The volcanic mountain eventually erodes away, but the magma that solidified in the pipe is commonly more resistant to weathering and erosion and is often left as an erosional remnant, a **volcanic neck** (Fig. 3.18). A number of volcanic necks are present in the southwestern United States, especially in Arizona and New Mexico (◆ Fig. 3.20), and others are recognized elsewhere.

Batholiths and Stocks

Batholiths are the largest intrusive bodies. By definition they must have at least 100 km² of surface area, and

◆ FIGURE 3.20 A volcanic neck in northern Arizona.

most are much larger than this (Fig. 3.18). **Stocks** have the same general features as batholiths but are smaller, although some stocks are simply the exposed parts of much larger intrusions. Batholiths are generally discordant, and most consist of multiple intrusions. In other words, a batholith is a large composite body produced by repeated, voluminous intrusions of magma in the same area. The coastal batholith of Peru, for example, was emplaced over 60 to 70 million years and consists of perhaps as many as 800 individual plutons.

The igneous rocks composing batholiths are mostly granitic, although diorite may also occur (◆ Fig. 3.21). Most batholiths are emplaced along convergent plate margins. For example, the Coast Range batholith of British Colombia, Canada, was emplaced over a period of millions of years. Later uplift and erosion exposed this huge composite pluton at the Earth's surface. Other large batholiths in North America include the Idaho batholith and the Sierra Nevada batholith in California.

A number of mineral resources occur in rocks of batholiths and stocks and in the country rocks adjacent to them. For example, silica-rich igneous rocks, such as granite, are the primary source of gold, which forms from mineral-rich solutions moving through cracks and fractures of the igneous body. The copper deposits at

◆ FIGURE 3.21 Granitic rocks intruded by a basalt dike exposed in southern Ontario, Canada.

Butte, Montana, are in rocks near the margins of the granitic rocks of the Boulder batholith. Near Salt Lake City, Utah, copper is mined from the mineralized rocks adjacent to the Bingham stock, a composite pluton composed of granite and granite porphyry.

◥ MECHANICS OF BATHOLITH EMPLACEMENT

Geologists realized long ago that the emplacement of batholiths posed a space problem; that is, what happened to the rock that formerly occupied the space now occupied by a granite batholith? One solution to this space problem was to propose that no displacement had occurred, but rather that the granite had been formed in place by alteration of the country rock through a process called *granitization.* According to this view, granite did not originate as a magma, but rather from hot, ion-rich solutions that simply altered the country rock and transformed it into granite. Granitization is a solid-state phenomenon so it is essentially an extreme type of metamorphism (see Chapter 6).

Many granites show clear evidence of an intrusive origin. For example, the contact between these granites and the adjacent country rock is sharp rather than gradational, and elongated mineral crystals are commonly oriented parallel with the contact. Some granitic rocks lack sharp contacts, however, and gradually change in character until they resemble the adjacent country rocks. Some of these have likely been altered by granitization. Most geologists think that only small quantities of granite are formed by this process, and

that it cannot account for the huge granite batholiths of the world. These geologists think an igneous origin for granite is clear, but then they must deal with the space problem. One solution is that these large igneous bodies melted their way into the crust. In other words, they simply assimilated the country rock as they moved upward (Fig. 3.9). The presence of inclusions, especially near the tops of such intrusive bodies, indicates that assimilation does occur. Nevertheless, as we noted previously, assimilation is a limited process because magma is cooled as country rock is assimilated; calculations indicate that far too little heat is available in a magma to assimilate the necessarily huge quantities of country rock.

Most geologists now agree that batholiths were emplaced as magma and that the magma, being less dense than the rock from which it was derived, moved upward toward the surface. Recall, however, that granite is derived from viscous felsic magma and, thus, it rises slowly. It appears that the magma deforms and shoulders aside the country rock, and as it rises further, some of the country rock fills the space beneath the magma (◆ Fig. 3.22). A somewhat analogous situation was discovered in which large masses of sedimentary rock called rock salt rise through the overlying rocks to form salt domes.

◆ FIGURE 3.22 Emplacement of a hypothetical batholith. As the magma rises, it shoulders aside and deforms the country rock.

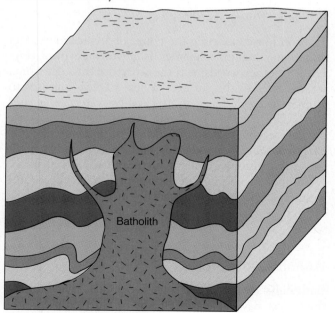

Batholith

Salt domes are recognized in several areas of the world, including the Gulf Coast of the United States. Layers of rock salt exist at some depth, but salt is less dense than most other types of rock materials. Thus, when under pressure, it rises toward the surface even though it remains solid, and as it moves upward, it pushes aside and deforms the country rock (◆ Fig. 3.23). Natural examples of rock salt flowage are known, and it can easily be demonstrated experimentally. For example, in the arid Middle East, salt moving upward in the manner described actually flows out at the surface.

Some batholiths do indeed show evidence of having been emplaced forcefully by shouldering aside and deforming the country rock. This mechanism probably occurs in the deeper parts of the crust where temperature and pressure are high and the country rocks are easily deformed in the manner described. At shallower depths, however, the crust is more rigid and tends to deform by fracturing. In this environment, batholiths may be emplaced by **stoping,** a process involving detaching and engulfing pieces of country rock by rising magma (◆ Fig. 3.24). According to this concept, magma moves upward along fractures and the planes separating layers of country rock. Eventually, pieces of country rock are detached and settle into the magma. No new room is created during stoping; the magma simply fills the space formerly occupied by country rock (Fig. 3.24).

▼ PEGMATITES

Pegmatite is a very coarsely crystalline igneous rock, commonly associated with plutons (◆ Fig. 3.25). Such rocks contain minerals measuring at least 1 cm across, and many crystals are much larger. The name pegmatite reflects texture rather than a specific composition, but most are composed largely of quartz, potassium feldspar, and sodium-rich plagioclase, and thus correspond to the composition of granite. Many pegmatites are associated with granite batholiths and appear to represent the minerals that formed from the remaining fluid and vapor phases that existed after most of the granite crystallized.

◆ FIGURE 3.24 Emplacement of a batholith by stoping. (*a*) The magma is injected into the country rock along fractures and planes between layers. (*b*) Blocks of country rock are detached and engulfed in the magma by stoping. Some of these blocks may be assimilated.

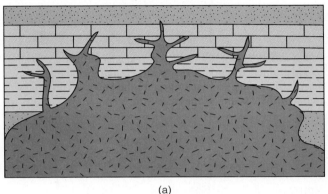

(a)

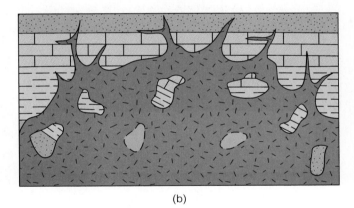

(b)

◆ FIGURE 3.23 Three stages in the origin of a salt dome. Rock salt is a low-density sedimentary rock that (*a*) when deeply buried (*b*) tends to rise toward the surface, (*c*) pushing aside and deforming the country rock and forming a dome. Salt domes are thought to originate in much the same manner as batholiths are intruded into the Earth's crust.

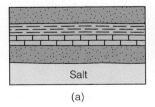 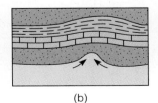 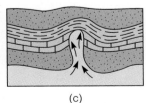

(a)   (b)   (c)

The water-rich vapor phase that exists after most of a magma has crystallized as granite has properties that differ from the magma from which it separated. It has a lower density and viscosity, and thus commonly invades the country rock where it crystallizes as dikes and, more rarely, as sills. The water-rich vapor phase ordinarily contains a number of elements that rarely enter into the common minerals that form granite. Pegmatites crystallizing to form very coarsely crystalline granite are simple pegmatites, whereas those with minerals containing elements such as lithium, beryllium, cesium,

◆ FIGURE 3.25 Pegmatite is a textural term for very coarse-grained igneous rock; most pegmatites have a composition close to that of granite, however. The mineral grains in this specimen measure 2 to 3 cm.

tin, and several others are complex pegmatites (see Perspective 3.2). Some complex pegmatites contain 300 different mineral species, a few of which are important economically. In addition, several gem minerals such as emerald and aquamarine, both of which are varieties of the silicate mineral beryl, and tourmaline are found in some pegmatites. Many rare minerals of lesser value and well-formed crystals of common minerals, such as quartz, are also mined and sold to collectors and museums.

The formation and growth of mineral-crystal nuclei in pegmatites are similar to those processes in magma, but with one critical difference: the vapor phase from which pegmatites crystallize inhibits the formation of nuclei. However, some nuclei do form, and because the appropriate ions in the liquid can move easily and attach themselves to a growing crystal, individual mineral grains have the opportunity to grow to very large sizes (see Perspective 3.2).

⬇ PLATE TECTONICS AND PLUTONISM

Recall that most of the magma generated within the Earth forms plutons rather than being erupted onto the surface. For example, mafic magma originates beneath spreading ridges, and some is erupted at the surface as lava flows and/or pyroclastic materials; however, much of it is simply emplaced at depth as vertical dikes and gabbro plutons (◆ Fig. 3.26). In fact, the oceanic crust is composed largely of such mafic rock, and only the

◆ FIGURE 3.26 Intrusive igneous activity at a spreading ridge. The oceanic crust is composed largely of vertical dikes of basaltic composition and gabbro that appears to have crystallized in the upper part of a magma chamber.

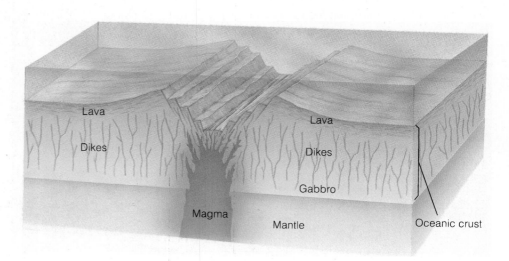

COMPLEX PEGMATITES

Even though complex pegmatites vary considerably, they share some common features. For example, they all possess concentric zones that differ from one another in texture and composition. Ideally, four zones are present; these are generally designated as the border, wall, intermediate, and core zones, the last of which is usually composed almost entirely of quartz. The zonation in complex pegmatites is the subject of continuing debate. One hypothesis holds that crystallization occurred from the border zone inward, with the fluid phase changing composition as successive layers of minerals crystallized from it. According to another hypothesis, a simple pegmatite formed first and was later partially or completely replaced by hydrothermal (hot water) solutions. At present, no compelling evidence supports one hypothesis to the exclusion of the other.

Another feature of complex pegmatites is the occurrence of giant mineral crystals in the inner zones. Numerous large tourmaline crystals have been recovered from the Dunton pegmatite in Maine (◆Fig. 1). Crystals of muscovite measuring 2.44 m across have been recovered from pegmatites in Ontario, Canada, and pegmatites in California have yielded huge feldspar crystals. Giant quartz crystals have been found in many pegmatites.

Some pegmatites have been mined for gemstones as well as for tin, industrial minerals such as feldspars, micas, and quartz, and minerals containing elements such as cesium, rubidium, lithium, and beryllium. Some of the largest reserves of lithium in the Western Hemisphere are in pegmatites of the tin-spodumene belt of North Carolina. Pegmatites are the only known commercial sources of some of the rare earth elements.

Pegmatites are common in the Black Hills of South Dakota, where more than 20,000 have been identified in the country rock adjacent to the Harney Peak Granite. The stone faces of presidents Washington, Jefferson, Lincoln, and Theodore Roosevelt on Mount Rushmore were carved into rocks of the Harney Peak Granite (◆Fig. 2). These pegmatites formed about 1.7 billion years ago, during the Late Proterozoic Eon, when the granite was emplaced as a composite pluton consisting of numerous sills and dikes. Subsequent uplift during the Late Cretaceous Period resulted in erosion of the overlying rocks, thus exposing the granite and its associated pegmatites.

Most of the Black Hills pegmatites are simple, with compositions closely resembling that of the Harney Peak Granite; about 1% are complex pegmatites.

Many of these complex pegmatites have been mined for various resources. One of the best known is the Etta pegmatite, which contains crystals of spodumene, a lithium-bearing silicate mineral, that commonly measure 1 to 3 m long; the larger spodumene crystals are the size of large logs, and one was more than 14 m long! Micas and tin were originally mined from the Etta pegmatite, and for many years it was a major producer of lithium. It closed in 1960, however, because more economical sources of lithium are available from arid region lake deposits.

◆ FIGURE 1 Tourmaline from the Dunton mine in Maine.

◆ FIGURE 2 The presidents' faces at Mount Rushmore, South Dakota are carved into the Harney Peak Granite.

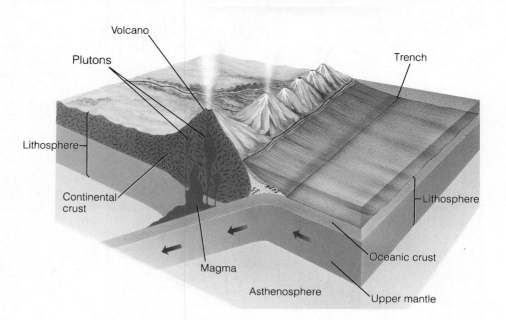

Volcano

Plutons

Trench

Lithosphere

Continental crust

Lithosphere

Oceanic crust

Magma

Asthenosphere

Upper mantle

◆ FIGURE 3.27 Generation of magma and the emplacement of plutons at a convergent plate margin.

upper part of the oceanic crust is composed of volcanic rock, mostly in the form of pillow lavas (see Fig. 4.9).

Magmas generated by partial melting of mafic oceanic crust and silica-rich continental margin sediments at convergent plate margins where subduction takes place are mostly intermediate and felsic in composition. Some of this magma is erupted at the surface and forms the large composite volcanoes that characterize such plate margins. Much of it, however, simply cools at depth as large plutons, especially batholiths (◆ Fig. 3.27). Considerably more on the relationship between plate tectonics and igneous activity is discussed in Chapter 4.

Chapter Summary

1. Magma is molten rock material below the Earth's surface, whereas lava is magma that reaches the surface. The silica content of magmas varies and serves to differentiate felsic, intermediate, and mafic magmas.
2. The viscosity of lava flows depends mostly on their temperature and composition. Silica-rich (felsic) lava is more viscous than silica-poor (mafic) lava.
3. Minerals crystallize from magma and lava when small crystal nuclei form and grow.
4. Volcanic rocks generally have aphanitic textures because of rapid cooling, whereas slow cooling and phaneritic textures characterize plutonic rocks. Igneous rocks with a porphyritic texture have mineral crystals of markedly different sizes. Other igneous rock textures include vesicular and pyroclastic.
5. The composition of igneous rocks is determined largely by the composition of the parent magma. It is possible, however, for an individual magma to yield igneous rocks of differing compositions.

6. Under ideal cooling conditions, a mafic magma yields a sequence of different minerals that are stable within specific temperature ranges. This sequence, called Bowen's reaction series, consists of a discontinuous branch and a continuous branch.
   a. The discontinuous branch contains only ferromagnesian minerals, each of which reacts with the melt to form the next mineral in the sequence.
   b. The continuous branch involves changes only in plagioclase feldspar as sodium replaces calcium in the crystal structure.
7. The ferromagnesian minerals that form first in Bowen's reaction series can settle and become concentrated near the base of a magma chamber or intrusive body. Such settling of iron- and magnesium-rich minerals causes a chemical change in the remaining melt.
8. A magma can be changed compositionally when it assimilates country rock, but this process usually has only a

limited effect. Magma mixing may also bring about compositional changes in magmas.

9. Most igneous rocks are classified on the basis of their textures and composition. Two fundamental groups of igneous rocks are recognized: volcanic or extrusive rocks, and plutonic or intrusive rocks.
   a. Common volcanic rocks include tuff, rhyolite, andesite, and basalt.
   b. Common plutonic rocks include granite, diorite, and gabbro.
10. Plutons are igneous bodies that formed in place or were intruded into the Earth's crust. Various types of plutons are classified by their geometry and whether they are concordant or discordant.
11. Common plutons include dikes (tabular geometry, discordant); sills (tabular geometry, concordant); volcanic necks (cylindrical geometry, discordant); laccoliths (mushroom shaped, concordant); and batholiths and stocks (irregular geometry, discordant).
12. By definition batholiths must have at least 100 km² of surface area; stocks are similar to batholiths but smaller. Many batholiths are large composite bodies consisting of many plutons emplaced over a long period of time.
13. Most batholiths appear to have formed in the cores of mountain ranges during episodes of mountain building.
14. Some geologists think that granite batholiths are emplaced when felsic magma moves upward and shoulders aside and deforms the country rock. The upward movement of rock salt and the formation of salt domes provide a somewhat analogous situation.
15. Pegmatites are very coarse-grained igneous rocks, most of which have an overall composition similar to that of granite. Crystallization from a vapor-rich phase left over after the crystallization of granite accounts for the very large mineral crystals in pegmatites.
16. Most plutons form in areas where volcanism occurs, such as at spreading ridges and above subduction zones.

Important Terms

aphanitic	laccolith	porphyritic
assimilation	lava	pyroclastic materials
batholith	lava flow	pyroclastic (fragmental) texture
Bowen's reaction series	mafic magma	sill
concordant	magma	stock
crystal settling	magma mixing	stoping
dike	pegmatite	vesicle
discordant	phaneritic	viscosity
felsic magma	phenocryst	volcanic neck
igneous rock	pluton	volcanic pipe
intermediate magma	plutonic (intrusive igneous) rock	volcanic (extrusive igneous) rock

Review Questions

1. The first minerals to crystallize from a mafic magma are:
   a. ___ quartz and potassium feldspar; b. ___ calcium-rich plagioclase and olivine; c. ___ biotite and muscovite; d. ___ amphibole and pyroxene; e. ___ andesite and basalt.
2. The most common aphanitic igneous rock is:
   a. ___ basalt; b. ___ granite; c. ___ pumice; d. ___ obsidian; e. ___ rhyolite.
3. Volcanic rocks can usually be distinguished from plutonic rocks by:
   a. ___ color; b. ___ composition; c. ___ iron-magnesium content; d. ___ the size of their mineral grains; e. ___ specific gravity.
4. An example of a concordant pluton having a tabular geometry is a:
   a. ___ sill; b. ___ batholith; c. ___ volcanic neck; d. ___ lava flow; e. ___ dike.
5. Most pegmatites are essentially:
   a. ___ light-colored gabbro; b. ___ thick accumulations of pyroclastic materials; c. ___ very coarse-grained granite; d. ___ rhyolite porphyry; e. ___ cylindrical plutons.
6. An igneous rock possessing a combination of mineral grains with markedly different sizes is:
   a. ___ a natural glass; b. ___ the product of very rapid cooling; c. ___ formed by explosive volcanism; d. ___ a porphyry; e. ___ a tuff.
7. Which of the following minerals is likely to be separated from a mafic magma by crystal settling?

a. ___ sodium-rich plagioclase; b. ___ muscovite;
c. ___ quartz; d. ___ olivine; e. ___ potassium feldspar.

8. The process whereby a magma reacts with and incorporates preexisting rock is:
   a. ___ crystal differentiation; b. ___ granitization; c. ___ plutonism; d. ___ magma mixing; e. ___ assimilation.

9. Igneous rocks composed largely of ferromagnesian minerals are characterized as:
   a. ___ pyroclastic; b. ___ ultramafic; c. ___ intermediate; d. ___ felsic; e. ___ mafic.

10. Which of the following pairs of igneous rocks have the same mineral composition?
    a. ___ granite-tuff; b. ___ andesite-rhyolite;
    c. ___ pumice-diorite; d. ___ basalt-gabbro;
    e. ___ peridotite-andesite.

11. Which of the following is a concordant pluton?
    a. ___ sill; b. ___ stock; c. ___ volcanic neck; d. ___ dike;
    e. ___ batholith.

12. Batholiths are composed mostly of what type of rock?
    a. ___ granitic; b. ___ gabbro; c. ___ basalt; d. ___ andesite; e. ___ peridotite.

13. An igneous rock possessing mineral grains large enough to be seen without magnification is said to have a _____ texture.
    a. ___ porphyritic; b. ___ aphanitic; c. ___ fragmental;
    d. ___ phaneritic; e. ___ vesicular.

14. What are the two major kinds of igneous rocks? How do they differ?

15. Describe the process whereby mineral crystals form and grow. Why are volcanic rocks generally aphanitic?

16. What is a natural glass, and how does it form?

17. In terms of composition, how are granite and diorite similar and dissimilar?

18. Compare the continuous and discontinuous branches of Bowen's reaction series.

19. Describe how the composition of a magma can be changed by crystal settling; by assimilation. Cite evidence indicating that both of these processes occur.

20. What is a welded tuff?

21. How do dikes and sills differ? How is each emplaced?

22. Describe the sequence of events in the formation of a volcanic neck.

23. Briefly explain where and how batholiths form.

24. What are pegmatites? Explain why some pegmatites contain very large mineral crystals.

25. Are extrusive and intrusive igneous activity related, or are these completely separate phenomena? Explain.

26. In what plate tectonic settings does intrusive igneous activity occur?

27. Why are felsic lava flows so much more viscous than mafic lava flows?

28. Why is silica the major component of most magmas?

Additional Readings

Baker, D. S. 1983. *Igneous rocks.* Englewood Cliffs, N.J.: Prentice-Hall.

Best, M. G. 1982. *Igneous and metamorphic petrology.* San Francisco: W. H. Freeman.

Dietrich, R. V., and B. J. Skinner. 1979. *Rocks and rock minerals.* New York: Wiley.

Dietrich, R. V. and R. Wicander. 1983. *Minerals, rocks, and fossils.* New York: Wiley.

Ernst, W. G. 1969. *Earth materials.* Englewood Cliffs, N.J.: Prentice-Hall.

Hall, A. 1987. *Igneous petrology.* Essex, England: Longman Scientific and Technical.

Hess, P. C. 1989. *Origins of igneous rocks.* Cambridge, Mass.: Harvard Univ. Press.

McBirney, A. R. 1984. *Igneous petrology.* San Francisco: Freeman, Cooper.

MacKenzie, W. S., C. H. Donaldson, and C. Guilford. 1982. *Atlas of igneous rocks and their textures.* New York: Halsted.

Middlemost, E. A. K. 1985. *Magma and magmatic rocks.* London: Longman.

CHAPTER
4

Chapter Outline

Mount Lassen in northern California erupted numerous times between 1914 and 1921. This eruption occurred in 1915.

Volcanism

Prologue

During the summer of 1914, Mount Lassen in northern California began erupting without warning and culminated with the "Great Hot Blast," a huge steam explosion on May 22, 1915. Fortunately, the area was sparsely settled, and little property damage and no deaths resulted, even though a large area of forest on the volcano's eastern and northeastern flanks was devastated. Activity largely ceased by 1921, but hot springs, boiling mud pots, and gas vents known as *fumaroles* remind us that a source of heat still exists beneath the surface.

Mount Lassen is one of 15 large volcanoes in the Cascade Range of northern California, Oregon, Washington, and southern British Columbia, Canada (◆ Fig. 4.1). After the 1914–1921 eruptions, the Cascade volcanoes remained quiet for 63 years. Then, on March 16, 1980, following an inactive period of 123 years, Mount St. Helens in southern Washington (◆ Fig. 4.2) showed signs of renewed activity, and on May 18 it erupted violently, causing the worst volcanic disaster in U.S. history.

The awakening of Mount St. Helens came as no surprise to geologists of the U.S. Geological Survey (USGS) who warned in 1978 that Mount St. Helens is ". . . an especially dangerous volcano because [of] its past behavior and [its] relatively high frequency of eruptions during the last 4,500 years . . . ."* Although no one could predict precisely when Mount St. Helens would erupt, the USGS report included maps showing areas in which damage from an eruption could be expected. Forewarned with such data, local

. . . .

* D. R. Crandell and D. R. Mullineaux, "Potential Hazards from Future Eruptions of Mt. St. Helens Volcano, Washington," *United States Geological Survey Bulletin 1383-C,* (1978):C1.

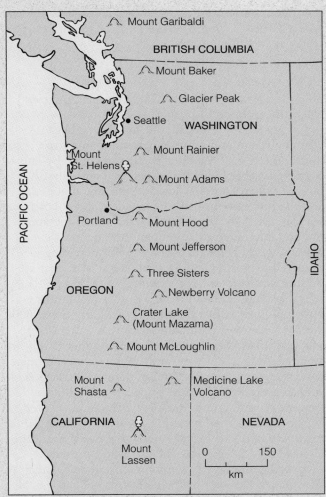

◆ FIGURE 4.1 The major volcanoes of the Cascade Range. Mount Lassen and Mount St. Helens have erupted during this century, and several others have been active during the last two hundred years.

◆ FIGURE 4.2 View of Mount St. Helens from the southwest in 1978.

◆ FIGURE 4.3 The eruption of Mount St. Helens on May 18, 1980. The lateral blast occurred when the bulge on the north face of the mountain collapsed and reduced the pressure on the molten rock within the mountain.

officials were better prepared to formulate policies when the eruption did occur.

On March 27, 1980, Mount St. Helens began erupting steam and ash and continued to do so during the rest of March and most of April. By late March, a visible bulge had developed on its north face as molten rock was injected into the mountain, and the bulge continued to expand at about 1.5 m per day. On May 18, an earthquake shook the area, the unstable bulge collapsed, and the pent-up volcanic gases below expanded rapidly, creating a tremendous northward-directed lateral blast that blew out the north side of the mountain (◆ Fig. 4.3). The lateral blast accelerated from 350 to 1,080 km/hr, obliterating virtually everything in its path. Some 600 km² of forest were completely destroyed; trees were snapped off at their bases and strewn about the countryside, and trees as far as 30 km from the bulge were seared by the intense heat. Tens of thousands of animals were killed; roads, bridges, and buildings were destroyed; and 63 people perished.

Shortly after the lateral blast, volcanic ash and steam erupted and formed a cloud above the volcano 19 km high (◆ Fig. 4.4). The ash cloud drifted east-northeast, and the resulting ash fall at Yakima, Washington, 130 km to the east, caused almost total darkness at midday. Detectable amounts of ash were deposited over a huge area. Flows of hot gases and volcanic ash raced down the north flank of the

◆ FIGURE 4.4 Shortly after the lateral blast of May 18, 1980, Mount St. Helens erupted a steam and ash cloud that rose about 19 km high.

mountain, causing steam explosions when they encountered bodies of water or moist ground. Steam explosions continued for weeks, and at least one occurred a year later.

Snow and glacial ice on the upper slopes of Mount St. Helens melted and mixed with ash and other surface debris to form thick, pasty volcanic mudflows. The largest and most destructive mudflow surged down the valley of the North Fork of the Toutle River. Ash and mudflows displaced water in lakes and streams and flooded downstream areas. Ash and other particles carried by the flood waters were deposited in stream channels; many kilometers from Mount St. Helens, the navigation channel of the Columbia River

was reduced from 12 m to less than 4 m as a result of such deposition.

Although the damage resulting from the eruption of Mount St. Helens was significant and the deaths were tragic, it was not a particularly large or deadly eruption compared with some historic eruptions. For example, the 1902 eruption of Mount Pelée on the island of Martinique killed 28,000 people, and the 1815 eruption of Tambora in Indonesia resulted in an estimated 92,000 deaths (• Table 4.1). The Tambora eruption is the greatest volcanic eruption in recorded history in terms of both casualties and the amount of material erupted; it produced at least 80 times more ash than the 0.9 km^3 that spewed forth from Mount St. Helens.

• **TABLE 4.1** Some Notable Volcanic Eruptions

Date	Volcano	Deaths
Aug. 24, 79	Mt. Vesuvius, Italy	3,360 killed in Pompeii and Herculaneum.
1586	Kelut, Java	Mudflows kill 10,000.
Dec. 16, 1631	Mt. Vesuvius, Italy	3,500 killed.
Aug. 4, 1672	Merapi, Java	3,000 killed by mudflows and pyroclastic flows.
Dec. 10, 1711	Awu, Indonesia	3,000 killed by pyroclastic flows.
Sept. 22, 1760	Makian, Indonesia	Eruption kills 2,000; island evacuated for seven years.
June 8, 1783	Lakagigar, Iceland	Largest historic lava flows: 12 km^3; 9,350 die.
July 26, 1783	Asama, Japan	Pyroclastic flows and floods kill 1,200+.
May 21, 1782	Unzen, Japan	14,500 die in debris avalanche and tsunami.
Apr. 10, 1815	Tambora, Indonesia	92,000 killed; another 80,000 reported to have died from famine and disease.
Oct. 8, 1822	Galunggung, Java	4,011 die in pyroclastic flows and mudflows.
Mar. 2, 1856	Awu, Indonesia	Pyroclastic flows kill 2,806.
Aug. 27, 1883	Krakatau, Indonesia	36,417 killed; most by tsunami.
June 7, 1892	Awu, Indonesia	1,532 die in pyroclastic flows.
May 8, 1902	Mt. Pelée, Martinique	St. Pierre destroyed by pyroclastic flow; 28,000 killed.
Oct. 24, 1902	Santa María, Guatemala	5,000 killed.
June 6, 1912	Novarupta, Alaska	Largest 20th-century eruption: about 33 km^3 of pyroclastic materials erupted; no fatalities.
May 19, 1919	Kelut, Java	Mudflows kill 5,110, devastate 104 villages.
Jan. 21, 1951	Lamington, New Guinea	2,942 killed by pyroclastic flows.
Mar. 17, 1963	Agung, Indonesia	1,148 killed.
Aug. 12, 1976	Soufrière, Guadeloupe	74,000 residents evacuated.
May 18, 1980	Mt. St. Helens, Washington	63 killed; 600 km^2 of forest devastated.
Mar. 28, 1982	El Chichón, Mexico	Pyroclastic flows kill 1,877.
Nov. 13, 1985	Nevado del Ruiz, Colombia	Mudflows kill 23,000.
Aug. 21, 1986	Oku volcanic field, Cameroon	1,746 asphyxiated by cloud of CO_2 released from Lake Nyos.
June 1991	Unzen, Japan	~43 killed; at least 8,500 fled.
June 1991	Mt. Pinatubo, Philippines	~281 killed during initial eruption; 83 killed by later mudflows; 358 died of illness.
Feb. 2, 1993	Mt. Mayon, Phillipines	At least 70 killed; 60,000 evacuated.

SOURCE: American Geological Institute Data Sheets, except for last three entries.

▼ INTRODUCTION

Erupting volcanoes are the most impressive manifestations of the dynamic processes operating within the Earth. During many eruptions, molten rock rises to the surface and flows as incandescent streams or is ejected into the atmosphere in fiery displays that are particularly impressive at night (◆ Fig. 4.5). In some parts of the world, volcanic eruptions are commonplace events. The residents of the Philippines, Iceland, Hawaii, and Japan are fully cognizant of volcanoes and their effects. In the United States, other than Alaska and Hawaii, volcanic eruptions are not particularly common and are currently localized in the Cascade Range of the northwest (see the Prologue).

◆ FIGURE 4.5 Lava fountains such as this one in Hawaii are particularly impressive at night.

Ironically, eruptions of volcanoes are constructive processes when considered in the context of Earth history. The Hawaiian Islands and Iceland, for example, owe their existence to volcanism. The oceanic crust is continually produced by volcanism at spreading ridges, and volcanic eruptions during the early history of the Earth released gases that are thought to have formed the atmosphere and the surface waters.

▼ VOLCANISM

Volcanism refers to the processes whereby magma and its associated gases rise through the Earth's crust and are extruded onto the surface or into the atmosphere. Currently, more than 500 volcanoes are *active*—that is, they have erupted during historic time. Well-known examples of active volcanoes include Mauna Loa and Kilauea on the island of Hawaii, Mount Etna on Sicily, Fujiyama in Japan, and Mount St. Helens in Washington (Fig. 4.4). Only two other bodies in the solar system are believed to possess active volcanoes: the Jovian moon Io, and possibly the Neptunian moon Triton.

In addition to active volcanoes, numerous *dormant volcanoes* exist that have not erupted recently but may do so again. For example, Mount Vesuvius in Italy had not erupted in human memory until A.D. 79 when it erupted and destroyed the cities of Herculaneum and Pompeii. Some volcanoes have not erupted during recorded history and show no evidence of doing so again; thousands of these *extinct* or *inactive* volcanoes are known.

Volcanic Gases

Samples of gases taken from present-day volcanoes indicate that 50 to 80% of all volcanic gases are water vapor. Lesser amounts of carbon dioxide, nitrogen, sulfur gases, especially sulfur dioxide and hydrogen sulfide, and very small amounts of carbon monoxide, hydrogen, and chlorine are also commonly emitted. In areas of recent volcanism, such as Lassen Volcanic National Park in California, gases continue to be emitted. One cannot help but notice the rotten-egg odor of hydrogen sulfide gas in such areas.

When magma rises toward the surface, the pressure is reduced and the contained gases begin to expand. However, in felsic magmas, which are highly viscous, expansion is inhibited and gas pressure increases. Eventually, the pressure may become great enough to cause an explosion and produce pyroclastic materials

VOLCANIC GASES AND CLIMATE

· · · · · · · · · · ·   · · · · · · · · · · ·

Most volcanic gases quickly dissipate in the atmosphere and pose little danger to humans, but on several occasions such gases have caused numerous fatalities. In 1783, toxic gases, probably sulfur dioxide, erupted from Laki fissure in Iceland had devastating effects. About 75% of the nation's livestock died, and the haze resulting from the gas caused lower temperatures and crop failures; about 24% of Iceland's population died as a result of the ensuing Blue Haze Famine.

Obviously, large volcanic eruptions can devastate local areas, but they can also affect climate over much larger regions—in some cases worldwide. The 1783 Laki fissure eruption produced what Benjamin Franklin called a "dry fog" that was responsible for dimming the intensity of sunlight in Europe. The severe winter of 1783–1784 in Europe and eastern North America is attributed to the presence of this "dry fog" in the upper atmosphere. In Iceland, the winter temperature was 4.8°C below the long-term average; the country suffered its coldest winter in 225 years.

More recently, in 1986, in the African nation of Cameroon 1,746 people died when a cloud of carbon dioxide engulfed them. The gas accumulated in the waters of Lake Nyos, which occupies a volcanic crater. No agreement exists on what caused the gas to suddenly burst forth from the lake, but once it did, it flowed downhill along the surface because it was denser than air. In fact, the density and velocity of the gas cloud were great enough to flatten vegetation, including trees, a few kilometers from the lake. Unfortunately, thousands of animals and many people, some as far as 23 km from the lake, were asphyxiated.

Volcanic ash erupted into the upper atmosphere has some effect on climate, but all particles except the smallest settle quickly and produce no long-lasting effect. Sulfur gases emitted during large eruptions have more important effects; small gas molecules remain in the upper atmosphere for years, absorbing incoming solar radiation and reflecting it back into space. In 1816, a persistent "dry fog" caused unusually cold spring and summer weather in Europe, the eastern United States, and eastern Canada. In North America, 1816 was called "The Year Without a Summer" or "Eighteen Hundred and Froze to Death." Killing frosts occurred through the summer in New England, resulting in crop failures and food shortages.

The particularly cold spring and summer of 1816 are attributed to the 1815 eruption of Tambora in Indonesia, the largest and most deadly eruption during historic time. The eruption of Mayon volcano in the Philippines during the previous year may have contributed to the cool spring and summer of 1816 as well. Another large historic eruption that had widespread climatic effects was the eruption of Krakatau in 1883.

In comparison with Tambora and Krakatau, the 1980 Mount St. Helens eruption was small. Furthermore, it did not emit much sulfur gas, and its explosion was directed laterally so that most of the particulate matter did not enter the upper atmosphere. In fact, the much smaller 1982 eruption of El Chichón in Mexico had a greater effect on the climate, because it erupted so much sulfur gas and its gases and ash were ejected vertically so that much of them entered the upper atmosphere.

---

such as ash. In contrast, low-viscosity mafic magmas allow gases to expand and escape easily. Accordingly, mafic magmas generally erupt rather quietly.

The amount of gases contained in magmas varies; it is rarely more than a few percent by weight. Even though volcanic gases constitute a small proportion of a magma, they can be dangerous, and, in some cases, have had far-reaching climatic effects (see Perspective 4.1).

Lava Flows and Pyroclastic Materials

Lava flows are frequently portrayed in movies and on television as fiery streams of incandescent rock material posing a great danger to humans. Actually, lava flows are the least dangerous manifestation of volcanism, although they may destroy buildings and cover agricultural land. Most lava flows do not move particularly fast, and because they are fluid, they follow existing low areas. Thus, once a flow erupts from a volcano, determining the path it will take is fairly easy, and anyone in areas likely to be affected can be evacuated.

The surfaces of lava flows may be marked by such features as pressure ridges and spatter cones. **Pressure ridges** are buckled areas on the surface of a lava flow (◆ Fig. 4.6a) that form because of pressure on the partly solid crust of a moving flow. **Spatter cones** form when gases escaping from a flow hurl globs of molten lava into the air. These globs fall back to the surface and adhere to one another, forming these small, steep-sided cones (◆ Fig. 4.6b).

Two types of lava flows, both of which were named for Hawaiian flows, are generally recognized. A **pahoe-**hoe (pronounced pah-hoy-hoy) flow has a ropy surface almost like taffy (◆ Fig. 4.7a). The surface of an **aa** (pronounced ah-ah) flow is characterized by rough, jagged angular blocks and fragments (◆ Fig. 4.7b). Pahoehoe flows are less viscous than aa flows; indeed, the latter are viscous enough to break up into blocks and move forward as a wall of rubble.

Columnar joints are common in many lava flows, especially mafic flows, but they also occur in other kinds of flows and in some intrusive igneous rocks (◆ Fig. 4.8). A lava flow contracts as it cools and thus produces forces that cause fractures called *joints* to open up. On the surface of a flow, these joints commonly form polygonal (often six-sided) cracks. These cracks also extend downward into the flow, thus form-

◆ FIGURE 4.6 (*a*) Pressure ridge on a 1982 lava flow in Hawaii. (*b*) A row of spatter cones formed on February 25, 1983, on a flow at Kilauea volcano, Hawaii.

◆ FIGURE 4.7 (*a*) Pahoehoe flow in the east rift zone of Kilauea volcano in 1972. (*b*) An aa flow in the east rift zone of Kilauea volcano, Hawaii in 1983. The flow front is about 2.5 m high.

(a)

(a)

(b)

(b)

ing parallel columns with their long axes perpendicular to the principal cooling surface. Excellent examples of columnar joints can be seen at Devil's Postpile National Monument in California (Fig. 4.8), Devil's Tower National Monument in Wyoming (see Chapter 3 Prologue), the Giant's Causeway in Ireland, and many other areas.

Much of the igneous rock in the upper part of the oceanic crust is of a distinctive type; it consists of bulbous masses of basalt resembling pillows, hence the name **pillow lava**. It was long recognized that pillow lava forms when lava is rapidly chilled beneath water, but its formation was not observed until 1971. Divers near Hawaii saw pillows form when a blob of lava broke through the crust of an underwater lava flow and cooled almost instantly, forming a glassy exterior. Remaining fluid inside then broke through the crust of the pillow, resulting in an accumulation of interconnected pillows (◆ Fig. 4.9).

Much pyroclastic material is erupted as **ash**, a designation for pyroclastic particles measuring less than 2.0 mm (◆ Fig. 4.10). Ash may be erupted in two ways: an ash fall or an ash flow. During an ash fall, ash is ejected into the atmosphere and settles to the surface over a wide area. In 1947, ash that erupted from Mount Hekla in Iceland fell 3,800 km away on Helsinki, Finland. Ash is also erupted in ash flows, which are coherent clouds of ash and gas that commonly flow along or close to the land surface. Such flows can move at more than 100 km per hour, and some of them cover vast areas.

Pyroclastic materials larger than ash are also erupted by explosive volcanoes. Particles measuring from 2 to 64 mm are known as *lapilli,* and any particle larger than 64 mm is called a *bomb* or *block* depending on its shape. Bombs have twisted, streamlined shapes that indicate they were erupted as globs of fluid that cooled and solidified during their flight through the air (◆ Fig. 4.11). Blocks are angular pieces of rock ripped from a volcanic conduit or pieces of a solidified crust of a magma. Because of their large size, volcanic bomb and

◆ FIGURE 4.8 (*a*) Columnar joints in a lava flow at Devil's Postpile National Monument, California. (*b*) Surface view of the same columnar joints showing their polygonal pattern. The straight lines and polish resulted from glacial ice moving over this surface.

(a)

(b)

◆ FIGURE 4.9 These bulbous masses of pillow lava form when magma is erupted under water.

◆ FIGURE 4.10 Mount Pinatubo in the Philippines is one of many volcanoes in a belt nearly encircling the Pacific Ocean basin. It is shown here erupting on June 12, 1991. A huge, thick cloud of ash and steam rises above Clark Air Force Base, from which about 15,000 people had already been evacuated to Subic Bay Naval Base. Following this eruption, the remaining 900 people at the base were also evacuated.

block accumulations are not nearly as widespread as ash deposits; instead, they are confined to the immediate area of eruption.

Volcanoes

Conical mountains formed around a vent where lava and pyroclastic materials are erupted are **volcanoes**. Volcanoes, which are named for *Vulcan*, the Roman deity of fire, come in many shapes and sizes, but geologists recognize several major categories, each of which has a distinctive eruptive style. One must realize, however, that each volcano is unique in terms of its overall history of eruptions and development. One of

◆ FIGURE 4.11 Pyroclastic materials: volcanic bombs collected in Hawaii.

the duties of the U.S. Geological Survey is monitoring active volcanoes and developing methods of forecasting eruptions (see Perspective 4.2).

Most volcanoes have a circular depression or **crater** at their summit. Craters form as a result of the extrusion of gases and lava from a volcano and are connected via a conduit to a magma chamber below the surface. It is not unusual, however, for magma to erupt from vents on the flanks of large volcanoes where smaller, parasitic cones develop. For example, Mount Etna on Sicily has some 200 smaller vents on its flanks.

Some volcanoes are characterized by a **caldera** rather than a crater. Craters are generally less than 1 km in diameter, whereas calderas exceed this dimension and have steep sides. One of the best-known calderas in this country is the misnamed Crater Lake in Oregon—Crater Lake is actually a caldera (◆ Fig. 4.12). It formed about 6,600 years ago after voluminous eruptions partially drained the magma chamber. This drainage left the summit of the mountain, Mount Mazama, unsupported, and it collapsed into the magma chamber, forming a caldera more than 1,200 m deep and measuring 9.7 by 6.5 km. Many calderas have probably formed when a summit has collapsed during particularly large, explosive eruptions as in the case of Crater Lake, but a few have apparently formed when the top of the original volcano was blasted away.

Shield Volcanoes

Shield volcanoes resemble the outer surface of a shield lying on the ground with the convex side up (◆ Fig. 4.13). They have low, rounded profiles with gentle

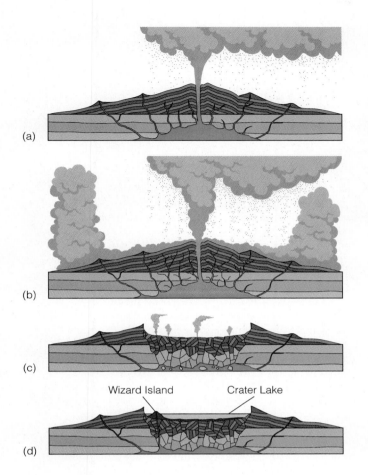

(a)

(b)

(c)

Wizard Island   Crater Lake

(d)

(e)

◆ FIGURE 4.12 The sequence of events leading to the origin of Crater Lake, Oregon. (*a–b*) Ash clouds and ash flows partly drain the magma chamber beneath Mount Mazama. (*c*) The collapse of the summit and formation of the caldera. (*d*) Post-caldera eruptions partly cover the caldera floor, and the small volcano known as Wizard Island forms. (*e*) View from the rim of Crater Lake showing Wizard Island.

◆ FIGURE 4.13 A shield volcano. Each layer shown consists of numerous, thin basalt lava flows.

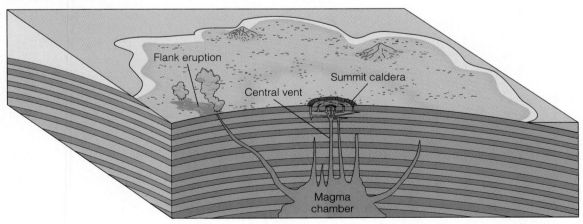

Flank eruption

Summit caldera

Central vent

Magma chamber

MONITORING VOLCANOES
AND FORECASTING ERUPTIONS

Two facilities in this country staffed by geologists of the U.S. Geological Survey (USGS) are devoted to volcano monitoring; Hawaiian Volcano Observatory on the rim of the crater of Kilauea volcano and the David A. Johnston Cascades Volcano Observatory in Vancouver, Washington. The latter was established in 1981 and named in memory of the USGS geologist killed during the 1980 Mount St. Helens eruption. This facility is responsible for monitoring the various Cascade Range volcanoes (Fig. 4.1).

Numerous volcanoes on the margins of the Earth's tectonic plates have erupted explosively during historic time and have the potential to do so again. As a matter of fact, volcanic eruptions are not as unusual as one might think; 376 separate outbursts occurred between 1975 and 1985. Fortunately, none of these compared to the 1815 eruption of Tambora; nevertheless, fatalities occurred in several instances, the worst being in 1985 in Colombia where about 23,000 perished in mudflows generated by an eruption (Table 4.1). Only a small minority of these potentially dangerous volcanoes are monitored, including some in Italy, Japan, New Zealand, the former Soviet Union, and the Cascade Range.

Many of the methods for monitoring active volcanoes were developed at the Hawaiian Volcano Observatory. These methods involve recording and analyzing various changes in both the physical and chemical attributes of volcanoes. Tilt-meters are used to detect changes in the slopes of a volcano when it inflates as magma is injected into it, while a geo-dimeter uses a laser beam to measure horizontal distances, which also change when a volcano inflates (◆ Fig. 1). Geologists also monitor gas emissions and changes in the local magnetic and electrical fields of volcanoes.

Of critical importance in volcano monitoring and eruption forecasting are a sudden increase in earthquake activity and the detection of *harmonic tremor*. Harmonic tremor is continuous ground motion as opposed to the sudden jolts produced by earthquakes. It precedes all eruptions of Hawaiian volcanoes and also preceded the eruption of Mount St. Helens. Such activity indicates that magma is moving below the surface.

The analysis of data gathered during monitoring is not by itself sufficient to forecast eruptions; the past history of a particular volcano must also be known. To determine the eruptive history of a volcano, the record of previous eruptions as preserved in rocks must be studied and analyzed. Indeed, prior to 1980, Mount St. Helens was considered one of the most likely Cascade volcanoes to erupt because detailed studies indicated that it has had a record of explosive activity for the past 4,500 years.

For the better monitored volcanoes, such as those in Hawaii, it is now possible to make accurate short-term forecasts of eruptions. For example, in 1960 the warning signs of an eruption of Kilauea were recognized soon enough to evacuate the residents of a small village that was subsequently buried by lava flows. Unfortunately, forecasting for more than a few months cannot be done at present.

For some volcanoes little or no information is available for making predictions. For example, on January 14, 1993, Colombia's Galeras volcano erupted without warning, killing 6 of 10 volcanologists on a field trip and 3 Colombian tourists. Ironically, the volcanologists were attending a conference on improving methods for predicting volcanic eruptions.

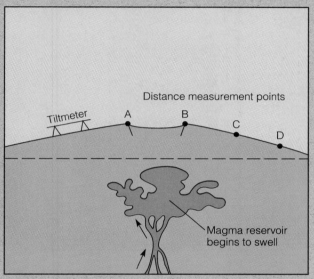

(a) Stage 1

Distance measurement points

Tiltmeter

A B

C

D

Magma reservoir
begins to swell

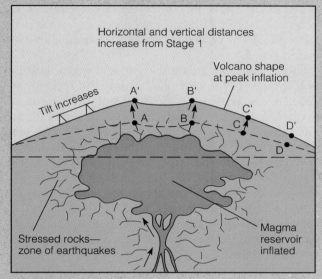

(b) Stage 2

Horizontal and vertical distances
increase from Stage 1

Tilt increases

A' B'

Volcano shape
at peak inflation

C'

A B

C

D'

D

Stressed rocks—
zone of earthquakes

Magma
reservoir
inflated

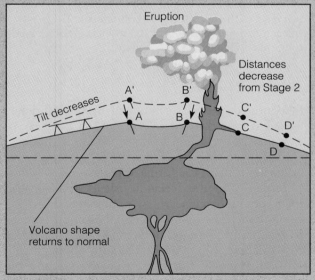

(c) Stage 3

Eruption

Distances
decrease
from Stage 2

A' B'

C'

Tilt decreases

A B

C

D'

C

D

Volcano shape
returns to normal

◆ FIGURE 1 Volcano monitoring. These diagrams show three stages in a typical eruption of a Hawaiian volcano: (*a*) The volcano begins to inflate; (*b*) inflation reaches its peak; (*c*) the volcano erupts and then deflates, returning to its normal shape.

slopes ranging from about 2 to 10 degrees. Their low slopes reflect the fact that they are composed mostly of mafic flows that had low viscosity, so the flows spread out and formed thin layers. Eruptions from shield volcanoes, sometimes called *Hawaiian-type volcanoes*, are quiet compared to those of volcanoes such as Mount St. Helens; lavas most commonly rise to the surface with little explosive activity, so they usually pose little danger to humans. Lava fountains, some up to 400 m high, contribute some pyroclastic materials to shield volcanoes (Fig. 4.5), but otherwise these volcanoes are composed largely of basalt lava flows; flows comprise more than 99% of the Hawaiian volcanoes above sea level.

Shield volcanoes are most common in oceanic areas, such as those of the Hawaiian Islands and Iceland, but some are also present on the continents—for example, in east Africa. The island of Hawaii consists of five huge shield volcanoes, two of which, Kilauea and Mauna Loa, are active much of the time. These Hawaiian volcanoes are the largest volcanoes in the world. Mauna Loa is nearly 100 km across at the base and stands more than 9.5 km above the surrounding sea floor. Its volume is estimated at about 50,000 km³. By contrast, the largest volcano in the continental United States, Mount Shasta in northern California, has a volume of only about 205 km³.

Shield volcanoes have a summit crater or caldera and a number of smaller cones on their flanks through which lava is erupted (Fig. 4.13). For example, a vent opened on the flank of Kilauea and grew to more than 250 m high between June 1983 and September 1986.

Cinder Cones

Volcanic peaks composed of pyroclastic materials that resemble cinders are known as **cinder cones** (◆ Fig. 4.14). They form when pyroclastic materials are ejected into the atmosphere and fall back to the surface to accumulate around the vent, thus forming small, steep-sided cones. The slope angle may be as much as 33 degrees, depending on the angle that can be maintained by the irregularly shaped pyroclastic materials. Cinder cones are rarely more than 400 m high, and many have a large, bowl-shaped crater.

Many cinder cones form on the flanks or within the calderas of larger volcanic mountains and appear to represent the final stages of activity, particularly in areas formerly characterized by basalt lava flows. Wizard Island in Crater Lake, Oregon, is a small cinder cone

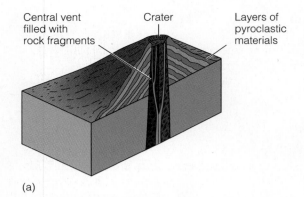

(a)

(b)

◆ FIGURE 4.14 (*a*) Cinder cones are composed of layers of angular pyroclastic materials. (*b*) The town of Vestmannaeyjar in Iceland was threatened by lava flows from Eldfell, a cinder cone, that formed in 1973. Within two days of the initial eruption on January 23, the new volcano had grown to about 100 m high. Another cinder cone called Helgafell is also visible.

that formed after the summit of Mount Mazama collapsed to form the caldera (Fig. 4.12). Cinder cones are common in the southern Rocky Mountain states, particularly New Mexico and Arizona, and many others occur in northern California, Oregon, and Washington.

In 1973, on the Icelandic island of Heimaey, the town of Vestmannaeyjar was threatened by a new cinder cone. The initial eruption began on January 23, and within two days a cinder cone, later named Eldfell, rose to about 100 m above the surrounding area (Fig. 4.14). Pyroclastic materials from the volcano buried parts of the town, and by February a massive aa lava flow was advancing toward the town. The flow's leading edge ranged from 10 to 20 m thick, and its central part was as much as 100 m thick. By spraying the leading edge of the flow with sea water, which caused it to cool and solidify, the residents of Vestmannaeyjar successfully diverted the flow before it did much damage to the town.

Composite Volcanoes

Composite volcanoes, also called *stratovolcanoes*, are composed of both pyroclastic layers and lava flows (◆ Fig. 4.15). Typically, both materials have an intermediate composition, and the flows cool to form andesite. Recall that lava of intermediate composition is more viscous than mafic lava. In addition to lava flows and pyroclastic layers, a significant proportion of a composite volcanoe is made up of **lahars** (volcanic mudflows). Some lahars form when rain falls on layers of loose pyroclastic materials and creates a muddy slurry that moves downslope. On November 13, 1985, mudflows resulting from a rather minor eruption of Nevado del Ruiz in Colombia killed about 23,000 people. In the Philippines, 83 of the 722 victims of the June 1991 eruptions of Mount Pinatubo were killed by lahars (Table 4.1).

Composite volcanoes are steep sided near their summits, perhaps as much as 30 degrees, but the slope decreases toward the base where it is generally less than 5 degrees. Mayon volcano in the Philippines is one of the most perfectly symmetrical composite volcanoes on Earth. Its concave slopes rise ever steeper to the summit with its central vent through which lava and pyroclastic materials are periodically erupted (Fig. 4.15). Mayon erupted for the twelfth time this century in February 1993.

Composite volcanoes are the typical large volcanoes of the continents and island arcs. Familiar examples include Fujiyama in Japan and Mount Vesuvius in Italy as well as Mount St. Helens and many of the other volcanic peaks in the Cascade Range of western North America (Fig. 4.2).

Lava Domes

If the upward pressure in a volcanic conduit is great enough, the most viscous magmas move upward and form bulbous, steep-sided **lava domes** (◆ Fig. 4.16). Lava domes are generally composed of felsic lavas although some are of intermediate composition. Because such magma is so viscous, it moves upward very slowly; the lava dome that formed in Santa María volcano in Guatemala in 1922 took two years to grow to 500 m high and 1,200 m across. Lava domes contribute significantly to many composite volcanoes. Beginning in 1980, a number of lava domes were emplaced in the crater of Mount St. Helens; most of these were destroyed during subsequent eruptions. Since 1983, Mount St. Helens has been characterized by sporadic dome growth.

◆ FIGURE 4.15 (*a*) Composite volcanoes are the typical, large volcanic mountains on continents. They are composed of lava flows, pyroclastic layers, and volcanic mudflows. (*b*) Mayon volcano in the Philippines is one of the most nearly symmetrical composite volcanoes in the world.

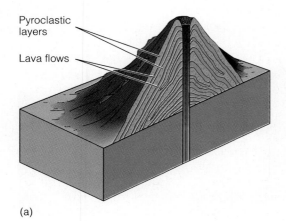

(a)

(b)

In June 1991, a dome in Japan's Unzen volcano collapsed, causing a flow of debris and hot ash that killed 43 people in a nearby town. Lava domes are also often responsible for extremely explosive eruptions. In 1902, viscous magma accumulated beneath the summit of Mount Pelée on the island of Martinique. Eventually, the pressure within the mountain increased to the point that it could no longer be contained, and the side of the mountain blew out in a tremendous explosion. When this occurred, a mobile, dense cloud of pyroclastic materials and gases called a **nuée ardente** (French for "glowing cloud") was ejected and raced downhill at about 100 km/hr, engulfing the city of St. Pierre (◆ Fig. 4.17). This nuée ardente had internal temperatures of 700°C and incinerated everything in its path. Of the 28,000 residents of St. Pierre, only 2 survived, a prisoner in a cell below the ground surface and a man on the surface who was terribly burned by the nuée ardente.

Fissure Eruptions

During the Miocene and Pliocene epochs (between about 17 million and 5 million years ago), some 164,000 km^2 of eastern Washington and parts of Oregon and Idaho were covered by overlapping basalt lava flows. These Columbia River basalts, as they are called, are now well exposed in the walls of the canyons eroded by the Snake and Columbia rivers (◆ Fig. 4.18). These lavas, which were erupted from long fissures, were so fluid that volcanic cones failed to develop. Such **fissure eruptions** yield flows that spread out over large areas and form **basalt plateaus** (◆ Fig. 4.19). The Columbia River basalt flows have an aggregate thickness of about 1,000 m, and some individual flows cover huge areas—for example, the Roza flow, which is 30 m thick, advanced along a front about 100 km wide and covered 40,000 km^2.

◆ FIGURE 4.16 A cross section showing the internal structure of a lava dome. Lava domes form when a viscous mass of magma, generally of felsic composition, is forced up through a volcanic conduit.

◆ FIGURE 4.17 St. Pierre, Martinique after it was destroyed by a nuée ardente erupted from Mount Pelée in 1902. Only 2 of the city's 28,000 inhabitants survived.

Fissure eruptions and basalt plateaus are not common, although several large areas of such features are known. The only area where such activity is currently occurring is in Iceland. A number of volcanic mountains are present in Iceland, but the bulk of the island is composed of basalt flows erupted from fissures. Two major fissure eruptions, one in A.D. 930 and the other in 1783, account for about half of the magma erupted in Iceland during historic time. The 1783 eruption occurred along the Laki fissure, which is 25 km long; lava flowed several tens of kilometers from the fissure and in one place filled a valley to a depth of about 200 m.

◆ FIGURE 4.18 The Columbia River basalts.

◆ FIGURE 4.19 A block diagram showing fissure eruptions and the origin of a basalt plateau.

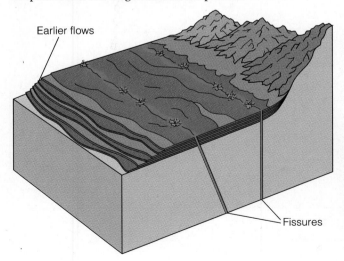

Earlier flows

Fissures

Pyroclastic Sheet Deposits

More than 100 years ago, geologists were aware of vast areas covered by felsic volcanic rocks a few meters to hundreds of meters thick. It seemed improbable that these could have formed as vast lava flows, but it also seemed equally unlikely that they were ash fall deposits. Based on observations of historic pyroclastic flows, such as the nuée ardente erupted by Mount Pelée in 1902, it now seems probable that these ancient rocks originated as pyroclastic flows. They cover far greater areas than any observed during historic time, however, and apparently erupted from long fissures rather than from a central vent. The pyroclastic materials of many of these flows were so hot that they fused together to form *welded tuff* (tuff is a volcanic rock composed of consolidated ash).

It now appears that major pyroclastic flows issue from fissures formed during the origin of calderas. For example, the Yellowstone Tuff was erupted during the formation of a large caldera in the area of present-day Yellowstone National Park in Wyoming (◆ Fig. 4.20). Similarly, the Bishop Tuff of eastern California appears to have been erupted shortly before the formation of the Long Valley caldera. Interestingly, earthquake activity in the Long Valley caldera and nearby areas beginning in 1978 may indicate that magma is moving upward beneath part of the caldera. Thus, the possibility of future eruptions in that area cannot be discounted.

◆ FIGURE 4.20 The Yellowstone Tuff in the walls of the Grand Canyon of the Yellowstone, Yellowstone National Park, Wyoming. Tuff is a volcanic rock composed of consolidated ash.

♼ DISTRIBUTION OF VOLCANOES

Rather than being distributed randomly around the Earth, volcanoes occur in well-defined zones or belts. More than 60% of all active volcanoes are in the **circum-Pacific belt** that nearly encircles the margins of the Pacific Ocean basin (♦ Fig. 4.21). This belt includes the volcanoes along the west coast of South America, those in Central America, Mexico, and the Cascade Range, and the Alaskan volcanoes in the Aleutian Island arc. The belt continues on the western side of the Pacific Ocean basin where it extends through Japan, the Philippines, Indonesia, and New Zealand. Mount Pinatubo and Mayon volcano, two Philippine volcanoes that have erupted since June 1991, are in this belt. The circum-Pacific belt also includes the southernmost active volcano, Mount Erebus in Antarctica,

and a large caldera at Deception Island that erupted during 1970 (Fig. 4.21).

About 20% of all active volcanoes are in the **Mediterranean belt** (Fig. 4.21). Included in this belt are the famous Italian volcanoes such as Mount Etna, Stromboli, and Mount Vesuvius.

Most of the large volcanoes in the circum-Pacific and Mediterranean belts are composite volcanoes, but a number of them have had lava domes emplaced in their craters or calderas. The fact that most of the volcanoes in these two belts are composite volcanoes is significant. Recall that such volcanoes are composed of lava flows and pyroclastic layers of intermediate and felsic composition whereas those within the ocean basins are composed primarily of mafic lavas.

Most of the rest of the active volcanoes are at or near the mid-oceanic ridges (Fig. 4.21). The longest of these

♦ FIGURE 4.21 Most volcanoes are at or near plate boundaries. Two major volcano belts are recognized: the circum-Pacific belt contains about 60% of all active volcanoes, about 20% are in the Mediterranean belt, and most of the rest are located along mid-oceanic ridges.

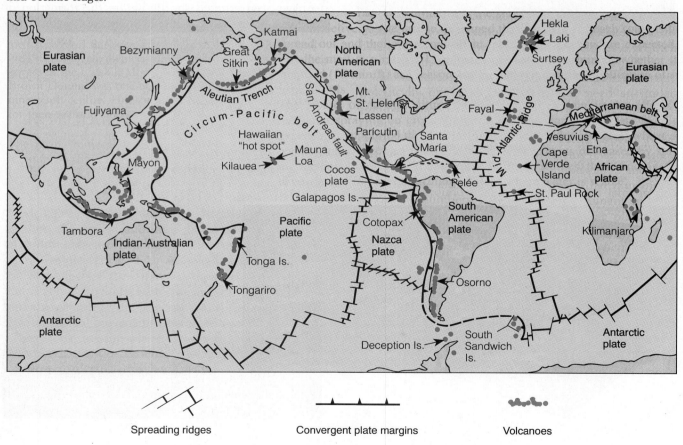

ridges is the Mid-Atlantic Ridge, which is near the middle of the Atlantic Ocean basin and curves around the southern tip of Africa where it continues as the Indian Ridge. Branches of the Indian Ridge extend into the Red Sea and East Africa. Mount Kilimanjaro in Africa is on this latter branch (Fig. 4.21). Most of the volcanism along the mid-oceanic ridges is submarine, and much of it goes undetected; but in a few places, such as Iceland, it occurs above sea level.

Volcanism is occurring in a few other areas at present, most notably on and near the island of Hawaii (Fig. 4.21). Only two volcanoes are currently active on the island, Mauna Loa and Kilauea, although a submarine volcano named Loihi exists about 32 km to the south; Loihi rises more than 3,000 m above the sea floor, but its summit is still about 940 m below sea level.

▼ PLATE TECTONICS AND IGNEOUS ACTIVITY

At this point, two questions might be raised regarding volcanoes: (1) What accounts for the alignment of volcanoes in belts? (2) Why do magmas erupted within ocean basins and magmas erupted at or near continental margins have different compositions? Recall from Chapter 1 that the outer part of the Earth is divided into large plates, which are sections of the lithosphere. Lithosphere can consist of upper mantle and oceanic crust or upper mantle and continental crust, called oceanic and continental lithosphere, respectively. Most volcanism occurs at spreading ridges where plates diverge or along subduction zones where plates converge.

Volcanism at Spreading Ridges

Spreading ridges are areas where new oceanic lithosphere is produced by volcanism as plates diverge from one another (Fig. 4.21). Most spreading ridges are in the ocean basins, but some extend into continents, as in east Africa.

Some of the volcanism at spreading ridges is apparent because it occurs above sea level, but, as previously noted, much of it is submarine and goes undetected. However, research submarines have descended into the rifts at the crests of spreading ridges where scientists have observed pillow lavas that formed during submarine eruptions (Fig. 4.9).

The fact that volcanism occurs at spreading ridges is undisputed, but how magma originates beneath the ridges is not fully understood. One explanation is

related to the manner in which the Earth's temperature increases with depth. We know from deep mines and deep drill holes that a temperature increase, called the *geothermal gradient*, does occur and that, on average, the gradient is about 25°C/km. Accordingly, rocks at depth are hot, but remain solid because their melting temperature rises with increasing pressure.

Rising hot rock beneath spreading ridges maintains a geothermal gradient well above average, and locally the temperature exceeds the melting temperature, at least in part, because pressure effects are overcome (◆ Fig. 4.22). That is, rifting at spreading ridges probably causes a decrease in pressure on the hot rocks at depth, thus initiating melting. Rifting is unlikely to be the sole cause of melting, however, because melting also occurs in some areas where there appears to be no rifting, such as beneath the Hawaiian Islands.

Another explanation for spreading ridge volcanism is that localized, cylindrical plumes of hot mantle material, called **mantle plumes** rise beneath spreading ridges and spread outward in all directions (◆ Fig. 4.23). Perhaps localized concentrations of radioactive minerals within the crust and upper mantle decay and

◆ FIGURE 4.22 Calculated geothermal gradients under the continents and ocean basins (somewhat speculative). The melting of dry basalt has been experimentally investigated only in the pressure-temperature region indicated by the solid line. Melting of dry peridotite occurs between 100 and 125 km beneath the ocean basins. However, should the pressure be reduced, as occurs at spreading ridges, melting might occur at even shallower depths.

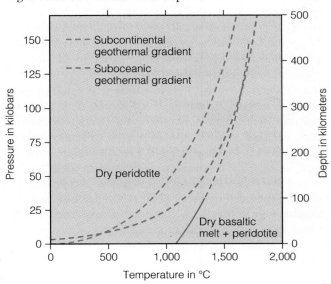

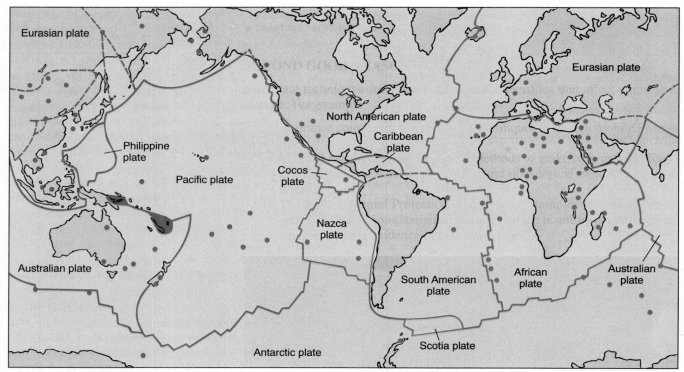

◆ FIGURE 4.23 Some of the "hot spots" in the Earth's crust that are thought to overlie rising mantle plumes.

generate the heat responsible for the melting associated with these hot mantle plumes.

The lavas erupted at spreading ridges are invariably mafic and cool to form the igneous rock basalt. However, the upper mantle, from which these lavas are derived, is composed of ultramafic rock, probably peridotite, which consists largely of ferromagnesian silicates and lesser amounts of nonferromagnesian silicates. To explain how mafic magma (45-52% silica) originates from ultramafic rock (≤ 45% silica), geologists propose that the magma is formed from source rock that only partially melts. This phenomenon of *partial melting* occurs because various minerals have different melting temperatures (see Fig. 3.7). When ultramafic rock begins to melt, the minerals richest in silica melt first followed by those containing less silica. Accordingly, if melting is not complete, a mafic magma containing proportionately more silica than the source rock results. Once this mafic magma is formed, some of it rises to the surface where it is erupted, cools, and crystallizes to form the extrusive igneous rock basalt (see Table 3.1).

Volcanism at Subduction Zones

Three types of plate convergence are recognized: convergence between two oceanic plates, convergence between an oceanic plate and a continental plate, and convergence between two continental plates (Fig. 1.7). In the first two types of convergence, an oceanic plate is subducted beneath another plate, and volcanism occurs near the leading margin of the overriding plate. Continental lithosphere is not dense enough to be subducted, so little volcanism occurs where two such plates converge.

To illustrate how volcanism is related to subduction, let us consider what occurs along the west coast of South America. As a consequence of spreading at the East Pacific Rise, the Nazca plate moves east and collides with the South American plate which is moving west (◆ Fig. 4.24). Thus, a collision occurs between oceanic and continental lithosphere, and because the Nazca plate is denser, it plunges beneath the South American plate (Fig. 4.24).

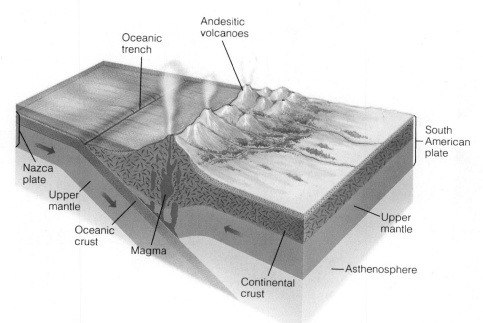

Oceanic trench

Andesitic volcanoes

South American plate

Nazca plate

Upper mantle

Oceanic crust

Magma

Upper mantle

Asthenosphere

Continental crust

◆ FIGURE 4.24 The subduction of the Nazca plate beneath the South American plate produces magma, some of which is erupted to form andesitic volcanoes.

The belt of large composite volcanoes near the western margin of South America formed from the magma created by partial melting of the subducted plate. As the Nazca plate descends toward the asthenosphere, it is heated by the Earth's geothermal gradient. When the descending plate reaches a depth where the temperature is high enough, partial melting occurs and magma is generated (Fig. 4.24). Additionally, the wet oceanic crust descends to a depth at which dewatering occurs. As the water rises into the overlying mantle, it enhances melting, and a magma may be generated.

Partial melting is one phenomenon accounting for the fact that magmas generated at subduction zones are intermediate and felsic in composition. Recall that partial melting of ultramafic rock of the upper mantle yields mafic magma. Likewise, partial melting of oceanic crust, which has a mafic composition, may yield magma richer in silica than the source rock. Additionally, some of the silica-rich sediments and sedimentary rocks of continental margins are probably carried downward with the subducted plate and contribute their silica to the magma. In addition, mafic magma rising through the lower continental crust may be contaminated with felsic materials, which change its composition.

Intermediate and felsic magmas are typically produced at convergent plate margins where subduction occurs. The intermediate magma that is erupted is more viscous than mafic magma and tends to form composite volcanoes. Much felsic magma is intruded into the continental crust where it forms various intrusive igneous bodies (see Chapter 3), but some is erupted as pyroclastic materials or emplaced as lava domes, thus accounting for the explosive eruptions that characterize convergent plate margins.

Intraplate Volcanism

Mauna Loa and Kilauea on the island of Hawaii and Loihi just to the south are within the interior of a rigid plate far from any spreading ridge or subduction zone (Fig. 4.21). It is postulated that a mantle plume creates a local "hot spot" beneath Hawaii. The magma is mafic and thus relatively fluid, so it builds up shield volcanoes (◆ Fig. 4.25).

Even though these Hawaiian volcanoes are unrelated to spreading ridges or subduction zones, the evolution of the Hawaiian Islands is related to plate tectonics. Notice in Figure 4.25 that the ages of the rocks composing the islands in the Hawaiian chain increase toward the northwest; Kauai formed 3.0 to 5.6 million years ago, whereas Hawaii began forming less than one million years ago, and Loihi began forming even more recently. Continuous motion of the Pacific plate over the "hot spot," now beneath Hawaii, has created the various islands in succession.

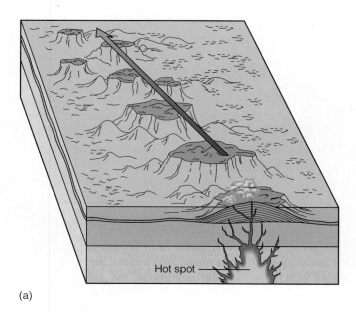

(a)

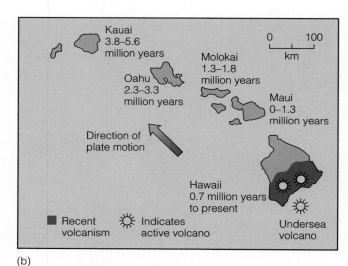

(b)

◆ FIGURE 4.25 (a) Generalized diagram showing the origin of the Hawaiian Islands. As the lithospheric plate moves over a hot spot, a succession of volcanoes forms. The only present-day volcanism occurs on Hawaii and beneath the sea just to the south. (b) Map showing the age of the islands in the Hawaiian chain.

Mantle plumes and "hot spots" have also been proposed to explain volcanism in a few other areas. A mantle plume may be beneath Yellowstone National Park in Wyoming. Some source of heat at depth is responsible for the present-day hot springs and geysers such as Old Faithful, but many geologists think that the source of heat is a body of intruded magma that has not yet completely cooled rather than a mantle plume.

Chapter Summary

1. Volcanism is the process whereby magma and its associated gases erupt at the surface. Some magma erupts as lava flows, and some is ejected explosively as pyroclastic materials.

2. Only a few percent by weight of a magma consists of gases, most of which is water vapor. Sulfur gases emitted during large eruptions can have far-reaching climatic effects.

3. Many lava flows are characterized by pressure ridges and spatter cones. Columnar joints form in some lava flows when they cool. Pillow lavas are erupted under water and consist of interconnected bulbous masses.

4. Volcanoes are conical mountains built up around a vent where lava flows and/or pyroclastic materials are erupted.

5. Shield volcanoes have low, rounded profiles and are composed mostly of mafic flows that have cooled and formed basalt. Cinder cones form where pyroclastic materials that resemble cinders are erupted and accumulate as small, steep-sided cones. Composite volcanoes are composed of lava flows of intermediate composition, layers of pyroclastic materials, and volcanic mudflows.

6. Viscous masses of lava, generally of felsic composition, are forced up through the conduits of some volcanoes and form bulbous, steep-sided lava domes. Volcanoes with lava domes are dangerous because they erupt explosively and frequently eject nuée ardentes.

7. The summits of volcanoes are characterized by a circular or oval crater or a much larger caldera. Many calderas form by summit collapse when an underlying magma chamber is partly drained.

8. Fluid mafic lava erupted from long fissures (fissure eruptions) spreads over large areas to form basalt plateaus.

9. Pyroclastic flows erupted from fissures formed during the origin of calderas cover vast areas. Such eruptions of pyroclastic materials form sheetlike deposits.

10. Most active volcanoes are distributed in linear belts. The circum-Pacific belt and Mediterranean belt contain more than 80% of all active volcanoes.

11. Volcanism in the circum-Pacific and Mediterranean belts is at convergent plate margins where subduction occurs. Partial melting of the subducted plate generates intermediate and felsic magmas.

12. Magma derived by partial melting of the upper mantle beneath spreading ridges accounts for the mafic lavas of ocean basins. Melting in these areas may be caused by reduction in pressure and/or hot mantle plumes.

13. The two active volcanoes on the island of Hawaii and one just to the south are thought to lie above a hot mantle plume. The Hawaiian Islands developed as a series of volcanoes formed on the Pacific plate as it moved over the mantle plume.

Important Terms

aa	crater	pahoehoe
ash	fissure eruption	pillow lava
basalt plateau	lahar	pressure ridge
caldera	lava dome	shield volcano
cinder cone	mantle plume	spatter cone
circum-Pacific belt	Mediterranean belt	volcanism
columnar joint	nuée ardent	volcano
composite volcano (stratovolcano)		

Review Questions

1. Which of the following is most dangerous to humans?
   a. ___ nuée ardente; b. ___ lava flows; c. ___ volcanic bombs; d. ___ pahoehoe; e. ___ pillow lava.

2. A lava flow with a surface of jagged blocks is termed:
   a. ___ lapilli; b. ___ vesicular; c. ___ aa; d. ___ obsidian; e. ___ pyroclastic sheet deposit.

3. Most calderas form by:
   a. ___ summit collapse; b. ___ explosions; c. ___ fissure eruptions; d. ___ forceful injection; e. ___ erosion of lava domes.

4. Basalt plateaus form as a result of:
   a. ___ repeated eruptions of cinder cones; b. ___ widespread ash falls; c. ___ accumulation of thick layers of pyroclastic materials; d. ___ the origin of lahars on composite volcanoes; e. ___ eruptions of fluid lava from long fissures.

5. One other Cascade Range volcano besides Mount St. Helens has erupted during this century. It is:
   a. ___ Mount Hood, Oregon; b. ___ Mount Lassen, California; c. ___ Mount Garibaldi, British Columbia; d. ___ Mount Adams, Washington; e. ___ Mount Mazama, Oregon.

6. Volcanic or extrusive igneous rocks form by the cooling and crystallization of lava flows and the:
   a. ___ crystallization of magma beneath the surface; b. ___ consolidation of pyroclastic materials; c. ___ reaction of volcanic gases with the atmosphere; d. ___ heating of sedimentary rocks beneath lava flows; e. ___ all of these.

7. The most commonly emitted volcanic gas is:
   a. ___ carbon dioxide; b. ___ hydrogen sulfide; c. ___ nitrogen; d. ___ chlorine; e. ___ water vapor.

8. Small, steep-sided cones that form on the surfaces of lava flows where gases escape are:
a. ___ lava tubes; b. ___ spatter cones; c. ___ columnar joints; d. ___ pahoehoe; e. ___ volcanic bombs.

9. Much of the upper part of the oceanic crust is composed of interconnected bulbous masses of igneous rock called:
a. ___ pillow lava; b. ___ lapilli; c. ___ pyroclastic material; d. ___ parasitic cones; e. ___ blocks.

10. Shield volcanoes have low slopes because they are composed of:
a. ___ mostly pyroclastic layers; b. ___ lahars and viscous lava flows; c. ___ fluid mafic lava flows; d. ___ felsic magma; e. ___ pillow lavas.

11. Crater Lake in Oregon is an excellent example of a:
a. ___ caldera; b. ___ cinder cone; c. ___ shield volcano; d. ___ basalt plateau; e. ___ lava dome.

12. The volcanic conduit of a lava dome is most commonly plugged by:
a. ___ mafic magma; b. ___ columnar joints; c. ___ viscous, felsic magma; d. ___ volcanic mudflows; e. ___ spatter cones.

13. Most active volcanoes are in:
a. ___ the Mediterranean belt; b. ___ the Hawaiian Islands; c. ___ Iceland; d. ___ the circum-Pacific belt; e. ___ the oceanic ridge belt.

14. The magma generated beneath spreading ridges is mostly:
a. ___ mafic; b. ___ felsic; c. ___ intermediate; d. ___ all of these; e. ___ answers (a) and (b) only.

15. The volcanoes of _____ are unrelated to either a divergent or a convergent plate margin.
a. ___ East Africa; b. ___ the mid-oceanic ridges; c. ___ the Cascade Range; d. ___ the Hawaiian Islands; e. ___ Iceland.

16. The largest volcano in the world is:
a. ___ Mount St. Helens, Washington; b. ___ Mount Etna, Sicily; c. ___ Fujiyama, Japan; d. ___ Mount Vesuvius, Italy; e. ___ Mauna Loa, Hawaii.

17. The only area where fissure eruptions are currently occurring is:
a. ___ the Red Sea; b. ___ western South America; c. ___ the Pacific Northwest; d. ___ Iceland; e. ___ Japan.

18. Explain how pyroclastic materials and volcanic gases can affect climate.

19. How do spatter cones and columnar joints form?

20. What accounts for the fact that volcanic ash can cover vast areas, whereas pyroclastic materials such as cinders are not very widely distributed?

21. Explain how most calderas form.

22. What kinds of warning signs enable geologists to forecast eruptions?

23. Why do shield volcanoes have such low slopes?

24. How do pahoehoe and aa lava flows differ?

25. Draw a cross section of a composite volcano. Indicate its constituent materials, and show how and where a flank eruption might occur.

26. Why do composite volcanoes occur in belts near convergent plate margins? Are such volcanoes present at all convergent plate margins?

27. Why are lava domes so dangerous?

28. Compare basalt plateaus and pyroclastic sheet deposits.

29. Give a brief summary of the origin and development of the Hawaiian Islands.

Additional Readings

Aylesworth, T. G., and V. Aylesworth. 1983. *The Mount St. Helens disaster: What we've learned.* New York: Franklin Watts.

Bullard, F. M. 1984. *Volcanoes of the Earth.* 2d ed. Austin, Tex.: Univ. of Texas Press.

Coffin, M. F. and O. Eldholm. 1993. Large Igneous Provinces. *Scientific American* 269, no. 4: 42–49.

Decker, R. W., and Decker, B. B. 1991. *Mountains of fire: The nature of volcanoes.* New York: Cambridge Univ. Press.

Erickson, J. 1988. *Volcanoes & earthquakes.* Blue Ridge Summit, Pa.: Tab Books.

Grove, N. 1992. Volcanoes: Crucibles of creation. *National Geographic* 182, no. 6: 5–41.

Lipman, P. W., and D. R. Mullineaux, eds. 1981. The 1980 eruptions of Mount St. Helens, Washington. *United States Geological Survey Professional Paper 1250.*

McClelland, L., T. Simkin, M. Summers, E. Nielsen, and T. C. Stein, eds. 1989. *Global volcanism 1975–1985.* Englewood Cliffs, N.J.: Prentice-Hall.

Rampino, M. R., S. Self, and R. B. Strothers. 1988. Volcanic winters. *Annual Review of Earth and Planetary Sciences* 16:73–99.

Simkin, T. *et al.* 1981. *Volcanoes of the world: A regional gazetteer, and chronology of volcanism during the last 10,000 years.* Stroudsburg, Pa.: Hutchison Ross.

Tilling, R. I. 1987 *Eruptions of Mount St. Helens: Past, present, and future.* U.S. Geological Survey.

Tilling, R. I., C. Heliker, and T. L. Wright. 1987. *Eruptions of Hawaiian volcanoes: Past, present, and future.* U. S. Geological Survey.

Volcanoes and the Earth's interior. 1982. Readings from Scientific American. San Francisco: W. H. Freeman.

Wenkam, R. 1987. *The edge of fire: Volcano and earthquake country in western North America and Hawaii.* San Francisco: Sierra Club Books.

CHAPTER
5

Sedimentary rocks exposed in the Sheep Rock area of John Day Fossil Beds National Monument, Oregon. This small hill is capped by the remnants of a lava flow.

Weathering, Soil, Sediment, and Sedimentary Rocks

Prologue

About 50 million years ago, two large lakes existed in what are now parts of Wyoming, Utah, and Colorado. Sand, mud, and dissolved minerals were carried into these lakes where they accumulated as layers of sediment that were subsequently converted into sedimentary rock. These sedimentary rocks, called the Green River Formation, contain the fossilized remains of millions of fish, plants, and insects and are a potential source of large quantities of oil, combustible gases, and other substances.

Thousands of fossilized fish skeletons are found on single surfaces within the Green River Formation, indicating that mass mortality must have occurred repeatedly (♦ Fig. 5.1) The cause of these events is not known with certainty, but some geologists have suggested that blooms of blue-green algae produced toxic substances that killed the fish. Others propose that rapidly changing water temperature or excessive salinity at times of increased evaporation was responsible. Whatever the cause, the fish died by the thousands and settled to the lake bottom where their decomposition was inhibited because the water contained little or no oxygen. One area of the formation in Wyoming where fossil plants are particularly abundant has been designated as Fossil Butte National Monument.

The Green River Formation is also well known for its huge deposits of oil shale (♦ Fig. 5.2). Oil shale consists of small clay particles and an organic substance known as *kerogen*. When the appropriate extraction processes are used, liquid oil and combustible gases can be produced from the kerogen

♦ FIGURE 5.1 Fossil fish from the Green River Formation of Wyoming. (Photo courtesy of Sue Monroe.)

♦ FIGURE 5.2 Exposures of oil shales of the Green River Formation.

of oil shale. To be designated as a true oil shale, however, the rock must yield a minimum of 10 gallons of oil per ton of rock. The use of oil shale as a source of fuel is not new. During the Middle Ages, people in Europe used oil shale as solid fuel for domestic purposes, and during the 1850s, small oil shale industries existed in the eastern United States; the latter were discontinued, however, when drilling and pumping of oil began in 1859. Oil shales occur on all continents, but the Green River Formation contains the most extensive deposits and has the potential to yield huge quantities of oil.

Oil can be produced from oil shale by a process in which the rock is heated to nearly 500°C in the absence of oxygen, and hydrocarbons are driven off as gases and recovered by condensation. During this process, 25 to 75% of the organic matter of oil shale can be converted to oil and combustible gases. The Green River Formation oil shales yield between 10 and 140 gallons of oil per ton of rock processed, and the total amount of oil recoverable with present processes is estimated at 80 billion barrels. Currently, however, no oil is produced from oil shale in the United States, because conventional drilling and pumping is less expensive. Nevertheless, the Green River oil shale constitutes one of the largest untapped sources of oil in the world. If more effective processes are developed, it could eventually yield even more than the currently estimated 80 billion barrels.

One should realize, however, that at the current and expected consumption rates of oil in the United States, oil production from oil shale will not solve all of our energy needs. Furthermore, large-scale mining that would be necessary would have considerable environmental impact. What would be done with the billions of tons of processed rock? Can such large-scale mining be conducted with minimal disruption of wildlife habitats and groundwater systems? Where will the huge volumes of water necessary for processing come from—especially in an area where water is already in short supply?

These and other questions are currently being considered by scientists and industry. Perhaps at some future time, the Green River Formation will provide some of our energy needs.

⩒ INTRODUCTION

Weathering is the physical breakdown (disintegration) and chemical alteration (decomposition) of rocks and minerals at or near the Earth's surface. It is the process whereby rocks and minerals are physically and chemically altered such that they are more nearly in equilibrium with a new set of environmental conditions. For example, many rocks form within the Earth's crust where little or no water or oxygen is present and where temperatures, pressures, or both are high. At or near the surface, however, the rocks are exposed to low temperatures and pressures and are attacked by atmospheric gases, water, acids, and organisms.

Geologists are interested in the phenomenon of weathering because it is an essential part of the rock cycle (◆ Fig. 5.3). The *parent material*, or rock being weathered, is broken down into smaller pieces, and some of its constituent minerals are dissolved or altered and removed from the weathering site. The removal of the weathered materials is known as **erosion.** Running water, wind, or glaciers commonly **transport** the weathered materials elsewhere where they are deposited as sediment, which may become sedimentary rock (Fig. 5.3). Whether they are eroded or not, weathered rock materials can also be further modified to form a soil. Thus, weathering provides the raw materials for both sedimentary rocks and soils. Weathering is also important in the origin of some mineral resources such as aluminum ores, and it is responsible for the enrichment of other deposits of economic importance.

Two types of weathering are recognized, *mechanical* and *chemical.* Both types occur simultaneously at the weathering site, during erosion and transport, and even in the environments where weathered materials are deposited.

◆ FIGURE 5.3 The rock cycle, with emphasis on weathering, sediments, and sedimentary rocks.

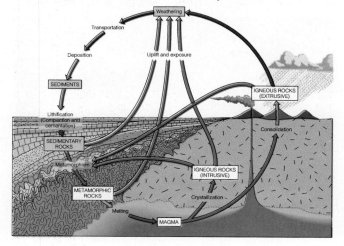

MECHANICAL WEATHERING

Mechanical weathering occurs when physical forces break rock materials into smaller pieces that retain the chemical composition of the parent material. For example, granite may be mechanically weathered to yield smaller pieces of granite, or disintegration may liberate individual mineral grains from it (◆ Fig. 5.4). The physical processes responsible for mechanical weathering include *frost action, pressure release, thermal expansion and contraction,* and the *activities of organisms.*

Frost action involves the repeated freezing and thawing of water in cracks and crevices in rocks. When water seeps into a crack and freezes, it expands by about 9% and exerts great force on the walls of the crack, thus widening and extending it by *frost wedging.* As a consequence of repeated freezing and thawing, pieces of rock are eventually detached from the parent material (◆ Fig. 5.5). The debris produced by frost wedging in mountains commonly accumulates as large cones of **talus** lying at the bases of slopes (◆ Fig. 5.6). Frost action is most effective in areas where temperatures commonly fluctuate above and below freezing, as in the high mountains of the western United States and Canada.

In the phenomenon known as *frost heaving,* a mass of sediment or soil undergoes freezing, expansion, and actual lifting, followed by thawing, contraction, and lowering of the mass. Frost heaving is particularly evident where water freezes beneath roadways and sidewalks.

Pressure release is a mechanical weathering process especially evident in rocks that formed as deeply buried intrusive bodies such as batholiths, but it occurs in other types of rocks as well. When a batholith forms, the magma crystallizes under tremendous pressure (the weight of the overlying rock) and is stable under these pressure conditions. When the batholith is uplifted and the overlying rock is stripped away by erosion, the pressure is reduced. However, the rock contains energy

◆ FIGURE 5.5 Frost wedging occurs when water seeps into cracks and expands as it freezes. Repeated freezing and thawing pry loose angular pieces of rock.

◆ FIGURE 5.4 Mechanically weathered granite. The sandy material consists of small pieces of granite (rock fragments) and minerals such as quartz and feldspars liberated from the parent material.

◆ FIGURE 5.6 Talus in the Canadian Rocky Mountains.

that is released by expansion and the formation of *sheet joints,* large fractures that more or less parallel the rock surface. Slabs of rock bounded by sheet joints may slip, slide, or spall (break) off of the host rock—a process called *exfoliation*—and accumulate as talus. The large rounded domes of rock resulting from this process are **exfoliation domes;** examples are found in Yosemite National Park in California (◆ Fig. 5.7) and Stone Mountain in Georgia. Sheet-jointing and exfoliation constitute an engineering problem in many areas (see Perspective 5.1).

During **thermal expansion and contraction** the volume of solids, such as rocks, changes in response to heating and cooling. In a desert, where the temperature may vary as much as 30°C in one day, rocks expand when heated and contract as they cool. Expansion and contraction do not occur uniformly throughout rocks, however. For one thing, a rock is a poor conductor of heat, so its outside heats up more than its inside. Consequently, the surface expands more than the interior, causing stresses that may cause fracturing. Furthermore, dark minerals absorb heat faster than light-colored minerals, so differential expansion occurs even between the mineral grains of some rocks. Experiments in which rocks are heated and cooled repeatedly to simulate years of such activity indicate that thermal expansion and contraction is not an important agent of mechanical weathering. But thermal expansion and contraction may be a significant mechanical weathering process on the Moon, where extreme temperature changes occur quickly.

Animals, plants, and bacteria all participate in the mechanical and chemical alteration of rocks. Burrowing animals, such as worms, reptiles, rodents, and many others, constantly mix soil and sediment particles and bring material from depth to the surface where further weathering may occur. The roots of plants, especially large bushes and trees, wedge themselves into cracks in rocks and further widen them (◆ Fig. 5.8).

◆ CHEMICAL WEATHERING

Chemical weathering is the process whereby rock materials are decomposed by chemical alteration of the parent material. Such weathering is accomplished by the action of atmospheric gases, especially oxygen, and by water and acids. Organisms also play an important role in chemical weathering. Rocks that have lichens (composite organisms consisting of fungi and algae) growing on their surfaces undergo more extensive chemical alteration than lichen-free rocks. Plants remove ions from soil water and reduce the chemical stability of soil minerals, and their roots release organic acids.

During **solution** the ions of a substance become dissociated from one another in a liquid, and the solid substance dissolves. Water is a remarkable solvent because its molecules have an asymmetric shape; they consist of one oxygen atom with two hydrogen atoms arranged such that the angle between the two hydrogens is about 104 degrees (◆ Fig. 5.9). Because of this asymmetry, the oxygen end of the molecule retains a slight negative electrical charge, whereas the hydrogen end retains a slight positive charge. When a soluble substance such as the mineral halite (NaCl) comes in contact with a water molecule, the positively charged sodium ions are attracted to the negative end of the water molecule, and the negatively charged chloride ions are attracted to the positively charged end of the water molecule (Fig. 5.9). Thus, ions are liberated from the crystal structure, and the solid dissolves.

Most minerals are not very soluble in pure water because the attractive forces of water molecules are not sufficient to overcome the forces between particles in minerals. For example, the mineral calcite ($CaCO_3$), the major constituent of the sedimentary rock limestone and the metamorphic rock marble, is practically insoluble in pure water, but rapidly dissolves if a small amount of acid is present. An easy way to make water acidic is by dissociating the ions of carbonic acid as follows:

$$H_2O + CO_2 \rightleftharpoons H_2CO_3 \rightleftharpoons H^+ + HCO_3^-$$

water   carbon   carbonic   hydrogen   bicarbonate
dioxide   acid   ion   ion

◆ FIGURE 5.7 Exfoliation domes in Yosemite National Park, California.

Perspective 5.1

BURSTING ROCKS AND SHEET JOINTS

The fact that solid rock can expand and produce fractures is a well-known phenomenon. In deep mines, for example, masses of rock suddenly detach from the sides of the excavation, often with explosive violence. Such *rock bursts* generally occur below depths of about 600 m; spectacular rock bursts have been recorded in deep gold mines in South Africa and Canada and in zinc mines in Idaho. Obviously, rock bursts and related phenomena, such as less violent *popping,* pose a danger to mine workers. In South Africa, about 20 miners are killed by rock bursts every year.

In some quarrying operations,* the removal of surface materials to a depth of only 7 or 8 m has led to the formation of sheet joints in the underlying rock (◆ Fig. 1). At quarries in Vermont and Tennessee, for example, the excavation of marble exposed rocks that were formerly buried under great pressure. When the overlying rock was removed, the marble expanded and sheet joints formed. Some slabs of rock that were bounded by sheet joints burst so violently that quarrying machines weighing more than a ton were thrown from their

. . . .

*A quarry is a surface excavation, generally for the extraction of building stone.

tracks, and some quarries had to be abandoned because fracturing rendered the stone useless.

Sheet joints paralleling the walls of the Vaiont River Valley in Italy contributed to the worst reservoir disaster in history. On October 9, 1963, more than 240 million m^3 of rock slid into the Vaiont Reservoir. Although several factors contributed to this slide, it moved partly along a system of sheet joints. The slide displaced water in the reservoir, causing a large wave to overtop the dam and flood the downstream area where nearly 3,000 people drowned.

The Sierra Nevada of California are partly composed of granitic rocks, many of which contain numerous sets of sheet joints parallel to the canyon walls. Large slabs of granite bounded by sheet joints lie on steeply inclined surfaces above highways and railroad tracks where they pose a danger to the road or trackway below (◆ Fig. 2). Occasionally, a mass of this unsupported rock slides or falls, blocking highways and railroad tracks.

◆ FIGURE 2 Sheet joints in granite of the Sierra Nevada in California.

◆ FIGURE 1 Sheet joints formed by expansion in the Mount Airy Granite in North Carolina. (Photo courtesy of W. D. Lowry.)

◆ FIGURE 5.8 The contribution of organisms to mechanical weathering. Tree roots enlarge cracks in rocks.

$$CaCO_3 + H_2O + CO_2 \rightleftharpoons Ca^{++} + 2HCO_3 -$$
calcite water carbon calcium bicarbonate
dioxide ion ion

The dissolution of the calcite in limestone and marble has had dramatic effects in many places ranging from small cavities to large caverns such as Mammoth Cave in Kentucky and Carlsbad Caverns in New Mexico (see Chapter 13).

Oxidation refers to reactions with oxygen to form oxides or, if water is present, hydroxides. For example, iron rusts when it combines with oxygen to form the iron oxide hematite:

$$4Fe + 3O_2 \rightarrow 2Fe_2O_3$$
iron oxygen iron oxide
(hematite)

Of course, atmospheric oxygen is abundantly available for oxidation reactions, but oxidation is generally a slow process unless water is present. Thus, most oxidation is carried out by oxygen dissolved in water.

Oxidation is very important in the alteration of ferromagnesian minerals such as olivine, pyroxenes, amphiboles, and biotite. Iron in these minerals combines with oxygen to form the reddish iron oxide hematite (Fe_2O_3) or the yellowish or brown hydroxide limonite ($FeO(OH) \cdot nH_2O$). The yellow, brown, and red colors of many soils and sedimentary rocks are caused by the presence of small amounts of hematite or limonite.

Hydrolysis is the chemical reaction between the hydrogen (H^+) ions and hydroxyl (OH^-) ions of water and a mineral's ions. In hydrolysis, hydrogen ions actually replace positive ions in minerals. Such replacement changes the composition of minerals and liberates iron that then may be oxidized.

As an illustration of hydrolysis, consider the chemical alteration of feldspars. All feldspars are framework silicates, but when altered, they yield soluble salts and clay minerals, such as kaolinite, which are sheet silicates. The chemical weathering of potassium feldspar by hydrolysis occurs as follows:

$$2KAlSi_3O_8 + 2H^+ + 2HCO_3^- + H_2O \rightarrow$$
orthoclase hydrogen bicarbonate water
ion ion

$$Al_2Si_2O_5(OH)_4 + 2K^+ + 2HCO_3^- + 4SiO_2$$
clay(kaolinite) Potassium bicarbonate silica
ion ion

According to this chemical equation, water and carbon dioxide combine to form *carbonic acid,* a small amount of which dissociates to yield hydrogen and bicarbonate ions. The concentration of hydrogen ions determines the acidity of a solution; the more hydrogen ions present, the stronger the acid.

There are several sources of carbon dioxide that may combine with water and react to form acid solutions. The atmosphere is mostly nitrogen and oxygen, but about 0.03% is carbon dioxide, causing rain to be slightly acidic. Carbon dioxide is also produced in soil by the decay of organic matter and the respiration of organisms, so groundwater is also generally slightly acidic. Climate affects the acidity, however, with arid regions tending to have alkaline groundwater (that is, it has a low concentration of hydrogen ions).

Whatever the source of carbon dioxide, once an acidic solution is present, calcite rapidly dissolves according to the following reaction:

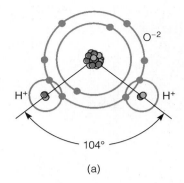

(a)

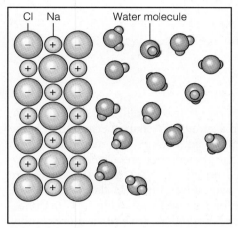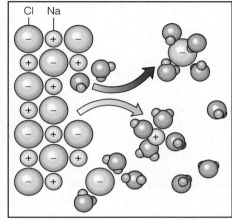

(b)

◆ FIGURE 5.9 (a) The structure of a water molecule. The asymmetric arrangement of the hydrogen atoms causes the molecule to have a slight positive electrical charge at its hydrogen end and a slight negative charge at its oxygen end. (b) The dissolution of sodium chloride (NaCl) in water.

In this reaction hydrogen ions attack the ions in the orthoclase structure, and some liberated ions are incorporated in a developing clay mineral. The potassium and bicarbonate ions go into solution and combine to form a soluble salt. On the right side of the equation is excess silica that would not fit into the crystal structure of the clay mineral.

Factors Controlling the Rate of Chemical Weathering

Chemical weathering processes operate on the surfaces of particles; that is, chemically weathered rocks or minerals are altered from the outside inward. Several factors including particle size, climate, and parent material control the rate of chemical weathering.

Because chemical weathering affects particle surfaces, the greater the surface area, the more effective is the weathering. It is important to realize that small particles have larger surface areas compared to their volume than do large particles. Notice in ◆ Figure 5.10 that a block measuring 1 m on a side has a total surface area of 6 m², but when the block is broken into particles measuring 0.5 m on a side, the total surface area increases to 12 m². And if these particles are all reduced to 0.25 m on a side, the total surface area increases to 24 m². Note that while the surface area in this example increases, the total volume remains the same at 1 m³.

Because chemical weathering is a surface process, the fact that small objects have proportionately more surface area compared to volume than do large objects has profound implications. We can conclude that mechanical weathering, which reduces the size of particles, contributes to chemical weathering by exposing more surface area.

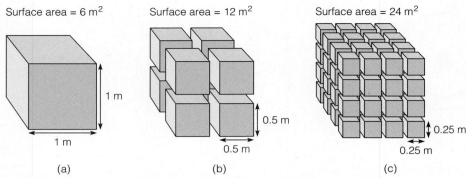

Surface area = 6 m²   Surface area = 12 m²   Surface area = 24 m²

1 m   1 m   0.5 m   0.5 m   0.25 m   0.25 m

(a)   (b)   (c)

◆ FIGURE 5.10 Particle size and chemical weathering. As a rock is reduced into smaller and smaller particles, its surface area increases but its volume remains the same. Thus, in (a) the surface area is 6 m², in (b) it is 12 m², and in (c) 24 m², but the volume remains the same at 1 m³. Accordingly, small particles have more surface area in proportion to their volume than do large particles.

Most chemical processes occur more rapidly at high temperatures and in the presence of liquids. Accordingly, it is not surprising that chemical weathering is more effective in tropical regions than in arid and arctic regions because temperatures and rainfall are high and evaporation rates are low. In addition, vegetation and animal life are much more abundant in the tropics than in arid or cold regions. Consequently, the effects of weathering extend to depths of several tens of meters in the tropics, but commonly extend only centimeters to a few meters deep in arid and arctic regions.

Some rocks are chemically more stable than others and thus are not altered as rapidly by chemical processes. For example, the metamorphic rock quartzite, composed of quartz, is an extremely stable substance that alters very slowly compared to most other rock types. In contrast, rocks such as granite, which contain large amounts of feldspar minerals, decompose rapidly because feldspars are chemically unstable. Ferromagnesian minerals are also chemically unstable and, when chemically weathered, yield clays, iron oxides, and ions in solution. In fact, the stability of common minerals is just the opposite of their order of crystallization in Bowen's reaction series (◆ Fig. 5.11): the minerals that form last in this series are chemically stable, whereas those that form early are easily altered by chemical processes because they are most out of equilibrium with their conditions of formation.

One manifestation of chemical weathering is **spheroidal weathering** (◆ Fig. 5.12). In spheroidal weather-

ing, a stone, even one that is rectangular to begin with, weathers to form a spheroidal shape because that is the most stable shape it can assume. The reason is that on a rectangular stone the corners are attacked by weathering processes from three sides, and the edges are attacked from two sides, but the flat surfaces are weathered more or less uniformly (Fig. 5.12). Consequently, the corners and edges are altered more rapidly, the material sloughs off them, and a more spherical shape develops (◆ Fig. 5.13). Once a spherical shape is present, all surfaces weather at the same rate.

⏷ SOIL

In most places the land surface is covered by a layer of unconsolidated rock and mineral fragments called *regolith*. Regolith may consist of volcanic ash, sediment deposited by wind, streams, or glaciers, or weathered rock material formed in place as a residue. Regolith that consists of weathered material, water, air, and organic matter and can support plant growth is recognized as **soil.**

Weathered rock material including sand, silt, and clay makes up about 45% of a good, fertile soil for gardening or farming, but an essential constituent of such soils is *humus*. Humus, which gives many soils their dark color, is derived by bacterial decay of organic matter. It contains more carbon and less nitrogen than the original material and is nearly resistant to further bacterial decay.

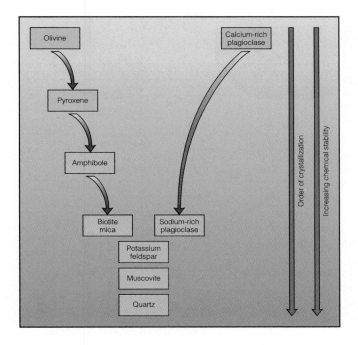

Olivine

Pyroxene

Amphibole

Biotite mica

Calcium-rich plagioclase

Sodium-rich plagioclase

Potassium feldspar

Muscovite

Quartz

Order of crystallization

Increasing chemical stability

◆ FIGURE 5.11 Bowen's reaction series and chemical stability. The minerals forming first in this series are most out of equilibrium with their conditions of formation and are thus most chemically unstable.

Some weathered materials in soils are simply sand- and silt-sized mineral grains, especially quartz, but other weathered materials may be present as well. Such solid particles are important because they hold soil particles apart, allowing oxygen and water to circulate more freely. Clay minerals are also important constituents of soils and aid in the retention of water as well as supplying nutrients to plants. Soils with excess clay minerals, however, drain poorly and are sticky when wet and hard when dry.

Residual soils are formed where parent material has weathered. For example, if a body of granite weathers, and the weathering residue accumulates over the granite and is converted to soil, the soil thus formed is residual. In contrast, *transported soils* are developed on weathered material eroded and transported from the weathering site to a new location.

The Soil Profile

Soil-forming processes begin at the surface and work downward, so the upper layer of soil is more altered from the parent material than the layers below. Observed in vertical cross section, a soil consists of distinct layers or **soil horizons** that differ from one another in texture, structure, composition, and color (◆ Fig. 5.14). Starting from the top, the horizons typical of soils are designated O, A, B, and C, but the boundaries between horizons are transitional rather than sharp.

The O horizon, which is generally only a few centimeters thick, consists of organic matter. The remains of plant materials are clearly recognizable in the upper part of the O horizon, but its lower part consists of humus.

Horizon A, called *top soil,* contains more organic matter than those below. It is also characterized by intense biological activity because plant roots, bacteria, fungi, and animals such as worms are abundant. Threadlike soil bacteria give freshly plowed soil its earthy aroma. In soils developed over a long period of time, the A horizon consists mostly of clays and chemically stable minerals such as quartz. Water percolating down through horizon A dissolves the soluble minerals that were originally present and carries them away or downward to lower levels in the soil by a process called *leaching* (Fig. 5.14).

Horizon B, or *subsoil,* contains fewer organisms and less organic matter than horizon A (Fig. 5.14). Horizon B is also called the *zone of accumulation,* because

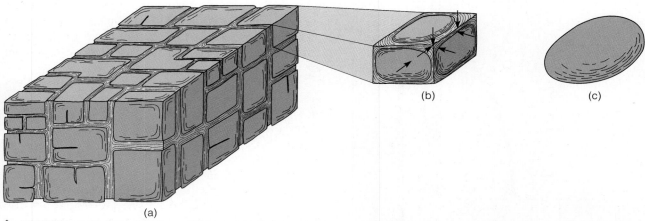

◆ FIGURE 5.12 Spheroidal weathering. (a) The rectangular blocks outlined by joints are attacked by chemical weathering processes, (b) but the corners and edges are weathered most rapidly. (c) When a block has been weathered so that it is spherical, its entire surface is weathered evenly, and no further change in shape occurs.

◆ FIGURE 5.13 Spheroidal weathering of granite in Australia.

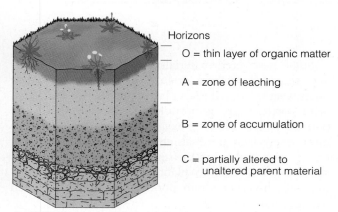

Horizons
O = thin layer of organic matter
A = zone of leaching
B = zone of accumulation
C = partially altered to unaltered parent material

◆ FIGURE 5.14 The soil horizons in a fully developed or mature soil.

soluble minerals leached from horizon A accumulate as irregular masses. If horizon A is stripped away by erosion leaving horizon B exposed, plants do not grow as well, and if horizon B is clayey, it is harder when dry and stickier when wet than other soil horizons.

Horizon C, the lowest soil layer, consists of partially altered to unaltered parent material (Fig. 5.14). In horizons A and B, the composition and texture of the parent material have been so thoroughly altered that the parent material is no longer recognizable. In contrast, rock fragments and mineral grains of the parent material retain their identity in horizon C. Horizon C contains little organic matter.

Factors Controlling Soil Formation

It has long been acknowledged that climate is the single most important factor in soil origins. A very general classification recognizes three major soil types characteristic of different climatic settings. Soils that develop in humid regions such as the eastern United States and much of Canada are **pedalfers,** a name derived from the Greek word *pedon,* meaning soil, and from the chemical symbols for aluminum (Al) and iron (Fe). Because these soils form where abundant moisture is present, most of the soluble minerals have been leached from horizon A. Although it may be gray, horizon A is generally dark colored because of abundant organic matter, and aluminum-rich clays and iron oxides tend to accumulate in horizon B.

Pedocals are soils characteristic of arid and semiarid regions and are found in much of the western United States, especially the southwest. Pedocal derives its name in part from the first three letters of calcite. Such soils contain less organic matter than pedalfers, so horizon A is generally lighter colored and contains more unstable minerals because of less intense chemical weathering. As soil water evaporates, calcium carbonate leached from above commonly precipitates in horizon B where it forms irregular masses of *caliche*. Precipitation of sodium salts in some desert areas where soil water evaporation is intense yields *alkali soils* that are so alkaline that they cannot support plants.

Laterite is a soil formed in the tropics where chemical weathering is intense and leaching of soluble minerals is complete. Such soils are red, commonly extend to depths of several tens of meters, and are composed largely of aluminum hydroxides, iron oxides, and clay minerals; even quartz, a chemically stable mineral, is generally leached out (◆ Fig. 5.15).

Although laterites support lush vegetation, they are not very fertile. The native vegetation is sustained by nutrients derived mostly from the surface layer of organic matter, but little humus is present in the soil itself because bacterial action destroys it. When such soils are cleared of their native vegetation, the existing surface accumulation of organic matter is rapidly oxidized, and there is little to replace it. Consequently, when societies practicing slash-and-burn agriculture

◆ FIGURE 5.15 Laterite, shown here in Madagascar, is a deep, red soil that forms in response to intense chemical weathering in the tropics.

clear these soils, they can raise crops for only a few years at best. Then the soil is completely depleted of plant nutrients, the clay-rich laterite bakes brick hard in the tropical sun, and the farmers move on to another area where the process is repeated.

One aspect of laterites is of great economic importance. If the parent material is rich in aluminum, aluminum hydroxides may accumulate in horizon B as *bauxite,* the ore of aluminum. Because such intense chemical weathering currently does not occur in North America, we are almost totally dependent on foreign sources for aluminum ores. Some aluminum ores do exist in Arkansas, Alabama, and Georgia, which had a tropical climate about 50 million years ago, but currently it is cheaper to import aluminum ore than to mine these deposits.

The same rock type can yield different soils in different climatic regimes, and in the same climatic regime the same soils can develop on different rock types. Thus, it seems that climate is more important than parent material in determining the type of soil that develops. Nevertheless, rock type does exert some control. For example, the metamorphic rock quartzite will have a thin soil over it because it is chemically stable, whereas an adjacent body of granite will have a much deeper soil.

Soils depend on organisms for their fertility, and in return they provide a suitable habitat for many organisms. Earthworms—as many as one million per acre— ants, sowbugs, termites, centipedes, millipedes, and nematodes, along with various types of fungi, algae, and single-celled animals, make their homes in the soil. All of these contribute to the formation of soils and provide humus when they die and are decomposed by bacterial action.

Much humus in soils is provided by grasses or leaf litter that microorganisms decompose to obtain food. In so doing, they break down organic compounds within plants and release nutrients back into the soil. Additionally, organic acids produced by decaying soil organisms are important in further weathering of parent materials and soil particles.

Burrowing animals constantly churn and mix soils, and their burrows provide avenues for gases and water. Soil organisms, especially some types of bacteria, are extremely important in changing atmospheric nitrogen into a form of soil nitrogen suitable for use by plants.

Relief is the difference in elevation between high and low points in a region. Because climate changes with elevation, relief affects soil-forming processes largely through elevation. *Slope* affects soils in two ways. One is simply *slope angle:* the steeper the slope, the less opportunity for soil development because weathered material is eroded faster than soil-forming processes can work. The other slope control is the direction the slope faces. In the Northern Hemisphere, north-facing slopes receive less sunlight than south-facing slopes. If a north-facing slope is steep, it may receive no sunlight at all. Consequently, north-facing slopes have soils with cooler internal temperatures, may support different vegetation, and, if in a cold climate, remain frozen longer.

The properties of a soil are determined by the factors of climate and organisms altering parent material through time; the longer these processes have operated, the more fully developed the soil will be. If a soil is weathered for extended periods of time, however, its fertility decreases as plant nutrients are leached out, unless new materials are delivered.

How much time is needed to develop a centimeter of soil or a fully developed soil a meter or so deep? No definitive answer can be given because weathering proceeds at vastly different rates depending on climate and parent material, but an overall average might be about 2.5 cm per century. However, a lava flow a few centuries old in Hawaii may have a well-developed soil on it, whereas a flow the same age in Iceland will have considerably less soil. Given the same climatic conditions, soil will develop faster on unconsolidated sediment than it will on solid bedrock.*

Under optimum conditions of soil formation, the soil-forming process occurs at a rapid rate in the context of geologic time. From the human perspective, however, soil formation is a slow process; consequently, soil is regarded as a nonrenewable resource.

Soil Erosion

The Soil Conservation Service of the U.S. Department of Agriculture has determined that soil losses exceeding five tons per acre per year adversely affect the productivity of the soil. Most soils in the United States are being eroded at rates less than this maximum. This same agency estimates that 13% of all agricultural land accounts for 71% of the erosion. In some parts of the world, however, soil erosion is a much more serious problem. Madagascar, for example, has lost a large

. . . .

*Bedrock is a general term for the rock underlying soil or unconsolidated sediment.

percentage of its soil to poor farming practices, overgrazing, and deforestation.

Most soil erosion occurs by the action of wind and water. When the natural vegetation is removed and a soil is pulverized by plowing, the fine particles are easily blown away. (see Perspective 5.2). Falling rain disrupts soil particles, and when it runs off at the surface, it carries soil with it. This is particularly devastating on steep slopes from which the vegetative cover has been removed by overgrazing or deforestation. Two types of erosion by water are recognized: sheet erosion and rill erosion.

Sheet erosion is more or less evenly distributed over the surface and removes thin layers of soil. *Rill erosion,* on the other hand, occurs when running water scours small channels (♦ Fig. 5.16). If these *rills* become too deep to be eliminated by plowing (about 30 cm), they are *gullies.* Where gullying becomes extensive, croplands can no longer be tilled and must be abandoned.

If the rate of soil erosion is less than five tons per year per acre—as is the case in most parts of the United States—soil-forming processes can keep pace, and the soil remains productive. If the maximum is exceeded, however, the upper layers of soil—the most productive layers—are removed first, thus exposing horizon B. Such losses are problems, of course, but there are additional consequences. For one thing, the eroded soil is transported elsewhere, perhaps onto neighboring cropland, onto roads, or into channels. Sediment accumulates in canals and irrigation ditches, and agricultural fertilizers and insecticides are carried into streams and lakes.

♦ FIGURE 5.16 Rill erosion in a field during a rainstorm. This rill was later plowed over.

Problems experienced during the past, particularly during the 1930s, have stimulated the development of methods to minimize soil erosion on agricultural lands. Various practices including crop rotation, contour plowing, and the construction of terraces have all proved helpful (◆ Fig. 5.17). Other practices include no-till planting in which harvested crop residue is left on the ground to protect the surface from the ravages of wind and water.

▼ SEDIMENT

Weathered materials are commonly eroded and transported to another location and deposited as sediment. Thus, all **sediment** is derived from preexisting rocks and can be characterized in two ways:

1. *Detrital sediment,* which consists of rock fragments and mineral grains.
2. *Chemical sediment,* which consists of the minerals precipitated from solution by inorganic chemical processes or extracted from solution by organisms.

In any case, sediment is deposited as an aggregate of loose solids. Much accumulated sediment settled from a fluid, such as mud in a lake, or from the atmosphere as dust. The term *sediment* is derived from the Latin *sedimentum,* meaning settling.

◆ FIGURE 5.17 One soil conservation practice is contour plowing, which involves plowing parallel to the contours of the land. The furrows and ridges are perpendicular to the direction that water would otherwise flow downhill and thus inhibit erosion.

Most **sedimentary rocks** formed from sediment that was transformed into solid rock, but a few sedimentary rocks skipped the unconsolidated sediment stage. For example, coral reefs form as solids when the reef organisms extract dissolved mineral matter from seawater for their skeletons. However, if a reef is broken apart during a storm, the solid pieces of reef material deposited on the sea floor are sediment.

One important criterion for classifying detrital sediments and the rocks formed from them is the size of the particles (● Table 5.1). *Gravel* refers to any sedimentary particle measuring more than 2.0 mm, whereas *sand* is any particle, regardless of composition, that measures 1/16 to 2.0 mm. Gravel- and sand-sized particles are large enough to be observed with the unaided eye or with low-power magnification, but silt- and clay-sized particles are too small to be observed except with very high magnification. We should note that *clay* has two meanings: in textural terms, clay refers to sedimentary grains less than 1/256 mm in size, and in compositional terms, clay refers to certain types of sheet silicate materials.

Sediment Transport and Deposition

Detrital sediment can be transported by any geologic agent possessing enough energy to move particles of a given size. Glaciers are very effective agents of transport and can move any sized particle. Wind, on the other hand, can transport only sand-sized and smaller sediment. Waves and marine currents also transport sediment, but by far the most effective way to erode sediment from the weathering site and transport it elsewhere is by streams.

During transport, *abrasion* reduces the size of sedimentary particles. The sharp corners and edges are abraded the most as the particles, especially gravel and sand, collide with one another and become **rounded**

● **TABLE 5.1** Classification of Sedimentary Particles

Size	Sediment Name	
>2 mm	Gravel	
1/16–2 mm	Sand	
1/256–1/16 mm	Silt	} Mud*
<1/256 mm	Clay	

*Mixtures of silt and clay are generally referred to as mud.

THE DUST BOWL

The stock market crash of 1929 ushered in the Great Depression, a time when millions of people were unemployed and many had no means to provide food and shelter. Urban areas were affected most severely by the depression, but rural areas suffered as well, especially during the great drought of the 1930s. Prior to the 1930s, farmers had enjoyed a degree of success unparalleled in U.S. history. During World War I, the price of wheat soared, and after the war when Europe was recovering, the government subsidized wheat prices. High prices and mechanized farming practices resulted in more and more land being tilled. Even the weather cooperated, and land in the western United States that would otherwise have been marginally productive was plowed. Deep-rooted prairie grasses that held the soil in place were replaced by shallow-rooted wheat.

Beginning in about 1930, drought conditions prevailed throughout the country; only two states—Maine and Vermont—were not drought-stricken. Drought conditions varied from moderate to severe, and the consequences of the drought were particularly severe in the southern Great Plains. Some rain fell, but in amounts insufficient to maintain agricultural production. And since the land, even marginal land, had been tilled, the native vegetation was no longer available to keep the soil from blowing away. And blow away it did—in huge quantities. Nothing stopped the wind from removing large quantities of top soil.

A large region in the southern Great Plains that was particularly hard hit by the drought, dust storms, and soil erosion came to be known as the Dust Bowl. Although its boundaries were not well defined, it included parts of Kansas, Colorado, and New Mexico as well as the panhandles of Oklahoma and Texas (◆ Fig. 1); together the Dust Bowl and its less affected fringe area covered more than 400,000 km²!

Dust storms became common during the 1930s, and some reached phenomenal sizes (◆ Fig. 2). One of the largest storms occurred in 1934 and covered more than 3.5 million km². It lifted dust nearly 5 km into the air, obscured the sky over large parts of six states, and blew hundreds of millions of tons of soil eastward where it settled on New York City, Washington, D.C., and other eastern cities as well as on ships some 480 km out in the Atlantic Ocean. The Soil Conservation Service reported that dust storms of regional extent occurred on 140 occasions during 1936 and 1937. Dust was everywhere. It seeped into houses, suffocated wild animals and livestock, and adversely affected human health.

The dust was, of course, the material derived from the tilled lands; in other words, much of the soil in many regions was simply blown away. Blowing dust was not the only problem; sand piled up along fences, drifted against houses and farm machinery, and covered what otherwise might have been productive soils. Agricultural production fell precipitously in the Dust Bowl, farmers could not meet their mortgage payments, and by 1935 tens of thousands were leaving. Many of these people went west to California and became the migrant farm workers immortalized in John Steinbeck's novel *The Grapes of Wrath.*

(◆ Fig. 5.18a). Another sediment property modified during transport is sorting. **Sorting** refers to the size distribution in an aggregate of sediment; if all the particles are approximately the same size, the sediment is characterized as well sorted, but if a wide range of grain sizes occur, the sediment is poorly sorted. (◆ Fig. 5.18b). Sorting results from processes that selectively transport and deposit particles by size. Windblown

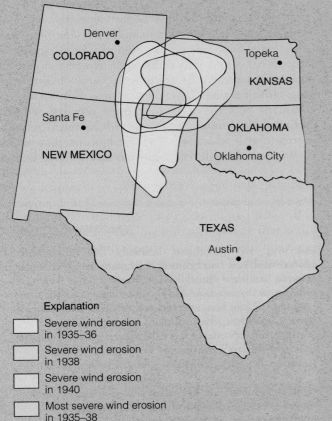

The Dust Bowl was an economic disaster of great magnitude. Droughts had stricken the southern Great Plains before, and have done so since, but the drought of the 1930s was especially severe. Political and economic factors also contributed to the disaster. Due in part to the artificially inflated wheat prices, many farmers were deeply in debt—mostly because they had purchased farm machinery in order to produce more and benefit from the high prices. Feeling economic pressure because of their huge debts, they tilled marginal land, and employed few, if any, soil conservation measures.

If the Dust Bowl has a bright side, it is that the government, farmers, and the public in general no longer take soil for granted or regard it as a substance that needs no nurturing. In addition, a number of soil conservation methods developed then have now become standard in agriculture.

◆ FIGURE 2 The huge dust storm of April 14, 1935, also known as Black Sunday, photographed at Hugoton, Kansas.

Explanation

Severe wind erosion in 1935–36

Severe wind erosion in 1938

Severe wind erosion in 1940

Most severe wind erosion in 1935–38

◆ FIGURE 1 The Dust Bowl of the 1930s. Drought conditions extended far beyond the boundaries shown here, but this area was particularly hard hit by drought and wind erosion.

dunes are composed of well-sorted sand; the deposits of mud flows tend to be poorly sorted.

Sediment may be transported a considerable distance from its source area, but eventually it is deposited. Some of the sand and mud being deposited at the mouth of the Mississippi River at the present time came from such distant places as Ohio, Minnesota, and Wyoming. Any geographic area in which sediment is

(a)

(b)

◆ FIGURE 5.18 Rounding and sorting of sedimentary particles. (*a*) A deposit consisting of well-sorted and well-rounded gravel. (*b*) Poorly sorted, angular gravel. (Photos courtesy of R. V. Dietrich.)

deposited is a *depositional environment.* Although no completely satisfactory classification of depositional environments exists, geologists generally recognize three major depositional settings: continental, transitional, and marine (◆ Fig. 5.19, • Table 5.2).

Lithification: Sediment to Sedimentary Rock

At present, calcium carbonate mud (chemical sediment) is accumulating in the shallow waters of Florida Bay, and detrital sand is being deposited in river channels, on beaches, and in sand dunes. Such deposits of sediment may be compacted and/or cemented and thereby converted into sedimentary rock; the process by which sediment is transformed into sedimentary rock is **lithification.**

When sediment is deposited, it consists of solid particles and *pore spaces,* which are the voids between particles (◆ Fig. 5.20a). The amount of pore space varies depending on the depositional process, the size of the sediment grains, and sorting. When sediment is buried, *compaction,* resulting from the pressure exerted by the weight of overlying sediments, reduces the amount of pore space, and thus the volume of the deposit (◆ Fig. 5.20b). When deposits of mud, which can have as much as 80% water-filled pore space, are buried and compacted, water is squeezed out, and the volume can be reduced by up to 40%. Sand may have up to 50% pore space, although it is generally somewhat less, and it, too, can be compacted so that the sand grains fit more tightly together.

Compaction alone is generally sufficient for lithification of mud, but for sand and gravel deposits *cementation* is necessary (◆ Fig. 5.20c). Recall that calcium carbonate ($CaCO_3$) readily dissolves in water containing a small amount of carbonic acid, and that chemical weathering of feldspars and other silicate minerals yields silica (SiO_2) in solution. These dissolved compounds may be precipitated in the pore spaces of sediments, where they act as a cement that effectively binds the sediment together (Fig 5.20c).

Calcium carbonate and silica are the most common cements in sedimentary rocks, but iron oxides and hydroxides, such as hematite (Fe_2O_3) and limonite [$FeO(OH) \cdot nH_2O$], respectively, also form a chemical cement in some rocks. Much of the iron oxide cement is derived from the oxidation of iron in ferromagnesian minerals present in the original deposit, although some is carried in by circulating groundwater. Yellow, brown, and red sedimentary rocks are colored by small amounts of iron oxide and hydroxide cement.

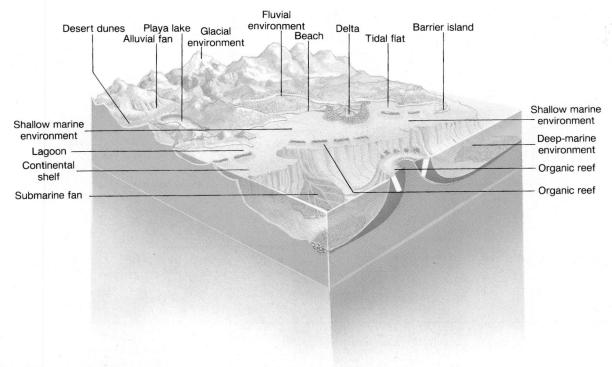

FIGURE 5.19 Major depositional environments are shown in this generalized diagram.

SEDIMENTARY ROCKS

Even though about 95% of the Earth's crust is composed of igneous and metamorphic rocks, sedimentary rocks are the most common at or near the surface. About 75% of the surface exposures on continents consist of sediments or sedimentary rocks, and they cover most of the sea floor. Sedimentary rocks are generally classified as *detrital* or *chemical* (◆ Tables 5.3 and ◆ 5.4).

Detrital Sedimentary Rocks

Detrital sedimentary rocks consist of *detritus,* the solid particles of preexisting rocks. Such rocks have a *clastic texture,* meaning that they are composed of fragments or particles also known as *clasts.* Several varieties of detrital sedimentary rocks are recognized, each of which is characterized by the size of its constituent particles (Table 5.3).

Both *conglomerate and sedimentary breccia* consist of gravel-sized particles (Table 5.3; ◆ Fig. 5.21a and b, page 114). The only difference between conglomerate and sedimentary breccia is the shape of the gravel particles; conglomerate consists of rounded gravel, whereas sedimentary breccia is composed of angular gravel called *rubble.*

Conglomerate is a fairly common rock type, but sedimentary breccia is rather rare. The reason is that gravel-sized particles become rounded very quickly during transport. Thus, if a sedimentary breccia is encountered, one can conclude that the rubble that composes it was not transported very far. High-energy transport agents such as rapidly flowing streams and waves are needed to transport gravel, so gravel tends to be deposited in high-energy environments such as stream channels and beaches.

The term *sand* is simply a size designation, so *sandstone* may be composed of grains of any type of mineral or rock fragment. However, most sandstones consist primarily of the mineral quartz (◆ Fig. 5.21c, page 114) with small amounts of a number of other minerals. Geologists recognize several types of sandstones, each characterized by its composition. *Quartz sandstone,* composed mostly of quartz, is the most common, but *arkose,* which contains more than 25% feldspars, is also fairly common (Table 5.3).

• TABLE 5.2 A Summary of Depositional Environments

CONTINENTAL ENVIRONMENTS	Fluvial	Braided Stream Meandering Stream
	Desert	Sand Dunes Alluvial Fans Playa Lakes
	Glacial	Ice Deposition (Moraines) Fluvial Deposition (Outwash) Glacial Lakes
TRANSITIONAL ENVIRONMENTS	Delta	Stream-Dominated Wave-Dominated Tide-Dominated
	Beach	
	Barrier Island	Beach Sand Dune Lagoon
MARINE ENVIRONMENTS	Continental Shelf	Detrital Deposition Carbonate Deposition
	Carbonate Platform	
	Continental Slope and Rise	Submarine Fans (Turbidites)
	Deep Ocean Basin	Oozes Brown Clay
	Evaporite Environments*	

*Evaporites may be deposited in a variety of environments including playa lakes, saline lakes, lagoons and tidal flats in arid regions, and in marine environments.

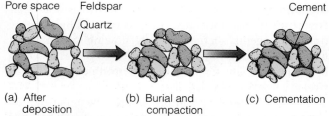

◆ FIGURE 5.20 Lithification of sand. (*a*) When initially deposited, sand has considerable pore space between grains. (*b*) Compaction resulting from the weight of overlying sediments reduces the amount of pore space. (*c*) Sand is converted to sandstone as cement is precipitated in pore spaces from groundwater.

The *mudrocks* include all detrital sedimentary rocks composed of silt- and clay-sized particles (◆ Fig. 5.21d). Among the mudrocks we can differentiate between *siltstone, mudstone,* and *claystone*. Siltstone, as the name implies, is composed of silt-sized particles; mudstone contains a mixture of silt- and clay-sized particles; and claystone is composed mostly of clay (Table 5.3). Some mudstones and claystones are designated as *shale* if they are fissile, which means they break along closely spaced parallel planes (Fig. 5.21d).

Chemical Sedimentary Rocks

Chemical sedimentary rocks originate from the ions and salts taken into solution in the weathering environ-

● **TABLE 5.3** Classification of Detrital Sedimentary Rocks

Sediment Name and Size	Description	Rock Name
Gravel (>2 mm)	Rounded gravel Angular gravel	Conglomerate Sedimentary breccia
Sand (1/16–2 mm)	Mostly quartz Quartz with >25% feldspar	Quartz sandstone Arkose
Mud (<1/16 mm)	Mostly silt Silt and clay Mostly clay	Siltstone Mudstone* ⎱ Mudrocks Claystone* ⎰

*Mudrocks possessing the property of fissility, meaning they break along closely spaced, parallel planes, are commonly called *shale*.

● **TABLE 5.4** Classification of Chemical and Biochemical Sedimentary Rocks

Texture	Composition	Rock Name
Clastic or crystalline	Calcite ($CaCO_3$)	Limestone (includes coquina, chalk, and oolitic limestone) ⎱ Carbonates
	Dolomite [$CaMg (CO_3)_2$]	Dolostone ⎰
Crystalline	Gypsum ($CaSO_4 \cdot 2H_2O$) Halite (NaCl)	Rock gypsum ⎱ Evaporites Rock salt ⎰
Usually crystalline	Microscopic SiO_2 shells	Chert
	Altered plant remains	Coal

ment (Table 5.4). Such dissolved materials are transported to lakes and the oceans where they become concentrated. Inorganic chemical processes remove these substances from solution, and they accumulate as solid minerals. *Biochemical sedimentary rocks,* which constitute a subcategory of chemical sedimentary rocks, result from the chemical processes of organisms.

Calcite (the main component of limestone) and dolomite (the mineral comprising the rock dolostone) are both carbonate minerals; calcite is a calcium carbonate ($CaCO_3$), whereas dolomite [$CaMg(CO_3)_2$] is a calcium magnesium carbonate. Thus, limestone and dolostone are *carbonate rocks*. Recall that calcite readily dissolves in water containing a small amount of acid, but the chemical reaction leading to dissolution is reversible, so solid calcite can be precipitated from

solution. Accordingly, some limestone, although probably not very much, results from inorganic chemical reactions.

Most limestones are conveniently classified as *biochemical sedimentary rocks,* because organisms with calcium carbonate shells play such a significant role in their origin (◆ Fig. 5.22a). For example, the limestone known as *coquina* (◆ Fig. 5.22b) consists entirely of broken shells cemented by calcium carbonate, and *chalk* is a soft variety of biochemical limestone composed largely of microscopic shells of organisms.

One distinctive type of limestone contains small spherical grains called *ooids*. Ooids have a small nucleus, a sand grain or shell fragment perhaps—around which concentric layers of calcite precipitate; lithified deposits of ooids form *oolitic limestones* (◆ Fig. 5.22c).

(a)

(b)

(c)

(d)

◆ FIGURE 5.21 Detrital sedimentary rocks: (*a*) conglomerate; (*b*) sedimentary breccia; (*c*) sandstone; and (*d*) the mudrock shale. (Photos courtesy of Sue Monroe.)

The near-absence of recent dolostone and evidence from chemistry and studies of rocks indicate that most dolostone was originally limestone that has been changed to dolostone. Many geologists think most dolostones originated through the replacement of some of the calcium in calcite by magnesium. For example, in a restricted environment, such as a lagoon, where evaporation rates are high, much of the calcium in solution is extracted as it goes into calcite ($CaCO_3$) and gypsum ($CaSO_4 \cdot 2H_2O$). Magnesium (Mg), on the other hand, becomes concentrated in the water, which then becomes denser and permeates preexisting limestone and converts it to dolostone by the addition of magnesium.

Evaporites include such rocks as *rock salt* and *rock gypsum,* which form by inorganic chemical precipita-

tion of minerals from solution (Table 5.4; ◆ Fig. 5.23). We noted earlier that some minerals are dissolved during chemical weathering, but a solution can hold only a certain volume of dissolved mineral matter. If the volume of a solution is reduced by evaporation, the amount of dissolved mineral matter increases in proportion to the volume of the solution and eventually reaches the saturation limit, the point at which precipitation must occur.

Rock salt, composed of the mineral halite (NaCl), is simply sodium chloride that was precipitated from seawater or, more rarely, lake water (Fig. 5.23a). Rock gypsum, the most common evaporite rock, is composed of the mineral gypsum ($CaSO_4 \cdot H_2O$), which also precipitates from evaporating solutions (Fig. 5.23b). A

(a)

(b)

(c)

◆ FIGURE 5.22 Three types of limestones.
(*a*) Fossiliferous limestone. (*b*) Coquina is composed
of the broken shells of organisms. (Photos a and b courtesy
of Sue Monroe.) (*c*) Present-day oolites from the Bahamas.

number of other evaporite rocks and minerals are known, but most of these are rare.

Chert is a hard rock composed of microscopic crystals of quartz (SiO_2) (Table 5.4; Fig. 5.23c). Various color varieties of chert are known as *flint,* which is black because of inclusions of organic matter, and *jasper,* which is red or brown because of iron oxide inclusions. Because chert lacks cleavage and can be shaped to form sharp cutting edges, many cultures have used it for the manufacture of tools, spear points, and arrowheads. Chert occurs as irregular masses or *nodules* in other rocks, especially limestones, and as distinct layers of rock called *bedded chert.*

Coal is a biochemical sedimentary rock composed of the compressed, altered remains of organisms, especially land plants (Table 5.4; ◆ Fig. 5.24). It forms in swamps and bogs where the water is deficient in oxygen or where organic matter accumulates faster than it decomposes. The bacteria that decompose vegetation in swamps can exist without oxygen, but their wastes must be oxidized, and because no oxygen is present, the wastes accumulate and kill the bacteria. Thus, bacterial decay ceases and plant materials are not completely destroyed. These partly altered plant remains accumulate as layers of organic muck, which commonly smell of hydrogen sulfide (the rotten-egg odor of swamps). When buried, this organic muck becomes *peat,* which looks rather like coarse pipe tobacco. Where peat is abundant, as in Ireland and Scotland, it is burned as a fuel. Peat that is buried more deeply and compressed, especially if it is also heated, is altered to a type of dark brown coal called *lignite,* in which plant remains are still clearly visible. During the change from organic muck to coal, such volatile elements of the vegetation as oxygen, hydrogen, and nitrogen are partly vaporized and driven off, thus enriching the residue in carbon; lignite contains about 70% carbon as opposed to about 50% in peat.

Bituminous coal, which contains about 80% carbon, is a higher grade coal than lignite. It is dense and black and has been so thoroughly altered that plant remains can only rarely be seen. The highest grade coal is *anthracite,* which is a metamorphic type of coal (see Chapter 6). It contains up to 98% carbon and, when burned, yields more heat per unit volume than other types of coal.

▼ SEDIMENTARY FACIES

If a layer of sediment or sedimentary rock is traced laterally, it generally changes in composition, texture, or

(a)

(b)

(c)

◆ FIGURE 5.23 (*a*) Core of rock salt from a well in Michigan. (*b*) Rock gypsum. (*c*) Chert. (Photos courtesy of Sue Monroe.)

◆ FIGURE 5.24 Coal is a biochemical sedimentary rock composed of the altered remains of land plants. (Photo courtesy of Sue Monroe.)

both. It changes by lateral gradation resulting from the simultaneous operation of different depositional processes in adjacent depositional environments. For example, sand may be deposited in a high-energy nearshore environment while mud and carbonate sediments accumulate simultaneously in the laterally adjacent low-energy offshore environments (◆ Fig. 5.25). Deposition in each of these environments produces bodies of sediment, each of which is characterized by a distinctive set of physical, chemical, and biological attributes. Such distinctive bodies of sediment, or sedimentary rock, are **sedimentary facies.**

Any aspect of sedimentary rocks that makes them recognizably different from adjacent rocks of the same

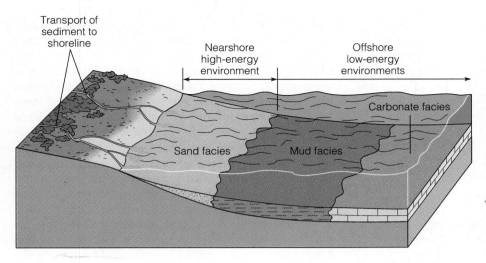

Transport of sediment to shoreline

Nearshore high-energy environment

Offshore low-energy environments

Carbonate facies

Sand facies

Mud facies

◆ FIGURE 5.25 Deposition in adjacent environments yields distinct bodies of sediment, each of which is designated as a sedimentary facies.

age, or approximately the same age, can be used to establish a sedimentary facies. Figure 5.25 illustrates three sedimentary facies: a sand facies, a mud facies, and a carbonate facies. If these sediments become lithified, they are called sandstone, mudstone (or shale), and limestone facies, respectively.

Marine Transgressions and Regressions

Many sedimentary rocks in the interiors of continents show clear evidence of having been deposited in marine environments. The strata in ◆ Figure 5.26, for example, consist of a sandstone facies that was deposited in a nearshore marine environment overlain by shale and limestone facies that were deposited in offshore environments. Such a vertical sequence of facies can be explained by deposition occurring during a time when sea level rose with respect to the continents. When sea level rises with respect to a continent, the shoreline moves inland, giving rise to a **marine transgression** (Fig. 5.26). As the shoreline advances inland, the depositional environments parallel to the shoreline do likewise. Remember that each laterally adjacent environment in Figure 5.26 is the depositional site of a different sedimentary facies. As a result of a marine transgression, the facies that formed in the offshore environments are superimposed over the facies deposited in the nearshore environment, thus accounting for the vertical succession of sedimentary facies (Fig. 5.26).

Another important aspect of marine transgressions is that an individual facies can be deposited over a huge geographic area (Fig. 5.26). Even though the nearshore environment is long and narrow at any particular time, deposition occurs continuously as the environment migrates landward during a marine transgression. The sand deposited under these conditions may be tens to hundreds of meters thick, but have horizontal dimensions of length and width measured in hundreds of kilometers.

The opposite of a marine transgression is a **marine regression.** If sea level falls with respect to a continent, the shoreline and environments that parallel the shoreline move in a seaward direction. The vertical sequence produced by a marine regression has facies of the nearshore environment superposed over facies of offshore environments. Marine regressions can also account for the deposition of a facies over a large geographic area.

⋎ ENVIRONMENTAL ANALYSIS

When geologists investigate sedimentary rocks in the field, they are observing the products of processes that operated during the past. The only record of these processes is preserved in the rocks, so geologists must evaluate those aspects of sedimentary rocks that allow inferences to be made about the original processes and the environment of deposition. Sedimentary textures such as sorting and rounding can give clues to the depositional process. Windblown dune sands, for example, tend to be well sorted and well rounded. The geometry or three-dimensional shape of rock bodies is another important criterion in environmental interpretation. Marine transgressions and regressions yield sediment bodies with a blanket or sheetlike geometry, whereas deposits in stream channels tend to be long and narrow and are therefore described as having a shoestring geometry. Other aspects of sedimentary rocks that are important in environmental analysis include *sedimentary structures* and *fossils*.

Sedimentary Structures

When sediments are deposited, they contain **sedimentary structures** that form as a result of the physical and

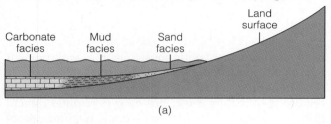

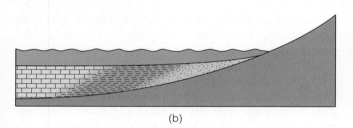

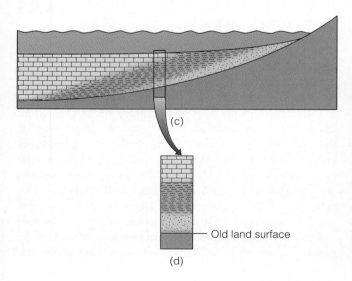

◆ FIGURE 5.26 (*a*), (*b*), and (*c*) Three stages of a marine transgression. (*d*) Diagrammatic representation of the vertical sequence of facies resulting from the transgression.

biological processes operating in the depositional environment. Accordingly, sedimentary structures are useful for interpreting ancient depositional environments. One of the most common sedimentary structures is layering, or what is called **bedding** or *stratification* (◆ Fig. 5.27). Individual layers are referred to as *laminae* if they are less than 1 cm thick, and *beds* if they are thicker. Individual beds are separated from one another by *bedding planes* and are distinguished from one another by differences in composition, grain size, color, or a combination of features. Almost all sedimentary rocks show some kind of bedding; a few, such as limestones that formed as coral reefs, lack this feature.

Graded bedding involves an upward decrease in grain size within a single bed (◆ Fig. 5.28). Most graded bedding appears to have formed from turbidity current deposition, although some forms in stream channels during the waning stages of floods. *Turbidity currents* are underwater flows of sediment-water mixtures that are denser than sediment-free water. Such flows move downslope along the bottom of the sea or a lake until they reach the relatively level sea or lake floor (Fig. 5.28). There, they rapidly slow down and begin depositing transported sediment, the coarsest first followed by progressively smaller particles.

Many sedimentary rocks are characterized by **cross-bedding;** cross-beds are arranged such that they are at an angle to the surface upon which they have accumulated (◆ Fig. 5.29). Cross-bedding is common in desert dunes and in sediments in stream channels and shallow marine environments. Invariably, cross-beds result from transport by wind or water currents, and the cross-beds

◆ FIGURE 5.27 Most sedimentary rocks show some kind of layering or bedding as these sandstones in Montana.

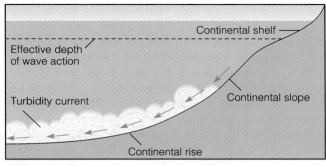

(a)

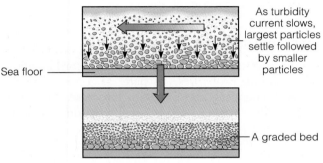

As turbidity current slows, largest particles settle followed by smaller particles

Sea floor

A graded bed

(b)

◆ FIGURE 5.28 (*a*) Turbidity currents flow downslope along the sea floor (or lake bottom) because of their density. (*b*) Graded bedding formed by deposition from a turbidity current.

◆ FIGURE 5.29 Cross-bedding forms when the beds are inclined with respect to the surface upon which they accumulate. Cross-beds indicate ancient current directions by their dip, to the left in this case.

are inclined downward, or dip, in the direction of flow. Because their orientation depends on the direction of flow, cross-beds are good indicators of ancient current directions or *paleocurrents* (Fig. 5.29).

In sand deposits one can commonly observe small-scale, ridgelike **ripple marks** on bedding planes. Two common types of ripple marks are recognized. One type is asymmetrical in cross section, having a gentle upstream slope and a steep downstream slope. They form as a result of currents that move in one direction, as in a stream channel. These are *current ripple marks* (◆ Fig. 5.30a) and, like cross-bedding, are good paleocurrent indicators. In contrast, ripples that tend to be symmetrical in cross section are produced by the to-and-fro motion of waves and are known as *wave-formed ripple marks* (◆ Fig. 5.30b).

◆ FIGURE 5.30 (*a*) Current ripples on the bed of a stream in California. (*b*) Wave-formed ripples on Heron Island, Australia.

(a)

(b)

Weathering, Soil, Sediment, and Sedimentary Rocks   119

Mud cracks are found in clay-rich sediment that has dried out (◆ Fig. 5.31). When such sediment dries, it shrinks and forms intersecting fractures (mud cracks). Such features in ancient sedimentary rocks indicate that the sediment was deposited where periodic drying was possible as on a river floodplain, near a lake shore, or where muddy deposits are exposed on marine shorelines at low tide.

It is important to realize that no single sedimentary structure is unique to a particular depositional environment. Current ripples, for example, are common in stream channels but may also be found in tidal channels, on the seafloor, or near lake margins. Associations of sedimentary structures are particularly useful in environmental analyses, especially when considered with other aspects of a rock unit such as composition, texture, and fossils (◆ Fig. 5.32).

Fossils

Fossils, the remains or traces of ancient organisms, are also important in determining depositional environment. Studies of a fossil's structure and its living descendants, if there are any, are helpful in this endeavor. For example, clams with heavily constructed shells typically live in shallow, turbulent water. In contrast, organisms dwelling in low-energy environments commonly have thin, fragile shells. Also, organisms that carry on photosynthesis are restricted to the zone of sunlight penetration in the seas, which is generally less than 200 m. The amount of suspended sediment is also a limiting factor on the distribution of organisms. Many corals, for example, live in shallow, clear seawater, because suspended sediment clogs their respiratory and food-gathering organs, and some have photosynthesizing algae living in their tissues.

Microfossils are particularly useful for environmental studies because the remains of numerous individual organisms can be recovered from small rock samples. In oil-drilling operations, for example, small rock chips called *well cuttings* are brought to the surface. Such samples rarely contain entire macrofossils but may contain numerous microfossils that aid in age and environmental interpretations.

Environment of Deposition

The sedimentary rocks in the geologic record acquired their various properties, in part, as a result of the physical, chemical, and biological processes that operated in the original depositional environment (Fig.

◆ FIGURE 5.32 Associations of sedimentary structures are useful for environmental interpretations, especially when considered along with other rock properties such as textures and fossils. This example shows an idealized vertical sequence deposited by a turbidity current. The flute casts are produced by erosion, but as a turbidity current slows down, units A through D are deposited in succession. The deep-sea mud (unit E) forms as particles settle from seawater. It commonly contains the *Nereites* trace fossil association consisting of crawling traces made by organisms.

◆ FIGURE 5.31 Mud cracks form in clay-rich sediments when they dry and shrink.

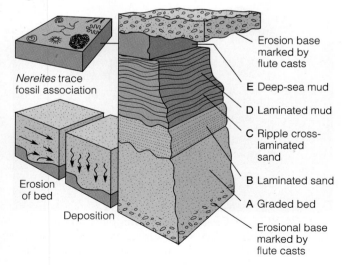

Nereites trace fossil association

Erosion of bed

Deposition

Erosion base marked by flute casts

E Deep-sea mud

D Laminated mud

C Ripple cross-laminated sand

B Laminated sand

A Graded bed

Erosional base marked by flute casts

5.19; Table 5.2). One of geologists' major tasks is to determine the specific depositional environment of sedimentary rocks. Based on their knowledge of sedimentary structures and present-day processes, such as sediment transport and deposition by streams, geologists can make inferences regarding the depositional environments of ancient sedimentary rocks.

While conducting field studies, geologists commonly make some preliminary interpretations. For example, some sedimentary particles such as ooids in limestones most commonly form in shallow marine environments where currents are vigorous. Large-scale cross-bedding is typical of, but not restricted to, desert dunes. Fossils of land plants and animals can be washed into transitional environments, but most of them are preserved in deposits of continental environments. Fossil shells of such marine-dwelling animals as corals obviously indicate marine depositional environments.

Much environmental interpretation is done in the laboratory where the data and rock samples collected during field work can be more fully analyzed. Such analyses might include microscopic and chemical examination of rock samples, identification of fossils, and graphic representations showing the three-dimensional shapes of rock units and their relationships to other

rock units. In addition, the features of sedimentary rocks are compared with those of sediments from present-day depositional environments; the contention is that features in ancient rocks, such as ripple marks, formed during the past in response to the same processes responsible for them now. Finally, when all data have been analyzed, an environmental interpretation is made.

The following examples illustrate how environmental interpretations are made. The Green Pond Conglomerate, Shawangunk Conglomerate, and Tuscarora Sandstone, three ancient formations* in the eastern United States, possess characteristic grain sizes, rock types, and sedimentary structures that indicate deposition in a continental environment, particularly a system of streams that flowed generally westward (the paleocurrent direction was determined by the orientation of cross-beds) (◆ Fig. 5.33). As supporting evidence for this interpretation, these ancient deposits possess textures and sedimentary structures very similar to those of the present-day deposits of the Platte River in Colorado and Nebraska.

. . . .

*A formation is a body of rock with distinctive upper and lower boundaries that is extensive enough to be depicted on a geologic map. The term is generally applied to sedimentary rocks, but can be used for some igneous and metamorphic rocks as well.

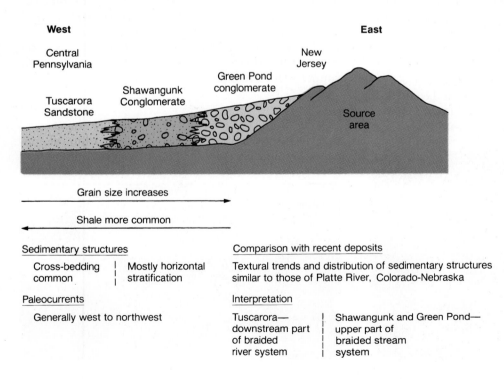

◆ FIGURE 5.33 A simplified cross section showing the lateral relationships for three rock units in the eastern United States.

West

Central Pennsylvania

East

New Jersey

Green Pond conglomerate

Shawangunk Conglomerate

Tuscarora Sandstone

Source area

Grain size increases →

← Shale more common

Sedimentary structures

Cross-bedding common | Mostly horizontal stratification

Paleocurrents

Generally west to northwest

Comparison with recent deposits

Textural trends and distribution of sedimentary structures similar to those of Platte River, Colorado-Nebraska

Interpretation

Tuscarora— downstream part of braided river system | Shawangunk and Green Pond— upper part of braided stream system

The composition of the sedimentary particles indicates they were derived from a source region in the area of the present-day Appalachian Mountains.

In the Grand Canyon of Arizona, a number of formations are exposed, and many of these can be traced for great distances. Three of these, the Tapeats Sandstone, the Bright Angel Shale, and the Mauv Limestone, occur in vertical sequence and contain features, including fossils, that clearly indicate that they were deposited in transitional and marine environments. In fact, all three were forming simultaneously, but a marine transgression caused them to be superposed in the order now observed (◆ Fig. 5.34). Similar sequences of rocks of approximately the same age in Utah, Colorado, Wyoming, Montana, and South Dakota indicate that this marine transgression was widespread indeed.

▼ WEATHERING AND MINERAL RESOURCES

In a preceding section, we discussed intense chemical weathering in the tropics and the origin of *bauxite,* the chief ore of aluminum. Such an accumulation of valuable minerals formed by the selective removal of soluble substances is a *residual concentration.* In addition to bauxite, a number of other residual concentrations are economically important; for example, ore

◆ FIGURE 5.34 View of the Tapeats Sandstone (lower cliff), Bright Angel Shale (forming the slope in the middle distance), and Mauv Limestone (upper cliff) in the Grand Canyon in Arizona. These formations were deposited during a widespread marine transgression.

deposits of iron, manganese, clays, nickel, phosphate, tin, diamonds, and gold.

Some of the sedimentary iron deposits of the Lake Superior region were enriched by chemical weathering when the soluble constituents that were originally present were carried away.

A number of kaolinite deposits in the southern United States were formed by the chemical weathering of feldspars in pegmatites and of clay-bearing limestones and dolostones. Kaolinite is a type of clay mineral used in the manufacture of paper and ceramics.

Gossans, oxidized ores, and *supergene enrichment* of ores are interrelated, and all result from chemical weathering (◆ Fig. 5.35). A gossan is a yellow to reddish deposit composed largely of hydrated iron oxides that formed by the oxidation and leaching of sulfide minerals such as pyrite (FeS_2). The dissolution of such sulfide minerals forms sulfuric acid, which causes other metallic minerals to dissolve, and these tend to be carried downward toward the groundwater table (Fig. 5.35). Oxidized ores form just above the groundwater table as a result of chemical reactions with these descending solutions. Some of the minerals formed in this zone contain copper, zinc, and lead.

Supergene enrichment of ores occurs where metal-bearing solutions penetrate below the water table (Fig. 5.35). Such deposits are characterized by the replacement of sulfide minerals of the primary deposit with sulfide minerals introduced by the descending solutions. For example, the iron in iron sulfides may be replaced by other metals such as lead, zinc, nickel, and copper that have a greater affinity for sulfur.

◆ FIGURE 5.35 A cross section showing a gossan and the origin of oxidized ores and the supergene enrichment of ores.

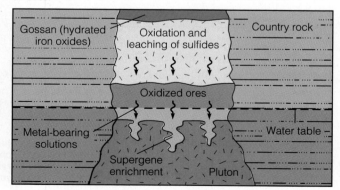

Gossans have been used occasionally as sources of iron, but they are far more important as indicators of underlying ore deposits. One of the oldest known underground mines exploited such ores about 3,400 years ago in what is now southern Israel. Supergene enriched ore bodies are generally small but extremely rich sources of various metals. The largest copper mine in the world, at Bingham, Utah, was originally mined for supergene ores, but currently only primary ores are being mined.

⩗ SEDIMENTS, SEDIMENTARY ROCKS, AND NATURAL RESOURCES

The uses of sediments and sedimentary rocks or the materials they contain vary considerably. Sand and gravel are essential to the construction industry, pure clay deposits are used for ceramics, and limestone is used in the manufacture of cement and in blast furnaces where iron ore is refined to make steel. Evaporites are the source of common table salt as well as a number of chemical compounds, and rock gypsum is used to manufacture wallboard.

Anthracite coal is an especially desirable resource because it burns hot with a smokeless flame. Unfortunately, it is the least common type of coal, so most coal used for heating buildings and for generating electrical energy is bituminous. Bituminous coal is also used to make *coke,* a hard, gray substance consisting of the fused ash of bituminous coal; coke is prepared by heating the coal and driving off the volatile matter. Coke is used to fire blast furnaces during the production of steel. Synthetic oil and gas and a number of other products are also made from bituminous coal and lignite.

Petroleum and Natural Gas

Both petroleum and natural gas are *hydrocarbons,* meaning that they are composed of hydrogen and carbon. Hydrocarbons form from the remains of microscopic organisms that exist in the seas and in some large lakes. When these organisms die, their remains settle to the sea or lake floor where little oxygen is available to decompose them. They are then buried under layers of sediment, and as the depth of burial increases, they are heated and transformed into petroleum and natural gas. The rock in which the hydrocarbons formed is called the *source rock.*

For petroleum and natural gas to occur in economic quantities, they must migrate from the source rock into some kind of rock in which they can be trapped. If there were no trapping mechanism, both would migrate upward and eventually seep out at the surface. The rock in which petroleum and natural gas accumulate is known as *reservoir rock* (◆ Fig. 5.36). Effective reservoir rocks contain a considerable amount of pore space so that appreciable quantities of hydrocarbons can accumulate. Furthermore, the reservoir rocks must possess high *permeability,* or the capacity to transmit fluids; otherwise hydrocarbons cannot be extracted in reasonable quantities. In addition, some kind of impermeable *cap rock* must be present over the reservoir rock to prevent upward migration of the hydrocarbons (Fig. 5.36).

Many hydrocarbon reservoirs consist of nearshore marine sandstones in proximity with fine-grained, organic-rich source rocks. Such oil and gas traps are called *stratigraphic traps* because they owe their existence to variations in the strata (Fig. 5.36a). Ancient coral reefs are also good stratigraphic traps. Indeed, some of the oil in the Persian Gulf region is trapped in ancient reefs. *Structural traps* result when rocks are deformed by folding, fracturing, or both. In areas where sedimentary rocks have been deformed into a series of folds, hydrocarbons migrate to the high parts of such structures (Fig. 5.36b). Displacement of rocks along faults (fractures along which movement has occurred) also yields situations conducive to trapping hydrocarbons (Fig. 5.36b).

Other sources of petroleum that will probably become increasingly important in the future include *oil shales* and *tar sands.* The United States has about two-thirds of all known oil shales, although large deposits also occur in South America, and all continents have some oil shale. The richest deposits in the United States are in the Green River Formation of Colorado, Utah, and Wyoming (see the Prologue).

Tar sand is a type of sandstone in which viscous, asphaltlike hydrocarbons fill the pore spaces. This substance is the sticky residue of once-liquid petroleum from which the volatile constituents have been lost. Liquid petroleum can be recovered from tar sand, but to do so, large quantities of rock must be mined and processed. Since the United States has few tar sand deposits, it cannot look to this source as a significant future energy resource. The Athabaska tar sands in Alberta, Canada, however, are one of the largest deposits of this type. These deposits are currently being mined, and it is estimated that they contain several hundred billion barrels of recoverable petroleum.

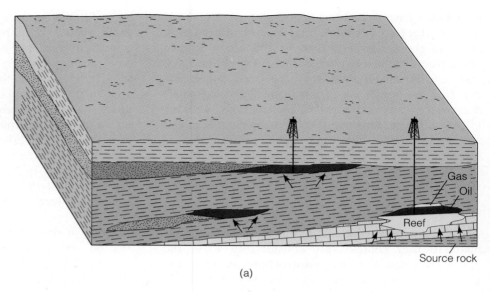

(a)

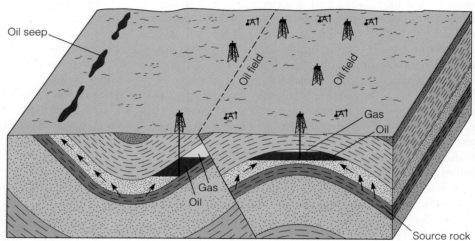

◆ FIGURE 5.36 Oil and natural gas traps. The arrows in both diagrams indicate the migration of hydrocarbons. (a) Two examples of stratigraphic traps. (b) Two examples of structural traps, one formed by folding, the other by faulting.

(b)

Uranium

Most of the uranium used in nuclear reactors in North America comes from the complex potassium-, uranium-, vanadium-bearing mineral *carnotite* found in some sedimentary rocks. Some uranium is also derived from *uraninite* (UO_2), a uranium oxide that occurs in granitic rocks and hydrothermal veins. Uraninite is easily oxidized and dissolved in groundwater, transported elsewhere, and chemically reduced and precipitated in the presence of organic matter.

The richest uranium ores in the United States are widespread in the Colorado Plateau area of Colorado and adjoining parts of Wyoming, Utah, Arizona, and New Mexico. These ores, consisting of fairly pure masses and encrustations of carnotite, are associated with plant remains in sandstones that formed in ancient stream channels. Although most of these ores are associated with fragmentary plant remains, some petrified trees also contain large quantities of uranium.

Large reserves of low-grade uranium ore also occur in the Chattanooga Shale. The uranium is finely disseminated in this black, organic-rich mudrock that underlies large parts of several states including Illinois, Indiana, Ohio, Kentucky, and Tennessee.

Canada remains the world's largest producer and exporter of uranium. For 1987, it is estimated that 13,612 metric tons of uranium valued at more than $1 billion was exported, much of it to the United States.

Banded Iron Formation

Banded iron formation is a chemical sedimentary rock of great economic importance. Such rocks consist of alternating thin layers of chert and iron minerals, mostly the iron oxides hematite and magnetite (◆ Fig. 5.37). Banded iron formations are present on all the continents and account for most of the iron ore mined in the world today. Vast banded iron formations are present in the Lake Superior region of the United States and Canada and in the Labrador Trough of eastern Canada. The origin of banded iron formations is considered in Chapter 20.

◆ FIGURE 5.37 Outcrop of banded iron formation in northern Michigan.

Chapter Summary

1. Mechanical and chemical weathering are processes whereby parent material is disintegrated and decomposed so that it is more nearly in equilibrium with new physical and chemical conditions. The products of weathering include solid particles, soluble salts, and ions in solution.

2. Mechanical weathering includes such processes as frost action, pressure release, thermal expansion and contraction, and the activities of organisms. Particles liberated by mechanical weathering retain the chemical composition of the parent material.

3. Solution, oxidation, and hydrolysis are chemical weathering processes; they result in a chemical change of the weathered products. Clay minerals, various ions in solution, and soluble salts are formed during chemical weathering.

4. Mechanical weathering aids chemical weathering by breaking parent material into smaller pieces, thereby exposing more surface area.

5. Mechanical and chemical weathering produce regolith, some of which is soil if it consists of solids, air, water, and humus and supports plant growth.

6. Soils are characterized by horizons that are designated, in descending order, as O, A, B, and C; soil horizons differ from one another in texture, structure, composition, and color.

7. Soils called pedalfers develop in humid regions such as the eastern United States and much of Canada. Arid and semiarid region soils are pedocals, many of which con-

tain irregular masses of caliche in horizon B. Laterite is a soil resulting from intense chemical weathering as in the tropics. Such soils are deep, red, and sources of aluminum ores if derived from aluminum-rich parent material.

8. Soil erosion, caused mostly by sheet and rill erosion, is a problem in some areas. Human practices such as construction, agriculture, and deforestation can accelerate losses of soil to erosion.

9. Sedimentary particles are designated in order of decreasing size as gravel, sand, silt, and clay.

10. Sedimentary particles are rounded and sorted during transport although the degree of rounding and sorting depends on particle size, transport distance, and depositional process.

11. Any area in which sediment is deposited is a depositional environment. Major depositional settings are continental, transitional, and marine, each of which includes several specific depositional environments.

12. Compaction and cementation are the processes of sediment lithification in which sediment is converted into sedimentary rock. Silica and calcium carbonate are the most common chemical cements, but iron oxide and iron hydroxide cements are important in some rocks.

13. Detrital sedimentary rocks consist of solid particles derived from preexisting rocks. Chemical sedimentary rocks are derived from ions in solution by inorganic chemical processes or the biochemical activities of organ-

isms. A subcategory called biochemical sedimentary rocks is recognized.

14. Sedimentary facies are bodies of sediment or sedimentary rock that are recognizably different from adjacent sediments or rocks.

15. Some sedimentary facies are geographically widespread because they were deposited during marine transgressions or marine regressions.

16. Sedimentary structures such as bedding, cross-bedding, and ripple marks commonly form in sediments when or shortly after they are deposited. Such features preserved in sedimentary rocks aid geologists in determining ancient current directions and depositional environments.

17. Depositional environments of ancient sedimentary rocks are determined by studying sedimentary textures and structures, examining fossils, and making comparisons with present-day depositional processes.

18. Intense chemical weathering is responsible for the origin of residual concentrations, many of which contain valuable minerals such as iron, lead, copper, and clay.

19. Many sediments and sedimentary rocks including sand, gravel, evaporites, coal, and banded iron formations are important natural resources. Most oil and natural gas are found in sedimentary rocks.

Important Terms

bedding
carbonate rock
chemical sedimentary rock
chemical weathering
cross-bedding
depositional environment
detrital sedimentary rock
erosion
evaporite
exfoliation dome
fossil
frost action
graded bedding
hydrolysis

laterite
lithification
marine regression
marine transgression
mechancial weathering
mud crack
oxidation
parent material
pedalfer
pedocal
pressure release
regolith
ripple mark
rounding

sediment
sedimentary facies
sedimentary rock
sedimentary structure
soil
soil horizon
solution
sorting
spheroidal weathering
talus
thermal expansion and contraction
transport
weathering

Review Questions

1. Which mechanical weathering process forms exfoliation domes?
   a. ___ heating and cooling; b. ___ expansion and contraction; c. ___ the activities of organisms; d. ___ oxidation and reduction; e. ___ pressure release.

2. When the ions in a substance become dissociated, the substance has been:
   a. ___ weathered mechanically; b. ___ altered to clay; c. ___ dissolved; d. ___ oxidized; e. ___ converted to soil.

3. The process whereby hydrogen and hydroxyl ions of water replace ions in minerals is:
   a. ___ supergene enrichment; b. ___ oxidation; c. ___ laterization; d. ___ hydrolysis; e. ___ carbonization.

4. Granite weathers more rapidly than quartzite because it contains abundant:
   a. ___ feldspars; b. ___ quartz; c. ___ ferromagnesian minerals; d. ___ carbonate minerals; e. ___ caliche.

5. Horizon B of a soil is also known as the:
   a. ___ top soil; b. ___ humus layer; c. ___ alkali zone; d. ___ zone of accumulation; e. ___ organic-rich layer.

6. The removal of thin layers of soil by water over a more or less continuous surface is:
   a. ___ gullying; b. ___ sheet erosion; c. ___ weathering; d. ___ leaching; e. ___ exfoliation.

7. Which of the following is detrital sediment?
   a. ___ broken sea shells; b. ___ ions in solution; c. ___ quartz sand; d. ___ conglomerate; e. ___ graded bedding.

8. If an aggregate of sediment consists of particles that are all about the same size, it is said to be:
   a. ___ well sorted; b. ___ poorly rounded; c. ___ completely abraded; d. ___ sandstone; e. ___ lithified.

9. The process whereby dissolved mineral matter precipitates in the pore spaces of sediment and binds it together is:

a. ___ compaction; b. ___ rounding; c. ___ bedding;
d. ___ weathering; e. ___ cementation.

10. Most limestones have a large component of calcite that was originally extracted from seawater by:
a. ___ inorganic chemical reactions; b. ___ organisms; c. ___ evaporation; d. ___ chemical weathering; e. ___ lithification.

11. The most common evaporite rock is:
a. ___ rock gypsum; b. ___ chert; c. ___ bituminous coal; d. ___ rock salt; e. ___ siltstone.

12. The superposition of offshore facies over nearshore facies occurs when sea level rises and the shoreline migrates inland during a marine:
a. ___ superposition; b. ___ regression; c. ___ facies; d. ___ invasion; e. ___ transgression.

13. Which of the following can be used to determine paleocurrent direction?
a. ___ mud cracks; b. ___ graded bedding; c. ___ cross-bedding; d. ___ turbidity currents; e. ___ grain size.

14. Traps for petroleum and natural gas resulting from variations in the properties of sedimentary rocks are ___ traps.

a. ___ reservoir; b. ___ stratigraphic; c. ___ cap rock; d. ___ structural; e. ___ salt dome.

15. How does the gravel in sedimentary breccia differ from the gravel in conglomerate?

16. What are the two meanings of the term "clay"?

17. Explain why the sediment in windblown sand dunes is better sorted than that in glacial deposits.

18. What are the common chemical cements in sedimentary rocks, and how do they form?

19. Explain how sheet joints and exfoliation domes originate.

20. Describe the process whereby soluble minerals such as halite (NaCl) are dissolved.

21. What is an acid solution, and why are acid solutions important in chemical weathering?

22. Explain why particle size is an important factor in chemical weathering.

23. Draw a soil profile and list the characteristics of each soil horizon.

24. How do organisms contribute to soil formation?

25. Compare and contrast pedalfer, pedocal, and laterite.

>=o=o=o=o=o=o=o=o=o=o=o=o=o=o=o=o=o=o=o=o=o=o=o=o=o=o=o=o=o=o=o=o=o=o=o=o=o=o=o=o=o=o=o=o=o=o=o=o=o=o=o=<

Additional Readings

Bear, F. E. 1986. *Earth: The stuff of life.* 2d revised ed. Norman, Okla.: Univ. of Oklahoma Press.

Boggs, S., Jr. 1987. *Principles of sedimentology and stratigraphy.* Columbus, Ohio: Merrill.

Buol, S. W., F. D. Hole, and R. J. McCracken. 1980. *Soil genesis and classification.* Ames, Iowa: Iowa State Univ. Press.

Carroll, D. 1970. *Rock weathering.* New York: Plenum Press.

Collinson, J. D., and D. B. Thompson. 1989. *Sedimentary structures.* 2d ed. London: Allen & Unwin.

Coughlin, R. C. 1984. *State and local regulations for reducing agricultural erosion.* American Planning Association, Planning Advisory Service Report No. 386.

Friedman, G. M., J. E. Sanders, and D. C. Kopaska-Merkel. 1992. *Principles of sedimentary deposits.* New York: Macmillan.

Gibbons, B. 1984. Do we treat our soil like dirt? *National Geographic* 166, no. 3:350–89.

LaPorte, L. F. 1979. *Ancient environments.* 2d ed. Englewood Cliffs, N.J.: Prentice-Hall.

Parfit, M. 1989. The dust bowl. *Smithsonian* 20, no. 3:44–54, 56–57.

Selley, R. C. 1982. *An introduction to sedimentology.* 2d ed. New York: Academic Press.

CHAPTER
6

Chapter Outline

Marble quarry, northcentral Vermont. (Photo courtesy of R. V. Dietrich.)

Metamorphism and Metamorphic Rocks

Prologue

Marble is a metamorphic rock formed from limestone or dolostone. Because of its homogeneity, softness, and various textures, marble has been a favorite rock of sculptors throughout history. As the value of authentic marble sculptures has increased through the years, the number of forgeries has also increased. With the price of some marble sculptures in the millions of dollars, private collectors and museums need some means of assuring the authenticity of the work they are buying. Aside from the monetary considerations, it is important that such forgeries do not become part of the historical and artistic legacy of human endeavor.

Experts have traditionally relied on the artistic style of the object as well as its weathering characteristics to determine whether a marble sculpture is authentic or a forgery. Because marble is not very resistant to weathering, forgers have had to resort to a variety of methods to produce the weathered appearance of an authentic ancient work. Now, however, with new techniques of analyzing marble, geologists can differentiate a naturally weathered surface from one that has been artificially altered.

Although marbles result when the agents of metamorphism (heat, pressure, and fluid activity) are applied to carbonate rocks, the type of marble formed depends, in part, on the original composition of the parent carbonate rock as well as the type and intensity of metamorphism. Therefore, one way to authenticate a marble sculpture is to determine the origin of the marble itself. The major quarrying localities of the Preclassical, Greek, and Roman periods include the islands of Naxos, Thasos, and Paros in the Aegean Sea as well as the Greek mainland, Turkey, and Italy (◆ Fig. 6.1a).

In order to determine where the marble in various sculptures has come from, geologists have employed a wide variety of analytical techniques. These include hand specimen and thin-section analysis of the marble, trace element analysis by X-ray fluorescence, stable isotopic ratio analysis for carbon and oxygen, and other more esoteric techniques. Currently, however, carbon and oxygen isotopic analysis has proven to be the most powerful and reliable method for source area determination. This is because each quarry yields marble with a distinctive set of carbon and oxygen isotope values (◆ Fig. 6.1b).

Recall that isotopes are forms of individual elements with different atomic mass numbers (see Fig. 2.3). If the carbon and oxygen isotope ratios of a sculpture fall outside the typical range of the locality from which the marble supposedly comes, then it is probably a forgery. Using this technique, geologists showed that a marble head of Achilles owned by the J. Paul Getty Museum in Malibu, California, was a forgery. When the carbon and oxygen isotope ratios from the Getty Museum specimen were compared with those obtained from another marble head of known authenticity, they did not match, indicating that the two sculptures were carved from marbles that came from two different quarries.

Norman Herz of the Geology Department of the University of Georgia has sampled all of the major and many of the minor ancient marble quarries in the Aegean Sea region and assembled a large isotopic data base for these quarries. Using this data base for comparative purposes, Herz has been able to

(a)

◆ FIGURE 6.1 (*a*) The principal quarries of the Aegean Sea region that supplied much of the marble used for sculpturing during the Preclassical, Greek, and Roman periods. (*b*) Carbon 13 and oxygen 18 isotope values for marbles from various geographic localities in the Aegean region. Each delta value represents a 0.1% change from a standardized sample. The lower the delta value, the higher the concentration of the lighter carbon 12 and oxygen 16 isotopes.

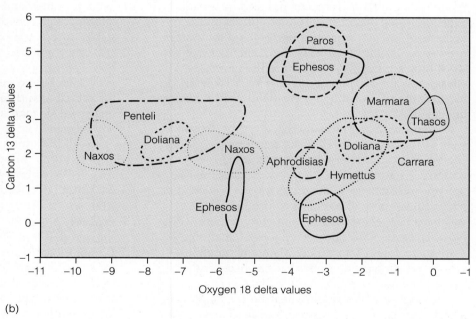

(b)

determine the source area of many marble pieces, as well as demonstrating that some marble sculptures have been reassembled from marbles that came from different localities and therefore were not part of the original piece.

In one especially interesting case, Herz was able to determine that the five fragments composing the Antonia Minor portrait in the Fogg Museum at Harvard University (◆ Fig. 6.2) are not all the same marble. The portrait was purchased by the earl of

◆ FIGURE 6.2 Carbon and oxygen isotopic analysis of the 53-cm tall Antonia Minor portrait showed that the head is authentic, but unrelated to the other four pieces that compose it.

Pembroke in 1678 and its restoration was completed in 1758. Since that time, art historians have debated the portrait's authenticity and method of restoration, with some claiming the portrait was assembled from completely different statues.

The five fragments composing the portrait are the head, the end of the ponytail, the right shoulder and breast, the lower left shoulder, and the upper left shoulder and breast. Carbon and oxygen isotopic analysis of the five fragments revealed that three of the pieces were of Parian marble and two were Carrara marble. It was concluded that the head is authentic, but unrelated to the other pieces, with the right shoulder and breast and the upper left shoulder and breast being comparatively recent additions.

Many museums are now making geological testing to authenticate marble sculptures an important part of their curatorial functions. In addition, a large body of data about the characteristics and origin of marble is being amassed as more sculptures and quarries are analyzed.

✦ INTRODUCTION

Metamorphic rocks (from the Greek *meta* meaning change and *morpho* meaning shape) are the third major group of rocks. They result from the transformation of other rocks by metamorphic processes that usually occur beneath the Earth's surface (◆ Fig. 6.3). During metamorphism, rocks are subjected to sufficient heat, pressure, and fluid activity to change their mineral composition and/or texture, thus forming new rocks. These transformations take place in the solid state, and the type of metamorphic rock formed depends on the original composition and texture of the parent rock, the agents of metamorphism, and the amount of time the parent rock was subjected to the effects of metamorphism.

A large portion of the Earth's continental crust is composed of metamorphic and igneous rocks. Together, they form the crystalline basement rocks that underlie the sedimentary rocks of a continent's surface. This basement rock is widely exposed in regions of the continents known as *shields,* which have been very stable during the past 600 million years (◆ Fig. 6.4). Metamorphic rocks also constitute a sizable portion of the crystalline core of large mountain ranges. Some of the oldest known rocks, dated at 3.96 billion years from the Canadian Shield, are metamorphic, indicating they formed from even older rocks.

◆ FIGURE 6.3 The rock cycle, showing how metamorphic, igneous, and sedimentary rocks are interrelated.

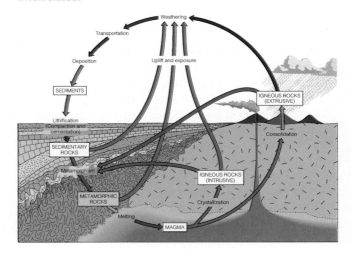

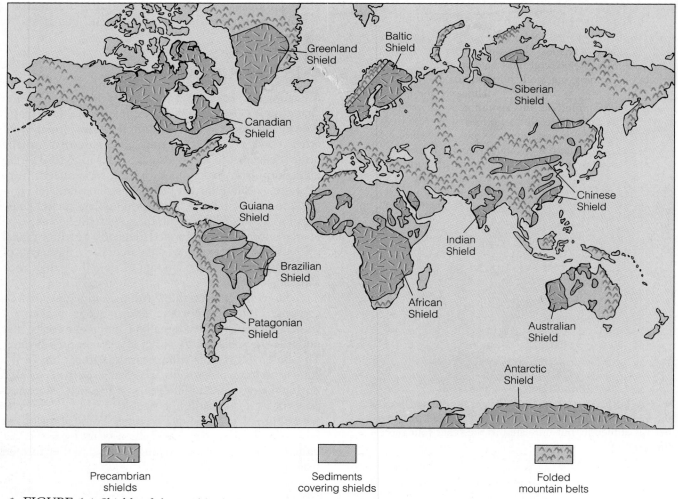

FIGURE 6.4 Shields of the world. Shields are the exposed portion of the crystalline basement rocks that underlie each continent; these areas have been very stable during the past 600 million years.

Precambrian shields

Sediments covering shields

Folded mountain belts

Why is it important to study metamorphic rocks? For one thing, they provide information about geological processes operating within the Earth and about the way these processes have varied through time. From the presence of certain minerals in metamorphic rocks, geologists can determine the approximate temperatures and pressures that parent rocks were subjected to during metamorphism and thus gain insights into the physical and chemical changes occurring at different depths within the Earth's crust. Furthermore, metamorphic rocks such as marble and slate are used as building materials, and certain metamorphic minerals are eco-

nomically important. For example, garnets are used as gemstones or abrasives; talc is used in cosmetics, in the manufacture of paint, and as a lubricant; asbestos is used for insulation and fireproofing (see Perspective 6.1); and kyanite is used in the production of refractory materials used in sparkplugs.

▼ THE AGENTS OF METAMORPHISM

The three agents of metamorphism are heat, pressure, and fluid activity. During metamorphism, the original rock undergoes change so as to achieve equilibrium with its new environment. The changes may result in the formation of new minerals and/or a change in the texture of the rock by the reorientation of the original minerals. In some instances the change is minor, and features of the parent rock can still be recognized. In other cases the rock changes so much that the identity of the parent rock can be determined only with great difficulty, if at all.

Heat

Heat is an important agent of metamorphism because it increases the rate of chemical reactions that may produce new mineral assemblages different from those in the original rock. The heat may come from intrusive magmas or result from deep burial in the Earth's crust such as occurs during subduction along a convergent plate boundary.

When rocks are intruded by bodies of magma, they are subjected to intense heat that affects the surrounding rock; the most intense heating usually occurs adjacent to the magma body and gradually decreases with distance from the intrusion. The zone of metamorphosed rocks that forms in the country rock adjacent to an intrusive igneous body is usually rather distinct and easy to recognize.

Recall from Chapter 4 that temperature increases with depth and that the Earth's geothermal gradient averages about 25°C/km. Rocks forming at the Earth's surface may be transported to great depths by subduction along a convergent plate boundary and thus subjected to increasing temperature and pressure. During subduction, some minerals may be transformed into other minerals that are more stable under the higher temperature and pressure conditions.

Pressure

When rocks are buried, they are subjected to increasingly greater **lithostatic pressure;** this pressure, which results from the weight of the overlying rocks, is applied equally in all directions (◆ Fig. 6.5a). A similar situation occurs when an object is immersed in water. For example, the deeper a styrofoam cup is submerged in the ocean, the smaller it gets because pressure increases with depth and is exerted on the cup equally in all directions, thereby compressing the styrofoam (◆ Fig. 6.5b).

Just as in the styrofoam cup example, rocks are subjected to increasing lithostatic pressure with depth

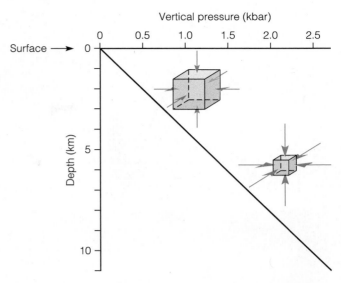

1 kilobar (kbar) = 1,000 bars
Atmospheric pressure at sea level = 1 bar

(a)

(b)

◆ FIGURE 6.5 (*a*) Lithostatic pressure is applied equally in all directions in the Earth's crust due to the weight of the overlying rocks. Thus, pressure increases with depth. (*b*) A similar situation occurs when 200 ml styrofoam cups are lowered to ocean depths of approximately 750 m and 1,500 m. Increased pressure is exerted equally in all directions on the cups, and they consequently decrease in volume, while still maintaining their general shape. (Styrofoam cups courtesy of David J. Matty and Jane M. Matty. Photo courtesy of Sue Monroe.)

such that the mineral grains within a rock may become more closely packed. Under such conditions, the minerals may *recrystallize;* that is, they may form smaller and denser minerals that either are the same chemical composition or are of different mineral assemblages.

Perspective 6.1

ASBESTOS

Asbestos (from the Latin, meaning unquenchable) is a general term applied to any silicate mineral that easily separates into flexible fibers (♦ Fig. 1).

♦ FIGURE 1 Hand specimen of chrysotile from Thetford, Quebec, Canada. Chrysotile is the fibrous form of serpentine asbestos.

The combination of such features as non-combustibility and flexibility makes asbestos an important industrial material of considerable value. In fact, asbestos has more than 3,000 known uses. These include brake linings and clutch facings, fireproof fabrics, heat insulators, cements, shingles, acid and chemical equipment, insulation, and binders for various plasters, porcelains, and electrical insulators to name only a few.

The unique properties of asbestos were certainly known in the ancient world. The Romans used it to make lamp wicks that never burned out and also wove it into cremation clothes for the nobility.

Asbestos can be divided into two broad groups, serpentine and amphibole asbestos. *Chrysotile,* which is a hydrous magnesium silicate with the chemical formula $Mg_3Si_2O_5(OH)_4$, is the fibrous form of serpentine asbestos; it is the most valuable type and constitutes the bulk of all commercial asbestos. Chrysotile's strong, silky fibers are easily spun and can withstand temperatures up to 2,750°C.

The vast majority of chrysotile asbestos occurs in serpentine that has been altered from such ultramafic igneous rocks as peridotite under low- and medium-grade metamorphic conditions. Serpentine is believed to form from the alteration of olivine by hot, chemically active, residual fluids emanating from cooling magma. The chrysotile asbestos forms veinlets of fiber within the

---

In addition to the lithostatic pressure resulting from burial, rocks may also experience **differential pressures.** In this case, the pressures are not equal on all sides, and the rock is consequently distorted. Differential pressures typically occur during deformation associated with mountain building and can produce distinctive metamorphic textures and features (♦ Fig. 6.6).

Fluid Activity

In almost every region where metamorphism occurs, water and carbon dioxide (CO_2) are present in varying amounts along mineral grain boundaries or in the pore spaces of rocks. This water, which may contain ions in solution, enhances metamorphism by increasing the

serpentine and may comprise up to 20% of the rock. Other chrysotile results when the metamorphism of magnesium limestone or dolostone produces discontinuous serpentine bands within the carbonate beds.

At least five varieties of amphibole asbestos are known, but *crocidolite,* a sodium-iron amphibole with the chemical formula $Na_2(Fe^{+3})_2(Fe^{+2})_3Si_8O_{22}(OH)_2$, is the most common. Crocidolite, which is also known as blue asbestos, is a long, coarse, spinning fiber that is stronger but more brittle than chrysotile and also less resistant to heat. The other varieties of amphibole asbestos have little commercial value and are used chiefly for insulation.

Crocidolite is found in such metamorphic rocks as slates and schists. It is thought that crocidolite forms by the solid-state alteration of other minerals within the high temperature and high pressure environment that results from deep burial. Unlike chrysotile, crocidolite is rarely found associated with igneous intrusions.

In spite of its widespread use, the federal Environmental Protection Agency has enacted a gradual ban on all new asbestos products. The ban was imposed because asbestos can cause cancer and scarring of the lungs if its fibers are inhaled. The threat of lung cancer has resulted in legislation mandating the removal of asbestos already in place in many buildings, including all public and private schools. Important questions have recently been raised, however, concerning the health threat posed by asbestos.[*]

Central to the debate is whether all varieties of asbestos should be lumped together. Chrysotile, whose fibers tend to be curly, does not become lodged in the lungs. Furthermore, the fibers are generally soluble and disappear in tissue. In contrast, crocidolite has long, straight, thin fibers that penetrate the lungs and stay there. Thus, crocidolite, not chrysotile, is overwhelmingly responsible for asbestos-related lung cancer. Because about 95% of the asbestos in place in the United States is chrysotile, many people are questioning whether the dangers from asbestos have been somewhat exaggerated.

Removing asbestos from buildings where it has been installed might cost as much as $150 billion, and some recent studies have indicated that the air in buildings containing asbestos has essentially the same amount of airborne asbestos fibers as the air outdoors. In fact, unless the material containing the asbestos is disturbed, asbestos does not shed fibers. Furthermore, improper removal of asbestos can lead to contamination. In most cases of improper removal, the concentration of airborne asbestos fibers is far higher than if the asbestos had been left in place.

. . . .

[*]P. H. Abelson, "The Asbestos Removal Fiasco," *Science* 247 no. 4946 (1990): 1017.

rate of chemical reactions. Under dry conditions, most minerals react very slowly, but when even small amounts of fluid are introduced, reaction rates increase, mainly because ions can move readily through the fluid and thus enhance chemical reactions and the formation of new minerals.

The following reaction provides a good example of how new minerals can be formed by **fluid activity.** Here, seawater moving through hot basaltic rock transforms olivine into the metamorphic mineral serpentine:

$$2Mg_2SiO_4 + 2H_2O \rightarrow Mg_3Si_2O_5(OH)_4 + MgO$$

olivine   water   serpentine   carried away in solution

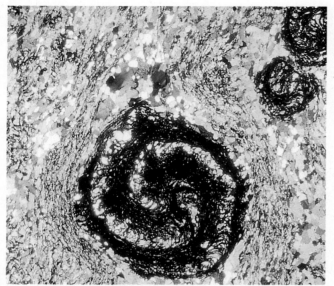

◆ FIGURE 6.6 Differential pressure is pressure that is unequally applied to an object. Rotated garnets are a good example of the effects of differential pressure applied to a rock during metamorphism. These rotated garnets come from a calcareous schist of the Waits River Formation, north of Springfield, Vermont. (Photo courtesy of John L. Rosenfeld, University of California, Los Angeles.)

The chemically active fluids that are part of the metamorphic process come primarily from three sources. The first is water trapped in the pore spaces of sedimentary rocks as they form; as these rocks are subjected to heat and pressure, the water is heated, thus accelerating the various chemical reaction rates. A second source is the volatile fluid within magma; as these hot fluids disperse through the surrounding rock, they frequently react with and alter the mineralogy of the country rock by adding or removing ions. The third source is the dehydration of water-bearing minerals such as gypsum ($CaSO_4 \cdot 2H_2O$) and some clays; when these minerals, which contain water as part of their crystal chemistry, are subjected to heat and pressure, the water may be driven off and enhance metamorphism.

▼ TYPES OF METAMORPHISM

Three major types of metamorphism are recognized: *contact metamorphism* in which magmatic heat and fluids act to produce change; *dynamic metamorphism,* which is principally the result of high differential pressures associated with intense deformation; and *regional metamorphism,* which occurs within a large area and is caused primarily by mountain-building forces. Even though we will discuss each type of metamorphism separately, the boundary between them is not always distinct and depends largely on which of the three metamorphic agents was dominant.

Contact Metamorphism

Contact metamorphism takes place when a body of magma alters the surrounding country rock. At shallow depths an intruding magma raises the temperature of the surrounding rock, causing thermal alteration. Furthermore, the release of hot fluids into the country rock by the cooling intrusion can also aid in the formation of new minerals.

Important factors in contact metamorphism are the initial temperature and size of the intrusion as well as the fluid content of the magma and/or country rock. The initial temperature of an intrusion is controlled, in part, by its composition: mafic magmas are hotter than felsic magmas (see Chapter 4) and hence have a greater thermal effect on the rocks directly surrounding them. The size of the intrusion is also important. In the case of small intrusions, such as dikes and sills, usually only those rocks in immediate contact with the intrusion are affected. Because large intrusions, such as batholiths, take a long time to cool, the increased temperature in the surrounding rock may last long enough for a larger area to be affected.

Fluids also play an important role in contact metamorphism. Many magmas are wet and contain hot, chemically active fluids that may emanate into the surrounding rock. These fluids can react with the rock and aid in the formation of new minerals. In addition, the country rock may contain pore fluids that, when heated by the magma, also increase reaction rates.

Temperatures can reach nearly 900°C adjacent to an intrusion, but they gradually decrease with distance. The effects of such heat and the resulting chemical reactions usually occur in concentric zones known as **aureoles** (◆ Fig. 6.7). The boundary between an intrusion and its aureole may be either sharp or transitional (◆ Fig. 6.8).

Metamorphic aureoles vary in width depending on the size, temperature, and composition of the intrusion as well as the mineralogy of the surrounding country rock. Typically, large intrusive bodies have several metamorphic zones, each characterized by distinctive

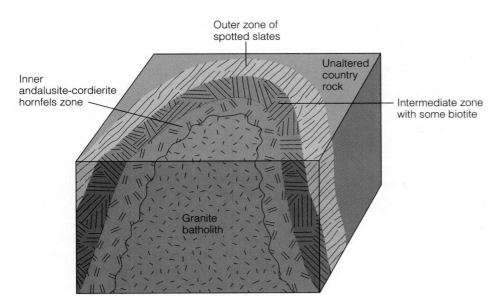

Outer zone of
spotted slates

Inner
andalusite-cordierite
hornfels zone

Unaltered
country
rock

Intermediate zone
with some biotite

Granite
batholith

◆ **FIGURE 6.7** A metamorphic aureole typically surrounds many igneous intrusions. The metamorphic aureole associated with this idealized granite batholith contains three zones of mineral assemblages reflecting the decreases in temperature with distance from the intrusion. An andalusite-cordierite hornfels forms the inner zone adjacent to the batholith. This is followed by an intermediate zone of extensive recrystallization in which some biotite develops, and farthest from the intrusion is the outer zone, which is characterized by spotted slates.

◆ FIGURE 6.8 A sharp and clearly defined boundary occurs between the intruding light-colored igneous rock on the left and the dark-colored metamorphosed country rock on the right. The intrusion is part of the Peninsular Ranges Batholith, east of San Diego, California. (Photo courtesy of David J. Matty.)

mineral assemblages indicating the decrease in temperature with distance from the intrusion (Fig. 6.7). The zone closest to the intrusion, and hence subject to the highest temperatures, may contain high-temperature metamorphic minerals (that is, minerals in equilibrium with the higher temperature environment) such as sillimanite. The outer zones may be characterized by lower temperature metamorphic minerals such as chlorite, talc, and epidote.

The formation of new minerals by contact metamorphism depends not only on proximity to the intrusion, but also on the mineralogy of the country rock. Shales, mudstones, impure limestones, and impure dolostones, for example, are particularly susceptible to the formation of new minerals by contact metamorphism, whereas pure sandstones or pure limestones typically are not.

Two types of contact metamorphic rocks are generally recognized: those resulting from baking of country rock and those altered by hot solutions. Many of the rocks resulting from contact metamorphism have the texture of porcelain; that is, they are hard and fine grained. This is particularly true for rocks with a high clay content, such as shale. Such texture results because the clay minerals in the rock are baked, just as a clay pot is baked when fired in a kiln.

During the final stages of cooling when an intruding magma begins to crystallize, large amounts of hot, watery solutions are often released. These solutions may react with the country rock and produce new metamorphic minerals. This process, which usually occurs near the Earth's surface, is called *hydrothermal alteration,* and may result in valuable mineral deposits such as the Kuroko sulfide deposit in Japan (see Chapter 9). Geologists think that many of the world's ore deposits result from the migration of metallic ions in hydrothermal solutions. Examples include copper, gold, iron ores, tin, and zinc in various localities including Australia, Canada, China, Cyprus, Finland, the former Soviet Union, and the western United States.

Dynamic Metamorphism

Most **dynamic metamorphism** is associated with fault zones where rocks are subjected to high differential pressures. The metamorphic rocks resulting from pure dynamic metamorphism are called *mylonites* and they are typically restricted to narrow zones adjacent to faults (fractures along which movement has occurred). Mylonites are hard, dense, fine-grained rocks, many of which are characterized by thin laminations (◆ Fig. 6.9). Tectonic settings where mylonites occur include the Moine Thrust Zone in northwest Scotland and portions of the San Andreas fault in California.

Regional Metamorphism

Most metamorphic rocks result from **regional metamorphism,** which occurs over a large area and is usually caused by tremendous temperatures, pressures, and deformation within the deeper portions of the Earth's crust. Regional metamorphism is most obvious along convergent plate margins where rocks are intensely deformed and recrystallized during convergence and subduction. Within these metamorphic rocks, there is usually a gradation of metamorphic intensity from areas that were subjected to the most intense pressures and/or highest temperatures to areas of lower pressures and temperatures. Such a gradation in metamorphism can be recognized by the metamorphic minerals that are present.

Regional metamorphism is not just confined to convergent margins. It also occurs in areas where plates diverge, though usually at much shallower depths because of the high geothermal gradient associated with these areas.

From field studies and laboratory experiments, certain minerals are known to form only within specific temperature and pressure ranges. Such minerals are known as **index minerals** because their presence allows geologists to recognize low-, intermediate-, and high-grade metamorphic zones (● Table 6.1). For example, in clay-rich rocks such as shale, the mineral chlorite is produced under relatively low temperatures of about 200°C, so its presence indicates low-grade metamorphism. At the high-grade end of the metamorphic spectrum for clay-rich rocks, sillimanite may form in environments where the temperature exceeds 500°C. A typical progression of index minerals from chlorite to sillimanite involves the sequential formation of the following minerals:

chlorite → biotite → amphibole → staurolite → sillimanite

Different rock compositions develop different index minerals. The sequence of index minerals just listed forms primarily in rocks that were originally clay rich. When sandy dolomites are metamorphosed, for example, they produce an entirely different set of index minerals. Thus, a specific set of index minerals commonly forms in specific rock types as metamorphism progresses.

Although such common minerals as mica, quartz, and feldspar can occur in both igneous and metamorphic rocks, other minerals such as andalusite, sillimanite, and kyanite generally occur only in metamorphic rocks derived from clay-rich sediments. Although these three minerals all have the same chemical formula (Al_2SiO_5), they differ in crystal structure and other physical properties because each forms under a different range of pressures and temperatures (◆ Fig. 6.10). Thus, these polymorphs are sometimes used as index minerals for metamorphic rocks formed from clay-rich sediments.

◆ FIGURE 6.9 This light-colored, 15-cm thick mylonite unit is part of the Carthage-Colton Mylonite Zone exposed along Route 3, south of Harrisville, New York. (Photo courtesy of Eric Johnson.)

Metamorphic Grade	Metamorphic Zone for Clay-Rich Rocks	Mineral Assemblage Produced for Different Country Rocks		
		Mudrocks	Limestones	Mafic Igneous Rocks
Increasing Low	Chlorite	Chlorite,* quartz, muscovite, plagioclase	Chlorite,* calcite or dolomite, plagioclase	Chlorite,* plagioclase
	Biotite	Biotite,* quartz, plagioclase		
Medium	Garnet	Garnet,* mica, quartz, plagioclase	Garnet,* epidote, hornblende, calcite	Garnet,* chlorite, epidote, plagioclase
	Staurolite	Staurolite,* mica, garnet, quartz, plagioclase	Garnet, hornblende,* plagioclase	
High	Kyanite	Kyanite,* mica, garnet, quartz, plagioclase		
metamorphism				Hornblende,* plagioclase
	Sillimanite	Sillimanite,* garnet, mica, quartz, plagioclase	Garnet, augite,* plagioclase	

*Index mineral.

⩔ CLASSIFICATION OF METAMORPHIC ROCKS

For purposes of classification, metamorphic rocks are commonly divided into two groups: those exhibiting a *foliated texture* and those with a *nonfoliated texture* (● Table 6.2).

Foliated Metamorphic Rocks

Rocks subjected to heat and differential pressure during metamorphism typically have minerals arranged in a parallel fashion that gives them a **foliated texture** (◆ Fig. 6.11). The size and shape of the mineral grains determine whether the foliation is fine or coarse. If the foliation is such that the individual grains cannot be recognized without magnification, the rock is said to be slate (◆ Fig. 6.12a). A coarse foliation results when granular minerals such as quartz and feldspar are segregated into roughly parallel and streaky zones that differ in composition and color as in a gneiss (Fig. 6.14). Foliated metamorphic rocks can be arranged in order of increasingly coarse grain size and perfection of foliation.

Slate is a very fine-grained metamorphic rock that commonly exhibits *slaty cleavage* (◆ Fig. 6.12b). Slate is

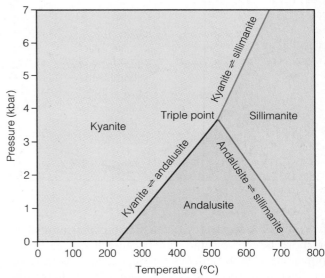

◆ FIGURE 6.10 Andalusite, sillimanite, and kyanite all have the same chemical composition (Al_2SiO_5) but very different physical properties and crystal structures because they form under different pressure-temperature conditions. The double arrows between the mineral combinations mean that the transformations are reversible; that is, they can go in either direction. The triple point where the three lines meet is the pressure-temperature value where, in theory, all three polymorphs can exist in equilibrium.

TABLE 6.2 Classification of Common Metamorphic Rocks

Texture	Metamorphic Rock	Typical Minerals	Metamorphic Grade	Characteristics of Rocks	Parent Rock
Foliated	Slate	Clays, micas, chlorite	Low	Fine grained, splits easily into flat pieces	Shale, claystones, volcanic ash
	Phyllite	Fine-grained quartz, micas, chlorite	Low to medium	Fine grained, glossy or lustrous sheen	Mudrocks
	Schist	Micas, chlorite, quartz, talc, hornblende, garnet, staurolite, graphite	Low to high	Distinct foliation, minerals visible	Variable mudrocks, impure carbonates, mafic igneous rocks
	Gneiss	Quartz, feldspars, hornblende, micas	High	Segregated light and dark bands visible	Mudrocks, sandstones, felsic igneous rocks
	Amphibolite	Hornblende, plagioclase	Medium to high	Dark colored, weakly foliated	Mafic igneous rocks
	Migmatite	Quartz, feldspars, hornblende, micas	High	Streaks or lenses of granite intermixed with gneiss	Felsic igneous rocks mixed with sedimentary rocks
Nonfoliated	Marble	Calcite, dolomite	Low to high	Interlocking grains of calcite or dolomite, reacts with HCl	Limestone or dolostone
	Quartzite	Quartz	Medium to high	Interlocking quartz grains, hard, dense	Quartz sandstone
	Greenstone	Chlorite, epidote, hornblende	Low to high	Fine grained, green color	Mafic igneous rocks
	Hornfels	Micas, garnets, andalusite, cordierite, quartz	Low to medium	Fine grained, equidimensional grains, hard, dense	Shale
	Anthracite	Carbon	High	Black, lustrous, subconcoidal fracture	Low-grade coal

the result of low-grade regional metamorphism of shale or, more rarely, volcanic ash. Because it can easily be split along cleavage planes into flat pieces, slate is an excellent rock for roofing and floor tiles, billiard and pool table tops, and blackboards. The different colors of most slates are caused by minute amounts of graphite (black), iron oxide (red and purple), and/or chlorite (green).

Phyllite is similar in composition to slate, but is coarser grained. However, the minerals are still too small to be identified without magnification. Phyllite can be distinguished from slate by its glossy or lustrous sheen. It represents an intermediate grain size between slate and schist.

Schist is most commonly produced by regional metamorphism. The type of schist formed depends on the intensity of metamorphism and the character of the parent rock (◆ Fig. 6.13). Metamorphism of many rock types can yield schist, but most schist appears to have formed from clay-rich sedimentary rocks (Table 6.2).

All schists contain more than 50% platy and elongated minerals, all of which are large enough to be clearly visible. Their mineral composition imparts a *schistosity* or *schistose foliation* to the rock that com-

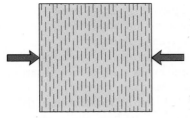

Random arrangement of elongated minerals before pressure is applied to two sides

Elongated minerals arranged in a parallel fashion as a result of pressure applied to two sides

(a)

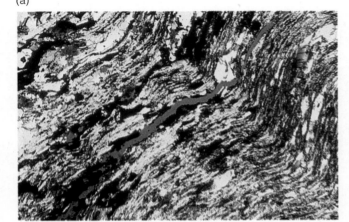

(b)

◆ FIGURE 6.11 (*a*) When rocks are subjected to differential pressure, the mineral grains are typically arranged in a parallel fashion, producing a foliated texture. (*b*) Photomicrograph of a metamorphic rock with a foliated texture showing the parallel arrangement of mineral grains.

monly produces a wavy type of parting when split. Schistosity is common in low- to high-grade metamorphic environments. Because a schist's mineral grains can be readily identified, each type is known by its most conspicuous mineral or minerals, for example, mica schist, chlorite schist, and talc schist.

Gneiss is a metamorphic rock that is streaked or has segregated bands of light and dark minerals. Gneisses are composed mostly of granular minerals such as quartz and/or feldspar with lesser percentages of platy or elongated minerals such as micas or amphiboles (◆ Fig. 6.14). Quartz and feldspar are the principal light-colored minerals, while biotite and hornblende are the typical dark-colored minerals. Most gneiss breaks in an irregular manner, much like coarsely crystalline nonfoliated rocks.

(a)

(b)

◆ FIGURE 6.12 (*a*) Hand specimen of slate. (*b*) This panel of Arvonia Slate from Albemarne Slate Quarry, Virginia, shows bedding (upper right to lower left) at an angle to the slaty cleavage. (Photo (*a*) courtesy of Sue Monroe; photo (*b*) courtesy of R. V. Dietrich.)

Most gneiss probably results from recrystallization of clay-rich sedimentary rocks during regional metamorphism (Table 6.2). Gneiss also can form from crystalline igneous rocks such as granite or older metamorphic rocks.

Another fairly common foliated metamorphic rock is *amphibolite*. It is dark in color and composed mainly of hornblende and plagioclase. The alignment of the hornblende crystals produces a slightly foliated texture. Many amphibolites result from medium- to high-grade metamorphism of such ferromagnesian mineral-rich igneous rocks as basalt.

In some areas of regional metamorphism, exposures of "mixed rocks" having both igneous and high-grade metamorphic characteristics are present. These rocks,

called *migmatites,* usually consist of streaks or lenses of granite intermixed with high-grade ferromagnesian-rich metamorphic rocks (◆ Fig. 6.15).

Most migmatites are thought to be the product of extremely high-grade metamorphism, and several models for their origin have been proposed. Part of the problem in determining the origin of migmatites is explaining how the granitic component formed. According to one model, the granitic magma formed in place by the partial melting of rock during intense metamorphism. Such an origin is possible providing

◆ FIGURE 6.14 Gneiss is characterized by segregated bands of light and dark minerals. This folded gneiss crops out at Wawa, Ontario, Canada.

◆ FIGURE 6.13 Schist. (*a*) Garnet-mica schist. (*b*) Hornblende-mica-garnet schist. (Photos courtesy of Sue Monroe.)

(a)

(b)

◆ FIGURE 6.15 Migmatites consist of high-grade metamorphic rock intermixed with streaks or lenses of granite. This Precambrian(?) migmatite crops out at Thirty Thousand Islands of Georgian Bay, Lake Huron, Ontario, Canada. (Photo by Ed Bartram, courtesy of R. V. Dietrich.)

that the host rocks contained quartz and feldspars and that water was present. Another possibility is that the granitic components formed by the redistribution of minerals by recrystallization in the solid state, that is, by pure metamorphism.

Nonfoliated Metamorphic Rocks

Some metamorphic rocks do not show discernible preferred orientation of their mineral grains. Instead, they generally consist of a mosaic of roughly equidimensional minerals and are characterized as **nonfoliated** (◆ Fig. 6.16). Most nonfoliated metamorphic rocks result from contact or regional metamorphism of rocks in which no platy or prismatic minerals are present. Frequently, the only indication that a granular rock has been metamorphosed is the large grain size resulting from recrystallization. Nonfoliated metamorphic rocks are generally of two types: those composed mainly of only one mineral, for example, marble or quartzite; and those in which the different mineral grains are too small to be seen without magnification, such as greenstone and hornfels.

Marble is a relatively well-known metamorphic rock composed predominantly of calcite or dolomite; its grain size ranges from fine to coarsely granular (Figs. 6.2 and ◆ 6.17). Marble results from either contact or regional metamorphism of limestones or dolostones (Table 6.2). Pure marble is snowy white or bluish, but varieties of all colors exist because of the presence of mineral impurities in the parent sedimentary rock. The softness of marble, its uniform texture, and its various colors have made it the favorite rock of builders and sculptors throughout history (see the Prologue).

Quartzite is a hard, compact rock typically formed from quartz sandstone under medium-to-high-grade metamorphic conditions during contact or regional metamorphism (◆ Fig. 6.18). Because recrystallization is so complete, metamorphic quartzite is of uniform strength and therefore usually breaks across the component quartz grains rather than around them when it is struck. Pure quartzite is white, but iron and other impurities commonly impart a reddish or other color to it. Quartzite is commonly used as foundation material for road and railway beds.

The name *greenstone* is applied to any compact, dark-green, altered, mafic igneous rock that formed under low-to-high-grade metamorphic conditions. The green color results from the presence of chlorite, epidote, and hornblende.

◆ FIGURE 6.16 Nonfoliated textures are characterized by a mosaic of roughly equidimensional minerals as in this photomicrograph of marble.

◆ FIGURE 6.17 Marble results from the metamorphism of the sedimentary rocks limestone and dolostone. (Photos courtesy of Sue Monroe.)

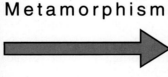

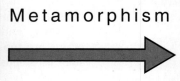

◆ FIGURE 6.18 Quartzite results from the metamorphism of quartz sandstone. (Photos courtesy of Sue Monroe.)

Hornfels is a fine-grained, nonfoliated metamorphic rock resulting from contact metamorphism; it is composed of various equidimensional mineral grains. The composition of hornfels is directly dependent upon the composition of the parent rock, and many compositional varieties are known. However, the majority of hornfels are apparently derived from contact metamorphism of clay-rich sedimentary rocks or impure dolostones.

Anthracite is a black, lustrous, hard coal that contains a high percentage of fixed carbon and a low percentage of volatile matter. It usually forms from the metamorphism of lower grade coals by heat and pressure and is thus considered by many geologists to be a metamorphic rock.

❦ METAMORPHIC ZONES AND FACIES

The first systematic study of metamorphic zones was conducted during the late 1800s by George Barrow and other British geologists working in the Dalradian schists of the southwestern Scottish Highlands. Here, clay-rich sedimentary rocks have been subjected to regional metamorphism, and the resulting metamorphic rocks can be divided into different zones based on the presence of distinctive silicate mineral assemblages. These mineral assemblages, each recognized by the presence of one or more index minerals, reflect different degrees of metamorphism. The index minerals Barrow and his associates chose to represent increasing metamorphic intensity were, in order, chlorite, biotite, garnet, staurolite, kyanite, and sillimanite (Table 6.1). Note that these are the metamorphic minerals produced from clay-rich sediments. Other mineral assem-

blages and index minerals are produced from rocks with different original compositions (Table 6.1).

The successive appearance of metamorphic index minerals reflects gradually increasing or decreasing intensity of metamorphism. Going from lower toward higher grade zones, the first appearance of a particular index mineral indicates the location of the minimum temperature and pressure conditions needed for the formation of that mineral. When the locations of the first appearances of that index mineral are connected on a map, the result is a line of equal metamorphic intensity or an **isograd.** The region between isograds is known as a **metamorphic zone.** The rocks within each zone represent a **metamorphic grade.** By noting the occurrence of metamorphic index minerals, geologists can construct a map showing the metamorphic zones of an entire area (◆ Fig. 6.19).

Numerous studies of different metamorphic rocks have demonstrated that while the texture and mineralogy of any rock may be altered by metamorphism, the overall chemical composition may be little changed. Thus, the different mineral assemblages found in increasingly higher grade metamorphic rocks derived from the same parent rock result from changes in temperature and pressure (Table 6.1).

A **metamorphic facies** is a group of metamorphic rocks that are characterized by particular mineral assemblages formed under the same broad temperature-pressure conditions (◆ Fig. 6.20). Each facies is named after its most characteristic rock or mineral. For example, the green metamorphic mineral chlorite, which forms under relatively low temperatures and pressures, yields rocks said to belong to the *greenschist facies.*

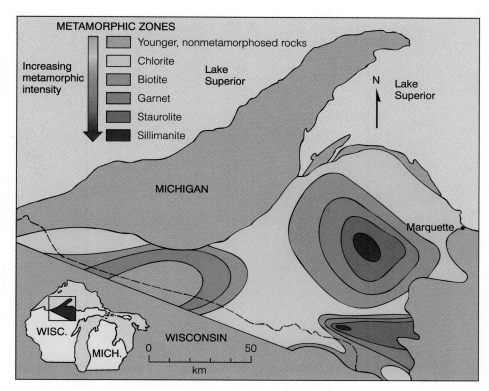

◆ FIGURE 6.19 Metamorphic zones in the upper peninsula of Michigan. The zones in this region are based on the appearance of distinctive silicate mineral assemblages resulting from the metamorphism of sedimentary rocks during an interval of mountain building and minor granitic intrusion during the Proterozoic Eon, about 1.5 billion years ago.

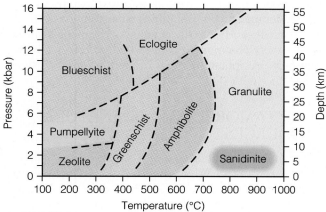

◆ FIGURE 6.20 A pressure-temperature diagram showing where various metamorphic facies occur. A facies is characterized by a particular mineral assemblage that formed under the same broad temperature-pressure conditions. Each facies is named after its most characteristic rock or mineral.

Under increasingly higher temperatures and pressures, other metamorphic facies, such as the *amphibolite* and *granulite facies* develop.

Although usually applied to areas where the original rocks were clay rich, the concept of metamorphic facies can be used with modification in other situations. It cannot, however, be used in areas where the original rocks were pure quartz sandstones or pure limestones or dolostones. Such rocks would yield only quartzites and marbles, respectively.

▼ METAMORPHISM AND PLATE TECTONICS

Although metamorphism is associated with all three types of plate boundaries (see Fig. 1.7), it is most common along convergent plate margins. Metamorphic rocks form at convergent plate boundaries because temperature and pressure increase as a result of plate collisions.

◆ Figure 6.21 illustrates the various temperature-pressure regimes that are produced along an oceanic-continental convergent plate boundary and the type of

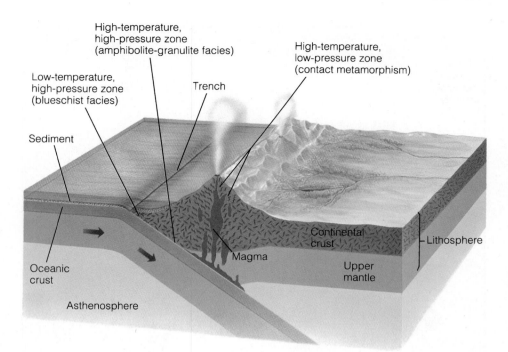

High-temperature,
high-pressure zone
(amphibolite-granulite facies)

Low-temperature,
high-pressure zone
(blueschist facies)

High-temperature,
low-pressure zone
(contact metamorphism)

Trench

Sediment

Continental
crust

Lithosphere

◆ FIGURE 6.21 Metamorphic facies resulting from various temperature-pressure conditions produced along an oceanic-continental convergent plate boundary.

Oceanic
crust

Magma

Upper
mantle

Asthenosphere

metamorphic facies and rocks that can result. When an oceanic plate collides with a continental plate, tremendous pressure is generated as the oceanic plate is subducted. Because rock is a poor heat conductor, the cold descending oceanic plate heats very slowly, and metamorphism is caused mostly by increasing pressure with depth. Metamorphism in such an environment produces rocks typical of the *blueschist facies* (low temperature, high pressure), which is characterized by the blue-colored amphibole mineral glaucophane (Fig. 6.20). Thus, geologists use the occurrence of blueschist facies rocks as evidence of ancient subduction zones. An excellent example of blueschist metamorphism can be found in the California Coast Ranges. Here rocks of the Franciscan Group were metamorphosed under low-temperature, high-pressure conditions that clearly indicate the presence of a former subduction zone (◆ Fig. 6.22).

As subduction along the oceanic-continental plate boundary continues, both temperature and pressure increase with depth and can result in high-grade metamorphic rocks. Eventually, the descending plate begins to melt and generates a magma that moves upward. This rising magma may alter the surrounding rock by contact metamorphism, producing migmatites in the deeper portions of the crust and hornfels at shallower depths. Such an environment is characterized by high temperatures and low to medium pressures.

While metamorphism is most common along convergent plate margins, many divergent plate boundaries are characterized by contact metamorphism. Rising magma at mid-oceanic ridges heats the adjacent rocks, producing contact metamorphic minerals and textures. In addition to contact metamorphism, fluids emanating from the rising magma—and from the reaction of the magma and sea water—very commonly produce hydrothermal solutions that may precipitate minerals of great economic value.

⩔ METAMORPHISM AND NATURAL RESOURCES

Many metamorphic rocks and minerals are valuable natural resources. While these resources include various types of ore deposits, the two most familiar and widely used metamorphic rocks, as such, are marble and slate, which, as previously discussed, have been used for centuries in a variety of ways.

Many ore deposits result from contact metamorphism in which hot, ion-rich fluids migrate from igneous intrusions into the surrounding rock, thereby producing rich ore deposits. The most common sulfide ore minerals associated with contact metamorphism are bornite, chalcopyrite, galena, pyrite, and sphalerite, while two common oxide ore minerals are hematite and magnetite.

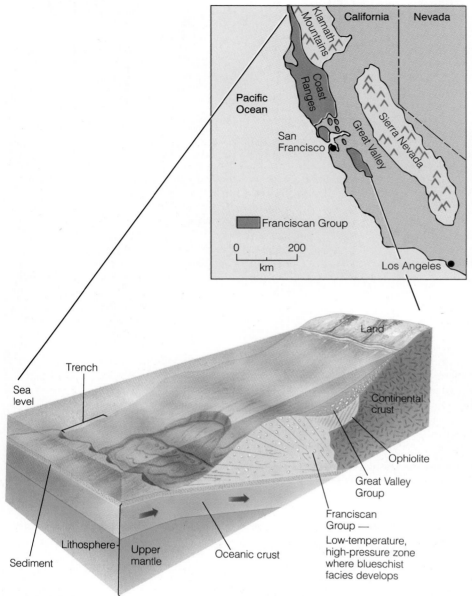

◆ FIGURE 6.22 Index map of California showing the location of the Franciscan Group and a diagrammatic reconstruction of the environment in which it was regionally metamorphosed under low-temperature, high-pressure subduction conditions approximately 150 million years ago.

Tin and tungsten are also important ores associated with contact metamorphism (● Table 6.3).

Other economically important metamorphic minerals include talc for talcum powder; graphite for pencils and dry lubricants (see Perspective 6.2); garnets and corundum, which are used as abrasives or gemstones, depending on their quality; and andalusite, kyanite, and sillimanite, all of which are used in the manufacture of high-temperature porcelains and refractives for products such as sparkplugs and the linings of furnaces.

Perspective 6.2

GRAPHITE

· · · · · · · · · · ·   · · · · · · · · · ·

Graphite (from the Greek *grapho,* meaning write) is a gray to black soft mineral that has a greasy feel and is composed of the element carbon. Graphite occurs in two varieties: crystalline, which consists of thin, flat, nearly pure black flakes; and massive, an impure variety found in compact masses.

Graphite has the same composition as diamond (see Perspective 2.2), but its carbon atoms are strongly bonded together in sheets, with the sheets weakly held together by van der Waals bonds (see Fig. 2.5). Because the sheets are loosely held together, they easily slide over one another, giving graphite its ability to mark paper and serve as a dry lubricant.

Graphite occurs mainly in metamorphic rocks produced by contact and regional metamorphism. It is found in marble, quartzite, schist, gneiss, and even in anthracite. Contact metamorphism of impure limestones by igneous intrusions produces some of the graphite found in marbles. The graphite resulting from regional metamorphism of sedimentary rocks probably came from organic matter present in the sediments. Some evidence, however, indicates that the graphite in Precambrian aged rocks (>570 million years) may be the result of the reduction of calcium carbonate ($CaCO_3$) by an inorganic process. Graphite is also found in igneous rocks, pegmatite dikes, and veins; it is thought to have formed in these environments from the primary constituents of the magma or from the hot fluids and vapors released by the cooling magma.

The major producers of graphite are Mexico, the former Soviet Union, Ceylon, Madagascar, Korea, and Canada. In the United States, graphite has been mined in 27 states, but production is now generally limited to Alabama and New York.

Graphite is used for many purposes. The oldest use is in pencil leads, where it is finely ground, mixed with clay, and baked. The amount of clay and the baking time give pencil leads their desired hardness. Other important uses include batteries, brake linings, carbon brushes, crucibles, foundry facings, lubricants, refractories, and steel making.

Synthetic graphite can be produced from anthracite coal or petroleum coke and now accounts for most graphite production. Its extreme purity (99% to 99.5% pure) makes it especially valuable where high purity is required such as in the rods that slow down the reaction rates in nuclear reactors.

● **TABLE 6.3** The Main Ore Deposits Resulting from Contact Metamorphism

Ore Deposit	Major Mineral	Formula	Use
Copper	Bornite Chalcopyrite	Cu_5FeS_4 $CuFeS_2$	Important sources of copper, which is used in various aspects of manufacturing, transportation, communications, and construction.
Iron	Hematite Magnetite	Fe_2O_3 Fe_3O_4	Major sources of iron for manufacture of steel, which is used in nearly every form of construction, manufacturing, transportation, and communications.
Lead	Galena	PbS	Chief source of lead, which is used in batteries, pipes, solder, and elsewhere where resistance to corrosion is required.
Tin	Cassiterite	SnO_2	Principal source of tin, which is used for tin plating, solder, alloys, and chemicals.
Tungsten	Scheelite Wolframite	$CaWO_4$ $(Fe, Mn) WO_4$	Chief sources of tungsten, which is used in hardening metals and manufacturing carbides.
Zinc	Sphalerite	$(Zn, Fe)S$	Major source of zinc, which is used in batteries and in galvanizing iron and making brass.

Chapter Summary

1. Metamorphic rocks result from the transformation of other rocks, usually beneath the Earth's surface, as a consequence of one or a combination of three agents: heat, pressure, and fluid activity.
2. Heat for metamorphism comes from intrusive magmas or deep burial. Pressure is either lithostatic or differential. Fluids trapped in sedimentary rocks or emanating from intruding magmas can enhance chemical changes and the formation of new minerals.
3. The three major types of metamorphism are contact, dynamic, and regional.
4. Metamorphic rocks are classified primarily according to their texture. In a foliated texture, platy minerals have a preferred orientation. A nonfoliated texture does not exhibit any discernible preferred orientation of the mineral grains.
5. Foliated metamorphic rocks can be arranged in order of grain size and/or perfection of their foliation. Slate is very fine grained, followed by phyllite and schist; gneiss displays segregated bands of minerals. Amphibolite is another fairly common foliated metamorphic rock.
6. Common nonfoliated metamorphic rocks are marble, quartzite, greenstone, and hornfels.
7. Metamorphic rocks can be arranged into metamorphic zones based on the conditions of metamorphism. Individual metamorphic facies are characterized by particular minerals that formed under specific metamorphic conditions. Such facies are named for a characteristic rock or mineral.
8. Metamorphism can occur along all three kinds of plate boundaries. It most commonly occurs, however, at convergent plate margins.
9. Metamorphic rocks formed near the Earth's surface along an oceanic-continental plate boundary result from low-temperature, high-pressure conditions. As a subducted oceanic plate descends, it is subjected to increasingly higher temperatures and pressures that result in higher grade metamorphism.
10. Many metamorphic rocks and minerals, such as marble, slate, graphite, talc, and asbestos, are valuable natural resources.

Important Terms

aureole
contact metamorphism
differential pressure
dynamic metamorphism
fluid activity
foliated texture

heat
index mineral
isograd
lithostatic pressure
metamorphic facies

metamorphic grade
metamorphic rock
metamorphic zone
nonfoliated texture
regional metamorphism

Review Questions

1. The metamorphic rock formed from limestone is:
   a. ___ quartzite; b. ___ hornfels; c. ___ marble; d. ___ slate; e. ___ greenstone.
2. From which of the following rock groups can metamorphic rocks form?
   a. ___ plutonic; b. ___ sedimentary; c. ___ metamorphic; d. ___ volcanic; e. ___ all of these.
3. Which of the following is not an agent of metamorphism?
   a. ___ gravity; b. ___ heat; c. ___ pressure; d. ___ fluid activity; e. ___ none of these.
4. Pressure exerted equally in all directions on an object is:
   a. ___ differential; b. ___ directional; c. ___ lithostatic; d. ___ shear; e. ___ none of these.
5. In which type of metamorphism are magmatic heat and fluids the primary agents of change?
   a. ___ contact; b. ___ dynamic; c. ___ regional; d. ___ local; e. ___ thermodynamic.
6. Concentric zones surrounding an igneous intrusion are:
   a. ___ metamorphic layers; b. ___ thermodynamic rings; c. ___ aureoles; d. ___ hydrothermal regions; e. ___ none of these.
7. Metamorphic rocks resulting from pure dynamic metamorphism are:
   a. ___ fault breccias; b. ___ quartzites; c. ___ greenstones; d. ___ mylonites; e. ___ hornfels.
8. Which type of metamorphism produces the majority of metamorphic rocks?

a. ___ contact; b. ___ dynamic; c. ___ regional; d. ___ lithostatic; e. ___ lithospheric.

9. Which of the following metamorphic rocks displays a foliated texture?
a. ___ marble; b. ___ quartzite; c. ___ greenstone; d. ___ schist; e. ___ hornfels.

10. What is the correct metamorphic sequence of increasingly coarser grain size?
a. ___ phyllite → slate → gneiss → schist;
b. ___ slate → phyllite → schist → gneiss;
c. ___ gneiss → phyllite → slate → schist;
d. ___ schist → gneiss → phyllite → slate;
e. ___ slate → schist → gneiss → phyllite.

11. An excellent rock for billiard table tops, floor and roofing tiles, and blackboards is:
a. ___ marble; b. ___ gneiss; c. ___ phyllite; d. ___ hornfels; e. ___ slate.

12. Mixed rocks containing the characteristics of both igneous and high-grade metamorphic rocks are:
a. ___ mylonites; b. ___ migmatites; c. ___ amphibolites; d. ___ hornfels; e. ___ greenstones.

13. Metamorphic zones:
a. ___ are characterized by distinctive mineral assemblages; b. ___ are separated from each other by isograds; c. ___ reflect a metamorphic grade; d. ___ all of these; e. ___ none of these.

14. To which metamorphic facies do metamorphic rocks formed under low-temperature, low-pressure conditions belong?
a. ___ granulite; b. ___ greenschist; c. ___ amphibolite; d. ___ blueschist; e. ___ eclogite.

15. Along what type of plate boundary is metamorphism most common?
a. ___ convergent; b. ___ divergent; c. ___ transform; d. ___ mantle plume; e. ___ static.

16. Which of the following is not a metamorphic mineral?
a. ___ graphite; b. ___ asbestos; c. ___ talc; d. ___ garnet; e. ___ gypsum.

17. Which of the following is the dangerous variety of asbestos?
a. ___ chrysotile; b. ___ crocidolite; c. ___ tremolite; d. ___ actinolite; e. ___ anthophyllite.

18. Metamorphic rocks form a significant proportion of:
a. ___ shields; b. ___ the cores of mountain ranges; c. ___ oceanic crust; d. ___ answers (a) and (b); e. ___ answers (b) and (c).

19. What are metamorphic rocks, and how do they form?

20. Name the three agents of metamorphism, and explain how each contributes to metamorphism.

21. What are the two types of pressure? What type of metamorphic textures does each produce?

22. Where does contact metamorphism occur, and what type of changes does it produce?

23. What are aureoles? How can they be used to determine the effects of metamorphism?

24. What is regional metamorphism, and under what conditions does it occur?

25. Describe the two types of metamorphic texture, and explain how they may be produced.

26. Starting with a shale, what metamorphic rocks would be produced by increasing heat and pressure?

27. Name the three common nonfoliated rocks, and describe their characteristics.

28. What is the difference between a metamorphic zone and a metamorphic facies?

29. What types of metamorphic rocks and facies are produced along a convergent plate margin?

30. Name some common metamorphic rocks or minerals that are economically valuable, and describe their uses.

Additional Readings

Best, M. G. 1982. *Igneous and metamorphic petrology.* San Francisco: W. H. Freeman.

Bowes, D. R., ed. 1989. *The encyclopedia of igneous and metamorphic petrology.* New York: Van Nostrand Reinhold.

Gillen, C. 1982. *Metamorphic geology.* London: George Allen & Unwin.

Hyndman, D. W. 1985. *Petrology of igneous and metamorphic rocks.* 2d ed. New York: McGraw-Hill.

Margolis, S. V. 1989. Authenticating ancient marble sculpture. *Scientific American* 260, no. 6:104–11.

Turner, F. J. 1981. *Metamorphic petrology.* 2d ed. New York: McGraw-Hill.

CHAPTER
7

Chapter Outline

Serpentine Fence west of Yellowstone National Park along Highway 287 in Montana. This fence was bent when seismic waves passed through the ground during the August 17, 1959 earthquake (magnitude 7.1) at Hebgen Lake, Montana.

Earthquakes and the Earth's Interior

Prologue

In the early evening of October 17, 1989, millions of baseball fans around the country turned on their television sets expecting to see the third game of the World Series between the Oakland Athletics and the San Francisco Giants. Instead of the baseball game, viewers witnessed the results of another, far more important event that was taking place 100 km south of San Francisco's Candlestick Park. At a few minutes past 5 P.M., near Loma Prieta peak in the Santa Cruz Mountains, a 40-km-long segment of the San Andreas fault ruptured beneath the Earth's surface, triggering a major earthquake (◆ Fig. 7.1).

Within seconds of the break, southward-moving shock waves demolished the downtown area of Santa Cruz. The shock waves also damaged or destroyed much of the town of Watsonville as well as damaging several other nearby communities.

The northward-racing shock waves shattered homes and businesses in Los Gatos, and as 50 million stunned viewers watched on television, Candlestick Park and 62,000 fans shook and swayed when the seismic waves passed beneath it. Fortunately, the stadium was built on solid bedrock, and thus the shaking resulted in only minor damage.

Those districts of the San Francisco-Oakland Bay Area that were built on artificial fill or reclaimed bay mud were not so fortunate, however. The soft fill amplified the shaking effects of the waves with devastating results.

In the Marina district of San Francisco, numerous buildings were destroyed, and a fire, fed by broken gas lines, lit up the night sky (Fig. 7.1a). A 15 m section of the upper deck of the San Francisco–

(a)

(b)

◆ FIGURE 7.1 Damage caused by the 1989 Loma Prieta earthquake. (*a*) Marina district fire in San Francisco caused by broken gas lines. (*b*) Aerial view looking west at part of the collapsed two-tiered Interstate 880 in Oakland. Only 1 of the 51 double-deck spans did not collapse.

Oakland Bay Bridge collapsed when bolts holding it in place snapped because of the swaying. The failure of the columns supporting a portion of the two-tiered Interstate 880 freeway in Oakland sent it crashing down, killing 42 unfortunate motorists (Fig. 7.1b).

The shaking lasted less than 15 seconds but resulted in 63 deaths, 3,800 injuries, at least 12,000 people left homeless, and property damage totaling almost $6 billion (approximately 28,000 buildings were damaged or destroyed). If the shaking had lasted even a few seconds longer, it is very likely that many more buildings and freeways would have failed, and the losses from property damage and human suffering would have been much higher.

Although the Loma Prieta earthquake was a major one in terms of energy released and damage done, it was not the "Big One" that Californians have long been expecting. That is not to say, however, that it was totally unexpected. Some sections along the San Andreas fault have not experienced any significant movement for many years and can be thought of as "locked." When a portion of a fault is locked, instead of slipping and releasing energy by small earthquakes, the fault essentially sticks. Potential energy builds up in the rocks adjacent to the fault until it finally snaps, releasing the energy as a major earthquake. Several segments of the San Andreas fault are currently locked and have the potential of producing the Big One.

In anticipation of such an earthquake, what lessons can be learned from the Loma Prieta earthquake? As was so dramatically demonstrated, the underlying geology and type of building construction are probably the two most important factors determining the amount of damage that can occur. Furthermore, the importance of careful planning and preparation in earthquake-prone areas was strongly reinforced. For instance, none of the structures in San Francisco that were constructed in compliance with current building codes collapsed.

Within hours after the earthquake, shelters were open and emergency relief services were in place and operating smoothly. This was due, in part, to the numerous rehearsals that various agencies conducted in preparation for just such an emergency. Certainly, more can be done to prepare for the Big One. However, Loma Prieta demonstrated that California is putting into practice what has been learned from a long history of dealing with earthquakes.

▼ INTRODUCTION

Earthquakes are violent and usually unpredictable; typically, they produce a feeling of helplessness. As one of nature's most frightening and destructive phenomena, they have always aroused a sense of fear. Even when an earthquake begins, there is no way to tell how strong the shaking will be or how long it will last.

It is estimated that more than 13 million people have died as a result of earthquakes during the past 4,000 years, and approximately 1 million of these deaths occurred during the last century (● Table 7.1).

How do geologists define an earthquake? An **earthquake** is the vibration of the Earth caused by the sudden release of energy, usually as a result of displacement of rocks along fractures, or faulting, beneath the Earth's surface.

Following an earthquake, adjustments along a fault commonly generate a series of earthquakes referred to as **aftershocks.** Most of these are smaller than the main shock, but they can cause considerable damage to already weakened structures. Indeed, much of the destruction from the 1755 earthquake in Lisbon, Portugal, was caused by aftershocks. After a small earthquake, aftershock activity usually ceases within a few days, but it may persist for months following a large earthquake.

Early humans and cultures had much more imaginative and colorful explanations of earthquakes than this scientific explanation. For example, many cultures believed that the Earth rested on some type of organism whose movements caused the Earth to shake. In Japan, it was a giant catfish; in Mongolia, a giant frog; in China, an ox; in India, a giant mole; in parts of South America, a whale; and to the Algonquin Indians of North America, an immense tortoise.

Many people believed earthquakes were a punishment or warning to the unrepentant. This view was strongly reinforced by the great Lisbon earthquake that occurred on November 1, 1755 (All Saints' Day), when many people were attending church services. So strong was this earthquake that buildings shook all over Europe and chandeliers rattled in parts of the United States. Approximately 70,000 people were killed by a combination of collapsing buildings, a giant seismic sea wave that devastated the waterfront, and a fire that burned throughout the city.

The Greek philosopher Aristotle offered what he considered to be a natural explanation for earthquakes. He believed that atmospheric winds were drawn into the Earth's interior where they caused fires and swept

● **TABLE 7.1** Significant Earthquakes of the World

Year	Location	Magnitude (estimated before 1935)	Deaths (estimated)
893	India		180,000
1138	Syria		100,000
1556	China (Shanxi Province)	8.0	1,000,000
1668	China (Shandong Province)	8.5	50,000
1730	Japan (Hokkaidō Prefecture)		137,000
1737	India (Calcutta)		300,000
1755	Portugal (Lisbon)	8.6	70,000
1811–12	USA (New Madrid, Missouri)	7.5	20
1835	Chile (Concepción)	8.5	
1857	USA (Fort Tejon, California)	8.3	1
1868	Chile and Peru	8.5	25,000
1872	USA (Owens Valley, California)	8.5	27
1886	USA (Charleston, South Carolina)	7.0	60
1905	India (Punjab-Kashmir region)	8.6	19,000
1906	USA (San Francisco, California)	8.3	700
1908	Italy (Messina)	7.5	83,000
1920	China (Gansu)	8.6	100,000
1923	Japan (Tokyo)	8.3	143,000
1950	India (Assam) and Tibet	8.6	1,530
1960	Chile	8.5	4,000+
1964	USA (Alaska)	8.6	131
1970	Peru (Chimbote)	7.8	25,000
1971	USA (San Fernando, California)	6.6	65
1975	China (Haicheng)	7.3	Some
1976	Guatamala	7.5	23,000
1976	China (Tangshan)	8.0	242,000
1985	Mexico (Mexico City)	8.1	9,500
1988	Armenia	7.0	25,000
1989	USA (Loma Prieta, California)	7.1	63
1990	Iran	7.3	40,000
1992	Turkey	6.8	570
1992	USA (California)	Five earthquakes varying from 4.6 to 7.5 and numerous aftershocks occurred from April through June.	At least 150 injured, but only 1 death.
1993	India	6.4	20,000+

around the various subterranean cavities trying to escape. It was this movement of underground air that caused earthquakes and occasional volcanic eruptions. Today, geologists know that the majority of earthquakes result from faulting associated with plate movements.

▼ ELASTIC REBOUND THEORY

Based on studies conducted after the 1906 San Francisco earthquake, H. F. Reid of Johns Hopkins University formulated the **elastic rebound theory** to explain how earthquakes occur. Reid studied three sets of measurements taken across the portion of the San Andreas fault that had broken during the 1906 earthquake. The measurements revealed that points on opposite sides of the fault had moved 3.2 m during the 50-year period prior to breakage in 1906, with the west side moving northward (◆ Fig. 7.2).

According to Reid, any straight line such as a fence or road that crossed the San Andreas fault would

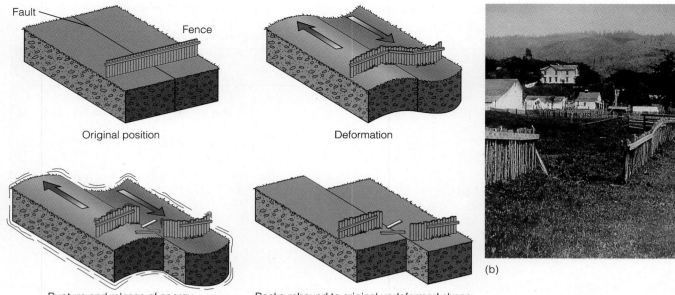

Fault

Fence

Original position

Deformation

Rupture and release of energy

Rocks rebound to original undeformed shape

(a)

(b)

◆ FIGURE 7.2 (a) According to the elastic rebound theory, when rocks are deformed, they store energy and bend. When the inherent strength of the rocks is exceeded, they rupture, releasing the energy in the form of earthquake waves that radiate outward in all directions. Upon rupture, the rocks rebound to their former undeformed shape. (b) During the 1906 San Francisco earthquake, this fence in Marin County was displaced 2.5 m.

gradually be bent, as rocks on one side of the fault moved relative to rocks on the other side (Fig. 7.2). Eventually, the strength of the rocks was exceeded, the rocks on opposite sides of the fault rebounded or "snapped back" to their former undeformed shape, and the energy stored was released as earthquake waves radiating outward from the break (Fig. 7.2). Additional field and laboratory studies conducted by Reid and others have confirmed that elastic rebound is the mechanism by which earthquakes are generated.

▼ SEISMOLOGY

Seismology, the study of earthquakes, began emerging as a true science around 1880 with the development of instruments that effectively recorded earthquake waves. The earliest earthquake detector was invented by the Chinese scholar Chang Heng in about A.D. 132 (◆ Fig. 7.3). Over the succeeding centuries, other instruments were invented to study earthquakes, but it was not until the late nineteenth century that the first effective seis-

◆ FIGURE 7.3 The world's first earthquake detector was invented by Chang Heng sometime around A.D. 132. Movement of the vase dislodged a ball from a dragon's mouth into the waiting mouth of a frog below.

mograph was developed. A *seismograph* is an instrument that detects, records, and measures the various vibrations produced by an earthquake (◆ Fig. 7.4). The record made by a seismograph is a *seismogram*.

When an earthquake occurs, energy in the form of *seismic waves* radiates outward in all directions from the point of release (◆ Fig. 7.5). Most earthquakes result when rocks in the Earth's crust rupture along a fault because of the buildup of excessive pressure, which is usually caused by plate movement. Once a rupture begins, it moves along the fault at a velocity of several km/sec for as long as conditions for failure exist. The length of the fault along which rupture occurs can range from a few meters to several hundred kilometers. The

longer the rupture, the more time it takes for all of the stored energy in the rocks to be released, and therefore the longer the ground will shake.

The location within the crust where rupture initiates, and thus where the energy is released, is referred to as the **focus** or *hypocenter*. The point on the Earth's surface vertically above the focus is the **epicenter** which is the location that is usually given in news reports on earthquakes (Fig. 7.5).

Seismologists recognize three categories of earthquakes based on the depth of their foci. *Shallow-focus* earthquakes have a focal depth of less than 70 km. Earthquakes with foci between 70 and 300 km are referred to as *intermediate focus*, and those with foci

(a)

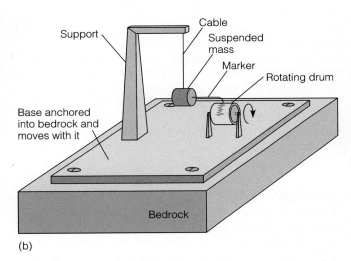

(b)

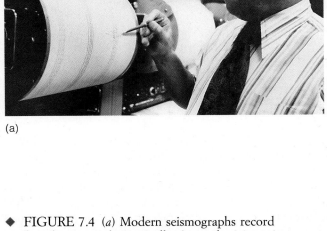

◆ FIGURE 7.4 (*a*) Modern seismographs record earthquake waves electronically. A geophysicist points to the trace of an earthquake recorded by a seismograph at the National Earthquake Information Service, Golden, Colorado. (*b*) A horizontal-motion seismograph. Because of its inertia, the heavy mass that contains the marker will remain stationary while the rest of the structure moves along with the ground during an earthquake. As long as the length of the arm is not parallel to the direction of ground movement, the marker will record the earthquake waves on the rotating drum. (*c*) A vertical-motion seismograph. This seismograph operates on the same principle as a horizontal-motion instrument and records vertical ground movement.

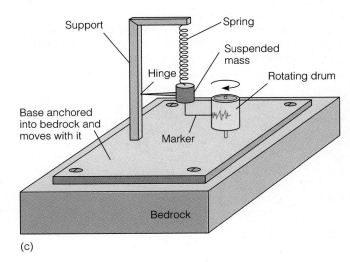

(c)

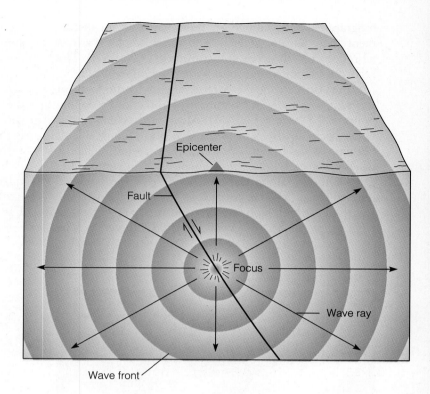

◆ FIGURE 7.5 The focus of an earthquake is the location where rupture begins and energy is released. The place on the Earth's surface vertically above the focus is the epicenter. Seismic wave fronts move outward in all directions from their source, the focus of an earthquake. Wave rays are lines drawn perpendicular to wave fronts.

greater than 300 km are called *deep focus*. Earthquakes are not evenly distributed among these three categories. Approximately 90% of all earthquake foci occur at a depth of less than 100 km. Shallow-focus earthquakes are, with few exceptions, the most destructive.

There is an interesting relationship between earthquake foci and plate margins. Earthquakes generated along divergent or transform plate boundaries are always shallow focus, while almost all intermediate- and deep-focus earthquakes occur within the circum-Pacific belt along convergent margins (◆ Fig. 7.6). Furthermore, a pattern emerges when the focal depths of earthquakes near island arcs and their adjacent ocean trenches are plotted. Notice in ◆ Figure 7.7 that the focal depth increases beneath the Tonga Trench in a narrow, well-defined zone that dips approximately 45°. Dipping seismic zones, called *Benioff zones*, are a feature common to island arcs and deep ocean trenches. Such zones indicate the angle of plate descent along a convergent plate boundary.

⩔ THE FREQUENCY AND DISTRIBUTION OF EARTHQUAKES

Most earthquakes (almost 95%) occur in seismic belts that correspond to plate boundaries where stresses develop as plates converge, diverge, and slide past each other. Earthquake activity distant from plate margins is minimal, but can be devastating when it occurs. The relationship between plate margins and the distribution of earthquakes is readily apparent when the locations of earthquake epicenters are superimposed on a map showing the boundaries of the Earth's plates (Fig. 7.6).

The majority of all earthquakes (approximately 80%) occur in the *circum-Pacific belt*, a zone of seismic activity that encircles the Pacific Ocean basin. Most of these earthquakes are a result of convergence along plate margins.

The second major seismic belt is the *Mediterranean-Asiatic belt* where approximately 15% of all earthquakes occur. This belt extends westerly from Indonesia through the Himalayas, across Iran and Turkey, and

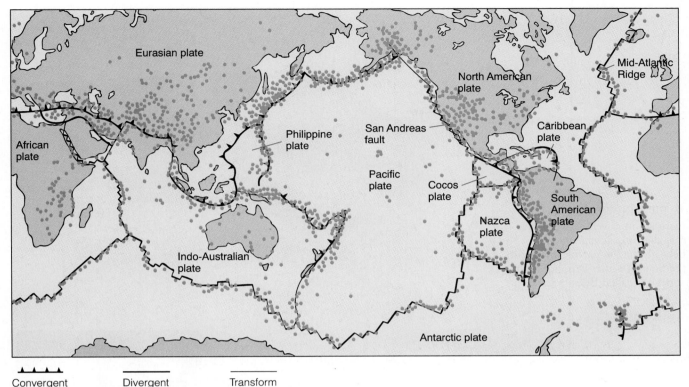

Convergent boundary   Divergent boundary   Transform boundary

◆ FIGURE 7.6 The relationship between the distribution of earthquake epicenters and plate boundaries. Approximately 80% of earthquakes occur within the circum-Pacific belt, 15% within the Mediterranean-Asiatic belt, and the remaining 5% within the interiors of plates or along oceanic spreading ridge systems. Each dot represents a single earthquake epicenter.

westerly through the Mediterranean region of Europe. The devastating earthquake that struck Armenia in 1988 killing 25,000 people and the 1990 earthquake in Iran that killed 40,000 are recent examples of the destructive earthquakes that strike this region (Table 7.1).

The remaining 5% of earthquakes occur mostly in the interiors of plates and along oceanic spreading ridge systems. Most of these earthquakes are not very strong although there have been several major intraplate earthquakes that are worthy of mention. For example, the 1811 and 1812 earthquakes near New Madrid, Missouri killed approximately 20 people and nearly destroyed the town of New Madrid. So strong were these earthquakes that they were felt from the Rocky Mountains to the Atlantic Ocean and from the Canadian border to the Gulf of Mexico. Another major intraplate earthquake struck Charleston, South Carolina, on August 31, 1886, killing 60 people and causing $23 million in property damage (◆ Fig. 7.8).

The cause of intraplate earthquakes is not well understood, but geologists believe they arise from localized stesses caused by the compression that most plates experience along their margins. The release of these stresses and hence the resulting intraplate earthquakes are due to local factors. Interestingly, many intraplate earthquakes are associated with very ancient and presumed inactive faults that are reactivated at various intervals.

More than 150,000 earthquakes strong enough to be felt by someone are recorded every year by the

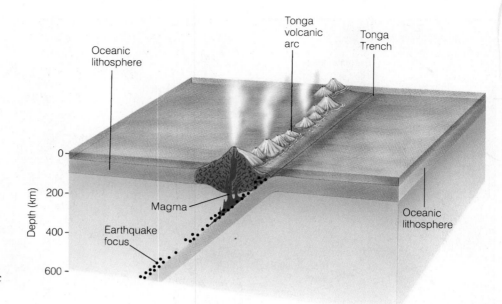

◆ FIGURE 7.7 Focal depth increases in a well-defined zone that dips approximately 45° beneath the Tonga volcanic arc in the South Pacific. Dipping seismic zones are called Benioff zones.

◆ FIGURE 7.8 Damage done to Charleston, South Carolina, by the earthquake of August 31, 1886. This earthquake is the largest reported in the eastern United States.

world-wide network of seismograph stations. In addition, it has been estimated that about 900,000 earthquakes occur annually that are recorded by seismographs, but are too small to be individually cataloged. These small earthquakes result from the energy released as continual adjustments between the Earth's various plates occur.

⍦ SEISMIC WAVES

The shaking and destruction resulting from earthquakes are caused by two different types of seismic waves: *body waves*, which travel through the Earth and are somewhat like sound waves; and *surface waves*, which travel only along the ground surface and are analogous to ocean waves.

An earthquake generates two types of body waves: P-waves and S- waves (◆ Fig 7.9). **P-waves** or *primary waves* are the fastest seismic waves and can travel through solids, liquids, and gases. P-waves are compressional, or push-pull, waves and are similar to sound waves in that they move material forward and backward along a line in the same direction that the waves themselves are moving (Fig. 7.9b). Thus, the material through which P-waves travel is expanded and compressed as the wave moves through it and returns to its original size and shape after the wave passes by.

S-waves or *secondary waves* are somewhat slower than P-waves and can only travel through solids. S-waves are *shear waves* because they move the material perpendicular to the direction of travel, thereby producing shear stresses in the material they move through (Fig. 7.9c). Because liquids (as well as gases) are not rigid, they have no shear strength and S-waves cannot be transmitted through them.

The velocities of P- and S-waves are determined by the density and elasticity of the materials through which they travel. For example, seismic waves travel more slowly through rocks of greater density, but more rapidly through rocks with greater elasticity. *Elasticity* is a property of solids, such as rocks, and means that once they have been deformed by an applied force, they return to their original shape when the force is no longer present. Because P-wave velocity is greater than S-wave velocity in all materials, however, P-waves always arrive at seismic stations first.

Surface waves travel along the surface of the ground, or just below it, and are slower than body waves. Unlike the sharp jolting and shaking that body waves cause, surface waves generally produce a rolling or swaying motion, much like the experience of being on a boat.

Several types of surface waves are recognized. The two most important are **Rayleigh waves** (**R-waves**) and **Love waves** (**L-waves**), named after the British scientists who discovered them, Lord Rayleigh and A. E. H. Love. Rayleigh waves are generally the slower of the two and behave like water waves in that they move forward while the individual particles of material move in an elliptical path within a vertical plane oriented in the direction of wave movement (Fig. 7.9d).

The motion of a Love wave is similar to that of an S-wave, but the individual particles of the material only move back and forth in a horizontal plane perpendicular to the direction of wave travel (Fig. 7.9e). This type of lateral motion can be particularly damaging to building foundations.

▼ LOCATING AN EARTHQUAKE

The various seismic waves travel at different speeds and thus arrive at a seismograph at different times. As ◆ Figure 7.10 illustrates, the first waves to arrive, and thus the fastest, are the P-waves, which travel at nearly twice the velocity of the S-waves that follow. Both the P- and S-waves travel directly from the focus to the seismograph through the interior of the Earth. The last waves to arrive are the L- and R-waves, which are the slowest and also travel the longest route along the Earth's surface.

By accumulating a tremendous amount of data over the years, seismologists have determined the average travel times of P- and S-waves for any specific distance. These P- and S-wave travel times are published as *time-distance graphs* and illustrate that the difference between the arrival times of the P- and S-waves is a function of the distance of the seismograph from the focus (◆ Fig. 7.11).

As ◆ Figure 7.12 (page 164) demonstrates, the epicenter of any earthquake can be determined by using a time-distance graph and knowing the arrival times of the P- and S-waves at any three seismograph locations. Subtracting the arrival time of the first P-wave from the arrival time of the first S-wave gives the time interval between the arrivals of the two waves for each seismograph location. Each time interval is then plotted on the time-distance graph, and a line is drawn straight down to the distance axis of the graph, thus indicating how far away each station is from the focus of the earthquake. Then a circle whose radius equals the distance shown on the time-distance graph from each of the three seismograph locations is drawn on a map (Fig. 7.12). The intersection of the three circles is the location of the earthquake's epicenter. A minimum of three locations is needed because two locations will provide two possible epicenters and one location will provide an infinite number of possible epicenters.

▼ MEASURING EARTHQUAKE INTENSITY AND MAGNITUDE

Geologists measure the strength of an earthquake in two different ways. The first, *intensity*, is a qualitative assessment of the kinds of damage done by an earthquake. The second, *magnitude*, is a quantitative measurement of the amount of energy released by an

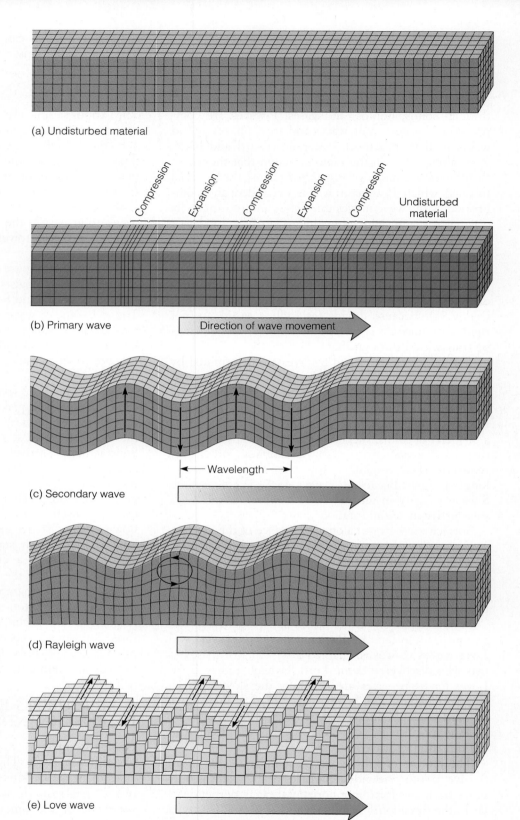

(a) Undisturbed material

Compression Expansion Compression Expansion Compression

Undisturbed material

(b) Primary wave

Direction of wave movement

◄——— Wavelength ———►

(c) Secondary wave

(d) Rayleigh wave

(e) Love wave

◆ FIGURE 7.9 Seismic waves. (a) Undisturbed material. (b) Primary waves (P-waves) compress and expand material in the same direction as the wave movement. (c) Secondary waves (S-waves) move material perpendicular to the direction of wave movement. (d) Rayleigh waves (R-waves) move material in an elliptical path within a vertical plane oriented parallel to the direction of wave movement. (e) Love waves (L-waves) move material back and forth in a horizontal plane perpendicular to the direction of wave movement.

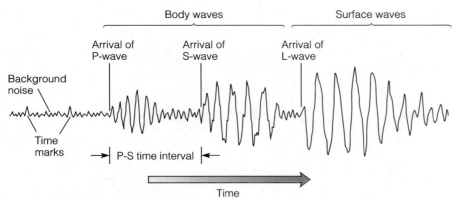

◆ FIGURE 7.10 A schematic seismogram showing the arrival order and pattern produced by P-, S-, and L-waves. When an earthquake occurs, body and surface waves radiate outward from the focus at the same time. Because P-waves are the fastest, they arrive at a seismograph first, followed by S-waves, and then by surface waves, which are the slowest waves. The difference between the arrival times of the P- and the S-waves is the P-S time interval; it is a function of the distance of the seismograph station from the focus.

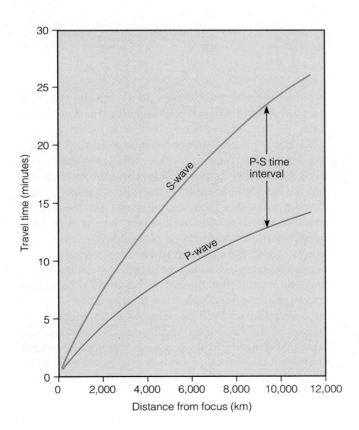

◆ FIGURE 7.11 A time-distance graph showing the average travel times for P- and S-waves. The farther away a seismograph station is from the focus of an earthquake, the longer the interval between the arrivals of the P- and S-waves, and hence the greater the distance between the curves on the time-distance graph as indicated by the P-S time interval.

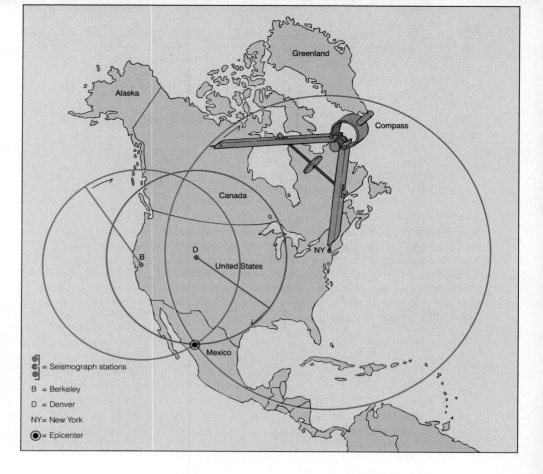

◆ FIGURE 7.12 Three seismograph stations are needed to locate the epicenter of an earthquake. The P-S time interval is plotted on a time-distance graph for each seismograph station to determine the distance that station is from the epicenter. A circle with that radius is drawn from each station, and the intersection of the three circles is the epicenter of the earthquake.

earthquake. Each method provides geologists with important data about earthquakes and their effects. This information can then be used to prepare for future earthquakes.

Intensity is a subjective measure of the kind of damage done by an earthquake as well as people's reaction to it. Since the mid-nineteenth century, geologists have used intensity as a rough approximation of the size and strength of an earthquake. The most common intensity scale used in the United States is the **Modified Mercalli Intensity Scale**, which has values ranging from I to XII (● Table 7.2).

After an assessment of the earthquake damage is made, *isoseismal lines* (lines of equal intensity) are drawn on a map, dividing the affected region into various intensity zones. The intensity value given for each zone is the maximum intensity that the earthquake produced for that zone. Even though intensity maps are not precise because of the subjective nature of the measurements, they do provide geologists with a rough approximation of the location of the earthquake, the

kind and extent of the damage done, and the effects of local geology and types of building construction (◆ Fig. 7.13).

While it is generally true that a large earthquake will produce greater intensity values than a small earthquake, many other factors besides the amount of energy released by an earthquake affect its intensity. These include the distance from the epicenter, the focal depth of the earthquake, the population density and local geology of the area, the type of building construction employed, and the duration of shaking.

If earthquakes are to be compared quantitatively, we must use a scale that measures the amount of energy released and is independent of intensity. Such a scale was developed in 1935 by Charles F. Richter, a seismologist at the California Institute of Technology. The **Richter Magnitude Scale** measures earthquake **magnitude**, which is the total amount of energy released by an earthquake at its source. It is an open-ended scale with values beginning at 1. The largest magnitude recorded has been 8.6, and though values greater than 9 are

● **TABLE 7.2** Modified Mercalli Intensity Scale

I	Not felt except by a very few under especially favorable circumstances.
II	Felt only by a few people at rest, especially on upper floors of buildings.
III	Felt quite noticeably indoors, especially on upper floors of buildings, but many people do not recognize it as an earthquake. Standing automobiles may rock slightly.
IV	During the day felt indoors by many, outdoors by few. At night some awakened. Sensation like heavy truck striking building, standing automobiles rocked noticeably.
V	Felt by nearly everyone, many awakened. Some dishes, windows, etc. broken, a few instances of cracked plaster. Disturbance of trees, poles, and other tall objects sometimes noticed.
VI	Felt by all, many frightened and run outdoors. Some heavy furniture moved, a few instances of fallen plaster or damaged chimneys. Damage slight.
VII	Everybody runs outdoors. Damage negligible in buildings of good design and construction; slight to moderate in well-built ordinary structures; considerable in poorly built or badly designed structures; some chimneys broken. Noticed by people driving automobiles.
VIII	Damage slight in specially designed structures; considerable in normally constructed buildings with possible partial collapse; great in poorly built structures. Fall of chimneys, monuments, walls. Heavy furniture overturned. Sand and mud ejected in small amounts.
IX	Damage considerable in specially designed structures. Buildings shifted off foundations. Ground noticeably cracked. Underground pipes broken.
X	Some well-built wooden structures destroyed; most masonry and frame structures with foundations destroyed; ground badly cracked. Rails bent. Landslides considerable from river banks and steep slopes. Water splashed over river banks.
XI	Few, if any (masonry) structures remain standing. Bridges destroyed. Broad fissures in ground. Underground pipelines completely out of service.
XII	Damage total. Waves seen on ground surfaces. Objects thrown upward into the air.

SOURCE: United States Geological Survey.

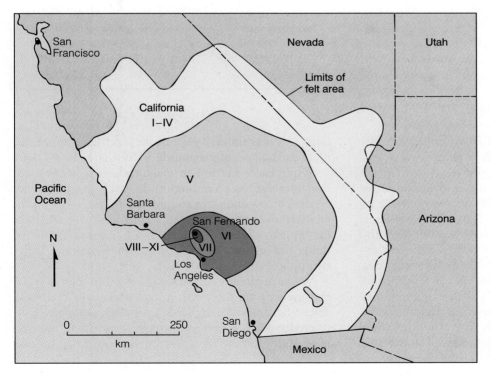

◆ FIGURE 7.13 Modified Mercalli Intensity map for the 1971 San Fernando Valley, California, earthquake showing the region divided into intensity zones based on the kind of damage done. This earthquake had a magnitude of 6.6.

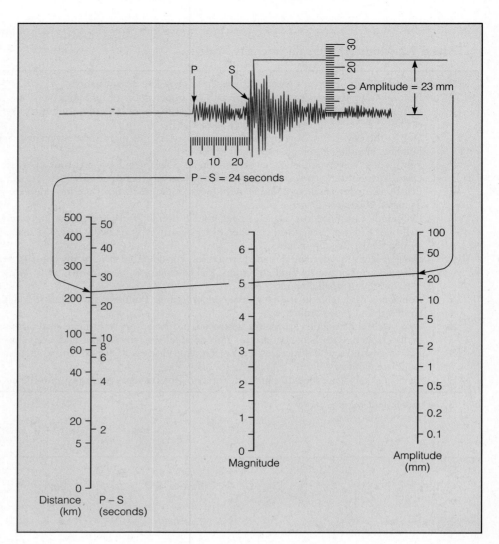

◆ FIGURE 7.14 The Richter Magnitude Scale measures magnitude, which is the total amount of energy released by an earthquake at its source. The magnitude is determined by measuring the maximum amplitude of the largest seismic wave and marking it on the right-hand scale. The difference between the arrival times of the P- and S-waves (recorded in seconds) is marked on the left-hand scale. When a line is drawn between the two points, the magnitude of the earthquake is the point at which the line crosses the center scale.

theoretically possible, they are highly improbable because rocks are not able to store the energy necessary to generate earthquakes of this magnitude.

The magnitude of an earthquake is determined by measuring the amplitude of the largest seismic wave as recorded on a seismogram (◆ Fig. 7.14). To avoid large numbers, Richter used a conventional base-10 logarithmic scale to convert the amplitude of the largest recorded seismic wave to a numerical magnitude value (Fig. 7.14). Therefore, each integer increase in magnitude represents a 10-fold increase in wave amplitude. For example, the amplitude of the largest seismic wave for an earthquake of magnitude 6 is 10 times that produced by an earthquake of magnitude 5, 100 times as large as a magnitude 4 earthquake, and 1,000 times that of an earthquake of magnitude 3 ($10 \times 10 \times 10 = 1,000$).

While each increase in magnitude represents a 10-fold increase in wave amplitude, each magnitude increase corresponds to a roughly 30-fold increase in the amount of energy released. Thus, the 1964 Alaska earthquake with a magnitude of 8.6 released about 900 times the energy of the 1971 San Fernando Valley earthquake of magnitude 6.6.

We have already mentioned that more than 900,000 earthquakes are recorded around the world each year. These figures can be placed in better perspective by reference to ● Table 7.3, which shows that the vast majority of earthquakes have a Richter magnitude of

● **TABLE 7.3** Average Number of Earthquakes of Various Magnitudes per Year Worldwide

Magnitude	Effects	Average Number per Year
<2.5	Typically not felt, but recorded.	900,000
2.5–6.0	Usually felt. Minor to moderate damage to structures.	31,000
6.1–6.9	Potentially destructive, especially in populated areas.	100
7.0–7.9	Major earthquakes. Serious damage results.	20
>8.0	Great earthquakes. Usually result in total destruction.	1 every 5 years

SOURCE: Data from *Earthquake Information Bulletin* and Gutenberg and Richter (1949).

less than 2.5, and that great earthquakes (those with a magnitude greater than 8.0) occur, on average, only once every five years.

▼ THE DESTRUCTIVE EFFECTS OF EARTHQUAKES

The destructive effects of earthquakes are many and varied and include such phenomena as ground shaking, fire, seismic sea waves, and landslides, as well as disruption of vital services, panic, and psychological shock. The amount of property damage, loss of life, and injury depends on the time of day an earthquake occurs, its magnitude, the distance from the epicenter, the geology of the area, the type of construction of the various structures, the population density, and the duration of shaking. Generally speaking, earthquakes occurring during working and school hours in densely populated urban areas are the most destructive.

Ground shaking usually causes more damage and results in more loss of life and injuries than any other earthquake hazard. Structures built on solid bedrock generally suffer less damage than those built on poorly consolidated material such as water-saturated sediments or artificial fill (◆ Fig. 7.15; see Perspective 7.1). Structures on poorly consolidated or water-saturated material are subjected to ground shaking of longer duration and greater S-wave amplitude than those on bedrock. In addition, fill and water-saturated sediments tend to liquefy, or behave as a fluid, a process known as *liquefaction*. When shaken, the individual grains lose cohesion and the ground flows. Dramatic examples of damage resulting from liquefaction include Niigata, Japan, where large apartment buildings were tipped to

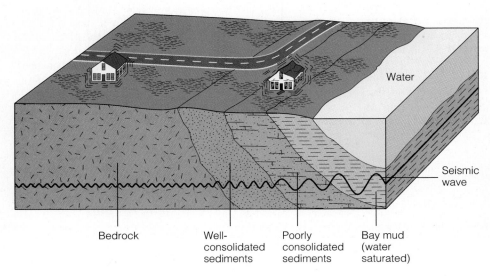

Bedrock

Well-consolidated sediments

Poorly consolidated sediments

Bay mud (water saturated)

Water

Seismic wave

◆ FIGURE 7.15 The amplitude, duration, and period of seismic waves vary in different types of materials. The amplitude and duration of the waves generally increase as they pass from bedrock to poorly consolidated or water-saturated material. Thus, structures built on weaker material typically suffer greater damage than similar structures built on bedrock.

DESIGNING EARTHQUAKE-RESISTANT STRUCTURES

One way to reduce property damage, injuries, and loss of life is to design and build structures as earthquake-resistant as possible. Many things can be done to improve the safety of current structures and of new buildings.

California has a Uniform Building Code that sets minimum standards for building earthquake-resistant structures that is used as a model around the world. The California code is far more stringent than federal earthquake building codes and requires that structures be able to withstand a 25-second main shock. Unfortunately, many earthquakes are of far longer duration. For example, the main shock of the 1964 Alaskan earthquake lasted approximately three minutes and was followed by numerous aftershocks. While many of the extensively damaged buildings in this earthquake had been built according to the California code, they were not designed to withstand shaking of such long duration. Nevertheless, in California and elsewhere in the world, structures built since the California code went into effect have fared much better during moderate to major earthquakes than those built before its implementation.

The major objective in designing earthquake-resistant structures is minimizing the loss of life, injuries, and damage. To achieve this goal, engineers must understand the dynamics and mechanics of earthquakes including the type and duration of the ground motion and how rapidly the ground accelerates during an earthquake. An understanding of the area's geology is also very important because certain ground materials such as water-saturated sediments or landfill can lose their strength and cohesiveness during an earthquake. Such materials should be avoided if at all possible. Finally, engineers must be aware of how different structures behave under different earthquake conditions.

With the level of technology currently available, a well-designed, properly constructed building should be able to withstand small, short-duration earthquakes of less than 5.5 magnitude with little or no damage. In moderate earthquakes (5.5 to 7.0 magnitude), the damage suffered should not be serious and should be repairable. In a major earthquake of greater than 7.0 magnitude, the building should not collapse, although it may later have to be demolished.

Many factors enter into the design of an earthquake-resistant structure, but the most important is that the building be tied together; that is, the foundation, walls, floors, and roof should all be joined together to create a structure that can withstand both horizontal and vertical shaking (◆ Fig. 1). Almost all of the structural failures that have resulted from earthquake ground movement have occurred at weak connections, where the various parts of a structure were not securely tied together.

The size and shape of a building can also affect its resistance to earthquakes. Rectangular box-shaped buildings are inherently stronger than those of irregular size or shape because different parts of an irregular building may sway at different rates, increasing the stress and likelihood of structural failure (◆ Fig. 2b). Buildings with open or unsupported first stories are particularly susceptible to damage. Some reinforcement must be done or collapse is a distinct possibility.

Tall buildings, such as skyscrapers, must be designed so that a certain amount of swaying or flexing can occur, but not so much that they touch neighboring buildings during swaying (◆ Fig. 2d). If a building is brittle and does not give, it will crack and fail. In addition to designed flexibility, engineers must make sure that a building does not resonate at the same frequency as the ground does during an earthquake. When that happens, the force applied by the seismic waves at ground level is multiplied several times by the time they reach the top of the building (◆ Fig. 2c). This condition is particularly troublesome in areas of poorly consolidated sediment (◆ Fig. 3). Fortunately, buildings can be designed so that they will sway at a different frequency from the ground.

What about structures built many years ago? Almost every city and town has older single and multistory

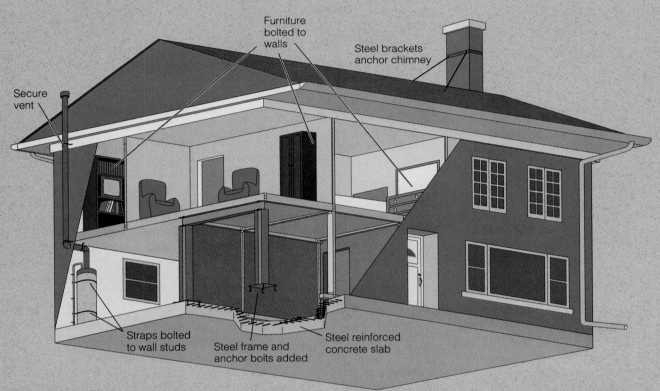

Furniture
bolted to
walls

Steel brackets
anchor chimney

Secure
vent

Straps bolted
to wall studs

Steel frame and
anchor bolts added

Steel reinforced
concrete slab

◆ FIGURE 1 This diagram shows some of the things a homeowner can do to reduce the potential damage to a building because of ground shaking during an earthquake.

structures, constructed of unreinforced brick masonry, poor-quality concrete, and rotting or decaying wood. Just as in new buildings, the most important thing that can be done to increase the stability and safety is to tie the different components of each building together. This can be done by adding a steel frame to unreinforced parts of a building such as a garage, bolting the walls to the foundation, adding reinforced beams to the exterior, and using beam and joist connectors whenever possible. Although such modifications are expensive, they are usually cheaper than having to replace a building that was destroyed by an earthquake.

continued on next page

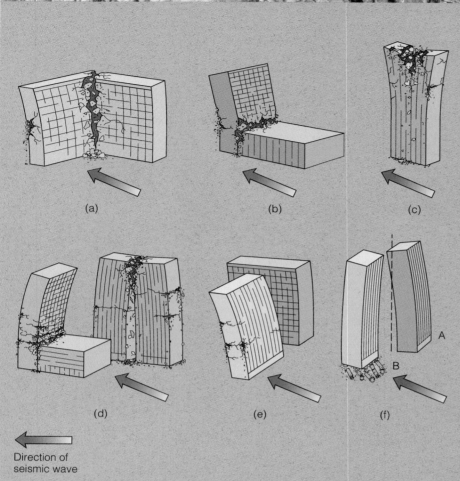

◆ FIGURE 2 The effects of ground shaking on various tall buildings of differing shapes. (*a*) Damage will occur if two wings of a building are joined at right angles and experience different motions. (*b*) Buildings of different heights will sway differently leading to damage at the point of connection. (*c*) Shaking increases with height and is greatest at the top of a building. (*d*) Closely spaced buildings may crash into each other due to swaying. (*e*) A building whose long axis is parallel to the direction of the seismic waves will sway less than a building whose axis is perpendicular. (*f*) Two buildings of different design will behave differently even when subjected to the same shaking conditions. Building A sways as a unit and remains standing while building B whose first story is composed of only tall columns collapses because most of the swaying takes place in the "soft" first story.

◆ FIGURE 3 This 15-story reinforced concrete building collapsed due to the ground shaking that occurred during the 1985 Mexico City earthquake. The soft lake bed sediments on which Mexico City is built enhanced the seismic waves as they passed through.

(a)　　(b)　　(c)

(d)　　(e)　　(f)

A
B

Direction of
seismic wave

◆ FIGURE 7.16 The effects of ground shaking on water-saturated soil are dramatically illustrated by the collapse of these buildings in Niigata, Japan, during a 1964 earthquake. The buildings were designed to be earthquake-resistant and fell over on their sides intact.

◆ FIGURE 7.17 Many of the approximately 242,000 people who died in the 1976 earthquake in Tangshan, China, were killed by collapsing structures. Many of the buildings were constructed from unreinforced brick, which has no flexibility, and quickly fell down during the earthquake. A few tents and temporary shelters can be seen in this oblique aerial view of a part of Tangshan.

their sides after the water-saturated soil of the hillside collapsed (◆ Fig. 7.16), and Turnagain Heights, Alaska, where many homes were destroyed when the Bootlegger Cove Clay lost all of its strength when shaken by the 1964 earthquake (see Fig. 11.9).

In addition to the magnitude of an earthquake and the regional geology, the material used and the type of construction also affect the amount of damage done (see Perspective 7.1). Adobe and mud-walled structures are the weakest of all and almost always collapse during an earthquake. Unreinforced brick structures and poorly built concrete structures are also particularly susceptible to collapse. For example, the 1976 earthquake in Tangshan, China, completely leveled the city because almost none of the structures were built to resist seismic forces. In fact, most of them had unreinforced brick walls, which have no flexibility, and consequently they collapsed during the shaking (◆ Fig. 7.17).

In many earthquakes, particularly in urban areas, fire is a major hazard. Almost 90% of the damage done in the 1906 San Francisco earthquake was caused by fire. The shaking severed many of the electrical and gas lines, which touched off flames and started numerous fires all over the city. Because water mains were ruptured by the earthquake, there was no effective way to fight the fires. Hence, they raged out of control for three days, destroying much of the city.

During the September 1, 1923 earthquake in Japan, fires destroyed 71% of the houses in Tokyo and practically all the houses in Yokohama. In all, a total of 576,262 houses were completely destroyed by fire, and 143,000 people died, many as a result of the fire.

Seismic sea waves or **tsunami** are destructive sea waves that are usually produced by earthquakes but can also be caused by submarine landslides or volcanic eruptions. Tsunami are popularly called tidal waves, although they have nothing to do with tides. Instead, most tsunami result from the sudden movement of the sea floor, which sets up waves within the water that travel outward, much like the ripples that form when a stone is thrown into a pond.

Tsunami travel at speeds of several hundred km/hr and are commonly not felt in the open ocean because their wave height is usually less than 1 m and the distance between wave crests is typically several hundred kilometers. However, when tsunami approach shorelines, the waves slow down and water piles up to heights of up to 65 m (◆ Fig. 7.18).

Following a 1946 tsunami that killed 159 people and caused $25 million in property damage in Hawaii, the U.S. Coast and Geodetic Survey established a Tsunami Early Warning System in Honolulu, Hawaii, in an attempt to minimize tsunami devastation. This system combines seismographs and instruments that can detect earthquake-generated waves. Whenever a strong earthquake occurs anywhere within the Pacific basin, its location is determined, and instruments are checked to see if a tsunami has been generated. If it has, a warning is sent out to evacuate people from low-lying areas that may be affected.

Earthquake-triggered landslides are particularly dangerous in mountainous regions and have been responsible for tremendous amounts of damage and many

◆ FIGURE 7.18 A tsunami destroying the pier at Hilo, Hawaii, in 1946. This tsunami was generated by an earthquake in the Aleutian Islands. The man in the path of the waves was never seen again.

deaths. For example, the 1970 Peru earthquake caused an avalanche that completely destroyed the town of Yungay, resulting in 25,000 deaths.

▼ EARTHQUAKE PREDICTION

Can earthquakes be predicted? A successful prediction must include a time frame for the occurrence of the earthquake, its location, and its strength. In spite of the tremendous amount of information geologists have gathered about the cause of earthquakes, successful predictions are still quite rare. Nevertheless, if reliable predictions can be made, they can greatly reduce the number of deaths and injuries.

From an analysis of historic records and the distribution of known faults, *seismic risk maps* can be constructed that indicate the likelihood and potential severity of future earthquakes based on the intensity of past earthquakes (◆ Fig. 7.19). Although such maps cannot predict when the next major earthquake will occur, they are useful in helping people plan for future earthquakes.

Studies conducted over the past several decades indicate that most earthquakes are preceded by both short-term and long-term changes within the Earth. Such changes are called *precursors*.

One long-range prediction technique used in seismically active areas involves plotting the location of major earthquakes and their aftershocks to detect areas that

◆ FIGURE 7.19 A 1969 seismic risk map for the United States based on intensity data collected by the U.S. Coast and Geodetic Survey.

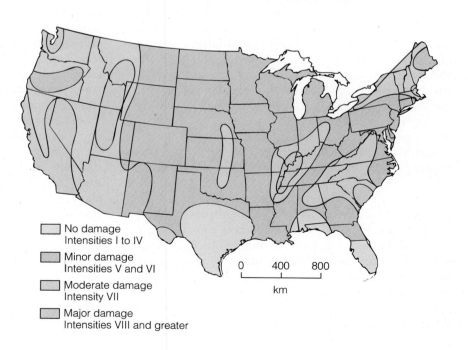

No damage
Intensities I to IV

Minor damage
Intensities V and VI

Moderate damage
Intensity VII

Major damage
Intensities VIII and greater

have had major earthquakes in the past but are currently inactive. Such regions are locked and not releasing energy, making these *seismic gaps* prime locations for future earthquakes. Several seismic gaps along the San Andreas fault have the potential for future major earthquakes.

Changes in elevation and tilting of the land surface have frequently preceded earthquakes and may be warnings of impending quakes. Extremely slight changes in the angle of the ground surface can be measured by tiltmeters. Tiltmeters have been placed on both sides of the San Andreas fault to measure tilting of the ground surface that is believed to result from increasing pressure in the rocks. Data from measurements in central California indicate significant tilting occurred immediately preceding small earthquakes. Furthermore, extensive tiltmeter work performed in Japan prior to the 1964 Niigata earthquake clearly showed a relationship between increased tilting and the main shock. While more research is needed, such changes appear to be useful in making short-term earthquake predictions.

Other earthquake precursors include fluctuations in the water level of wells and changes in the Earth's magnetic field and the electrical resistance of the ground. These fluctuations are believed to result from changes in the amount of pore space in rocks due to increasing pressure. A change in animal behavior prior to an earthquake also is frequently mentioned. It may be that animals are sensing small and subtle changes in the Earth prior to a quake that humans simply do not sense.

Many of the precursors just discussed can be related to the *dilatancy model,* which is based on changes occurring in rocks subjected to very high pressures. Laboratory experiments have shown that rocks undergo an increase in volume, known as dilatancy, just before rupturing. As pressure builds in rocks along faults, numerous small cracks are produced that alter the physical properties of the rocks. Water enters the cracks and increases the fluid pressure; this further increases the volume of the rocks and decreases their inherent strength until failure eventually occurs, producing an earthquake.

Currently, only four nations—the United States, Japan, Russia, and China—have government-sponsored earthquake prediction programs. These programs include laboratory and field studies of the behavior of rocks before, during, and after major earthquakes as well as monitoring activity along major active faults. Most earthquake prediction work in the United States

is done by the United States Geological Survey (USGS) and involves a variety of research into all aspects of earthquake-related phenomena.

The Chinese have perhaps one of the most ambitious earthquake prediction programs anywhere in the world, which is understandable considering their long history of destructive earthquakes. The Chinese program on earthquake prediction was initiated soon after two large earthquakes occurred at Xingtai (300 km southwest of Beijing) in 1966. The Chinese program includes extensive study and monitoring of all possible earthquake precursors. In addition, the Chinese also emphasize changes in phenomena that can be observed by seeing and hearing without the use of sophisticated instruments. The Chinese have had remarkable success in predicting earthquakes, particularly in the short term, such as the 1975 Haicheng earthquake. They failed, however, to predict the devastating 1976 Tangshan earthquake that killed at least 242,000 people.

Great strides are being made toward dependable, accurate earthquake predictions, and studies are underway to assess public reactions to long-, medium-, and short-term earthquake warnings. However, unless short-term warnings are actually followed by an earthquake, most people will probably ignore the warnings as they frequently do now for hurricanes, tornadoes, and tsunami.

▼ EARTHQUAKE CONTROL

If earthquake prediction is still in the future, can anything be done to control earthquakes? Because of the tremendous forces involved, humans are certainly not going to be able to prevent earthquakes. However, there may be ways to dissipate the destructive energy of major earthquakes by releasing it in small amounts that will not cause extensive damage.

During the early to mid-1960s, Denver experienced numerous small earthquakes. This was surprising because Denver had not been prone to earthquakes in the past. In 1962, David M. Evans, a geologist, suggested that the earthquakes in Denver were directly related to the injection of contaminated waste water into a disposal well 3,674 m deep at the Rocky Mountain Arsenal, northeast of Denver. The U.S. Army initially denied that there was any connection, but a USGS study concluded that the pumping of waste fluids into the disposal well was the cause of the earthquakes.

◆ Figure 7.20 shows the relationship between the average number of earthquakes in Denver per month and the average amount of contaminated waste fluids

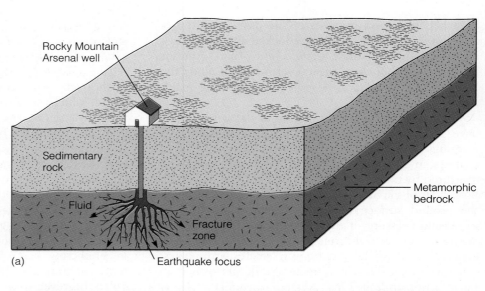

(a)

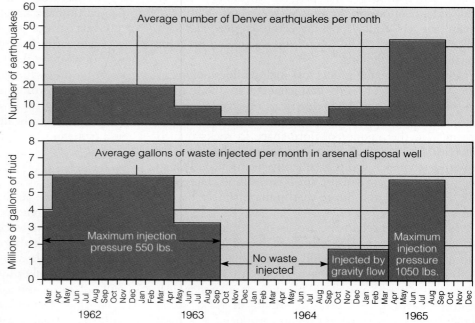

(b)

◆ FIGURE 7.20 (a) A block diagram of the Rocky Mountain Arsenal well and the underlying geology. (b) A graph showing the relationship between the amount of waste injected into the well per month and the average number of Denver earthquakes per month.

injected into the disposal well per month. Obviously, a high degree of correlation between the two exists, and the correlation is particularly convincing considering that during the time when no waste fluids were injected, earthquake activity decreased dramatically. The geology of the area consists of highly fractured gneiss overlain by sedimentary rocks. When water was pumped into these fractures, it decreased the friction on opposite sides of the fractures and, in essence, lubricated them so that movement occurred, causing the earthquakes that Denver experienced.

Experiments conducted in 1969 at an abandoned oil field near Rangely, Colorado, confirmed the arsenal hypothesis. Water was pumped in and out of abandoned oil wells, the pore-water pressure in these wells was measured, and seismographs were installed in the area to measure any seismic activity. Monitoring showed that small earthquakes were occurring in the area when fluid was injected and that earthquake activity declined when the fluids were pumped out. What the geologists were doing was starting and stopping earthquakes at will, and the relationship between pore-water pressures and earthquakes was established.

Based upon these results, some geologists have proposed that fluids be pumped into the locked segments of active faults to cause small- to moderate-sized earthquakes. They believe that this would relieve the pressure on the fault and prevent a major earthquake from occurring. While this plan is intriguing, it also has many potential problems. For instance, there is no guarantee that only a small earthquake might result. Instead a major earthquake might occur, causing tremendous property damage and loss of life. Who would be responsible? Certainly, a great deal more research is needed before such an experiment is performed, even in an area of low population density.

It appears that until such time as earthquakes can be accurately predicted or controlled, the best means of defense is careful planning and preparation.

☙ THE EARTH'S INTERIOR

The Earth's interior has always been an inaccessible, mysterious realm. During most of historic time, it was perceived as an underground world of vast caverns, heat, and sulfur gases, populated by demons. By the 1860s, scientists knew what the average density of the Earth was and that pressure and temperature increase with depth. And even though the Earth's interior is hidden from direct observation, scientists have a reasonably good idea of its internal structure and composition.

Scientists have known for more than 200 years that the Earth's interior is not homogeneous. Sir Isaac Newton (1642–1727) noted in a study of the planets that the Earth's average density is 5.0 to 6.0 g/cm³ (water has a density of 1 g/cm³). In 1797, Henry Cavendish calculated a density value very close to the 5.5 g/cm³ now accepted. The Earth's average density is considerably greater than that of surface rocks, most of which range from 2.5 to 3.0 g/cm³. Thus, in order for the average density to be 5.5 g/cm³, much of the interior must consist of materials with a density greater than the Earth's average density.

The Earth is generally depicted as consisting of concentric layers that differ in composition and density separated from adjacent layers by rather distinct boundaries (◆ Fig. 7.21). Recall that the outermost layer, or the **crust,** is the very thin skin of the Earth. Below the crust and extending about halfway to the Earth's center is the **mantle,** which comprises more than 80% of the Earth's volume (◉ Table 7.4). The central part of the Earth consists of a **core,** which is divided into a solid inner core and a liquid outer part (Fig. 7.21).

The behavior and travel times of P- and S-waves within the Earth provide geologists with much information about its internal structure. Seismic waves travel outward as wave fronts from their source areas, although it is most convenient to depict them as *wave rays,* which are lines showing the direction of movement of small parts of wave fronts (Fig. 7.5). Any disturbance, such as a passing train or construction equipment, can cause seismic waves, but only those generated by large earthquakes, explosive volcanism, asteroid impacts, and nuclear explosions can travel completely through the Earth.

As we noted earlier, the velocities of P- and S-waves are determined by the density and elasticity of the materials through which they travel. Both the density and elasticity of rocks increase with depth, but elasticity increases faster than density, resulting in a general increase in the velocity of seismic waves. P-waves travel faster than S-waves through all materials. However, unlike P-waves, S-waves cannot be transmitted through a liquid because liquids have no shear strength (rigidity)—they simply flow in response to a shear stress.

As a seismic wave travels from one material into another of different density and elasticity, its velocity and direction of travel change. That is, the wave is bent, a phenomenon known as **refraction** (◆ Fig. 7.22). Because seismic waves pass through materials of differing density and elasticity, they are continually refracted so that their paths are curved; the only exception is that wave rays are not refracted if their direction of travel is perpendicular to a boundary (Fig. 7.22). In that case they travel in a straight line.

In addition to refraction, seismic rays are also **reflected,** much as light is reflected from a mirror. Some of the energy of seismic rays that encounter a boundary separating materials of different density or elasticity within the Earth is *reflected* back to the Earth's surface (Fig. 7.22). If we know the wave velocity and the time

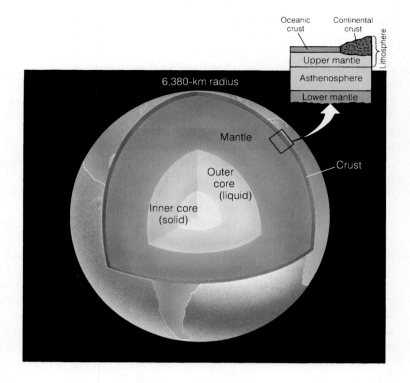

6,380-km radius

Oceanic crust — Continental crust — Lithosphere

Upper mantle
Asthenosphere
Lower mantle

Mantle

Outer core (liquid)

Inner core (solid)

Crust

◆ FIGURE 7.21 The internal structure of the Earth.

● **TABLE 7.4** Data on the Earth

	Volume (thousands of km³)	Percentage of the Total	Mass (trillions of metric tons)	Percentage of the Total
Inner core	7,512,800	0.70%	19,000,000,000	31.79%
Outer core	169,490,000	15.68		
Mantle	896,990,000	83.02	40,500,000,000	67.77
Continental crust	4,760,800	0.44	250,000,000	0.42
Oceanic crust	1,747,200	0.16		
Atmosphere, water, ice			14,351,000	0.02

required for it to travel from its source to the boundary and back to the surface, we can calculate the depth of the reflecting boundary. Such information is useful in determining not only the depths of the various layers within the Earth, but also the depths of sedimentary rocks that may contain petroleum.

Although changes in seismic wave velocity occur continuously with depth, P-wave velocity increases suddenly at the base of the crust and decreases abruptly at a depth of about 2,900 km (◆ Fig. 7.23). Such marked changes in the velocity of seismic waves indicate a boundary called a **discontinuity** across which a

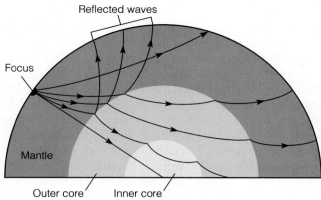

◆ FIGURE 7.22 Refraction and reflection of P-waves. When seismic waves pass through a boundary separating Earth materials of different density or elasticity, they are refracted, and some of their energy is reflected back to the surface.

significant change in Earth materials or their properties occurs. Such discontinuities are the basis for subdividing the Earth's interior into concentric layers.

▼ THE EARTH'S CORE

In 1906, R. D. Oldham of the Geological Survey of India discovered that seismic waves arrived later than expected at seismic stations more than 130° from an earthquake focus. He postulated the existence of a core that transmits seismic waves at a slower rate than shallower Earth materials. We now know that P-wave velocity decreases markedly at a depth of 2,900 km, thus indicating a major discontinuity now recognized as the core-mantle boundary (Fig. 7.23).

The sudden decrease in P-wave velocity at the core-mantle boundary causes P-waves entering the core to be refracted in such a way that very little P-wave energy reaches the Earth's surface in the area between 103° and 143° from an earthquake focus (◆ Fig. 7.24). This area in which little P-wave energy is recorded by seismometers is a **P-wave shadow zone.**

The P-wave shadow zone is not a perfect shadow zone. That is, some weak P-wave energy reaches the surface within the zone. Several hypotheses were proposed to explain this phenomenon, all of which were rejected by the Danish seismologist Inge Lehmann, who in 1936 postulated that the core is not entirely liquid. She demonstrated that reflection from a solid inner core could account for the arrival of weak

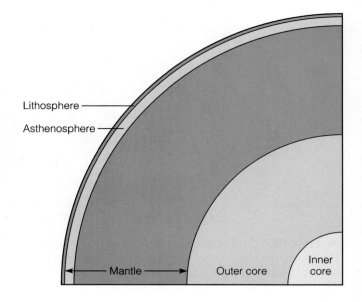

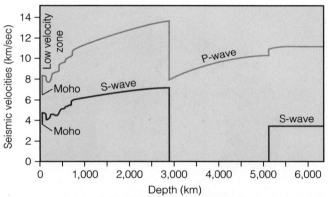

◆ FIGURE 7.23 Profiles showing seismic wave velocities versus depth. Several discontinuities are shown across which seismic wave velocities change rapidly.

P-waves in the P-wave shadow zone. Lehmann's proposal of a solid inner core was quickly accepted by seismologists.

In 1926, the British physicist Harold Jeffreys realized that S-waves were not simply slowed by the core, but were completely blocked by it. Thus, in addition to a P-wave shadow zone, a much larger and more complete **S-wave shadow zone** exists (◆ Fig. 7.25). At locations greater than 103° from an earthquake focus, no S-waves are recorded, indicating that S-waves cannot be transmitted through the core, S-waves will not pass

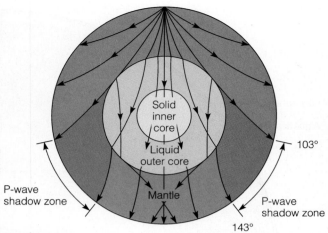

◆ FIGURE 7.24 P-waves are refracted so that no direct P-wave energy reaches the Earth's surface in the P-wave shadow zone.

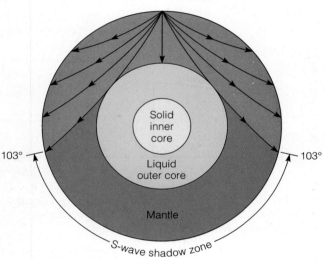

◆ FIGURE 7.25 The presence of an S-wave shadow zone indicates that S-waves are being blocked within the Earth.

through a liquid, so it seems that the outer core must be liquid or behave as a liquid.

The core constitutes 16.4% of the Earth's volume and nearly one-third of its mass (Table 7.4). We can estimate the core's density and composition by using seismic evidence and laboratory experiments. Furthermore, meteorites, which are thought to represent rem-

nants of the material from which the solar system formed, can be used to make estimates of density and composition. For example, meteorites composed of iron and nickel alloys may represent the differentiated interiors of large asteroids and approximate the density and composition of the Earth's core. The density of the outer core varies from 9.9 to 13.0 g/cm³ (● Table 7.5). At the Earth's center, the pressure is equivalent to about 3.5 million times normal atmospheric pressure.

The core cannot be composed of the minerals most common at the Earth's surface, because even under the tremendous pressures at great depth they would still not be dense enough to yield an average density of 5.5 g/cm³ for the Earth. Both the outer and inner core are thought to be composed largely of iron, but pure iron is too dense to be the sole constituent of the outer core. Thus, it must be "diluted" with elements of lesser density. Laboratory experiments and comparisons with iron meteorites indicate that about 12% of the outer core may consist of sulfur, and perhaps some silicon and small amounts of nickel and potassium (Table 7.5).

In contrast, pure iron is not dense enough to account for the estimated density of the inner core. Most geologists think that perhaps 10 to 20% of the inner core also consists of nickel. These metals form an iron-nickel alloy that under the pressure at that depth is thought to be sufficiently dense to account for the density of the inner core. When the core formed during early Earth history, it was probably completely molten and has since cooled to the point that its interior has crystallized.

▼ THE MANTLE

Another significant discovery about the Earth's interior was made in 1909 when the Yugoslavian seismologist Andrija Mohorovičić detected a discontinuity at a depth of about 30 km. While studying arrival times of seismic waves from Balkan earthquakes, Mohorovičić noticed that two distinct sets of P- and S-waves were recorded at seismic stations a few hundred kilometers from an earthquake's epicenter. He reasoned that one set of waves traveled directly from the epicenter to the seismic station, whereas the other waves had penetrated a deeper layer where they were refracted (◆ Fig. 7.26).

From his observations Mohorovičić concluded that a sharp boundary separating rocks with different properties exists at a depth of about 30 km. He postulated that P-waves below this boundary travel at 8 km/sec,

● **TABLE 7.5** Composition and Density of the Earth

	Composition	Density (g/cm³)
Inner core	Iron with 10 to 20% nickel	12.6–13.0
Outer core	Iron with perhaps 12% sulfur; also silicon and small amounts of nickel	9.9–12.2
Mantle	Peridotite (composed mostly of ferrogmagnesian silicates)	3.3–5.7
Oceanic crust	Upper part basalt, lower part gabbro	~3.0
Continental crust	Average composition of granodiorite	~2.7

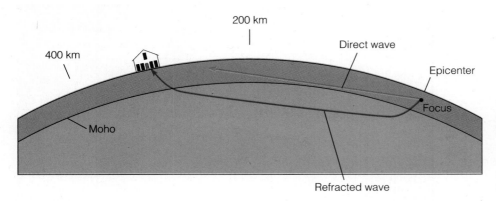

◆ FIGURE 7.26 Andrija Mohorovičić studied seismic waves and detected a seismic discontinuity at a depth of about 30 km. The deeper, faster seismic waves arrive at seismic stations first, even though they travel farther.

whereas those above the boundary travel at 6.75 km/sec. Thus, when an earthquake occurs, some waves travel directly from the focus to the seismic station, while others travel through the deeper layer and some of their energy is refracted back to the surface (Fig. 7.26). Waves traveling through the deeper layer travel farther to a seismic station but they do so more rapidly than those in the shallower layer. The boundary identified by Mohorovičić separates the crust from the mantle and is now called the **Mohorovičić discontinuity,** or simply the **Moho.** It is present everywhere except beneath spreading ridges, but its depth varies: beneath the continents it averages 35 km, but ranges from 20 to 90 km; beneath the sea floor it is 5 to 10 km deep (◆ Fig. 7.27).

Although seismic wave velocity in the mantle generally increases with depth, several discontinuities also exist. Between depths of 100 and 250 km, both P- and S-wave velocities decrease markedly (◆ Fig. 7.28). This layer between 100 and 250 km deep is the **low-velocity zone;** it corresponds closely to the *asthenosphere,* a layer in which the rocks are close to their melting point and thus are less elastic; this decrease in elasticity accounts for the observed decrease in seismic wave velocity. The asthenosphere is an important zone because it may be where some magmas are generated. Furthermore, it lacks strength and flows plastically and is thought to be the layer over which the outer, rigid *lithosphere* moves.

Other discontinuities have been detected at deeper levels within the mantle that probably represent structural changes in minerals rather than compositional changes. In other words, geologists believe the mantle is composed of the same material throughout, but the structural states of minerals such as olivine change with depth. At depths of 400 km, 640 to 720 km, and 1,050 km, seismic wave velocity increases slightly as a consequence of such changes in mineral structure. These

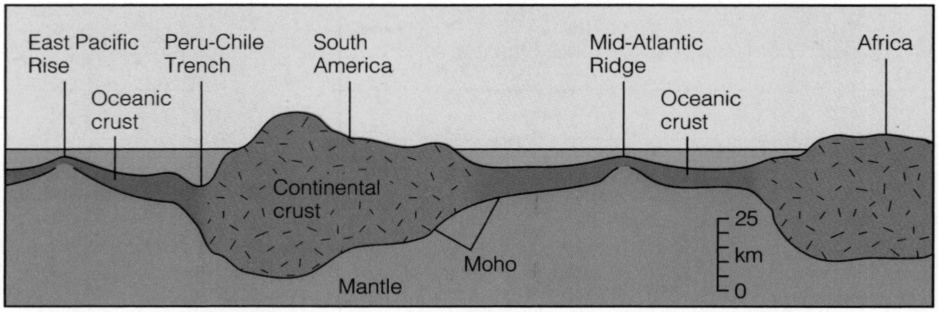

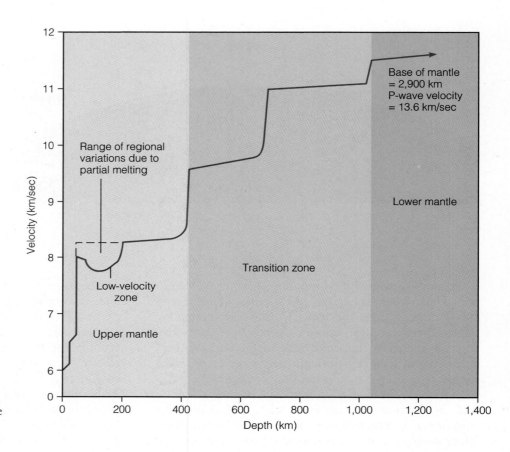

◆ FIGURE 7.27 The Moho is present everywhere except beneath spreading ridges such as the East Pacific Rise and the Mid-Atlantic Ridge. However, the depth of the Moho varies considerably.

◆ FIGURE 7.28 Variations in P-wave velocity in the upper mantle and transition zone.

three discontinuities are within what is called a *transition zone* separating the upper mantle from the lower mantle (Fig. 7.28).

Although the mantle's density, which varies from 3.3 to 5.7 g/cm³, can be inferred rather accurately from seismic waves, its composition is less certain. The igneous rock *peridotite* containing mostly ferromagne-sian minerals, is considered the most likely component. Laboratory experiments indicate that it possesses physical properties that would account for the mantle's density and observed rates of seismic wave transmissions. Peridotite also forms the lower parts of igneous rock sequences believed to be fragments of the oceanic crust and upper mantle emplaced on land. In addition, peridotite occurs as inclusions in volcanic rock bodies such as *kimberlite pipes* that are known to have come from great depths. These inclusions are thought to be pieces of the mantle (see Perspective 7.2).

THE EARTH'S INTERNAL HEAT

During the nineteenth century, scientists realized that the Earth's temperature in deep mines increases with depth. More recently, the same trend has been observed in deep drill holes, but even in these we can measure temperatures directly down to a depth of only a few kilometers. The temperature increase with depth, or **geothermal gradient,** near the surface is about 25°C/km, although it varies from area to area. For example, in areas of active or recently active volcanism, the geothermal gradient is greater than in adjacent nonvolcanic areas, and temperature rises faster beneath spreading ridges than elsewhere beneath the sea floor.

Unfortunately, the geothermal gradient is not useful for estimating temperatures deep in the Earth. If we were simply to extrapolate from the surface downward, the temperature at 100 km would be so high that in spite of the great pressure, all known rocks would melt. Yet except for pockets of magma, it appears that the mantle is solid rather than liquid because it transmits S-waves. Accordingly, the geothermal gradient must decrease markedly.

Current estimates of the temperature at the base of the crust are 800° to 1,200°C. The latter figure seems to be an upper limit: if it were any higher, melting would be expected. Furthermore, fragments of mantle rock in kimberlite pipes (see Perspective 7.2), thought to have come from depths of about 100 to 300 km, appear to have reached equilibrium at these depths and at a temperature of about 1,200°C. At the core-mantle boundary, the temperature is probably between 3,500° and 5,000°C; the wide spread of values indicates the uncertainties of such estimates. If these figures are reasonably accurate, the geothermal gradient in the mantle is only about 1°C/km.

Considering that the core is so remote and so many uncertainties exist regarding its composition, only very general estimates of its temperature can be made. The maximum temperature at the center of the core is thought to be about 6,500°C, very close to the estimated temperature for the surface of the Sun!

SEISMIC TOMOGRAPHY

The model of the Earth's interior consisting of an iron-rich core and a rocky mantle is probably accurate but is also rather imprecise. Recently, however, geophysicists have developed a new technique called *seismic tomography* that allows them to develop three-dimensional models of the Earth's interior. In seismic tomography numerous crossing seismic waves are analyzed in much the same way radiologists analyze CAT (computerized axial tomography) scans. In CAT scans, X-rays penetrate the body, and a two-dimensional image of the inside of a patient is formed. Repeated CAT scans, each from a slightly different angle, are computer analyzed and stacked to produce a three-dimensional picture.

In a similar fashion geophysicists use seismic waves to probe the interior of the Earth. From its time of arrival and distance traveled, the velocity of a seismic ray is computed at a seismic station. Only average velocity is determined, however, rather than variations in velocity. In seismic tomography numerous wave rays are analyzed so that "slow" and "fast" areas of wave travel can be detected (◆ Fig. 7.29, page 184). Recall that seismic wave velocity is controlled partly by elasticity; cold rocks have greater elasticity and therefore transmit seismic waves faster than hot rocks.

Using this technique, geophysicists have detected areas within the mantle at a depth of about 150 km where seismic velocities are slower than expected. These anomalously hot regions lie beneath volcanic areas and beneath the mid-oceanic ridges, where convection cells of rising hot mantle rock are thought to exist. In contrast, beneath the older interior parts of continents, where tectonic activity ceased hundreds of millions or billions of years ago, anomalously cold spots are recognized. In effect, tomographic maps and three-dimensional diagrams show heat variations within the Earth.

Seismic tomography has also yielded additional and sometimes surprising information about the core. For example, the core-mantle boundary is not a smooth surface, but has broad depressions and rises extending several kilometers into the mantle. Of course, the base of the mantle possesses the same features in reverse; geophysicists have termed these features "anticontinents" and "antimountains." It appears that the surface of the core is continually deformed by sinking and rising masses of mantle material.

As a result of seismic tomography, a much clearer picture of the Earth's interior is emerging. It has already given us a better understanding of complex convection within the mantle, including upwelling convection currents thought to be responsible for the movement of the Earth's lithospheric plates.

KIMBERLITE PIPES — WINDOWS TO THE MANTLE

Diamonds have been economically important throughout history, yet prior to 1870, they had been found only in river gravels, where they occur as the result of weathering, transport, and deposition. In 1870, however, the source of diamonds in South Africa was traced to cone-shaped igneous bodies called *kimberlite pipes* found near the town of Kimberly (◆ Fig. 1). Kimberlite pipes are the source rocks for most diamonds.

◆ FIGURE 1 Generalized cross section of a kimberlite pipe. Most kimberlite pipes measure less than 500 m in diameter at the surface.

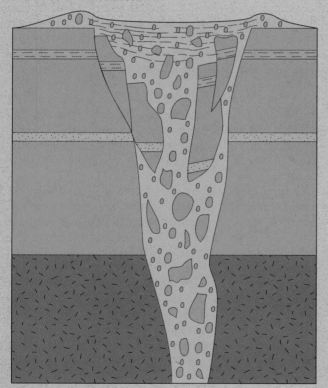

The greatest concentrations of kimberlite pipes are in southern Africa and Siberia, but they occur in many other areas as well. In North America they have been found in the Canadian Arctic, Colorado, Wyoming, Missouri, Montana, Michigan, and Virginia, and one at Murfreesboro, Arkansas, was briefly worked for diamonds. Diamonds discovered in glacial deposits in some midwestern states indicate that kimberlite pipes are present father north. The precise source of these diamonds has not been determined, although some kimberlite pipes have recently been identified in northern Michigan.

Kimberlite pipes are composed of dark gray or blue igneous rock called *kimberlite,* which contains olivine, a potassium- and magnesium-rich mica, serpentines, and calcite and silica. Some of these rocks contain inclusions of *peridotite* that are thought to represent pieces of the mantle brought to the surface during the explosive volcanic eruptions that form kimberlite pipes.

If peridotite inclusions are, in fact, pieces of the mantle, they indicate that the magna in kimberlite pipes originated at a depth of at least 30 km. Indeed, the presence of diamonds and the structural form of the silica in the kimberlite can be used to establish both minimum and maximum depths for the origin of the magma. Diamond and graphite are different crystalline forms of carbon (see Fig. 2.5), but diamond forms only under high-pressure, high-temperature conditions. The presence of diamond and the absence of graphite in kimberlite indicate that such conditions existed where the magma originated.

The calculated geothermal gradient and the pressure increase with depth beneath the continents are shown in ◆ Figure 2. Laboratory experiments have established a diamond-graphite inversion curve showing the pressure-temperature conditions at which graphite is favored over diamond (Fig. 2). According to the data in Figure 2, the intersection of the diamond-graphite inversion curve with the geothermal gradient indicates that kimberlite magma came from a minimum depth of about 100 km.

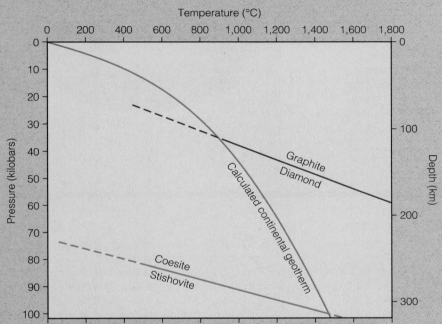

◆ FIGURE 2 The forms of carbon and silica in kimberlite pipes provide information on the depth at which the magma formed. The presence of diamond and coesite in kimberlite indicates that the magma probably formed between 100 and 300 km as shown by the intersection of the calculated continental geotherm with the graphite-diamond and coesite-stishovite inversion curves.

Diamond can establish only a minimum depth for kimberlite because it is stable at any pressure greater than that occurring at a depth of 100 km. The silica found in kimberlite, on the other hand, is a form that indicates a maximum depth of about 300 km. Quartz is the form of silica found under low-pressure, low-temperature conditions. Under great pressure, however, the crystal structure of quartz changes to its high-pressure equivalent called coesite, and at even greater pressure it changes to stishovite.* Kimberlite pipes contain coesite but no stishovite, indicating that the kimberlite magma must have come from a depth of less than 300 km as indicated by the intersection of the coesite-stishovite inversion curve with the geothermal gradient (Fig. 2).

. . . .

*Coesite and stishovite are also known from other high-pressure environments such as meteorite impact sites.

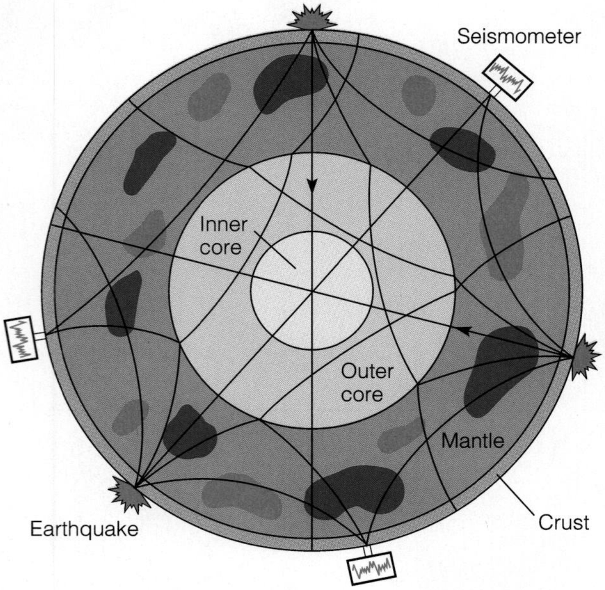

◆ FIGURE 7.29 Numerous earthquake waves are analyzed to detect areas within the Earth that transmit seismic waves faster or slower than adjacent areas. Areas of fast wave travel correspond to "cold" regions (blue), whereas "hot" regions (red) transmit seismic waves more slowly.

Chapter Summary

1. Earthquakes are vibrations of the Earth caused by the sudden release of energy, usually along a fault.

2. The elastic rebound theory states that pressure builds in rocks on opposite sides of a fault until the inherent strength of the rocks is exceeded and rupture occurs. When the rocks rupture, stored energy is released as they snap back to their original position.

3. Seismology is the study of earthquakes. Earthquakes are recorded on seismographs, and the record of an earthquake is a seismogram.

4. The focus of an earthquake is the point where energy is released. Vertically above the focus on the Earth's surface is the epicenter.

5. Most earthquakes occur within seismic belts. Approximately 80% of all earthquakes occur in the circum-Pacific belt, 15% within the Mediterranean-Asiatic belt, and the remaining 5% mostly in the interior of the plates or along oceanic spreading ridge systems.

6. The two types of body waves are P-waves and S-waves. Both travel through the Earth, although S-waves do not travel through liquids, P-waves are the fastest waves and are compressional, while S-waves are shear. Surface waves travel along or just below the Earth's surface. The two types of surface waves are Rayleigh and Love waves.

7. The epicenter of an earthquake can be located by the use of a time-distance graph of the P- and S-waves from any given distance. Three seismographs are needed to locate the epicenter of an earthquake.

8. Intensity is a measure of the kind of damage done by an earthquake and is expressed by values from I to XII in the Modified Mercalli Intensity Scale.

9. Magnitude measures the amount of energy released by an earthquake and is expressed in the Richter Magnitude Scale. Each increase in the magnitude number represents about a 30-fold increase in energy released.

10. Ground shaking is the most destructive of all earthquake hazards. The amount of damage done by an earthquake depends upon the geology of the area, the type of building construction, the magnitude of the earthquake, and the duration of shaking. Tsunami are seismic sea waves that are usually produced by earthquakes. They can do a tremendous amount of damage to coastlines.

11. The Earth is concentrically layered into an iron-rich core with a solid inner core and a liquid outer part, a rocky mantle, and an oceanic crust and continental crust.

12. Much of the information about the Earth's interior has been derived from studies of P- and S-waves that travel through the Earth. Laboratory experiments, comparisons with meteorites, and studies of inclusions in volcanic rocks provide additional information.

13. Density and elasticity of Earth materials determine the velocity of seismic waves. Seismic waves are refracted when

their direction of travel changes. Wave reflection occurs at boundaries across which the properties of rocks change.

14. The behavior of P- and S-waves within the Earth and the presence of P- and S-wave shadow zones allow geologists to estimate the density and composition of the Earth's interior and to estimate the size and depth of the core and mantle.

15. The Earth's inner core is thought to be composed of iron and nickel, whereas the outer core is probably composed mostly of iron with 10 to 20% sulfur and other substances in less quantities. Peridotite is the most likely component of the mantle.

16. The geothermal gradient of 25°C/km cannot continue to great depths, otherwise most of the Earth would be molten. The geothermal gradient for the mantle and core is probably about 1°C/km. The temperature at the Earth's center is estimated to be 6,500°C.

Important Terms

aftershock
core
crust
discontinuity
earthquake
elastic rebound theory
epicenter
focus
geothermal gradient
intensity

Love wave (L-wave)
low-velocity zone
magnitude
mantle
Modified Mercalli Intensity Scale
Mohorovičić discontinuity (Moho)
P-wave
P-wave shadow zone
Rayleigh wave (R-wave)
reflection

refraction
Richter Magnitude Scale
seismogram
seismograph
seismology
S-wave
S-wave shadow zone
time-distance graph
tsunami

Review Questions

1. According to the elastic rebound theory:
   a. ___ earthquakes originate deep within the Earth; b. ___ earthquakes originate in the asthenosphere where rocks are plastic; c. ___ earthquakes occur where the strength of the rock is exceeded; d. ___ rocks are elastic and do not rebound to their former position; e. ___ none of these.

2. With few exceptions, the most destructive earthquakes are:
   a. ___ shallow focus; b. ___ intermediate focus; c. ___ deep focus; d. ___ answers (a) and (b); e. ___ answers (b) and (c).

3. The majority of all earthquakes occur in the:
   a. ___ Mediterranean-Asiatic belt; b. ___ interior of plates; c. ___ circum-Atlantic belt; d. ___ circum-Pacific belt; e. ___ along spreading ridges.

4. The fastest of the four seismic waves are:
   a. ___ P; b. ___ S; c. ___ Rayleigh; d. ___ Love; e. ___ tsunami.

5. An epicenter is:
   a. ___ the location where rupture begins; b. ___ the point on the Earth's surface vertically above the focus; c. ___ the same as the hypocenter; d. ___ the location where energy is released; e. ___ none of these.

6. A qualitative assessment of the kinds of damage done by an earthquake is expressed by:
   a. ___ seismicity; b. ___ dilatancy; c. ___ magnitude; d. ___ intensity; e. ___ none of these.

7. How much more energy is released by a magnitude 5 earthquake than by one of magnitude 2?
   a. ___ 2.5 times; b. ___ 3 times; c. ___ 30 times; d. ___ 1,000 times; e. ___ 27,000 times.

8. A tsunami is a:
   a. ___ measure of the energy released by an earthquake; b. ___ seismic sea wave; c. ___ precursor to an earthquake; d. ___ locked portion of a fault; e. ___ seismic gap.

9. The average density of the Earth is _____ g/cm³.
   a. ___ 12.0; b. ___ 5.5; c. ___ 6.75; d. ___ 1.0; e. ___ 2.5.

10. A line showing the direction of movement of a small part of a wave front is a:
    a. ___ seismic discontinuity; b. ___ P-wave reflection; c. ___ wave ray; d. ___ particle beam; e. ___ seismic gradient.

11. When seismic waves travel through materials having different properties, their direction of travel changes. This phenomenon is wave:
    a. ___ elasticity; b. ___ energy dissipation; c. ___ refraction; d. ___ deflection; e. ___ reflection.

12. A major seismic discontinuity at a depth of 2,900 km is the:
a. ___ core-mantle boundary; b. ___ oceanic crust-continental crust boundary; c. ___ Moho; d. ___ inner core-outer core boundary; e. ___ lithosphere-asthenosphere boundary.

13. The Earth's core is probably composed mostly of:
a. ___ sulfur; b. ___ silica; c. ___ nickel; d. ___ potassium; e. ___ iron.

14. Which of the following provides evidence for the composition of the core?
a. ___ inclusions in volcanic rocks; b. ___ diamonds; c. ___ meteorites; d. ___ peridotite; e. ___ S-wave shadow zone.

15. The seismic discontinuity at the base of the crust is the:
a. ___ magnetic anomaly; b. ___ Moho; c. ___ geothermal gradient; d. ___ high-velocity zone; e. ___ transition zone.

16. How does the elastic rebound theory explain how energy is released during an earthquake?

17. What is the difference between body waves and surface waves?

18. How do P-waves differ from S-waves? How do Rayleigh waves differ from Love waves?

19. How is the epicenter of an earthquake determined?

20. What is the relationship between plate boundaries and earthquakes?

21. What is the relationship between plate boundaries and focal depth?

22. Explain the difference between intensity and magnitude and between the Modified Mercalli Intensity Scale and the Richter Magnitude Scale.

23. Explain how tsunami are produced and why they are so destructive.

24. How can earthquake precursors be used to predict earthquakes?

25. What determines the velocity of P- and S-waves?

26. Explain how seismic waves are refracted and reflected.

27. What is the significance of the S-wave shadow zone?

28. Why is the inner core thought to be composed of iron and nickel whereas the outer core is probably composed of iron and sulfur?

29. Several seismic discontinuities exist within the mantle. What accounts for these discontinuities?

30. Explain the reasoning used by Mohorovičić to determine that a discontinuity, now called the Moho, exists between the crust and the mantle.

Additional Readings

Anderson, D. L., and A. M. Dziewonski. 1984. Seismic tomography. *Scientific American* 251, no. 4:60–68.

Bolt, B. A. 1982. *Inside the Earth: Evidence from earthquakes.* San Francisco: W. H. Freeman.

———. 1988. *Earthquakes.* New York: W. H. Freeman.

Canby, T. Y. 1990. California earthquake—prelude to the big one? *National Geographic* 177, no. 5:76–105.

Dawson, J. 1993. CAT scanning the Earth. *Earth* 2, no. 3:36–41.

Fischman, J. 1992. Falling into the gap: A new theory shakes up earthquake predictions. *Discover* October 1992: 56–63.

Fowler, C. M. R. 1990. *The solid Earth.* New York: Cambridge Univ. Press.

Frohlich, C. 1989. Deep earthquakes. *Scientific American* 260, no. 1:48–55.

Hanks, T. C. 1985. *National earthquake hazard reduction program: Scientific status.* U.S. Geological Survey Bulletin 1659.

Jeanloz, R. 1983. The Earth's core. *Scientific American* 249, no. 3:56–65.

———. 1990. The nature of the Earth's core. *Annual Review of Earth and Planetary Sciences,* 18:357–386.

——— and T. Lay. 1993. The core-mantle boundary. *Scientific American* 268, no. 5:48–55.

Johnston, A. C., and L. R. Kanter. 1990. Earthquakes in stable continental crust. *Scientific American* 262, no. 3:68–75.

McKenzie, D. P. 1983. The Earth's mantle. *Scientific American* 249, no. 3:66–78.

Wesson, R. L., and R. E. Wallace. 1985. Predicting the next great earthquake in California. *Scientific American* 252, no. 2:35–43.

CHAPTER
8

Chapter Outline

Pillow lava on the floor of the Pacific Ocean near the Galápagos Islands.

The Sea Floor

Prologue

In 1979, researchers aboard the submersible *Alvin* descended about 2,500 m to the Galapagos Rift in the eastern Pacific Ocean basin and observed hydrothermal vents on the sea floor (◆ Fig. 8.1). Such vents occur near spreading ridges where seawater seeps down into the oceanic crust through cracks and fissures, is heated by the hot rocks, and then rises and is discharged onto the sea floor as hot springs. During the 1960s, hot metal-rich brines apparently derived from hydrothermal vents were detected and sampled in the Red Sea. These dense brines were concentrated in pools along the axis of the sea; beneath them thick deposits of metal-rich sediments were found. During the early 1970s, researchers observed hydrothermal vents on the Mid-Atlantic Ridge about 2,900 km east of Miami, Florida, and in 1978 moundlike mineral deposits were sampled from the East Pacific Rise just south of the Gulf of California.

When the submersible *Alvin* descended to the Galapagos Rift in 1979, mounds of metal-rich sediments were observed. Near these mounds the researchers saw what they called black smokers (chimneylike vents) discharging plumes of hot, black water (Fig. 8.1). Since 1979 similar vents have been observed at or near spreading ridges in several other areas.

Submarine hydrothermal vents are interesting for several reasons. Near the vents live communities of organisms, including bacteria, crabs, mussels, starfish, and tubeworms, many of which had never been seen before (Fig. 8.1). In most biological communities, photosynthesizing organisms form the base of the food chain and provide nutrients for the herbivores and carnivores. In vent communities, however, no

◆ FIGURE 8.1 The submersible *Alvin* sheds light on hydrothermal vents at the Galapagos Rift, a branch of the East Pacific Rise. Seawater seeps down through the oceanic crust, becomes heated, and then rises and builds chimneys on the sea floor. The plume of "black smoke" is simply heated water saturated with dissolved minerals. Communities of organisms, including tubeworms, giant clams, crabs, and several types of fish, live near the vents.

sunlight is available for photosynthesis, and the base of the food chain consists of bacteria that practice chemosynthesis; they oxidize sulfur compounds from the hot vent waters, thus providing their own nutrients and the nutrients for other members of the food chain.

Another interesting aspect of these submarine hydrothermal vents is their economic potential. When seawater circulates downward through the oceanic crust, it is heated to as much as 400°C. The hot water then reacts with the crust and is transformed into a metal-bearing solution. As the hot solution rises and discharges onto the sea floor, it cools, precipitating iron, copper, and zinc sulfides and other minerals that accumulate to form a chimneylike vent (Fig. 8.1). These vents are ephemeral, however; one observed in 1979 was inactive six months later. When their activity ceases, the vents eventually collapse and are incorporated into a moundlike mineral deposit.

The economic potential of hydrothermal vent deposits is tremendous. The deposits in the Atlantis II Deep of the Red Sea contain an estimated 100 million tons of metals, including iron, copper, zinc, silver, and gold. Many of these sulfide deposits now on land are believed to have formed on the sea floor by hydrothermal vent activity.

Hydrothermal vent sulfide deposits have formed throughout geologic time. None are currently being mined, but the technology to exploit them exists. In fact, the Saudi Arabian and Sudanese governments have determined that it is feasible to recover such deposits from the Red Sea and are making plans to do so.

▼ INTRODUCTION

Although the oceans are distinct enough to be designated by separate names such as Pacific, Atlantic, and Indian, a single interconnected body of salt water covers more than 70% of the Earth's surface. During most of historic time, people knew little of the oceans and, until fairly recently, believed that the sea floor was flat and featureless. Although the ancient Greeks had determined the size of the Earth rather accurately, Western Europeans were not aware of the vastness of the oceans until the fifteenth and sixteenth centuries when various explorers sought new trade routes to the Indies. When Christopher Columbus set sail on August 3, 1492, in an attempt to find a route to the Indies, he greatly underestimated the width of the Atlantic Ocean.

During these and subsequent voyages, Europeans sailed to the Americas, the Pacific Ocean, Australia, New Zealand, the Hawaiian Islands, and many other islands previously unknown to them. Continuing exploration of the oceans revealed that the sea floor is not flat and featureless as formerly believed. Indeed, scientists discovered that the sea floor possesses varied topography, including oceanic trenches, submarine ridges, broad plateaus, hills, and vast plains. Some people have suggested that some of these features are remnants of the mythical lost continent of Atlantis (see Perspective 8.1).

▼ OCEANOGRAPHIC RESEARCH

The Deep Sea Drilling Project, an international program sponsored by several oceanographic institutions and funded by the National Science Foundation, began in 1968. Its first research vessel, the *Glomar Challenger,* could drill in water more than 6,000 m deep and recover long cores of sea-floor sediment and the oceanic crust. During the next 15 years, the *Glomar Challenger* drilled more than 1,000 holes in the sea floor.

The Deep Sea Drilling Project came to an end in 1983 when the *Glomar Challenger* was retired. However, an international project, the Ocean Drilling Program, continued where the Deep Sea Drilling Project left off, and a larger, more advanced research vessel, the JOIDES* *Resolution,* made its first voyage in 1985.

In addition to surface vessels, submersibles, both remotely controlled and those carrying scientists, have been added to the research arsenal of oceanographers. In 1985, for example, the *Argo,* towed by a surface vessel and equipped with sonar and television systems, provided the first views of the British ocean liner R.M.S. *Titanic* since it sank in 1912. The U.S. Geological Survey is using a towed device to map the sea floor (♦ Fig. 8.2). The system uses sonar to produce images resembling aerial photographs. Researchers aboard the submersible *Alvin* have observed submarine hydrothermal vents (see the Prologue) and have explored parts of the oceanic ridge system.

The first measurements of the oceanic depths were made by lowering a weighted line to the sea floor and measuring the length of the line. Now, however, an instrument called an *echo sounder* is used. Sound waves from a ship are reflected from the sea floor and

· · · · · · · · · ·

*JOIDES is an acronym for Joint Oceanographic Institutions for Deep Earth Sampling.

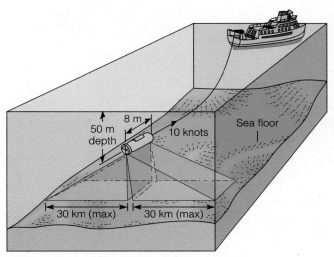

◆ FIGURE 8.2 The sonar system used by the U.S. Geological Survey for sea-floor mapping.

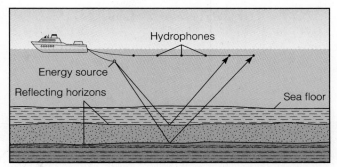

◆ FIGURE 8.3 Diagram showing how seismic profiling is used to detect buried layers at sea. Some of the energy generated at the energy source is reflected from various horizons back to the surface where it is detected by hydrophones.

detected by instruments on the ship, thus yielding a continuous profile of the sea floor. Depth is determined by knowing the velocity of sound waves in water and the time it takes for the waves to reach the sea floor and return to the ship.

Seismic profiling is similar to echo sounding but even more useful. Strong waves are generated at an energy source, the waves penetrate the layers beneath the sea floor, and some of the energy is reflected from various geologic horizons back to the surface (◆ Fig. 8.3). Seismic profiling has been particularly useful in mapping the structure of the oceanic crust beneath sea-floor sediments.

Although scientific investigations of the oceans have been yielding important information for more than two hundred years, much of our current knowledge has been acquired since World War II. This statement is particularly true with respect to the sea floor, because only in recent decades has instrumentation been available to study this largely hidden domain.

▼ CONTINENTAL MARGINS

Continental margins are zones separating the part of a continent above sea level from the deep-sea floor. The continental margin consists of a gently sloping *continental shelf,* a more steeply inclined *continental slope,* and, in some cases, a deeper, gently sloping *continental rise* (◆ Fig. 8.4, page 194). Seaward of the continental margin is the deep-ocean basin. Thus, the continental margin extends to increasingly greater depths until it merges with the deep-sea floor.

Most people perceive continents as land areas outlined by sea level. However, the true geologic margin of a continent—that is, where continental crust changes to oceanic crust—is below sea level, generally somewhere beneath the continental slope. Accordingly, marginal parts of continents are submerged.

The Continental Shelf

The **continental shelf** is an area where the sea floor slopes very gently in a seaward direction (Fig. 8.4). The outer edge of the continental shelf is generally taken to correspond to the point at which the inclination of the sea floor increases rather abruptly to several degrees; this *shelf-slope break* occurs at an average depth of about 135 m (Fig. 8.4). Continental shelves vary considerably in width, ranging from a few tens of meters to more than 1,000 km.

At times during the Pleistocene Epoch (1,600,000 to 10,000 years ago), sea level was as much as 130 m lower than at present. As a consequence of lower sea level, much of the sediment on continental shelves accumulated in stream channels and floodplains. In fact, in areas such as northern Europe and parts of North America, glaciers extended onto the exposed shelves and deposited gravel, sand, and mud. Sea level has risen since the Pleistocene Epoch, submerging the shelf sediments, which are now being reworked by marine processes.

Perspective 8.1

LOST CONTINENTS

Most people have heard of the mythical lost continent of Atlantis, but few are aware of the source of the Atlantis legend or the evidence that is cited for the former existence of this continent. Only two known sources of the Atlantis legend exist, both written in about 350 B.C. by the Greek philosopher Plato. In two of his philosophical dialogues, the *Timaeus* and the *Critias*, Plato tells of Atlantis, a large island continent that, according to him, was located in the Atlantic Ocean west of the Pillars of Hercules, which we now call the Strait of Gibraltar (◆Fig. 1). Plato also wrote that following the conquest of Atlantis by Athens, the continent disappeared:

... there were violent earthquakes and floods and one terrible day and night came when ... Atlantis ... disappeared beneath the sea. And for this reason even

◆ FIGURE 1 According to Plato, Atlantis was a large continent west of the Pillars of Hercules, which we now call the Strait of Gibraltar.

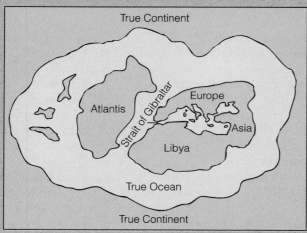

now the sea there has become unnavigable and unsearchable, blocked as it is by the mud shallows which the island produced as it sank.*

If one assumes that the destruction of Atlantis was a real event, rather than one conjured up by Plato to make a philosophical point, he nevertheless lived long after it was supposed to have occurred. According to Plato, Solon, an Athenian who lived about 200 years before Plato, heard the story from Egyptian priests who claimed the event had occurred 9,000 years before their time. Solon told the story to his grandson, Critias, who in turn told it to Plato.

Present-day proponents of the Atlantis legend generally cite two types of evidence to support their claim that Atlantis did indeed exist. First, they point to supposed cultural similarities on opposite sides of the Atlantic Ocean basin, such as the similar shapes of the pyramids of Egypt and Central and South America. They contend that these similarities are due to cultural diffusion from the highly developed civilization of Atlantis. According to archaeologists, however, few similarities actually exist, and those that do can be explained as the independent development of analogous features by different cultures.

Secondly, supporters of the legend assert that remnants of the sunken continent can be found. No "mud shallows" exist in the Atlantic as Plato claimed, but the Azores, Bermuda, the Bahamas, and the Mid-Atlantic Ridge are alleged to be remnants of Atlantis. If a continent had actually sunk in the Atlantic, however, it could be easily detected by a gravity survey. Recall that continental crust has a granitic composition and a lower density than oceanic crust. Thus, if a continent were actually present beneath the Atlantic Ocean, there would be a huge negative gravity anomaly, but no such anomaly has been

*From the *Timaeus.* Quoted in E. W. Ramage, ed., *Atlantis: Fact or Fiction?* (Bloomington, Ind.: Indiana University Press, 1978), p. 13.

detected. Furthermore, the crust beneath the Atlantic has been drilled in many places, and all the samples recovered indicate that its composition is the same as that of oceanic crust elsewhere.

In short, there is no geological evidence for Atlantis. Nevertheless, some archaeologists think that the legend may be based on a real event. About 1390 B.C., a huge volcanic eruption destroyed the island of Thera in the Mediterranean Sea, which was an important center of early Greek civilization. The eruption was one of the most violent during historic time, and much of the island disappeared when it subsided to form a caldera (◆ Fig. 2). Most of the island's inhabitants escaped, but the eruption probably contributed to the demise of the Minoan culture on Crete. At least 10 cm of ash fell on parts of Crete, and the coastal areas of the island were probably devastated by tsunami. It is possible that Plato used an account of the destruction of Thera, but fictionalized it for his own purposes, thereby giving rise to the Atlantis legend.

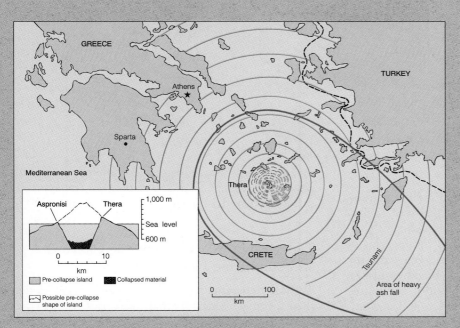

◆ FIGURE 2 The island of Thera was destroyed by a huge eruption about 1390 B.C. Ash was carried more than 950 km to the southeast, and tsunami probably devastated nearby coastal areas. The inset shows the possible profile of the island before the eruption and its shape immediately after the caldera formed.

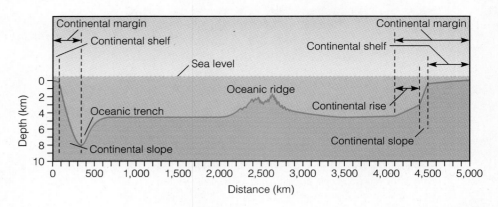

◆ FIGURE 8.4 A generalized profile of the sea floor showing features of the continental margins. The vertical dimensions of the features in this profile are greatly exaggerated because the vertical and horizontal scales differ.

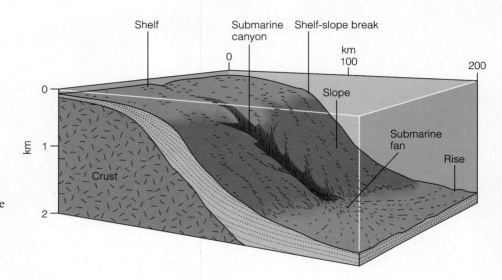

◆ FIGURE 8.5 Submarine fans formed by the deposition of sediments carried down submarine canyons by turbidity currents. Much of the continental rise is composed of overlapping submarine fans.

The Continental Slope and Rise

The seaward margin of the continental shelf is marked by the shelf-slope break (at an average depth of 135 m) where the relatively steep **continental slope** begins (Fig. 8.4). In many places, especially around the margins of the Atlantic, the continental slope merges with the more gently sloping **continental rise.** In other places, such as around the Pacific Ocean, slopes commonly descend directly into an oceanic trench, and a continental rise is absent (Fig. 8.4).

The shelf-slope break is a very important feature in terms of sedimentation. Landward from the break, the shelf is affected by waves and tidal currents. Seaward of the break, the bottom sediments are completely unaffected by surface processes, and their transport onto the slope and rise is controlled by gravity. The continental slope and rise system is the area where most of

the sediment derived from continents is eventually deposited. Much of this sediment is transported by turbidity currents through submarine canyons.

Turbidity Currents, Submarine Canyons, and Submarine Fans

Turbidity currents are sediment-water mixtures denser than normal seawater that flow downslope to the deep-sea floor (see Fig. 5.28a). An individual turbidity current flows onto the relatively flat sea floor where it slows and begins depositing sediment; the coarsest particles are deposited first, followed by progressively smaller particles, thus forming graded bedding (see Fig. 5.28b). These deposits accumulate as a series of overlapping **submarine fans,** which constitute a large part of the continental rise (◆ Fig. 8.5). At their seaward margins, these fans grade into the deposits of the deep-ocean basins.

No one has ever observed a turbidity current in progress, so for many years there was considerable debate about their existence. Perhaps the most compelling evidence for their existence is the pattern of trans-Atlantic cable breaks that occurred south of Newfoundland on November 18, 1929 (◆ Fig. 8.6). Initially, it was assumed that an earthquake that occurred on that date had ruptured several trans-Atlantic telephone and telegraph cables. However, while the breaks on the continental shelf near the epicenter occurred when the earthquake struck, cables farther seaward were broken later and in succession. The last cable to break was 720 km from the source of the earthquake, and it did not snap until 13 hours after the first break occurred (Fig. 8.6). In 1949, geologists realized that the earthquake had generated a turbidity current that moved downslope, breaking the cables in succession. The precise time at which each cable broke was known, so it was a simple matter to calculate the velocity of the turbidity current. It apparently moved at about 80 km/hr on the continental slope, but slowed to about 27 km/hr when it reached the continental rise.

Deep, steep-sided **submarine canyons** occur on the continental shelves, but they are best developed on continental slopes (Fig. 8.5). Some submarine canyons can be traced across the shelf to associated streams on land and apparently formed as stream valleys when sea level was lower during the Pleistocene. However, many have no such association, and their origin is not fully understood. It is known that strong currents move through submarine canyons and perhaps play some role in their origin. Furthermore, turbidity currents periodi-cally move through these canyons and are now thought to be the primary agent responsible for their erosion.

▼ TYPES OF CONTINENTAL MARGINS

Two types of continental margins are generally recognized, *passive* and *active*. An **active continental margin** develops at the leading edge of a continental plate where oceanic lithosphere is subducted (◆ Fig. 8.7a). The west coast of South America is a good example. Here, the continental margin is characterized by seismicity, a geologically young mountain range, and andesitic volcanism. Additionally, the continental shelf is narrow, and the continental slope descends directly into the Peru-Chile Trench.

The configuration and geologic activity of the continental margins of eastern North and South America differ considerably from their western margins. These **passive continental margins** are on the trailing edge of a continental plate (◆ Fig. 8.7b). They possess broad continental shelves and a continental slope and rise; vast, flat *abyssal plains* are commonly present adjacent to the rises (Fig. 8.7b). Furthermore, passive continental margins lack the intense seismic and volcanic activity characteristic of active continental margins.

Active and passive continental margins share some features, but in other respects they differ markedly (Fig. 8.7). Active continental margins obviously lack a continental rise because the slope descends directly into an oceanic trench. Just as on passive continental margins, sediment is transported down the slope by turbidity currents, but it simply fills the trench rather than forming a rise. The proximity of the trench to the continent also explains why the continental shelf is so

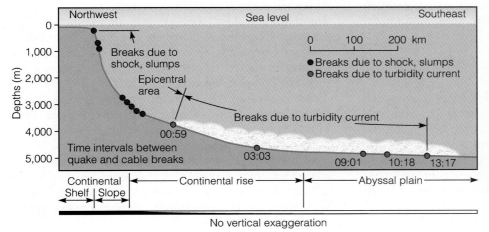

◆ FIGURE 8.6 Submarine cable breaks caused by an earthquake-generated turbidity current south of Newfoundland. This profile of the sea floor shows the locations of the cables and the times at which they were severed. The vertical dimension in this profile is highly exaggerated. The profile labeled "no vertical exaggeration" shows what the sea floor actually looks like in this area.

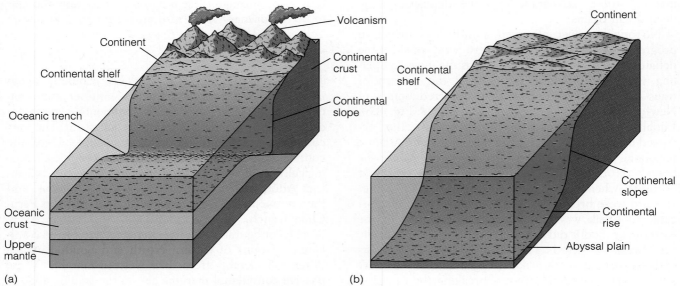

◆ FIGURE 8.7 Diagrammatic views of (a) an active continental margin and (b) a passive continental margin.

narrow. In contrast, the continental shelf of a passive continental margin is much wider because land-derived sedimentary deposits build outward into the ocean.

▼ THE DEEP-OCEAN BASIN

Submersibles have carried scientists to the greatest oceanic depths, so some of the sea floor has been observed directly. Nevertheless, much of the deep-ocean floor has been studied only by echo sounding, seismic profiling, and remote devices that have descended in excess of 11,000 m. Although oceanographers know considerably more about the deep-ocean basins than they did even a few years ago, many questions remain unanswered.

Abyssal Plains

Beyond the continental rises of passive continental margins are **abyssal plains,** flat surfaces covering vast areas of the sea floor. In some areas they are interrupted by peaks rising more than 1 km, but in general they are the flattest, most featureless areas on Earth (◆ Fig. 8.8). The flat topography is a consequence of sediment deposition covering the rugged topography of the oceanic crust.

Abyssal plains are invariably found adjacent to the continental rises, which are composed mostly of overlapping submarine fans that owe their origin to deposition by turbidity currents. Along active continental margins, sediments derived from the shelf and slope are trapped in an oceanic trench, and abyssal plains fail to develop. Accordingly, abyssal plains are common in the Atlantic Ocean basin, but rare in the Pacific Ocean basin (Fig. 8.8).

Oceanic Trenches

Although **oceanic trenches** constitute a small percentage of the sea floor, they are very important, for it is here that lithospheric plates are consumed by subduction (see Chapter 9). Oceanic trenches are long, narrow features restricted to active continental margins; thus, they are common around the margins of the Pacific Ocean basin (Fig. 8.8). Oceanic trenches are the sites of the greatest oceanic depths; a depth of more than 11,000 m has been recorded in the Challenger Deep of the Marianas Trench.

Oceanic trenches show anomalously low heat flow compared to the rest of the oceanic crust; thus, it appears that the crust here is cooler and slightly denser than elsewhere. Furthermore, gravity surveys reveal

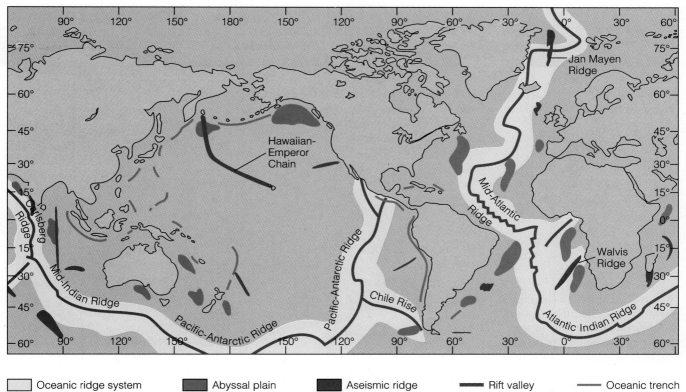

♦ FIGURE 8.8 The distribution of oceanic trenches, abyssal plains, aseismic ridges, and the oceanic ridge system.

Legend: Oceanic ridge system | Abyssal plain | Aseismic ridge | Rift valley | Oceanic trench

that trenches show a huge negative gravity anomaly, indicating that the crust is held down and is not in equilibrium. Seismic activity also occurs at or near trenches. In fact, trenches are characterized by Benioff zones in which earthquake foci become progressively deeper in a landward direction (see Fig. 7.7). Most of the Earth's intermediate and deep earthquakes occur in such zones. Finally, oceanic trenches are associated with volcanoes, either as an arcuate chain of volcanic islands (island arc) or as a chain of volcanoes on land (volcanic arc) adjacent to a trench along the margin of a continent, as in western South America (Fig. 8.7a).

Oceanic Ridges

A feature called the Telegraph Plateau was discovered in the Atlantic Ocean basin during the late nineteenth century when the first submarine cable was laid between North America and Europe. Following the 1925-1927 voyage of the German research vessel *Me-*

teor, scientists proposed that this plateau was actually a continuous feature extending the length of the Atlantic Ocean basin. Subsequent investigations revealed that this proposal was correct, and we now call this feature the Mid-Atlantic Ridge (Fig. 8.8).

The Mid-Atlantic Ridge is more than 2,000 km wide and rises about 2.5 km above the sea floor adjacent to it. It is, in fact, part of a much larger **oceanic ridge** system of submarine mountainous topography at least 65,000 km long (Fig. 8.8). Its length surpasses that of the largest mountain range on land. However, the latter ranges are typically composed of granitic and metamorphic rocks and sedimentary rocks that have been folded and fractured by compressional forces. The oceanic ridges, on the other hand, are composed of volcanic rocks (mostly basalt) and have features produced by tensional forces.

Running along the crests of some ridges is a rift that appears to have opened up in response to tensional forces (♦ Fig. 8.9), although portions of the East Pacific

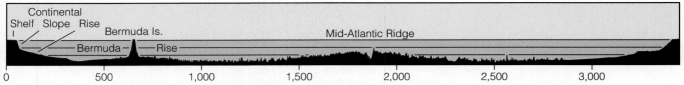

◆ FIGURE 8.9 Profile across the North Atlantic Ocean showing the Mid-Atlantic Ridge with its well-developed central rift.

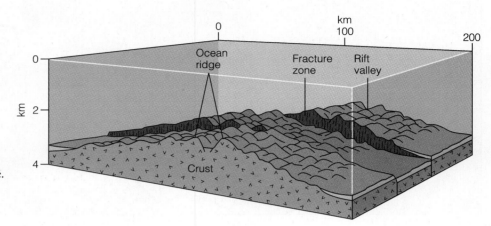

◆ FIGURE 8.10 Diagrammatic view of a fracture offsetting a ridge. Earthquakes occur only in the segments between offset ridge crests.

Rise lack such a feature. These rifts are commonly one to two kilometers deep and several kilometers wide. Such rifts open as sea-floor spreading occurs (discussed in Chapter 9); ridges are characterized by shallow-focus earthquakes, basaltic volcanism, and high heat flow.

Fractures in the Sea Floor

Oceanic ridges are not continuous features winding without interruption around the globe. They abruptly terminate where they are offset along major fractures oriented more or less at right angles to ridge axes (◆ Fig. 8.10). Such large-scale fractures run for hundreds of kilometers, although they are difficult to trace where they are buried beneath sea-floor sediments. Many geologists are convinced that some geologic features on the continents can best be accounted for by the extension of such fractures into continents.

Where these fractures offset oceanic ridges, they are characterized by shallow seismic activity only in the area between the displaced ridge segments (Fig. 8.10). Furthermore, because ridges are higher than the sea floor adjacent to them, the offset segments yield vertical relief on the sea floor. For example, nearly vertical escarpments 3 or 4 km high develop, as illustrated in

Figure 8.10. We will have more to say about such fractures, called transform faults, in Chapter 9.

Seamounts, Guyots, and Aseismic Ridges

As noted previously, the sea floor is not a flat, featureless plain, except for the abyssal plains, and even these are underlain by rugged topography. In fact, a large number of volcanic hills, seamounts, and guyots rise above the sea floor. Such features are present in all ocean basins, but are particularly abundant in the Pacific. All are of volcanic origin and differ from one another mostly in size. **Seamounts** rise more than one kilometer above the sea floor; if they are flat topped, they are called **guyots** rather than seamounts (◆ Fig. 8.11). Guyots are volcanoes that originally extended above sea level. However, as the plate upon which they were situated continued to move, they were carried away from a spreading ridge, and the oceanic crust cooled and descended to greater oceanic depths. Thus, what was once an island slowly sank beneath the sea, where it was eroded by waves, giving it the typical flat-topped appearance (Fig. 8.11).

Many other volcanic features exist on the sea floor; most of these are much smaller than seamounts, but

Seamount

Guyot

◆ FIGURE 8.11 Submarine volcanoes may build up above sea level to form seamounts. As the plate upon which these volcanoes rest moves away from a spreading ridge, the volcanoes sink beneath sea level and become guyots.

probably originated in the same way. These so-called *abyssal hills* average only about 250 m high. They are common on the sea floor and underlie thick sediments on the abyssal plains.

Other common features in the ocean basins are long, linear ridges and broad plateaulike features rising as much as 2 to 3 km above the surrounding sea floor. They are known as **aseismic ridges** because they lack seismic activity. A few of these ridges are thought to be small fragments separated from continents during rifting. Such fragments, referred to as *microcontinents,* are represented by such features as the Jan Mayen Ridge in the North Atlantic (Fig. 8.8).

Most aseismic ridges form as a linear succession of hot spot volcanoes. These may develop at or near an oceanic ridge, but each volcano so formed is carried laterally with the plate upon which it originated. The net result of such activity is a sequence of seamounts/guyots extending from an oceanic ridge (Fig. 8.11); the Walvis Ridge in the South Atlantic is a good example (Fig. 8.8). Aseismic ridges also form over hot spots unrelated to ridges. The Hawaiian-Emperor chain in the Pacific formed in such a manner (Fig. 8.8).

▼ DEEP-SEA SEDIMENTATION

Deep-sea sediments consist mostly of fine-grained windblown dust and volcanic ash from the continents and oceanic islands and the shells of microscopic organisms that live in the near-surface waters of the oceans. Other sources of sediment include cosmic dust

and deposits resulting from chemical reactions in seawater. The manganese nodules that are fairly common in all the ocean basins are a good example of the latter (◆ Fig. 8.12). These nodules are composed mostly of manganese and iron oxides, but also contain copper, nickel, and cobalt. Such nodules may be an important source of some metals in the future; the United States, which imports most of its manganese and cobalt, is particularly interested in this potential resource. The contribution of cosmic dust to deep-sea sediment is negligible.

The bulk of the sediments on the deep-sea floor are *pelagic,* meaning that they settled from suspension far from land. Two categories of pelagic sediment are

◆ FIGURE 8.12 Manganese nodules on the sea floor south of Australia.

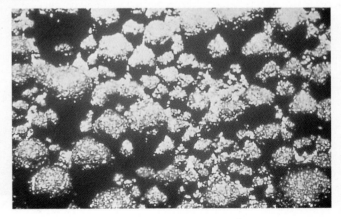

recognized: pelagic clay and ooze. **Pelagic clay** covers most of the deeper parts of the ocean basins. It is generally brown or reddish and is composed of clay-sized particles derived from the continents and oceanic islands. **Ooze,** on the other hand, is composed mostly of shells of microscopic marine animals and plants. It is characterized as *calcareous ooze* if it contains mostly calcium carbonate ($CaCO_3$) skeletons of tiny marine organisms such as foraminifera. *Siliceous ooze* is composed of the silica (SiO_2) skeletons of such single-celled organisms as radiolarians (animals) and diatoms (plants).

⩒ REEFS

Reefs are moundlike, wave-resistant structures composed of the skeletons of organisms (◆ Fig. 8.13). Commonly they are called coral reefs, but many other organisms in addition to corals make up reefs. A reef consists of a solid framework of skeletons of corals, clams, and such encrusting organisms as algae and sponges. Reefs grow to a depth of about 45 or 50 m and are restricted to shallow tropical seas where the water is clear, and the temperature does not fall below about 20°C.

Three types of reefs are recognized: fringing, barrier, and atoll (◆ Fig. 8.14). *Fringing reefs* are solidly attached to the margins of an island or continent. They have a rough, tablelike surface, are as much as one kilometer wide, and, on their seaward side, slope steeply down to the sea floor. *Barrier reefs* are similar to fringing reefs, except that they are separated from the mainland by a lagoon. Probably the best-known barrier reef in the world is the 2,000-km-long Great Barrier Reef of Australia.

An *atoll* is a circular to oval reef surrounding a lagoon. Such reefs form around volcanic islands that subside below sea level as the plate upon which they rest is carried progressively farther from an oceanic ridge (Fig. 8.14). As subsidence occurs, the reef organisms construct the reef upward so that the living part of the reef remains in shallow water. However, the island eventually subsides below sea level, leaving a circular lagoon surrounded by a more-or-less continuous reef. Such reefs are particularly common in the western Pacific Ocean basin (◆ Fig. 8.15). Many of these began as fringing reefs, but as subsidence occurred, they evolved first to barrier reefs and finally to atolls.

◆ FIGURE 8.13 Reefs such as this one fringing an island in the Pacific are wave-resistant structures composed of the skeletons of organisms.

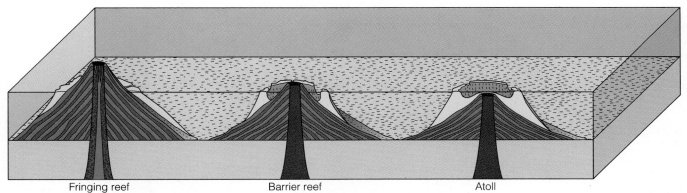

◆ FIGURE 8.14 Three-stage development of an atoll. In the first stage, a fringing reef forms, but as the island sinks, a barrier reef becomes separated from the island by a lagoon. As the island disappears beneath the sea, the barrier reef continues to grow upward, thus forming an atoll. An oceanic island carried into deeper water by plate movement can account for this sequence.

◆ FIGURE 8.15 View of an atoll in the Pacific Ocean.

⩗ RESOURCES FROM THE SEA

Seawater contains many elements in solution, some of which are extracted for various industrial and domestic uses. For example, in many places sodium chloride (table salt) is produced by the evaporation of seawater, and a large proportion of the world's magnesium is produced from seawater. Numerous other elements and compounds can be extracted from seawater, but for many, such as gold, the cost is prohibitive.

In addition to substances in seawater, deposits on the sea floor or within sea-floor sediments are becoming increasingly important. Many of these potential re-sources lie well beyond the margins of the continents, so their ownership is a political and legal problem that has not yet been resolved. Most nations bordering the ocean claim those resources occurring within their adjacent continental margin. The United States, for example, by a presidential proclamation issued on March 10, 1983, claims sovereign rights over an area designated as the **Exclusive Economic Zone (EEZ).** The EEZ extends seaward 200 nautical miles (371 km) from the coast, giving the United States jurisdiction over an area about 1.7 times larger than its land area (◆ Fig. 8.16).* Also included within the EEZ are the areas adjacent to U.S. territories, such as Guam, American Samoa, Wake Island, and Puerto Rico. In short, the United States claims a huge area of the sea floor and any resources on or beneath it.

Numerous resources occur within the EEZ, some of which have been exploited for many years. For example, sand and gravel for construction are mined from the continental shelf in several areas. About 17% of U.S. oil and natural gas production comes from wells on the continental shelf. Some 30 sedimentary basins occur within the EEZ, several of which are known to contain hydrocarbons whereas others are areas of potential hydrocarbon production. Ancient shelf deposits in the Persian Gulf region contain the world's largest reserves of oil.

. . . .

*A number of other nations also claim sovereign rights to resources within 200 nautical miles of their coasts.

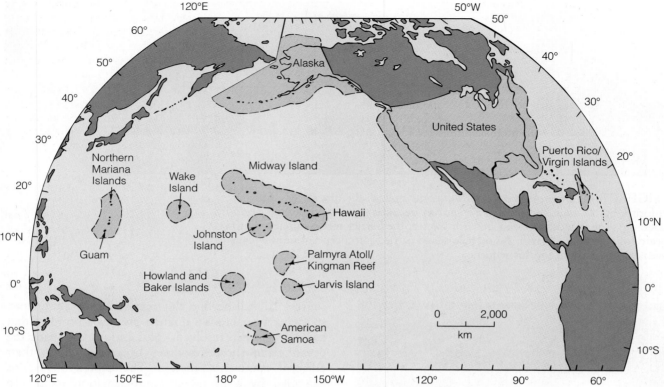

◆ FIGURE 8.16 The Exclusive Economic Zone (EEZ) includes a vast area adjacent to the United States and its possessions.

Other resources of interest include the massive sulfide deposits that form by submarine hydrothermal activity at spreading ridges (see the Prologue). Such deposits containing iron, copper, zinc, and other metals have been identified within the EEZ at the Gorda Ridge off the coasts of California and Oregon; similar deposits occur at the Juan de Fuca Ridge within the Canadian EEZ.

Other potential resources include the manganese nodules discussed previously, and metalliferous oxide crusts found on seamounts. Manganese nodules contain manganese, cobalt, nickel, and copper; the United States is heavily dependent on imports of the first three of these elements. Within the EEZ, manganese nodules occur near Johnston Island in the Pacific Ocean and on the Blake Plateau off the east coast of South Carolina and Georgia. In addition, seamounts and seamount chains within the EEZ in the Pacific are known to have metalliferous oxide crusts several centimeters thick from which cobalt and manganese could be mined.

Chapter Summary

1. Continental margins separate the continents above sea level from the deep ocean basin. They consist of a continental shelf, continental slope, and in some cases a continental rise.

2. Continental shelves slope gently in a seaward direction and vary from a few tens of meters to more than 1,000 km wide.

3. The continental slope begins at an average depth of 135 m where the inclination of the sea floor increases rather abruptly from less than 1° to several degrees.
4. Submarine canyons are characteristic of the continental slope, but some of them extend well up onto the shelf and lie offshore from large streams. Some submarine canyons were probably eroded by turbidity currents.
5. Turbidity currents commonly move through submarine canyons and deposit an overlapping series of submarine fans that constitutes a large part of the continental rise.
6. Active continental margins are characterized by a narrow shelf and a slope that descends directly into an oceanic trench with no rise present. Such margins are also characterized by seismic activity and volcanism.
7. Passive continental margins lack volcanism and exhibit little seismic activity. The continental shelf along such margins is broad, and the slope merges with a continental rise. Abyssal plains are commonly present seaward beyond the rise.
8. Oceanic trenches are long, narrow features where oceanic crust is subducted. They are characterized by low heat flow, negative gravity anomalies, and the greatest oceanic depths.
9. Oceanic ridges consisting of mountainous topography are composed of volcanic rocks, and many ridges possess a large rift caused by tensional forces. Basaltic volcanism and shallow-focus earthquakes occur at ridges. Oceanic ridges nearly encircle the globe, but they are interrupted and offset by large fractures in the sea floor.
10. Other important features on the sea floor include seamounts that rise more than a kilometer high and guyots, which are flat-topped seamounts. Many aseismic ridges are oriented more-or-less perpendicular to oceanic ridges and consist of a chain of seamounts and/or guyots.
11. Deep-sea sediments consist mostly of fine-grained particles derived from continents and oceanic islands and the microscopic shells of organisms. The primary types of deep-sea sediments are pelagic clay and ooze.
12. Reefs are wave-resistant structures composed of animal skeletons, particularly corals. Three types of reefs are recognized: fringing, barrier, and atoll.
13. The United States has claimed rights to all resources occurring within 200 nautical miles (371 km) of its shorelines. Numerous resources including various metals occur within this Exclusive Economic Zone.

Important Terms

abyssal plain
active continental margin
aseismic ridge
continental margin
continental rise
continental shelf
continental slope

Exclusive Economic Zone
guyot
oceanic ridge
oceanic trench
ooze
passive continental margin

pelagic clay
reef
seamount
submarine canyon
submarine fan
turbidity current

Review Questions

1. Much of the continental rise is composed of:
   a. ___ calcareous ooze; b. ___ submarine fans;
   c. ___ fringing reefs; d. ___ sheeted dikes; e. ___ ophiolite.
2. The greatest oceanic depths occur at:
   a. ___ aseismic ridges; b. ___ guyots; c. ___ the shelf-slope break; d. ___ oceanic trenches; e. ___ passive continental margins.
3. Abyssal plains are most common:
   a. ___ around the margins of the Atlantic; b. ___ adjacent to the East Pacific Rise; c. ___ along the west coast of South America; d. ___ in the rift valley of the Mid-Atlantic Ridge; e. ___ on continental shelves.
4. A circular reef enclosing a lagoon is a(n):
   a. ___ barrier reef; b. ___ seamount; c. ___ aseismic ridge; d. ___ guyot; e. ___ atoll.
5. Submarine canyons are most characteristic of the:
   a. ___ continental shelves; b. ___ abyssal plains;
   c. ___ continental slopes; d. ___ rift valleys; e. ___ fractures in the sea floor.
6. Continental shelves:
   a. ___ are composed of pelagic sediments; b. ___ lie between continental slopes and rises; c. ___ descend to an average depth of 1,500 m; d. ___ slope gently from the shoreline to the shelf-slope break; e. ___ are widest along active continental margins.
7. The flattest, most featureless areas on Earth are the:
   a. ___ oceanic ridges; b. ___ abyssal plains; c. ___ continental slopes; d. ___ aseismic ridges; e. ___ continental margins.

8. Sediment that settles from suspension far from land is:
   a. ___ abyssal; b. ___ pelagic; c. ___ volcanic; d. ___ generally coarse grained; e. ___ characterized by graded bedding.

9. Which of the following statements is correct?
   a. ___ most of the continental margins around the Atlantic are passive; b. ___ oceanic ridges are composed largely of deformed sedimentary rocks; c. ___ the deposits of turbidity currents consist of calcareous ooze; d. ___ most of the Earth's intermediate and deep earthquakes occur at or near oceanic ridges; e. ___ oceanic crust is thicker than continental crust.

10. Massive sulfide deposits form:
    a. ___ on passive continental margins; b. ___ as accumulations of microscopic shells on the sea floor; c. ___ by precipitation of minerals near hydrothermal vents; d. ___ from sediments derived from continents; e. ___ in oceanic trenches.

11. The most useful method of determining the structure of the oceanic crust beneath continental shelf sediments is:
    a. ___ echo sounding; b. ___ observations from submersible research vessels; c. ___ dredging; d. ___ seismic profiling; e. ___ underwater photography.

12. Which of the following is *not* characteristic of an active continental margin?
    a. ___ volcanism; b. ___ earthquakes; c. ___ oceanic trench; d. ___ volcanic arc; e. ___ continental rise.

13. Graded bedding is a characteristic of:
    a. ___ continental shelves; b. ___ turbidity current deposits; c. ___ pelagic clay; d. ___ siliceous ooze; e. ___ manganese nodules.

14. How do sulfide mineral deposits form on the sea floor?

15. What is an echo sounder, and how is it used to study the sea floor?

16. What are the characteristics of a passive continental margin? How does such a continental margin originate?

17. Describe the continental rise, and explain why a rise occurs at some continental margins and not at others.

18. Summarize the evidence indicating that turbidity currents transport sediment from the continental shelf onto the slope and rise.

19. Where do abyssal plains most commonly develop? Describe their composition.

20. What is the significance of oceanic trenches, and where are they found?

21. How do mid-oceanic ridges differ from mountain ranges on land?

22. Describe how an aseismic ridge forms.

23. What are the sources of deep-sea sediments?

24. Describe the sequence of events leading to the origin of an atoll.

25. What is the Exclusive Economic Zone? What types of metal deposits occur within it?

Additional Readings

Anderson, R. N. 1986. *Marine geology.* New York: Wiley.

Davis, R. A. 1987. *Oceanography: An introduction to the marine environment.* Dubuque, Iowa: William C. Brown.

Edmond, J. M., and K. Von Damm. 1983. Hot springs on the ocean floor. *Scientific American* 248, no. 4:78–93.

Gass, I. G. 1982. Ophiolites. *Scientific American* 247, no. 2:122–31.

Kennett, J. P. 1982. *Marine geology.* Englewood Cliffs, N.J.: Prentice-Hall.

Mark, K. 1976. Coral reefs, seamounts, and guyots. *Sea Frontiers* 22, no. 3:143–49.

Pinet, P. 1992. *Oceanography: An introduction to the planet oceanus.* St. Paul, Minn.: West Publishing.

Rona, P. A. 1986. Mineral deposits from sea-floor hot springs. *Scientific American* 254, no. 1:84–93.

Rona, P. A. 1992. Deep-sea geysers of the Atlantic. *National Geographic,* 182, no. 4: 105–109.

Ross, D. A. 1988. *Introduction to oceanography.* Englewood Cliffs, N.J.: Prentice-Hall.

Thurman, H. V. 1988. *Introductory oceanography.* 5th ed. Columbus, Ohio: Merrill.

CHAPTER
9

Vertical view of the Himalayas, the youngest and highest mountain system in the world. The Himalayas began forming when India collided with Asia 40 to 50 million years ago.

Plate Tectonics:
A Unifying Theory

Prologue

Two tragic events that occurred during 1985 serve to remind us of the dangers of living near a convergent plate margin. On September 19, a magnitude 8.1 earthquake killed more than 9,000 people in Mexico City. Two months later and 3,200 km to the south, a minor eruption of Colombia's Nevado del Ruiz volcano partially melted its summit glacial ice, causing a mudflow that engulfed Armero and several other villages and killed more than 23,000 people. These two tragedies resulted in more than 32,000 deaths, tens of thousands of injuries, and billions of dollars in property damage.

Both of these events occurred along the eastern portion of the Ring of Fire, a chain of intense seismic and volcanic activity that encircles the Pacific Ocean basin (◆ Fig. 9.1). Some of the world's greatest disasters occur along this ring because of volcanism and earthquakes generated by plate convergence. Although earthquakes and volcanic eruptions are very different geologic phenomena, both are related to the activities occurring at convergent plate margins. The Mexico City earthquake resulted from subduction of the Cocos plate at the Middle America Trench (Fig. 9.1). Sudden movement of the Cocos plate beneath Central America generated seismic waves that traveled outward in all directions. The violent shaking experienced in Mexico City, 350 km away, and elsewhere was caused by these seismic waves.

The strata underlying Mexico City consist of unconsolidated sediment deposited in a large ancient lake. Such sediment amplifies the shaking during earthquakes with the unfortunate consequence that buildings constructed there are commonly more heavily damaged than those built on solid bedrock (see Perspective 7.1, Fig. 3).

Less than two months after the Mexico City earthquake, Colombia experienced its greatest recorded natural disaster. Nevado del Ruiz is one of several active volcanoes resulting from the rise of magma generated where the Nazca plate is subducted beneath South America (Fig. 9.1). A minor eruption of Nevado del Ruiz partially melted the glacial ice on the mountain; the meltwater rushed down the valleys, mixed with the sediment, and turned it into a deadly viscous mudflow.

The city of Armero, Colombia, lies in the valley of the Lagunilla River, one of several river valleys inundated by mudflows. Twenty thousand of the city's 23,000 inhabitants died, and most of the city was destroyed (◆ Fig. 9.2). Another 3,000 people were killed in nearby valleys. A geologic hazard assessment study completed one month before the eruption showed that Armero was in a high-hazard mudflow area!

These two examples vividly illustrate some of the dangers of living in proximity to a convergent plate boundary. Subduction of one plate beneath another repeatedly triggers large earthquakes, the effects of which are frequently felt far from their epicenters. Since 1900, earthquakes have killed more than 112,000 people in Central and South America alone. While volcanic eruptions in this region have not caused nearly as many casualties as earthquakes, they have, nevertheless, caused tremendous property damage and have the potential for triggering devastating events such as the 1985 Colombian mudflow.

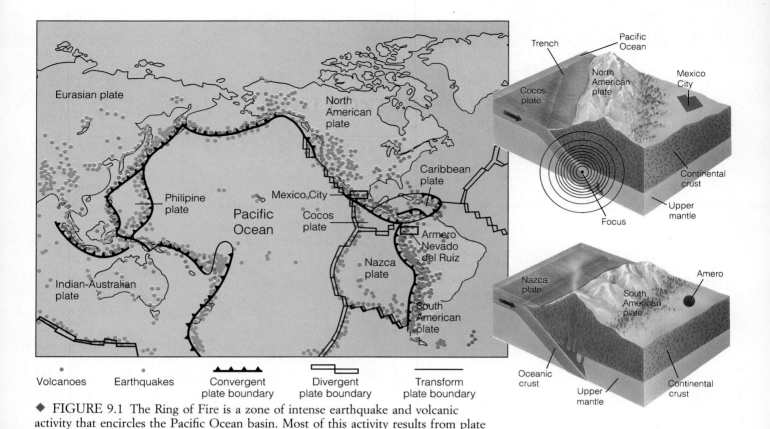

Volcanoes   Earthquakes   Convergent plate boundary   Divergent plate boundary   Transform plate boundary

◆ FIGURE 9.1 The Ring of Fire is a zone of intense earthquake and volcanic activity that encircles the Pacific Ocean basin. Most of this activity results from plate convergence as illustrated by the two insets.

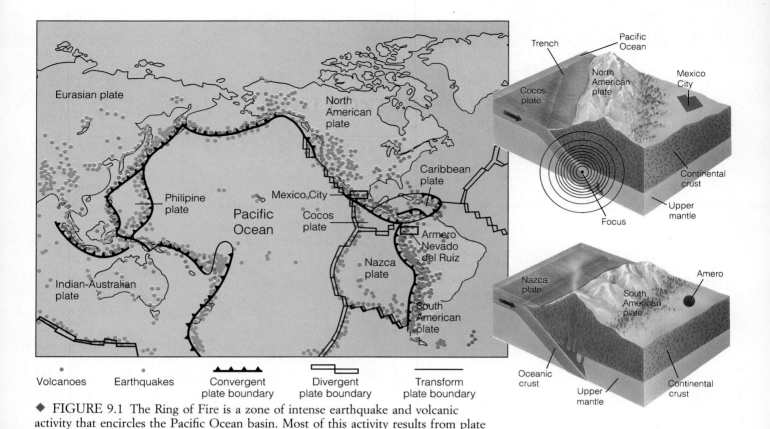

◆ FIGURE 9.2 The 1985 eruption of Nevado del Ruiz in Colombia melted some of its glacial ice. The meltwater mixed with sediments and formed a huge mudflow that destroyed the city of Armero and killed 20,000 of its inhabitants.

⯆ INTRODUCTION

The recognition that the Earth's geography has changed continuously through time has led to a revolution in the geological sciences, forcing geologists to greatly modify the way they view the Earth. Although many people have only a vague notion of what plate tectonic theory is, plate tectonics has a profound effect on all of our lives. It is now realized that most earthquakes and volcanic eruptions occur near plate margins and are not merely random occurrences. Furthermore, the formation and distribution of many important natural resources, such as metallic ores, are related to plate boundaries, and geologists are now incorporating plate tectonic theory into their prospecting efforts.

The interaction of plates determines the location of continents, ocean basins, and mountain systems, which in turn affects the atmospheric and oceanic circulation patterns that ultimately determine global climates. Furthermore, plate movements have profoundly influenced the geographic distribution, evolution, and extinction of plants and animals.

Plate tectonic theory is now almost universally accepted among geologists, and its application has led to a greater understanding of how the Earth has evolved and continues to do so. This powerful, unifying theory accounts for many apparently unrelated geologic events, allowing geologists to view such phenomena as part of a continuing story rather than as a series of isolated incidents.

We will first review the various hypotheses that preceded plate tectonic theory, examining the evidence that led some people to accept the idea of continental movement and others to reject it. Because plate tectonic theory has evolved from numerous scientific inquiries and observations, only the more important ones will be covered.

⩔ EARLY IDEAS ABOUT CONTINENTAL DRIFT

The idea that the Earth's geography was different during the past is not new. During the fifteenth century, Leonardo da Vinci observed that "above the plains of Italy where flocks of birds are flying today fishes were once moving in large schools." In 1620, Sir Francis Bacon commented on the similarity of the shorelines of western Africa and eastern South America but did not make the connection that the Old and New Worlds might once have been sutured together. Alexander von Humboldt made the same observation in 1801, although he attributed these similarities to erosion rather than the splitting apart of a larger continent.

One of the earliest specific references to continental drift is in Antonio Snider-Pellegrini's 1858 book *Creation and Its Mysteries Revealed.* He suggested that all of the continents were linked together during the Pennsylvanian Period and later split apart. He based his conclusions on the similarities between plant fossils in the Pennsylvanian-aged coal beds of Europe and North America. He envisioned that continental separation was a consequence of the biblical deluge.

During the late nineteenth century, the Austrian geologist Edward Suess noted the similarities between the Late Paleozoic plant fossils of India, Australia, Africa, Antarctica, and South America as well as evidence of glaciation in the rock sequences of these southern continents. In 1885 he proposed the name Gondwanaland (or **Gondwana** as we will use here) for a supercontinent composed of these southern land-masses. Gondwana is a province in east-central India

where, along with evidence of extensive glaciation, abundant fossils of the ***Glossopteris* flora** occur (◆ Fig. 9.3). Suess believed the distribution of plant fossils and glacial deposits was a consequence of extensive land bridges that once connected the continents and later sank beneath the ocean.

In 1910 the American geologist Frank B. Taylor published a pamphlet in which he presented his own theory of continental drift. In it he explained the formation of mountain ranges as a result of the lateral movement of continents. He also envisioned the present-day continents as parts of larger polar continents that

◆ FIGURE 9.3 Representative members of the *Glossopteris* flora. Fossils of these plants are found on all five of the Gondwana continents. *Glossopteris* leaves from (*a*) the Upper Permian Dunedoo Formation and (*b*) the Upper Permian Illawarra Coal Measures, Australia. (Photos courtesy of Patricia G. Gensel, University of North Carolina.)

(a)

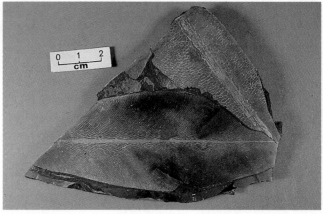

(b)

eventually broke apart and migrated toward the equator because of a slowing of the Earth's rotation due to gigantic tidal forces. According to Taylor, these tidal forces were generated when the Earth captured the Moon about 100 million years ago.

Although we now know that Taylor's mechanism is incorrect, one of his most significant contributions was his suggestion that the Mid-Atlantic Ridge, discovered by the 1872–1876 British H.M.S. *Challenger* expeditions, might mark the site along which an ancient continent broke apart to form the present-day Atlantic Ocean.

▼ ALFRED WEGENER AND THE CONTINENTAL DRIFT HYPOTHESIS

Alfred Wegener, a German meteorologist (◆ Fig. 9.4), is generally credited with developing the hypothesis of **continental drift.** In his monumental book, *The Origin of Continents and Oceans,* first published in 1915, Wegener proposed that all of the landmasses were originally united into a single supercontinent that he named **Pangaea,** from the Greek meaning "all land." Wegener portrayed his grand concept of continental movement in a series of maps showing the breakup of Pangaea and the movement of the various continents to their present-day locations. Wegener had amassed a tremendous amount of geological, paleontological, and

◆ FIGURE 9.4 Alfred Wegener, a German meteorologist, proposed the continental drift hypothesis in 1912 based on a tremendous amount of geological, paleontological, and climatological evidence. He is shown here waiting out the Arctic winter in an expedition hut.

climatological evidence in support of continental drift, but the initial reaction of scientists to his then-heretical ideas can best be described as mixed.

Opposition to Wegener's ideas became particularly widespread in North America after 1928 when the American Association of Petroleum Geologists held an international symposium to review the hypothesis of continental drift. After each side had presented its arguments, the opponents of continental drift were clearly in the majority, even though the evidence in support of continental drift, most of which came from the Southern Hemisphere, was impressive and difficult to refute. One problem with the hypothesis, however, was its lack of a mechanism to explain how continents, composed of granitic rocks, could seemingly move through the denser basaltic oceanic crust.

Nevertheless, the eminent South African geologist Alexander du Toit further developed Wegener's arguments and gathered more geological and paleontological evidence in support of continental drift. In 1937, du Toit published *Our Wandering Continents,* in which he contrasted the glacial deposits of Gondwana with coal deposits of the same age found in the continents of the Northern Hemisphere. To resolve this apparent climatological paradox, du Toit moved the Gondwana continents to the South Pole and brought the northern continents together such that the coal deposits were located at the equator. He named this northern landmass **Laurasia.** It consisted of present-day North America, Greenland, Europe, and Asia (except for India).

In spite of what seemed to be overwhelming evidence, most geologists still refused to accept the idea that continents moved. It was not until the 1960s when oceanographic research provided convincing evidence that the continents had once been joined together and subsequently separated that the hypothesis of continental drift finally became widely accepted.

▼ THE EVIDENCE FOR CONTINENTAL DRIFT

The evidence used by Wegener, du Toit, and others to support the hypothesis of continental drift includes the fit of the shorelines of continents; the appearance of the same rock sequences and mountain ranges of the same age on continents now widely separated; the matching of glacial deposits and paleoclimatic zones; and the similarities of many extinct plant and animal groups whose fossil remains are found today on widely separated continents.

Continental Fit

Wegener, like some before him, was impressed by the close resemblance between the coastlines of continents on opposite sides of the Atlantic Ocean, particularly between South America and Africa. He cited these similarities as partial evidence that the continents were at one time joined together as a supercontinent that subsequently split apart. As his critics pointed out, however, the configuration of coastlines results from erosional and depositional processes and therefore is continually being modified. Thus, even if the continents had separated during the Mesozoic Era, as Wegener proposed, it is not likely that the coastlines would fit exactly.

A more realistic approach is to fit the continents together along the continental slope where erosion would be minimal. Recall from Chapter 8 that the true margin of a continent—that is, where continental crust changes to oceanic crust—is beneath the continental slope. In 1965 Sir Edward Bullard, an English geophysicist, and two associates showed that the best fit between the continents occurs along the continental slope at a depth of about 2,000 m (◆ Fig. 9.5). Since then, other reconstructions using the latest ocean basin

◆ FIGURE 9.5 The best fit between continents occurs along the continental slope at a depth of about 2,000 m.

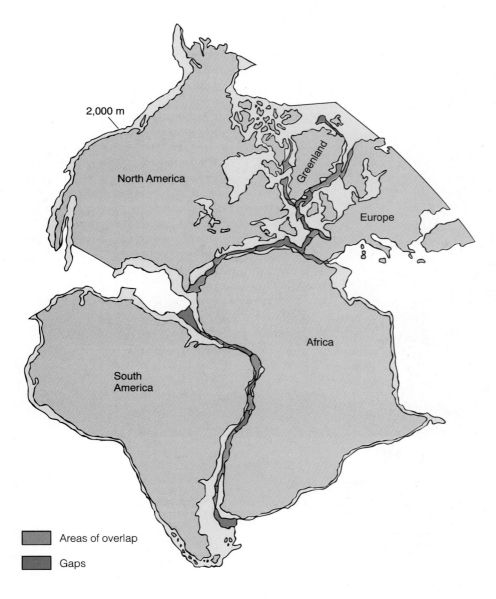

data have confirmed the close fit between continents when they are reassembled to form Pangaea.

Similarity of Rock Sequences and Mountain Ranges

If the continents were at one time joined together, then the rocks and mountain ranges of the same age in adjoining locations on the opposite continents should closely match. Such is the case for the Gondwana continents (◆ Fig. 9.6). Marine, nonmarine, and glacial rock sequences of Pennsylvanian to Jurassic age are almost identical for all five Gondwana continents, strongly indicating that they were at one time joined together.

The trends of several major mountain ranges also support the hypothesis of continental drift. These mountain ranges seemingly end at the coastline of one continent only to apparently continue on another continent across the ocean. For example, the folded Appalachian Mountains of North America trend northeastward through the eastern United States and Canada and terminate abruptly at the Newfoundland coastline (◆ Fig. 9.7a). Mountain ranges of the same age and deformational style occur in eastern Greenland, Ireland, Great Britain, and Norway. Even though these mountain ranges are currently separated by the Atlantic Ocean, they form an essentially continuous mountain range when the continents are positioned next to each other (◆ Fig. 9.7b).

◆ FIGURE 9.6 Marine, nonmarine, and glacial rock sequences of Pennsylvanian to Jurassic age are nearly the same for all Gondwana continents. Such close similarity strongly suggests that they were at one time joined together. The range indicated by G is that of the *Glossopteris* flora.

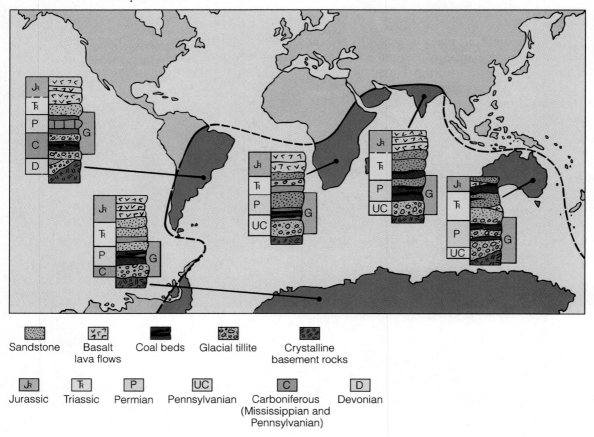

Sandstone   Basalt lava flows   Coal beds   Glacial tillite   Crystalline basement rocks

Jᴿ Jurassic   Tᴿ Triassic   P Permian   UC Pennsylvanian   C Carboniferous (Mississippian and Pennsylvanian)   D Devonian

Glacial Evidence

Massive glaciers covered large continental areas of the Southern Hemisphere during the Late Paleozoic Era. Evidence for this glaciation includes layers of till (sediments deposited by glaciers) and striations (scratch marks) in the bedrock beneath the till. Fossils and sedimentary rocks of the same age from the Northern Hemisphere, however, give no indication of glaciation. Fossil plants found in coals indicate that the Northern Hemisphere had a tropical climate during the time that the Southern Hemisphere was glaciated.

All of the Gondwana continents except Antarctica are currently located near the equator in subtropical to tropical climates. Mapping of glacial striations in bedrock in Australia, India, and South America indicates that the glaciers moved from the areas of the present-day oceans onto land (◆ Fig. 9.8a). However, this would be impossible because large continental glaciers (such as occurred on the Gondwana continents during the Late Paleozoic Era) flow outward from their central area of accumulation toward the sea.

If the continents did not move during the past, one would have to explain how glaciers moved from the oceans onto land and how large-scale continental glaciers formed near the equator. But if the continents are reassembled as a single landmass with South Africa located at the south pole, the direction of movement of Late Paleozoic continental glaciers makes sense. Furthermore, this geographic arrangement places the northern continents nearer the tropics, which is consistent with the fossil and climatological evidence from Laurasia (◆ Fig. 9.8b).

Fossil Evidence

Some of the most compelling evidence for continental drift comes from the fossil record. Fossils of the *Glossopteris* flora are found in equivalent Pennsylvanian- and Permian-aged coal deposits on all five Gondwana continents. The *Glossopteris* flora is characterized by the seed fern *Glossopteris* (Fig. 9.3) as well as by many other distinctive and easily identifiable plants. Pollen and spores of plants can be dispersed over great distances by wind, but *Glossopteris*-type plants produced seeds that are too large to have been carried by winds. Even if the seeds had floated across the ocean, they probably would not have remained viable for any length of time in salt water.

The present-day climates of South America, Africa, India, Australia, and Antarctica range from tropical to

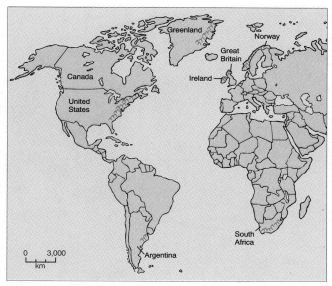

(a)

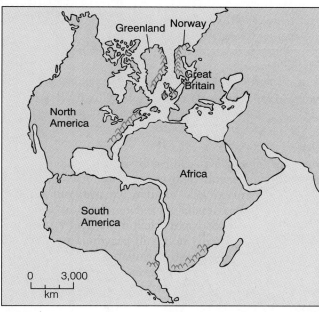

(b)

◆ FIGURE 9.7 (*a*) Various mountain ranges of the same age and style of deformation are currently widely separated by oceans. (*b*) When the continents are brought together, however, a single continuous mountain range is formed. Such evidence indicates the continents were at one time joined together and were subsequently separated.

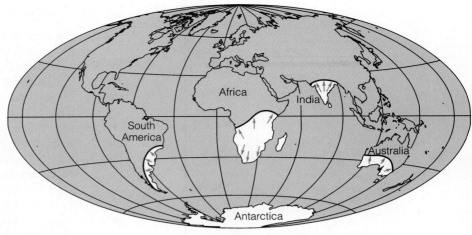

(a)

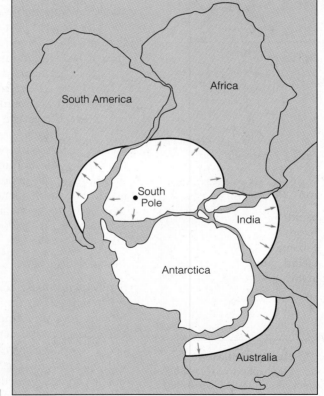

Glaciated area
Arrows indicate the direction of glacial movement based on striations preserved in bedrock.

◆ FIGURE 9.8 (a) If the continents did not move in the past, then Late Paleozoic glacial striations preserved in bedrock in Australia, India, and South America indicate that glacial movement for each continent was from the oceans onto land within a subtropical to tropical climate. Such an occurrence is highly unlikely. (b) If the continents are brought together, such that South Africa is located at the South Pole, then the glacial movement indicated by the striations makes sense. In this situation, the glacier, located in a polar climate, moved radially outward from a thick central area toward its periphery.

(b)

polar and are much too diverse to support the type of plants that compose the *Glossopteris* flora. Wegener reasoned therefore that these continents must once have been joined such that these widely separated localities were all in the same latitudinal climatic belt (◆ Fig. 9.9).

The fossil remains of animals also provide strong evidence for continental drift. One of the best examples is *Mesosaurus,* a freshwater reptile whose fossils are found in Permian-aged rocks in certain regions of Brazil and South Africa and nowhere else in the world

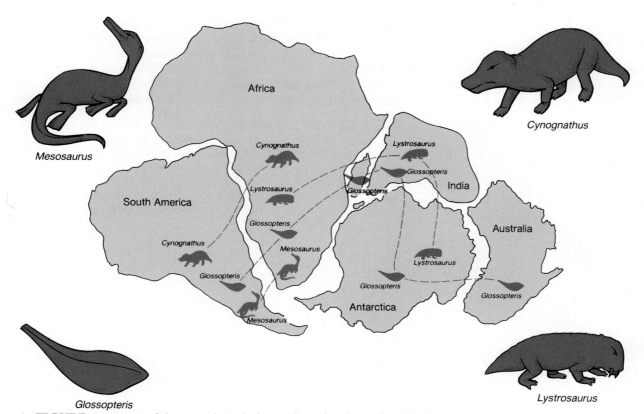

◆ FIGURE 9.9 Some of the animals and plants whose fossils are found today on the widely separated continents of South America, Africa, India, Australia, and Antarctica. These continents were joined together during the Late Paleozoic to form Gondwana, the southern landmass of Pangaea. *Glossopteris* and similar plants are found in Pennsylvanian- and Permian-aged deposits on all five continents. *Mesosaurus* is a freshwater reptile whose fossils are found in Permian-aged rocks in Brazil and South Africa. *Cynognathus* and *Lystrosaurus* are land reptiles who lived during the Early Triassic Period. Fossils of *Cynognathus* are found in South America and Africa, while fossils of *Lystrosaurus* have been recovered from Africa, India, and Antarctica.

(Fig. 9.9). Because the physiology of freshwater and marine animals is completely different, it is hard to imagine how a freshwater reptile could have swum across the Atlantic Ocean and found a freshwater environment nearly identical to its former habitat. Moreover, if *Mesosaurus* could have swum across the ocean, its fossil remains should be widely dispersed. It is more logical to assume that *Mesosaurus* lived in lakes in what are now adjacent areas of South America and Africa, but were then united into a single continent.

Lystrosaurus and *Cynognathus* are both land-dwelling reptiles that lived during the Triassic Period; their fossils are found only on the present-day conti-

nental fragments of Gondwana (Fig. 9.9). Because they are both land animals, they certainly could not have swum across the oceans currently separating the Gondwana continents. Therefore, the continents must once have been connected.

The evidence favoring continental drift seemed overwhelming to Wegener and his supporters, yet the lack of a suitable mechanism to explain continental movement prevented its widespread acceptance. Not until new evidence from studies of the Earth's magnetic field and oceanographic research showed that the ocean basins were geologically young features did renewed interest in continental drift occur.

PALEOMAGNETISM AND POLAR WANDERING

Interest in continental drift revived in the 1950s as a result of new evidence from paleomagnetic studies of the Earth. **Paleomagnetism** is the remanent magnetism in ancient rocks recording the direction of the Earth's magnetic poles at the time of the rock's formation. The Earth can be thought of as a giant dipole magnet in which the magnetic poles correspond closely to the location of the geographic poles (◆ Fig. 9.10). Such an arrangement means that the strength of the magnetic field is not con-

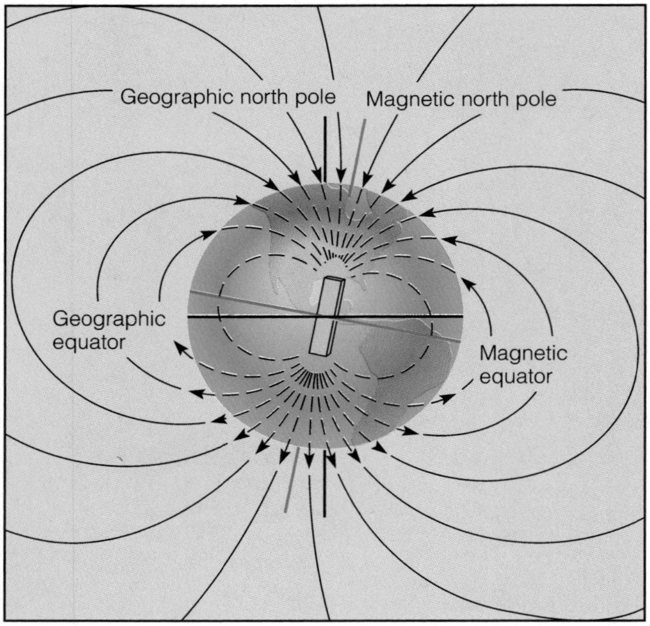

(a)

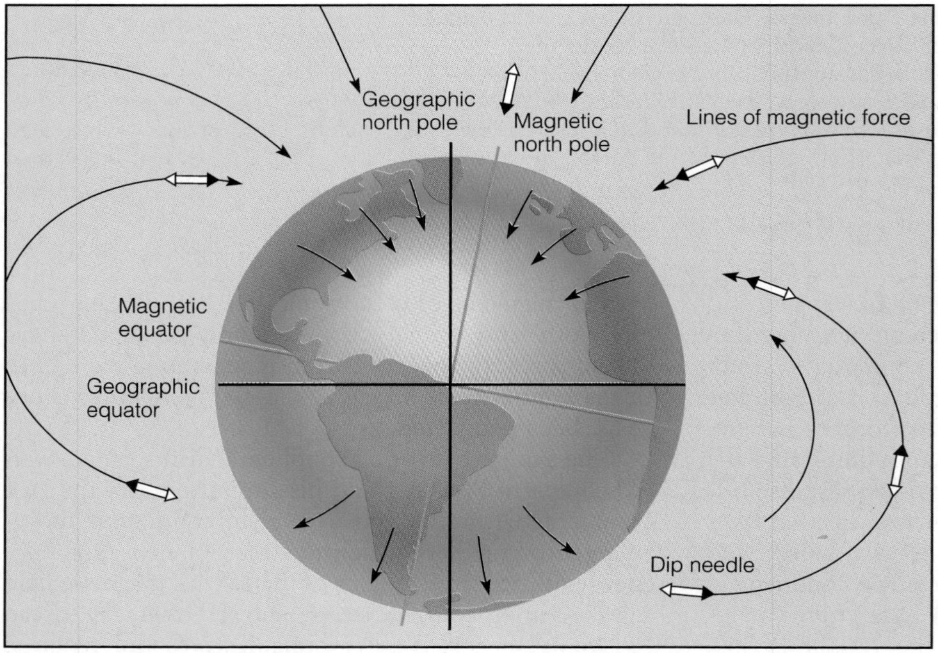

(b)

◆ FIGURE 9.10 (*a*) The magnetic field of the Earth has lines of force just like those of a bar magnet. (*b*) The strength of the magnetic field changes uniformly from the magnetic equator to the magnetic poles. This change in strength causes a dip needle to parallel the Earth's surface only at the magnetic equator, whereas its inclination with respect to the surface increases to 90° at the magnetic poles.

stant, but varies, being weakest at the equator and strongest at the poles. The Earth's magnetic field is thought to result from convection within the liquid outer core.

When a magma cools, the magnetic iron-bearing minerals align themselves with the Earth's magnetic field, recording both its direction and strength. The temperature at which iron-bearing minerals gain their magnetization is called the **Curie point**. As long as the rock is not subsequently heated above the Curie point, it will preserve that remanent magnetism. Thus, an ancient lava flow provides a record of the orientation and strength of the Earth's magnetic field at the time the lava flow cooled.

As paleomagnetic research in the 1950s progressed, some unexpected results emerged. When geologists measured the paleomagnetism of recent rocks, they found it was generally consistent with the Earth's current magnetic field. The paleomagnetism of ancient rocks, however, showed different orientations. For example, paleomagnetic studies of Silurian lava flows in North America indicated that the north magnetic pole was located in the western Pacific Ocean at that time, while the paleomagnetic evidence from Permian lava flows indicated a pole in Asia, and that of Cretaceous lava flows pointed to yet another location in northern Asia. When plotted on a map, the paleomagnetic readings of numerous lava flows from all ages in North America trace the apparent movement of the magnetic pole through time (◆ Fig. 9.11). Thus, the paleomagnetic evidence from a single continent could be interpreted in three ways: the continent remained fixed and the north magnetic pole moved; the north magnetic pole stood still and the continent moved; or both the continent and the north magnetic pole moved.

Upon analysis, magnetic minerals from European Silurian and Permian lava flows pointed to a different magnetic pole location than those of the same age from North America (Fig. 9.11). Furthermore, analysis of lava flows from all continents indicated each continent had its own series of magnetic poles. Does this mean there were different north magnetic poles for each continent? That would be highly unlikely and difficult to reconcile with the theory accounting for the Earth's magnetic field. The best explanation for such data is that the magnetic poles have remained at their present locations near the geographic north and south poles and the continents have moved. When the continental margins are fitted together so that the paleomagnetic data point to only one magnetic pole, we find, just as Wegener did, that the rock sequences and glacial deposits match, and that the fossil evidence is consistent with the reconstructed paleogeography.

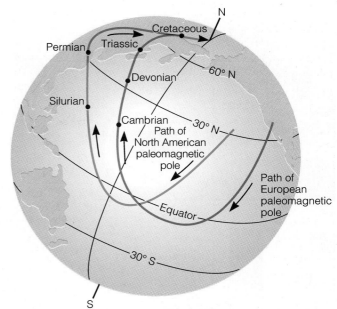

◆ FIGURE 9.11 The apparent paths of polar wandering for North America and Europe. The apparent location of the north magnetic pole is shown for different periods on each continent's polar wandering path.

▼ MAGNETIC REVERSALS AND SEA-FLOOR SPREADING

Geologists refer to the Earth's present magnetic field as being normal, that is, with the north and south magnetic poles located approximately at the north and south geographic poles. At numerous times in the geologic past, the Earth's magnetic field has completely reversed. The existence of such **magnetic reversals** was discovered by dating and determining the orientation of the remanent magnetism in lava flows on land (◆ Fig. 9.12). Once their existence was well established, magnetic reversals were also discovered in ocean basalts as part of the extensive mapping of the ocean basins that took place during the 1960s. Although the cause of magnetic reversals is still uncertain, their occurrence in the geologic record is well documented.

In addition to the discovery of magnetic reversals, mapping of the ocean basins also revealed a ridge system 65,000 km long, constituting the most extensive mountain range in the world. Perhaps the best-known part of the ridge system is the Mid-Atlantic Ridge, which divides the Atlantic Ocean basin into two nearly equal parts (◆ Fig. 9.13).

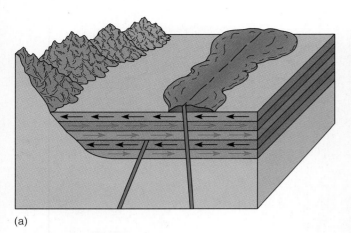

(a)

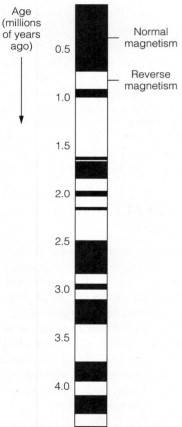

Age
(millions
of years
ago)

0.5 — Normal
magnetism

1.0 — Reverse
magnetism

1.5

2.0

2.5

3.0

3.5

4.0

(b)   4.5

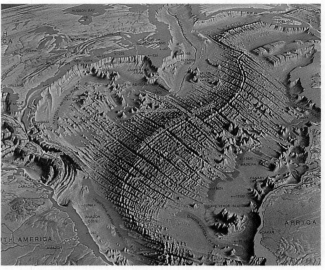

◆ FIGURE 9.13 Artistic view of what the Atlantic Ocean basin would look like without water. The major feature is the Mid-Atlantic Ridge.

◆ FIGURE 9.12 (*a*) Magnetic reversals recorded in a succession of lava flows are shown diagrammatically by red arrows, whereas the record of normal polarity events is shown by black arrows. The lava flows containing a record of such magnetic-polarity events can be radiometrically dated so that a magnetic time scale as in (*b*) can be constructed. (*b*) Magnetic reversals for the last 4.5 million years as determined from lava flows on land. Black bands represent normal magnetism and blue bands represent reverse magnetism.

In 1962, as a result of the oceanographic research conducted in the 1950s, Harry Hess of Princeton University proposed the theory of **sea-floor spreading** to account for continental movement. Hess suggested that continents do not move across oceanic crust, but rather that the continents and oceanic crust move together. He suggested that sea floor separates at oceanic ridges where new crust is formed by upwelling magma. As the magma cools, the newly formed oceanic crust moves laterally away from the ridge, thus explaining how volcanic islands that formed at or near ridge crests later became guyots (see Fig. 8.11). As a mechanism to drive this system, Hess revived the idea of **thermal convection cells** in the mantle; that is, hot magma rises from the mantle, intrudes along rift zone fractures defining oceanic ridges, and thus forms new crust. Cold crust is subducted back into the mantle at deep-sea trenches, where it is heated and recycled, thus completing a thermal convection cell.

How could Hess's hypothesis be confirmed? Magnetic surveys of the oceanic crust revealed striped **magnetic anomalies** (deviations from the average strength of the Earth's magnetic field) in the rocks that were both parallel to and symmetrical with the oceanic ridges (◆ Fig. 9.14). Furthermore, the pattern of oceanic magnetic anomalies matched the pattern of magnetic reversals already known from studies of continental lava flows (Fig. 9.12). When magma wells up and cools along a ridge summit, it records the Earth's

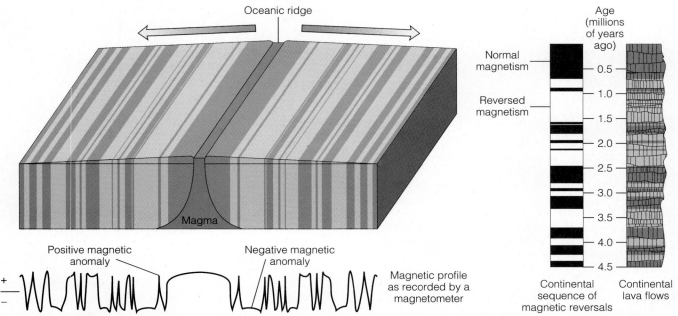

Oceanic ridge

Normal magnetism

Reversed magnetism

Age (millions of years ago)

Magma

Positive magnetic anomaly

Negative magnetic anomaly

Magnetic profile as recorded by a magnetometer

Continental sequence of magnetic reversals

Continental lava flows

◆ FIGURE 9.14 The sequence of magnetic anomalies preserved within the oceanic crust on both sides of an oceanic ridge is identical to the sequence of magnetic reversals already known from continental lava flows. Magnetic anomalies are formed when basaltic magma intrudes into oceanic ridges; when the magma cools below the Curie point, it records the Earth's magnetic polarity at the time. Subsequent sea-floor spreading split the previously formed crust in half, so that is moves laterally away from the oceanic ridge. Repeated intrusions produce a symmetrical series of magnetic anomalies that reflect periods of normal and reversed polarity. The magnetic anomalies are recorded by a magnetometer, which measures the strength of the magnetic field.

magnetic field at that time as either normal or reversed. As new crust forms at the summit, the previously formed crust moves laterally away from the ridge. These magnetic stripes, representing times of normal or reversed polarity, are parallel to and symmetrical around oceanic ridges (where upwelling magma forms new oceanic crust), conclusively confirming Hess's theory of sea-floor spreading.

One of the consequences of the sea-floor spreading theory is its confirmation that ocean basins are geologically young features whose openings and closings are partially responsible for continental movement (◆ Fig. 9.15, page 222). Radiometric dating reveals that the oldest oceanic crust is less than 180 million years old, whereas the oldest continental crust is 3.96 billion years old.

▼ PLATE TECTONIC THEORY

Most geologists accept **plate tectonic theory,** in part because the evidence for it is overwhelming, and also because it is a unifying theory that accounts for a variety of apparently unrelated geologic features and events. Consequently, geologists now view many geologic processes, such as mountain building, seismicity, and volcanism, from the perspective of plate tectonics (see Perspective 9.1). Furthermore, because all of the terrestrial planets have had a similar origin and early history, geologists are interested in determining whether plate tectonics is unique to Earth or whether it operates in the same way on the other terrestrial planets.

Plate tectonic theory is based on a simple model of the Earth. The rigid lithosphere, consisting of both oceanic and continental crust, as well as the underlying upper mantle, consists of numerous variable-sized pieces called **plates** (◆ Fig. 9.16, page 223). The plates vary in thickness; those composed of upper mantle and continental crust are as much as 250 km thick, whereas those of upper mantle and oceanic crust are up to 100 km thick.

The lithosphere overlies the hotter and weaker semi-plastic asthenosphere. It is believed that movement re-

THE SUPERCONTINENT CYCLE

At the end of the Paleozoic Era, all continents were amalgamated into the supercontinent Pangaea. Pangaea began fragmenting during the Triassic Period and continues to do so, thus accounting for the present distribution of continents and oceans. It now appears that another supercontinent existed at the end of the Proterozoic Eon, and there is some evidence for even earlier supercontinents (see Chapter 20). Recently, it has been proposed that supercontinents consisting of all or most of the Earth's landmasses form, break up, and re-form in a cycle spanning about 500 million years.

The supercontinent cycle hypothesis is an expansion on the ideas of the Canadian geologist J. Tuzo Wilson. During the early 1970s, Wilson proposed a cycle (now known as the Wilson cycle) that includes continental fragmentation, the opening and closing of an ocean basin, and finally reassembly of the continent. According to the supercontinent cycle hypothesis, heat accumulates beneath a supercontinent because rocks of continents are poor conductors of heat. As a consequence of the heat accumulation, the supercontinent domes upward and fractures. Basaltic magma rising from below fills the fractures. As a basalt-filled fracture widens, it begins subsiding and forms a long narrow ocean such as the present-day Red Sea. Continued rifting eventually forms an expansive ocean basin such as the Atlantic.

According to proponents of the supercontinent cycle, one of the most convincing arguments for their hypothesis is the "surprising regularity" of mountain building caused by compression during continental collisions. Such mountain-building episodes occur about every 400 to 500 million years, and are followed by an episode of rifting about 100 million years later. In other words, a supercontinent fragments and its individual plates disperse following a rifting episode, an interior ocean forms, and then the dispersed fragments reassemble to form another supercontinent (◆ Fig. 1).

In addition to explaining the dispersal and reassembly of supercontinents, the supercontinent cycle hypothesis also explains two distinct types of orogens (linear parts of the Earth's crust that were deformed during mountain building): *interior orogens* resulting from compression induced by continental collisions and *peripheral orogens* that form at continental margins in response to subduction and igneous activity. Following the fragmentation of a supercontinent, an interior ocean forms and widens as plates separate. After about 200 million years, however, plate separation ceases as the oceanic crust becomes cooler and denser and begins to subduct at the margins of the interior ocean basin, thus transforming inward-facing passive continental margins into active continental margins. Rising magma resulting from subduction forms plutons and volcanoes along these continental margins, and an interior orogeny develops when plates converge, causing compression-induced deformation, crustal thickening, and metamorphism. The present-day Appalachian Mountains of the eastern United States and Canada originally formed as an interior orogen (see Chapter 21).

Peripheral orogenies are caused by convergence of oceanic plates with the margins of a continent, igneous activity resulting from subduction, and collisions of island arcs with the continent. Subduction-related volcanism in peripheral orogens is rather continuous, but the collisions of island arcs with the supercontinent are episodic. Peripheral orogenies can develop at any time during the supercontinent cycle. A good example of a presently active peripheral orogeny is the Andes of western South America where the Nazca plate is being subducted beneath the South American plate (see Fig. 10.24).

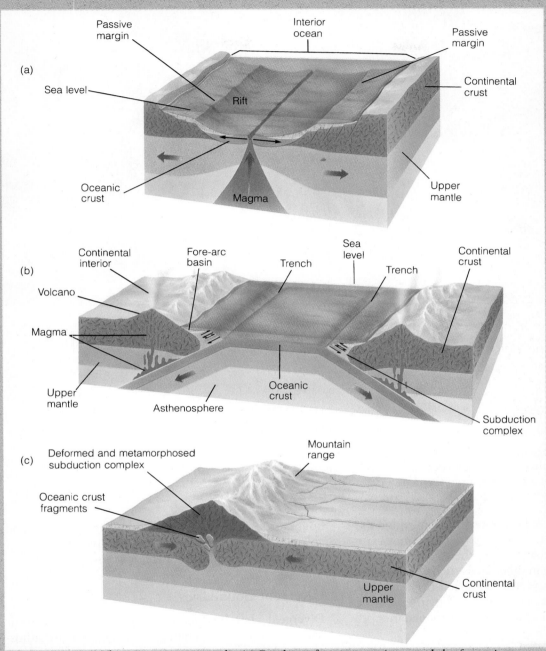

(a)

Passive margin

Interior ocean

Passive margin

Sea level

Continental crust

Rift

Oceanic crust

Magma

Upper mantle

(b)

Continental interior

Fore-arc basin

Trench

Sea level

Trench

Continental crust

Volcano

Magma

Upper mantle

Asthenosphere

Oceanic crust

Subduction complex

(c)

Deformed and metamorphosed subduction complex

Mountain range

Oceanic crust fragments

Upper mantle

Continental crust

◆ FIGURE 1 The supercontinent cycle. (*a*) Breakup of a supercontinent and the formation of an interior ocean basin. (*b*) Subduction along the margins of the interior ocean basin begins approximately 200 million years later, resulting in volcanic activity and deformation along an active oceanic-continental plate boundary. (*c*) Continental collisions and the formation of a new supercontinent result when all of the oceanic crust of the interior ocean basin is subducted.

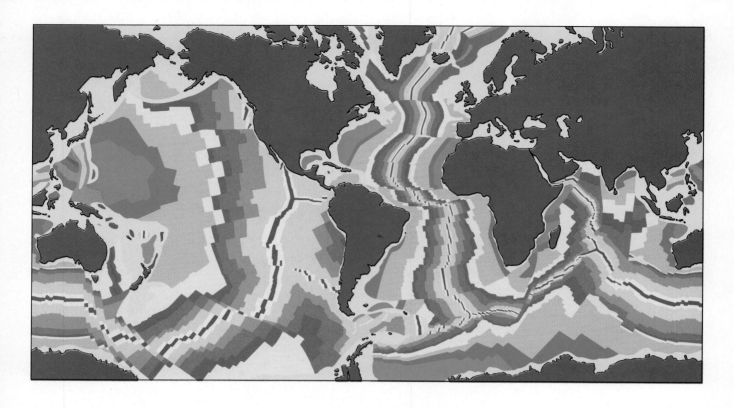

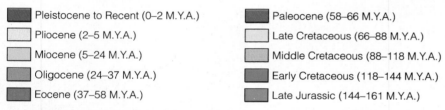

■ Pleistocene to Recent (0–2 M.Y.A.)		■ Paleocene (58–66 M.Y.A.)	
□ Pliocene (2–5 M.Y.A.)		□ Late Cretaceous (66–88 M.Y.A.)	
□ Miocene (5–24 M.Y.A.)		□ Middle Cretaceous (88–118 M.Y.A.)	
■ Oligocene (24–37 M.Y.A.)		■ Early Cretaceous (118–144 M.Y.A.)	
■ Eocene (37–58 M.Y.A.)		□ Late Jurassic (144–161 M.Y.A.)	

◆ FIGURE 9.15 The age of the world's ocean basins established from magnetic anomalies demonstrates that the youngest oceanic crust is adjacent to the spreading ridges and that its age increases away from the ridge axis.

sulting from some type of heat transfer system within the asthenosphere causes the overlying plates to move. As plates move over the asthenosphere, they separate, mostly at oceanic ridges, while in other areas such as at oceanic trenches, they collide and are subducted back into the mantle. Individual plates are recognized by the types of geological phenomena occurring at their boundaries.

▼ PLATE BOUNDARIES

Plates move relative to one another such that their boundaries can be characterized as *divergent, convergent,* and *transform.* Interaction of plates at their boundaries accounts for most of the Earth's seismic and volcanic activity and, as will be apparent in the next chapter, the origin of mountain systems.

Divergent Boundaries

Divergent plate boundaries or **spreading ridges** occur where plates are separating and new oceanic lithosphere is forming. Divergent boundaries are places where the crust is being extended, thinned, and fractured as magma, derived from the partial melting of the mantle, rises to the surface. The magma is almost entirely basaltic and intrudes into vertical fractures to

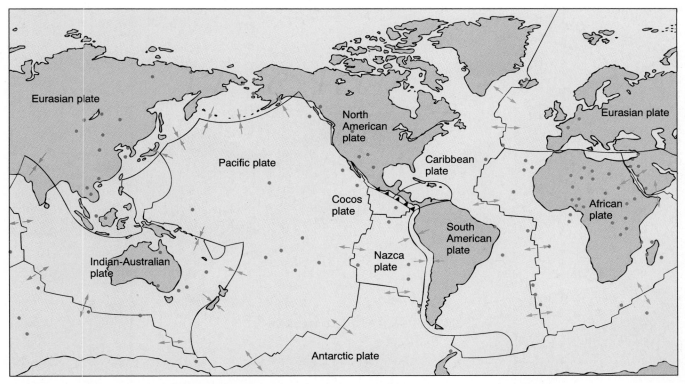

● Hot spot   → Direction of movement

◆ FIGURE 9.16 A map of the world showing the plates, their boundaries, direction of movement, and hot spots.

form dikes and lava flows (◆ Fig. 9.17). As successive injections of magma cool and solidify, they form new oceanic crust and record the intensity and orientation of the Earth's magnetic field (Fig. 9.14). Divergent boundaries most commonly occur along the crests of oceanic ridges, for example, the Mid-Atlantic Ridge. Oceanic ridges are thus characterized by rugged topography with high relief resulting from displacement of rocks along large fractures, shallow-focus earthquakes, high heat flow, and basaltic flows or pillow lavas.

Divergent plate boundaries also occur under continents during the early stages of continental breakup (◆ Fig. 9.18). When magma wells up beneath a continent, the crust is initially elevated, stretched, and thinned. Such stretching eventually produces fractures and rift valleys. During this stage, magma typically intrudes into the fractures, forming sills and dikes, as well as flowing onto the rift floor. The East African rift valleys are an excellent example of this stage of conti-

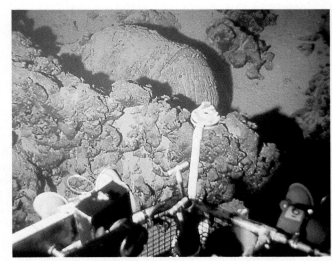

◆ FIGURE 9.17 Pillow lavas forming along the Mid-Atlantic Ridge. Their distinctive bulbous shape is the result of underwater eruption.

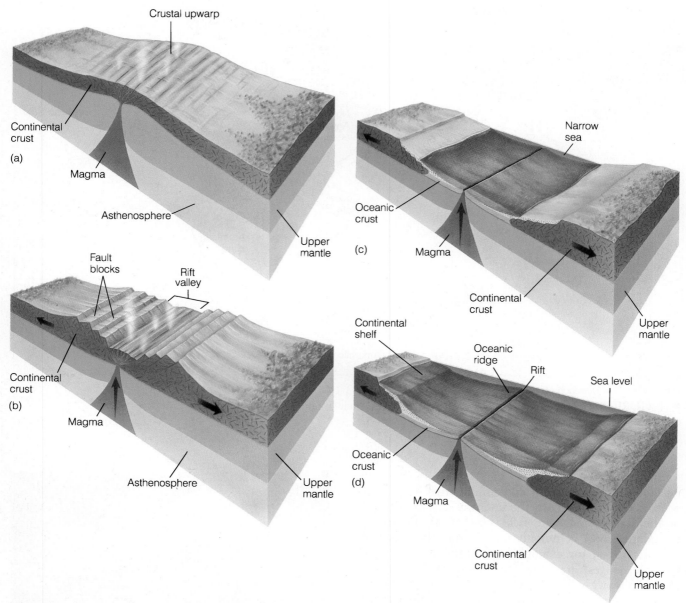

◆ FIGURE 9.18 History of a divergent plate boundary. (*a*) Rising magma beneath a continent pushes the crust up, producing numerous cracks and fractures. (*b*) As the crust is stretched and thinned, rift valleys develop, and lava flows onto the valley floors. (*c*) Continued spreading further separates the continent until a narrow seaway develops. (*d*) As spreading continues, an oceanic ridge system forms, and an ocean basin develops and grows.

nental breakup (◆ Fig. 9.19). As rifting proceeds, the continental crust eventually breaks. If magma continues welling up, the two parts of the continent will move away from each other, as is happening today beneath the Red Sea. As this newly formed narrow ocean basin continues to enlarge, it will eventually become an expansive ocean basin such as the Atlantic or Pacific Ocean basins are today.

An Example of Ancient Rifting

What features in the rock record can geologists use to recognize ancient rifting? Associated with regions of continental rifting are normal faults, dikes, sills, lava flows, and thick sedimentary sequences within rift valleys. The Triassic fault basins of the eastern United States are a good example of ancient continental rifting (see Fig. 23.8). These fault basins mark the zone of rifting that occurred when North America split apart from Africa. They contain thousands of meters of continental sediment and are riddled with dikes and sills (see Chapter 23).

Convergent Boundaries

Whereas new crust forms at divergent plate boundaries, old crust is destroyed at many **convergent plate boundaries.** One plate is subducted under another and eventually is resorbed in the asthenosphere. When we talk about convergent plate boundaries, we are really talking about three different types of boundaries: *oceanic-oceanic, oceanic-continental,* and *continental-continental.* The basic processes are the same for all three types of boundaries, but because different types of crust are involved, the results are different.

Oceanic-Oceanic Boundaries

When two oceanic plates converge, one of them is subducted under the other along an **oceanic-oceanic plate boundary** (◆ Fig. 9.20). The subducting plate bends downward to form the outer wall of an oceanic trench. A *subduction complex,* composed of wedge-shaped slices of highly folded and faulted marine sediments and oceanic lithosphere scraped off the descending plate, forms along the inner wall. This subduction complex is elevated as a result of uplift along faults as subduction continues.

As the subducting plate descends into the mantle, it is heated and partially melted, generating a magma commonly of andesitic composition. This magma is less dense than the surrounding mantle rocks and rises to the surface of the nonsubducted plate, forming a curved chain of volcanic islands called a **volcanic island arc.** This arc is nearly parallel to the oceanic trench and is separated from it by a distance of up to several hundred kilometers—the distance depends on the angle of dip of the subducting plate.

Located between the volcanic island arc and the subduction complex of the oceanic trench is a **fore-arc basin** (Fig. 9.20). It typically contains a diverse assort-

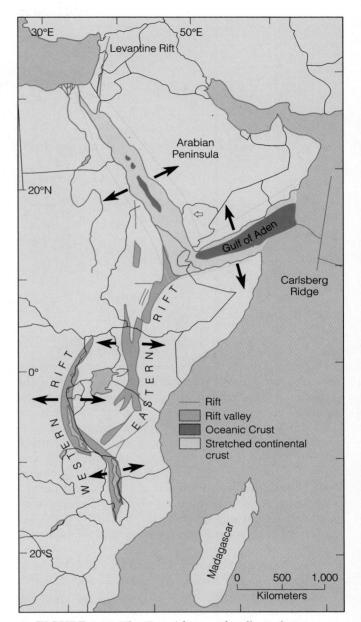

◆ FIGURE 9.19 The East African rift valley is being formed by the separation of eastern Africa from the rest of the continent along a divergent plate boundary. The Red Sea represents a more advanced stage of rifting in which two continental blocks are separated by a narrow sea.

ment of generally flat-lying detrital sediments up to 5 km thick. These sediments are derived from the weathering and erosion of the island arc volcanoes and reflect a progressive shallowing as the basin fills up.

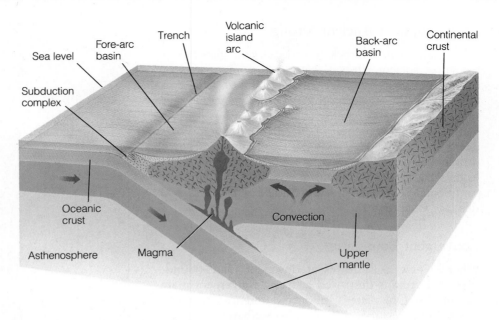

♦ FIGURE 9.20 Oceanic-oceanic plate boundary. An oceanic trench forms where one oceanic plate is subducted beneath another. On the nonsubducted oceanic plate, a volcanic island arc forms from the rising magma generated from the subducting plate.

In those areas where the rate of subduction is faster than the forward movement of the overriding plate, the lithosphere on the landward side of the volcanic island arc may be subjected to tensional stress and stretched and thinned, resulting in the formation of a **back-arc basin.** This back-arc basin may grow by spreading if magma breaks through the thin crust and forms new oceanic crust (Fig. 9.20). In any case, the back-arc basin will fill with a mixture of volcanic rocks and detrital sediments. A good example of a back-arc basin associated with an oceanic-oceanic plate boundary is the Sea of Japan between the Asian continent and the islands of Japan.

Most present-day active volcanic island arcs are in the Pacific Ocean basin and include the Aleutian Islands, the Kermadec-Tonga arc, and the Japanese and Philippine Islands. The Scotia and Antillean (Caribbean) island arcs are present in the Atlantic Ocean basin.

Oceanic-Continental Boundaries

When oceanic crust is subducted under continental crust along an **oceanic-continental plate boundary,** a subduction complex, consisting of wedge-shaped slices of complexly folded and faulted rocks, forms the inner wall of the trench. Between it and the continent is a fore-arc basin containing detrital sediments derived

from the erosion of the continent (♦ Fig. 9.21). These sediments are typically flat-lying or only mildly deformed. The andesitic magma generated by subduction rises beneath the continent and either crystallizes as plutons before reaching the surface or erupts at the surface, producing a chain of andesitic volcanoes (also called a volcanic arc). A back-arc basin may be filled with continental detrital sediments, pyroclastic materials, and lava flows, derived from and thickening toward the volcanic arc. An excellent example of an oceanic-continental plate boundary is the Pacific coast of South America where the oceanic Nazca plate is presently being subducted under South America. The Peru-Chile Trench marks the site of subduction, and the Andes Mountains are the resulting volcanic mountain chain on the nonsubducting plate (Fig. 9.16).

Continental-Continental Boundaries

Two continents approaching each other will initially be separated by an ocean floor that is being subducted under one continent. The edge of that continent will display the features characteristic of oceanic-continental convergence. As the ocean floor continues to be subducted, the two continents will come closer together until they eventually collide. Because continental lithosphere, which consists of continental crust and the upper mantle, is less dense than oceanic lithosphere

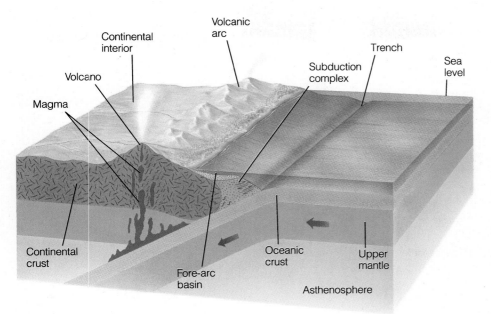

Volcanic arc

Continental interior

Volcano

Magma

Trench

Subduction complex

Sea level

Continental crust

Fore-arc basin

Oceanic crust

Upper mantle

Asthenosphere

◆ FIGURE 9.21 Oceanic-continental plate boundary. When an oceanic plate is subducted beneath a continental plate, an andesitic volcanic mountain range is formed on the continental plate as a result of rising magma.

(oceanic crust and the upper mantle), it cannot sink into the asthenosphere. Although one continent may partly slide under the other, it cannot be pulled or pushed down into a subduction zone (◆ Fig. 9.22).

When two continents collide, they are welded together along a zone marking the former site of subduction. At this **continental-continental plate boundary,** an interior mountain belt is formed consisting of thrusted and folded sediments, igneous intrusions, metamorphic rocks, and fragments of oceanic crust. In addition, the entire region is subjected to numerous earthquakes. The Himalayas in central Asia are the result of a continental-continental collision between India and Asia that began about 40 to 50 million years ago and is still continuing.

Ancient Convergent Plate Boundaries

How can former subduction zones be recognized in the rock record? One clue is provided by igneous rocks. The magma that erupts on the Earth's surface, forming island arc volcanoes and continental volcanoes, is of andesitic composition. Another clue can be found in the zone of intensely deformed rocks that occurs between the deep-sea trench where subduction is taking place and the area of igneous activity. Here, sediments and submarine rocks are folded, faulted, and metamor-

phosed into a chaotic mixture of rocks termed a *melange.*

During subduction, pieces of oceanic lithosphere are sometimes incorporated into the melange and accreted onto the edge of the continent. Such slices of oceanic crust and upper mantle are called **ophiolites** (◆ Fig. 9.23). They consist of a layer of deep-sea sediments that include graywackes (poorly sorted sandstones containing abundant feldspars and rock fragments, usually in a clay-rich matrix), black shales, and cherts. These deep-sea sediments are underlain by pillow basalts, a sheeted dike complex, massive gabbro, and layered gabbro, all of which form the oceanic crust. Beneath the gabbro is peridotite, which probably represents the upper mantle. Ophiolites are key features in recognizing plate convergence along a subduction zone (see Perspective 20.1).

Elongate belts of folded and thrust-faulted marine sediments, andesites, and ophiolites are found in the Appalachians, Alps, Himalayas, and Andes mountains. The combination of such features is good evidence that these mountain ranges resulted from deformation along convergent plate boundaries.

Transform Boundaries

The third type of plate boundary is a **transform plate boundary.** These occur along fractures in the sea floor

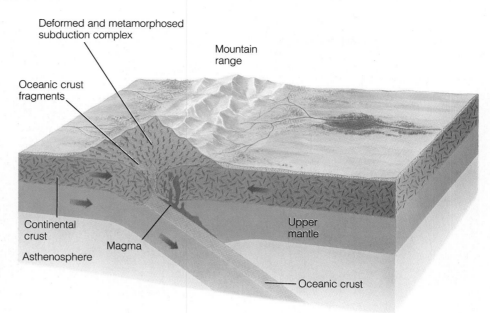

Deformed and metamorphosed subduction complex

Mountain range

Oceanic crust fragments

Continental crust

Asthenosphere

Magma

Upper mantle

Oceanic crust

◆ FIGURE 9.22 Continental-continental plate boundary. When two continental plates converge, neither is subducted because of their great thickness and low and equal densities. As the two plates collide, a mountain range is formed in the interior of a new and larger continent.

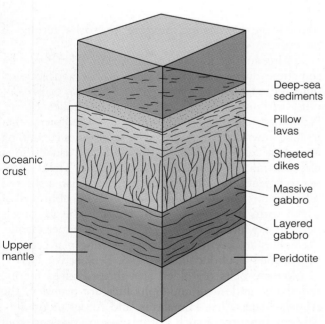

Oceanic crust

Upper mantle

Deep-sea sediments

Pillow lavas

Sheeted dikes

Massive gabbro

Layered gabbro

Peridotite

◆ FIGURE 9.23 Ophiolites are sequences of rock on land consisting of deep-sea sediments, oceanic crust, and upper mantle.

(see Fig. 8.10) known as *transform faults* where plates slide laterally past one another roughly parallel to the direction of plate movement. Although lithosphere is neither created nor destroyed along a transform boundary, the movement between plates results in a zone of intensely shattered rock and numerous shallow-focus earthquakes.

Transform faults are particular types of faults that "transform" or change one type of motion between plates into another type of motion. The majority of transform faults connect two oceanic ridge segments, but they can also connect ridges to trenches and trenches to trenches (◆ Fig. 9.24). While the majority of transform faults occur in oceanic crust and are marked by distinct fracture zones, they may also extend into continents.

One of the best-known transform faults is the San Andreas fault in California. It separates the Pacific plate from the North American plate and connects spreading ridges in the Gulf of California and the ridge separating the Juan de Fuca and Pacific plates off the coast of northern California (◆ Fig. 9.25). The many earthquakes that affect California are the result of movement along this fault.

Unfortunately, transform faults generally do not leave any characteristic or diagnostic features except for the obvious displacement of the rocks with which they are associated. This displacement is commonly large, on the order of tens to hundreds of kilometers. Such large displacements in ancient rocks can sometimes be related to transform fault systems.

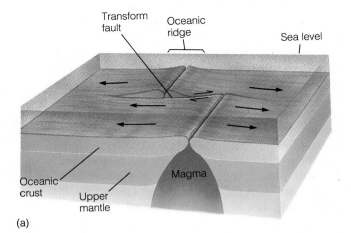

(a)

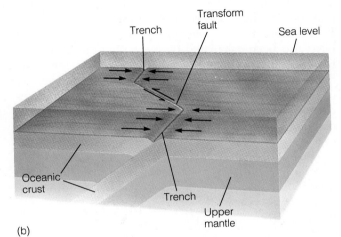

(b)

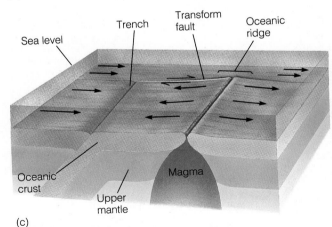

(c)

◆ FIGURE 9.24 Horizontal movement between plates occurs along a transform fault. (*a*) The majority of transform faults connect two oceanic ridge segments. Note that relative motion between the plates only occurs between the two ridges. (*b*) A transform fault connecting two trenches. (*c*) A transform fault connecting a ridge and a trench.

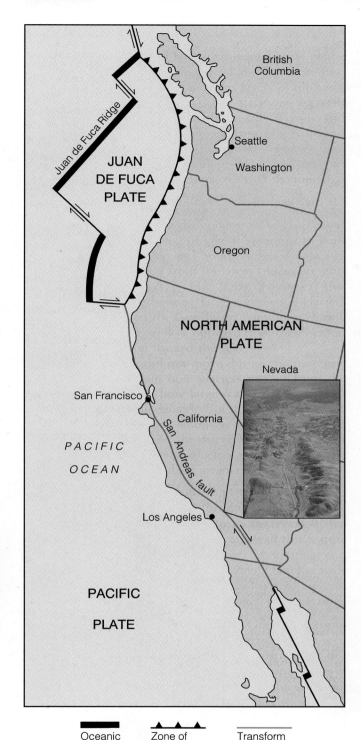

▬▬▬	▲▲▲	———
Oceanic ridge	Zone of subduction	Transform fault

◆ FIGURE 9.25 Transform plate boundary. The San Andreas fault is a transform fault separating the Pacific plate from the North American plate. Movement along this fault has caused numerous earthquakes. The photograph shows a segment of the San Andreas fault as it cuts through the Leona Valley, California. (Photo courtesy of Eleanora I. Robbins, U.S. Geological Survey.)

❦ PLATE MOVEMENT AND MOTION

How fast and in what direction are the Earth's various plates moving, and do they all move at the same rate? Rates of plate movement can be calculated in several ways. The least accurate method is to determine the age of the sediments immediately above any portion of the oceanic crust and divide that age by the distance from the spreading ridge. Such calculations give an average rate of movement.

A more accurate method of determining both the average rate of movement and relative motion is by dating the magnetic reversals in the crust of the sea floor. The distance from an oceanic ridge axis to any magnetic reversal indicates the width of new sea floor that formed during that time interval. Thus, for a given interval of time, the wider the strip of sea floor, the faster the plate has moved. In this way not only can the present average rate of movement and relative motion be determined (◆ Fig. 9.26), but the average rate of movement during the past can also be calculated by dividing the distance between reversals by the amount of time elapsed between reversals.

From the information in Figure 9.26, it is obvious that the rate of movement varies among plates. The southeastern part of the Pacific plate and the Cocos plates are the two fastest moving plates, while the Arabian and southern African plates are the slowest.

The average rate of movement as well as the relative motion between any two plates can also be determined by satellite laser ranging techniques. Laser beams from a station on one plate are bounced off a satellite (in geosynchronous orbit) and returned to a station on a different plate. As the plates move away from each other, there is an increase in the length of time that the laser beam takes to go from the sending station to the stationary satellite and back to the receiving station. This difference in elapsed time is used to calculate the rate of movement and relative motion between plates. In addition, rates of movement and relative motion have also been calculated by measuring the difference between arrival times of radio signals from the same quasar to receiving stations on different plates. The rate of plate movement determined by these two techniques correlates closely with those determined from magnetic reversals.

◆ FIGURE 9.26 This map shows the average rate of movement in centimeters per year and relative motion of the Earth's plates.

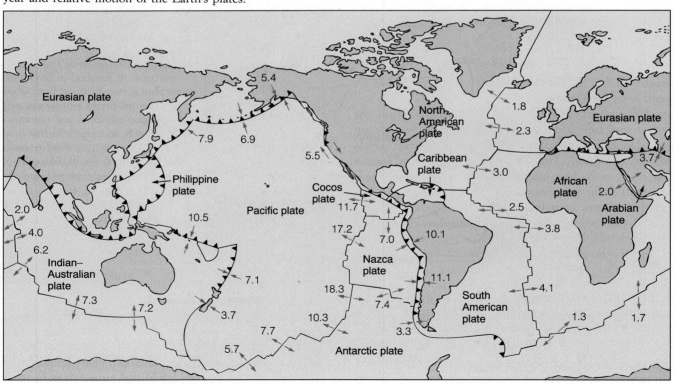

Hot Spots and Absolute Motion

Plate motions derived from magnetic reversals, satellites, and lasers give only the relative motion of one plate with respect to another. To determine absolute motion, we must have a fixed reference from which the rate and direction of plate movement can be determined. **Hot spots,** which may provide reference points, are locations where stationary columns of magma, originating deep within the mantle (mantle plumes), slowly rise to the Earth's surface and form volcanoes or flood basalts (Fig. 9.16).

One of the best examples of hot spot activity is that over which the Emperor Seamount–Hawaiian Island chain formed (◆ Fig. 9.27). Currently, the only active volcanoes in this island chain are on the island of Hawaii and the Loihi Seamount. The rest of the islands and seamounts of the chain are also of volcanic origin and are progressively older west-northwestward along the Hawaiian chain and north-northwestward along the Emperor Seamount chain.

The reason these islands and seamounts are progressively older as one moves toward the north and northwest is that the Pacific plate has moved over an apparently stationary mantle plume. Thus, a line of volcanoes was formed near the middle of the Pacific plate, marking the direction of the plate's movement. In the case of the Emperor Seamount–Hawaiian Island chain, the Pacific plate moved first north-northwesterly and then west-northwesterly over a single mantle plume.

Mantle plumes and hot spots are useful to geologists in helping to explain some of the geologic activity occurring within plates as opposed to that occurring at or near plate boundaries. In addition, if mantle plumes are essentially fixed with respect to the Earth's rotational axis—although some evidence suggests they might not be (see Chapter 7)—they may prove useful as reference points for determining paleolatitude.

▼ THE DRIVING MECHANISM OF PLATE TECTONICS

A major obstacle to the acceptance of continental drift was the lack of a driving mechanism to explain continental movement. When it was shown that continents and ocean floors moved together and not separately and that new crust formed at spreading ridges by rising magma, most geologists accepted some type of convective heat system as the basic process responsible for

◆ FIGURE 9.27 The Emperor Seamount–Hawaiian Island chain formed as a result of movement of the Pacific plate over a hot spot. The line of the volcanic islands traces the direction of plate movement. The numbers indicate the age of the islands in millions of years.

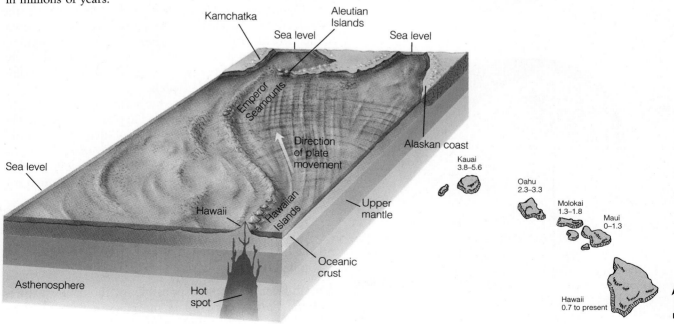

plate motion. The question still remains, however, as to what exactly drives the plates.

Two models involving thermal convection cells have been proposed to explain plate movement (◆ Fig. 9.28). In one model, thermal convection cells are restricted to the asthenosphere, whereas in the second model the entire mantle is involved. In both models spreading ridges mark the ascending limbs of adjacent convection cells, while trenches occur where the convection cells descend back into the Earth's interior. Thus, the locations of spreading ridges and trenches are determined by the convection cells themselves. Furthermore, in both models, the lithosphere is considered to be the top of the thermal convection cell, and each plate therefore corresponds to a single convection cell.

Although most geologists agree that the Earth's internal heat plays an important role in plate movement, problems are inherent in both models. The major problem associated with the first model is the difficulty in explaining the source of heat for the convection cells and why they are restricted to the asthenosphere. In the second model, the source of heat comes from the Earth's outer core, but it is still not known how heat is transferred from the outer core to the mantle. Nor is it clear how convection can involve both the lower mantle and the asthenosphere.

Some geologists believe that in addition to being generated by thermal convection within the Earth, plate movement also occurs, in part, because of a mechanism involving "slab-pull" or "ridge-push" (◆ Fig. 9.29). Both of these mechanisms are gravity driven, but still depend on thermal differences within the Earth. In "slab-pull," the subducting cold slab of lithosphere is denser than the surrounding warmer asthenosphere and thus pulls the rest of the plate along with it as it descends into the asthenosphere. As the lithosphere moves downward, there is a corresponding upward flow back into the spreading ridge.

Operating in conjunction with "slab-pull" is the "ridge-push" mechanism. As a result of rising magma, the oceanic ridges are higher than the surrounding oceanic crust. It is believed that gravity pushes the oceanic lithosphere away from the higher spreading ridges and toward the trenches.

Currently, geologists are fairly certain that some type of convective system is involved in plate movement. The extent to which other mechanisms such as "slab-pull" and "ridge-push" are involved, however, is still unresolved. Consequently, a comprehensive theory of plate movement has not as yet been developed, and much still remains to be learned about the Earth's interior.

▼ PLATE TECTONICS AND THE DISTRIBUTION OF NATURAL RESOURCES

In addition to being responsible for the major features of the Earth's crust, plate movements also affect the formation and distribution of the Earth's natural re-

◆ FIGURE 9.28 Two models involving thermal convection cells have been proposed to explain plate movement. (a) In one model, thermal convection cells are restricted to the asthenosphere. (b) In the other model, thermal convection cells involve the entire mantle.

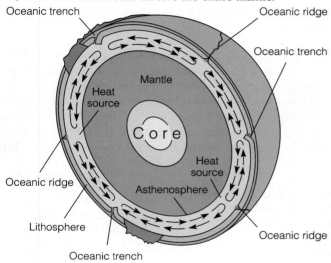

(a)

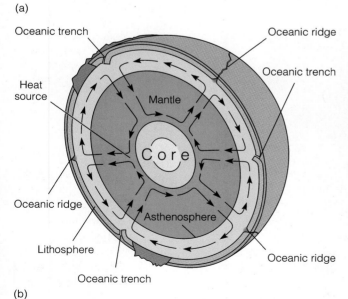

(b)

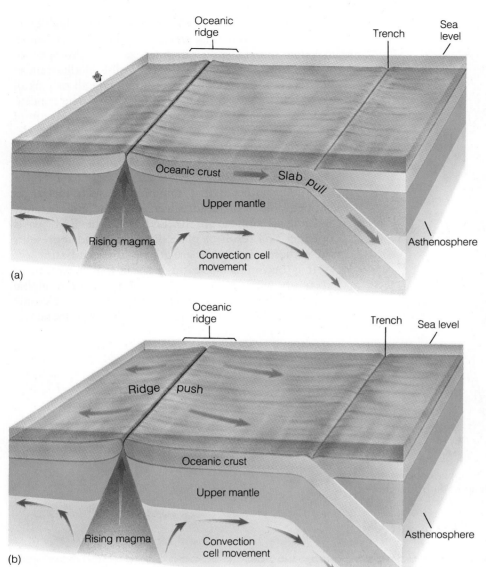

Oceanic ridge

Trench

Sea level

Oceanic crust → Slab pull

Upper mantle

Rising magma

Convection cell movement

Asthenosphere

(a)

Oceanic ridge

Trench

Sea level

Ridge push

Oceanic crust

Upper mantle

Rising magma

Convection cell movement

Asthenosphere

(b)

◆ FIGURE 9.29 Plate movement is also thought to occur because of gravity-driven "slab-pull," or "ridge-push" mechanisms. (*a*) In "slab-pull," the edge of the subducting plate descends into the Earth's interior, and the rest of the plate is pulled downward. (*b*) In "ridge-push," rising magma pushes the oceanic ridges higher than the rest of the oceanic crust. Gravity thus pushes the oceanic lithosphere away from the ridges and toward the trenches.

sources. Consequently, geologists are using plate tectonic theory in their search for new mineral deposits and in explaining the occurrence of known deposits (see Chapter 25).

Many metallic mineral deposits such as copper, gold, lead, silver, tin, and zinc are related to igneous and associated hydrothermal activity, so it is not surprising that a close relationship exists between plate boundaries and the occurrence of these valuable deposits.

The magma generated by partial melting of a subducting plate rises toward the Earth's surface, and as it cools, it precipitates and concentrates various metallic sulfide ores. Many of the world's major metallic ore deposits, such as the porphyry copper deposits of western North and South America, are excellent examples of the relationship between convergent plate boundaries and the distribution, concentration, and exploitation of metallic ores (◆ Fig. 9.30).

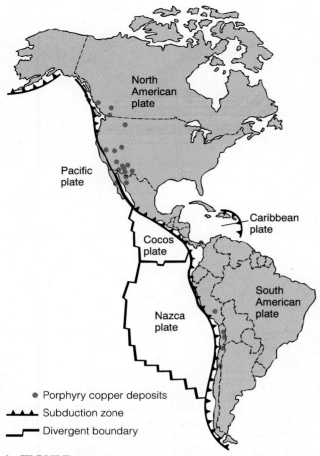

Divergent plate boundaries also yield valuable resources. Hydrothermal vents are the sites of much metallic mineral precipitation (see the Prologue in Chapter 8). The island of Cyprus in the Mediterranean is rich in copper and has been supplying all or part of the world's needs for the last 3,000 years. The concentration of copper on Cyprus formed as a result of precipitation adjacent to hydrothermal vents. This deposit was brought to the surface when the copper-rich sea floor collided with the European plate, warping the sea floor and forming Cyprus.

Studies indicate that minerals of such metals as copper, gold, iron, lead, silver, and zinc are currently forming as sulfides in the Red Sea. The Red Sea is opening as a result of plate divergence and represents the earliest stage in the growth of an ocean basin.

It is becoming increasingly clear that if we are to keep up with the continuing demands of a global industrialized society, the application of plate tectonic theory to the origin and distribution of mineral resources is essential.

◆ FIGURE 9.30 Important porphyry copper deposits are located along the west coasts of North and South America.

Chapter Summary

1. The concept of continental movement is not new. The earliest maps showing the similarity between the east coast of South America and the west coast of Africa provided people with the first evidence that the continents may once have been united and subsequently drifted away from each other.
2. Alfred Wegener is generally credited with developing the hypothesis of continental drift. He provided abundant geological and paleontological evidence to show that the continents were once united into one supercontinent he named Pangaea. Unfortunately, Wegener could not explain how the continents moved, and most geologists ignored his ideas.
3. The hypothesis of continental drift was revived during the 1950s when paleomagnetic studies revealed that there apparently were multiple magnetic north poles instead of just one as there is today. This paradox was resolved by moving the continents into different positions. When this was done, the paleomagnetic data were consistent with a single magnetic north pole.
4. Magnetic surveys of the oceanic crust reveal magnetic anomalies in the rocks indicating that the Earth's magnetic field has reversed itself in the past. Because the anomalies are parallel and form symmetric belts adjacent to the oceanic ridges, new oceanic crust must have formed as the seafloor was spreading.

5. Sea-floor spreading has been confirmed by radiometric dating of rocks on oceanic islands. Such dating reveals that the oceanic crust becomes older with distance from spreading ridges.
6. Plate tectonic theory became widely accepted by the 1970s because of the overwhelming evidence supporting it and because it provides geologists with a powerful theory for explaining such phenomena as volcanism, seismicity, mountain building, global climatic changes, past and present animal and plant distribution, and the distribution of mineral resources.
7. Three types of plate boundaries are recognized: divergent boundaries, where plates move away from each other; convergent boundaries, where two plates collide; and transform boundaries, where two plates slide past each other.
8. Ancient plate boundaries can be recognized by their associated rock assemblages and geologic structures. For divergent boundaries, these may include rift valleys with thick sedimentary sequences and numerous dikes and sills. For convergent boundaries, ophiolites and andesitic rocks are two characteristic features. Transform faults generally do not leave any characteristic or diagnostic features in the rock record.
9. The average rate of movement and relative motion of plates can be calculated several ways. These different methods all yield similar average rates of plate movement and indicate that the plates move at different average velocities.
10. Absolute motion of plates can be determined by the movement of plates over mantle plumes. A mantle plume is an apparently stationary column of magma that rises to the Earth's surface where it becomes a hot spot and forms a volcano.
11. Although a comprehensive theory of plate movement has yet to be developed, geologists think that some type of convective heat system is involved.
12. A close relationship exists between the formation of some mineral deposits and plate boundaries. Furthermore, the formation and distribution of the Earth's natural resources are related to plate movements.

Important Terms

back-arc basin
continental-continental plate boundary
continental drift
convergent plate boundary
Curie point
divergent plate boundary
fore-arc basin
Glossopteris flora
Gondwana

hot spot
Laurasia
magnetic anomaly
magnetic reversal
oceanic-continental plate boundary
oceanic-oceanic plate boundary
ophiolite
paleomagnetism
Pangaea

plate
plate tectonic theory
sea-floor spreading
spreading ridge
thermal convection cell
transform plate boundary
transform fault
volcanic island arc

Review Questions

1. The man who is credited with developing the continental drift hypothesis is:
   a. ___ Wilson; b. ___ Hess; c. ___ Vine; d. ___ Wegener; e. ___ du Toit.
2. The southern part of Pangaea, consisting of South America, Africa, India, Australia, and Antarctica, is called:
   a. ___ Gondwana; b. ___ Laurasia; c. ___ Atlantis; d. ___ Laurentia; e. ___ Pacifica.
3. Which of the following has been used as evidence for continental drift?
   a. ___ continental fit; b. ___ fossil plants and animals; c. ___ similarity of rock sequences; d. ___ paleomagnetism; e. ___ all of these.
4. Magnetic surveys of the ocean basins indicate that:
   a. ___ the oceanic crust is oldest adjacent to spreading ridges; b. ___ the oceanic crust is youngest adjacent to the continents; c. ___ the oceanic crust is youngest adjacent to spreading ridges; d. ___ the oceanic crust is the same age in all ocean basins; e. ___ answers (a) and (b).
5. Plates:
   a. ___ are the same thickness everywhere; b. ___ vary in thickness; c. ___ include the crust and upper mantle; d. ___ answers (a) and (c); e. ___ answers (b) and (c).
6. Divergent boundaries are the areas where:
   a. ___ new continental lithosphere is forming; b. ___ new oceanic lithosphere is forming; c. ___ two plates come

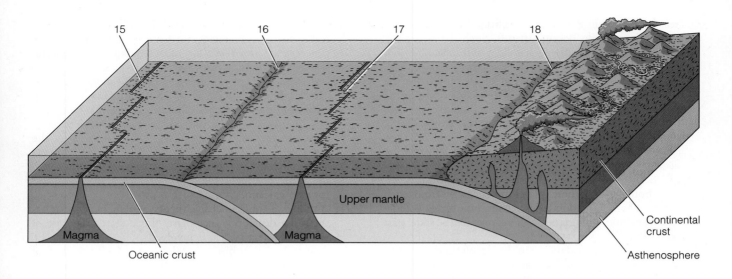

15   16   17   18

Upper mantle

Magma   Magma

Oceanic crust

Continental crust

Asthenosphere

together; d. ___ two plates slide past each other;
e. ___ answers (b) and (d).

7. Along what type of plate boundary does subduction occur?
a. ___ divergent;   b. ___ transform;   c. ___ convergent;
d. ___ answers (a) and (b); e. ___ answers (b) and (c).

8. The west coast of South America is an example of a(n) _____ plate boundary.
a. ___ divergent; b. ___ continental-continental;
c. ___ oceanic-oceanic; d. ___ oceanic-continental;
e. ___ transform.

9. Back-arc basins are associated with _____ plate boundaries.
a. ___ divergent; b. ___ convergent; c. ___ transform;
d. ___ answers (a) and (b); e. ___ answers (b) and (c).

10. The San Andreas fault is an example of a(n) _____ boundary.
a. ___ divergent; b. ___ convergent; c. ___ transform;
d. ___ oceanic-continental; e. ___ continental-continental.

11. Which of the following will allow you to determine the absolute motion of plates?
a. ___ hot spots; b. ___ the age of the sediment directly above any portion of the ocean crust; c. ___ magnetic reversals in the sea floor crust; d. ___ satellite laser ranging techniques; e. ___ all of these.

12. The formation of the island of Hawaii and the Loihi Seamount are the result of:
a. ___ oceanic-oceanic plate boundaries; b. ___ hot spots; c. ___ divergent plate boundaries; d. ___ transform boundaries; e. ___ oceanic-continental plate boundaries.

13. The driving mechanism of plate movement is thought to be:
a. ___ composition;   b. ___ magnetism;   c. ___ thermal convection cells; d. ___ rotation of the Earth; e. ___ none of these.

14. The formation and distribution of copper deposits are associated with _____ boundaries.
a. ___ divergent;   b. ___ convergent;   c. ___ transform;
d. ___ answers (a) and (b); e. ___ answers (b) and (c).

Name the type of plate boundary indicated in the illustration found at the top of this page.

15. _____ .

16. _____ .

17. _____ .

18. _____ .

19. What evidence convinced Wegener that the continents were once joined together and subsequently broke apart?

20. Why can't the similarity between the coastlines of continents alone be used to prove they were once joined together?

21. What is the significance of polar wandering in relation to continental drift?

22. How can magnetic anomalies be used to show that the sea floor has been spreading?

23. What evidence besides magnetic anomalies convinced geologists of sea-floor spreading?

24. Why is plate tectonics such a powerful unifying theory?

25. Summarize the geologic features characterizing the three different types of plate boundaries.

26. Explain how rates of movement of plates can be determined.

27. What are mantle plumes and hot spots? How can they be used to determine the direction and rate of movement of plates?

28. What are some of the positive and negative features of the various models proposed to explain plate movement?

29. What features would an astronaut look for on the Moon or another planet to find out if plate tectonics is currently active or if it was active during the past?

30. Briefly discuss how a geologist could use plate tectonic theory to help locate mineral deposits.

Additional Readings

Allégre, C. 1988. *The behavior of the Earth.* Cambridge, Mass.: Harvard Univ. Press.

Bonatti, E. 1987. The rifting of continents. *Scientific American* 256, no. 3:96–103.

Brimhall, G. 1991. The genesis of ores. *Scientific American* 264, no. 5:84–91.

Condie, K. 1989. *Plate tectonics and crustal evolution.* 3d ed. New York: Pergamon.

Cox, A., and R. B. Hart. 1986. *Plate tectonics: How it works.* Palo Alto, Calif.: Blackwell Scientific.

Cromie, W. J. 1989. The roots of midplate volcanism. *Mosaic* 20, no. 4:19–25.

Gass, I. G. 1982. Ophiolites. *Scientific American* 247, no. 2:122–31.

Jordan, T. H., and J. B. Minster. 1988. Measuring crustal deformation in the American west. *Scientific American* 259, no. 2:48–59.

Kearey, P., and F. J. Vine. 1990. *Global tectonics.* Palo Alto, Calif.: Blackwell Scientific.

Murphy, J. B., and R. D. Nance. 1992. Mountain belts and the supercontinent cycle. *Scientific American* 266, no. 4:84–91.

Nance, R. D., T. R. Worsley, and J. B. Moody. 1988. The supercontinent cycle. *Scientific American* 259, no. 1:72–79.

Vink, G. E., W. J. Morgan, and P. R. Vogt. 1985. The Earth's hot spots. *Scientific American* 252, no. 4:50–57.

CHAPTER
10

Chapter Outline

View of the Teton Range in Wyoming. The Grand Teton is the highest peak visible.

Deformation, Mountain Building, and the Continents

Prologue

Of the many scenic mountain ranges in the continental United States, few compare in grandeur to the Teton Range of northwestern Wyoming. The Native Americans of the region called these mountains Teewinot, meaning many pinnacles. This is an appropriate name indeed, for the Teton Range consists of numerous jagged peaks, the loftiest of which, the Grand Teton, rises to 4,190 m above sea level. There are higher and larger mountain ranges in the United States, but none rise so abruptly as the Tetons. They ascend nearly vertically more than 2,100 m above the floor of Jackson Hole, the valley to their east. This range and the surrounding region comprise Grand Teton National Park.

Mountains began forming in this region about 90 million years ago. These early mountains were quite different from the present ones in that the long axes of these ranges were oriented northwest-southeast, and they originated as the Earth's crust was contorted and folded. The present-day Teton Range, which runs north-south, began forming about 10 million years ago when part of the crust was uplifted along a large fracture called the Teton fault (◆ Fig. 10.1).

Most of the rocks exposed in the Teton Range are Precambrian-aged metamorphic and plutonic rocks formed at great depth beneath sedimentary rocks. Movement on the Teton fault resulted in uplift of the Teton block relative to the block to the east; the total displacement on this fault is about 6,100 m. As the Teton block rose, the overlying sedimentary rocks were eroded, exposing the underlying metamorphic and plutonic rocks (Fig. 10.1). The fault is along the east side of the Teton block, so as uplift occurred, the block has been tilted ever more steeply toward the west (Fig. 10.1). Displacement of recent sedimentary deposits along the east flank of the Teton Range shows that uplift is continuing today.

The spectacular, rugged topography of the Teton Range developed rather recently. Currently, the range supports about a dozen small glaciers, but periodically during the last 200,000 years it was more heavily glaciated. Glaciers are particularly effective agents of erosion; they scoured out valleys and intricately sculpted the uplifted Teton block, producing excellent examples of glacial landforms. The Grand Teton, which is a horn peak, is one of the most prominent of these.

▼ INTRODUCTION

Many ancient rocks are fractured or highly contorted—an indication that forces within the Earth caused deformation during the past. Seismic activity is a manifestation of forces continuing to operate within the Earth, as is the Teton Range uplift in Wyoming. Colliding plates generate forces causing deformation and mountain building along convergent plate margins, and in so doing, they add material to the margins of continents. Mountain systems within continents form when two continents collide and become sutured, thereby forming a larger landmass. Mountains also form when continents are stretched during rifting events. In short, deformation, mountain building, and the evolution of continents are interrelated, although not all deformation results in the origin of mountains.

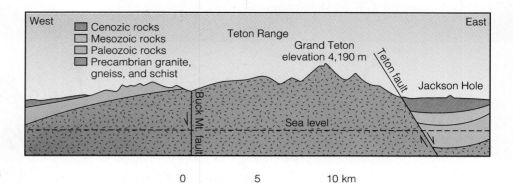

◆ FIGURE 10.1 A cross section of the Teton Range, Wyoming.

The study of deformed rocks has several applications. For example, the geologic structures produced during deformation, such as folds and faults, provide a record of the kinds and intensities of forces that operated during the past. Interpretations of such structures allow geologists to make inferences regarding Earth history. Understanding the nature of the local geologic structures also helps geologists find and recover natural resources. For example, several geologic structures form traps for petroleum and natural gas (see Fig. 5.36). Furthermore, geologic structures are considered when sites are selected for dams, large bridges, and nuclear power plants, especially if such sites are in areas of active deformation.

DEFORMATION

Fractured and contorted rock layers such as those in ◆ Figure 10.2 are said to be *deformed*; that is, their original shape or volume or both have been altered by **stress**, which is the result of force applied to a given area of rock. If the intensity of the stress is greater than the internal strength of the rock, it will undergo **strain**, which is deformation caused by stress.

Three types of stress are recognized: compressional, tensional, and shear. **Compression** results when rocks are squeezed or compressed by external forces directed toward one another. Rock layers subjected to compression are commonly shortened in the direction of stress by folding or faulting (◆ Fig. 10.3a). **Tension** results from forces acting in opposite directions along the same line (◆ Fig. 10.3b). Such stress tends to lengthen rocks or pull them apart. In **shear stress**, forces act parallel to one another but in opposite directions,

◆ FIGURE 10.2 Deformed layers of rock. The folded rock layers are considered to be ductile because they show considerable plastic deformation, whereas the fractured rocks are brittle.

resulting in deformation by displacement of adjacent layers along closely spaced planes (◆ Fig. 10.3c).

Strain is characterized as **elastic** if the deformed rocks return to their original shape when the stresses are relaxed. Squeezing a tennis ball, for example, causes strain, but once the stress is released, the tennis ball returns to its original shape. Rocks that are strained beyond their elastic limit cannot recover their original shape, however, and retain the configuration produced by the stress. Such rocks either deform by **plastic strain**, as when they are folded, or behave as brittle solids and are **fractured** (Fig. 10.2).

The type of strain that occurs depends on the kind of stress applied, the amount of pressure, the tempera-

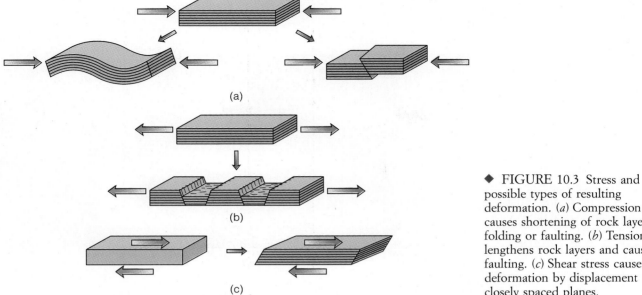

(a)

(b)

(c)

◆ FIGURE 10.3 Stress and possible types of resulting deformation. (*a*) Compression causes shortening of rock layers by folding or faulting. (*b*) Tension lengthens rock layers and causes faulting. (*c*) Shear stress causes deformation by displacement along closely spaced planes.

ture, the rock type, and the length of time the rock is subjected to the stress. For example, a small stress applied over a long period of time, as on the slab shown in ◆ Figure 10.4, will cause plastic deformation. By contrast, a large stress applied rapidly to the same object, as when it is struck by a hammer, will probably result in fracture. Rock type is important because not all rocks respond to stress in the same way. Rocks are

◆ FIGURE 10.4 This marble slab in the Rock Creek Cemetery, Washington, D.C., bent under its own weight in about 80 years.

considered to be either *ductile* or *brittle* depending on the amount of plastic strain they exhibit. Brittle rocks show no plastic strain before fracture, whereas ductile rocks exhibit a great deal (Fig. 10.2).

Many rocks show the effects of plastic deformation that must have occurred deep within the Earth's crust where the temperature and pressure are high. Recall that rock materials behave very differently under these conditions compared to their behavior near the surface. At or near the surface, they behave as brittle solids, whereas under conditions of high temperature and high pressure, they are more commonly ductile and deform plastically rather than fracture.

Strike and Dip

A basic principle in geology holds that when sediments are deposited, they accumulate in nearly horizontal layers. Thus, sedimentary rock layers that are steeply inclined must have been tilted following deposition and lithification (Fig. 10.2). Some igneous rocks, especially ash falls and many lava flows, also form nearly horizontal layers. To describe the orientation of deformed rock layers, geologists use the concept of *strike and dip*.

Strike is the direction of a line formed by the intersection of a horizontal plane with an inclined

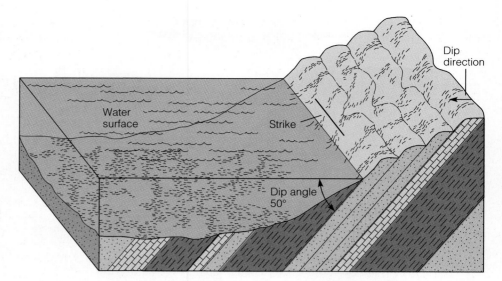

◆ FIGURE 10.5 Strike and dip. The strike is formed by the intersection of a horizontal plane (the water surface) with the surface of an inclined plane (the surface of the rock layer). The dip is the maximum angular deviation of the inclined plane from horizontal.

plane, such as a rock layer. For example, in ◆ Figure 10.5, the surface of any of the tilted rock layers constitutes an inclined plane. The intersection of a horizontal plane with any of these inclined planes forms a line, the direction of which is the strike. The strike line's orientation is determined by using a compass to measure its angle with respect to north. **Dip** is a measure of the maximum angular deviation of an inclined plane from horizontal, so it must be measured perpendicular to the strike direction (Fig. 10.5).

Geologic maps indicate strike and dip by using a long line oriented in the strike direction and a short line perpendicular to the strike line and pointing in the dip direction (◆ Fig. 10.6a). The number adjacent to the

◆ FIGURE 10.6 (a) Strike and dip symbol. The long bar is oriented in the strike direction, and the short bar points in the dip direction. The number indicates the dip angle. (b) The symbol used to indicate horizontal rock layers. (c) The symbol for vertical rock layers.

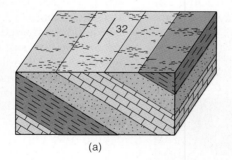

(a)

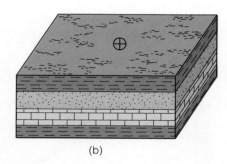

(b)

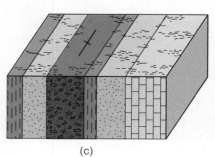

(c)

strike and dip symbol indicates the dip angle. A circled cross is used to indicate horizontal strata, and a strike symbol with a short crossbar indicates layers dipping vertically (◆ Fig. 10.6b and c).

Folds

If you place your hands on a tablecloth and move them toward one another, the tablecloth is deformed by compression into a series of up- and down-arched folds. Similarly, rock layers within the Earth's crust commonly respond to compression by folding. As opposed to the tablecloth, however, folding in rock layers is permanent; that is, the rocks have been strained plastically. Most folding probably occurs deep within the crust because rocks at or near the surface are brittle and generally deform by fracturing rather than by folding.

Monoclines, Anticlines, and Synclines

A **monocline** is a simple bend or flexure in otherwise horizontal or uniformly dipping rock layers (◆ Fig. 10.7). An **anticline** is an up-arched fold, while a **syncline** is a down-arched fold (◆ Fig. 10.8). Both anticlines and synclines are characterized by an *axial plane* that divides them into halves; the part of a fold on opposite sides of the axial plane is a *limb* (◆ Fig. 10.9). Because folds most commonly occur as a series of

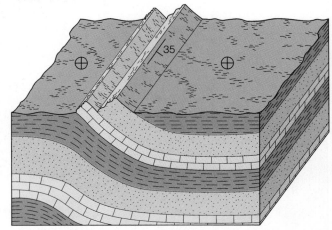

◆ FIGURE 10.7 A monocline. Notice the strike and dip symbols and the symbol for horizontal layers.

anticlines alternating with synclines, a limb is generally shared by an anticline and an adjacent syncline.

Even where the exposed view has been eroded, anticlines and synclines can easily be distinguished from each other by strike and dip and by the relative ages of the folded strata. As ◆ Figure 10.10 shows, in an eroded anticline, the strata of each limb dip outward or away from the center, where the oldest strata are located. In eroded synclines, on the other hand, the

◆ FIGURE 10.8 Anticline and syncline in the Calico Mountains of southeastern California.

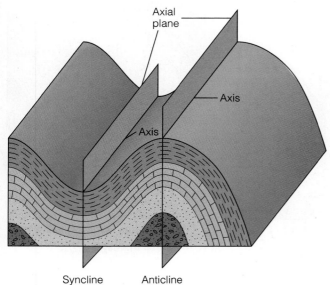

◆ FIGURE 10.9 Syncline and anticline showing the axial plane, axis, and fold limbs.

strata in each limb dip inward toward the center, and the youngest strata coincide with the center of the fold.

Thus far, we have described *symmetrical*, or upright, folds in which the axial plane is vertical, and each fold limb dips at the same angle (Fig. 10.9). However, if the axial plane is inclined, the limbs dip at different angles,

and the fold is characterized as *asymmetrical* (◆ Fig. 10.11a). In an *overturned fold*, both limbs dip in the same direction. In other words, one fold limb has been rotated more than 90 degrees from its original position such that it is now upside down (◆ Fig. 10.11b). Folds in which the axial plane is horizontal are referred to as *recumbent* (◆ Fig. 10.11c). Overturned and recumbent folds are particularly common in many mountain ranges (discussed later in this chapter).

Plunging Folds

Folds may be further characterized as *nonplunging* or *plunging*. In the former, the fold *axis*, a line formed by the intersection of the axial plane with the folded beds, is horizontal (Fig. 10.9). However, it is much more common for the axis to be inclined so that it appears to plunge beneath the surrounding strata; folds possessing an inclined axis are **plunging folds** (◆ Fig. 10.12). To differentiate plunging anticlines from plunging synclines, geologists use exactly the same criteria used for nonplunging folds: that is, all strata dip away from the fold axis in plunging anticlines, whereas in plunging synclines all strata dip inward toward the axis. The oldest exposed strata are in the center of an eroded plunging anticline, whereas the youngest exposed strata are in the center of an eroded plunging syncline (Fig. 10.12b).

In Chapter 5 we noted that anticlines form a type of structural trap for petroleum and natural gas (see Fig.

◆ FIGURE 10.10 Identifying eroded anticlines and synclines.

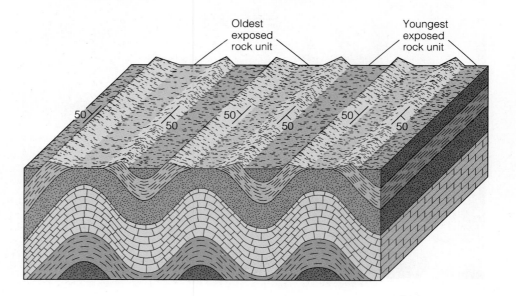

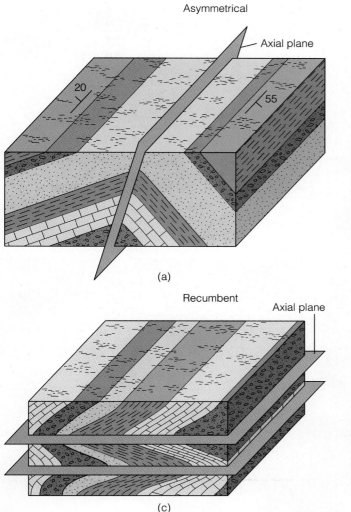

Asymmetrical

Axial plane

20

55

(a)

Overturned

Axis

Axial plane

75   25   75   25

(b)

Recumbent

Axial plane

(c)

◆ FIGURE 10.11 (*a*) An asymmetrical fold. The axial plane is not vertical, and the fold limbs dip at different angles. (*b*) Overturned folds. Both fold limbs dip in the same direction, but one limb is inverted. Notice the special strike and dip symbol to indicate overturned beds. (*c*) Recumbent folds.

5.36). As a matter of fact, most of the world's petroleum production comes from anticlinal traps, although several other types are important as well. Accordingly, geologists are particularly interested in correctly identifying the geologic structures in areas of potential petroleum and natural gas production.

Domes and Basins

Anticlines and synclines are elongate structures; that is, they tend to be long and narrow. **Domes** and **basins**, on the other hand, are the circular to oval equivalents of anticlines and synclines (◆ Fig. 10.13). In an eroded dome, the oldest exposed rock is at the center, whereas in a basin the opposite is true. All of the strata in a

dome dip away from a central point (as opposed to dipping away from a fold axis, which is a line). By contrast, all the strata in a basin dip inward toward a central point (Fig. 10.13).

Many domes and basins are of such large proportions that they can be visualized only on geologic maps or aerial photographs. The Black Hills of South Dakota, for example, are a large oval dome. One of the best-known large basins in the United States is the Michigan basin. Most of the Michigan basin is buried beneath younger strata so it is not directly observable at the surface. Nevertheless, strike and dip of exposed strata near the basin margin and thousands of drill holes for oil and gas clearly show that the strata are deformed into a large structural basin.

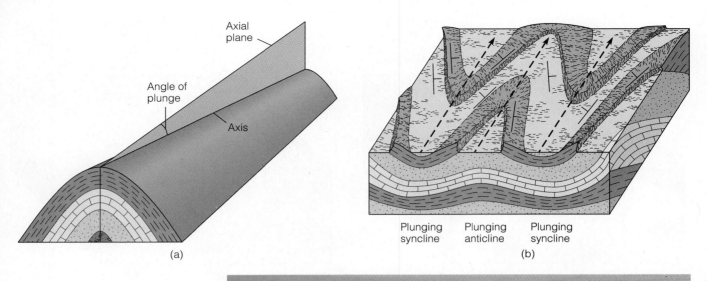

(a)

(b)

Plunging   Plunging   Plunging
syncline   anticline   syncline

◆ FIGURE 10.12 Plunging folds.
(*a*) A schematic illustration of a
plunging fold. (*b*) A block diagram
showing surface and cross-sectional
views of plunging folds. The long
arrow at the center of each fold
shows the direction of plunge.
(*c*) Surface view of the eroded,
plunging Sheep Mountain anticline
in Wyoming.

(c)

Joints

Joints are fractures along which no movement has
occurred, or where movement has been perpendicular
to the fracture surface. In other words, the fracture may
open up, but no relative movement of the masses of
rock on opposite sides of the fracture occurs parallel to
the fracture. The term "joint" was originally used by
coal miners long ago for cracks in rocks that appeared
to be surfaces where adjacent blocks were "joined"
together. Joints are the commonest structures in rocks;

almost all near-surface rocks are jointed to some degree
(◆ Fig. 10.14). The lack of any movement parallel to
joint surfaces is what distinguishes them from *faults,*
which do show movement parallel with the fracture
surface.

Joints can form under a variety of conditions. For
example, anticlines are produced by compression, but
the rock layers are arched such that tension occurs
perpendicular to fold crests, and joints form parallel to
the long axis of the fold in the upper part of a folded
layer (◆ Fig. 10.15a, page 250). Joints also form in

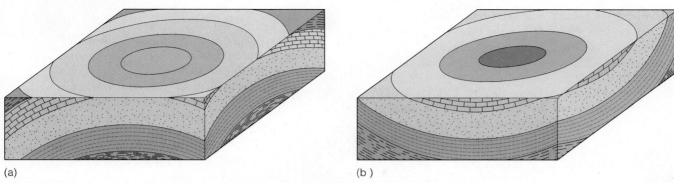

(a)   (b)

◆ FIGURE 10.13 (*a*) A block diagram of a dome. (*b*) A block diagram of a basin.

◆ FIGURE 10.14 Jointed strata on the northeast flank of the Salt Valley anticline, Arches National Park, Utah.

response to tension when rock layers are simply stretched (◆ Fig. 10.15b). Compressive stresses can also produce joints, as shown in ◆ Figure 10.15c.

Joints vary from minute fractures to those of regional extent (Fig. 10.14). Furthermore, they are often arranged in parallel or nearly parallel sets, and it is common for a region to have two or perhaps three prominent sets. Regional mapping reveals that joints and joint sets are usually related to other geologic structures such as large folds. Weathering and erosion of jointed rocks in Utah has produced the spectacular scenery of Arches National Park (see Perspective 10.1).

One type of joint pattern that we have already discussed consists of columnar joints that form in some lava flows and in some intrusive igneous bodies. Recall from Chapters 3 and 4 that as cooling lava contracts, it develops tensional stresses that form polygonal fracture patterns (see Fig. 4.8). Another type of jointing previously discussed is sheet jointing that forms in response to unloading (see Fig. 5.7).

Faults

Faults are fractures along which movement has occurred parallel to the fracture surface. A *fault plane* is the fracture surface along which blocks of rock on opposite sides have moved relative to one another. Notice in ◆ Figure 10.16 (page 250) that the blocks adjacent to the fault plane are labeled *hanging wall block* and *footwall block*. The **hanging wall block** is the block that overlies the fault, whereas the **footwall block** lies beneath the fault plane. Hanging wall and footwall blocks can be defined with respect to any fault plane except those that are vertical. Understanding the concept of hanging wall and footwall blocks is important because geologists use the movement of the hanging wall block relative to the footwall block to distinguish between two different types of faults.

Like layers of rock, fault planes can be characterized by their strike and dip (Fig. 10.16). Two basic types of faults are distinguished on the basis of whether the blocks on opposite sides of the fault plane have moved parallel to the direction of dip or along the direction of strike.

Dip-Slip Faults

Dip-slip faults are those on which all movement is parallel with the dip of the fault plane (◆ Fig. 10.17, page 251). In other words, all movement is such that one block moves up or down relative to the block on the opposite side of the fault. Although it is not

Perspective 10.1

FOLDING, JOINTS, AND ARCHES

Arches National Park in eastern Utah is noted for its panoramic vistas, which include such landforms as Delicate Arch, Double Arch, Landscape Arch, and many others (◆ Fig. 1). Unfortunately, the term *arch* is used for a variety of geologic features of different origin, but here we will restrict the term to mean an opening through a wall of rock that is formed by weathering and erosion.

The arches of Arches National Park continue to form as a result of weathering and erosion of the folded and jointed Entrada Sandstone, the rock underlying much of the park. Accordingly, geologic structures play a significant

role in the origin of arches. Where the Entrada Sandstone was folded into anticlines, it was stretched so that parallel, vertical joints formed. Weathering and erosion occur most vigorously along joints because these processes can attack the exposed rock from both the top and the sides, whereas only the top is attacked in unjointed strata (Fig. 10.14).

Erosion along joints causes them to enlarge, thereby forming long slender fins of rock between adjacent joints. Many such fins are clearly visible in Figure 10.14. Some parts of these fins are more susceptible to weathering and erosion than others, and as the sides are attacked, a recess

◆ FIGURE 1 Delicate Arch in Arches National Park, Utah formed by weathering and erosion of jointed sedimentary rocks, as shown in Figure 3.

◆ FIGURE 2 Baby Arch shows the early development of an arch. (Photo courtesy of Sue Monroe.)

possible to tell how the blocks actually moved, it is usually easy to determine which block appears to have moved up or down in relation to the other. Thus, geologists refer to *relative movement* on faults. For example, in Figure 10.17a, the hanging wall block appears to have moved downward relative to the footwall block. Such faults are called **normal faults**,

whereas those where the hanging wall block moved up relative to the footwall block are **reverse faults** (Fig. 10.17b). A type of reverse fault involving a fault plane with a dip of less than 45° is a **thrust fault** (Fig. 10.17c).

Normal faults are caused by tensional forces, such as those that occur when the Earth's crust is stretched and thinned by rifting. The mountain ranges of a large area

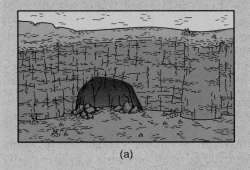

(a)

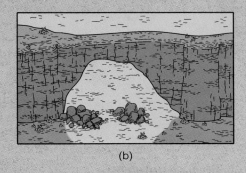

(b)

(c)

◆ FIGURE 3 (*a*) Weathering and erosion of a fin form a recess. (*b*) These recesses expand and eventually develop into arches. (*c*) The arches continue to enlarge until they finally collapse.

may form. If it does, eventually pieces of the unsupported rock above the recess will fall away, forming an arch as the original recess is enlarged (◆ Figs. 2 and 3). Thus, arches are remnants of fins formed by weathering and erosion along joints.

Historical observations show that arches continue to form today. For example, in 1940, Skyline Arch was enlarged when a large block fell from its underside. The park also contains many examples of arches that collapsed during prehistoric time. When arches collapse, they leave isolated pinnacles and spires. Arches National Park is well worth visiting; the pinnacles, spires, and arches areimpressive features indeed.

called the Basin and Range Province in the western United States are bounded on one or both sides by major normal faults. A large normal fault is present along the east side of the Sierra Nevada in California; these mountains have been uplifted along this normal fault so that they now stand more than 3,000 m above the lowlands to the east. Continued normal faulting is also found along the eastern margin of the Teton Range in Wyoming (see the Prologue).

Unlike normal faults, reverse (and thrust) faults are caused by compression (◆ Fig. 10.18, page 252). Many large reverse and thrust faults are present in mountain ranges that form at convergent plate margins (discussed later in the chapter). A well-known thrust fault is the

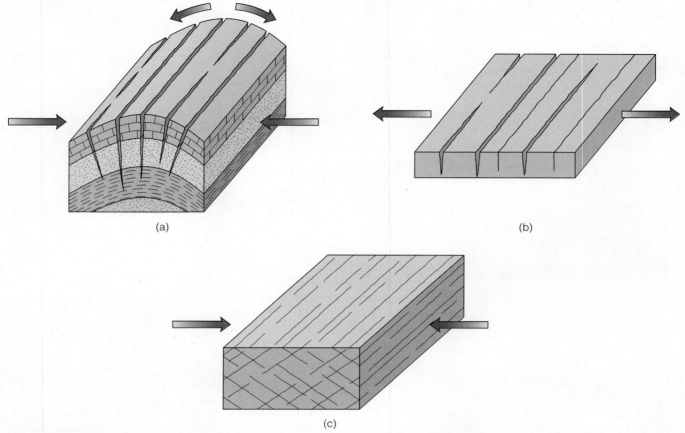

(a)

(b)

(c)

◆ FIGURE 10.15 (*a*) Folding and the formation of joints parallel to the crest of an anticline. (*b*) Joints produced by tension. (*c*) Joints formed in response to compression.

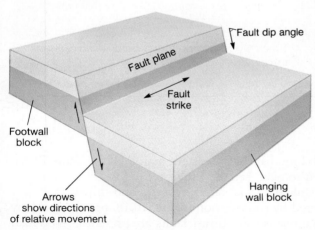

◆ FIGURE 10.16 Fault terminology.

Lewis overthrust of Montana. A large slab of Precambrian-aged rocks moved at least 75 km eastward on this fault and now rests upon much younger rocks of Cretaceous age (see Chapter 23).

Strike-Slip Faults

Shearing forces are responsible for **strike-slip faulting**, a type of faulting involving horizontal movement in which blocks on opposite sides of a fault plane slide sideways past one another (Fig. 10.17d). In other words, all movement is in the direction of the fault plane's strike. One of the best-known strike-slip faults is the San Andreas fault of California.*

. . . .

*Recall from Chapter 9 that the San Andreas fault is also called a transform fault in plate tectonics terminology.

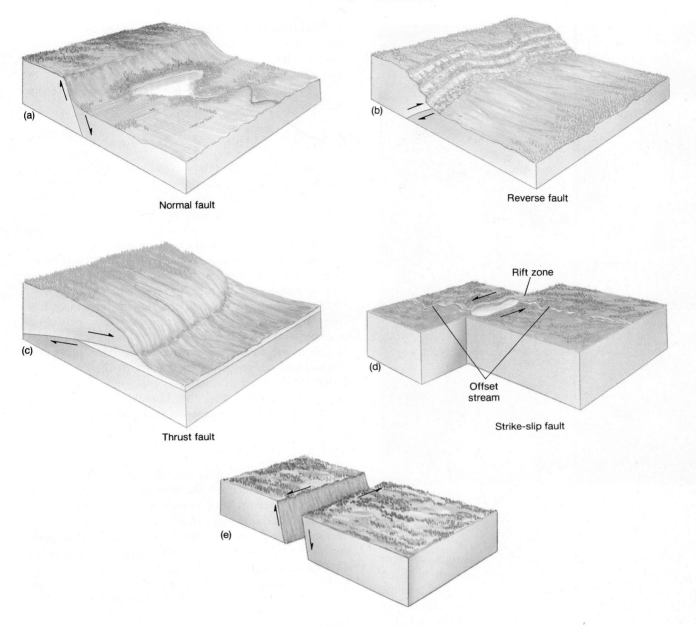

(a)

Normal fault

(b)

Reverse fault

(c)

Thrust fault

(d)

Rift zone

Offset
stream

Strike-slip fault

(e)

Oblique-slip fault

◆ FIGURE 10.17 Types of faults. (*a*), (*b*), and (*c*) are dip-slip faults. (*a*) Normal
fault—hanging wall block down relative to footwall block. (*b*) and (*c*) Reverse and
thrust faults—hanging wall block up relative to footwall block. (*d*) Strike-slip fault—
all movement parallel to strike of fault. (*e*) Oblique-slip fault—combination of dip-slip
and strike-slip.

◆ FIGURE 10.18 Reverse fault in welded tuff, Mojave Desert, California.

Strike-slip faults can be characterized as right-lateral or left-lateral, depending on the apparent direction of offset. In Figure 10.17d, for example, an observer looking at the block on the opposite side of the fault determines whether it appears to have moved to the right or to the left. In this example, movement appears to have been to the left, so the fault is characterized as a *left-lateral strike-slip fault*. Had this been a *right-lateral strike-slip fault*, the block across the fault from the observer would appear to have moved to the right. The San Andreas fault is a right-lateral strike-slip fault (◆ Fig. 10.19).

Oblique-Slip Faults

It is possible for movement on a fault to show components of both dip-slip and strike-slip. For example,

strike-slip movement may be accompanied by a dip-slip component giving rise to a combined movement that includes left-lateral and reverse, or right-lateral and normal (Fig. 10.17e). Faults having components of both dip-slip and strike-slip movement are **oblique-slip faults**.

⯆ MOUNTAINS

The term *mountain* refers to any area of land that stands significantly higher than the surrounding country. Some mountains are single, isolated peaks, but much more commonly they are parts of a linear association of peaks and/or ridges called *mountain ranges* that are related in age and origin. A *mountain system* is a complex mountainous region consisting of several or many mountain ranges; the Rocky Mountains and Appalachians are examples of mountain systems.

Major mountain systems are indeed impressive features and represent the effects of dynamic processes operating within the Earth. They are large-scale manifestations of tremendous forces that have produced folded, faulted, and thickened parts of the crust. Furthermore, in some mountain systems, such as the Andes of South America and the Himalayas of Asia, the mountain-building processes remain active today.

Types of Mountains

Mountainous topography can develop in a variety of ways, some of which involve little or no deformation of the Earth's crust. For example, a single volcanic mountain can develop over a hot spot, but more commonly a series of volcanoes develops as a plate moves over the hot spot, as in the case of the Hawaiian Islands (see Fig. 9.27).

Mountainous topography also forms where the crust has been intruded by batholiths that are subsequently uplifted and eroded (◆ Fig. 10.20). The Sweetgrass Hills of northern Montana consist of resistant plutonic rocks exposed following uplift and erosion of the softer overlying sedimentary rocks.

Yet another way to form mountains—*block-faulting*—involves considerable deformation (◆ Fig. 10.21). Block-faulting involves movement on normal faults so that one or more blocks are elevated relative to adjacent areas. A classic example is the large-scale block-faulting currently occurring in the Basin and Range Province of the western United States, a large area centered on Nevada but extending into several

◆ FIGURE 10.19 Right-lateral offset of a gully by the San Andreas fault in southern California. The gully is offset about 21 m.

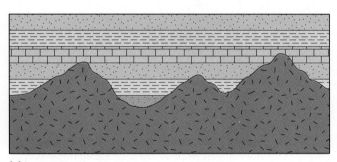

(a)

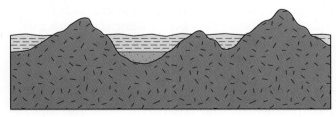

(b)

◆ FIGURE 10.20 (*a*) Pluton intruded into sedimentary rocks. (*b*) Erosion of the softer overlying rocks reveals the pluton and forms small mountains.

adjacent states and northern Mexico. In the Basin and Range Province, the Earth's crust is being stretched in an east-west direction; thus, tensional stresses produce north-south oriented, range-bounding faults. Differential movement on these faults has yielded uplifted blocks called *horsts* and down-dropped blocks called *grabens* (Fig. 10.21). Horsts and grabens are bounded on both sides by parallel normal faults. Erosion of the horsts has yielded the mountainous topography now present, and the grabens have filled with sediments eroded from the horsts (Fig. 10.21).

The processes discussed above can certainly yield mountains. However, the truly large mountain systems of the continents, such as the Alps of Europe and the Appalachians in North America, were produced by compression along convergent plate margins.

▼ MOUNTAIN BUILDING: OROGENESIS

An **orogeny** is an episode of mountain building during which intense deformation occurs, generally accompanied by metamorphism and the emplacement of plutons, especially batholiths. Mountain building, called

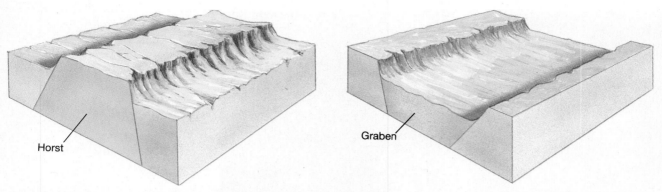

◆ FIGURE 10.21 Block-faulting and the origin of a horst and a graben.

orogenesis, is still not completely understood, but it is known to be related to plate movements. In fact, the advent of plate tectonic theory has completely changed the way geologists view the origin of mountain systems. Any theory accounting for orogenesis must adequately explain the characteristics of mountain systems such as their long, narrow geometry and their location at or near plate margins. The intensity of deformation increases from the continental interior into mountain systems where overturned and recumbent folds and reverse and thrust faults indicating compression are common. Furthermore, both shallow and deep marine sedimentary rocks in mountain systems have been elevated far above sea level—in some cases as high as 9,000 m!

Plate Boundaries and Orogenesis

Most of the Earth's geologically recent and present-day orogenic activity is concentrated in two major zones or belts: the *Alpine-Himalayan orogenic belt* and the *circum-Pacific orogenic belt* (◆ Fig. 10.22). Both belts are composed of a number of smaller segments called *orogens*; each orogen is a zone of deformed rocks, many of which have been metamorphosed and intruded by plutons.

Most orogenies occur at convergent plate boundaries where one plate is subducted beneath another or where two continents collide. Subduction-related orogenies are those involving oceanic-oceanic and oceanic-continental plate boundaries.

Orogenesis at Oceanic-Oceanic Plate Boundaries

Orogenies occurring where oceanic lithosphere is subducted beneath oceanic lithosphere are characterized by the formation of a volcanic island arc and by deformation and igneous activity. Deformation occurs when sediments derived from the volcanic island arc are compressed along a convergent plate boundary. These sediments are deposited on the adjacent sea floor and in the back-arc basin. Those on the sea floor, including sediments deposited in the oceanic trench, are deformed and scraped off against the landward side of the trench (◆ Fig. 10.23), thus forming a subduction complex, or *accretionary wedge*, of intricately folded rocks cut by numerous compression-induced thrust faults. In addition, orogenesis generated by plate convergence results in low-temperature, high-pressure metamorphism characteristic of the blueschist facies (see Fig. 6.20).

Deformation of sedimentary rocks also occurs in the island arc system where it is caused largely by the emplacement of plutons, and many rocks show evidence of high-temperature, low-pressure metamorphism. The overall effect of island arc orogenesis is the origin of two more-or-less parallel orogenic belts consisting of a landward volcanic island arc underlain by batholiths and a seaward belt of deformed trench rocks (Fig. 10.23).

Orogenesis at Oceanic-Continental Plate Boundaries

Many major mountain systems including the Alps of Europe and the Andes of South America formed at oceanic-continental plate boundaries. The Andes are perhaps the best example of such continuing orogeny (Fig. 10.22). Among the ranges of the Andes are the highest mountain peaks in the Americas and many active volcanoes. Furthermore, the west coast of South

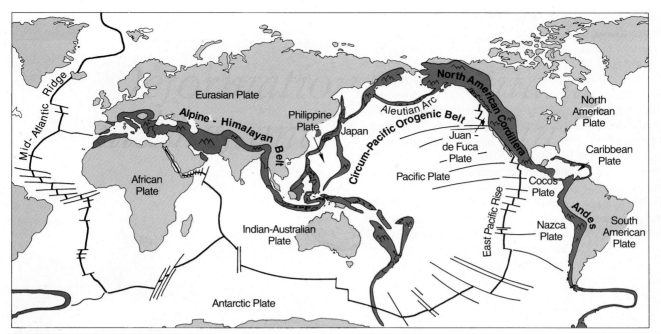

◆ FIGURE 10.22 Most of the Earth's geologically recent and current orogenic activity is concentrated in the circum-Pacific and Alpine-Himalayan orogenic belts.

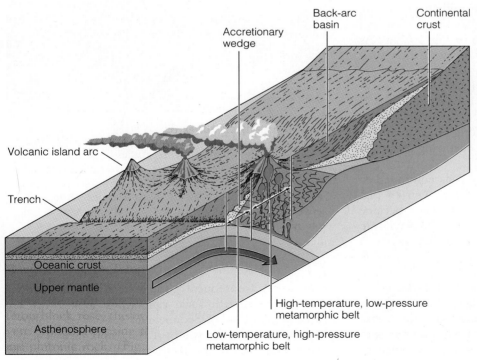

◆ FIGURE 10.23 Orogenesis and the origin of a volcanic island arc at an oceanic-oceanic plate boundary.

Deformation, Mountain Building, and the Continents   255

America is an extremely active segment of the circum-Pacific earthquake belt. One of the Earth's great oceanic trench systems, the Peru-Chile Trench, lies just off the west coast.

Prior to 200 million years ago, the western margin of South America was a site where sediments accumulated on the continental shelf, slope, and rise much as they currently do along the east coast of North America. However, when Pangaea split apart in response to rifting along what is now the Mid-Atlantic Ridge, the South American plate moved westward. As a consequence, the oceanic lithosphere west of South America began subducting beneath the continent (◆ Fig. 10.24). As subduction proceeded, sedimentary rocks of the passive continental margin were folded and faulted and are now part of the accretionary wedge along the west coast of South America. Subduction also resulted in partial melting of the descending plate producing a volcanic arc, and numerous large plutons were emplaced beneath the arc (Fig. 10.24).

Orogenesis at Continental-Continental Plate Boundaries

The best example of orogenesis at a continental-continental plate boundary is provided by the Himalayas. The Himalayas of Asia began forming when India collided with Asia about 40 to 50 million years ago. Prior to that time, India was far south of Asia and separated from it by an ocean basin (◆ Fig. 10.25a). As the Indian plate moved northward, a subduction zone formed along the southern margin of Asia where oceanic lithosphere was consumed. Partial melting generated magma, which rose to form a volcanic arc, and large granite plutons were emplaced into what is now Tibet. At this stage, the activity along Asia's southern margin was similar to what is now occurring along the west coast of South America.

The ocean separating India from Asia continued to close, and India eventually collided with Asia (◆ Fig. 10.25b). As a result, two continental plates became welded, or sutured, together. Thus, the Himalayas are now located within a continent rather than along a continental margin. The exact time of India's collision with Asia is uncertain, but between 40 and 50 million years ago, India's rate of northward drift decreased abruptly—from 15 to 20 cm per year to about 5 cm per year. Because continental lithosphere is not dense enough to be subducted, this decrease in rate seems to mark the time of collision and India's resistance to

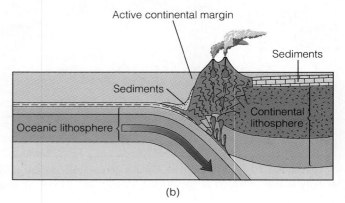

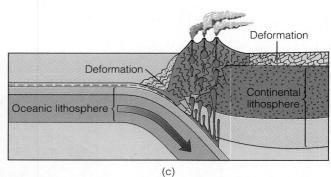

◆ FIGURE 10.24 Generalized diagrams showing three stages in the development of the Andes of South America. (*a*) Prior to 200 million years ago, the west coast of South America was a passive continental margin. (*b*) Orogenesis began when the west coast of South America became an active continental margin. (*c*) Continued deformation, volcanism, and plutonism.

subduction. Consequently, the leading margin of India was thrust beneath Asia, causing crustal thickening, thrusting, and uplift. Sedimentary rocks that had been deposited in the sea south of Asia were thrust northward, and two major thrust faults carried rocks of

Asian origin onto the Indian plate (◆ Fig. 10.25c and d). Rocks deposited in the shallow seas along India's northern margin now form the higher parts of the Himalayas. Since its collision with Asia, India has been underthrust about 2,000 km beneath Asia. Currently, India is moving north at a rate of about 5 cm per year.

◆ FIGURE 10.25 Simplified cross sections showing the collision of India with Asia and the origin of the Himalayas. (*a*) The northern margin of India before its collision with Asia. Subduction of oceanic lithosphere beneath southern Tibet as India approached Asia. (*b*) About 40 to 50 million years ago, India collided with Asia, but since India was too light to be subducted, it was underthrust beneath Asia. (*c*) Continued convergence accompanied by thrusting of rocks of Asian origin onto the Indian Subcontinent. (*d*) Since about 10 million years ago, India has moved beneath Asia along the main boundary fault. Shallow marine sedimentary rocks that were deposited along India's northern margin now form the higher parts of the Himalayas. Sediment eroded from the Himalayas has been deposited on the Ganges Plain.

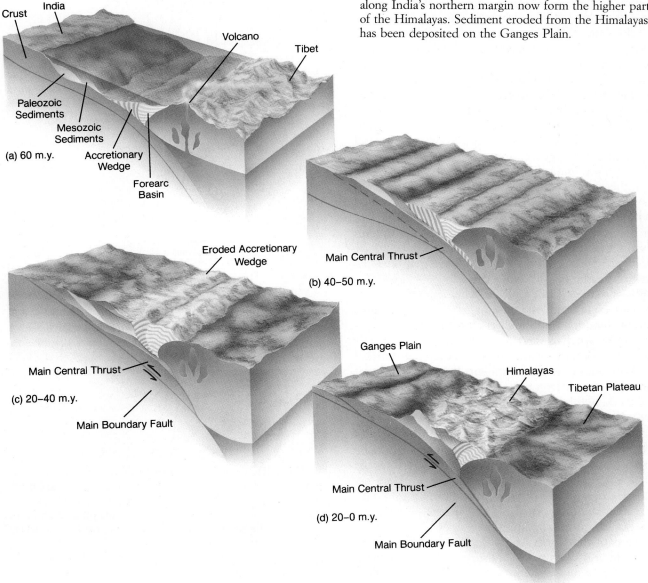

▼ MICROPLATE TECTONICS AND MOUNTAIN BUILDING

In the preceding sections, we discussed orogenies along convergent plate boundaries resulting in continental accretion, the addition of material to a continent along its margin. Much of the material accreted to continents during such events is simply eroded older continental crust, but a significant amount of new material is added to continents as well—igneous rocks that formed as a consequence of subduction and partial melting, for example. While subduction is the predominant influence on the tectonic history in many regions of orogenesis, other processes are also involved in mountain building and continental accretion, especially the accretion of microplates.

During the late 1970s and 1980s, geologists discovered that portions of many mountain systems are composed of small accreted lithospheric blocks that are clearly of foreign origin. These *microplates* differ completely in their fossil content, stratigraphy, structural trends, and paleomagnetic properties from the rocks of the surrounding mountain system and adjacent craton. In fact, these microplates are so different from adjacent rocks that most geologists think that they formed elsewhere and were carried great distances as parts of other plates until they collided with other microplates or continents.

Geologic evidence indicates that more than 25% of the entire Pacific coast from Alaska to Baja California consists of accreted microplates. The accreting microplates are composed of volcanic island arcs, oceanic ridges, seamounts, and small fragments of continents that were scraped off and accreted to the continent's margin as the oceanic plate with which they were carried was subducted under the continent. It is estimated that more than 100 different-sized microplates have been added to the western margin of North America during the last 200 million years (◆ Fig. 10.26).

The basic plate tectonic reconstruction of orogenies and continental accretion remains unchanged, but the details of such reconstructions are decidedly different in view of microplate tectonics. Furthermore, these accreted microplates are often new additions to a continent, rather than reworked older continental material.

So far, most microplates have been identified in mountains of the North American Pacific coast region, but a number of such plates are suspected to be present in other mountain systems as well. They are more difficult to recognize in older mountain systems, such

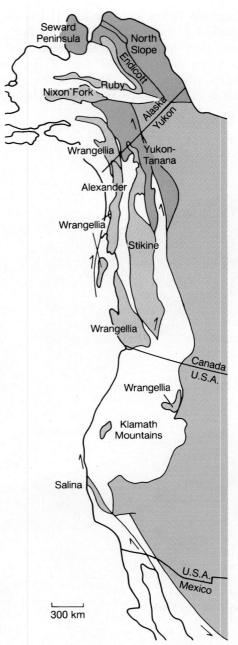

◆ FIGURE 10.26 Some of the accreted lithospheric blocks called microplates that form the western margin of the North American craton. The light brown blocks probably originated as parts of continents other than North America. The reddish brown blocks are possibly displaced parts of North America.

as the Appalachians, however, because of greater deformation and erosion. Nevertheless, about a dozen microplates have been identified in the Appalachians, but their boundaries are hard to identify. Thus, microplate tectonics provides a new way of viewing the Earth and of gaining a better understanding of the geologic history of the continents.

COMPOSITION AND STRUCTURE OF THE EARTH'S CRUST

The Earth's crust is the most accessible and best studied of its concentric layers, but it is also the most complex both chemically and physically. Whereas the core and mantle seem to vary mostly in a vertical dimension, the crust shows considerable vertical and lateral variation, although more lateral variation exists in the mantle than was once thought. The crust along with that part of the upper mantle above the low-velocity zone constitutes the lithosphere of plate tectonic theory.

Two types of crust are recognized—continental crust and oceanic crust—both of which are less dense than the underlying mantle. **Continental crust** is the more complex, consisting of a wide variety of igneous, sedimentary, and metamorphic rocks. It is generally described as "granitic," meaning that its overall composition is similar to that of granitic rocks. Continental crust varies in density depending on rock type, but with the exception of metal-rich rocks, such as iron ore deposits, most rocks have densities of 2.0 to 3.0 g/cm^3, and the overall density is about 2.70 g/cm^3 (see Table 7.5).

The continental crust varies considerably in thickness. It averages about 35 km thick, but is much thinner in such areas as the Rift Valleys of East Africa and the Basin and Range Province in the western United States. The crust in these areas is being stretched and thinned in what appear to be the early stages of rifting; it is as little as 20 km thick in these areas. In contrast, continental crust beneath mountain ranges is much thicker and projects deep into the mantle. For example, beneath the Himalayas of Asia, the continental crust is as much as 90 km thick. Crustal thickening beneath mountain ranges is an important point that will be discussed in "The Principle of Isostasy" later in the chapter.

Each continent is characterized by one or more areas of exposed ancient rocks called a **Precambrian shield** (◆ Fig. 10.27). Extending outward from these shields are broad platforms of ancient rocks that are buried beneath younger sediments and sedimentary rocks. The

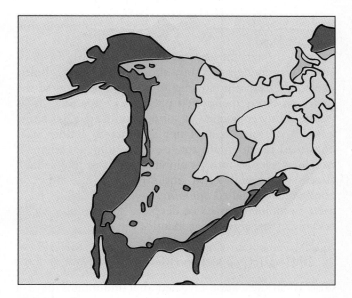

◻ Canadian Shield

◼ Other exposed Precambrian rocks

◻ Covered Precambrian rocks

◆ FIGURE 10.27 The North American craton. The Canadian Shield is a large area of exposed Precambrian-aged rocks. Extending from the shield are platforms of buried Precambrian rocks. The shield and platforms collectively make up the craton.

shields and buried platforms are collectively called **cratons,** so shields are simply the exposed parts of cratons. Cratons are considered to be the stable interior parts of continents.

In North America, the *Canadian Shield* includes much of Canada; a large part of Greenland; parts of the Lake Superior region in Minnesota, Wisconsin, and Michigan; and parts of the Adirondack Mountains of New York (Fig. 10.27). In general, the Canadian Shield is a vast area of subdued topography, numerous lakes, and exposed ancient metamorphic, volcanic, plutonic, and sedimentary rocks. The geologic history of the cratons will be discussed in some detail in Chapters 20 through 24.

Oceanic crust is less complex than continental crust. Sampling and direct observations of the oceanic ridges reveal that the upper part of the oceanic crust is composed of basalt. Much of this basalt is in the form of pillow lavas (see Fig. 4.9), but sheet flows are also present. Deep-sea drill holes have penetrated through the upper oceanic crust into a sheeted dike complex, a zone consisting almost entirely of vertical basaltic dikes.

What lies below this sheeted dike complex has not been sampled.

Even though the oceanic crust is 5 to 10 km thick and can be penetrated only about 1 km by drill holes, geologists have a good idea of the composition of the entire crust. As mentioned previously, oceanic crust is continuously consumed at subduction zones, but a tiny amount of this crust is not subducted. Rather it is emplaced in mountain ranges on continents, where it usually arrives by moving along large thrust faults. Such slivers of oceanic crust and upper mantle now on continents are called *ophiolites* (see Fig. 9.23). A complete ophiolite consists of deep-sea sedimentary rocks, oceanic crust, and upper mantle.

⩔ THE PRINCIPLE OF ISOSTASY

Gravity studies have revealed that mountains have a low-density "root" projecting deep into the mantle. If it were not for this low-density root, a gravity survey across a mountainous area would reveal a huge positive gravity anomaly. The fact that no such anomaly exists indicates that some of the dense mantle at depth must be displaced by lighter crustal rocks, as shown in ◆ Figure 10.28.

According to the **principle of isostasy,** the Earth's crust is in floating equilibrium with the more dense mantle below. This phenomenon is easy to understand by an analogy to an iceberg. Ice is slightly less dense than water, and thus it floats. However, according to Archimedes' principle of buoyancy, an iceberg will sink in the water until it displaces a volume of water that

equals its total weight. When the iceberg has sunk to an equilibrium position, only about 10% of its volume is above water level. If some of the ice above water level should melt, the iceberg will rise in order to maintain the same proportion of ice above and below water.

The Earth's crust is similar to the iceberg, or a ship, in that it sinks into the mantle to its equilibrium level. Where the crust is thickest, as beneath mountain ranges, it sinks further down into the mantle but also rises higher above the equilibrium surface (Fig. 10.28). Continental crust being thicker and less dense than oceanic crust stands higher than the ocean basins. Should the crust be loaded, as where widespread glaciers accumulate, it responds by sinking further into the mantle to maintain equilibrium. In Greenland and Antarctica, for example, the surface of the crust has been depressed below sea level by the weight of glacial ice. The crust also responds isostatically to widespread erosion and sediment deposition (◆ Fig. 10.29).

Unloading of the Earth's crust causes it to respond by rising upward until equilibrium is again attained. This phenomenon, known as **isostatic rebound,** occurs in areas that are deeply eroded and in areas that were formerly glaciated. Scandinavia, for example, which was covered by a vast ice sheet until about 10,000 years ago, is still rebounding isostatically at a rate of up to 1 m per century. Isostatic rebound has also occurred in eastern Canada, where the land has risen as much as 100 m during the last 6,000 years (◆ Fig. 10.30).

The principle of isostasy implies that the mantle behaves as a liquid. In preceding discussions, however, we said that the mantle must be solid because it

◆ FIGURE 10.28 (*a*) Gravity measurements along the line shown would indicate a positive gravity anomaly over the excess mass of the mountains if the mountains were simply thicker crust resting on denser material below. (*b*) An actual gravity survey across a mountain region shows no departure from the expected and thus no gravity anomaly. Such data indicate that the mass of the mountains above the surface must be compensated for at depth by low-density material displacing denser material.

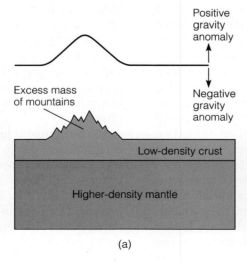

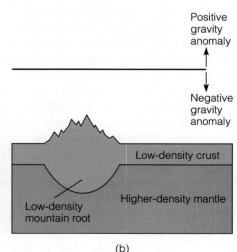

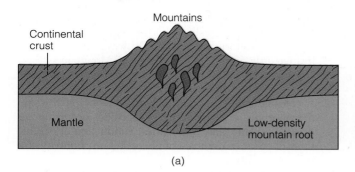

(a)

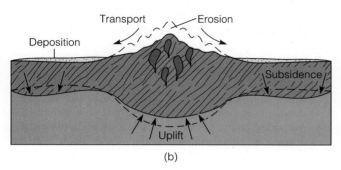

(b)

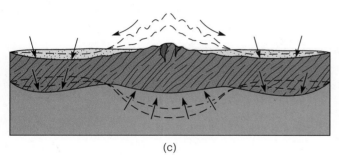

(c)

◆ FIGURE 10.29 A diagrammatic representation showing the isostatic response of the crust to erosion (unloading) and widespread deposition (loading). Notice in (b) and (c) that isostatic rebound occurs as the mountain range is eroded, and subsidence of the crust occurs where deposition occurs.

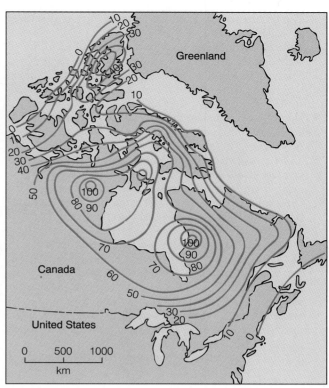

◆ FIGURE 10.30 Isostatic rebound in eastern Canada in meters during the last 6,000 years.

transmits S-waves, which will not move through a liquid. How can this apparent paradox be resolved? When considered in terms of the short time necessary for S-waves to pass through it, the mantle is indeed solid. However, when subjected to stress over long periods of time, it will yield by flowage and thus at these time scales can be considered a viscous liquid. A familiar substance that has the properties of a solid or a liquid depending on how rapidly deforming forces are applied is silly putty. It will flow under its own weight if given sufficient time, but shatters as a brittle solid if struck a sharp blow.

Chapter Summary

1. Contorted and fractured rocks have been deformed or strained by applied stresses.
2. Stresses are characterized as compressional, tensional, or shear. Elastic strain is not permanent, meaning that when the stress is removed, the rocks return to their original shape or volume. Plastic strain and fracture are both permanent types of deformation.

3. The orientation of deformed layers of rock is described by strike and dip.
4. Rock layers that have been buckled into up- and down-arched folds are anticlines and synclines, respectively. They can be identified by the strike and dip of the folded rocks and by the relative age of the rocks in the center of eroded folds.
5. Domes and basins are the circular to oval equivalents of anticlines and synclines, but are commonly much larger structures.
6. Two types of structures resulting from fracturing are recognized: joints are fractures along which the only movement, if any, is perpendicular to the fracture surface, and faults are fractures along which the blocks on opposite sides of the fracture move parallel to the fracture surface.
7. Joints, which are the commonest geologic structures, form in response to compression, tension, and shear.
8. On dip-slip faults, all movement is in the dip direction of the fault plane. Two varieties of dip-slip faults are recognized: normal faults form in response to tension, while reverse faults are caused by compression.
9. Strike-slip faults are those on which all movement is in the direction of strike of the fault plane. They are characterized as right-lateral or left-lateral depending on the apparent direction of offset of one block relative to the other.
10. Some faults show components of both dip-slip and strike-slip; they are called oblique-slip faults.
11. Mountains can form in a variety of ways, some of which involve little or no folding or faulting. Mountain systems consisting of several mountain ranges result from deformation related to plate movements.
12. A volcanic island arc, deformation, igneous activity, and metamorphism characterize orogenies occurring at oceanic-oceanic plate boundaries. Subduction of oceanic lithosphere at an oceanic-continental plate boundary also results in orogeny.
13. Some mountain systems, such as the Himalayas, are within continents far from a present-day plate boundary. Such mountains formed when two continental plates collided and became sutured.
14. The oceanic and continental crusts are basaltic and granitic in composition, respectively.
15. A craton is the stable core of a continent. Broad areas in which the cratons of continents are exposed are called shields; each continent has at least one shield area.
16. According to the principle of isostasy, the Earth's crust is floating in equilibrium with the denser mantle below. Continental crust stands higher than oceanic crust because it is thicker and less dense.

Important Terms

anticline
basin
compression
continental crust
craton
dip
dip-slip fault
dome
elastic strain
fault
footwall block
fracture

hanging wall block
isostatic rebound
joint
monocline
normal fault
oblique-slip fault
oceanic crust
orogeny
plastic strain
plunging fold
Precambrian shield

principle of isostasy
reverse fault
shear stress
shield
strain
stress
strike
strike-slip fault
syncline
tension
thrust fault

Review Questions

1. Strain is characterized as _____ if deformed rocks regain their shape when they are no longer subjected to stress.
   a. ___ compression; b. ___ elastic; c. ___ tensional; d. ___ plastic; e. ___ shear.
2. Rocks that show a large amount of plastic strain are said to be:
   a. ___ brittle; b. ___ fractured; c. ___ ductile; d. ___ sheared; e. ___ all of these.
3. An elongate fold in which all the strata dip in toward the center is a(n):
   a. ___ dome; b. ___ monocline; c. ___ basin; d. ___ syncline; e. ___ anticline.

4. An overturned fold is one in which:
   a. ___ both limbs dip in the same direction; b. ___ the axial plain is vertical; c. ___ the axis is inclined; d. ___ the strata in one limb are horizontal; e. ___ the strata are faulted as well as folded.
5. An oval to circular fold with all strata dipping outward from a central point is a(n):
   a. ___ plunging anticline; b. ___ dome; c. ___ overturned syncline; d. ___ recumbent syncline; e. ___ basin.
6. A fault on which the hanging wall block appears to have moved down relative to the footwall block is a _____ fault.
   a. ___ thrust; b. ___ strike-slip; c. ___ normal; d. ___ reverse; e. ___ joint.
7. Faults on which both dip-slip and strike-slip movement has occurred are referred to as:
   a. ___ plunging; b. ___ recumbent; c. ___ oblique-slip; d. ___ nonplunging; e. ___ normal-slip.
8. A graben is a:
   a. ___ fold with a horizontal axial plane; b. ___ type of reverse fault with a very low dip; c. ___ fracture along which no movement has occurred; d. ___ down-dropped block bounded by normal faults; e. ___ type of structure resulting from compression.
9. Strike-slip faults:
   a. ___ are low-angle reverse faults; b. ___ have mainly vertical displacement; c. ___ have mainly horizontal movement; d. ___ are faults on which no movement has yet occurred; e. ___ are characterized by uplift of the footwall block.
10. Solids that have been deformed by movement along closely spaced slippage planes are said to have been:
    a. ___ sheared; b. ___ folded; c. ___ subjected to tension; d. ___ elastically strained; e. ___ overturned.
11. Fractures along which no movement has occurred are:
    a. ___ joints; b. ___ monoclines; c. ___ transform faults; d. ___ axial planes; e. ___ fold limbs.
12. The intersection of an inclined plane with a horizontal plane is the definition of:
    a. ___ horizontal strata; b. ___ dip-slip movement; c. ___ folded strata; d. ___ strike; e. ___ joint.
13. In mountain systems that form at continental margins:
    a. ___ the Earth's crust is thicker than average; b. ___ most deformation is caused by tensional stresses; c. ___ little or no volcanic activity occurs; d. ___ stretching and thinning of the continental crust occur; e. ___ most deformation results from rifting.
14. Continental crust has an overall composition corresponding closely to that of:
    a. ___ basalt; b. ___ sandstone; c. ___ granite; d. ___ iron-nickel alloy; e. ___ gabbro.
15. Oceanic crust is:
    a. ___ 20 to 90 km thick; b. ___ thinnest at spreading ridges; c. ___ granitic in composition; d. ___ less dense than continental crust; e. ___ the primary source of magma.
16. According to the principle of isostasy:
    a. ___ more heat escapes from oceanic crust than from continental crust; b. ___ the Earth's crust is floating in equilibrium with the more dense mantle below; c. ___ the Earth's crust behaves both as a liquid and a solid; d. ___ much of the asthenosphere is molten; e. ___ magnetic anomalies result when the crust is loaded by glacial ice.
17. What types of evidence indicate that stress remains active within the Earth?
18. How do compression, tension, and shear differ from one another?
19. Explain how the factors of rock type, time, temperature, and pressure influence the type of strain in rocks.
20. Domes and basins show the same patterns on geologic maps, but differ in two important ways. What are the two criteria for distinguishing between them?
21. How do joints differ from faults?
22. Draw a simple cross section showing the displacement on a normal fault.
23. Explain what is meant by an oblique-slip fault.
24. Draw a simple sketch map showing the displacement on a left-lateral strike-slip fault.
25. Cite two examples of mountain systems in which mountain-building processes remain active.
26. Explain why two roughly parallel orogenic belts develop where oceanic lithosphere is subducted beneath continental lithosphere.
27. How do geologists account for mountain systems within continents, such as the Himalayas of Asia?
28. What is the difference between a reverse fault and a thrust fault?
29. How do oceanic and continental crust differ in composition and thickness?
30. If the continental crust is deeply eroded in one area and loaded by widespread, thick sedimentary deposits in another, how will it respond isostatically at each location?

Additional Readings

Davis, G. H. 1984. *Structural geology of rocks and regions.* New York: Wiley.

Dennis, J. G. 1987. *Structural geology: An introduction.* Dubuque, Iowa: William C. Brown.

Hatcher, R. D., Jr. 1990. *Structural geology: Principles, concepts, and problems.* Columbus, Ohio: Merrill.

Howell, D. G. 1985. Terranes. *Scientific American* 253, no. 5:116–125.

_____. 1989. *Tectonics of suspect terranes: Mountain building and continental growth.* London: Chapman and Hall.

Jones, D. L., A. Cox, P. Coney, and M. Beck. 1982. The growth of western North America. *Scientific American* 247, no. 5:70–84.

Lisle, R. J. 1988. *Geological structures and maps: A practical guide.* New York: Pergamon.

Miyashiro, A., K. Aki, and A. M. C. Segnor. 1982. *Orogeny.* New York: Wiley.

Molnar, P. 1986. The geologic history and structure of the Himalaya. *American Scientist* 74, no. 2:144–154.

_____. 1986. The structure of mountain ranges. *Scientific American* 255, no. 1:70–79.

Spencer, E. W. 1988. *Introduction to the structure of the Earth.* New York: McGraw-Hill.

CHAPTER
11

Chapter Outline

Hong Kong's most destructive landslide occurred on Po Shan road on June 18, 1972. Sixty-seven people were killed when a 68-m wide portion of this steep hillside failed, destroying a four-story building and a 13-story apartment block.

Mass Wasting

Prologue

On May 31, 1970, a devastating earthquake occurred about 25 km west of Chimbote, Peru. High in the Peruvian Andes, about 65 km to the east, the violent shaking from the earthquake tore loose a huge block of snow, ice, and rock from the north peak of Nevado Huascarán (6,654 m), setting in motion one of this century's worst landslides. Free-falling for about 1,000 m, this block of material smashed to the ground, displacing thousands of tons of rock and generating a gigantic debris flow (◆ Fig. 11.1). Hurtling down the mountain's steep glacial valley at speeds up to 320 km per hour, the avalanche, consisting of more than 50,000,000 m^3 of mud, rock, and water, flowed over ridges 140 m high obliterating everything in its path.

About 3 km east of the town of Yungay, part of the debris flow overrode the valley walls and within seconds buried Yungay, instantly killing more than 20,000 of its residents (Fig. 11.1). The main mass of the flow continued down the valley, overwhelming the town of Ranrahirca and several other villages and burying about 5,000 more people. By the time the flow reached the bottom of the valley, its momentum carried it across the Rio Santa and some 60 m up the opposite bank. In a span of roughly four minutes from the time of the initial ground shaking, approximately 25,000 people died, and most of the area's transportation, power, and communication network was destroyed.

Ironically, the only part of Yungay that was not buried was Cemetery Hill, where 92 people survived by running to its top (◆ Fig. 11.2). A Peruvian geophysicist who was giving a French couple a tour

of Yungay provided a vivid eyewitness account of the disaster:

As we drove past the cemetery the car began to shake. It was not until I had stopped the car that I realized that we were experiencing an earthquake. We immediately got out of the car and observed the effects of the earthquake around us. I saw several homes as well as a small bridge crossing a creek near Cemetery Hill collapse. It was, I suppose, after about one-half to three-quarters of a minute when the earthquake shaking began to subside. At that time I heard a great roar coming from Huascarán. Looking up, I saw what appeared to be a cloud of dust and it looked as though a large mass of rock and ice was breaking loose from the north peak. My immediate reaction was to run for the high ground of Cemetery Hill, situated about 150 to 200 m away. I began running and noticed that there were many others in Yungay who were also running toward Cemetery Hill. About half to three-quarters of the way up the hill, the wife of my friend stumbled and fell and I turned to help her back to her feet.

The crest of the wave had a curl, like a huge breaker coming in from the ocean. I estimated the wave to be at least 80 m high. I observed hundreds of people in Yungay running in all directions and many of them toward Cemetery Hill. All the while, there was a continuous loud roar and rumble. I reached the upper level of the cemetery near the top just as the debris flow struck the base of the hill and I was probably only 10 seconds ahead of it.

At about the same time, I saw a man just a few meters down hill who was carrying two small children toward the hilltop. The debris flow caught him and he threw the two children toward the hilltop, out of the path of the flow, to safety, although the debris flow swept him down the valley, never to be seen again. I also remember two women who were no more than a few meters behind me and I never did

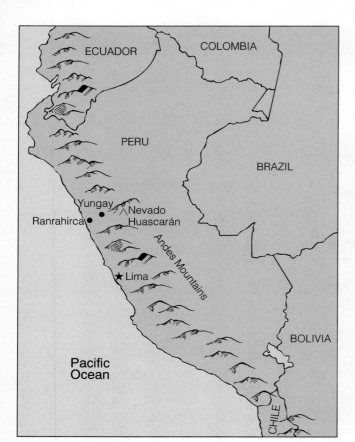

◆ FIGURE 11.1 An earthquake 65 km away triggered a landslide on Nevado Huascarán, Peru, that destroyed the towns of Yungay and Ranrahirca and killed more than 25,000 people.

◆ FIGURE 11.2 Cemetery Hill was the only part of Yungay to escape the 1970 landslide that destroyed the rest of the town. Only 92 people survived the destruction by running to the top of the hill.

● **TABLE 11.1** Selected Landslides, Their Cause, and the Number of People Killed

Date	Location	Type	Deaths
218 B.C.	Alps (European)	Avalanche—destroyed Hannibal's army	18,000
1512	Alps (Biasco)	Landslide—temporary lake burst	>600
1556	China (Hsian)	Landslides—earthquake triggered	1,000,000
1689	Austria (Montaton Valley)	Avalanche	>300
1806	Switzerland (Goldau)	Rock glide	457
1881	Switzerland (Elm)	Rockfall	115
1892	France (Haute-Savoie)	Icefall, mudflow	150
1903	Canada (Frank, Alberta)	Rock glide	70
1920	China (Kansu)	Landslides—earthquake triggered	~200,000
1936	Norway (Loen)	Rockfall into fiord	73
1941	Peru (Huaraz)	Avalanche and mudflow	7,000
1959	USA (Madison Canyon, Montana)	Landslide—earthquake triggered	26
1962	Peru (Mt. Huascarán)	Ice avalanche and mudflow	~4,000
1963	Italy (Vaiont Dam)	Landslide—subsequent flood	~2,000
1966	Brazil (Rio de Janeiro)	Landslides	279
1966	United Kingdom (Aberfan, South Wales)	Debris flow—collapse of mining waste tip	144
1970	Peru (Mt. Huascarán)	Rockfall and debris flow—earthquake triggered	25,000
1971	Canada (St. Jean-Vianney, Quebec)	Quick clays	31
1972	USA (West Virginia)	Landslide and mudflow—collapse of mining waste tip	400
1974	Peru (Mayunmarca)	Rock glide and debris flow	430
1978	Japan (Myoko Kogen Machi)	Mudflow	12
1979	Indonesia (Sumatra)	Landslide	80
1980	USA (Washington)	Avalanche and mudflow	63
1981	Indonesia (West Irian)	Landslide—earthquake triggered	261
1981	Indonesia (Java)	Mudflow	252
1983	Iran (Northern area)	Landslide and avalanche	90
1987	El Salvador (San Salvador)	Landslide	1,000
1988	Chile (Tupungatito area)	Mudflow	41
1989	Tadzhikistan	Mudflow—earthquake triggered	274
1989	Indonesia (West Irian)	Landslide—earthquake triggered	120
1991	Guatemala (Santa Maria)	Landslide	33

SOURCE: Data from J. Whittow, *Disasters: The Anatomy of Environmental Hazards* (Athens, Ga.: University of Georgia Press, 1979) and *Geotimes*.

see them again. Looking around, I counted 92 persons who had also saved themselves by running to the top of the hill. It was the most horrible thing I have ever experienced and I will never forget it.*

As tragic and devastating as this debris avalanche was, it was not the first time a destructive landslide had swept down the Rio Shacsha valley. In January 1962, another large chunk of snow, ice, and rock broke off from the main glacier and generated a large debris avalanche that buried several villages and killed about 4,000 people.

. . . .

*B. A. Bolt et al., *Geological Hazards* (New York: Springer-Verlag, 1977), pp. 37–39.

▼ INTRODUCTION

Geologists use the term *landslide* in a general sense to cover a wide variety of mass movements that may cause loss of life, property damage, or a general disruption of human activities. For example, in 218 B.C., avalanches in the European Alps buried 18,000 people; an earthquake-generated landslide in Hsian, China, killed an estimated 1,000,000 people in 1556; and 7,000 people died when mudflows and avalanches destroyed Huaraz, Peru, in 1941. What makes these mass movements so terrifying, and yet so fascinating, is that they almost always occur with little or no warning and are over in a very short time, leaving behind a legacy of death and destruction (● Table 11.1).

Every year about 25 people are killed by landslides in the United States alone, while the total annual cost of damages from them exceeds $1 billion. Almost all of the major landslides have natural causes, yet many of the smaller ones are the result of human activity and could have been prevented or their damage minimized.

Mass wasting (also called *mass movement*) is defined as the downslope movement of material under the direct influence of gravity. Most types of mass wasting are aided by weathering and usually involve surficial material. The material moves at rates ranging from almost imperceptible, as in the case of creep, to extremely fast as in a rockfall or slide. While water can play an important role, the relentless pull of gravity is the major force behind mass wasting.

Mass wasting is an important geologic process that can occur at any time and almost any place. While most people associate mass wasting with steep and unstable slopes, it can also occur on near-level land, given the right geologic conditions. Furthermore, while the rapid types of mass wasting, such as avalanches and mudflows, typically get the most publicity, the slow, imperceptible types, such as creep, usually do the greatest amount of property damage.

▼ FACTORS INFLUENCING MASS WASTING

When the gravitational force acting on a slope exceeds its resisting force, slope failure (mass wasting) occurs. The resisting forces helping to maintain slope stability include the slope material's strength and cohesion, the amount of internal friction between grains, and any external support of the slope (◆ Fig.

11.3). These factors collectively define a slope's **shear strength.**

Opposing a slope's shear strength is the force of gravity. Gravity operates vertically but has a component acting parallel to the slope, thereby causing instability (Fig. 11.3). The greater a slope's angle, the greater the component of force acting parallel to the slope, and the greater the chance for mass wasting. The steepest angle that a slope can maintain without collapsing is its *angle of repose.* At this angle, the shear strength of the slope's material exactly counterbalances the force of gravity. For unconsolidated material, the angle of repose normally ranges from 25° to 40°. Slopes steeper than 40° usually consist of unweathered solid rock.

All slopes are in a state of dynamic equilibrium, which means that they are constantly adjusting to new conditions. While we tend to view mass wasting as a disruptive and usually destructive event, it is one of the ways that a slope adjusts to new conditions. Whenever a building or road is constructed on a hillside, the equilibrium of that slope is affected. The slope must then adjust, perhaps by mass wasting, to this new set of conditions.

Many factors can cause mass wasting: a change in slope gradient, weakening of material by weathering, increased water content, changes in the vegetation cover, and overloading. Although most of these are interrelated, we will examine them separately for ease of discussion, but will also show how they individually and collectively affect a slope's equilibrium.

Slope Gradient

Slope gradient is probably the major cause of mass wasting. Generally speaking, the steeper the slope, the

◆ FIGURE 11.3 A slope's shear strength depends on the slope material's strength and cohesiveness, the amount of internal friction between grains, and any external support of the slope. These factors promote slope stability. The force of gravity operates vertically but has a component acting parallel to the slope. When this force, which promotes instability, exceeds a slope's shear strength, slope failure occurs.

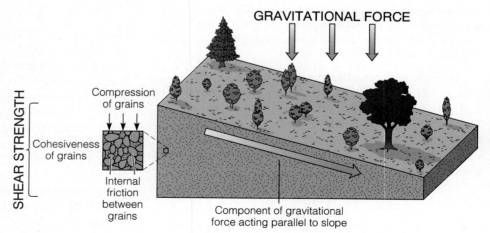

less stable it is. Therefore, steep slopes are more likely to experience mass wasting than gentle ones.

A number of processes can oversteepen a slope. One of the most common is undercutting by stream or wave action (◆ Fig. 11.4). This removes the slope's base, increases the slope angle, and thereby increases the gravitational force acting parallel to the slope. Wave action, especially during storms, often results in mass movements along the shores of oceans or large lakes.

Excavations for road cuts and hillside building sites are another major cause of slope failure (◆ Fig. 11.5). Grading the slope too steeply, or cutting into its side, increases the stress in the rock or soil to the point that it is no longer strong enough to remain at the steeper angle and mass movement ensues. Such action is analogous to undercutting by streams or waves and has the same result, thus explaining why so many mountain roads are plagued by frequent mass movements. Fortunately, many of the slope failures associated with hillside roadcuts and building construction can be avoided or greatly minimized by better understanding of the factors involved.

Weathering and Climate

Mass wasting is more likely to occur in loose or poorly consolidated slope material than in solid bedrock. As soon as solid rock is exposed at the Earth's surface, weathering begins to disintegrate and decompose it, thus reducing its shear strength and increasing its susceptibility to mass wasting. The deeper the weathering zone extends, the greater the likelihood of some type of mass movement.

Recall from Chapter 5 that some rocks are more susceptible to weathering than others and that climate plays an important role in the rate and type of weathering. In the tropics, for example, where temperatures are high and considerable rainfall occurs, the effects of weathering extend to depths of several tens of meters, and rapid mass movements most commonly occur in the deep weathering zone. In arid and semiarid regions, however, the weathering zone is usually considerably shallower. Nevertheless, localized and intense cloudbursts can drop large quantities of water on an area in a short time. With little vegetation to absorb this water, runoff is rapid and frequently results in mudflows.

Water Content

The amount of water in rock or soil influences slope stability. Large quantities of water from melting snow

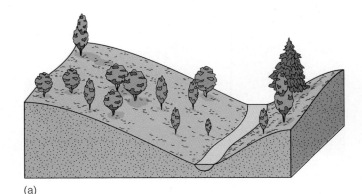
(a)

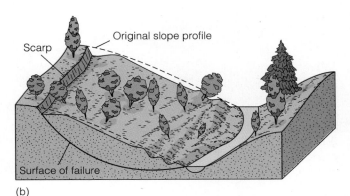
(b)

(c)

◆ FIGURE 11.4 Undercutting by stream erosion (*a*) removes a slope's base, which increases the slope angle and (*b*) can lead to slope failure. (*c*) Undercutting by stream erosion caused slumping along this stream near Weidman, Michigan.

or heavy storms greatly increase the likelihood of slope failure. The additional weight that water adds to a slope can be enough to cause mass movement. Furthermore,

◆ FIGURE 11.5 (*a*) Highway excavations disturb the equilibrium of a slope by (*b*) removing a portion of its support as well as oversteepening it at the point of excavation. (*c*) Such action can result in frequent landslides. (*d*) Cutting into the hillside to construct this portion of the Pan American Highway in Mexico resulted in a rockfall that completely blocked the road. (Photo courtesy of R. V. Dietrich.)

water percolating through a slope's material helps to decrease friction between grains, contributing to a loss of cohesion. For example, slopes composed of dry clay are usually quite stable, but when wetted, they quickly lose cohesiveness and internal friction and become an unstable slurry. This occurs because clay, which can hold large quantities of water, consists of platy particles that easily slide over each other when wet. For this reason, clay beds are frequently the slippery layer along which overlying rock units slide downslope (see Perspective 11.1).

Vegetation

Vegetation affects slope stability in several ways. By absorbing the water from a rainstorm, vegetation decreases water saturation of a slope's material and the resultant loss of shear strength that frequently leads to mass wasting. In addition, the root system of vegetation helps to stabilize a slope by binding soil particles together and holding the soil to bedrock.

The removal of vegetation by either natural or human activity is a major cause of many mass movements. For example, summer brush and forest fires in southern California frequently leave the hillsides bare of vegetation. Fall rainstorms saturate the ground causing mudslides that do tremendous damage and cost millions of dollars to clean up (◆ Fig. 11.6). The soils of many hillsides in New Zealand are sliding because deep-rooted native bushes have been replaced by shallow-rooted grasses used for sheep grazing. When heavy rains saturate the soil, the shallow-rooted grasses cannot hold the slope in place, and parts of it slide downhill.

Overloading

Overloading is almost always the result of human activity and typically results from dumping, filling, or piling up of material. Under natural conditions, a material's load is carried by its grain-to-grain contacts, and a slope is thus maintained by the friction between the grains. The additional weight created by overloading, however, increases the water pressure within the material, which in turn decreases its shear strength, thereby weakening the slope material. If enough material is added, the slope will eventually fail, sometimes with tragic consequences.

Geology and Slope Stability

The relationship between topography and the geology of an area is important in determining slope stability (◆ Fig. 11.7, page 276). If the rocks underlying a slope dip in the same direction as the slope, mass wasting is more likely to occur than if the rocks are horizontal or dip in the opposite direction. When the rocks dip in the same direction as the slope, water can percolate along the various bedding planes and decrease the cohesiveness and friction between adjacent rock units (Fig. 11.7a). This is particularly true when there are interbedded clay layers because clay becomes very slippery when wet.

Even if the rocks are horizontal or dip in a direction opposite to that of the slope, joints may dip in the same direction as the slope. Water migrating through them weathers the rock and expands these openings until the weight of the overlying rock causes it to fall (Fig. 11.7b).

Triggering Mechanisms

While the factors previously discussed all contribute to slope instability, most—though not all—rapid mass

◆ FIGURE 11.6 A California Highway Patrol officer stands on top of a 2-m-high wall of mud that rolled over a patrol car near the Golden State Freeway on October 23, 1987. Flooding and mudslides also trapped other vehicles and closed the freeway.

THE TRAGEDY AT ABERFAN, WALES

· · · · · · · · · · · · · · · · · · · ·

The debris brought out of underground coal mines in southern Wales typically consists of a wet mixture of various sedimentary rock fragments. This material is usually dumped along the nearest valley slope where it builds up into large waste piles called tips. A tip is fairly stable as long as the material composing it is relatively dry and its sides are not oversteepened.

Between 1918 and 1966, seven large tips composed of mine debris had been built at various elevations on the valley slopes above the small coal-mining village of Aberfan. Shortly after 9:00 A.M. on October 21, 1966, the 250 m high, rain-soaked Tip No. 7 collapsed, and a black sludge flowed down the valley with the roar of a loud train (◆ Fig. 1). Before it came to a halt 800 m from its starting place, the flow had destroyed two farm cottages, crossed a canal, and buried Pantglas Junior School, suffocating virtually all the children of Aberfan. A total of 144 people died in the flow, among them 116 children who had gathered for morning assembly in the school.

After the disaster, everyone asked, "Why did this tragedy occur and could it have been prevented?" The subsequent investigation revealed that no stability studies had ever been made on the tips and that repeated warnings about potential failure of the tips, as well as previous slides, had all been ignored.

In 1939, 8 km to the south, a tip constructed under conditions almost identical to those of Tip No. 7 collapsed. Luckily no one was injured, but unfortunately the failure was soon forgotten and the Aberfan tips continued to grow. In 1944 Tip No. 4 failed, and again no one was injured.

In 1958 Tip No. 7 was sited solely on the basis of available space, with no regard to the area's geology. In spite of previous tip failures and warnings of slope failure by tip workers and others, mine debris was being piled onto Tip No. 7 until the day of the disaster.

What exactly caused Tip No. 7 and the others to fail? The official investigation revealed that the foundation of the tips had become saturated with water from the springs over which they were built. In the case of the collapsed tips, pore pressure from the water exceeded the friction between grains, and the entire mass liquefied like a "quicksand." Behaving as a liquid, the mass quickly moved downhill spreading out laterally. As it flowed, water escaped from the mass, and the sedimentary particles regained their cohesion.

Following the inquiry, it was recommended that a National Tip Safety Committee be established to assess the dangers of existing tips and advise on the construction of new tip sites.

---

movements are triggered by a force that temporarily disturbs slope equilibrium. The most common triggering mechanisms are strong vibrations from earthquakes and excessive amounts of water from a winter snow melt or a heavy rainstorm.

Volcanic eruptions, explosions, and even loud claps of thunder may also be enough to trigger a landslide if the slope is sufficiently unstable. Many *avalanches,* which are rapid movements of snow and ice down steep mountain slopes, are triggered by the sound of a loud gunshot or, in rare cases, even a person's shout.

▼ TYPES OF MASS WASTING

Geologists recognize a variety of mass movements (● Table 11.2). Some are of one distinct type, while

◆ FIGURE 1 Location map and aerial view of the Aberfan tip disaster in which 144 people died.

others are a combination of different types. It is not uncommon for one type of mass movement to change into another along its course. For example, a landslide may start out as a slump at its head and, with the addition of water, become an earthflow at its base. Even though many slope failures are combinations of different materials and movements, it is still convenient to classify them according to their dominant behavior.

Mass movements are generally classified on the basis of three major criteria (Table 11.2): (1) rate of movement (rapid or slow); (2) type of movement (primarily falling, sliding, or flowing); and (3) type of material involved (rock, soil, or debris).

Rapid mass movements involve a visible movement of material. Such movements usually occur quite suddenly, and the material moves very quickly downslope.

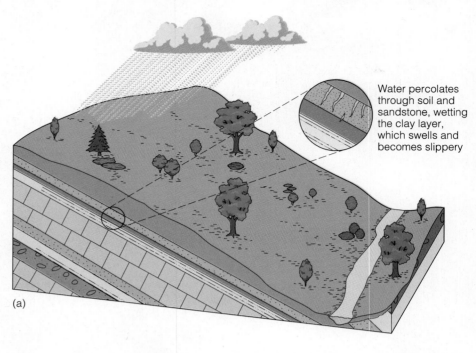

Water percolates through soil and sandstone, wetting the clay layer, which swells and becomes slippery

(a)

◆ FIGURE 11.7 (*a*) Rocks dipping in the same direction as a hill's slope are particularly susceptible to mass wasting. Undercutting of the base of the slope by a stream removes support and steepens the slope at the base. Water percolating through the soil and into the underlying rock increases its weight and, if clay layers are present, wets the clay making them slippery. (*b*) Fractures dipping in the same direction as a slope are enlarged by chemical weathering, which can remove enough material to cause mass wasting.

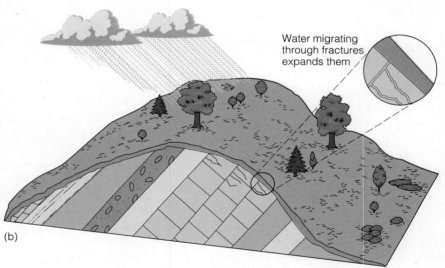

Water migrating through fractures expands them

(b)

Rapid mass movements are potentially dangerous and frequently result in loss of life and property damage. Most rapid mass movements occur on relatively steep slopes and can involve rock, soil, or debris.

Slow mass movements advance at an imperceptible rate and are usually only detectable by the effects of their movement such as tilted trees and power poles or cracked foundations. Although rapid mass movements are more dramatic, slow mass movements are respon-

sible for the downslope transport of a much greater volume of weathered material.

Falls

Rockfalls are a common type of extremely rapid mass movement in which rocks of any size fall through the air (◆ Fig. 11.8). Rockfalls occur along steep canyons, cliffs, and road cuts and build up accumulations of

● **TABLE 11.2** Classification of Mass Movements and Their Characteristics

Type of Movement	Subdivision	Characteristics	Rate of Movement
Falls	Rockfall	Rocks of any size fall through the air from steep cliffs, canyons, and road cuts	Extremely rapid
Slides	Slump	Movement occurs along a curved surface of rupture; most commonly involves unconsolidated or weakly consolidated material	Extremely slow to moderate
	Rock glide	Movement occurs along a generally planar surface	Rapid to very rapid
Flows	Mudflow	Consists of at least 50% silt- and clay-sized particles and up to 30% water	Very rapid
	Debris flow	Contains larger-sized particles and less water than mudflows	Rapid to very rapid
	Earthflow	Thick, viscous, tongue-shaped mass of wet regolith	Slow to moderate
	Quick clays	Composed of fine silt and clay particles saturated with water; when disturbed by a sudden shock, lose their cohesiveness and flow like a liquid	Rapid to very rapid
	Solifluction	Water-saturated surface sediment	Slow
	Creep	Downslope movement of soil and rock	Extremely slow
Complex movements		Combination of different movement types	Slow to extremely rapid

◆ FIGURE 11.8 Rockfalls result from failure along cracks, fractures, or bedding planes in the bedrock and are common features in areas of steep cliffs.

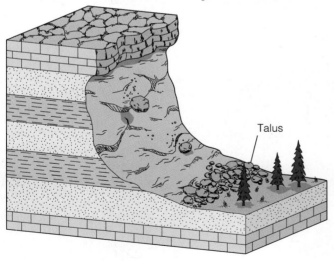

Talus

loose rocks and rock fragments at their base called *talus* (see Fig. 5.6).

Rockfalls result from failure along joints or bedding planes in the bedrock and are commonly triggered by natural or human undercutting of the slope, or by earthquakes. Many rockfalls in cold climates are the result of frost wedging (◆ Fig. 11.9). Chemical weathering caused by water percolating through the fissures in carbonate rocks (limestone, dolostone, and marble) is also responsible for many rockfalls.

Rockfalls range in size from small rocks falling from a cliff to massive falls involving millions of cubic meters of debris that destroy buildings, bury towns, and block highways (◆ Fig. 11.10). Rockfalls are a particularly common hazard in mountainous areas where roads have been built by blasting and grading through steep hillsides of bedrock. Anyone who has ever driven through the Appalachian Mountains, the Rocky Mountains, or the Sierra Nevada is familiar with the "Watch for Falling Rocks" signs posted to warn drivers of the danger. Slopes particularly prone to rockfalls are sometimes covered with wire mesh in an effort to prevent dislodged rocks from falling to the road below. Another tactic is to put up wire mesh fences along the base of the slope to catch or slow down bouncing or rolling rocks.

Slides

A **slide** involves movement of material along one or more surfaces of failure. The type of material may be soil, rock, or a combination of the two, and it may break apart during movement or remain intact. A slide's rate of movement can vary from extremely slow to very rapid (Table 11.2).

Two types of slides are generally recognized: (1) slumps or rotational slides, in which movement occurs along a curved surface; and (2) rock or block glides, which move along a more-or-less planar surface.

A **slump** involves the downward movement of material along a curved surface of rupture and is characterized by the backward rotation of the slump block (◆ Fig. 11.11). Slumps occur most commonly in unconsolidated or weakly consolidated material and range in size from small individual sets, such as occur along stream banks, to massive, multiple sets that affect large areas and cause considerable damage.

◆ FIGURE 11.9 Numerous rockfalls have resulted from frost wedging of these bedded and fractured rocks at Alberta Falls, Rocky Mountain National Park, Colorado. Accumulations of talus can be seen at the base of these outcrops.

◆ FIGURE 11.10 Rockfall in Jefferson County, Colorado. All eastbound traffic and part of the westbound lane of Interstate 70 were blocked by the rockfall. Heavy rainfall and failure along joints and foliation planes in Precambrian gneiss caused this rockfall.

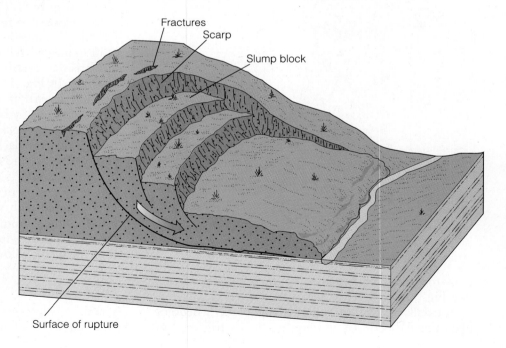

◆ FIGURE 11.11 In a slump, material moves downward along the curved surface of a rupture, causing the slump block to rotate backward. Most slumps involve unconsolidated or weakly consolidated material and are typically caused by erosion along the slope's base.

Slumps can be caused by a variety of factors, but the most common is erosion along the base of a slope, which removes support for the overlying material. This local steepening may be caused naturally by stream erosion along its banks (Fig. 11.11) or by wave action at the base of a coastal cliff. Slope oversteepening can also be caused by human activity, such as the construction of highways and housing developments. Slumps are particularly prevalent along highway cuts and fills where they are generally the most frequent type of slope failure observed.

While many slumps are merely a nuisance, large-scale slumps involving populated areas and highways can cause extensive damage. Such is the case in coastal southern California where slumping and sliding have been a con-stant problem. Many areas along the coast are underlain by poorly to weakly consolidated silts, sands, and gravels interbedded with clay layers, some of which are weath-ered ash falls. In addition, southern California is tectoni-cally active so that many of these deposits are cut by faults and joints, which allow the infrequent rains to percolate downward rapidly, wetting and lubricating the clay layers.

Southern California lies in a semiarid climate and is dry most of the year. When it does rain, typically be-tween November and March, large amounts of rain can fall in a short time. Thus, the ground quickly becomes saturated, leading to landslides along steep canyon walls as well as along coastal cliffs (◆ Fig. 11.12). Most of the

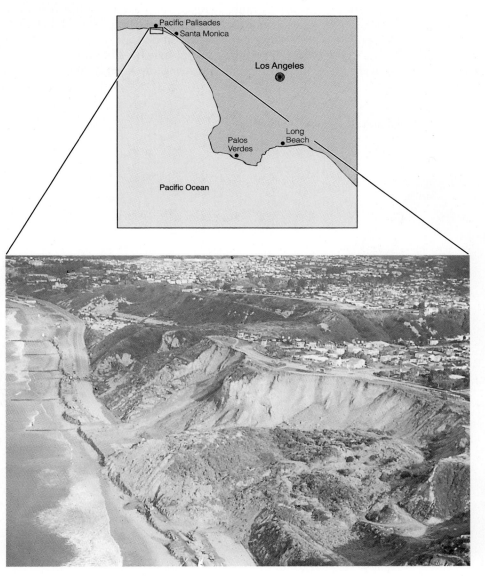

◆ FIGURE 11.12 Undercutting of steep sea cliffs by wave action resulted in massive slumping in the Pacific Palisades area of southern California on March 31 and April 3, 1958. Highway 1 was completely blocked. Note the heavy earth-moving equipment for scale.

slope failures along the southern California coast are the result of slumping. These slumps have destroyed many expensive homes and forced numerous roads to be closed and relocated.

A **rock** or *block* **glide** occurs when rocks move downslope along a more-or-less planar surface. Most rock glides occur because the local slopes and rock layers dip in the same direction (◆ Fig. 11.13), although they can also occur along fractures parallel to a slope. Rock glides are also common occurrences along the southern California coast. At Point Fermin, seaward-dipping rocks with interbedded slippery clay layers are undercut by waves causing numerous glides (◆ Fig. 11.14a).

Farther south in the town of Laguna Beach, startled residents watched as a rock glide destroyed or damaged 50 homes on October 2, 1978 (◆ Fig. 11.14b). Just as at Point Fermin, the rocks at Laguna Beach dip about 25° in the same direction as the slope of the canyon walls and contain clay beds that "lubricate" the overlying rock layers, causing the rocks and the houses built on them to glide. In addition, percolating water from the previous winter's heavy rains wet a subsurface clayey siltstone, thus reducing its shear strength and helping to activate the glide. Although the 1978 glide covered only about five acres, it was part of a larger ancient slide complex.

Not all rock glides are the result of rocks dipping in the same direction as a hill's slope. The rock glide at Frank, Alberta, Canada, on April 29, 1903, illustrates how nature and human activity can combine to create a situation with tragic results (◆ Fig. 11.15).

It would appear at first glance that the coal-mining town of Frank, lying at the base of Turtle Mountain, was in no danger from a landslide (Fig. 11.15). After all, many of the rocks dipped away from the mining valley. The joints in the massive limestone composing Turtle Mountain, however, dip steeply toward the val-

◆ FIGURE 11.13 Rock glides occur when material moves downslope along a generally planar surface.

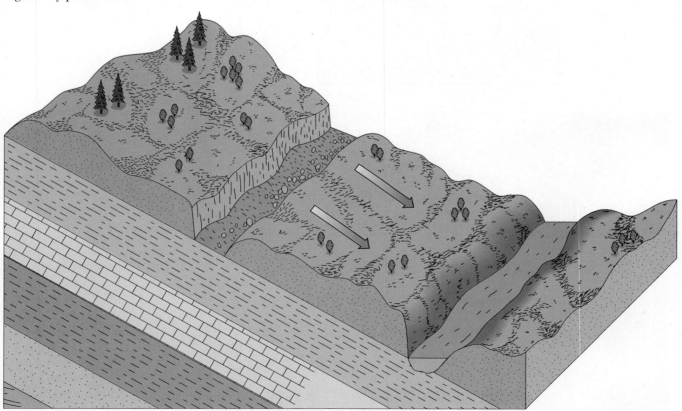

(a)

(b)

◆ FIGURE 11.14 (*a*) A combination of interbedded clay beds that become slippery when wet, rocks dipping in the same direction as the slope of the sea cliffs, and undercutting of the sea cliffs by wave action has caused numerous rock glides and slumps at Point Fermin, California. (*b*) The same combination of factors apparently activated a rock glide farther south at Laguna Beach that destroyed numerous homes and cars on October 2, 1978. (Photo (*a*) courtesy of Eleanora I. Robbins, U.S. Geological Survey.)

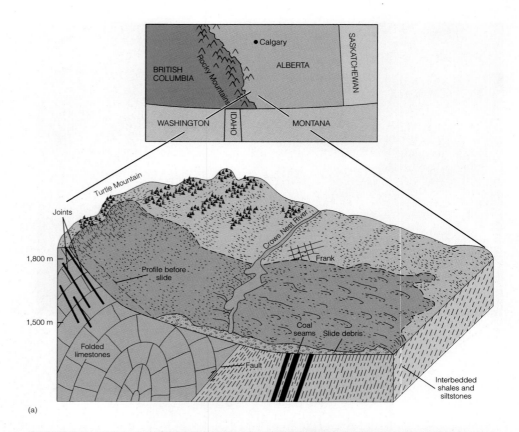

(a)

◆ FIGURE 11.15 (*a*) The tragic Turtle Mountain rock glide that killed 70 people and partially buried the town of Frank, Canada, on April 29, 1903, was caused by a combination of factors. These included joints that dipped in the same direction as the slope of Turtle Mountain, a fault partway down the mountain, weak shale and siltstone beds underlying the base of the mountain, and mined-out coal seams. (*b*) Results of the 1903 rock glide at Frank.

(b)

ley and are essentially parallel with the slope of the mountain itself. Furthermore, Turtle Mountain is supported by weak limestones, shales, and coal layers that underwent slow plastic deformation from the weight of the overlying massive limestone. Coal mining along the base of the valley also contributed to the stress on the rocks by removing some of the underlying support. All of these factors, as well as frost action and chemical weathering that widened the joints, finally resulted in a massive rock glide. Almost 40 million m^3 of rock slid down Turtle Mountain along joint planes, killing 70 people and partially burying the town of Frank.

Flows

Mass movements in which material flows as a viscous fluid or displays plastic movement are termed *flows*. Their rate of movement ranges from extremely slow to extremely rapid (Table 11.2). In many cases, mass movements may begin as falls, slumps, or slides and change into flows further downslope.

Mudflows are the most fluid of the major mass movement types. They consist of at least 50% silt- and clay-sized material combined with a significant amount of water (up to 30%). Mudflows are common in arid and semiarid environments where they are triggered by heavy rainstorms that quickly saturate the regolith, turning it into a raging flow of mud that engulfs everything in its path. Mudflows can also occur in mountain regions (◆ Fig. 11.16) and in areas covered by volcanic ash where they can be particularly destructive (see Chapter 4). Because mudflows are so fluid, they generally follow preexisting channels until the slope decreases or the channel widens, at which point they fan out.

Mudflows are very dangerous because they typically form quickly, usually move very rapidly (at speeds up to 80 km per hour), and are capable of transporting all different sizes of objects. As urban areas in arid and semiarid climates continue to expand, mudflows and the damage they create are becoming problems. For example, mudflows are very common in the steep hillsides around Los Angeles where they have damaged or destroyed many homes.

In addition to the damage they cause on hillsides, mudflows are also a hazard to structures built along the bases of steep mountain fronts. Any building, highway, or railroad tracks in the path of the mudflow will be quickly moved or buried. For example, a mudflow in

◆ FIGURE 11.16 Mudflow in Estes Park, Colorado.

Cajon Pass near Los Angeles carried a locomotive a distance of more than 600 m before burying it.

Debris flows are composed of larger-sized particles than mudflows and do not contain as much water. Consequently, they are usually more viscous than mudflows, typically do not move as rapidly, and rarely are confined to preexisting channels. Debris flows can, however, be just as damaging because they can transport large objects (◆ Fig. 11.17).

Earthflows move more slowly than either mudflows or debris flows. An earthflow slumps from the upper part of a hillside, leaving a scarp, and flows slowly downslope as a thick, viscous, tongue-shaped mass of wet regolith (◆ Fig. 11.18). Like mudflows and debris flows, earthflows can be of any size, and are frequently destructive. They occur, however, most commonly in humid climates on grassy soil-covered slopes following heavy rains.

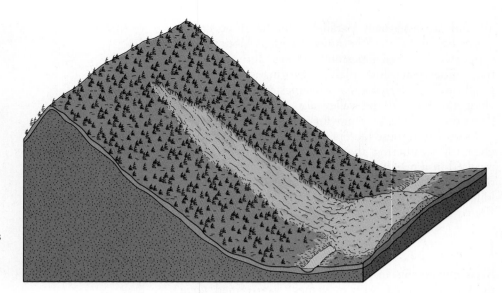

◆ FIGURE 11.17 Debris flows contain larger-sized particles than mudflows and are not as fluid. Debris flows can be very destructive in mountainous regions because of the steep slopes, loose material, and water available from melting snow.

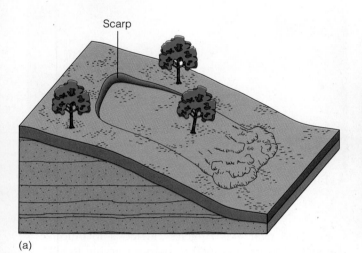

(a)

(b)

◆ FIGURE 11.18 (*a*) Earthflows form tongue-shaped masses of wet regolith that move slowly downslope. They occur most commonly in humid climates on grassy soil-covered slopes. (*b*) An earthflow near Baraga, Michigan.

Some clays spontaneously liquefy and flow like water when they are disturbed. Such **quick clays** have caused serious damage and loss of lives in Sweden, Norway, eastern Canada, and Alaska (Table 11.1). Quick clays are composed of fine silt and clay particles made by the grinding action of glaciers. Geologists believe these fine sediments were originally deposited in a marine environment where their pore space was filled with salt water. The ions in the salt water helped establish strong bonds between the clay particles, thus stabilizing and strengthening the clay. When the clays were subsequently uplifted above sea level, however, the salt water was flushed out by fresh groundwater, reducing the effectiveness of the ionic bonds between the clay particles and thereby reducing the overall strength and cohesiveness of the clay. Consequently, when the clay is disturbed by a sudden shock or shaking, it essentially turns to a liquid and flows.

An example of the damage that can be done by quick clays occurred in the Turnagain Heights area of Anchorage, Alaska, in 1964 (◆ Fig. 11.19). Underlying

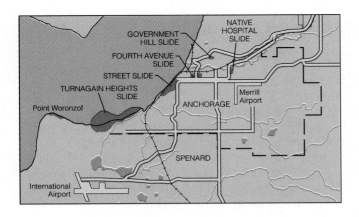

◆ FIGURE 11.19 (*a*) Groundshaking by the 1964 Alaska earthquake turned parts of the Bootlegger Cove Clay into a quick clay, causing numerous slides (*b*) that destroyed many homes in the Turnagain Heights subdivision of Anchorage.

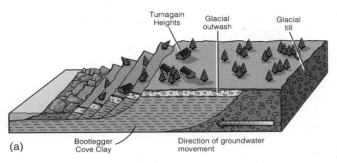

(a)

(b)

most of the Anchorage area is the Bootlegger Cove Clay, a massive clay unit of poor permeability. Because the Bootlegger Cove Clay forms a barrier preventing groundwater from flowing through the adjacent glacial deposits to the sea, considerable hydraulic pressure builds up behind the clay. Some of this water has flushed out the salt water in the clay and also has saturated the lenses of sand and silt associated with the clay beds. When the 8.5-magnitude Good Friday earthquake struck on March 27, 1964, the shaking turned parts of the Bootlegger Cove Clay into a quick clay and precipitated a series of massive slides in the coastal bluffs that destroyed most of the homes in the Turnagain Heights subdivision.

Solifluction is the slow downslope movement of water-saturated surface sediment. Solifluction can occur in any climate where the ground becomes saturated with water, but is most common in areas of permafrost.

Permafrost is ground that remains permanently frozen. It covers nearly 20% of the world's land surface (◆ Fig. 11.20a). During the warmer season when the upper portion of the permafrost thaws, water and surface sediment form a soggy mass that flows by solifluction and produces a characteristic lobate topography (◆ Fig. 11.20b).

As might be expected, many problems are associated with construction in a permafrost environment. A good example is what happens when an uninsulated building is constructed directly on permafrost. In this instance, heat escapes through the floor, thaws the ground below, and turns it into a soggy, unstable mush. Because the ground is no longer solid, the building settles unevenly into the ground, and numerous structural problems result (◆ Fig. 11.21).

Construction of the Alaska pipeline from the oil fields in Prudhoe Bay to the ice-free port of Valdez

◆ FIGURE 11.20 (a) Distribution of permafrost areas in the Northern Hemisphere. (b) Solifluction flows near Suslositna Creek, Alaska, show the typical lobate topography that is characteristic of solifluction conditions.

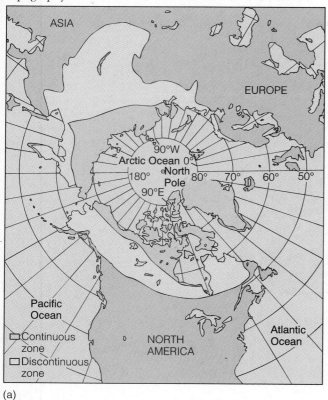

(a)

(b)

◆ FIGURE 11.21 This house south of Fairbanks, Alaska, has settled unevenly because the underlying permafrost in fine-grained silts and sands has thawed.

raised numerous concerns over the effect it might have on the permafrost and the potential for solifluction. Some thought that oil flowing through the pipeline would be warm enough to melt the permafrost, causing the pipeline to sink further into the ground and possibly rupture. After numerous studies were conducted, scientists concluded that the pipeline, completed in 1977, could safely be buried for more than half of its 1,280 km length; where melting of the permafrost might cause structural problems to the pipe, it was insulated and installed above ground.

Creep is the slowest type of flow. It is also the most widespread and significant mass wasting process in terms of the total amount of material moved downslope and the monetary damage that it does annually. Creep involves extremely slow downhill movement of soil or rock. Although it can occur anywhere and in any climate, it is most effective and significant as a geologic agent in humid rather than arid or semiarid climates. In fact, it is the most common form of mass wasting in the southeastern United States and the southern Appalachian Mountains.

Because the rate of movement is essentially imperceptible, we are frequently unaware of creep's existence until we notice its effects: tilted trees and power poles, broken streets and sidewalks, cracked retaining walls or foundations (◆ Fig. 11.22). Creep usually involves the whole hillside and probably occurs, to some extent, on any weathered or soil-covered, sloping surface.

Not only is creep difficult to recognize, it is difficult to control. Although engineers can sometimes slow or stabilize creep, many times the only course of action is to simply avoid the area if at all possible or, if the zone of creep is relatively thin, design structures that can be anchored into the solid bedrock.

Complex Movements

Recall that many mass movements are combinations of different movement types. When one type is dominant, the movement can be classified as one of the movements described thus far. If several types are more or less equally involved, however, it is called a **complex movement.**

The most common type of complex movement is the slide-flow in which there is sliding at the head and then some type of flowage farther along its course. Most slide-flow landslides involve well-defined slumping at the head, followed by a debris flow or earthflow (◆ Fig. 11.23). However, any combination of different mass movement types can be classified as a complex movement.

A **debris avalanche** is a complex movement that often occurs in very steep mountain ranges. Debris avalanches typically start out as rockfalls when large quantities of rock, ice, and snow are dislodged from a mountainside, frequently as a result of an earthquake. The material then slides or flows down the mountainside, picking up additional surface material and increasing in speed. The 1970 Peru earthquake set in motion the debris avalanche that destroyed the town of Yungay (see the Prologue).

▼ RECOGNIZING AND MINIMIZING THE EFFECTS OF MASS MOVEMENTS

The most important factor in eliminating or minimizing the damaging effects of mass wasting is a thorough geologic investigation of the region in question. In this way, former landslides and areas susceptible to mass movements can be identified and perhaps avoided (see Perspective 11.2, page 290). By assessing the risks of possible mass wasting before construction begins, steps can be taken to eliminate or minimize the effects of such events.

Identifying areas with a high potential for slope failure is important in any hazard assessment study; such studies include identifying former landslides as

◆ FIGURE 11.22 (*a*) Some evidence of creep: (A) curved tree trunks; (B) displaced monuments; (C) tilted power poles; (D) displaced and tilted fences; (E) roadways moved out of alignment; (F) hummocky surface. (*b*) Creep has bent these sandstone and shale beds of the Haymond Formation near Marathon, Texas. (*c*) Trees bent by creep, Wyoming. (*d*) Stone wall tilted due to creep, Champion, Michigan. (Photo courtesy of David J. Matty.)

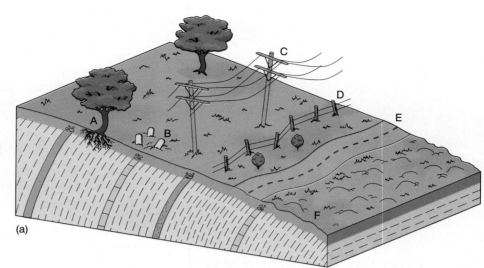

(a)

(b)

(c)

(d)

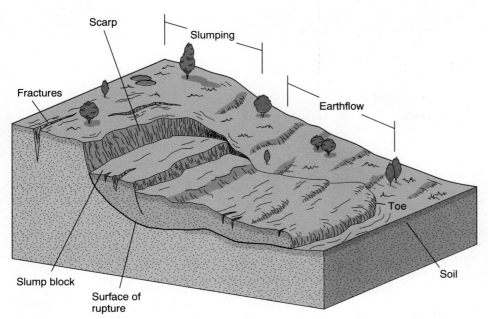

Scarp

Slumping

Fractures

Earthflow

Toe

Soil

Slump block

Surface of rupture

◆ FIGURE 11.23 A complex movement in which slumping occurs at the head followed by an earthflow.

well as sites of potential mass movement. Because of the effects of weathering, erosion, and vegetation, the evidence for previous mass wasting may be obscured. However, scarps, open fissures, displaced or tilted objects, a hummocky surface, and sudden changes in vegetation are some of the features indicating former landslides or an area susceptible to slope failure.

Soil and bedrock samples are also studied, both in the field and laboratory, to assess such characteristics as composition, susceptibility to weathering, cohesiveness, and ability to transmit fluids. Such studies help geologists and engineers predict slope stability under a variety of conditions.

The information derived from a hazard assessment study can be used to produce *slope stability maps* of the area. These maps allow planners and developers to make decisions about where to site roads, utility lines, and housing or industrial developments based on the relative stability or instability of a particular location. In addition, the maps also indicate how extensive an area's landslide problem is and the type of mass movement that may occur. This information is important for designing slopes or building structures to prevent or minimize slope failure damage.

Although most large mass movements usually cannot be prevented, geologists and engineers can employ various methods to minimize the danger and damage resulting from them. Because water plays such an important role in many landslides, one of the most effective and inexpensive ways to reduce the potential for slope failure or to increase existing slope stability is through surface and subsurface drainage of a hillside. Drainage serves two purposes. It reduces the weight of the material likely to slide and increases the shear strength of the slope material by lowering pore pressure.

Surface waters can be drained and diverted by ditches, gutters, or culverts designed to direct water away from slopes. Drainpipes perforated along one surface and driven into a hillside can help remove subsurface water (◆ Fig. 11.24, page 293). Finally, planting vegetation on hillsides helps stabilize slopes by holding the soil together and reducing the amount of water in the soil.

Another way to help stabilize a hillside is to reduce its slope. Recall that overloading or oversteepening by grading are common causes of slope failure. By reducing the gradient of a hillside, the potential for slope failure is decreased. Two common methods are generally employed to reduce a slope's gradient. In the *cut-and-fill* method, material is removed from the upper part of the slope and used as fill at the base, thus

Perspective 11.2

THE VAIONT DAM DISASTER

On October 9, 1963, a glacial valley in the Italian Alps was the site of the worst dam disaster in history. More than 240 million m³ of rock and soil slid into the Vaiont Reservoir, triggering a destructive flood that killed nearly 3,000 people (◆ Fig. 1). To fully appreciate the enormity of this catastrophe, consider the following: Within a period of 15 to 30 seconds, the slide filled the reservoir with a mass of debris 2 km long and as high as 175 m above the reservoir level. The impact of the debris created a wave of water that overflowed the dam by 100 m and was still more than 70 m high 1.6 km downstream. The slide also set off a blast of wind that shook houses, broke windows, and even lifted the roof off one house in the town of Casso, which is 260 m above the reservoir on the opposite side of the slide; it also set off shock waves recorded by seismographs throughout Europe. Considering the forces generated by the slide, it is a

tribute to the designer and construction engineer that the dam itself survived the disaster (◆ Fig. 2)!

The dam was built in a glacial valley underlain by a thick sequence of folded and faulted limestones and interbedded clay layers and marls that are further weakened by jointing (◆ Fig. 3). Signs of previous slides in the area were obvious, and the few boreholes in the valley slopes revealed clay seams and small-scale slide planes. In spite of the geological evidence of previous mass wasting in the area and objections to the site by some of the early investigators, construction of the 265.5 m high Vaiont Dam began.

A combination of both adverse geological features and conditions resulting from the dam construction contributed to the massive landslide. Among the geological causes were the rocks themselves, which were weak to begin with and dipped in the same direction as the valley walls of the

◆ FIGURE 1 Location of the Vaiont Dam disaster and features associated with the landslide.

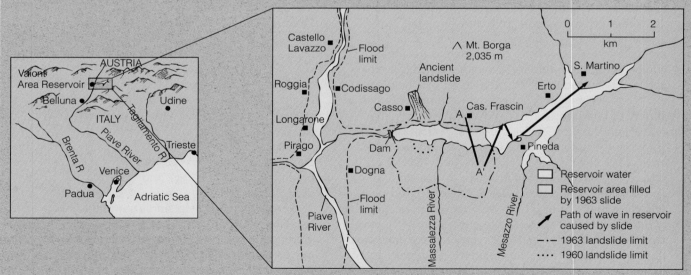

reservoir. Fractured limestones make up the bulk of the rocks and are interbedded with numerous clay beds that are particularly prone to slippage. Active solution of the limestones by slightly acid groundwater further weakened them by developing and expanding an extensive network of cracks, joints, fissures, and other openings.

During the two weeks before the slide occurred, heavy rains saturated the ground, adding extra weight and reducing the shear strength of the rocks. In addition to water from the rains, water from the reservoir infiltrated the rocks of the lower valley walls, further reducing their strength.

Soon after the dam was completed, a relatively small slide of one million m³ of material occurred on the south side of the reservoir. Following this slide, it was decided to limit the amount of water in the reservoir and to install monitoring devices throughout the potential slide area. Between 1960 and 1963, the eventual slide area moved an average of about 1 cm per week. On September 18, 1963, numerous monitoring stations reported movement had increased to about 1 cm per day. It was assumed that these were individual blocks moving, but it was actually the entire slide area!

Heavy rains fell between September 28 and October 9, increasing the amount of subsurface water. By October 8, the creep rate had increased to almost 39 cm per day. Engineers finally realized that the entire slide area was moving, and quickly began lowering the reservoir level. On October 9, the rate of movement in the slide area had increased still further, in some locations up to 80 cm per day, and there were reports that the reservoir level was actually rising. This was to be expected if the south bank was moving into the reservoir and displacing water. Finally, at 10:41 P.M. that night, during yet another rainstorm, the south bank of the Vaiont valley slid into the reservoir.

The lesson to be learned from this disaster is that before construction on any dam begins, a complete and systematic appraisal of an area must be conducted. Such a study should examine the geology of the area, identify past mass movements, assess their potential for recurrence, and evaluate the effects that the project will have on the rocks, including how it will alter their shear strength over time. Without these precautions, similar disasters will occur and lives will needlessly be lost.

◆ FIGURE 2 Aerial view of the Vaiont Dam.

continued on next page

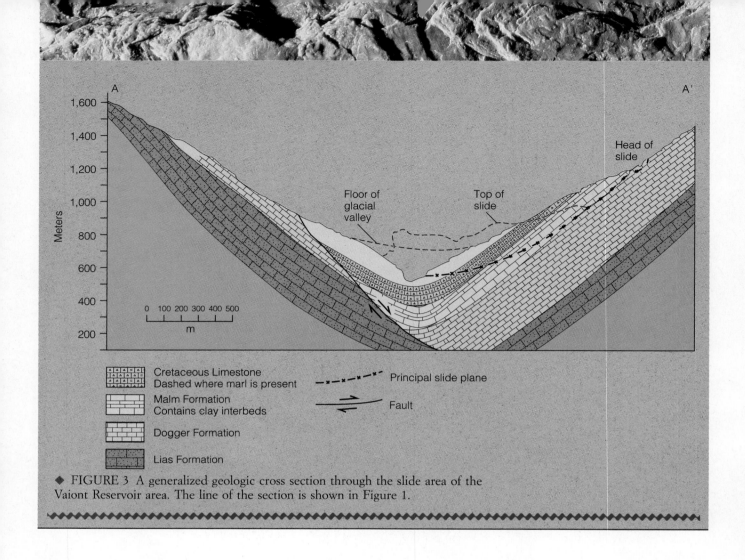

Cretaceous Limestone
Dashed where marl is present

Malm Formation
Contains clay interbeds

Dogger Formation

Lias Formation

Principal slide plane

Fault

◆ FIGURE 3 A generalized geologic cross section through the slide area of the Vaiont Reservoir area. The line of the section is shown in Figure 1.

providing a flat surface for construction and reducing the slope (◆ Fig. 11.25). The second method, which is called *benching,* involves cutting a series of benches or steps into a hillside. This process reduces the average slope, and the benches serve as collecting sites for small landslides or rockfalls that might occur. Benching is most commonly used on steep hillsides in conjunction with a system of surface drains to divert runoff.

In some situations, retaining walls can be constructed to provide support for the base of the slope

(◆ Fig. 11.26). These are usually anchored well into bedrock, backfilled with crushed rock, and provided with drain holes to prevent the buildup of water pressure in the hillside.

Rock bolts, similar to those employed in tunneling and mining, can sometimes be used to fasten potentially unstable rock masses into the underlying stable bedrock (◆ Fig. 11.27). This technique has been used successfully on the hillsides of Rio de Janeiro, Brazil, and to help secure the slopes at the Glen Canyon Dam on the Colorado River.

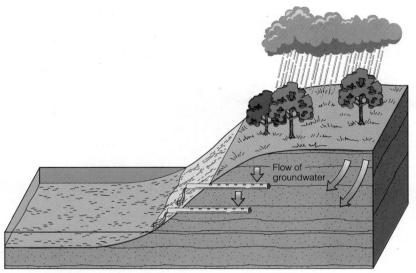

◆ FIGURE 11.24 (*a*) Driving drainpipes that are perforated on one side into a hillside with the perforated side up can remove some subsurface water and thus help stabilize a hillside. (*b*) A drainpipe driven into the hillside at Point Fermin, California, helps remove subsurface water and stabilize the slope.

Flow of groundwater

(a)

(b)

Recognition, prevention, and control of landslide-prone areas is expensive, but not nearly as expensive as the damage can be when such warning signs are ignored or not recognized. The Vaiont Dam disaster (see Perspective 11.2) and the collapse of Tip No.7 at Aberfan, Wales (see Perspective 11.1), are two tragic examples in which the warning signs of impending disaster were ignored.

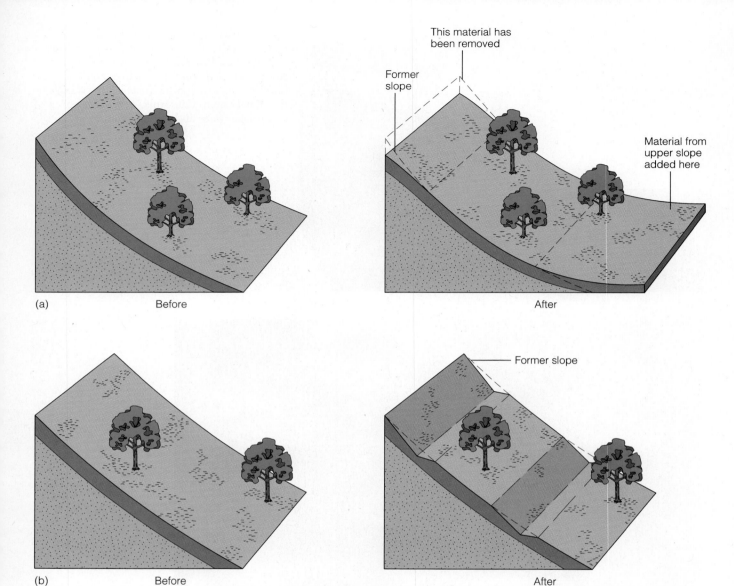

This material has been removed

Former slope

Material from upper slope added here

(a)   Before   After

Former slope

(b)   Before   After

◆ FIGURE 11.25 Two common methods used to help stabilize a hillside and reduce its slope. (*a*) In the cut-and-fill method, material from the steeper upper part of the hillside is removed, thereby reducing the slope angle, and is used to fill in the base. This provides some additional support at the base of the slope. (*b*) Benching involves making several cuts along a hillside to reduce the overall slope.

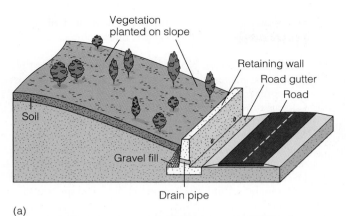

(a)

(b)

◆ FIGURE 11.26 (*a*) Retaining walls anchored into bedrock, backfilled with gravel, and provided with drainpipes can support a slope's base and reduce landslides. (*b*) Steel retaining wall built to stabilize the slope and keep falling and sliding rocks off of the highway.

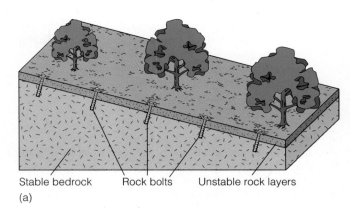

(a)

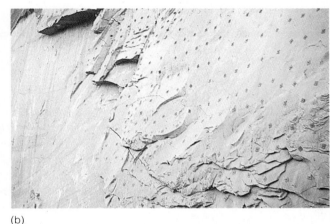

(b)

◆ FIGURE 11.27 (*a*) Rock bolts secured in bedrock can help stabilize a slope and reduce landslides. (*b*) Rock bolts are used to help secure rock above the outlet of the west diversion tunnel of the Glen Canyon Dam. As can be seen, however, some portions of rock still broke away.

Chapter Summary

1. Mass wasting is the downslope movement of material under the influence of gravity. It occurs when the gravitational force acting parallel to a slope exceeds the slope's strength.

2. Mass wasting frequently results in loss of life, as well as causing millions of dollars in damage annually.

3. Mass wasting can be caused by many factors including a slope's gradient, weathering of the slope material, the material's water content, overloading of the slope, and removal of vegetation. Usually, several of these factors in combination contribute to slope failure.

4. Mass movements are generally classified on the basis of their rate of movement, type of movement, and type of material.

5. Rockfalls are a common mass movement in which rocks free-fall.

6. Two types of slides are recognized. Slumps are rotational slides involving movement along a curved surface; they are most common in poorly consolidated or unconsolidated material. Rock glides occur when movement takes place along a more or less planar surface; they usually involve solid pieces of rock.

7. Several types of flows are recognized on the basis of their rate of movement, type of material, and amount of water.
8. Mudflows consist of mostly clay- and silt-sized particles and contain more than 30% water. They are most common in semiarid and arid environments and generally follow preexisting channels.
9. Debris flows are composed of larger-sized particles and contain less water than mudflows. They are more viscous and do not flow as rapidly as mudflows.
10. Earthflows move more slowly than either debris flows or mudflows; they move downslope as thick, viscous, tongue-shaped masses of wet regolith.
11. Quick clays are clays that spontaneously liquefy and flow like water when they are disturbed.
12. Solifluction is the slow downslope movement of water-saturated surface sediment and is most common in areas of permafrost.
13. Creep, the slowest type of flow, is the imperceptible downslope movement of soil or rock. Creep is the most widespread of all types of mass wasting.
14. Complex movements are combinations of different types of mass movements in which one type is not dominant. Most complex movements involve sliding and flowing.
15. The most important factor in reducing or eliminating the damaging effects of mass wasting is a thorough geologic investigation of the area including mapping, soil and rock analysis, and the construction of slope stability maps to outline areas susceptible to mass movements.

Important Terms

complex movement
creep
debris avalanche
debris flow
earthflow
mass wasting

mudflow
permafrost
quick clay
rapid mass movement
rockfall
rock glide

shear strength
slide
slow mass movement
slump
solifluction

Review Questions

1. Shear strength includes:
   a. ___ the strength and cohesion of material; b. ___ the amount of internal friction between grains; c. ___ gravity; d. ___ all of these; e. ___ answers (a) and (b).
2. Which of the following is not a factor influencing mass wasting?
   a. ___ gravity; b. ___ weathering; c. ___ slope gradient; d. ___ water content; e. ___ none of these.
3. Which of the following factors can actually enhance slope stability?
   a. ___ water content; b. ___ vegetation; c. ___ overloading; d. ___ rocks dipping in the same direction as the slope; e. ___ none of these.
4. Mass wasting can occur:
   a. ___ on gentle slopes; b. ___ on steep slopes; c. ___ in flat-lying areas; d. ___ all of these; e. ___ none of these.
5. A type of mass wasting common in mountainous regions in which talus accumulates is:
   a. ___ creep; b. ___ solifluction; c. ___ rockfalls; d. ___ slides; e. ___ mudflows.
6. Movement of material along a surface or surfaces of failure is a:
   a. ___ slide; b. ___ fall; c. ___ flow; d. ___ slip; e. ___ none of these.
7. Downslope movement along an essentially planar surface is a:
   a. ___ slump; b. ___ rockfall; c. ___ earthflow; d. ___ landslide; e. ___ rock glide.
8. Which of the following are the most fluid of mass movements?
   a. ___ earthflows; b. ___ debris flows; c. ___ mudflows; d. ___ solifluction; e. ___ slumps.
9. The most widespread and costly type of mass wasting in terms of total material moved and monetary damage is:
   a. ___ creep; b. ___ solifluction; c. ___ mudflow; d. ___ debris flow; e. ___ slumping.
10. Which of the following features indicate former landslides or areas susceptible to slope failure?
    a. ___ displaced objects; b. ___ scarps; c. ___ hummocky surfaces; d. ___ open fissures; e. ___ all of these.
11. Define mass wasting.
12. What are the forces that help to maintain slope stability?
13. What roles do climate and weathering play in mass wasting?
14. In what ways does the ground's water content affect slope stability? Give an example of how excessive water content has resulted in slope failure.
15. How does vegetation affect slope stability? Give several examples.

16. What is overloading and why is it dangerous?
17. Give several examples of how the relationship between topography and the underlying geology affects slope stability.
18. What are rockfalls? Where are they most common and why?
19. What is the difference between a slump and a rock glide?
20. Why are slumps particularly common along road cuts and fills?
21. Discuss and give examples of two different ways that rock glides might occur.
22. Differentiate between a mudflow, debris flow, and earthflow.
23. Why are quick clays so dangerous?
24. What precautions must be taken when building in permafrost areas?
25. Why is creep so prevalent, and why does it do so much damage?
26. How can creep be controlled once it has started?
27. What are complex movements, and how do they differ from other types of mass movements?
28. What are some of the indications of previous mass wasting? How can you recognize areas that are susceptible to mass movement?

Additional Readings

Brabb, E. E., and B. L. Harrod, eds. 1989. *Landslides: Extent and economic significance.* Brookfield, Va.: A. A. Balkema.

Crozier, M. J. 1989. *Landslides: Causes, consequences, and environment.* Dover, N. H.: Croom Helm.

Fleming, R. W., and F. A. Taylor. 1980. *Estimating the cost of landslide damage in the United States.* U.S. Geological Survey Circular 832.

Kiersch, G. A. 1964. Vaiont reservoir disaster. *Civil Engineering* 34:32–39.

McPhee, J. 1989. *The control of nature.* New York: Farrar, Straus & Giroux.

Parks, N. 1993. The fragile volcano. *Earth* 6, no. 4: 42–49.

Small, R. J., and M. J. Clark. 1982. *Slopes and weathering.* New York: Cambridge Univ. Press.

Zaruba, Q., and V. Mencl. 1982. *Landslides and their control.* 2d ed. Amsterdam, The Netherlands: Elsevier.

CHAPTER
12

Chapter Outline

Grand Sable Falls in Alger County, Michigan. (Photo courtesy of R. V. Dietrich.)

Running Water

Prologue

In 1877, the Italian astronomer Giovanni Virginio Schiaparelli viewed Mars through his telescope and was convinced that he saw straight lines. In his report he called these lines *canali,* the Italian word for channel, but it was translated into English as "canal." During the 1890s, Percival Lowell, who founded the Lowell Observatory near Flagstaff, Arizona, published numerous maps showing interconnecting canals on Mars. Many astronomers could not see Lowell's canals, but others thought they could. By the early 1900s, the public had accepted the idea that intelligent beings had constructed the canals to divert water from the polar icecaps to the equatorial regions of a planet that was becoming progressively drier. When *Mariner 4* flew past Mars on July 15, 1965, however, it sent back images indicating that no canals were present; apparently, they were simply an optical illusion.

Although no Martians or their works were discovered, *Mariner 4* and the subsequent *Viking* missions did reveal evidence of running water. Currently, liquid water cannot exist on the Martian surface because the atmospheric pressure is too low. If any water were present, it would rapidly vaporize. Mars does, however, show clear evidence of volcanism, although it is doubtful that any of the volcanoes are still active. Nevertheless, just as on Earth, Martian volcanoes no doubt emitted carbon dioxide, water vapor, and other gases. Furthermore, the gravitational attraction of Mars is sufficient to retain these gases, so most of the water that originated during its early history may still be there in some form. The polar icecaps of Mars are composed of frozen carbon dioxide and frozen water, and large quantities of water ice are probably present in the pore spaces of the surface deposits.

Studies of *Mariner* and *Viking* images reveal areas called *chaotic terrane* that appear to consist of loosely piled rubble. Winding valleys, termed *outflow channels,* extend from some of these areas of chaotic terrane (◆ Fig. 12.1). These channels are 10 to 100 km wide, some are more than 2,000 km long, and within them are a number of features indicating fluid flow. Apparently, the channels formed when huge quantities of subsurface ice suddenly melted, perhaps because of volcanic activity. The overlying rock then subsided, forming the chaotic terrane, and the water was released at the surface as flash floods. Judging from the size of these outflow channels, the flash floods probably exceeded any known on Earth.

What became of these flash flood waters? The outflow channels terminate at closed depressions where the water no doubt ponded and its surface froze. The water beneath very likely percolated downward and froze once again. The surface ice probably sublimated (vaporized without going through a liquid phase), fell as snow, melted, and percolated down into the surface deposits where it refroze.

▼ INTRODUCTION

Among the terrestrial planets, the Earth is unique in having abundant liquid water. Fully, 71% of the Earth's surface is covered by water, and a small but important quantity of water vapor is present in its atmosphere.

The volume of water on Earth is estimated at 1.36 billion km^3, most of which (97.2%) is in the oceans.

About 2% is frozen in glaciers, and the remaining 0.8% constitutes all the water in streams, lakes, swamps, groundwater, and the atmosphere (● Table 12.1). Thus, only a tiny portion of the total water on Earth is in streams, but running water is nevertheless the most important erosional agent modifying the Earth's surface. Even in most desert regions the effects of running water are manifest, although the channels are dry most of the time (◆ Fig. 12.2).

In addition to its significance as a geologic agent, running water is important for many other reasons as well. It is a source of fresh (nonsaline) water for industry, domestic use, and agriculture, and about 8% of the electricity used in North America is generated by falling water at hydroelectric stations. Streams have been, and continue to be, important avenues of commerce. Much of the interior of North America was first explored by following such large streams as the St. Lawrence, Mississippi, and Missouri rivers.

◆ FIGURE 12.1 Outflow channels extending from areas of chaotic terrain (CT) on Mars. Arrows show inferred directions of flow.

◆ FIGURE 12.2 The evidence for running water is common even in deserts as in Death Valley, California, but stream channels are dry most of the time.

● **TABLE 12.1** Water on Earth in km³

Location	Volume	Percent of Total
Oceans	1,327,500,000	97.20
Icecaps and glaciers	29,315,000	2.15
Groundwater	8,442,580	.625
Freshwater and saline lakes and inland seas	230,325	.017
Atmosphere at sea level	12,982	.001
Average in stream channels	1,255	.0001

❧ THE HYDROLOGIC CYCLE

Water is continually recycled from the oceans, through the atmosphere, to the continents, and back to the oceans. This continual recycling of water is called the **hydrologic cycle** (◆ Fig. 12.3). The hydrologic cycle, which is powered by solar radiation, is possible because water changes phases easily under Earth surface conditions. Huge quantities of water evaporate from the oceans each year as the surface waters are heated by solar energy. Approximately 85% of all water that enters the atmosphere is derived from the oceans; the remaining 15% comes from evaporation of water on land.

When water evaporates, the vapor rises into the atmosphere where the complex processes of condensation and cloud formation occur. About 80% of the precipitation falls directly into the oceans, in which case the hydrologic cycle is limited to a three-step process of evaporation, condensation, and precipitation.

About 20% of all precipitation falls on land as rain and snow, and the hydrologic cycle involves more steps: evaporation, condensation, movement of water vapor from the oceans to the continents, precipitation, and runoff and infiltration. Some of the precipitation evaporates as it falls and reenters the hydrologic cycle as vapor; water evaporated from lakes and streams also reenters the cycle as vapor as does moisture evaporated from plants by *transpiration* (Fig. 12.3).

Each year about 36,000 km³ of the precipitation falling on land returns to the oceans by **runoff,** the surface flow of streams. The water returning to the oceans by runoff enters the Earth's ultimate reservoir where it begins the hydrologic cycle again.

Some of the precipitation falling on land is temporarily stored in lakes, snow fields, and glaciers or seeps below the surface where it is temporarily stored as groundwater. This water is effectively removed from the system for up to thousands of years, but eventually, glaciers melt, lakes feed streams, and groundwater flows into streams or directly into the oceans (Fig. 12.3). Our concern here is with the comparatively small quantity returning to the oceans as runoff, for the

◆ FIGURE 12.3 The hydrologic cycle.

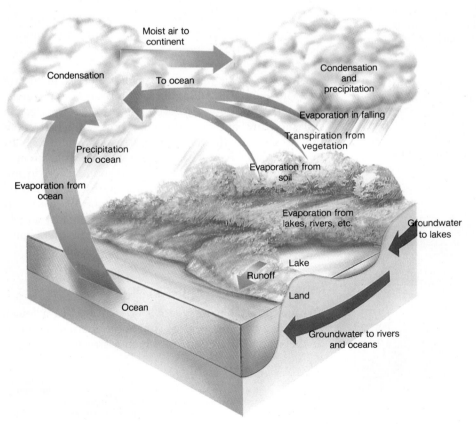

energy of running water is responsible for a great many surface features.

⩥ RUNNING WATER

Water possesses no strength so it will flow on any slope no matter how slight. The flow of water, or any other fluid, can be characterized as *laminar* or *turbulent*. In laminar flow, lines of flow called streamlines are all parallel with one another. In other words, all flow occurs in parallel layers with no mixing between layers (◆ Fig. 12.4a). By contrast, in turbulent flow, the streamlines intertwine, causing a complex mixing of the fluid (◆ Fig. 12.4b).

Laminar flow is most easily observed in viscous fluids such as cold motor oil or syrup. Turbulent flow, on the other hand, occurs in almost all streams. The primary control on the type of flow is velocity; roughness of the surface over which flow occurs also plays a role. Laminar flow occurs when water flows very slowly, as when groundwater moves through the tiny pores in sediments and soil. In streams, however, the flow is usually fast enough and the channel walls and bed rough enough so that flow is fully turbulent. Laminar flow is so slow, and generally so shallow, that it causes little or no erosion. Turbulent flow is much more energetic and thus is capable of considerable erosion and sediment transport.

◆ FIGURE 12.4 (*a*) In laminar flow, streamlines are all parallel to one another, and little or no mixing occurs between adjacent layers in the fluid. (*b*) In turbulent flow, streamlines are complexly intertwined, indicating mixing between adjacent layers in the fluid. Most flow in streams is turbulent.

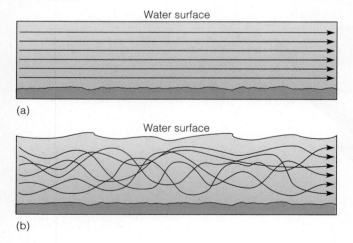

Water surface

(a)

Water surface

(b)

Sheet Flow versus Channel Flow

The amount of runoff in any area during a rainstorm depends on **infiltration capacity**, the maximum rate at which soil or other surface materials can absorb water. Infiltration capacity depends on several factors, including the intensity and duration of rainfall. Loosely packed, dry soils absorb water faster than tightly packed, wet soils.

If rain is absorbed as fast as it falls, no surface runoff occurs. However, should the infiltration capacity be exceeded, or should surface materials become saturated, excess water collects on the surface and, if a slope exists, moves downhill. Even on steep slopes such flow is initially slow, and hence little or no erosion occurs. As the water moves downslope, however, it accelerates and may move by *sheet flow,* a more-or-less continuous film of water flowing over the surface. Such flow is not confined to depressions, and it accounts for *sheet erosion,* a particular problem on some agricultural lands (see Chapter 5).

In *channel flow,* surface runoff is confined to long, troughlike depressions. Channels vary in size from rills containing a trickling stream of water to the Amazon River of South America, which is 6,450 km long and up to 2.4 km wide and 90 m deep. Channelized flow is described by various terms including rill, brook, creek, stream, and river, most of which are distinguished by size and volume. The term **stream** carries no connotation of size and is used here to refer to all runoff confined to channels regardless of size.

Streams receive water from several sources, including sheet flow and rain falling directly into stream channels. Far more important, however, is the water supplied by soil moisture and groundwater, both of which flow downslope and discharge into streams (Fig. 12.3). In humid areas where groundwater is plentiful, streams may maintain a fairly stable flow year round, even during dry seasons, because they are continuously supplied by groundwater. In contrast, the amount of water in streams of arid and semiarid regions fluctuates widely because such streams depend more on infrequent rainstorms and surface runoff for their water supply.

Stream Gradient, Velocity, and Discharge

Streams flow downhill from a source area to a lower elevation where they empty into another stream, a lake, or the sea.* The slope over which a stream flows is its

. . . .

*The flow in certain desert streams diminishes in a downstream direction by evaporation and infiltration until the streams disappear, and some streams in regions with numerous caverns may disappear below the ground.

gradient. For example, if the source (headwaters) of a stream is 1,000 m above sea level and the stream flows 500 km to the sea, it drops 1,000 m vertically over a horizontal distance of 500 km (◆ Fig. 12.5). Its gradient is calculated by dividing the vertical drop by the horizontal distance; in this example, it is 1,000 m/500 km = 2 m/km.

Gradients vary considerably, even along the course of a single stream. Generally, streams are steeper in their upper reaches where their gradients may be tens of meters per kilometer, but in their lower reaches the gradient may be as little as a few centimeters per kilometer.

Stream velocity and discharge are closely related variables. **Velocity** is simply a measure of the downstream distance traveled per unit of time. Velocity is usually expressed in feet per second (ft/sec) or meters per second (m/sec) and varies considerably among streams and even within the same stream.

Variations in flow velocity occur not only with distance along a stream channel but also across a channel's width. For example, flow velocity is slower and more turbulent near a stream bed or stream banks because of friction than it is farther from these boundaries (◆ Fig. 12.6). Other controls on velocity include channel shape and roughness. Broad, shallow channels

and narrow, deep channels have proportionally more water in contact with their perimeters than do channels with semicircular cross sections (◆ Fig. 12.7). Consequently the water in semicircular channels flows more rapidly because it encounters less frictional resistance. In sinuous (meandering) channels, the line of maximum flow velocity switches from one side of the channel to the other and corresponds to the channel center only along straight reaches (◆ Fig. 12.8).

Channel roughness is a measure of the frictional resistance within a channel. For example, frictional resistance to flow is greater in a channel containing large boulders than in one with banks and a bed composed of sand or clay. In channels with abundant vegetation, flow is slower than in barren channels of comparable size.

The most obvious control on velocity is gradient, and one might think that the steeper the gradient, the greater the flow velocity. In fact, the average velocity generally increases in a downstream direction, even though the gradient decreases in the same direction. Three factors contribute to this: First, velocity increases continuously, even as gradient decreases, in response to the acceleration of gravity unless other factors retard flow. Secondly, in their upstream reaches, streams commonly have boulder-strewn, broad, shallow channels,

◆ FIGURE 12.5 The average gradient of this stream is 2 m/km. Gradient can be calculated for any segment of a stream as shown in this example. Notice that the gradient is steepest in the headwaters area and decreases in a downstream direction.

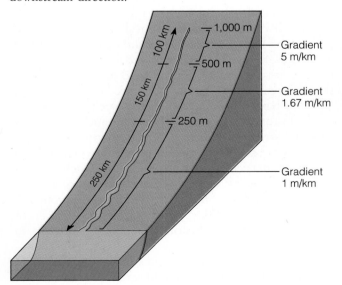

◆ FIGURE 12.6 In a stream, flow velocity varies as a result of friction with the banks and bed. The maximum flow velocity is near the center and top of a stream in a straight channel. The lengths of the arrows in this illustration are proportional to velocity.

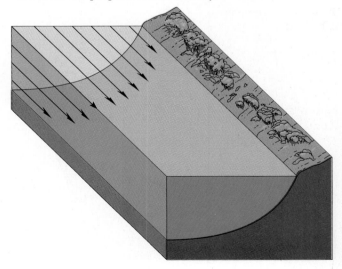

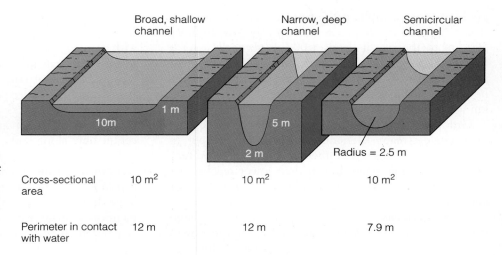

	Broad, shallow channel	Narrow, deep channel	Semicircular channel
Cross-sectional area	10 m²	10 m²	10 m²
Perimeter in contact with water	12 m	12 m	7.9 m

◆ FIGURE 12.7 All three of these channels have the same cross-sectional area, but each has a different shape. The semicircular channel has the least perimeter in contact with the water and thus causes the least frictional resistance to flow. If other variables, such as channel roughness, are the same in all of these channels, flow velocity will be greatest in the semicircular channel.

so flow resistance is high and velocity is correspondingly slower. Downstream, however, channels generally become more semicircular, and the bed and banks are usually composed of finer-grained materials, thus reducing the effects of friction. Thirdly, the number of tributary streams joining a larger stream increases in a downstream direction. Thus, the total volume of water

◆ FIGURE 12.8 In a sinuous (meandering) channel, flow velocity varies from one side of the channel to the other. As the water flows around curves, it flows fastest near the outer bank. The dashed line in this illustration follows the path of maximum flow velocity.

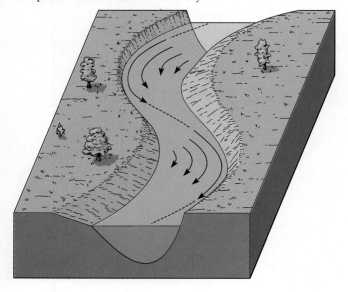

(discharge) increases, and increasing discharge results in increased velocity.

Discharge is the total volume of water in a stream moving past a particular point in a given period of time. To determine discharge, one must know the dimensions of a channel—that is, its cross-sectional area (A)—and its flow velocity (V). Discharge (Q) can then be calculated by the formula $Q = VA$, and it is generally expressed in cubic feet per second (ft³/sec) or cubic meters per second (m³/sec).

▼ STREAM EROSION

Most of the energy a stream possesses is dissipated as heat by fluid turbulence, but a small amount, perhaps 5%, is available to erode and transport sediment. Erosion involves the physical removal of dissolved substances and loose particles of soil and rock from a source area. Thus, the sediment transported in a stream consists of both dissolved materials and solid particles. Some of the *dissolved load* of a stream is acquired from the stream bed and banks where soluble rocks such as limestone and dolostone are exposed. However, much of it is carried into streams by sheet flow and by groundwater.

The solid sediment carried in streams ranges from clay-sized particles to large boulders. Much of this sediment finds its way into streams by mass wasting (◆ Fig. 12.9), but some is derived directly from the stream bed and banks. For example, the power of running water, called **hydraulic action,** is sufficient to set particles in motion.

Another process of erosion in streams is **abrasion,** in which exposed rock is worn and scraped by the impact of solid particles. If running water is transporting sand and gravel, the impact of these particles abrades exposed rock surfaces. One obvious manifestation of abrasion is the occurrence of *potholes* in the beds of streams (◆ Fig. 12.10). These circular to oval holes occur where eddying currents containing sand and gravel swirl around and erode depressions into solid rock.

▼ TRANSPORT OF SEDIMENT LOAD

Streams transport a solid load of sedimentary particles and a **dissolved load** consisting of ions taken into solution by chemical weathering. Sedimentary particles are transported either as suspended load or as bed load. **Suspended load** consists of the smallest particles, such as silt and clay, which are kept suspended by fluid turbulence (◆ Fig. 12.11).

Bed load consists of the coarser particles such as sand and gravel (Fig. 12.11). Fluid turbulence is insufficient to keep such large particles suspended, so they move along the stream bed. However, part of the bed load can be suspended temporarily. For example, an eddying current may swirl across a stream bed and lift sand grains into the water. These particles move forward at approximately the flow velocity, but at the same time they settle toward the stream bed where they come to rest, to be moved again later by the same process. This process of intermittent bouncing and skipping is *saltation* (Fig. 12.11).

Particles too large to be suspended even temporarily are transported by rolling or sliding (Fig. 12.11). Obviously, greater flow velocity is required to move particles of these sizes. The maximum-sized particles that a stream can carry define its *competence,* a factor related to flow velocity. *Capacity* is a measure of the total load a stream can carry. It varies as a function of discharge; with greater discharge, more sediment can be carried. A small, swiftly flowing stream may have the competence to move gravel-sized particles but not to transport a large volume of sediment. Thus, it has a low capacity. A large, slow-flowing stream, on the other hand, has a low competence, but may have a very large suspended load, and hence a large capacity.

▼ STREAM DEPOSITION

Streams can transport sediment a considerable distance from the source area. For example, some of the sediments deposited in the Gulf of Mexico by the Mississippi River came from such distant sources as Pennsylvania, Minnesota, and Wyoming. Along the way, deposition may

◆ FIGURE 12.9 Streams such as the Snake River in Idaho receive some of their sediment load by mass wasting processes. (Photo courtesy of R. V. Dietrich.)

◆ FIGURE 12.10 Potholes in the bed of the Chippewa River in Ontario, Canada.

◆ FIGURE 12.11 Methods of sediment transport by running water. The velocity profile at the right indicates that the water flows fastest near the surface and slowest along the stream bed.

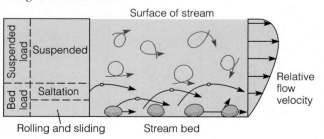

occur in a variety of environments, such as stream channels, the floodplains adjacent to channels, and the points where streams flow into lakes or the seas or flow from mountain valleys onto adjacent lowlands.

Streams do most of their erosion, sediment transport, and deposition when they flood. Consequently, stream deposits, collectively called **alluvium,** do not represent the continuous day-to-day activity of streams, but rather those periodic, large-scale events of sedimentation associated with flooding.

Braided Streams and Their Deposits

Braided streams possess an intricate network of dividing and rejoining channels (◆ Fig. 12.12). Braiding develops when a stream is supplied with excessive sediment, which over time is deposited as sand and gravel bars within its channel. During high-water stages, these bars are submerged, but during low-water stages, they are exposed and divide a single channel into multiple channels (Fig. 12.12). Braided streams have broad, shallow channels. They are generally characterized as bed load transport streams, and their deposits are composed mostly of sheets of sand and gravel.

Meandering Streams and Their Deposits

Meandering streams possess a single, sinuous channel with broadly looping curves called *meanders* (◆ Fig.

◆ FIGURE 12.12 A braided stream near Sante Fe, New Mexico.

12.13). Such stream channels are semicircular in cross section along straight reaches, but at meanders they are markedly asymmetric, being deepest near the outer bank, which commonly descends vertically into the channel. The outer bank is called the *cut bank* because flow velocity and turbulence are greatest on that side of the channel where it is eroded. In contrast, flow velocity is at a minimum near the inner bank, which slopes gently into the channel.

As a consequence of the unequal distribution of flow velocity across meanders, the cut bank is eroded and deposition occurs along the opposite side of the chan-

◆ FIGURE 12.13 Aerial view of a meandering stream. The broad, flat area adjacent to the stream channel is the floodplain. Notice the crescent-shaped lakes—these are cut-off meanders, or what are known as oxbow lakes.

nel. The deposit formed in this manner is a **point bar;** it consists of cross-bedded sand or, in some cases, gravel (◆ Fig. 12.14). Point bars are the characteristic deposits that accumulate within meandering stream channels.

It is not uncommon for meanders to become so sinuous that the thin neck of land separating adjacent meanders is eventually cut off during a flood. The valley floors of meandering streams are commonly marked by crescent-shaped **oxbow lakes,** which are actually cutoff meanders (Figs. 12.13, ◆ 12.15). These oxbow lakes may persist as lakes for some time, but are eventually filled with organic matter and fine-grained sediment carried by floods. Once filled, oxbow lakes are called *meander scars.*

Floodplain Deposits

Most streams periodically receive more water than their channel can carry, so they spread across low-lying, relatively flat areas called **floodplains** adjacent to their channels (Fig. 12.13; Perspective 12.1). Some flood-

◆ FIGURE 12.14 Two small point bars in a meandering stream.

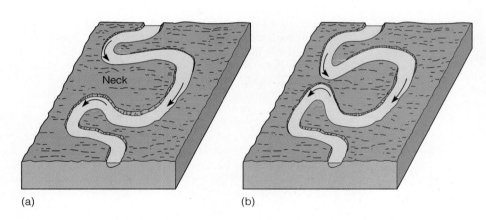

(a)

(b)

◆ FIGURE 12.15 Four stages in the origin of an oxbow lake. In (*a*) and (*b*), the meander neck becomes narrower. (*c*) The meander neck is cut off, and part of the channel is abandoned. (*d*) When it is completely isolated from the main channel, the abandoned meander is an oxbow lake.

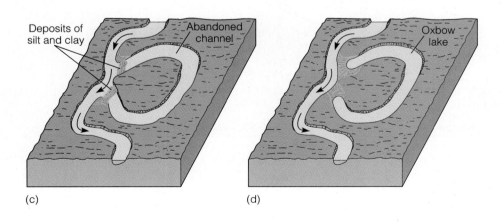

(c)

(d)

Perspective 12.1

PREDICTING AND CONTROLLING FLOODS

Occasionally, a stream receives more water than its channel can handle, and it floods, occupying part or all of its floodplain. To monitor stream behavior, the U.S. Geological Survey maintains more than 11,000 stream gauging stations, and various state agencies also monitor streams. Data collected at gauging stations can be used to construct a *hydrograph* showing how a stream's discharge varies over time (♦Fig. 1). Hydrographs are useful in planning irrigation and water supply projects, and they give planners a better idea of what to expect during flood events.

Stream gauge data are also used to construct *flood-frequency curves* (♦Fig. 2). To construct such a curve, the peak discharges are first arranged in order of volume; the flood with the greatest discharge has a magnitude rank of 1, the second largest is 2, and so on (● Table 1). The *recurrence interval*—that is, the time period during which a flood of a given magnitude or larger can be expected over an average of many years—is determined by the equation shown in Table 1. Once the recurrence interval has been calculated, it is plotted against discharge, and a line is drawn through the data points (Fig. 2).

According to Figure 2, the 10-year flood for the Rio Grande near Lobatos, Colorado, has a discharge of 245 m^3/sec. This means that, on average, we can expect one flood of this size or greater to occur within a 10-year interval. One cannot, however, predict that such a flood will occur in any particular year, only that it has a probability of 1 in 10 (1/10) of occurring in any year. Furthermore, 10-year floods are not necessarily separated by 10 years. That is, two such events could occur in the same year or in successive years, but over a period of centuries their average occurrence would be once every 10 years.

Unfortunately, stream gauge data in the United States have generally been available for only a few decades, and rarely for more than a century. Accordingly, we have a good idea of stream behavior over short periods, the 2-year and 5-year floods, for example, but our knowledge of long-term behavior is limited by the short period of record keeping. Thus, predictions of 50-year or 100-year floods from Figure 2 are unreliable. In fact, the largest magnitude flood shown in Figure 2 may have been a unique event for this stream that will never be repeated. On the other hand, it may actually turn out to be a magnitude 2 or 3 flood when data for a longer time are available.

Although flood-frequency curves have limited applicability, they are nevertheless helpful in making decisions regarding flood control. Careful mapping of floodplains can identify areas at risk for floods of a given magnitude. For a particular stream, planners must decide what magnitude of flood to protect against because the cost goes up faster than the increasing sizes of floods would indicate.

♦ FIGURE 1 Hydrograph for Sycamore Creek near Ashland City, Tennessee, for the February 1989 flood. (From U.S. Geological Survey Water-Resources Investigations Report 89–4207.)

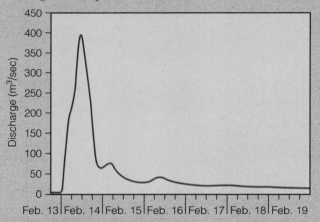

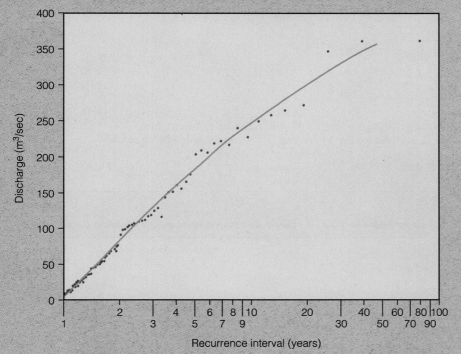

◆ FIGURE 2 Flood-frequency curve for the Rio Grande near Lobatos, Colorado. The curve was constructed from the data in Table 1.

Federal, state, and local agencies and land-use planners use flood-frequency analyses to develop recommendations and regulations concerning construction on and use of floodplains. Geologists and engineers are interested in such analyses for planning appropriate flood-control projects. For example, where should dams and basins be constructed to contain the excess water of floods?

When flood-control projects are well planned and constructed, they are functional. What many people fail to realize, however, is that such projects are designed to contain floods of a given size; should larger floods occur, streams spill onto floodplains anyway. Furthermore, dams occasionally collapse and reservoirs eventually fill with sediment unless dredged. In short, flood-control projects are not only initially expensive, but they require constant, costly maintenance. Such costs must be weighed against the cost of damage if no control projects were undertaken.

continued on next page

● **TABLE 1** Some of the Data and Recurrence Intervals for the Rio Grande near Lobatos, Colorado

Year	Discharge (m³/sec)	Rank	Recurrence Interval
1900	133	23	3.35
1901	103	35	2.20
1902	16	69	1.12
1903	362	2	38.50
1904	22	66	1.17
1905	371	1	77.00
1906	234	10	7.70
1907	249	7	11.00
1908	61	45	1.71
1909	211	13	5.92
.	The greatest yearly discharge is given a magnitude rank (m) ranging from 1 to N		
.	($N = 76$ in this example), and the recurrence interval (R) is calculated by the equation		
.	$R = (N + 1)/m$.		
1974	22	64	1.20
1975	68	43	1.79

SOURCE: U.S. Geological Survey Open-File Report 79–681.

plains are composed mostly of sand and gravel that were deposited as point bars. When a meandering stream erodes its cut bank and deposits on the opposite bank, it migrates laterally across its floodplain. As lateral migration occurs, a succession of point bars develops by *lateral accretion* (◆ Fig. 12.16). That is, the deposits build laterally as a consequence of repeated episodes of sedimentation on the inner banks of meanders.

Many floodplains are dominated by *vertical accretion* of fine-grained sediments. When a stream overflows its banks and floods, the velocity of the water spilling onto the floodplain diminishes rapidly because of greater frictional resistance to flow as the water spreads out as a broad, shallow sheet. In response to the diminished velocity, ridges of sandy alluvium called **natural levees** are deposited along the margins of the stream channel (◆ Fig. 12.17).

◆ FIGURE 12.16 Floodplain deposits forming by lateral accretion of point bars.

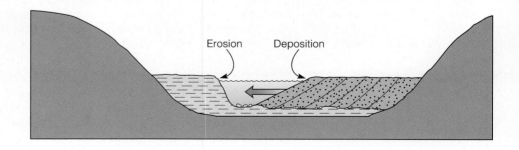

Erosion   Deposition

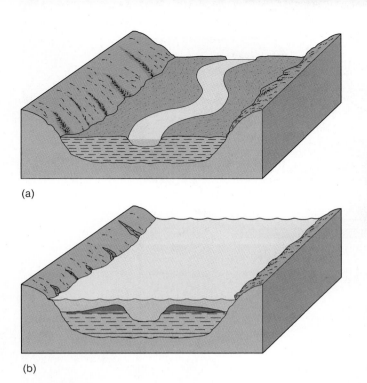

(a)

(b)

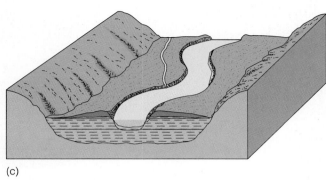

(c)

◆ FIGURE 12.17 Three stages in the formation of vertical accretion deposits on a floodplain. (*a*) Stream at low-water stage. (*b*) Flooding stream and deposition. Many such episodes of flooding form natural levees. (*c*) After flooding.

The flood waters spilling from a main channel carry large quantities of fine-grained sediment beyond the natural levees and onto the floodplain. During the waning stages of a flood, the flood waters may flow very slowly or not at all, and the suspended silt and clay eventually settle as layers of mud.

Deltas

When a stream flows into another body of water, its flow velocity decreases rapidly and deposition occurs. As a result of such deposition, a **delta** forms, causing the local shoreline to build out, or *prograde* (◆ Fig. 12.18). Deltas in lakes are common, but marine deltas are much larger and far more complex.

The simplest prograding deltas exhibit a characteristic vertical sequence in which *bottomset beds* are successsively overlain by *foreset beds* and *topset beds* (Fig. 12.18a). Such sequences develop when a stream enters another body of water, and the finest sediments are carried some distance beyond the river mouth, where they settle from suspension and form bottomset beds. Nearer the river mouth, foreset beds are formed as sand and silt are deposited in gently inclined layers.

The topset beds consist of coarse-grained sediments deposited in a network of *distributary channels* traversing the top of the delta. In effect, streams lengthen their channels as they extend across prograding deltas (Fig. 12.18).

Many small deltas in lakes have the three-part division described above, but large marine deltas are usually much more complex. Depending on the relative importance of stream, wave, and tidal processes, three major types of marine deltas are recognized (◆ Fig. 12.19). *Stream-dominated deltas,* such as the Mississippi River delta, consist of long fingerlike sand bodies, each deposited in a distributary channel that progrades far seaward. Such deltas are commonly called *bird's-foot deltas* because the projections resemble the toes of a bird. In contrast, the Nile delta of Egypt is *wave-dominated,* although it also possesses distributary channels; the seaward margin of the delta consists of a series of barrier islands formed by reworking of sediments by waves, and the entire margin of the delta progrades seaward. *Tide-dominated deltas,* such as the Ganges-Brahmaputra of Bangladesh, are continually modified into tidal sand bodies that parallel the direction of tidal flow.

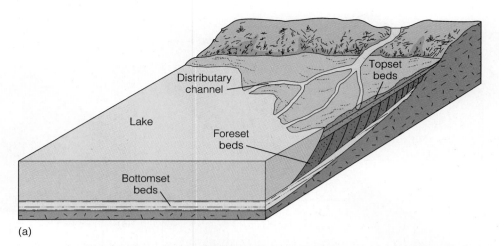

◆ FIGURE 12.18 (*a*) Internal structure of the simplest type of prograding delta. (*b*) A small delta in which bottomset, foreset, and topset beds are visible.

(b)

Alluvial Fans

Alluvial fans are lobate deposits on land (◆ Fig. 12.20). They form best on lowlands adjacent to highlands in arid and semiarid regions where little or no vegetation exists to stabilize surface materials. When periodic rainstorms occur, surface materials are quickly saturated and runoff begins. During a particularly heavy rain, all of the surface flow in a drainage area is funneled into a mountain canyon leading to an adjacent lowland. The stream is confined in the mountain canyon so that it cannot spread laterally. However, as it discharges from the canyon onto the lowland area, it quickly spreads out, its velocity diminishes, and deposition ensues.

The alluvial fans that develop by the process just described are mostly accumulations of sand and gravel, a large proportion of which is deposited by streams. In some cases, however, the water flowing through a mountain canyon picks up so much sediment that it becomes a viscous mudflow. Consequently, mudflow deposits make up a large part of many alluvial fans.

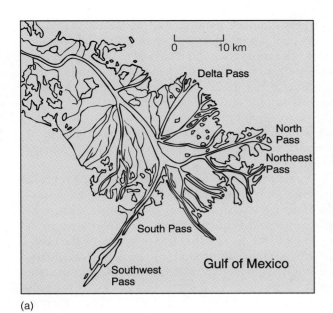

(a)

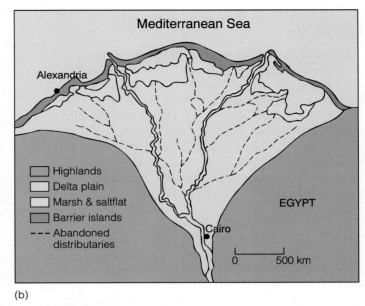

(b)

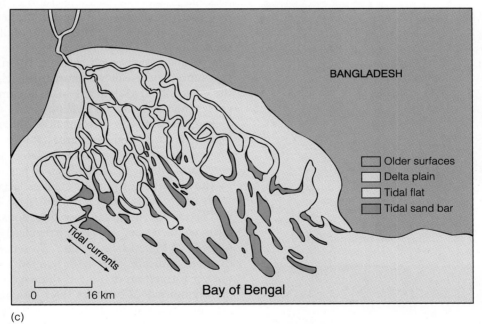

(c)

◆ FIGURE 12.19 (*a*) The Mississippi River delta of the U.S. Gulf Coast is stream-dominated, and (*b*) the Nile Delta of Egypt is wave-dominated. (*c*) The Ganges- Brahmaputra delta of Bangladesh is tide-dominated.

DRAINAGE BASINS AND DRAINAGE PATTERNS

Thousands of streams, most of which are parts of larger drainage systems, flow either directly or indirectly into the oceans. A stream such as the Mississippi River consists of a main stream and all of the smaller *tributary streams* that supply water to it. The Mississippi and all of its tributaries, or any other drainage system for that matter, carry surface runoff from an area known as the **drainage basin.** Individual drainage basins are separated from adjacent ones by topographically higher areas called **divides** (◆ Fig. 12.21).

◆ FIGURE 12.20 Alluvial fans form where a stream discharges from a mountain canyon onto an adjacent lowland. Alluvial fan adjacent to the Confusion Range, Utah.

◆ FIGURE 12.21 Small drainage basins separated from one another by divides.

Various **drainage patterns** are recognized based on the regional arrangement of channels in a drainage system. The most common is *dendritic drainage,* which consists of a network of channels resembling tree branching (◆ Fig. 12.22a). Dendritic drainage develops on gently sloping surfaces where the materials respond more or less homogeneously to erosion. Areas of flat-lying sedimentary rocks and some terrains of igneous or metamorphic rocks usually display a dendritic drainage pattern.

In marked contrast to dendritic drainage in which tributaries join larger streams at various angles, *rectangular drainage* is characterized by channels with right angle bends and tributaries that join larger streams at right angles (◆ Fig. 12.22b). The positions of the channels are strongly controlled by geologic structures, particularly regional joint systems that intersect at right angles.

In some parts of the eastern United States, such as Virginia and Pennsylvania, erosion of folded sedimentary rocks develops a landscape of alternating parallel ridges and valleys. The ridges consist of more resistant rocks, such as sandstone, whereas the valleys overlie less resistant rocks such as shale. Main streams follow the trends of the valleys. Short tributaries flowing from the adjacent ridges join the main stream at nearly right angles, hence the name *trellis drainage* (◆ Fig. 12.22c).

In *radial drainage,* streams flow outward in all directions from a central high area (◆ Fig. 12.22d). Radial drainage develops on large, isolated volcanic mountains, and where the Earth's crust has been arched up by the intrusion of plutons such as laccoliths.

In some areas streams flow in and out of swamps and lakes with irregular flow directions. Drainage patterns characterized by such irregularity are called *deranged* (◆ Fig. 12.22e). The presence of deranged drainage indicates that it developed recently and has not yet formed an organized drainage system. In areas of Minnesota, Wisconsin, and Michigan that were glaciated until about 10,000 years ago, the previously established drainage systems were obliterated by glacial ice. Following the final retreat of the glaciers, drainage systems became established, but have not yet become fully organized.

BASE LEVEL

Streams require a slope on which to flow, so they can erode downward only to the level of the body of water into which they flow. Thus, there exists a lower limit to

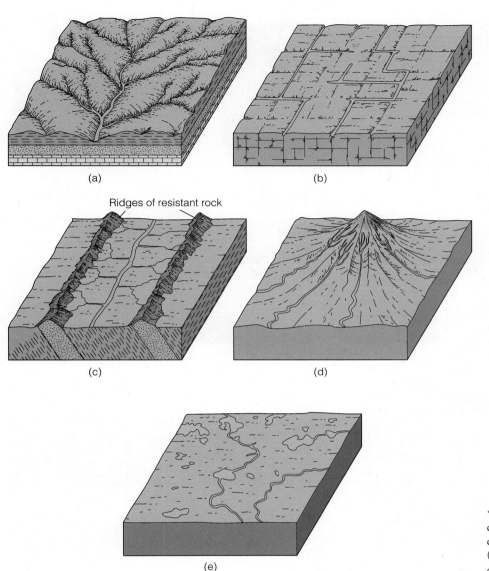

(a)

(b)

Ridges of resistant rock

(c)

(d)

(e)

◆ FIGURE 12.22 Examples of drainage patterns. (*a*) Dendritic drainage. (*b*) Rectangular drainage. (*c*) Trellis drainage. (*d*) Radial drainage. (*e*) Deranged drainage.

which streams can erode called **base level** (◆ Fig. 12.23). Theoretically, a stream could erode its entire valley to very near sea level, so sea level is commonly referred to as *ultimate base level.** Streams never reach ultimate base level, however, because they must have some gradient in order to maintain flow.

. . . .

*Streams flowing into depressions below sea level, such as Death Valley in California, have a base level corresponding to the lowest point of the depression and are not limited by sea level.

In addition to ultimate base level, streams have *local* or *temporary base levels.* For example, a lake or another stream can serve as a local base level for the upstream segment of a stream (Fig. 12.23). Likewise, where a stream flows across particularly resistant rock, a waterfall may develop, forming a local base level.

When sea level rises or falls with respect to the land, or the land over which a stream flows is uplifted or subsides, changes in base level occur. For example, during the Pleistocene Epoch when extensive glaciers were present on the Northern Hemisphere continents,

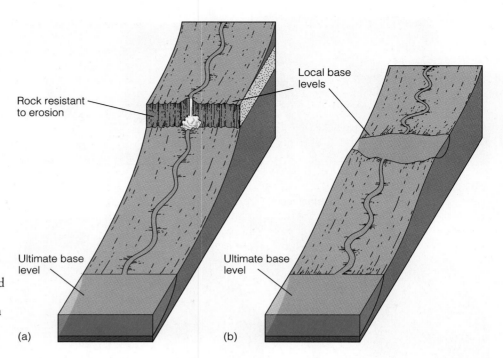

◆ FIGURE 12.23 In both (a) and (b), sea level is ultimate base level. In (a) a resistant rock layer forms a local base level, while in (b) a lake is a local base level.

Rock resistant to erosion

Local base levels

Ultimate base level

Ultimate base level

(a)

(b)

sea level was more than 100 m lower than at present. Accordingly, streams deepened their valleys by adjusting to a new, lower base level. Rising sea level at the end of the Pleistocene caused base level to rise, and the streams responded by depositing sediments and backfilling previously formed valleys.

Streams adjust to human intervention, but not always in anticipated or desirable ways. Geologists and engineers are well aware that the process of building a dam and impounding a reservoir creates a local base level (◆ Fig. 12.24a). Where a stream enters a reservoir, its flow velocity diminishes rapidly and deposition occurs; thus, reservoirs are eventually filled with sediment unless they are dredged. Another consequence of building a dam is that the water discharged at the dam is largely sediment free, but it still possesses energy to transport sediment. Commonly, such streams simply acquire a new sediment load by vigorously eroding downstream from the dam.

Draining a lake along a stream's course may seem like a small change that is well worth the time and expense to expose dry land for agriculture or commercial development. However, draining a lake lowers the base level for that part of the stream above the lake, and the stream will very likely respond by rapid downcutting (◆ Fig. 12.24b).

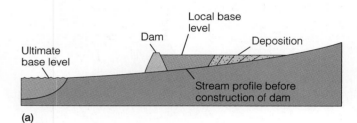

Local base level

Dam

Deposition

Ultimate base level

Stream profile before construction of dam

(a)

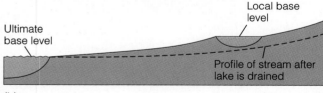

Local base level

Ultimate base level

Profile of stream after lake is drained

(b)

◆ FIGURE 12.24 (a) The process of constructing a dam and impounding a reservoir creates a local base level. A stream deposits much of its sediment load where it flows into a reservoir. (b) A stream adjusts to a lower base level when a lake is drained.

THE GRADED STREAM

A stream's *longitudinal profile* shows the elevations of a channel along its length as viewed in cross section (◆ Fig. 12.25). The longitudinal profiles of many streams show a number of irregularities such as lakes and waterfalls, which are local base levels (Fig. 12.25a). Over time such irregularities tend to be eliminated by stream processes; where the gradient is steep, erosion decreases it, and where the gradient is too low to maintain sufficient flow velocity for sediment transport, deposition occurs, steepening the gradient. In short, streams tend to develop a smooth, concave longitudinal profile of equilibrium, meaning that all parts of the system dynamically adjust to one another (Fig. 12.25b).

Streams possessing an equilibrium profile are said to be **graded streams;** that is, a delicate balance exists between gradient, discharge, flow velocity, channel characteristics, and sediment load such that neither significant erosion nor deposition occurs within the channel. Such a delicate balance is rarely attained; thus, the concept of a graded stream is an ideal. Nevertheless, many streams do indeed approximate the graded condition, although usually only temporarily.

Even though the concept of a graded stream is an ideal, we can generally anticipate the responses of a graded stream to changes altering its equilibrium. For example, a change in base level would cause a stream to adjust as previously discussed. Increased rainfall in a stream's drainage basin would result in greater discharge and flow velocity. In short, the stream would now possess greater energy—energy that must be dissipated within the stream system by, for example, a change in channel shape. A change from a semicircular to a broad, shallow channel would dissipate more energy by friction. On the other hand, the stream may respond by active downcutting in which it erodes a deeper valley and effectively reduces its gradient until it is once again graded.

DEVELOPMENT OF STREAM VALLEYS

Valleys are common landforms, and with few exceptions they form and evolve as a consequence of stream erosion, although other processes, especially mass wasting, also contribute. The shapes and sizes of valleys vary considerably; some are small, steep-sided *gullies,* whereas others are broad and have gently sloping valley walls. Some steep-walled, deep valleys of vast proportions are called *canyons,* such as the Grand Canyon of Arizona. Particularly narrow and deep valleys are *gorges* (◆ Fig. 12.26).

A valley may begin where runoff has sufficient energy to dislodge surface materials and excavate a small rill. Once formed, a rill collects more surface runoff and becomes deeper and wider until a full-fledged valley develops (◆ Fig. 12.27). Several pro-

◆ FIGURE 12.25 (*a*) An ungraded stream has irregularities in its longitudinal profile. (*b*) Erosion and deposition along the course of a stream eliminate irregularities and cause it to develop the smooth, concave profile typical of a graded stream.

Stream profile

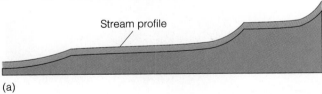

(a)

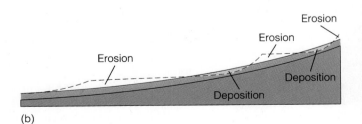

Erosion

Erosion

Erosion

Deposition

Deposition

(b)

◆ FIGURE 12.26 The Black Canyon of the Gunnison River in Colorado is a gorge because it is steep-walled and deep.

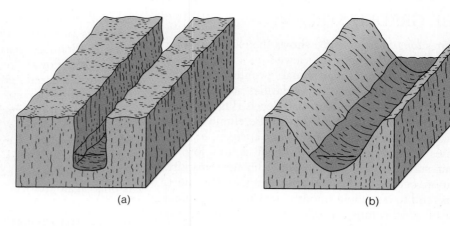

◆ FIGURE 12.27 Valley development. (*a*) If valleys formed mostly by downcutting, they would be narrow and steep sided. (*b*) Valleys are deepened by downcutting, but most of them are also widened by lateral erosion, mass wasting, and sheet wash.

(a)

(b)

cesses are involved in the origin and evolution of valleys, including downcutting, lateral erosion, mass wasting, sheet wash, and headward erosion.

Downcutting occurs when a stream possesses more energy than it requires to transport its sediment load, so some of its excess energy cuts its valley deeper. If downcutting were the only process operating, valleys would be narrow and steep sided, as in Figure 12.27a. In most cases, however, the valley walls are undercut by the stream. Such undermining, termed *lateral erosion,* creates unstable conditions so that part of a bank or valley wall may move downslope by any one or a combination of mass wasting processes (Fig. 12.27b). Furthermore, sheet wash and erosion of rill and gully tributaries carry materials from the valley walls into the main stream.

In addition to becoming deeper and wider, stream valleys are commonly lengthened as well. Valleys are lengthened in an upstream direction by *headward erosion* as drainage divides are eroded by entering runoff water (◆ Fig. 12.28a). In some cases headward erosion eventually breaches the drainage divide and diverts part of the drainage of another stream by a process called *stream piracy* (◆ Fig. 12.28b). Once stream piracy has occurred, both drainage systems must adjust; one now has more water, greater discharge, and greater potential to erode and transport sediment, whereas the other is diminished in all of these aspects.

⏷ SUPERPOSED STREAMS

Streams flow downhill in response to gravity, so their courses are determined by preexisting topography. Yet a number of streams seem, at first glance, to have defied this fundamental control. For example, the Delaware, Potomac, and Susquehanna rivers in the eastern United States have valleys that cut directly through ridges lying

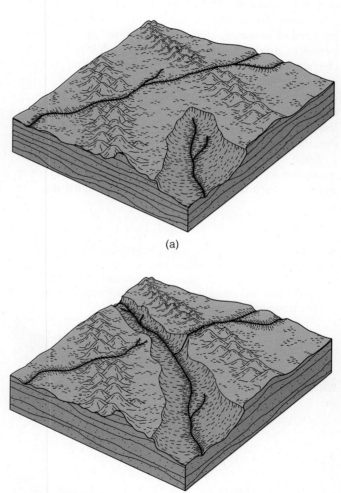

(a)

(b)

◆ FIGURE 12.28 Two stages in stream piracy. (*a*) In the first stage, the stream at the lower elevation extends its channel by headward erosion. In (*b*) it has captured some of the drainage of the stream flowing at the higher elevation.

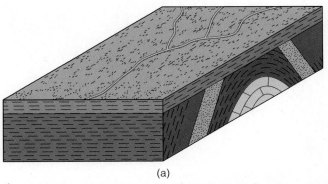
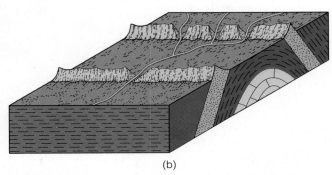

(a)

(b)

◆ FIGURE 12.29 The origin of a superposed stream. (*a*) A stream begins cutting down into horizontal strata. (*b*) A horizontal layer is removed by erosion, exposing the underlying structure. The stream flows across resistant beds that form the ridges.

in their paths. The Madison River in Montana meanders northward through a broad valley, then enters a narrow canyon cut into bedrock that leads to the next valley where the river resumes meandering.

All of the streams cited above are **superposed.** In order to understand how superposition occurs, it is necessary to know the geologic histories of these streams. In the case of the Madison River, the valleys it now occupies were once filled with sedimentary rocks so that the river flowed on a surface at a higher level (◆ Fig. 12.29). As the river eroded downward, it was superposed directly upon a preexisting knob of more resistant rock, and instead of changing its course, it cut a narrow, steep-walled canyon called a *water gap.*

Superposition also accounts for the fact that the Delaware, Potomac, and Susquehanna rivers flow through water gaps. During the Mesozoic Era, the Appalachian Mountain region was eroded to a sediment-covered plain across which numerous streams flowed generally eastward. During the Cenozoic Era, however, regional uplift commenced, and as a consequence of the uplift, the streams began eroding downward and were superposed on resistant strata, thus forming water gaps (Fig. 12.29).

≽ STEAM TERRACES

Adjacent to many streams are erosional remnants of floodplains formed when the streams were flowing at a higher level. These erosional remnants are **stream terraces.** They consist of a fairly flat upper surface and a steep slope descending to the level of the lower, present-day floodplain (◆ Fig. 12.30). In some cases, a stream has several steplike surfaces above its present-day floodplain, indicating that stream terraces formed several times.

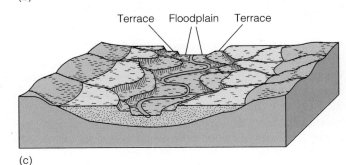
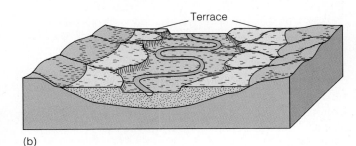
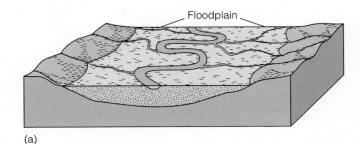

(a)

(b)

(c)

◆ FIGURE 12.30 Origin of stream terraces. (*a*) A stream has a broad floodplain adjacent to its channel. (*b*) The stream erodes downward and establishes a new floodplain at a lower level. Remnants of its old floodplain are stream terraces. (*c*) Another level of terraces originates as the stream erodes downward again.

Perspective 12.2

NATURAL BRIDGES

The term *natural bridge* has been used to describe a variety of features, including spans of rock resulting from wave erosion, the partial collapse of cavern roofs, and weathering and erosion along closely spaced, parallel joints as in Arches National Park in Utah (see Perspective 10.1). Here, however, we are concerned only with natural bridges that span a valley eroded by running water.

The best place to observe this type of natural bridge is in Natural Bridges National Monument in southwestern Utah. Three natural bridges are present within the monument, and all originated in the same way. Of these three, Sipapu Bridge is the largest (◆ Fig. 1); it stands 67 m above White Canyon and has a span of 81.5 m.

The process by which these natural bridges were formed is well understood, and, as a matter of fact, it is still going on. In the first stage, a meandering stream was incised into solid bedrock (◆ Fig. 2). In Natural Bridges National Monument, this rock unit is the Cutler Formation, which consists of sandstone formed from windblown sand deposited during the Permian Period. When local meandering streams became incised, lateral erosion created a thin wall of rock between adjacent meanders that was eventually breached (Fig. 2). As the breach was subsequently enlarged, the stream abandoned its old meander and the stream flow was diverted. As we discussed previously, oxbow lakes are formed by a similar process

(Fig. 12.15). The only significant difference is that the streams that form natural bridges are incised.

Natural bridges are temporary features. Once formed, they are destroyed by other processes. For example, rocks fall from the undersides of bridges, their surfaces are weathered and eroded, and eventually they collapse. The monument contains several examples of such collapsed bridges, but new ones are in the process of forming.

◆ FIGURE 1 Sipapu Bridge in Natural Bridges National Monument, Utah. (Photo courtesy of Sue Monroe.)

Although all stream terraces result from erosion, they are preceded by an episode of floodplain formation and deposition of sediment. Subsequent erosion causes the stream to cut downward until it is once again graded (Fig. 12.30). Once the stream again becomes graded, it begins eroding laterally and establishes a new floodplain at a lower level. Several such episodes account for the multiple terrace levels seen adjacent to some streams.

Renewed erosion and the formation of stream terraces are usually attributed to a change in base level. Either uplift of the land over which a stream flows or lowering of sea level yields a steeper gradient and increased flow velocity, thus initiating an episode of downcutting. When the stream reaches a level at which

it is once again graded, downcutting ceases. Although changes in base level no doubt account for many stream terraces, greater runoff in a stream's drainage basin can also result in the formation of terraces.

▼ INCISED MEANDERS

Some streams are restricted to deep, meandering canyons cut into solid bedrock, where they form features called **incised meanders.** For example, the San Juan River in Utah occupies a meandering canyon more than 390 meters deep (◆ Fig. 12.31). Such streams, being restricted by solid rock walls, are generally ineffective in eroding laterally; thus, they lack a floodplain and

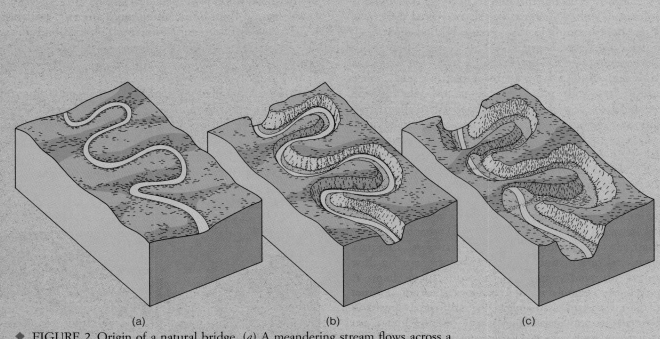

♦ FIGURE 2 Origin of a natural bridge. (*a*) A meandering stream flows across a gently sloping surface. (*b*) Incised meanders develop as the stream erodes down into solid rock. (*c*) A thin wall of rock between meanders is eventually breached, forming a natural bridge.

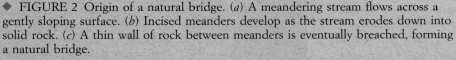

♦ FIGURE 12.31 Goose Necks of the San Juan River.

occupy the entire width of the canyon floor. Some incised meandering streams do erode laterally, thereby cutting off meanders and producing natural bridges (see Perspective 12.2).

It is not difficult to understand how a stream can cut downward into solid rock, but forming a meandering pattern in bedrock is another matter. Because lateral erosion is inhibited once downcutting begins, one must infer that the meandering course was established when the stream flowed across an area covered by alluvium. For example, suppose that a stream near base level has established a meandering pattern. If the land over which the stream flows is uplifted, erosion is initiated, and the meanders become incised into the underlying bedrock.

Chapter Summary

1. Water is continuously evaporated from the oceans, rises as water vapor, condenses, and falls as precipitation. About 20% of all precipitation falls on land and eventually returns to the oceans, mostly by surface runoff.

2. Running water moves by either laminar or turbulent flow. In the former, streamlines parallel one another, whereas in the latter they are complexly intertwined. Most flow in streams is turbulent.

3. Runoff can be characterized as either sheet flow or channel flow. Channels of all sizes are called streams.

4. Gradient generally varies from steep to gentle along the course of a stream, being steep in upper reaches and gentle in lower reaches.

5. Flow velocity and discharge are related. A change in one of these parameters causes the other to change as well.

6. A stream and its tributaries carry runoff from its drainage basin. Drainage basins are separated from one another by divides.

7. Streams erode by hydraulic action, abrasion, and dissolution of soluble rocks.

8. The coarser part of a stream's sediment load is transported as bed load, and the finer part as suspended load. Streams also transport a dissolved load of ions in solution.

9. Competence is a measure of the maximum-sized particles that a stream can carry and is related to velocity. Capacity is a function of discharge and is a measure of the total load transported by a stream.

10. Braided streams are characterized by a complex of dividing and rejoining channels. Braiding occurs whensediment transported by the stream is deposited within channels as sand and gravel bars.

11. Meandering streams have a single, sinuous channel with broad looping curves. Meanders migrate laterally as the cut bank is eroded and point bars form on the inner bank. Oxbow lakes are cutoff meanders in which fine-grained sediments and organic matter accumulate.

12. Floodplains are rather flat areas paralleling stream channels. They may be composed mostly of point bar deposits formed by lateral accretion or mud accumulated by vertical accretion during numerous floods.

13. Deltas are alluvial deposits at a stream's mouth. Many small deltas in lakes conform to the three-part division of bottomset, foreset, and topset beds, but large marine deltas are more complex.

14. Alluvial fans are lobate alluvial deposits on land that consist mostly of sand and gravel. They form best in arid regions where erosion rates are high.

15. Sea level is ultimate base level, the lowest level to which streams can erode. However, streams commonly have local base levels such as lakes, other streams, or the points where they flow across particularly resistant rocks.

16. Streams tend to eliminate irregularities in their channels so that they develop a smooth, concave profile of equilibrium. Such streams are graded. In a graded stream, a balance exists between gradient, discharge, flow velocity, channel characteristics, and sediment load so that little or no deposition occurs within the channel.

17. Stream valleys develop by a combination of processes including downcutting, lateral erosion, mass wasting, sheet wash, and headward erosion.

18. Many streams flowing through valleys cut into ridges directly in their paths are superposed, meaning that they once flowed on a higher surface and eroded downward into resistant rocks.

19. Renewed downcutting by a stream possessing a floodplain commonly results in the formation of stream terraces, which are remnants of an older floodplain at a higher level.

20. Incised meanders are generally attributed to renewed downcutting by a meandering stream such that it now occupies a deep, meandering valley.

Important Terms

abrasion
alluvial fan
alluvium
base level
bed load
braided stream
delta
discharge
dissolved load
divide

drainage basin
drainage pattern
floodplain
graded stream
gradient
hydraulic action
hydrologic cycle
incised meander
infiltration capacity
meandering stream

natural levee
oxbow lake
point bar
runoff
stream
stream terrace
superposed stream
suspended load
velocity

Review Questions

1. Trellis drainage develops on:
   a. ___ natural levees; b. ___ granite; c. ___ fractured basalt; d. ___ tilted sedimentary rock layers; e. ___ horizontal layers of volcanic rocks.
2. Mounds of sediment deposited on the margin of a stream are:
   a. ___ natural levees; b. ___ oxbow lakes; c. ___ bottomset beds; d. ___ incised meanders; e. ___ alluvial fans.
3. The direct impact of running water is:
   a. ___ bed load; b. ___ saltation; c. ___ hydraulic action; d. ___ meander cutoff; e. ___ base level.
4. Most of the fresh water on Earth is in:
   a. ___ the groundwater system; b. ___ the atmosphere; c. ___ lakes; d. ___ streams; e. ___ glaciers.
5. The vertical drop of a stream in a given horizontal distance is its:
   a. ___ discharge; b. ___ gradient; c. ___ velocity; d. ___ base level; e. ___ drainage pattern.
6. A _____ drainage pattern resembles the branching of a tree.
   a. ___ rectangular; b. ___ trellis; c. ___ dendritic; d. ___ deranged; e. ___ radial.
7. Sediment transport by intermittent bouncing and skipping along a stream bed is:
   a. ___ saltation; b. ___ dissolved load; c. ___ capacity; d. ___ suspended load; e. ___ alluvium.
8. The capacity of a stream is a measure of its:
   a. ___ volume of water; b. ___ velocity; c. ___ total load of sediment; d. ___ discharge; e. ___ ability to erode.
9. A meandering stream is one having:
   a. ___ numerous sand and gravel bars in its channel; b. a single, sinuous channel; c. ___ a broad, shallow channel; d. ___ a deep, narrow valley; e. ___ long, straight reaches and waterfalls.
10. In which of the following do foreset beds occur?
    a. ___ alluvial fans; b. ___ point bars; c. ___ floodplains; d. ___ deltas; e. ___ natural levees.
11. Which of the following is a local base level?
    a. ___ lake; b. ___ ocean; c. ___ floodplain; d. ___ point bar; e. ___ alluvial fan.
12. Erosional remnants of floodplains that are higher than the current level of a stream are:
    a. ___ oxbow lakes; b. ___ cut banks; c. ___ stream terraces; d. ___ incised meanders; e. ___ natural bridges.
13. All of the sediment carried by saltation and rolling and sliding along a stream bed is the:
    a. ___ suspended load; b. ___ drainage capacity; c. ___ stream profile; d. ___ bed load; e. ___ channel pattern.
14. A stream can lengthen its channel by:
    a. ___ runoff; b. ___ headward erosion; c. ___ vertical accretion; d. ___ hydraulic action; e. ___ downcutting.

15. Infiltration capacity is the:
    a. ___ rate at which a stream erodes; b. ___ distance a stream flows from its source to the ocean; c. ___ maximum rate at which surface materials can absorb water; d. ___ vertical distance a stream can erode below sea level; e. ___ variation in flow velocity across a stream channel.
16. A stream with a cross-sectional area of 250 m^2 and a flow velocity of 1.5 m/sec has a discharge of _____ m^3/sec.
    a. ___ 500; b. ___ 125; c. ___ 375; d. ___ 1,000; e. ___ 200.
17. Which of the following controls flow velocity in streams?
    a. ___ channel shape; b. ___ gradient; c. ___ channel roughness; d. ___ answers (a) and (b); e. ___ all of these.
18. The feature separating one drainage basin from another is a(an):
    a. ___ divide; b. ___ natural levee; c. ___ alluvial fan; d. ___ valley; e. ___ point bar.
19. A drainage pattern in which streams flow in and out of lakes with irregular flow directions is:
    a. ___ radial; b. ___ longitudinal; c. ___ deranged; d. ___ rectangular; e. ___ graded.
20. In which of the following would you expect to find mudflow deposits?
    a. ___ delta; b. ___ point bar; c. ___ incised meanders; d. ___ alluvial fan; e. ___ floodplain.
21. Why is the Earth the only planet that has abundant liquid water?
22. How do solar radiation, the changing phases of water, and runoff cause the recycling of water from the oceans to the atmosphere and back to the oceans?
23. What is the difference between laminar and turbulent flow, and why is flow in streams usually turbulent?
24. Explain what infiltration capacity is and why it is important in considering runoff.
25. A stream 2,000 m above sea level at its source flows 1,500 km to the sea. What is its gradient? Do you think the gradient that you calculated would be accurate for all segments of this stream?
26. How do channel shape and roughness control flow velocity?
27. Is the statement "the steeper the gradient, the greater the flow velocity" correct? Explain.
28. If a stream possesses rectangular drainage, what can you infer about the underlying rocks of the region?
29. How do streams erode and acquire a sediment load?
30. Explain the concepts of stream competence and capacity.
31. What do braided streams look like, and what do they transport and deposit?
32. How is it possible for a meandering stream to erode laterally yet maintain a more or less constant channel width?

33. How do oxbow lakes and meander scars form?
34. Explain how floodplains can develop by both lateral and vertical accretion.
35. How does a stream-dominated delta differ from a wave-dominated delta?
36. What are alluvial fans and where are they best developed?
37. What two depositional processes predominate on alluvial fans?
38. Sea level is ultimate base level for most streams. If sea level drops with respect to the land, how would a stream respond?
39. Why do streams tend to eliminate irregularities in their channels?
40. What is a graded stream, and why are streams rarely graded except temporarily?
41. How do headward erosion and stream piracy lengthen a stream channel?
42. Illustrate how a stream can be superposed and form a water gap.
43. How do stream terraces form?
44. How is it possible for a stream near base level to erode a deep meandering valley?

Additional Readings

Baker, V. R. 1982. *The channels of Mars.* Austin, Tex.: Univ. of Texas Press.

Beven, K., and P. Carling, eds. 1989. *Floods.* New York: Wiley.

Frater, A., ed. 1984. *Great rivers of the world.* Boston: Little, Brown.

Knighton, D. 1984. *Fluvial forms and processes.* London: Edward Arnold.

Leopold, L. B., M. G. Wolman, and J. P. Miller. 1964. *Fluvial processes in geomorphology.* San Francisco: W. H. Freeman.

McPhee, J. 1989. *The control of nature.* New York. Farrar, Straus, & Giroux.

Morisawa, M. 1968. *Streams: Their dynamics and morphology.* New York: McGraw-Hill.

Petts, G., and I. Foster. 1985. *Rivers and landscape.* London: Edward Arnold.

Rachocki, A. 1981. *Alluvial fans.* New York: Wiley.

Schumm, S. A. 1977. *The fluvial system.* New York: Wiley.

CHAPTER
13

Chapter Outline

The Leaning Tower of Pisa, Italy. The tilting is partly the result of subsidence due to the removal of groundwater.

Groundwater

Prologue

For more than two weeks in February 1925, Floyd Collins, an unknown farmer and cave explorer, became a household word (◆ Fig. 13.1). News about the attempts to rescue him from a narrow subsurface fissure near Mammoth Cave, Kentucky, captured the attention of the nation.

The saga of Floyd Collins is rooted in what is known as the Great Cave War of Kentucky. The western region of Kentucky is riddled with caves formed by groundwater weathering and erosion. Many of them were developed as tourist attractions to help supplement meager farm earnings. The largest and best known is Mammoth Cave (see Perspective 13.1). So spectacular is Mammoth, with its numerous caverns, underground rivers, and dramatic cave deposits, that it soon became the standard by which all other caves were measured.

As Mammoth Cave drew more and more tourists, rival cave owners became increasingly bold in attempting to lure visitors to their caves and curio shops. Signs pointing the way to Mammoth Cave frequently disappeared, while "official" cave information booths redirected unsuspecting tourists away from Mammoth Cave. It was in this environment that Floyd Collins grew up.

Seven years before his tragic death, Collins had discovered Crystal Cave on the family farm and opened it up for visitors. But like most of the caves in the area, Crystal Cave attracted few tourists—they visited Mammoth Cave instead. Perhaps it was the thought of discovering a cave rivaling Mammoth or even connecting to it that drove Collins to his fateful exploration of Sand Cave on January 30, 1925.

As Collins inched his way back up through the narrow fissure he had crawled down, he dislodged a small oblong piece of limestone from the ceiling that immediately pinned his left ankle. Try as he might, he was trapped in total darkness 17 m below ground. As he lay half on his left side, Collins's left arm was partially wedged under him, while his right arm was held fast by an overhanging ledge. During his struggles to free himself, he dislodged enough silt and small rocks to bury his legs, further immobilizing him and adding to his anguish.

The next day several neighbors reached Collins and were able to talk to him, feed him, encourage him, and try to make him more comfortable, but they could not get him out. Word of his plight quickly spread and the area soon swarmed with reporters. Eventually, volunteers were able to excavate an area around Collins's upper body, but could not free his pinned legs. While an anxious country waited, rescue attempts led by Floyd's brother Homer continued.

Three days after he had become trapped, a harness was put around Collins's chest and rescuers tried to pull him free. After numerous attempts to yank him out, workers had to abandon that plan because Collins was unable to bear the pain. Meanwhile at the surface, a carnival-like atmosphere had developed as hordes of up to 20,000 people converged on the scene, and the National Guard had to be called out to maintain order.

Two days after the attempt to pull Collins out of the fissure failed, part of the passageway used by rescuers collapsed, thus blocking passage to Collins (Fig. 13.1c). The only hope now was to dig a vertical

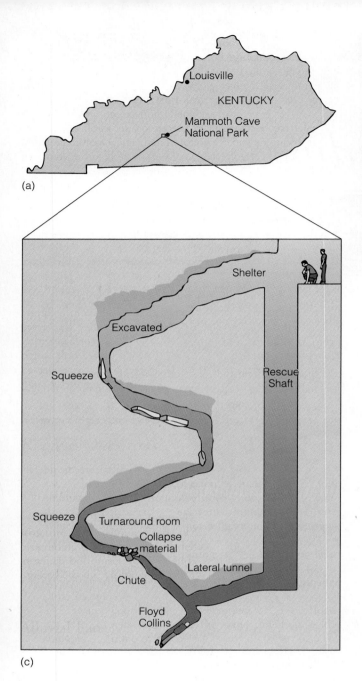

(a)

(c)

(b)

◆ FIGURE 13.1 (*a*) Location of the cave in which Floyd Collins was trapped. (*b*) Collins looking out of a fissure near the cave where he ultimately died. (*c*) Cross section showing the fissure where Collins was trapped, the rescue shaft that was sunk, and the lateral tunnel that finally reached him.

relief shaft from which a lateral tunnel could be dug to reach Collins. On February 16, rescuers finally reached the chamber where Collins's lifeless body lay entombed. After he was brought out, he was buried near Crystal Cave, where his grave is appropriately marked by a beautiful stalagmite and pink granite headstone.

▼ INTRODUCTION

Groundwater—the water stored in the open spaces within underground rocks and unconsolidated material—is a valuable natural resource that is essential to the lives of all people. Its importance to humans is not new. Groundwater rights have always been important in the western United States, and many legal battles have been fought over them. Groundwater also played a crucial role in the development of the U.S. railway system during the nineteenth century when railroads had to have a reliable source of water for their steam locomotives. Much of the water used by the locomotives came from groundwater tapped by wells.

Today, the study of groundwater and its movement has become increasingly important as the demand for fresh water by agricultural, industrial, and domestic users has reached an all-time high. More than 65% of the groundwater used in the United States each year goes for irrigation, with industrial use second, followed by domestic needs. Such demands have severely depleted the groundwater supply in many areas and led to such problems as ground subsidence and saltwater contamination. In other areas, pollution from landfills, toxic waste, and agriculture has rendered the groundwater supply unsafe.

As the world's population and industrial development expand, the demand for water, particularly groundwater, will increase. Not only is it important to locate new groundwater sources, but, once found, these sources must be protected from pollution and managed properly to ensure that users do not withdraw more water than can be replenished.

⩔ GROUNDWATER AND THE HYDROLOGIC CYCLE

Groundwater represents approximately 22% (8.4 million km^3) of the world's supply of fresh water. This amount is about 36 times greater than the total for all of the streams and lakes of the world (see Chapter 12) and equals about one-third the amount locked up in the world's ice caps (see Chapter 14). If the world's groundwater were spread evenly over the Earth's surface, it would be about 10 m deep.

Groundwater is one reservoir of the hydrologic cycle (see Fig. 12.3). The major source of groundwater is precipitation that infiltrates the ground and moves through the soil and pore spaces of rocks. Other sources include water infiltrating from lakes and streams, recharge ponds, and wastewater treatment systems. As the groundwater moves through soil, sediment, and rocks, many of its impurities, such as disease-causing microorganisms, are filtered out. Not all soils and rocks are good filters, however, and some serious pollutants are not removed. Groundwater eventually returns to the surface reservoir when it enters lakes, streams, or the ocean.

⩔ POROSITY AND PERMEABILITY

Porosity and *permeability* are important physical properties of rocks, sediment, and soil and are largely responsible for the amount, availability, and movement of groundwater. Water soaks into the ground because the soil, sediment, or rock has open spaces or pores. **Porosity** is the percentage of a material's total volume that is pore space. While porosity most often consists of the spaces between particles in soil, sediments, and sedimentary rocks, other types of porosity can include cracks, fractures, faults, and vesicles in volcanic rocks (◆ Fig. 13.2).

Porosity varies among different rock types and is dependent on the size, shape, and arrangement of the material composing the rock (● Table 13.1). Most igneous and metamorphic rocks as well as many limestones and dolostones have very low porosity because they are composed of tightly interlocking crystals. Their porosity can be increased, however, if they have been fractured or weathered by groundwater. This is particularly true for massive limestone and dolostone whose fractures can be enlarged by acidic groundwater.

◆ FIGURE 13.2 A rock's porosity is dependent on the size, shape, and arrangement of the material composing the rock. (*a*) A well-sorted sedimentary rock has high porosity while (*b*) a poorly sorted one has low porosity. (*c*) In soluble rocks such as carbonates, porosity can be increased by solution, while (*d*) crystalline rocks can be rendered porous by fracturing.

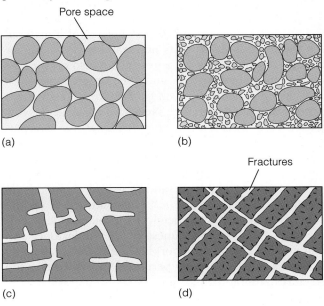

● TABLE 13.1 Porosity Values for Different Materials

Material	Percentage Porosity
Unconsolidated sediment	
Soil	55
Gravel	20–40
Sand	25–50
Silt	35–50
Clay	50–70
Rocks	
Sandstone	5–30
Shale	0–10
Solution activity in limestone, dolostone	10–30
Fractured basalt	5–40
Fractured granite	10

SOURCE: U.S. Geological Survey, Water Supply Paper 2220 (1983) and others.

By contrast, detrital sedimentary rocks composed of well-sorted and well-rounded grains can have very high porosity because any two grains touch only at a single point, leaving relatively large open spaces between the grains (Fig. 13.2a). Poorly sorted sedimentary rocks, on the other hand, typically have low porosity because finer grains fill in the space between the larger grains, further reducing porosity (Fig. 13.2b). In addition, the amount of cement between grains can also decrease porosity.

Although porosity determines the amount of groundwater a rock can hold, it does not guarantee that the water can be extracted. The capacity of a material for transmitting fluids is its **permeability**. Permeability is dependent not only on porosity, but also on the size of the pores or fractures and their interconnections. For example, deposits of silt or clay are typically more porous than sand or gravel. Nevertheless, shale has low permeability because the pores between its clay particles are very small, and the molecular attraction between the clay and the water is great, thereby preventing movement of the water. In contrast, the pore spaces between grains in sandstone and conglomerate are much larger, and the molecular attraction on the water is therefore low. Chemical and biochemical sedimentary rocks, such as limestone and dolostone, and many igneous and metamorphic rocks that are highly fractured can also be very permeable provided that the fractures are interconnected.

A permeable layer transporting groundwater is called an **aquifer**, from the Latin *aqua* meaning water. The most effective aquifers are deposits of well-sorted and well-rounded sand and gravel. Limestones in which fractures and bedding planes have been enlarged by solution are also good aquifers. Shales and many igneous and metamorphic rocks, however, are typically impermeable. Rocks such as these and any other materials that prevent the movement of groundwater are called **aquicludes**.

⩔ THE WATER TABLE

When precipitation occurs over land, some of it evaporates, some of it is carried away by runoff in streams, and the remainder seeps into the ground. As this water moves down from the surface, some of it adheres to the material that it is moving through and halts its downward progress. This region is the **zone of aeration,** and its water is called *suspended water* (◆ Fig. 13.3). The pore spaces in this zone contain both water and air. Extending irregularly upward a few centimeters to several meters from the base of the zone of aeration is the **capillary fringe**. Water moves upward in this region from the zone of saturation below because of surface

◆ FIGURE 13.3 The zone of aeration contains both air and water within its open space, while all of the open space in the zone of saturation is filled with groundwater. The water table is the surface separating the zones of aeration and saturation. Within the capillary fringe, water rises upward by surface tension from the zone of saturation into the zone of aeration.

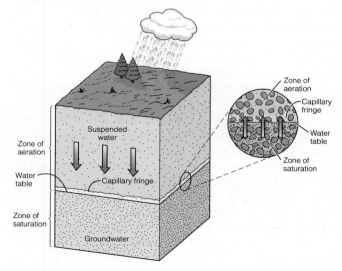

tension. Such movement is analogous to the upward movement of water through a paper towel.

Beneath the zone of aeration lies the **zone of saturation** in which all of the pore spaces are filled with groundwater (Fig. 13.3). The base of the zone of saturation varies from place to place, but usually extends to a depth where an impermeable layer is encountered or to a depth where confining pressure closes all open space.

The surface separating the zone of aeration from the underlying zone of saturation is the **water table** (Fig. 13.3). In general, the configuration of the water table is a subdued replica of the overlying land surface; that is, it has its highest elevations beneath hills and its lowest elevations in valleys. In most arid and semiarid regions, however, the water table is quite flat and is below the level of river valleys.

Several factors contribute to the surface configuration of a region's water table. These include regional differences in the amount of rainfall, permeability, and the rate of groundwater movement. During periods of high rainfall, groundwater tends to rise beneath hills because it cannot flow fast enough into the adjacent valleys to maintain a level surface. During droughts, however, the water table falls and tends to flatten out because it is not being replenished.

▼ GROUNDWATER MOVEMENT

Groundwater velocity varies greatly and depends on many factors. Velocities range from 250 m per day in some extremely permeable material to less than a few centimeters per year in nearly impermeable material. In most ordinary aquifers, however, the average velocity of groundwater is a few centimeters per day.

Gravity provides the energy for the downward movement of groundwater. Water entering the ground moves through the zone of aeration to the zone of saturation (◆ Fig. 13.4). When water reaches the water table, it continues to move through the zone of saturation from areas where the water table is high toward areas where it is lower, such as streams, lakes, or swamps (Fig. 13.4). Only some of the water follows the direct route along the slope of the water table. Most of it takes longer curving paths downward and then enters a stream, lake, or swamp from below. This occurs because groundwater moves from areas of high pressure toward areas of lower pressure within the saturated zone.

▼ SPRINGS, WATER WELLS, AND ARTESIAN SYSTEMS

Adding water to the zone of saturation is called **recharge**, and it causes the water table to rise. Water may be added by natural means, such as rainfall or melting snow, or artificially at recharge basins or wastewater treatment plants (◆ Fig. 13.5). If groundwater is discharged without sufficient replenishment, the water table drops. Groundwater discharges naturally whenever the water table intersects the ground surface as at a spring or along a stream, lake, or swamp. Groundwater can also be discharged artificially by pumping water from wells.

Springs

A **spring** is a place where groundwater flows or seeps out of the ground. Springs have always fascinated people because the water flows out of the ground for

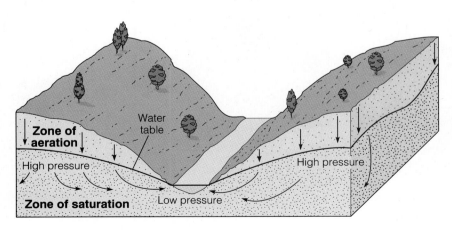

◆ FIGURE 13.4 Groundwater moves downward due to the force of gravity. It moves through the zone of aeration to the zone of saturation where some of it moves along the slope of the water table and the rest of it moves through the zone of saturation from areas of high pressure toward areas of low pressure.

◆ FIGURE 13.5 A recharge basin in Nassau County, New York.

no apparent reason and from no readily identifiable source. It is not surprising that springs have long been regarded with superstition and revered for their supposed medicinal value and healing powers. Nevertheless, there is nothing mystical or mysterious about springs.

Although springs can occur under a wide variety of geologic conditions, they all form in basically the same way (◆ Fig. 13.6). When percolating water reaches the water table or an impermeable layer, it flows laterally, and if this flow intersects the Earth's surface, the water discharges onto the surface as a spring (◆ Fig. 13.7). The Mammoth Cave area in Kentucky, for example, is underlain by fractured limestones that have been enlarged into caves by solution activity (see Perspective 13.1). In this geologic environment, springs occur where the fractures and caves intersect the ground surface allowing groundwater to exit onto the surface.

◆ FIGURE 13.6 Springs form wherever laterally moving groundwater intersects the Earth's surface. (a) Most commonly, they form when percolating water reaches an impermeable layer and migrates laterally until it seeps out at the surface. (b) Springs also can occur in areas underlain by fractured soluble rocks such as limestones where groundwater moves freely through underground cavities until it reaches the surface and flows out.

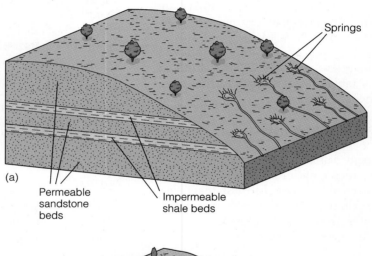

Springs

(a)

Permeable sandstone beds

Impermeable shale beds

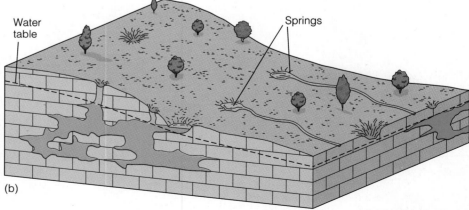

Water table

Springs

(b)

Springs most commonly occur along valley walls where streams have cut valleys below the regional water table.

Springs can also develop wherever a perched water table intersects the Earth's surface (◆ Fig. 13.8). A **perched water table** may occur wherever a local aquiclude occurs within a larger aquifer, such as a lens of shale within a sandstone. As water migrates through the zone of aeration, it is stopped by the local aquiclude, and a localized zone of saturation "perched" above the main water table is created. Water moving laterally along the perched water table may intersect the Earth's surface to produce a spring.

Water Wells

A **water well** is made by digging or drilling into the zone of saturation. Once the zone of saturation is reached, water percolates into the well and fills it to the level of the water table. Most wells must be pumped to bring the groundwater to the surface.

When a well is pumped, the water table in the area around the well is lowered, because water is removed from the aquifer faster than it can be replenished. A **cone of depression** thus forms around the well, varying in size according to the rate and amount of water being withdrawn (◆ Fig. 13.9). If water is pumped out of a well faster than it can be replaced, the cone of depression grows until the well goes dry. This lowering of the water table normally does not pose a problem for the average domestic well provided that the well is drilled sufficiently deep into the zone of saturation. The tremendous amounts of water used by industry and irrigation, however, may create a large cone of depres-

◆ FIGURE 13.7 Periodic Spring, near Afton, Wyoming.

◆ FIGURE 13.8 If a localized aquiclude, such as a shale layer, occurs within an aquifer, a perched water table may result with springs occurring where the perched water table intersects the Earth's surface.

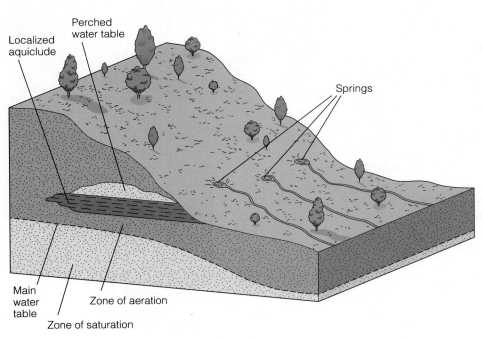

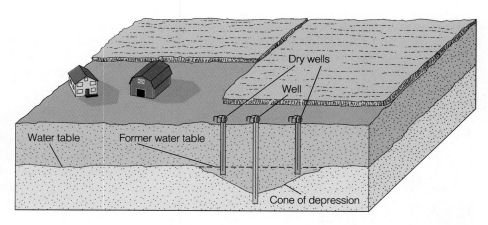

◆ FIGURE 13.9 A cone of depression forms whenever water is withdrawn from a well. If water is withdrawn faster than it can be replenished, the cone of depression will grow in depth and circumference, lowering the water table in the area and causing nearby shallow wells to go dry.

sion that lowers the water table sufficiently to cause shallow wells in the immediate area to go dry (Fig. 13.9). Such a situation is not uncommon and frequently results in lawsuits by the owners of the shallow dry wells. Furthermore, lowering of the regional water table is becoming a serious problem in many areas, particularly in the southwestern United States where rapid growth has placed tremendous demands on the groundwater system. Unrestricted withdrawal of groundwater cannot continue indefinitely, and the rising costs and decreasing supply of groundwater should soon limit the growth of this region of the United States.

Artesian Systems

The word *artesian* comes from the French town and province of Artois (called Artesium during Roman times) near Calais, where the first European artesian well was drilled in A.D. 1126 and is still flowing today. The term **artesian system** can be applied to any system in which groundwater is confined and builds up high hydrostatic (fluid) pressure. Water in such a system is able to rise above the level of the aquifer if a well is drilled through the confining layer, thereby reducing the pressure and forcing the water upward (◆ Fig. 13.10). For an artesian system to develop, three geologic conditions must be present (◆ Fig. 13.11): (1) the aquifer must be confined above and below by aquicludes to prevent water from escaping; (2) the rock sequence is usually tilted and exposed at the surface, enabling the aquifer to be recharged; and (3) there is sufficient precipitation in the recharge area to keep the aquifer filled.

◆ FIGURE 13.10 Artesian well at Deep Well Ranch, South Fork of the Madison River, Gallatin County, Montana.

The elevation of the water table in the recharge area and the distance of the well from the recharge area determine the height to which artesian water rises in a well. The surface defined by the water table in the recharge area, called the *artesian-pressure surface*, is indicated by the sloping dashed line in Figure 13.11. If there were no friction in the aquifer, well water from an artesian aquifer would rise exactly to the elevation of the artesian-pressure surface. Friction, however, slightly reduces the pressure of the aquifer water and conse-

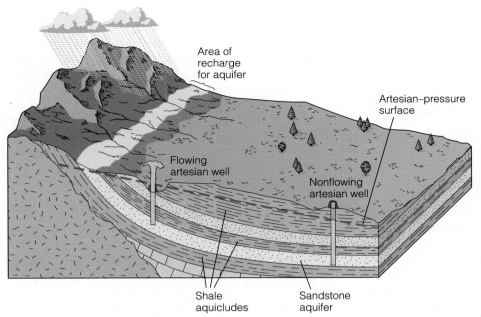

Area of
recharge
for aquifer

Artesian–pressure
surface

Flowing
artesian well

Nonflowing
artesian well

Shale
aquicludes

Sandstone
aquifer

◆ FIGURE 13.11 An artesian system must have an aquifer confined above and below by aquicludes, the aquifer must be exposed at the surface, and there must be sufficient precipitation in the recharge area to keep the aquifer filled. The elevation of the water table in the recharge area, which is indicated by a sloping dashed line (the artesian-pressure surface), defines the highest level to which well water can rise. If the elevation of a wellhead is below the elevation of the artesian-pressure surface, the well will be free-flowing because the water will rise toward the artesian-pressure surface, which is at a higher elevation than the wellhead. If the elevation of a wellhead is at or above that of the artesian-pressure surface, the well will be nonflowing.

quently the level to which artesian water rises. This is why the pressure surface slopes.

An artesian well will flow freely at the ground surface only if the wellhead is at an elevation below the artesian-pressure surface. In this situation, the water flows out of the well because it rises toward the artesian-pressure surface, which is at a higher elevation than the wellhead. In a nonflowing artesian well, the wellhead is above the artesian-pressure surface, and thus the water will rise in the well only as high as the artesian-pressure surface.

In addition to artesian wells, many artesian springs also exist. Such springs can occur if a fault or fracture intersects the confined aquifer allowing water to rise above the aquifer. Oases in deserts are commonly artesian springs.

Because the geologic conditions necessary for artesian water can occur in a variety of ways, artesian systems are quite common in many areas of the world underlain by sedimentary rocks. One of the best-known

artesian systems in the United States underlies South Dakota and extends southward to central Texas. The majority of the artesian water from this system is used for irrigation. The aquifer of this artesian system, the Dakota Sandstone, is recharged where it is exposed along the margins of the Black Hills of South Dakota. The hydrostatic pressure in this system was originally great enough to produce free-flowing wells and to operate waterwheels. The extensive use of water for irrigation over the years, however, has reduced the pressure in many of the wells so that they are no longer free-flowing and the water must be pumped.

▼ GROUNDWATER EROSION AND DEPOSITION

When rainwater begins seeping into the ground, it immediately starts to react with the minerals it contacts, weathering them chemically. In an area underlain by soluble rock, groundwater is the principal agent of erosion and thus is responsible for the formation of many major features of the landscape.

Limestone, a common sedimentary rock composed primarily of the mineral calcite ($CaCO_3$), underlies large areas of the Earth's surface (◆ Fig. 13.12). Although limestone is practically insoluble in pure water, it readily dissolves if a small amount of acid is present. Carbonic acid (H_2CO_3) is a weak acid that forms when carbon dioxide combines with water ($H_2O + CO_2 \rightarrow H_2CO_3$) (see Chapter 5). Because the atmosphere contains a small amount of carbon dioxide (0.03%), and carbon dioxide is also produced in soil by the decay of organic matter, most groundwater is slightly acidic. When groundwater percolates through the various openings in limestone, the slightly acidic water readily reacts with the calcite to dissolve the rock by forming soluble calcium bicarbonate, which is carried away in solution (see Chapter 5).

Sinkholes and Karst Topography

In regions underlain by soluble rock, the ground surface may be pitted with numerous depressions that vary in size and shape. These depressions, called **sinkholes** or merely *sinks*, mark areas where the underlying rock is soluble (◆ Fig. 13.13). Sinkholes usually form in one of two ways. The first is when the soluble rock below the soil is dissolved by seeping water. Natural openings in the rock are enlarged and filled in by the overlying soil. As the groundwater continues to dissolve the rock, the soil is eventually removed, leaving depressions that are typically shallow with gently sloping sides.

Sinkholes also form when a cave's roof collapses, usually producing a steep-sided crater. Sinkholes formed in this way are a serious hazard, particularly in populated areas. In regions prone to sinkhole formation, the depth and extent of underlying cave systems must be mapped before any development to ensure that the underlying rocks are thick enough to support planned structures.

◆ FIGURE 13.12 The distribution of the major limestone and karst areas of the world.

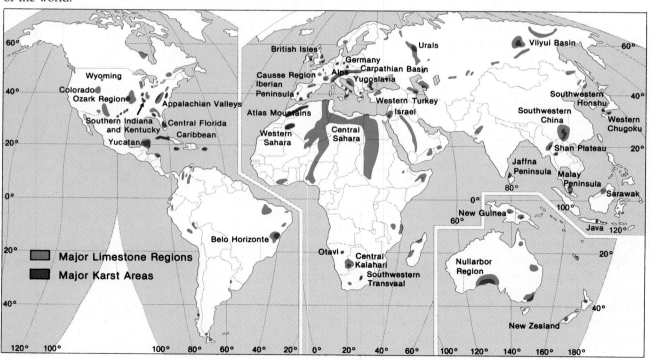

◆ FIGURE 13.13 This sinkhole formed on May 8 and 9, 1981, in Winter Park, Florida, due to a drop in the water table after prior dissolution of the underlying limestone. The sinkhole destroyed a house, numerous cars, and the municipal swimming pool. It has a diameter of 100 m and a depth of 35 m.

A **karst topography** is one that has developed largely by groundwater erosion (◆ Fig. 13.14). The name *karst* is derived from the plateau region of the border area between Yugoslavia and northeastern Italy where this type of topography is well developed. In the United States, regions of karst topography include large areas of southwestern Illinois, southern Indiana, Kentucky, Tennessee, northern Missouri, Alabama, and central and northern Florida (Fig. 13.12).

Karst topography is characterized by numerous caves, springs, sinkholes, solution valleys, and disappearing streams (Fig. 13.14). When adjacent sinkholes merge, they form a network of larger, irregular, closed depressions called *solution valleys. Disappearing streams* are another feature of areas of karst topography. They are so named because they typically flow only a short distance at the surface and then disappear into a sinkhole. The water continues flowing underground through various fractures or caves until it surfaces again at a spring or other stream.

◆ FIGURE 13.14 Some of the features of karst topography.

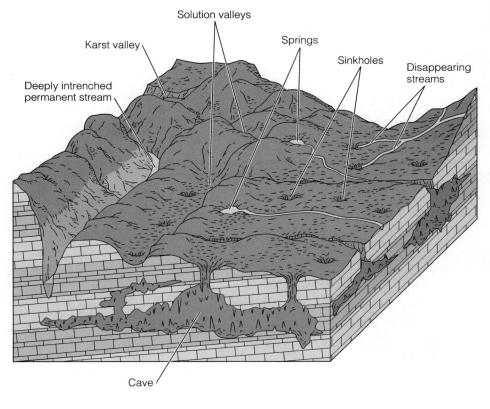

Karst topography can range from the spectacular high relief landscapes of China to the subdued and pockmarked landforms of Kentucky (♦ Fig. 13.15). What is common to all karst topography, however, is that thick-bedded, readily soluble rock is present at the surface or just below the soil, and enough water is present for solution activity to occur. Karst topography is, therefore, typically restricted to humid and temperate climates.

♦ FIGURE 13.15 (*a*) The Stone Forest, 126 km southeast of Kunming, People's Republic of China, is a high relief karst landscape formed by the dissolution of carbonate rocks. (*b*) Solution valleys, sinkholes, and sinkhole lakes dominate the subdued karst topography east of Bowling Green, Kentucky.

(a)

(b)

Caves and Cave Deposits

Caves are some of the most spectacular examples of the combined effects of weathering and erosion by groundwater. As groundwater percolates through carbonate rocks, it dissolves and enlarges original fractures and openings to form a complex interconnecting system of crevices, caves, caverns, and underground streams. A **cave** is usually defined as a naturally formed subsurface opening that is generally connected to the surface and is large enough for a person to enter. A *cavern* is a very large cave or a system of interconnected caves.

More than 17,000 caves are known in the United States. Most of them are small, but some are quite large and spectacular. Some of the more famous caves in the United States are Mammoth Cave, Kentucky (see Perspective 13.1); Carlsbad Caverns, New Mexico; Lewis and Clark Caverns, Montana; Wind Cave and Jewel Cave, South Dakota; Lehman Cave, Nevada; and Meramec Caverns, Missouri, which Jesse James and his outlaw band often used as a hideout. While the United States has many famous caves, the deepest known cave in North America is the 536 m–deep Arctomys Cave in Mount Robson Provincial Park, British Columbia, Canada.

Caves and caverns form as a result of the dissolution of carbonate rocks by weakly acidic groundwater (♦ Fig. 13.16). Groundwater percolating through the zone of aeration slowly dissolves the carbonate rock and enlarges its fractures and bedding planes. Upon reaching the water table, the groundwater migrates toward the region's surface streams (Fig. 13.4). As the groundwater moves through the zone of saturation, it continues to dissolve the rock and gradually forms a system of horizontal passageways through which the dissolved rock is carried to the streams. As the surface streams erode deeper valleys, the water table drops in response to the lower elevation of the streams. The water that flowed through the system of horizontal passageways now percolates down to the lower water table where a new system of passageways begins to form. The abandoned channelways now form an interconnecting system of caves and caverns that may continue to enlarge as groundwater percolates through them and dissolves the surrounding rock. As the caves increase in size, they may become unstable and collapse, littering the floor with fallen debris.

When most people think of caves, they think of the seemingly endless variety of colorful and bizarre-shaped deposits found in them. Although a great many

MAMMOTH CAVE NATIONAL PARK, KENTUCKY

Within the limestone region of western Kentucky lies the largest cave system in the world. In 1941, approximately 51,000 acres were set aside and designated as Mammoth Cave National Park. In 1981 it became a World Heritage Site. Recently, the National Park Service has been considering closing Mammoth Cave because of the health hazard created by raw sewage and contaminated groundwater in the area.

From ground level, the topography of the area is unimposing with numerous sinkholes, lakes, valleys, and disappearing streams. Beneath the surface, however, are more than 230 km of interconnecting passageways whose spectacular geologic features have been enjoyed by numerous cave explorers and tourists alike.

Based on carbon 14 dates from some of the many artifacts found in the cave (such as woven cord and wooden bowls), Mammoth Cave had been explored and used by Native Americans for more than 3,000 years prior to its rediscovery in 1799 by a bear hunter named Robert Houchins. During the War of 1812, approximately 180 metric tons of saltpeter (a potassium nitrate mineral), used in the manufacture of gunpowder, were mined from Mammoth Cave. At the end of the war, the saltpeter market collapsed, and Mammoth Cave was developed as a tourist attraction, easily overshadowing the other caves in the area. Over the next 150 years, the discovery of new passageways and caverns helped establish Mammoth Cave as the world's premier cave and the standard against which all others were measured (see the Prologue).

Mammoth Cave formed in much the same way as all other caves (Fig. 13.16). Groundwater flowing through the St. Genevieve Limestone eroded a complex network of openings, passageways, and caverns. Flowing through the various caverns is the Echo River, a system of subsurface streams that eventually joins the Green River at the surface.

The colorful cave deposits are the primary reason millions of tourists have visited Mammoth Cave. Here can be seen numerous stalactites, stalagmites, and columns, as well as spectacular travertine flowstone deposits (◆ Fig. 1). Other attractions include the Giant's Coffin, a 15 m collapse block of limestone, and giant rooms such as Mammoth Dome, which is about 58 m high (◆ Fig. 2). The cave is also home to more than 200 species of insects and other animals, including about 45 blind species; some of these can be seen on the Echo River Tour, which conveys visitors 5 km along the underground stream.

◆ FIGURE 1 Frozen Niagara is a spectacular example of massive travertine flowstone deposits.

◆ FIGURE 2 Looking up Mammoth Dome, the largest room in Mammoth Cave, Kentucky.

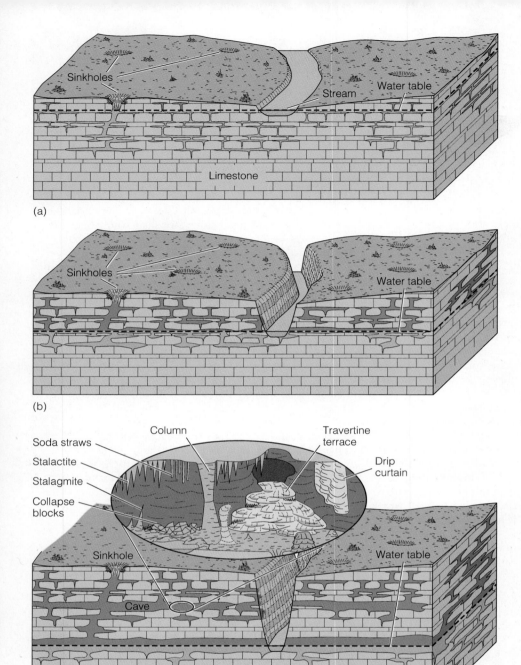

(a)

(b)

Soda straws
Stalactite
Stalagmite
Collapse
blocks

Column

Travertine
terrace

Drip
curtain

Sinkhole

Cave

Water table

(c)

◆ FIGURE 13.16 The formation of caves. (*a*) As groundwater percolates through
the zone of aeration and flows through the zone of saturation, it dissolves the carbonate
rocks and gradually forms a system of passageways. (*b*) Groundwater moves along the
surface of the water table, forming a system of horizontal passageways through which
dissolved rock is carried to the surface streams and thus enlarging the passageways.
(*c*) As the surface streams erode deeper valleys, the water table drops, and the
abandoned channelways form an interconnecting system of caves and caverns.

different types of cave deposits exist, most form in essentially the same manner and are collectively known as **dripstone**. As water seeps through a cave, some of the dissolved carbon dioxide in the water escapes, and a small amount of calcite is precipitated. In this manner, the various dripstone deposits are formed.

Stalactites are icicle-shaped structures hanging from cave ceilings that form as a result of precipitation from dripping water ($\blacklozenge$ Fig. 13.17). With each drop of water, a thin layer of calcite is deposited over the previous layer, forming a cone-shaped projection that grows downward from the ceiling. While many stalactites are solid, some are hollow and are appropriately called *soda straws*.

The water that drips from a cave's ceiling also precipitates a small amount of calcite when it hits the floor. As additional calcite is deposited, an upward growing projection called a **stalagmite** forms (Fig. 13.17). If a stalactite and stalagmite meet, they form a **column**. Groundwater seeping from a crack in a cave's ceiling may form a vertical sheet of rock called a *drip curtain*, while water flowing across a cave's floor may produce *travertine terraces* (Fig. 13.16).

▼ MODIFICATIONS OF THE GROUNDWATER SYSTEM AND THEIR EFFECTS

Groundwater is a valuable natural resource that is rapidly being exploited with little regard to the effects of overuse and misuse. Currently, about 20% of all water used in the United States is groundwater. This percentage is increasing, however, and unless this resource is used more wisely, sufficient amounts of clean groundwater will not be available in the future. Modifications of the groundwater system may have many consequences including (1) lowering of the water table, which causes wells to dry up; (2) loss of hydrostatic pressure, which causes once free-flowing wells to require pumping; (3) saltwater encroachment; (4) subsidence; and (5) contamination of the groundwater supply.

Lowering of the Water Table

Withdrawing groundwater at a significantly greater rate than it is replaced by either natural or artificial recharge can have serious effects. For example, the High Plains aquifer is one of the most important aquifers in the United States. Underlying most of Nebraska, large parts of Colorado and Kansas, portions of South Dakota, Wyoming, and New Mexico, as well as the panhandle regions of Oklahoma and Texas, it accounts for approximately 30% of the groundwater used for irrigation in the United States ($\blacklozenge$ Fig. 13.18). Irrigation from the High Plains aquifer is largely responsible for the high agricultural productivity of this region. A significant percentage of the nation's corn, cotton, and wheat is grown here, and half of our beef cattle are raised in this region. Large areas of land (more than 14

$\blacklozenge$ FIGURE 13.17 Stalactites are the icicle-shaped structures seen hanging from the ceiling, while the upward-pointing structures on the cave floor are stalagmites. Several columns are present where the stalactites and stalagmites have met in this chamber of Luray Caves, Virginia.

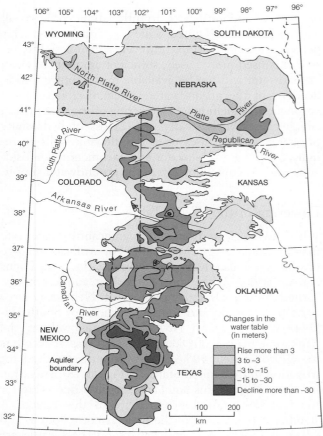

◆ FIGURE 13.18 Areal extent of the High Plains aquifer and changes in the water table, predevelopment to 1980.

million acres) are currently irrigated with water pumped from the High Plains aquifer. Irrigation is popular because yields from irrigated lands can be triple what they would be without irrigation.

While the High Plains aquifer has contributed to the high productivity of the region, it cannot continue providing the quantities of water that it has in the past. In some parts of the High Plains, from 2 to 100 times more water is being pumped annually than is being recharged. Consequently, water is being removed from the aquifer faster than it is being replenished, causing the water table to drop significantly in many areas (Fig. 13.18).

What will happen to this region's economy if long-term withdrawal of water from the High Plains aquifer greatly exceeds its recharge rate such that it can no longer supply the quantities of water necessary for irrigation? Solutions range from going back to farming without irrigation to diverting water from other regions

such as the Great Lakes. Farming without irrigation would result in greatly decreased yields and higher costs and prices for agricultural products, while the diversion of water from elsewhere would cost billions of dollars and the price of agricultural products would still rise.

Saltwater Incursion

The excessive pumping of groundwater in coastal areas can result in **saltwater incursion** such as occurred on Long Island, New York, during the 1960s. Along coastlines where permeable rocks or sediments are in contact with the ocean, the fresh groundwater, being less dense than seawater, forms a lens-shaped body above the underlying salt water (◆ Fig. 13.19a). The weight of the fresh water exerts pressure on the underlying salt water. As long as rates of recharge equal rates of withdrawal, the contact between the fresh groundwater and the seawater will remain the same. If excessive pumping occurs, however, a deep cone of depression forms in the fresh groundwater (◆ Fig. 13.19b). Because some of the pressure from the overlying fresh water has been removed, salt water migrates upward to fill the pore space that formerly contained fresh water. When this occurs, wells become contaminated with salt water and remain contaminated until recharge by fresh water restores the former level of the fresh groundwater water table.

Saltwater incursion is a major problem in many rapidly growing coastal communities. As the population in these areas grows, greater demand for groundwater creates an even greater imbalance between recharge and withdrawal.

To counteract the effects of saltwater incursion, recharge wells are often drilled to pump water back into the groundwater system (◆ Fig. 13.19c). Recharge ponds that allow large quantities of fresh surface water to infiltrate the groundwater supply may also be constructed (Fig. 13.5).

Subsidence

As excessive amounts of groundwater are withdrawn from poorly consolidated sediments and sedimentary rocks, the water pressure between grains is reduced, and the weight of the overlying materials causes the grains to pack closer together, resulting in subsidence of the ground. As more and more groundwater is pumped to meet the increasing needs of agriculture, industry, and population growth, subsidence is becoming more prevalent.

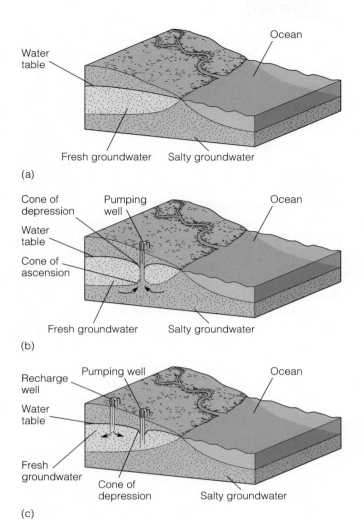

(a)

(b)

(c)

◆ FIGURE 13.19 Saltwater incursion. (*a*) Because fresh water is not as dense as salt water, it forms a lens-shaped body above the underlying salt water. (*b*) If excessive pumping occurs, a cone of depression develops in the fresh groundwater, and a cone of ascension forms in the underlying salty groundwater that may result in saltwater contamination of the well. (*c*) Pumping water back into the groundwater system through recharge wells can help lower the interface between the fresh groundwater and the salty groundwater and reduce saltwater incursion.

The San Joaquin Valley of California is a major agricultural region that relies largely on groundwater for irrigation. Between 1925 and 1975, groundwater withdrawals in parts of the valley caused subsidence of almost 9 m (◆ Fig. 13.20). Other examples of subsidence in the United States include New Orleans, Louisiana,

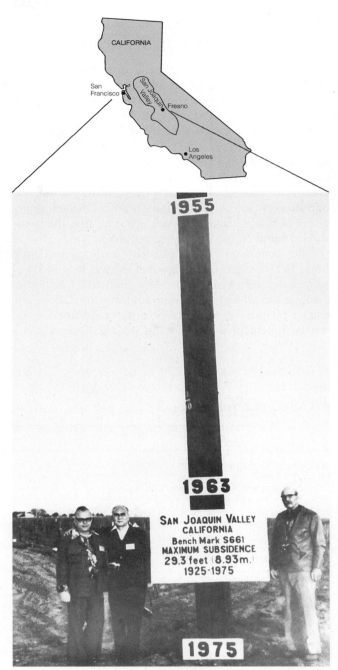

◆ FIGURE 13.20 The dates on this power pole dramatically illustrate the amount of subsidence the San Joaquin Valley has undergone since 1925. Due to withdrawal of groundwater for agricultural needs and the ensuing compaction of sediment, the ground subsided almost 9 m between 1925 and 1975.

and Houston, Texas, both of which have subsided more than 2 m, and Las Vegas, Nevada, which has subsided 8.5 m (● Table 13.2).

Elsewhere in the world, the tilt of the Leaning Tower of Pisa is partly due to groundwater withdrawal. The tower started tilting soon after construction began in 1173 because of differential compaction of the foundation. During the 1960s, the city of Pisa withdrew ever-larger amounts of groundwater, causing the ground to subside further; as a result, the tilt of the tower increased until it was considered in danger of falling over. However, strict control of groundwater withdrawal and stabilization of the foundation have reduced the amount of tilting to about 1 mm per year, ensuring that the tower should stand for several more centuries.

The extraction of oil can also cause subsidence. Long Beach, California, has subsided 9 m as a result of 34 years of oil production. More than $100 million of damage was done to the pumping, transportation, and harbor facilities in this area because of subsidence and encroachment of the sea (◆ Fig. 13.21). Once second-ary recovery wells began pumping water back into the oil reservoir and stabilizing it, subsidence virtually stopped.

Groundwater Contamination

A major problem facing our society is the safe disposal of the numerous pollutant by-products of an industrialized economy. We are becoming increasingly aware that our streams, lakes, and oceans are not unlimited reservoirs for waste, and that we must find new safe ways to dispose of pollutants.

The most common sources of contamination are sewage, landfills, toxic waste disposal sites (see Perspective 13.2), and agriculture. Once pollutants get into the groundwater system, they will spread wherever groundwater travels, which can make containment of the contamination difficult. Furthermore, because groundwater moves very slowly, it takes a very long time to cleanse a groundwater reservoir once it has become contaminated.

● **TABLE 13.2** Subsidence of Cities and Regions Due Primarily to Groundwater Removal

Location	Maximum Subsidence (m)	Area Affected (km²)
Mexico City, Mexico	8.0	25
Long Beach and Los Angeles, California	9.0	50
Taipei Basin, Taiwan	1.0	100
Shanghai, China	2.6	121
Venice, Italy	0.2	150
New Orleans, Louisiana	2.0	175
London, England	0.3	295
Las Vegas, Nevada	8.5	500
Santa Clara Valley, California	4.0	600
Bangkok, Thailand	1.0	800
Osaka and Tokyo, Japan	4.0	3,000
San Joaquin Valley, California	9.0	9,000
Houston, Texas	2.7	12,100

SOURCE: Data from R. Dolan and H.G. Goodell, "Sinking Cities," *American Scientist* 74 (1986): 38–47; and J. Whittow, *Disasters: The Anatomy of Environmental Hazards* (Athens, Ga.: University of Georgia Press, 1979).

TOTAL SUBSIDENCE
1928 TO 1968

◆ FIGURE 13.21 The withdrawal of petroleum from the oil field in Long Beach, California, resulted in up to 9 m of ground subsidence because of sediment compaction. It was not until secondary recovery wells began pumping water back into the reservoir to replace the petroleum that ground subsidence essentially ceased. (2 to 29 feet = 0.6 to 8.8 meters)

In many areas, septic tanks are the most common way of disposing of sewage. A septic tank slowly releases sewage into the ground where it is decomposed by oxidation and microorganisms and filtered by the sediment as it percolates through the zone of aeration. In most situations, by the time the water from the sewage reaches the zone of saturation, it has been cleansed of any impurities and is safe to use (◆ Fig. 13.22a, page 348). If, however, the water table is very close to the surface or if the rocks are very permeable, water entering the zone of saturation may still be contaminated and unfit to use.

Landfills are also potential sources of groundwater contamination (◆ Fig. 13.22b). Not only does liquid waste seep into the ground, but rainwater also carries dissolved chemicals and other pollutants downward into the groundwater reservoir. Unless the landfill is carefully designed and lined below by an impermeable layer such as clay, many toxic and cancer-causing compounds will find their way into the groundwater system. For example, paints, solvents, cleansers, pesticides, and battery acid are just a few of the toxic household items that end up in landfills and can pollute the groundwater supply.

Toxic waste sites in which dangerous chemicals are either buried or pumped underground are an increasing source of groundwater contamination. The United States alone must dispose of several thousand metric tons of hazardous chemical waste per year. Unfortu-

nately, much of this waste has been, and still is being, improperly dumped and is contaminating the surface water, soil, and groundwater.

Examples of indiscriminate dumping of dangerous and toxic chemicals can be found in every state. Perhaps the most famous is the Love Canal, near Niagara Falls, New York. During the 1940s, the Hooker Chemical Company dumped approximately 19,000 tons of chemical waste into the Love Canal. In 1953 it covered one of the dump sites with dirt and sold it for one dollar to the Niagara Falls Board of Education, which built an elementary school and playground on the site. Heavy rains and snow during the winter of 1976-1977 raised the water table and turned the area into a muddy swamp in the spring of 1977. Mixed with the mud were thousands of different toxic, noxious chemicals that formed puddles in the playground, oozed into people's basements, and covered gardens and lawns. Trees, lawns, and gardens began to die, and many of the residents of the area suffered from serious illnesses. The cost of cleaning up the Love Canal site and relocating its residents will eventually exceed $100 million, and the site and neighborhood are now vacant.

▼ HOT SPRINGS AND GEYSERS

The subsurface rocks in regions of recent volcanic activity usually stay hot for thousands of years.

RADIOACTIVE WASTE DISPOSAL

One of the problems of the nuclear age is finding safe storage sites for the radioactive waste from nuclear power plants, the manufacture of nuclear weapons, and the radioactive by-products of nuclear medicine. Radioactive waste can be grouped into two categories: low-level and high-level waste. Low-level wastes are low enough in radioactivity that, when properly handled, they do not pose a significant environmental threat. Most low-level wastes can be safely buried in controlled dump sites where the geology and groundwater system are well known and careful monitoring is provided.

High-level radioactive waste, such as the spent uranium fuel assemblies used in nuclear reactors and the material used in nuclear weapons, is extremely dangerous because of high amounts of radioactivity; it therefore presents a major environmental problem. Currently, more than 15,000 metric tons of spent uranium fuel are awaiting disposal, and the Department of Energy (DOE) estimates that by the year 2000 the nation will have produced almost 50,000 metric tons of highly radioactive waste that must be disposed of safely.

Near the end of 1987, Congress authorized the DOE to study the feasibility of using Yucca Mountain in southern Nevada as the nation's first high-level radioactive waste dump (◆Fig. 1). Such a facility must be able to isolate high-level waste from the environment for at least 10,000 years, which is the minimum time such waste will remain dangerous. The Yucca Mountain site will have a capacity of 70,000 metric tons of waste and will not be completely filled until around the year 2030, at which time its entrance shafts will be sealed and backfilled.

The canisters holding the waste are designed to remain leakproof for at least 300 years, so there is some possibility that leakage could occur over the next 10,000 years. The DOE believes, however, that the geology of the area will prevent radioactive isotopes from entering the groundwater system. Under an Environmental Protection Agency (EPA) regulation, a radioactive dump site must be located so that the travel time for groundwater from the site to the outside environment is at least 1,000 years.

The radioactive waste at the Yucca Mountain repository will be buried in a volcanic tuff at a depth of about 300 m. The water table in the area will be an additional 200 to 420 m below the dump site. Thus, the canisters will be stored in the zone of aeration, which was one of the reasons Yucca Mountain was selected. Only about 15 cm of rain fall in this area per year, and only a small amount of this percolates into the ground. Most of the water that does seep into the ground evaporates before it migrates very far. Thus, the rock at the depth the canisters are buried will be very dry, helping prolong the lives of the canisters.

Geologists believe that the radioactive waste at Yucca Mountain is most likely to contaminate the environment if it is in liquid form; if liquid, it could seep into the zone of saturation and enter the groundwater supply. But because of the low moisture in the zone of aeration, there is little water to carry the waste downward, and it will take well over 1,000 years to reach the zone of saturation. In fact, the DOE estimates that the waste will take longer than 10,000 years to move from the repository to the water table.

Groundwater percolating through these rocks is heated and, if returned to the surface, forms *hot springs* or *geysers*. Yellowstone National Park in the United States, Rotorua, New Zealand, and Iceland are all famous for their hot springs and geysers. They are all sites of recent volcanism, and consequently their subsurface rocks and groundwater are very hot.

A **hot spring** (also called a thermal spring or warm spring) is a spring in which the water temperature is warmer than the temperature of the human body (37°C) (◆ Fig. 13.23). Some hot springs, however, are much hotter, with temperatures ranging up to the boiling point in many instances. Of the approximately 1,100 known hot springs in the United States, more

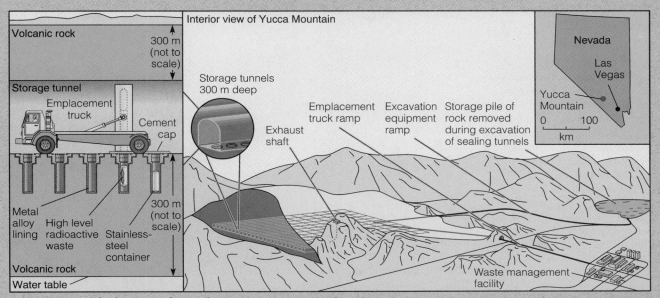

◆ FIGURE 1 The location of Nevada's Yucca Mountain and a schematic diagram of the proposed high-level radioactive waste dump.

One of the concerns of some geologists is that the climate will change during the next 10,000 years. If the region should become more humid, more water will percolate through the zone of aeration. This will increase the corrosion rate of the canisters and could cause the water table to rise, thereby decreasing the travel time between the repository and the zone of saturation. This area of the country was much more humid between 2 million and 10,000 years ago (see Chapter 14).

While it appears that Yucca Mountain meets all of the requirements for a safe high-level radioactive waste dump, the site is still controversial, and further studies must be conducted to ensure that the groundwater supply in this area is not rendered unusable by nuclear waste.

than 1,000 are in the Far West, while the rest are in the Black Hills of South Dakota, the Ouachita region of Arkansas, Georgia, and the Appalachian region.

Hot springs are also common in other parts of the world. One of the most famous is at Bath, England, where shortly after the Roman conquest of Britain in A.D. 43, numerous bathhouses and a temple were built around the hot springs (◆ Fig. 13.24).

The heat for most hot springs comes from magma or cooling igneous rocks. The geologically recent igneous activity in the western United States accounts for the large number of hot springs in that region. The water in some hot springs, however, is circulated deep into the Earth, where it is warmed by the normal increase in temperature, the geothermal gradient. For example, the spring water of Warm Springs, Georgia, is heated in

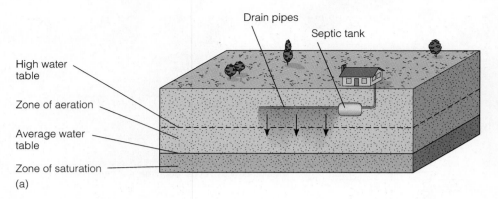

Drain pipes

Septic tank

High water
table

Zone of aeration

Average water
table

Zone of saturation

(a)

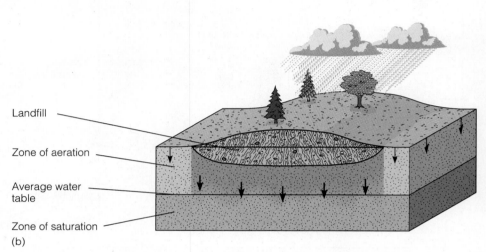

Landfill

Zone of aeration

Average water
table

Zone of saturation

(b)

◆ FIGURE 13.22 (a) A septic system slowly releases sewage into the zone of aeration. Oxidation, bacterial degradation, and filtering by the sediments usually remove all of the natural impurities before they reach the water table. If, however, the rocks are very permeable or the water table is too close to the septic system, contamination of the groundwater can result. (b) Unless there is an impermeable barrier between a landfill and the water table, pollutants can be carried into the zone of saturation and contaminate the groundwater supply.

◆ FIGURE 13.23 Hot springs are springs with a water temperature greater than 37°C. This hot spring is in West Thumb Geyser Basin, Yellowstone National Park, Wyoming.

this manner. This hot spring was a health and bathing resort long before the Civil War; later with the establishment of the Georgia Warm Springs Foundation, it was used to help treat polio victims.

Geysers are hot springs that intermittently eject hot water and steam with tremendous force. The word comes from the Icelandic *geysir* which means to gush or rush forth. One of the most famous geysers in the world is Old Faithful in Yellowstone National Park in Wyoming (◆ Fig. 13.25). With a thunderous roar, it erupts a column of hot water and steam every 30 to 90 minutes. Other well known geyser areas are found in Iceland and New Zealand.

Geysers are the surface expression of an extensive underground system of interconnected fractures within hot igneous rocks (◆ Fig. 13.26). Groundwater percolating down into the network of fractures is heated as it comes into contact with the hot rocks. Because the water near the bottom of the fracture system is under

◆ FIGURE 13.24 One of the many bathhouses in Bath, England, that were built around hot springs shortly after the Roman conquest in A.D. 43.

greater pressure than that near the top, it must be heated to a higher temperature before it will boil. Thus, when the deeper water is heated to very near the boiling point, a slight rise in temperature or a drop in pressure, such as from escaping gas, will cause it to instantly change to steam. The expanding steam quickly pushes the water above it out of the ground and into the air, thereby producing a geyser eruption. After the eruption, relatively cool groundwater starts to seep back into the fracture system where it is heated to near its boiling temperature and the eruption cycle begins again. Such a process explains how geysers can erupt with some regularity.

◆ FIGURE 13.25 Old Faithful Geyser in Yellowstone National Park, Wyoming, is one of the world's most famous geysers, erupting approximately every 30 to 90 minutes.

Hot spring and geyser water typically contains large quantities of dissolved minerals because most minerals dissolve more rapidly in warm water than in cold water. Due to this high mineral content, the waters of many hot springs are believed by some to have medicinal properties. Numerous spas and bathhouses have been built throughout the world at hot springs to take advantage of these supposed healing properties.

When the highly mineralized water of hot springs or geysers cools at the surface, some of the material in solution is precipitated, forming various types of deposits. The amount and type of precipitated mineral depend on the solubility and composition of the material through which the groundwater flows. If the groundwater contains dissolved calcium carbonate ($CaCO_3$), then *travertine* or *calcareous tufa* (both of which are varieties of limestone) are precipitated. Spectacular examples of hot spring travertine deposits are found at Mammoth Hot Springs in Yellowstone National Park (◆ Fig. 13.27) and at Pamukhale in Turkey. Groundwater containing dissolved silica will, upon reaching the surface, precipitate a soft, white, hydrated mineral called *siliceous sinter* or *geyserite*, which can accumulate around a geyser's opening.

Geothermal Energy

Energy that is harnessed from steam and hot water trapped within the Earth's crust is **geothermal energy**. It is a desirable and relatively nonpolluting alternate

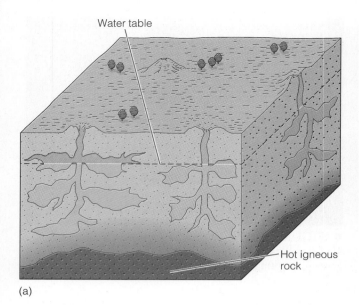

(a)

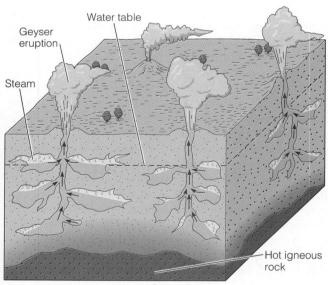

(b)

◆ FIGURE 13.27 Minerva Terrace at Mammoth Hot Springs in Yellowstone National Park, Wyoming, formed when calcium carbonate-rich hot spring water cooled, precipitating travertine deposits.

◆ FIGURE 13.26 The formation of a geyser. (*a*) Groundwater percolates downward into a network of interconnected openings and is heated by the hot igneous rocks. The water near the bottom of the fracture system is under greater pressure than that near the top and consequently must be heated to a higher temperature before it will boil. (*b*) Any rise in temperature of the water above its boiling point or a drop in pressure will cause the water to change to steam, which quickly pushes the water above it upward and out of the ground, producing a geyser eruption.

form of energy. Approximately 1 to 2% of the world's current energy needs could be met by geothermal energy. In those areas where it is plentiful, however, geothermal energy can supply most, if not all, of the energy needs, sometimes at a fraction of the cost of other types of energy. Some of the countries currently using geothermal energy in one form or another include Iceland, the United States, Mexico, Italy, New Zealand, Japan, the Philippines, and Indonesia.

Geothermal energy has been successfully used in Iceland since 1928. In Reykjavik, Iceland's capital, steam and hot water from wells drilled in geothermal areas are pumped into buildings for heating and hot water. Fruits and vegetables are grown year-round in hot houses heated from geothermal wells. Direct heating in this manner is significantly cheaper than fuel oil or electrical heating and much cleaner.

The city of Rotorua in New Zealand is world famous for its volcanoes, hot springs, geysers, and geothermal fields. Since the first well was sunk in the 1930s, more than 800 wells have been drilled to tap the hot water and steam below. Geothermal energy in Rotorua is used in a variety of ways, including home, commercial, and greenhouse heating.

In the United States, the first commercial geothermal electrical generating plant was built in 1960 at The Geysers, about 120 km north of San Francisco, California (◆ Fig. 13.28). Here, wells were drilled into the numerous near-vertical fractures underlying the region.

◆ FIGURE 13.28 The Geysers, Sonoma County, California. Plumes of steam can be seen rising from several steam-generating plants.

As pressure on the rising groundwater decreases, the water changes to steam that is piped directly to electricity-generating turbines. The present electrical generating capacity at The Geysers is about 2,000 megawatts, which is enough to supply about two-thirds of the electrical needs of the San Francisco Bay area.

As oil reserves decline, geothermal energy is becoming an attractive alternative, particularly in parts of the western United States, such as the Salton Sea area of southern California, where geothermal exploration and development have begun. While geothermally generated electricity is a generally clean source of power, it can also be expensive because most geothermal waters are acidic and very corrosive. Consequently, the turbines must either be built of expensive corrosion-resistant alloy metals or frequently replaced. Furthermore, geothermal power is not inexhaustible. The steam and hot water removed for geothermal power cannot be easily replaced, and eventually pressure in the wells drops to the point at which the geothermal field must be abandoned.

Chapter Summary

1. The water stored in the pore spaces of subsurface rocks and unconsolidated material is groundwater.

2. Groundwater is part of the hydrologic cycle and represents approximately 22% of the world's supply of fresh water.

3. Porosity is the percentage of a rock, sediment, or soil consisting of pore space. Permeability is the ability of a rock, sediment, or soil to transmit fluids. A material that transmits groundwater is an aquifer and one that prevents the movement of groundwater is an aquiclude.

4. The water table is the surface that separates the zone of aeration (in which pore spaces are filled with both air and water) from the zone of saturation (in which all pore spaces are filled with water).

5. Groundwater moves slowly through the pore spaces of rocks, sediment, or soil (zone of aeration) and moves through the zone of saturation to outlets such as streams, lakes, and swamps.

6. A spring occurs wherever the water table intersects the Earth's surface. Some springs are the result of a perched water table, that is, a localized aquiclude within an aquifer and above the regional water table.

7. Water wells are made by digging or drilling into the zone of saturation. When water is pumped out of a well, a cone of depression forms. If water is pumped out faster than it can be recharged, the cone of depression deepens and enlarges and may locally drop to the base of the well, resulting in a dry well.

8. Artesian systems are those in which confined groundwater builds up high hydrostatic pressure. Three conditions must generally be met before an artesian system can form: the aquifer must be confined above and below by aquicludes; the aquifer is usually tilted and exposed at the Earth's surface so it can be recharged; and precipitation must be sufficient to keep the aquifer filled.

9. Karst topography results from groundwater weathering and erosion, and is characterized by sinkholes, solution valleys, and disappearing streams.

10. Caves form when groundwater in the zone of saturation weathers and erodes soluble rock such as limestone. Cave deposits, called dripstone, result from the precipitation of calcite.

11. Modifications of the groundwater system can cause serious problems. Excessive withdrawal of groundwater can result in dry wells, loss of hydrostatic pressure, saltwater encroachment, and ground subsidence.

12. Groundwater contamination is becoming a serious problem and can result from sewage, landfills, and toxic waste.

13. Hot springs and geysers may occur where groundwater is heated by hot subsurface volcanic rocks. Geysers are hot springs that intermittently eject hot water and steam.

14. Geothermal energy comes from the steam and hot water trapped within the Earth's crust. It is a relatively nonpolluting form of energy that is used as a source of heat and to generate electricity.

Important Terms

aquiclude
aquifer
artesian system
capillary fringe
cave
column
cone of depression
dripstone
geothermal energy

geyser
groundwater
hot spring
karst topography
perched water table
permeability
porosity
recharge
saltwater incursion

sinkhole
spring
stalactite
stalagmite
water table
water well
zone of aeration
zone of saturation

Review Questions

1. What is the correct order, from highest to lowest, of groundwater usage in the United States?
   a. ___ agricultural, industrial, domestic; b. ___ industrial, domestic, agricultural; c. ___ domestic, agricultural, industrial; d. ___ agricultural, domestic, industrial; e. ___ industrial, agricultural, domestic.

2. What percentage of the world's supply of fresh water is represented by groundwater?
   a. ___ 5; b. ___ 18; c. ___ 22; d. ___ 43; e. ___ 50.

3. The capacity of a material to transmit fluids is:
   a. ___ porosity; b. ___ permeability; c. ___ solubility; d. ___ aeration quotient; e. ___ saturation.

4. The water table is a surface separating the:
   a. ___ zone of porosity from the underlying zone of permeability; b. ___ capillary fringe from the underlying zone of aeration; c. ___ capillary fringe from the underlying zone of saturation; d. ___ zone of aeration from the underlying zone of saturation; e. ___ zone of saturation from the underlying zone of aeration.

5. Groundwater:
   a. ___ moves slowly through the pore spaces of Earth materials; b. ___ moves fastest through the central area of a material's pore space; c. ___ can move upward against the force of gravity; d. ___ moves from areas of high pressure toward areas of low pressure; e. ___ all of these.

6. A perched water table:
   a. ___ occurs wherever there is a localized aquiclude within an aquifer; b. ___ is frequently the site of springs; c. ___ lacks a zone of aeration; d. ___ answers (a) and (b); e. ___ answers (b) and (c).

7. An artesian system is one in which:
   a. ___ water is confined; b. ___ water can rise above the level of the aquifer when a well is drilled; c. ___ there are no aquicludes; d. ___ answers (a) and (c); e. ___ answers (a) and (b).

8. Which of the following is not an example of groundwater erosion?
   a. ___ karst topography; b. ___ stalactites; c. ___ sinkholes; d. ___ caves; e. ___ caverns.

9. What percentage of the water used in the United States is provided by groundwater?
   a. ___ 50; b. ___ 40; c. ___ 30; d. ___ 20; e. ___ 10.

10. Rapid withdrawal of groundwater can result in:
    a. ___ a cone of depression; b. ___ ground subsidence; c. ___ saltwater incursion; d. ___ loss of hydrostatic pressure; e. ___ all of these.

11. The water in hot springs and geysers:
    a. ___ is believed to have curative properties; b. ___ is noncorrosive; c. ___ contains large quantities of dissolved minerals; d. ___ answers (a) and (b); e. ___ answers (a) and (c).

12. Which of the following is not a cave deposit?
    a. ___ stalagmite; b. ___ room; c. ___ dripstone; d. ___ stalactite; e. ___ none of these.

13. Discuss the role of groundwater in the hydrologic cycle.

14. How can a rock be porous and yet not be permeable?

15. What types of materials make good aquifers and aquicludes?

16. Why is the water table a subdued replica of the surface topography? What causes the water table level to fluctuate?

17. Why does groundwater move so much slower than surface water?
18. Where are springs likely to occur?
19. How does a perched water table differ from a regional water table?
20. What is a cone of depression and why is it so important?
21. Why are some artesian wells free-flowing while others must be pumped?
22. How does groundwater weather and erode?
23. List the surface features of karst topography and explain how they form.

24. How do caves and their various features form?
25. Discuss the various effects that excessive groundwater removal may have on a region. Give some examples.
26. Discuss the various ways that a groundwater system may become contaminated.
27. What is the difference between a thermal spring and a geyser?
28. In what ways has geothermal energy been used?

Additional Readings

Dolan, R., and H. G. Goodell. 1986. Sinking cities. *American Scientist* 74, no. 1:38–47.

Fetter, C. W. 1988. *Applied hydrogeology*. 2d ed. Columbus, Ohio: Merrill.

Fincher, J. 1990. Dreams of riches led Floyd Collins to a nightmarish end. *Smithsonian* 21, no. 2:137–49.

Freeze, R. A., and J. A. Cherry. 1979. *Groundwater*. Englewood Cliffs, N.J.: Prentice-Hall.

Jennings, J. N. 1983. Karst landforms. *American Scientist* 71, no. 6:578–86.

———. 1985. *Karst geomorphology*. 2d ed. Oxford, England: Basil Blackwell.

Monastersky, R. 1988. The 10,000-year test. *Science News* 133:139–41.

Price, M. 1985. *Introducing groundwater*. London: Allen & Unwin.

Rinehart, J. S. 1980. *Geysers and geothermal energy*. New York: Springer-Verlag.

Sloan, B., ed. 1977. *Caverns, caves, and caving*. New Brunswick, N.J.: Rutgers Univ. Press.

CHAPTER
14

Chapter Outline

Climbers ascending Ingraham Glacier on Mount Rainier, Washington.

Glaciers and Glaciation

Prologue

Following the Great Ice Age, which ended about 10,000 years ago, a general warming trend occurred that was periodically interrupted by short relatively cool periods. One such cool period, from about A.D. 1500 to the mid- to late-1800s, was characterized by the expansion of small glaciers in mountain valleys and the persistence of sea ice at high latitudes for longer periods than had occurred previously. This interval of nearly four centuries is known as the *Little Ice Age.*

The climatic changes leading to the Little Ice Age actually began by about A.D. 1300. During the preceding centuries, Europe had experienced rather mild temperatures, and the North Atlantic Ocean was warmer and more storm-free than it is at the present. During this time, the Vikings discovered and settled Iceland, and by A.D. 1200, about 80,000 people resided there. They also discovered Greenland and North America and established two colonies on the former and one on the latter. As the climate deteriorated, however, the North Atlantic became stormier, and sea ice occurred further south and persisted longer each year. As a consequence of poor sea conditions and political problems in Norway, all shipping across the North Atlantic ceased, and the colonies in Greenland and North America eventually disappeared.

During the Little Ice Age, many of the small glaciers in Europe and Iceland expanded and moved far down their valleys, reaching their greatest historic extent by the early 1800s. A small ice cap formed in Iceland where none had existed previously, and glaciers in Alaska and the mountains of the western United States and Canada also expanded to their greatest limits during historic time. Although glaciers caused some problems in Europe where they advanced across roadways and pastures, destroying some villages in Scandinavia and threatening villages elsewhere, their overall impact on humans was minimal. Far more important from the human perspective was that during much of the Little Ice Age the summers in northern latitudes were cooler and wetter.

Although worldwide temperatures were a little lower during this time, the change in summer conditions rather than cold winters or glaciers caused most of the problems. Particularly hard hit were Iceland and the Scandinavian countries, but at times much of northern Europe was affected (◆ Fig. 14.1). Growing seasons were shorter during many years, resulting in food shortages and a number of famines. Iceland's population declined from its high of 80,000 in 1200 to about 40,000 by 1700. Between 1610 and 1870, sea ice was observed near Iceland for as much as three months a year, and each time the sea ice persisted for long periods, poor growing seasons and food shortages followed.

Exactly when the Little Ice Age ended is debatable. Some authorities put the end at 1880, whereas others think it ended as early as 1850. In any case, during the late 1800s, the sea ice was retreating northward, glaciers were retreating back up their valleys, and summer weather became more stable.

▼ INTRODUCTION

Most people have some idea of what a glacier is, but many confuse glaciers with other masses of snow and

(a)

(b)

◆ FIGURE 14.1 (*a*) During the Little Ice Age, many of the glaciers in Europe, such as this one in Switzerland, extended much farther down their valleys than they do at present. *The Unterer Grindelwald* painted in 1826 by Samuel Birmann (1793–1847). (*b*) This mid-1600s painting by Jan-Abrahamsz Beerstraten titled *The Village of Nieukoop in Winter* shows the canals of Holland frozen. These canals rarely freeze today.

At the present time, glaciers cover nearly 15 million km², or about one-tenth of the Earth's land surface (● Table 14.1). Numerous glaciers exist in the mountains of the western United States, especially Alaska, western Canada, the Andes in South America, the Alps of Europe, the Himalayas of Asia, and other high mountains. Even Mount Kilimanjaro in Africa, although near the equator, is high enough to support glaciers. In fact, Australia is the only continent lacking glaciers. By far the largest existing glaciers are in Greenland and Antarctica; both areas are almost completely covered by glacial ice (Table 14.1).

Glaciers are particularly effective agents of erosion, transport, and deposition. They deeply scour the land over which they move, producing a number of easily recognized erosional landforms. Eventually, they deposit their sediment load just as do other agents of erosion and transport. Although numerous examples of landscapes that originated from recent glaciation can be found, most glacial landscapes developed during the Pleistocene Epoch, or what is commonly called the Ice Age (1.6 million to 10,000 years ago), a time during which glaciers covered much more extensive areas than they do now, particularly on the Northern Hemisphere continents (see Chapter 24).

● **TABLE 14.1** Present-Day Ice-Covered Areas

Antarctica	12,653,000km²
Greenland	1,802,600
Northeast Canada	153,200
Central Asian ranges	124,500
Spitsbergen group	58,000
Other Arctic islands	54,000
Alaska	51,500
South American ranges	25,000
West Canadian ranges	24,900
Iceland	11,800
Scandinavia	5,000
Alps	3,600
Caucasus	2,000
New Zealand	1,000
USA (other than Alaska)	650
Others	about 800
	14,971,550

Total volume of present ice: 28 to 35 million km³

SOURCE: C. Embleton and C. A. King, *Glacial Geomorphology* (New York: Halsted Press, 1975).

ice. A **glacier** is a mass of ice composed of compacted and recrystallized snow that flows under its own weight on land. Accordingly, sea ice as in, for example, the north polar region is not glacial ice, nor are drifting icebergs glaciers even though they may have derived from glaciers that flowed into the sea. Snow fields in high mountains may persist in protected areas for years, but these are not glaciers either because they are not actively moving.

GLACIERS AND THE HYDROLOGIC CYCLE

About 2% of the world's water is contained in glaciers and thus has been temporarily removed from the hydrologic cycle. However, many glaciers at high latitudes, as in Alaska, flow directly into the sea where they melt, or where icebergs break off by a process called *calving* and drift out to sea where they eventually melt. At lower latitudes where they can exist only at high elevations, glaciers flow to lower elevations where they melt and the water returns to the oceans by surface runoff. In areas of low precipitation, as in parts of the western United States, glaciers are important fresh water reservoirs that release water to streams during the dry season.

In addition to melting, glaciers also lose water by sublimation, a process in which ice changes directly to water vapor without an intermediate liquid phase. Water vapor thus derived rises into the atmosphere where it may condense and fall once again as snow or rain. In any case, it too is eventually returned to the oceans.

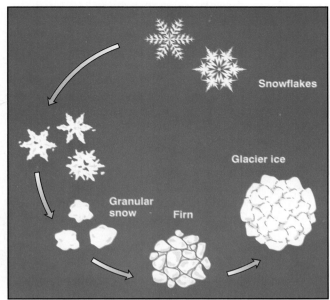

◆ FIGURE 14.2 The conversion of freshly fallen snow to firn and glacial ice.

THE ORIGIN OF GLACIAL ICE

Ice is a mineral in every sense of the word; it has a crystalline structure and possesses characteristic physical and chemical properties. Accordingly, geologists consider glacial ice to be rock, although it is a type of rock that is easily deformed. It forms in a fairly straightforward manner (◆ Fig. 14.2). When an area receives more winter snow than can melt during the spring and summer seasons, a net accumulation occurs. Freshly fallen snow consists of about 80% air and 20% solids, but it compacts as it accumulates, partly thaws, and refreezes; in the process, the original snow layer is converted to a granular type of ice called **firn.** Deeply buried firn is further compacted and is finally converted to **glacial ice,** consisting of about 90% solids (Fig. 14.2). When accumulated snow and ice reach a critical thickness of about 40 m, the pressure on the ice at depth is sufficient to cause deformation and flow, even though it remains solid. Once the critical thickness is reached and flow begins, the moving mass of ice becomes a *glacier.*

Plastic flow, which causes permanent deformation, occurs in response to pressure and is the primary way that glaciers move. They may also move by **basal slip,** which occurs when a glacier slides over the underlying

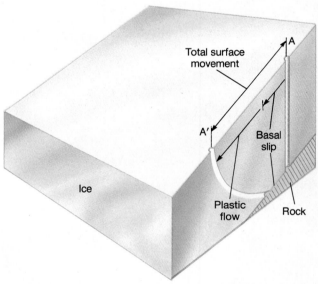

◆ FIGURE 14.3 Movement of a glacier by a combination of plastic flow and basal slip. If a glacier is solidly frozen to the underlying surface, it moves only by plastic flow.

surface (◆ Fig. 14.3). Basal slip is facilitated by the presence of meltwater that reduces frictional resistance between the underlying surface and the glacier.

▼ TYPES OF GLACIERS

Geologists generally recognize two basic types of glaciers: *valley* and *continental*. A **valley glacier,** as its name implies, is confined to a mountain valley or perhaps to an interconnected system of mountain valleys (◆ Fig. 14.4). Large valley glaciers commonly have several smaller tributary glaciers, much as large streams have tributaries. Valley glaciers flow from higher to lower elevations and are invariably small in comparison to continental glaciers.

Continental glaciers, also called *ice sheets,* cover vast areas (at least 50,000 km²) and are unconfined by topography (◆ Fig. 14.5). In contrast to valley glaciers, which flow downhill within the confines of a valley,

◆ FIGURE 14.4 A large valley glacier in Alaska. Notice the tributaries to the large glacier.

continental glaciers flow outward in all directions from a central area of accumulation. Currently, only two continental glaciers exist, one in Greenland and the other in Antarctica. Both are more than 3,000 m thick in their central areas, become thinner toward their margins, and cover all but the highest mountains (◆ Fig. 14.6). During the Pleistocene Epoch, such glaciers covered large parts of the Northern Hemisphere continents. Many of the erosional and depositional landforms in much of Canada and the northern tier of the United States formed as a consequence of Pleistocene glaciation.

Although valley and continental glaciers are easily differentiated by their size and location, an intermediate variety called an *ice cap* is also recognized. Ice caps are similar to, but smaller than, continental glaciers and cover less than 50,000 km². Some ice caps form when valley glaciers grow and overtop the divides and passes between adjacent valleys and coalesce to form a continuous ice cap. They also form on fairly flat terrain including some of the islands of the Canadian Arctic and Iceland.

▼ THE GLACIAL BUDGET

Just as a savings account grows and shrinks as funds are deposited and withdrawn, glaciers expand and contract in response to accumulation and wastage. Their behavior can be described in terms of a **glacial budget,** which is essentially a balance sheet of accumulation and wastage. The upper part of a valley glacier is a **zone of**

◆ FIGURE 14.5 The Antarctic ice sheet, one of two continental glaciers existing at present.

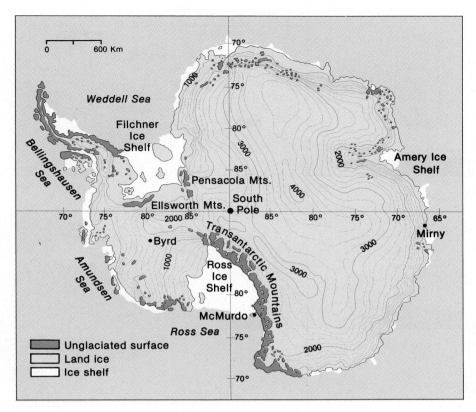

◆ FIGURE 14.6 Antarctica is almost completely covered by an ice sheet averaging about 2,160 m thick and reaching a maximum thickness of about 4,000 m.

accumulation where additions exceed losses, and the glacier's surface is perennially covered by snow. In contrast, the lower part of the same glacier is a **zone of wastage,** where losses from melting, sublimation, and calving of icebergs exceed the rate of accumulation (◆ Fig. 14.7).

At the end of winter, a glacier's surface is usually completely covered with the accumulated seasonal snowfall. During spring and summer, however, the snow begins to melt, first at lower elevations and then progressively higher up the glacier. The elevation to which snow recedes during a wastage season is called the **firn limit** (Fig. 14.7). One can easily identify the zones of accumulation and wastage by noting the position of the firn limit.

Observations of a single glacier reveal that the position of the firn limit usually changes from year to year. If it does not change or shows only minor fluctuations, the glacier is said to have a balanced budget; that is, additions in the zone of accumulation are exactly balanced by losses in the zone of wastage, and the distal end or *terminus* of the glacier remains

stationary (Fig. 14.7a). When the firn limit moves down the glacier, the glacier has a positive budget; its additions exceed its losses, and its terminus advances (Fig. 14.7b). If the budget is negative, the glacier recedes— its terminus retreats up the glacial valley (Fig. 14.7c). But even though a glacier's terminus may be receding, the glacial ice continues to move toward the terminus by plastic flow and basal slip. If a negative budget persists long enough, however, a glacier recedes and thins to the point at which it no longer flows, thus becoming a *stagnant glacier.*

Although we used a valley glacier as our example, the same budget considerations control the flow of continental glaciers as well. For example, the entire Antarctic ice sheet is in the zone of accumulation, but it flows into the ocean where wastage occurs.

⩔ RATES OF GLACIAL MOVEMENT

In general, valley glaciers move more rapidly than continental glaciers, but the rates for both vary, ranging from centimeters to tens of meters per day. Valley

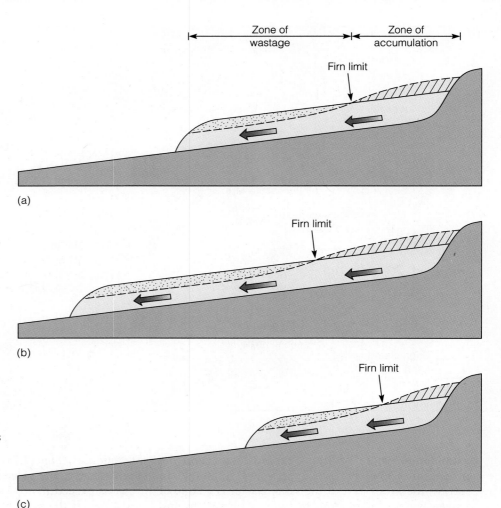

Zone of wastage | Zone of accumulation

Firn limit

(a)

Firn limit

(b)

Firn limit

(c)

◆ FIGURE 14.7 Response of a hypothetical glacier to changes in its budget. (*a*) If the losses in the zone of wastage, shown by stippling, equal additions in the zone of accumulation, shown by crosshatching, the terminus of the glacier remains stationary. (*b*) Gains exceed losses, and the glacier's terminus advances. (*c*) Losses exceed gains, and the glacier's terminus retreats, although the glacier continues to flow.

glaciers moving down steep slopes flow more rapidly than glaciers of comparable size on gentle slopes, assuming that all other variables are the same. The main glacier in a valley glacier system contains a greater volume of ice and thus has a greater discharge and flow velocity than its tributaries (Fig. 14.4). Temperature exerts a seasonal control on valley glaciers because although plastic flow remains rather constant year-round, basal slip is more important during warmer months when meltwater is more abundant.

Flow rates also vary within the ice itself. For example, flow velocity generally increases in the zone of accumulation until the firn limit is reached; from that point, the velocity becomes progressively slower toward the glacier's terminus. Valley glaciers are similar to streams, in that the valley walls and floor cause fric-

tional resistance to flow. Thus, the ice in contact with the walls and floor moves more slowly than the ice some distance away (◆ Fig. 14.8).

Notice in Figure 14.8 that the flow velocity increases upward until the top few tens of meters of ice are reached, but little or no additional increase occurs after that point. This upper ice constitutes the rigid part of the glacier that is moving as a consequence of basal slip and plastic flow below. The fact that this upper 40 m or so of ice behaves as a brittle solid is clearly demonstrated by large fractures called *crevasses* that develop when a valley glacier flows over a step in its valley floor where the slope increases or where it flows around a corner (◆ Fig. 14.9). In either case, the glacial ice is stretched (subjected to tension), and large crevasses develop, but they extend downward only to the zone of

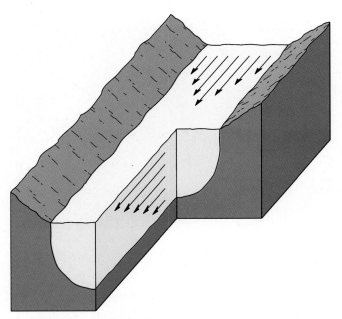

◆ FIGURE 14.8 Flow velocity in a valley glacier varies both horizontally and vertically. Velocity is greatest at the top-center of the glacier. Friction with the walls and floor of the glacial trough causes the flow to be slower adjacent to these boundaries. The length of the arrows in the figure is proportional to the velocity.

◆ FIGURE 14.9 Crevasses and an ice fall in a glacier in Alaska.

plastic flow. In some cases, a valley glacier descends over such a steep precipice that crevasses break up the ice into a jumble of blocks and spires, and an *ice fall* develops (Fig. 14.9).

The flow rates of valley glaciers are also complicated by *glacial surges,* which are bulges of ice that move through a glacier at a velocity several times faster than the normal flow. Although surges are best documented in valley glaciers, they occur in ice caps and continental glaciers as well. During a surge, a glacier's terminus may advance several kilometers during a year. The causes of surges are not fully understood, but some of them have occurred following a period of unusually heavy precipitation in the zone of accumulation.

One reason continental glaciers move comparatively slowly is that they exist at higher latitudes and are frozen to the underlying surface most of the time, which limits the amount of basal slip. Some basal slip does occur even beneath the Antarctic ice sheet, but most of its movement is by plastic flow. Nevertheless, some parts of continental glaciers manage to achieve extremely high flow rates. For example, near the margins of the Greenland ice sheet, the ice is forced between mountains in what are called *outlet glaciers.* In some of these outlets, flow velocities exceeding 100 m per day have been recorded.

▼ GLACIAL EROSION AND TRANSPORT

Glaciers are moving solids that can erode and transport huge quantities of materials, especially unconsolidated sediment and soil. In many areas of Canada and the northern United States, glaciers transported boulders, some of huge proportions, for long distances before depositing them. Such boulders are called **glacial erratics** (◆Fig. 14.10).

◆ FIGURE 14.10 A glacial erratic in Yellowstone National Park, Wyoming.

(a)

(b)

◆ FIGURE 14.11 (*a*) Glacial polish on quartzite near Marquette, Michigan. (*b*) Glacial striations in basalt at Devil's Postpile National Monument, California.

Important erosional processes associated with glaciers include bulldozing, plucking, and abrasion. Bulldozing, although not a formal geologic term, is fairly self-explanatory: a glacier simply shoves or pushes unconsolidated materials in its path. *Plucking,* also called *quarrying,* occurs when glacial ice freezes in the cracks and crevices of a bedrock projection and eventually pulls it loose.

Sediment-laden glacial ice can effectively erode by **abrasion.** For example, bedrock over which sediment-laden glacial ice has moved commonly develops a **glacial polish,** a smooth surface that glistens in reflected light (◆ Fig. 14.11a). Abrasion also yields **glacial striations,** consisting of rather straight scratches on rock surfaces (◆ Fig. 14.11b). Glacial striations are rarely more than a few millimeters deep, whereas **glacial grooves** are similar but much larger and deeper (◆ Fig. 14.12). Abrasion also thoroughly pulverizes rocks so that they yield an aggregate of clay- and silt-sized particles having the consistency of flour, hence the

◆ FIGURE 14.12 Glacial grooves on Kelly's Island in Lake Erie.

name *rock flour*. Rock flour is so common in streams discharging from glaciers that the water generally has a milky appearance.

Continental glaciers can derive sediment from mountains projecting through them, and windblown dust settles on their surfaces. Otherwise, most of their sediment is derived from the surface over which they move and is transported in the lower part of the ice sheet. In contrast, valley glaciers carry sediment in all parts of the ice, but it is concentrated at the base and along the margins (◆ Fig. 14.13). Some of the marginal sediment is derived by abrasion and plucking, but much of it is supplied by mass wasting processes. The sediments carried along the margins and center become *lateral* and *medial moraine* deposits, respectively, as discussed later in this chapter (Fig. 14.13).

Erosional Landforms of Valley Glaciers

Some of the world's most inspiring scenery is produced by valley glaciers. Many mountain ranges are scenic to begin with, but when modified by valley glaciers, they take on a unique aspect of jagged, angular peaks and ridges in the midst of broad valleys (◆ Fig. 14.14).

◆ FIGURE 14.13 Sediment is transported in all parts of a valley glacier. The sediment carried along the margins is lateral moraine; where two lateral moraines coalesce, they form a medial moraine.

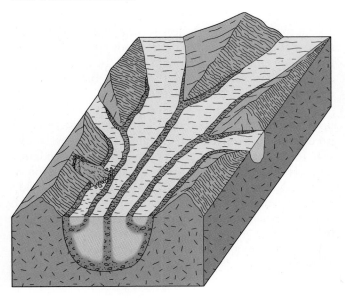

U-Shaped Glacial Troughs

A **U-shaped glacial trough** is one of the most distinctive features of valley glaciation (Fig 14.14c). Mountain valleys eroded by running water are typically V-shaped in cross section; that is, they have valley walls that descend steeply to a narrow valley bottom (Fig. 14.14a). In contrast, valleys scoured by glaciers are deepened, widened, and straightened such that they possess very steep or vertical walls, but have broad, rather flat valley floors; thus, they exhibit a U-shaped profile (◆ Fig. 14.15). Many glacial troughs contain triangular-shaped *truncated spurs,* which are cutoff or truncated ridges that extend into the preglacial valley (Fig. 14.14c).

During the Pleistocene, when glaciers were more extensive, sea level was about 130 m lower than at present, so glaciers flowing into the sea eroded their valleys to much greater depths than they do now. When the glaciers melted at the end of the Pleistocene, sea level rose, and the ocean filled the lower ends of the glacial troughs so that now they are long, steep-walled embayments called **fiords** (◆ Fig. 14.16).

Lower sea level during the Pleistocene was not responsible for the formation of all fiords. Unlike running water, glaciers can erode a considerable distance below sea level. In fact, a glacier 500 m thick can stay in contact with the sea floor and effectively erode it to a depth of about 450 m before the buoyant effects of water cause the glacial ice to float!

Hanging Valleys

Although waterfalls can form in several ways, some of the world's highest and most spectacular are found in recently glaciated areas. For example, Yosemite Falls in Yosemite National Park, California, plunges from a **hanging valley,** which is a tributary valley whose floor is at a higher level than that of the main valley (◆ Fig. 14.17). As Figure 14.14 shows, the large glacier in the main valley vigorously erodes, whereas the smaller glaciers in tributary valleys are less capable of large-scale erosion. When the glaciers disappear, the smaller tributary valleys remain as hanging valleys. Accordingly, streams flowing through hanging valleys plunge over vertical or very steep precipices.

Cirques, Arêtes, and Horns

Perhaps the most spectacular erosional landforms in areas of valley glaciation occur at the upper ends of glacial troughs and along the divides separating adjacent

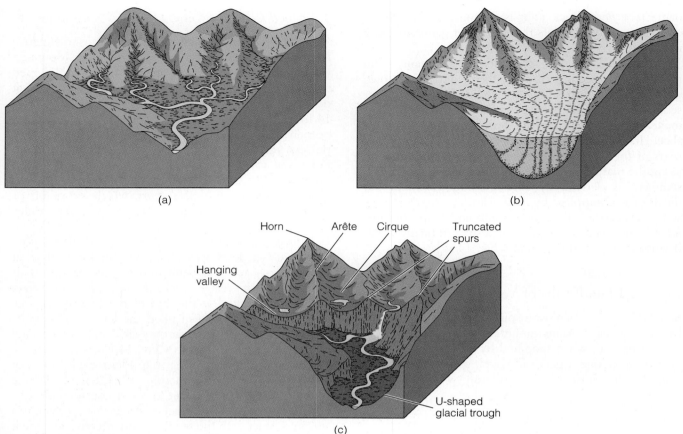

(a)

(b)

Horn   Arête   Cirque   Truncated spurs

Hanging valley

U-shaped glacial trough

(c)

◆ FIGURE 14.14 Erosional landforms produced by valley glaciers. (*a*) A mountain area before glaciation. (*b*) The same area during the maximum extent of the valley glaciers. (*c*) After glaciation.

◆ FIGURE 14.15 A U-shaped glacial trough in northwestern Montana.

glacial troughs. Valley glaciers form and move out from steep-walled, bowl-shaped depressions called **cirques** at the upper end of their troughs (Fig. 14.14c). Cirques are typically steep-walled on three sides, but one side is open and leads into the glacial trough.

Although the details of cirque origin are not fully understood, they apparently form by erosion of a preexisting depression on a mountain side. As snow and ice accumulate in the depression, frost wedging and plucking enlarge it until it takes on the typical cirque shape. Many cirques have a lip or threshold, indicating that the glacial ice not only moves outward but rotates as well, scouring out a depression rimmed by rock.

Cirques become wider and are cut deeper into mountain sides by headward erosion as a consequence

◆ FIGURE 14.16 Milford Sound, a fiord in New Zealand. (Photo courtesy of George and Linda Lohse.)

of abrasion, plucking, and several mass wasting processes. Thus, a combination of processes can erode a small mountain side depression into a large cirque; the largest one known is the Walcott Cirque in Antarctica, which is 16 km wide and 3 km deep.

The fact that cirques expand laterally and by headward erosion accounts for the origin of two other distinctive erosional features, *arêtes* and *horns*. **Arêtes**— narrow, serrated ridges—can form in two ways. In many cases, cirques form on opposite sides of a ridge, and headward erosion reduces the ridge until only a thin partition of rock remains (Fig. 14.14c). The same effect occurs when erosion in two parallel glacial troughs reduces the intervening ridge to a thin spine of rock (◆ Fig. 14.18).

The most majestic of all mountain peaks are **horns;** these steep-walled, pyramidal peaks are formed by headward erosion of cirques. In order for a horn to form, a mountain peak must have at least three cirques on its flanks, all of which erode headward (Fig. 14.14c). Excellent examples of horns include Mount Assiniboine in the Canadian Rockies, the Grand Teton in Wyoming (Fig. 10.1), and the most famous of all, the Matterhorn in Switzerland (◆ Fig. 14.19).

Erosional Landforms of Continental Glaciers

Areas eroded by continental glaciers tend to be smooth and rounded because such glaciers bevel and abrade high areas that projected into the ice. Rather than yielding the sharp, angular landforms typical of valley

◆ FIGURE 14.17 Yosemite Falls in Yosemite National Park, California plunges 435 m vertically, cascades down a steep slope for another 205 m, and then falls vertically 97 m, for a total descent of 737 m. (Photo courtesy of Sue Monroe.)

glaciation, they produce a landscape of rather flat, monotonous topography interrupted by rounded hills.

In a large part of Canada, particularly the vast Canadian Shield region, continental glaciation has stripped off the soil and unconsolidated surface sediment, revealing extensive exposures of striated and polished bedrock (◆ Fig. 14.20). Similar though smaller bedrock exposures are also widespread in the northern United States from Maine through Minnesota.

Another consequence of erosion in these areas is the complete disruption of drainage that has not yet become reestablished. Thus, much of the area is characterized by deranged drainage (see Fig. 12.22e), numerous lakes and swamps, low relief, extensive bedrock exposures, and little or no soil. Such areas are generally referred to as *ice-scoured plains* (Fig. 14.20).

◆ FIGURE 14.18 The knifelike ridges adjacent to these glaciers in the North Cascades of Washington are arêtes.

◆ FIGURE 14.19 The Matterhorn in Switzerland is a well-known horn.

◆ FIGURE 14.20 An ice-scoured plain in the Northwest Territories of Canada.

GLACIAL DEPOSITS

All sediment deposited as a consequence of glacial activity is **glacial drift.** Glacial deposits in several upper midwestern states are important sources of groundwater and rich soils, and in several states they are exploited for their sand and gravel.

Geologists generally recognize two distinct types of glacial drift, *till* and *stratified drift.* **Till** consists of sediment deposited directly by glacial ice. It is not sorted or stratified; that is, its particles are not separated by size or density, and it does not exhibit any layering. Till deposited by valley glaciers looks much like the till of continental glaciers except that the latter's deposits are much more extensive and have generally been transported much farther.

Stratified drift is sorted by size and density and, as its name implies, is layered. In fact, most of the sediments recognized as stratified drift are braided stream deposits; the streams in which they were deposited received their water and sediment load directly from melting glacial ice.

Landforms Composed of Till

Landforms composed of till include several types of *moraines* and elongated hills known as *drumlins.*

End Moraines

The terminus of either a valley or a continental glacier may become stabilized in one position for some period of time, perhaps a few years or even decades. Such stabilization of the ice front does not mean that the glacier has ceased flowing, only that it has a balanced budget. When an ice front is stationary, flow within the glacier continues, and the sediment transported within or upon the ice is dumped as a pile of rubble at the glacier's terminus. Such deposits are **end moraines,** which continue to grow as long as the ice front is stabilized (♦ Fig. 14.21). End moraines of valley glaciers are commonly crescent-shaped ridges of till spanning the valley occupied by the glacier. Those of continental glaciers similarly parallel the ice front, but are much more extensive.

♦ FIGURE 14.21 (*a*) The origin of an end moraine. (*b*) End moraines are described as terminal moraines or recessional moraines depending on their relative positions with respect to the glacier that produced them.

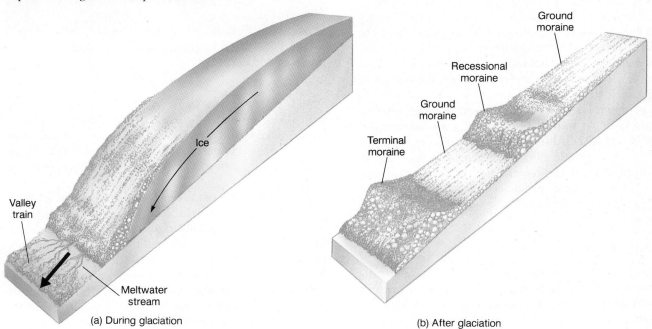

(a) During glaciation

(b) After glaciation

Following a period of stabilization, a glacier may advance or retreat, depending on changes in its budget. If it advances, the ice front overrides and modifies its former moraine. Should a negative budget occur, however, the ice front retreats toward the zone of accumulation. As the ice front recedes, till is deposited as it is liberated from the melting ice and forms a layer of **ground moraine** (Fig. 14.21b). Ground moraine has an irregular, rolling topography, whereas end moraine consists of long ridgelike accumulations of sediment.

After a glacier has retreated for some time, its terminus may once again stabilize, and it will deposit another end moraine. Because the ice front has receded, such moraines are called **recessional moraines** (Fig. 14.21b). During the Pleistocene, continental glaciers in the mid-continent region extended as far south as southern Ohio, Indiana, and Illinois. Their outermost end moraines, marking the greatest extent of the glaciers, go by the special name **terminal moraine** (valley glaciers also deposit terminal moraines). As the glaciers retreated from the positions at which their terminal moraines were deposited, they temporarily ceased retreating numerous times and deposited dozens of recessional moraines.

Lateral and Medial Moraines

As we previously discussed, valley glaciers transport considerable sediment along their margins. Much of this sediment is abraded and plucked from the valley walls, but a significant amount falls or slides onto the glacier's surface by mass wasting processes. In any case, when a glacier melts, this sediment is deposited as long ridges of till called **lateral moraines** along the margin of the glacier (◆ Fig. 14.22).

Where two lateral moraines merge, as when a tributary glacier flows into a larger glacier, a **medial moraine** forms (Fig. 14.22). Although medial moraines are identified by their position on a valley glacier, they are, in fact, formed from the coalescence of two lateral moraines. One can generally determine how many tributaries a valley glacier has by the number of its medial moraines.

Drumlins

In many areas where continental glaciers have deposited till, the till has been reshaped into elongated hills called **drumlins.** Some drumlins measure as much as 50 m high and 1 km long, but most are much smaller. From the side, a drumlin looks like an inverted spoon

◆ FIGURE 14.22 Lateral and medial moraines on a glacier in Alaska.

with the steep end on the side from which the glacial ice advanced, and the gently sloping end pointing in the direction of ice movement (◆ Fig. 14.23). Thus, drumlins can be used to determine the direction of ice movement.

Drumlins are most often found in areas of ground moraine that were overridden by an advancing ice sheet. Although no one has fully explained the origin of drumlins, it appears that they form in the zone of plastic flow as glacial ice modifies preexisting till into streamlined hills. Drumlins rarely occur as single, isolated hills; instead they occur in *drumlin fields* in which hundreds or thousands of drumlins are present. Drumlin fields are found in several states and Ontario, Canada, but perhaps the finest example is near Palmyra, New York.

Landforms Composed of Stratified Drift

Stratified drift is associated with both valley and continental glaciers, but as one would expect, it is more extensive in areas of continental glaciation.

Outwash Plains and Valley Trains

Glaciers discharge meltwater laden with sediment most of the time, except perhaps during the coldest months. Such meltwater forms a series of braided streams that radiate out from the front of continental glaciers over a wide region (◆ Fig. 14.24). So much sediment is sup-

plied to these streams that much of it is deposited within the channels as sand and gravel bars. The vast blankets of sediments so formed are **outwash plains** (◆ Fig. 14.25).

Valley glaciers discharge huge amounts of meltwater and, like continental glaciers, have braided streams extending from them. However, these streams are generally confined to the lower parts of glacial troughs, and their long, narrow deposits of stratified drift are known as **valley trains** (Fig. 14.24).

◆ FIGURE 14.23 These elongated hills in Antrim County, Michigan are drumlins. (Photo courtesy of B. M. C. Pape.)

◆ FIGURE 14.24 Braided streams discharging from a valley glacier, such as these in the Yukon Territory, Canada, deposit the stratified drift of a valley train.

◆ FIGURE 14.25 Two stages in the origin of kettles, kames, and eskers. (*a*) During glaciation. (*b*) After glaciation.

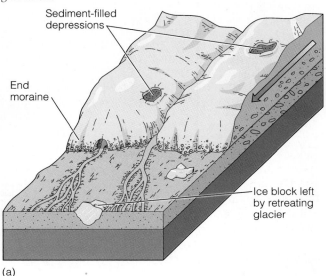

(a)

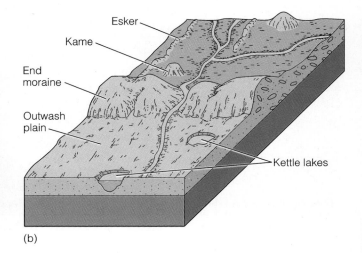

(b)

Perspective 14.1

GLACIAL LAKE MISSOULA
AND THE CHANNELED SCABLANDS

The term *scabland* is used in the Pacific Northwest to describe areas from which the surface deposits have been scoured, thus exposing the underlying rock. Such an area exists in a large part of eastern Washington where numerous deep and generally dry channels are present. Some of these channels, cut into basalt lava flows, are more than 70 m deep, and their floors are covered by gigantic "ripple marks" as much as 10 m high and 70 to 100 m apart. Additionally, a number of high hills in the area are arranged such that they appear to have been islands in a large braided stream.

In 1923, J Harlan Bretz proposed that the channeled scablands of eastern Washington were formed during a single, gigantic flood. Bretz's unorthodox explanation was rejected by most geologists who preferred a more traditional interpretation based on normal stream erosion over a long period of time. In contrast, Bretz held that the scablands were formed rapidly during a flood of glacial meltwater that lasted only a few days.

The problem with Bretz's hypothesis was that he could not identify an adequate source for his floodwater. He knew that the glaciers had advanced as far south as Spokane, Washington, but he could not explain how so much ice melted so rapidly. The answer to Bretz's dilemma came from western Montana where an enormous ice-dammed lake (Lake Missoula) had formed. Lake Missoula

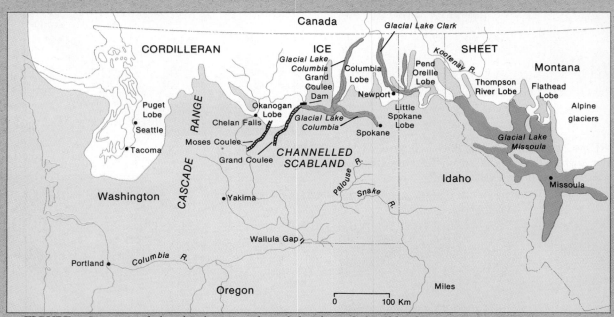

◆ FIGURE 1 Location of glacial Lake Missoula and the channeled scablands of eastern Washington.

formed when an advancing glacier plugged the Clark Fork Valley at Ice Cork, Idaho, causing the water to fill the valleys of western Montana (◆ Fig. 1). At its highest level, Lake Missoula covered about 7,800 km^2 and contained an estimated 2,090 km^3 of water (about 42% of the volume of present-day Lake Michigan). The shorelines of Lake Missoula are still clearly visible on the mountainsides around Missoula, Montana (◆ Fig. 2).

When the ice dam impounding Lake Missoula failed, the water rushed out at tremendous velocity and drained south and southwest across Idaho and into Washington. The maximum rate of flow is estimated to have been nearly 11 million m^3/sec, about 55 times greater than the average discharge of the Amazon River. When these raging floodwaters reached eastern Washington, they stripped away the soil and most of the surface sediment, carving out huge valleys in solid bedrock. The currents were so powerful and turbulent they plucked out and moved pieces of basalt measuring 10 m across. Within the channels, sand and gravel was shaped into huge ridges, the so-called giant ripple marks (◆ Fig. 3).

Bretz originally believed that one massive flood formed the channeled scablands, but geologists now know that Lake Missoula formed, flooded, and re-formed at least four times and perhaps as many as seven times. The largest lake formed 18,000 to 20,000 years ago, and its draining produced the last great flood.

How long did the flood last and did humans witness it? It has been estimated that approximately one month passed from the time the ice dam first broke and water rushed out onto the scablands to the time the scabland streams returned to normal flow. No one knows for sure if anyone witnessed the flood. The oldest known evidence of humans in the region is from the Marmes Man site in southeastern Washington dated at 10,130 years ago, nearly 2,000 years after the last flood from Lake Missoula. However, it is now generally accepted that Native Americans were present in North America at least 15,000 years ago.

◆ FIGURE 2 The horizontal lines on Sentinel Mountain at Missoula, Montana are wave-cut shorelines of glacial Lake Missoula.

◆ FIGURE 3 These gravel ridges are the so-called giant ripple marks that formed when glacial Lake Missoula drained across this area near Camas Hot Springs, Montana.

Outwash plains and valley trains commonly contain numerous circular to oval depressions, many of which contain small lakes. These depressions are *kettles;* they form when a retreating ice sheet or valley glacier leaves a block of ice that is subsequently partly or wholly buried (Fig. 14.25). When the ice block eventually melts, it leaves a depression; if the depression extends below the water table, it becomes the site of a small lake. So many kettles occur in some outwash plains that they are called *pitted outwash plains.* Although kettles are most common in outwash plains, they can also form in end moraines.

Kames and Eskers

Kames are conical hills as much as 50 m high composed of stratified drift (Figs. 14.25 and ◆ 14.26a). Many kames form when a stream deposits sediment in a depression on a glacier's surface; as the ice melts, the deposit is lowered to the surface. They also form in cavities within or beneath stagnant ice.

Long sinuous ridges of stratified drift, many of which meander and have tributaries, are called **eskers** (Figs. 14.25 and ◆ 14.26b). Some eskers are quite high, as much as 100 m, and can be traced for more than 100 km. Eskers occur most commonly in areas once covered by continental glaciers, but they are also associated with large valley glaciers. The sorting and stratification of the sediments within eskers clearly indicate deposition by running water. The physical properties of ancient eskers and observations of present-day glaciers indicate that they form in tunnels beneath stagnant ice and in meltwater channels on the surface of glaciers (Fig. 14.25).

(a)

(b)

◆ FIGURE 14.26 (*a*) This small, conical hill is a kame. (*b*) An area of ground moraine and an esker.

Glacial Lake Deposits

Numerous lakes exist in areas of glaciation. Some formed as a consequence of glaciers scouring out depressions; others occur where a stream's drainage was blocked; and others are the result of water accumulating behind moraines or in kettles. Regardless of how they formed, glacial lakes, like all lakes, are areas of deposition. Sediment may be carried into them and deposited as small deltas, but of special interest are the fine-grained deposits. Mud deposits in glacial lakes are commonly finely laminated, consisting of alternating light and dark layers. Each light-dark couplet is called a *varve* (◆ Fig. 14.27). Each varve represents an annual episode of deposition; the light layers form during the spring and summer and consist of silt and clay; the dark layers form during the winter when the smallest particles of clay and organic matter settle from suspension as the lake freezes over. The number of varves indicates how many years a glacial lake has existed.

Another distinctive feature of glacial lakes containing varved deposits is the presence of *dropstones* (Fig. 14.27). These are pieces of gravel, some of boulder size, in otherwise very fine-grained deposits. Most of them were probably carried into the lakes by icebergs that eventually melted and released sediment contained in the ice. One glacial lake of particular interest is glacial lake Missoula that existed in western Montana during the late Pleistocene (see Perspective 14.1).

◆ FIGURE 14.27 Glacial varves with a dropstone. (Photo courtesy of Canadian Geological Survey.)

⩔ GLACIERS AND ISOSTASY

In Chapter 10 we discussed the concept of isostasy and noted that loading or unloading of the Earth's crust causes it to respond isostatically to an increased or decreased load by subsiding and rising, respectively. There is no question that isostatic rebound has occurred in the areas formerly covered by continental glaciers.

When the Pleistocene ice sheets formed and increased in size, the weight of the ice caused the crust to respond by slowly subsiding deeper into the mantle. In some places, the Earth's surface was depressed as much as 300 m below preglacial elevations. As the ice sheets disappeared, the downwarped areas gradually rebounded to their former positions. As noted in Chapter 10, considerable isostatic rebound has occurred in Scandinavia and Canada (see Fig. 10.32).

The Great Lakes evolved as the glaciers retreated to the north. As one would expect, isostatic rebound began as the ice front retreated north. Rebound began first in the southern part of the region because that area was free of ice first. Furthermore, the greatest loading by glaciers, and hence the greatest crustal depression, occurred farther north in Canada in the zones of accumulation. For these reasons, rebound has not been evenly distributed over the entire glaciated area; it increases in magnitude from south to north. As a result of this uneven isostatic rebound, coastal features in the Great Lakes region, such as old shorelines, are now elevated higher above their former levels in the north and thus slope to the south.

⩔ CAUSES OF GLACIATION

Thus far we have examined the effects of glaciation, but have not addressed the central questions of what causes large-scale glaciation and why so few episodes of widespread glaciation have occurred. For more than a century, scientists have been attempting to develop a comprehensive theory explaining all aspects of ice ages, but have not yet been completely successful. One reason for their lack of success is that the climatic changes responsible for glaciation, the cyclic occurrence of glacial-interglacial episodes, and short-term events such as the Little Ice Age operate on vastly different time scales.

Only a few periods of glaciation are recognized in the geologic record, each separated from the others by long intervals of mild climate. Such long-term climatic changes probably result from slow geographic changes related to plate tectonic activity. Moving plates can carry continents to high latitudes where glaciers can exist, provided that they receive enough precipitation as snow. Plate collisions, the subsequent uplift of vast areas far above sea level, and the changing atmospheric and oceanic circulation patterns caused by the changing shapes and positions of plates also contribute to long-term climatic change.

A theory explaining ice ages must also address the fact that during the Pleistocene Ice Age (1.6 million to 10,000 years ago) several intervals of glacial expansion separated by warmer interglacial periods occurred. At least 4 major episodes of glaciation have been recognized in North America, and 6 or 7 glacial advances and retreats occurred in Europe. These intermediate-term climatic events occur on time scales of tens to hundreds of thousands of years. The cyclic nature of this most recent episode of glaciation has long been a problem in formulating a comprehensive theory of climatic change.

The Milankovitch Theory

A particularly interesting hypothesis for intermediate-term climatic events was put forth by the Yugoslavian

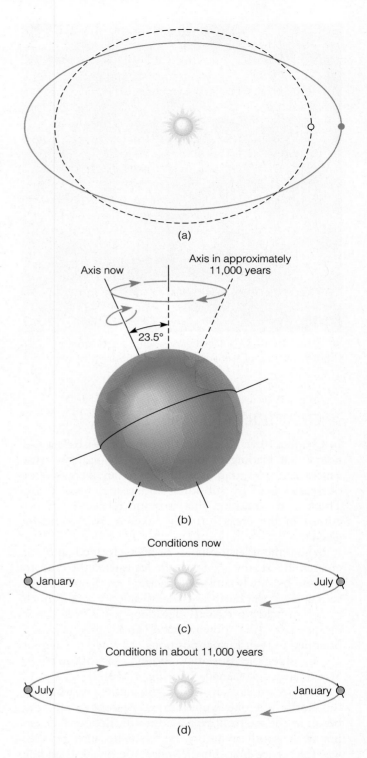

◆ FIGURE 14.28 (*a*) The Earth's orbit varies from nearly a circle (dashed line) to an ellipse (solid line) and back again in about 100,000 years. (*b*) The Earth moves around its orbit while spinning about its axis, which is tilted to the plane of the ecliptic at 23.5° and points toward the North Star. The Earth's axis of rotation slowly moves and traces out the path of a cone in space. (*c*) At present, the Earth is closest to the Sun in January when the Northern Hemisphere experiences winter. (*d*) In about 11,000 years, as a result of precession, the Earth will be closer to the Sun in July, when summer occurs in the Northern Hemisphere.

astronomer Milutin Milankovitch during the 1920s. He proposed that minor irregularities in the Earth's rotation and orbit are sufficient to alter the amount of solar radiation that the Earth receives at any given latitude and hence can affect climatic changes. Now called the **Milankovitch theory,** it was initially ignored, but has received renewed interest during the last 20 years.

Milankovitch attributed the onset of the Pleistocene Ice Age to variations in three parameters of the Earth's orbit (◆ Fig. 14.28). The first of these is orbital eccentricity, which is the degree to which the orbit departs from a perfect circle. Calculations indicate a roughly 100,000-year cycle between times of maximum eccentricity. This corresponds closely to 20 warm-cold climatic cycles that occurred during the Pleistocene. The second parameter is the angle between the Earth's axis and a line perpendicular to the plane of the ecliptic. This angle shifts about 1.5° from its current value of 23.5° during a 41,000-year cycle. The third parameter is the precession of the equinoxes, which causes the position of the equinoxes and solstices to shift slowly around the Earth's elliptical orbit in a 23,000-year cycle.

Continuous changes in these three parameters cause the amount of solar heat received at any latitude to vary slightly over time. The total heat received by the planet, however, remains little changed. Milankovitch proposed, and now many scientists agree, that the interaction of these three parameters provides the triggering mechanism for the glacial-interglacial episodes of the Pleistocene.

Short-Term Climatic Events

Climatic events having durations of several centuries, such as the Little Ice Age, are too short to be accounted for by plate tectonics or Milankovitch cycles. Several hypotheses have been proposed, including variations in solar energy and volcanism.

Variations in solar energy could result from changes within the Sun itself or from anything that would reduce the amount of energy the Earth receives from the Sun. The latter could result from the solar system passing through clouds of interstellar dust and gas or from substances in the Earth's atmosphere reflecting solar radiation back into space. Records kept over the past 75 years, however, indicate that during this time the amount of solar radiation has varied only slightly. Thus, although variations in solar energy may influence short-term climatic events, such a correlation has not been demonstrated.

During large volcanic eruptions, tremendous amounts of ash and gases are spewed into the atmosphere where they reflect incoming solar radiation and thus reduce atmospheric temperatures. Recall from Perspective 4.1 that small droplets of sulfur gases remain in the atmosphere for years and can have a significant effect on the climate. Several such large-scale volcanic events have been recorded, such as the 1815 eruption of Tambora, and are known to have had climatic effects. However, no relationship between periods of volcanic activity and periods of glaciation has yet been established.

Chapter Summary

1. Glaciers are masses of ice on land that move by plastic flow and basal slip. Glaciers currently cover about 10% of the land surface and contain 2% of all water on Earth.
2. Valley glaciers are confined to mountain valleys and flow from higher to lower elevations, whereas continental glaciers cover vast areas and flow outward in all directions from a zone of accumulation.
3. A glacier forms when winter snowfall in an area exceeds summer melt and therefore accumulates year after year. Snow is compacted and converted to glacial ice, and when the ice is about 40 m thick, pressure causes it to flow.
4. The behavior of a glacier depends on its budget, which is the relationship between accumulation and wastage. If a glacier possesses a balanced budget, its terminus remains stationary; a positive or negative budget results in advance or retreat of the terminus, respectively.

5. Glaciers move at varying rates depending on the slope, discharge, and season. Valley glaciers tend to flow more rapidly than continental glaciers.
6. Glaciers are powerful agents of erosion and transport because they are solids in motion. They are particularly effective at eroding soil and unconsolidated sediment, and they can transport any size sediment supplied to them.
7. Continental glaciers transport most of their sediment in the lower part of the ice, whereas valley glaciers may carry sediment in all parts of the ice.
8. Erosion of mountains by valley glaciers yields several sharp, angular landforms including cirques, arêtes, and horns. U-shaped glacial troughs, fiords, and hanging valleys are also products of valley glaciation.
9. Continental glaciers abrade and bevel high areas, producing a smooth, rounded landscape.

10. Depositional landforms include moraines, which are ridgelike accumulations of till. Several types of moraines are recognized, including terminal, recessional, lateral, and medial moraines.
11. Drumlins are composed of till that was apparently reshaped into streamlined hills by continental glaciers.
12. Stratified drift consists of sediments deposited in or by meltwater streams issuing from glaciers; it is found in outwash plains and valley trains. Ridges called eskers and conical hills called kames are also composed of stratified drift.
13. Loading of the Earth's crust by Pleistocene glaciers caused isostatic subsidence. When the glaciers disappeared, isostatic rebound began and continues in some areas.
14. Major glacial intervals separated by tens or hundreds of millions of years probably occur as a consequence of the changing positions of tectonic plates, which in turn cause changes in oceanic and atmospheric circulation patterns.
15. Currently, the Milankovitch theory is widely accepted as the explanation for glacial-interglacial intervals.
16. The reasons for short-term climatic changes, such as the Little Ice Age, are not understood. Two proposed causes for such events are changes in the amount of solar energy received by the Earth and volcanism.

Important Terms

abrasion	glacial drift	Milankovitch theory
arête	glacial erratic	outwash plain
basal slip	glacial groove	plastic flow
cirque	glacial ice	recessional moraine
continental glacier	glacial polish	stratified drift
drumlin	glacial striation	terminal moraine
end moraine	glacier	till
esker	ground moraine	U-shaped glacial trough
fiord	hanging valley	valley glacier
firn	horn	valley train
firn limit	lateral moraine	zone of accumulation
glacial budget	medial moraine	zone of wastage

Review Questions

1. Crevasses in glaciers extend down to:
   a. ___ about 300 m; b. ___ the base of the glacier; c. the zone of plastic flow; d. ___ variable depths depending on how thick the ice is; e. ___ the outwash layer.
2. If a glacier has a negative budget:
   a. ___ the terminus will retreat; b. ___ its accumulation rate is greater than its wastage rate; c. ___ all flow ceases; d. ___ the glacier's length increases; e. ___ crevasses will no longer form.
3. The bowl-shaped depression at the upper end of a glacial trough is a(an):
   a. ___ inselberg; b. ___ cirque; c. ___ lateral moraine; d. ___ drumlin; e. ___ till.
4. Which of the following is not an erosional landform?
   a. ___ horn; b. ___ arête; c. ___ lateral moraine; d. ___ U-shaped glacial trough; e. ___ cirque.
5. Headward erosion of a group of cirques on the flanks of a mountain may produce a(an):
   a. ___ tarn; b. ___ varve; c. ___ drumlin; d. ___ kettle; e. ___ horn.
6. Rocks abraded by glaciers develop a smooth surface that shines in reflected light. Such a surface is called glacial:
   a. ___ grooves; b. ___ polish; c. ___ flour; d. ___ striations; e. ___ till.
7. A small lake in a cirque is a:
   a. ___ pluvial lake; b. ___ proglacial lake; c. ___ tarn; d. ___ salt lake; e. ___ glacial trough lake.
8. Firn is:
   a. ___ freshly fallen snow; b. ___ a granular type of ice; c. ___ a valley train; d. ___ another name for the zone of wastage; e. ___ a type of glacial groove.
9. Pressure on ice at depth in a glacier causes it to move by:
   a. ___ rock creep; b. ___ fracture; c. ___ basal slip; d. ___ surging; e. ___ plastic flow.
10. A pyramid-shaped peak formed by glacial erosion is a:
    a. ___ fiord; b. ___ medial moraine; c. ___ horn; d. ___ cirque; e. ___ hanging valley.
11. Glacial drift is a general term for:
    a. ___ the erosional landforms of continental glaciers; b. ___ all the deposits of glaciers; c. ___ icebergs floating

at sea; d. ___ the movement of glaciers by plastic flow and basal slip; e. ___ the annual wastage rate of a glacier.

12. The number of medial moraines on a glacier generally indicates the number of its _____ .
    a. ___ tributary glaciers; b. ___ terminal moraines; c. ___ eskers; d. ___ outwash plains; e. ___ valley trains.

13. A knifelike ridge separating glaciers in adjacent valleys is a(an):
    a. ___ fiord; b. ___ horn; c. ___ arête; d. ___ cirque; e. ___ lateral moraine.

14. Which of the following is a glacial erratic?
    a. ___ deposit of unsorted, unstratified till; b. ___ glacially transported boulder far from its source; c. ___ sand and gravel deposited in a depression on a glacier; d. ___ U-shaped glacial trough; e. ___ deposits consisting of light and dark layers.

15. How does glacial ice form, and why is it considered to be a rock?

16. Other than size, how do valley glaciers differ from continental glaciers?

17. What is the relative importance of plastic flow and basal slip for glaciers at high and low latitudes?

18. Explain in terms of the glacial budget how a once active glacier becomes a stagnant glacier.

19. What is a glacial surge and what are the probable causes of surges?

20. Explain how glaciers erode by abrasion and plucking.

21. Why are glaciers more effective agents of erosion and transport than running water?

22. Describe the processes responsible for the origin of a cirque, U-shaped glacial trough, and hanging valley.

23. What is an arête and how does one form?

24. How do the erosional landforms of continental glaciers differ from those of valley glaciers?

25. Discuss the processes whereby terminal, recessional, and lateral moraines form.

26. How does a medial moraine form, and how can one determine the number of tributaries a valley glacier has by its medial moraines?

27. Describe drumlins, and explain how they form.

28. What are outwash plains and valley trains?

29. In a roadside outcrop, you observe a deposit of alternating light and dark laminated mud containing a few large boulders. Explain the sequence of events responsible for its deposition.

Additional Readings

Bell, M. 1994. Is our climate unstable? *Earth* 3, no. 1:24–31.

Broecker, W. S., and G. H. Denton. 1990. What drives glacial cycles? *Scientific American* 262, no. 1:49–56.

Carozzi, A. V. 1984. Glaciology and the ice age. *Journal of Geological Education* 32:158–70.

Covey, C. 1984. The Earth's orbit and the ice ages. *Scientific American* 250, no. 2:58–66.

Drewry, D. J. 1986. *Glacial geologic processes*. London: Edward Arnold.

Grove, J. M. 1988. *The Little Ice Age*. London: Methuen.

Imbrie, J., and K. P. Imbrie. 1979. *Ice ages: Solving the mystery*. Hillside, N.J.: Enslow Press.

John, B. S. 1977. *The ice age: Past and present*. London: Collins.

———. 1979. *The winters of the world*. London: David & Charles.

Kurten, B. 1988. *Before the Indians*. New York: Columbia Univ. Press.

Schneider, S. H. 1990. *Global warming: Are we entering the greenhouse century?* San Francisco: Sierra Club Books.

Sharp, R. P. 1988. *Living ice: Understanding glaciers and glaciation*. New York: Cambridge Univ. Press.

Williams, R. S., Jr. 1983. *Glaciers: Clues to future climate?* United States Geological Survey.

CHAPTER
15

Chapter Outline

Racetrack Playa, Death Valley, California, is famous for its "sliding rocks." Geologists think that strong winds push the rocks across a lake's exposed wet, slippery bed after a rainstorm. This limestone block was moved 24 m by the wind.

The Work of Wind and Deserts

Prologue

During the last few decades, deserts have been advancing across millions of acres of productive land, destroying rangelands, croplands, and even villages. Such expansion, estimated at 70,000 km² per year, has exacted a terrible toll in human suffering. Because of the relentless advance of deserts, hundreds of thousands of people have died of starvation or been forced to migrate as "environmental refugees" from their homelands to camps where the majority are severely malnourished. This expansion of deserts into formerly productive lands is **desertification.**

Most regions undergoing desertification lie along the margins of existing deserts. These margins have a delicately balanced ecosystem that serves as a buffer between the desert on one side and a more humid environment on the other. Their potential to adjust to increasing environmental pressures from natural causes or human activity is limited. Currently, such fringe areas include large regions in several parts of the world (◆ Fig. 15.1).

While natural processes such as climatic change result in gradual expansion and contraction of desert

◆ FIGURE 15.1 Desert areas of the world and areas threatened by desertification.

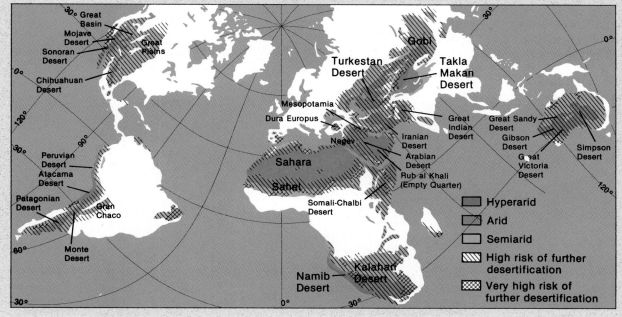

regions, much recent desertification has been greatly accelerated by human activities. In many areas, the natural vegetation has been cleared as crop cultivation has expanded into increasingly drier fringes to support the growing population. Because these areas are especially prone to droughts, crop failures are common occurrences, leaving the land bare and susceptible to increased wind and water erosion.

Because grasses constitute the dominant natural vegetation in most fringe areas, raising livestock is a common economic activity. Usually, these areas achieve a natural balance between vegetation and livestock as nomadic herders graze their animals on the available grasses. In many fringe areas, however, livestock numbers have been greatly increasing in recent years, and they now far exceed the land's capacity to support them. As a result, the vegetation cover that protects the soil has diminished, causing the soil to crumble. This leads to further drying of the soil and accelerated soil erosion by wind and water (♦ Fig. 15.2).

Drilling water wells also contributes to desertification because human and livestock activity around a well site strips away the vegetation. With its vegetation gone, the topsoil blows away, and the resultant bare areas merge with the surrounding desert. In addition, the water used for irrigation from these wells sometimes contributes to desertification by increasing the salt content of the soil. As the water

evaporates, a small amount of salt is deposited in the soil and is not flushed out as it would be in an area that receives more rain. Over time, the salt concentration becomes so high that plants can no longer grow. Desertification resulting from soil salinization is a major problem in North Africa, the Middle East, southwest Asia, and the western United States.

Collecting firewood for heating and cooking is another major cause of desertification, particularly in many less-developed countries where wood is the major fuel source. In the Sahel of Africa (a belt 300 to 1,100 km wide that lies south of the Sahara), the expanding population has completely removed all trees and shrubs in the areas surrounding many towns and cities. Journeys of several days on foot to collect firewood are common there. Furthermore, the use of dried animal dung to supplement firewood has exacerbated desertification because important nutrients in the dung are not returned to the soil.

The Sahel averages between 10 and 60 cm of rainfall per year, 90% of which evaporates when it falls. Because drought is common in the Sahel, the region can support only a limited population of livestock and humans. Traditionally, herders and livestock existed in a natural balance with the vegetation, following the rains north during the rainy season and returning south to greener rangeland during the dry seasons. Some areas were alternately

♦ FIGURE 15.2 A sharp line marks the boundary between pasture and an encroaching dune in Niger, Africa. As the goats eat the remaining bushes, the dune will continue to advance, and more land will be lost to desertification.

planted and left fallow to help regenerate the soil. During fallow periods, livestock fed off the stubble of the previous year's planting, and their dung helped fertilize the soil.

With the emergence of new nations and increased foreign aid to the Sahel during the 1950s and 1960s, nomads and their herds were restricted, and large areas of grazing land were converted to cash crops such as peanuts and cotton that have a short growing season. Expanding human and animal populations and more intensive agriculture put increasing demands on the land. These factors, combined with the worst drought of the century (1968–1973), brought untold misery to the people of the Sahel. Without rains, the crops failed and the livestock denuded the land of what little vegetation remained. As a result, nearly 250,000 people and 3.5 million cattle died of starvation, and the adjacent Sahara expanded southward as much as 150 km.

The tragedy of the Sahel and prolonged droughts in other desert fringe areas serve to remind us of the delicate equilibrium of ecosystems in such regions. Once the fragile soil cover has been removed by erosion, it will take centuries for new soil to form (see Chapter 5).

▼ INTRODUCTION

Most people associate the work of wind with deserts. Wind is an effective geologic agent in desert regions, but it also plays an important role wherever loose sediment can be eroded, transported, and deposited, such as along shorelines or the plains (see Perspective 5.2). Therefore, we will first consider the work of wind in general and then discuss the distribution, characteristics, and landforms of deserts.

▼ SEDIMENT TRANSPORT BY WIND

Wind is a turbulent fluid and therefore transports sediment in much the same way as running water. Although wind typically flows at a greater velocity than water, it has a lower density and, thus, can carry only clay- and silt-size particles as suspended load. Sand and larger particles are moved along the ground as bed load.

Bed Load

Sediments too large or heavy to be carried in suspension by water or wind are moved as *bed load* either by *saltation* or by rolling and sliding. As we discussed in Chapter 12, saltation is the process by which a portion of the bed load moves by intermittent bouncing along a stream bed. Saltation also occurs on land. The wind starts sand grains rolling and lifts and carries some grains short distances before they fall back to the surface. As the descending sand grains hit the surface, they strike other grains causing them to bounce along by saltation (◆ Fig. 15.3). Wind tunnel experiments have shown that once sand grains begin moving, they will continue to move, even if the wind drops below the speed necessary to start them moving! This happens because once saltation begins, it sets off a chain reaction of collisions between grains that keeps the sand grains in constant motion.

Saltating sand usually moves near the surface, and even when winds are strong, grains are rarely lifted higher than about a meter. If the winds are very strong, these wind-whipped grains can cause extensive abrasion. A car's paint can be removed by sandblasting in a short time, and its windshield will become completely frosted and translucent from pitting.

Particles larger than sand can also be moved along the ground by the process of *surface creep*. This type of movement occurs when saltating sand grains strike the larger particles and push them forward along the ground.

Suspended Load

Silt- and clay-sized particles constitute most of a wind's *suspended load*. Even though these particles are much smaller and lighter than sand-sized particles, wind usually starts the latter moving first. The reason for this phenomenon is that a very thin layer of motionless air lies next to the ground where the small silt and clay particles remain undisturbed. The larger sand grains,

◆ FIGURE 15.3 Most sand is moved near the ground surface by saltation. Sand grains are picked up by the wind and carried a short distance before falling back to the ground where they usually hit other grains, causing them to bounce and move in the direction of the wind.

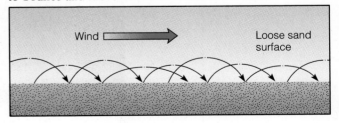

however, stick up into the turbulent air zone where they can be moved. Unless the stationary air layer is disrupted, the silt and clay particles remain on the ground providing a smooth surface. This phenomenon can be observed on a dirt road on a windy day. Unless a vehicle travels over the road, little dust is raised even though it is windy. When a vehicle moves over the road, it breaks the calm boundary layer of air and disturbs the smooth layer of dust, which is picked up by the wind and forms a dust cloud in the vehicle's wake.

In a similar manner, when a sediment layer is disturbed, silt- and clay-sized particles are easily picked up and carried in suspension by the wind, creating clouds of dust or even dust storms (◆ Fig. 15.4). Once these fine particles are lifted into the atmosphere, they may be carried thousands of kilometers from their source. For example, large quantities of fine dust from the southwestern United States were blown eastward and fell on New England during the Dust Bowl of the 1930s (see Perspective 5.2).

⋙ WIND EROSION

Recall that streams and glaciers are effective agents of erosion, much more so than wind. Even in deserts, where wind is most effective, running water is still responsible for most erosional landforms, although stream channels are typically dry (see Fig. 12.2). Nev-

ertheless, wind action can still produce many distinctive erosional features and is an extremely efficient sorting agent.

Abrasion

Wind erodes material in two ways: abrasion and deflation. **Abrasion** involves the impact of saltating sand grains on an object and is analogous to sandblasting. The effects of abrasion, however, are usually minor because sand, the most common agent of abrasion, is rarely carried more than 1 m above the surface. Rather than creating major erosional features, wind abrasion merely modifies existing features by etching, pitting, smoothing, or polishing. Thus, wind abrasion is most effective on soft sedimentary rocks.

Ventifacts are a common product of wind abrasion; these are stones whose surfaces have been polished, pitted, grooved, or faceted by the wind (◆ Fig. 15.5). If the wind blows from different directions, or if the stone is moved, the ventifact will have multiple facets. Ventifacts are most common in deserts, yet they can also form wherever stones are exposed to saltating sand grains, as on beaches in humid regions and some outwash plains in New England.

Yardangs are larger features than ventifacts and also result from wind erosion (◆ Fig. 15.6). They are elongated and streamlined ridges that look like an over-

◆ FIGURE 15.4 A dust storm in Death Valley, California.

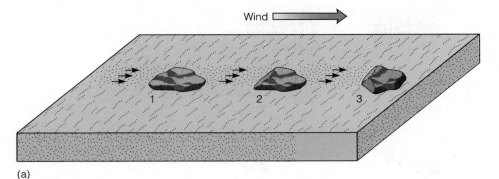

Wind

(a)

(b)

◆ FIGURE 15.5 (a) A ventifact forms when wind-borne particles (1) abrade the surface of a rock (2) forming a flat surface. If the rock is moved, (3) additional flat surfaces are formed. (b) A granite ventifact in the dune corridor along the Michigan shore of Lake Michigan. (Photo courtesy of Marion A. Whitney.)

turned ship's hull. They are typically found grouped in clusters aligned parallel to the prevailing winds. They probably form by differential erosion in which depressions, parallel to the direction of wind, are carved out of a rock body, leaving sharp, elongated ridges. These ridges may then be further modified by wind abrasion into their characteristic shape. Although yardangs are fairly common desert features, interest in them was renewed when images radioed back from Mars showed that they are also widespread features on the Martian surface (see Perspective 15.1).

Deflation

Another important mechanism of wind erosion is **deflation,** which is the removal of loose surface sediment by the wind. Among the characteristic features of deflation in many arid and semiarid regions are

◆ FIGURE 15.6 Profile view of a streamlined yardang in the Roman playa deposits of the Kharga Depression, Egypt. (Photo courtesy of Marion A. Whitney.)

Perspective 15.1

EVIDENCE OF WIND ACTIVITY ON MARS

Data gathered by the two *Viking* landers (July and September 1976) as well as by the *Viking* and *Mariner 9* orbiters reveal that Mars is a planet with a complex geologic history involving volcanism, running water, and wind (see Chapter 19 and the Prologue to Chapter 12). Although the majority of Martian topographic features formed early in the planet's history by volcanism and running water, many surface features were created by wind and, thus, are still forming.

One of the surprising features of Mars is the general lack of sand and sand dunes. While dunes do exist around the northern Martian ice caps, other areas of the Martian surface resemble a rocky volcanic desert. At the two *Viking* landing sites, the terrain varies from flat to rolling plains littered with rocks (◆Fig. 1). Evidence of wind activity can be seen in the form of linear deposits or streaks of fine-grained material on the lee (downwind) sides of most of the rocks, as well as small dunes and angular-faceted rocks similar to ventifacts on Earth (Fig. 1).

◆ FIGURE 2 A planetary dust storm obscured *Mariner 9's* view of the Martian surface for the first few weeks after it went into orbit around Mars in 1971.

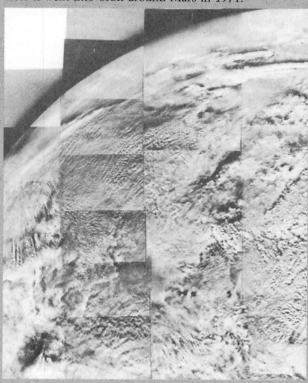

◆ FIGURE 1 Scattered rocks, streaks, and small dunes are visible in this ground-level panorama of the Martian surface taken by *Viking 2*.

Large-scale dust storms are seasonal occurrences on Mars (◆ Fig. 2) and occur during the southern summer when Mars is closest to the Sun. Once winds lift the fine particles into the atmosphere, they remain suspended for long periods. The dust raised by these storms may obscure the entire planet for weeks at a time. One result of these yearly dust storms is the seasonal change in the dust cover on the ground, which accounts for the seasonal color variations observed from Earth.

While the atmosphere of Mars is too thin to suspend anything larger than dust-sized particles, sand-sized particles can be moved along the ground by saltation. Large dune fields composed of sand-sized particles have been discovered surrounding the north polar ice cap (◆ Fig. 3). The origin of these dunes is still controversial. Geologists think that most of the debris on the northern plains and the dunes themselves consist of material eroded from the polar deposits. When the deposits of dust-sized particles were removed by the wind, the sand-sized particles were left behind and were transported by saltation to form dunes.

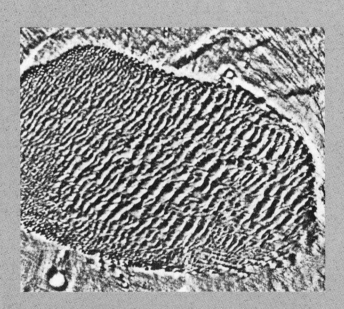

◆ FIGURE 3 Large dune fields surrounding the north polar ice cap are testimony to the incessant wind action occurring on Mars.

deflation hollows (also called *blowouts*). These shallow depressions of variable dimensions result from differential erosion of surface materials (◆ Fig. 15.7). Ranging in size from several kilometers in diameter and tens of meters deep to small depressions only a few meters wide and less than a meter deep, deflation hollows are common in the southern Great Plains region of the United States. In addition, large areas measuring hundreds of square kilometers, such as the Qattara Depression in northwestern Egypt and some of Australia's interior desert basins, may also have formed primarily by deflation.

In many dry regions, the removal of sand-sized and smaller particles by wind leaves a surface of pebbles, cobbles, and boulders. As the wind removes the fine-grained material from the surface, the effects of gravity and occasional floodwaters rearrange the remaining coarse particles into a mosaic of close-fitting rocks called **desert pavement** (◆ Fig. 15.8). Once a desert pavement forms, it protects the underlying material from further deflation.

▼ WIND DEPOSITS

Although wind is of minor importance as an erosional agent, wind deposits can form impressive structures. These deposits are primarily of two types. The first, called *dunes,* occur in several distinctive types, all of which consist of sand-sized particles that are usually deposited near their source. The second is *loess,* which consists of layers of windblown silt and clay that are deposited over large areas downwind and commonly far from their source.

The Formation and Migration of Dunes

The most characteristic features associated with sand-covered regions are **dunes,** which are mounds or ridges of wind-deposited sand. Dunes form when wind flows over and around an obstruction. This results in two zones of quiet air, called wind shadows, that form immediately in front of and behind the obstruction (◆ Fig. 15.9). As saltating sand grains settle in these wind shadows, they begin to accumulate and build up a deposit of sand. As they grow, these sand deposits become self-generating in that they form ever-larger wind barriers that further reduce the wind's velocity, forcing it to deposit any sand it carries.

Most dunes have an asymmetrical profile, with a gentle windward slope and a steeper downwind or leeward slope that is inclined in the direction of the prevailing wind (◆ Fig. 15.10a). Sand grains move up the gentle windward slope by saltation and accumulate on the leeward side forming an angle between 30° and 34° from the horizontal, which is the angle of repose of dry sand. When this angle is exceeded by accumulating sand, the slope collapses, and the sand slides down the leeward slope, coming to rest at its base. As sand moves from a dune's windward side and periodically slides down its leeward slope, the dune slowly migrates in the direction of the prevailing wind (◆ Fig. 15.10b). When preserved in the rock record, dunes help geologists determine the prevailing direction of ancient winds (◆ Fig. 15.11).

Dune Types

Four major dune types are generally recognized (*barchan, longitudinal, transverse,* and *parabolic*), although intermediate forms between the major types also exist. The size, shape, and arrangement of dunes result from the interaction of such factors as sand supply, the direction and velocity of the prevailing wind, and the amount of vegetation. While dunes are usually found in deserts, they can also occur wherever there is an abundance of sand such as along the upper parts of many beaches.

Barchan dunes are crescent-shaped dunes whose tips point downwind (◆ Fig. 15.12). They form in areas where there is a generally flat, dry surface with little vegetation, a limited supply of sand, and a nearly constant wind direction. Most barchans are small, with the largest reaching about 30 m in height. Barchans are

◆ FIGURE 15.7 A deflation hollow in Death Valley, California.

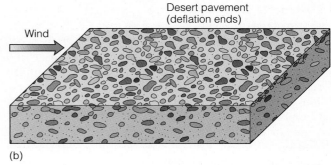

(a)

(b)

(c)

◆ FIGURE 15.8 (*a*) Desert pavement forms when deflation removes fine-grained material from the ground surface leaving behind larger-sized particles. (*b*) As deflation continues and more material is removed, the larger particles are concentrated and form a desert pavement, which protects the underlying material from additional deflation. (*c*) Desert pavement in the Mojave Desert, California. Several ventifacts can also be seen in the lower left of the photograph. (Photo courtesy of David J. Matty.)

the most mobile of the major dune types, moving at rates that can exceed 10 m per year.

Longitudinal dunes (also called *seif dunes*) are long, parallel ridges of sand aligned generally parallel to the direction of the prevailing winds; they form where the sand supply is somewhat limited (◆ Fig. 15.13). Longitudinal dunes result when winds converge from slightly different directions to produce the prevailing wind.

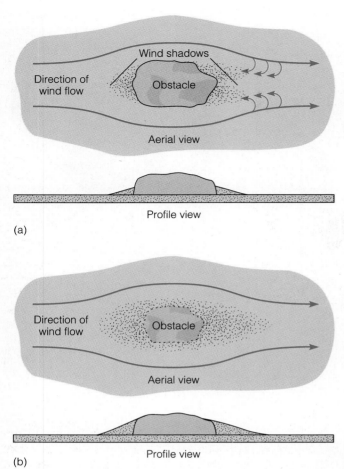

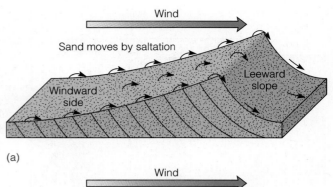

(a)

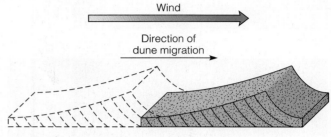

(b)

◆ FIGURE 15.10 (a) Profile view of a sand dune. (b) Dunes migrate when sand moves up the windward side and slides down the leeward slope. Such movement of the sand grains produces a series of cross beds that slope in the direction of wind movement.

◆ FIGURE 15.9 (a) When wind flows around an obstacle, two wind shadows form, one in front of the obstacle and the other behind it. Sand accumulates in both of these wind shadows. (b) The accumulating sand forms a mound that may develop into a dune.

◆ FIGURE 15.11 Cross-bedding in this sandstone in Zion National Park, Utah, helps geologists determine the prevailing direction of wind that formed these ancient sand dunes.

They range in size from about 3 m to more than 100 m high, and some stretch for more than 100 km. These dunes are especially well developed in central Australia, where they cover nearly one-fourth of the continent. They also cover extensive areas in Saudi Arabia, Egypt, and Iran.

Transverse dunes form long ridges perpendicular to the prevailing wind direction in areas where abundant sand is available and little or no vegetation exists (◆ Fig. 15.14). When viewed from the air, transverse dunes have a wavelike appearance and are therefore sometimes called sand seas. The crests of transverse dunes can be as high as 200 m, and the dunes may be as

much as 3 km wide. Some transverse dunes develop a clearly distinguishable barchan form and may separate into individual barchan dunes along the edges of the

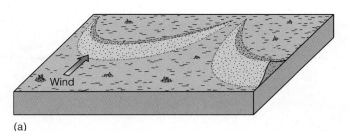
(a)

(b)

◆ FIGURE 15.12 (*a*) Barchan dunes form where there is a limited amount of sand, a nearly constant wind direction, and a generally flat, dry surface with little vegetation. The tips of barchan dunes point downwind. (*b*) Several barchan dunes west of the Salton Sea, California.

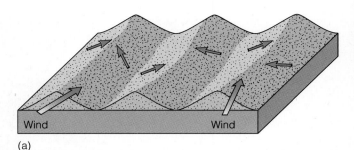
(a)

(b)

◆ FIGURE 15.13 (*a*) Longitudinal dunes form long, parallel ridges of sand aligned roughly parallel to the prevailing wind direction. They typically form where sand supplies are limited. (*b*) Aerial view of the great seif dune field near Glamis, southern California.

dune field where there is less sand. Such intermediate-form dunes are known as **barchanoid dunes** (◆ Fig. 15.15).

Parabolic dunes are most common in coastal areas with abundant sand, strong onshore winds, and a partial cover of vegetation (◆ Fig. 15.16, page 392). Although parabolic dunes have a crescent shape like barchan dunes, their tips point upwind. Parabolic dunes form when the vegetation cover is broken and deflation produces a deflation hollow or blowout. As the wind transports the sand out of the depression, it builds up on the convex downwind dune crest. The central part of the dune is excavated by the wind, while vegetation holds the ends and sides fairly well in place.

Loess

Windblown silt and clay deposits composed of angular quartz grains, feldspar, micas, and calcite are known as **loess.** The distribution of loess shows that it is derived from three main sources: deserts, Pleistocene glacial outwash deposits, and the floodplains of rivers in semiarid regions. It must be stabilized by moisture and vegetation in order to accumulate. Consequently, loess is not found in deserts, even though they provide much of its material. Because of its unconsolidated nature, loess is easily eroded, and as a result, eroded loess areas are characterized by steep cliffs and rapid lateral and headward stream erosion (◆ Fig. 15.17, page 392).

At present, loess deposits cover approximately 10% of the Earth's land surface and 30% of the United States (◆ Fig. 15.18, page 393). The most extensive and thickest loess deposits occur in northeast China where accumulations greater than 30 m are common. The extensive deserts in central Asia are the source for this loess. Other important loess deposits are on the North European Plain from Belgium eastward to the Ukraine, Central Asia, and the Pampas of Argentina. In the

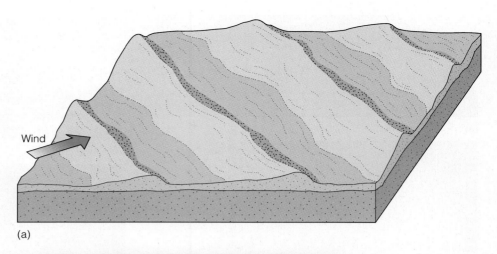

(a)

◆ FIGURE 15.14 (*a*) Transverse dunes form long ridges of sand that are perpendicular to the prevailing wind direction in areas of little or no vegetation and abundant sand. (*b*) Aerial view of transverse dunes, Great Sand Dunes National Monument, Colorado.

(b)

United States, they occur in the Great Plains, the Midwest, the Mississippi River Valley, and eastern Washington (see Perspective 14.1).

Loess-derived soils are some of the world's most fertile. It is therefore not surprising that the world's major grain-producing regions correspond to the distribution of large loess deposits such as the North European Plain, the Ukraine, and the Great Plains of the United States.

▼ AIR PRESSURE BELTS AND GLOBAL WIND PATTERNS

To understand the work of wind and the distribution of deserts, we need to consider the global pattern of air pressure belts and winds, which are responsible for the Earth's atmospheric circulation patterns. Air pressure is the density of air exerted on its surroundings (that is, its weight). When air is heated, it expands and rises, reducing its mass for a given volume and causing a decrease in air pressure. Conversely, when air is cooled, it contracts and air pressure increases. Therefore, those areas of the Earth's surface that receive the most solar radiation, such as the equatorial regions, have low air pressure, while the colder areas, such as the polar regions, have high air pressure.

Air flows from high-pressure zones to low-pressure zones. If the Earth did not rotate, winds would move in a straight line from one zone to another. Because the Earth rotates, however, winds are deflected to the right of their direction of motion (clockwise) in the Northern Hemisphere and to the left of their direction of motion

◆ FIGURE 15.15 Barchanoid dunes at White Sands National Monument, New Mexico.

(counterclockwise) in the Southern Hemisphere. Such a deflection of air between latitudinal zones resulting from the Earth's rotation is known as the **Coriolis effect.** Therefore, the combination of latitudinal pressure differences and the Coriolis effect produces a worldwide pattern of east-west–oriented wind belts (◆ Fig. 15.19, page 394).

The Earth's equatorial zone receives the most solar energy, which heats the surface air, causing it to rise. As the air rises, it cools and releases moisture that falls as rain in the equatorial region (Fig. 15.19). The rising air is now much drier as it moves northward and southward toward each pole. By the time it reaches 20° to 30° north and south latitude, the air has become cooler and denser and begins to descend. Compression of the atmosphere warms the descending air mass and produces a warm, dry, high-pressure area, providing the perfect conditions for the formation of the low-latitude

deserts of the Northern and Southern hemispheres (◆ Fig. 15.20, page 395).

▼ THE DISTRIBUTION OF DESERTS

Dry climates occur in the low and middle latitudes. In these climates, the potential loss of water by evaporation exceeds the yearly precipitation (Fig. 15.20). Dry climates cover 30% of the Earth's land surface and are subdivided into *semiarid* and *arid* regions. Semiarid regions receive more precipitation than arid regions, yet are moderately dry. Their soils are usually well developed and fertile and support a natural grass cover. Arid regions, generally described as **deserts,** are very dry; they receive less than 25 cm of rain per year, typically have poorly developed soils, and are mostly or completely devoid of vegetation.

The majority of the world's deserts are found in the dry climates of the low and middle latitudes (Fig. 15.20). In North America, most of the southwestern United States and northern Mexico are characterized by this hot, dry climate, while in South America this climate is primarily restricted to the Atacama Desert of coastal Chile and Peru. The Sahara in Northern Africa, the Arabian Desert in the Middle East, along with the majority of Pakistan and western India form the largest essentially unbroken desert environment in the Northern Hemisphere. More than 40% of Australia is desert, and most of the rest of it is semiarid. It is no wonder that it is called the "desert continent."

The remaining dry climates of the world are found in the middle and high latitudes, mostly within continental interiors in the Northern Hemisphere (Fig. 15.20). Many of these areas are dry because of their remoteness from moist maritime air and the presence of mountain ranges that produce a **rainshadow desert** (◆ Fig. 15.21, page 395). When moist marine air moves inland and meets a mountain range, it is forced upward. As it rises, it cools, forming clouds and producing precipitation that falls on the windward side of the mountains. The air that descends on the leeward side of the mountain range is much warmer and drier, producing a rainshadow desert.

Three widely separated areas are included within the mid-latitude dry climate zone (Fig. 15.20). The largest of these is the central part of Eurasia extending from just north of the Black Sea eastward to north-central China. The Gobi Desert in China is the largest desert in this region. The Great Basin area of North America, is

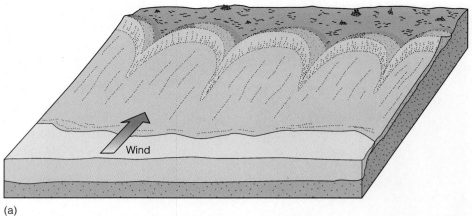

(a)

◆ FIGURE 15.16 (*a*) Parabolic dunes typically form in coastal areas where there is a partial cover of vegetation, a strong onshore wind, and abundant sand. (*b*) Parabolic dune developed along the Lake Michigan shoreline, west of St. Ignace, Michigan.

(b)

◆ FIGURE 15.17 These steep banks along the Yukon River, Yukon Territory, Canada are formed of loess.

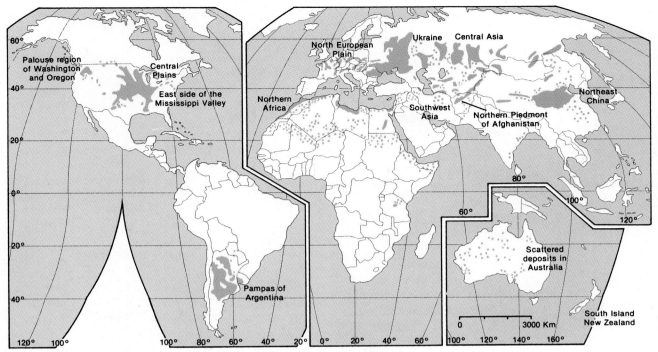

◆ FIGURE 15.18 The distribution of the Earth's major loess-covered areas.

the second largest mid-latitude dry climate zone and results from the rainshadow produced by the Sierra Nevada (see Perspective 15.2, pages 396–397). This region adjoins the southwestern deserts of the United States that formed as a result of the low-latitude subtropical high-pressure zone. The smallest of the mid-latitude dry climate areas is the Patagonian region of southern and western Argentina. Its dryness results from the rainshadow effect of the Andes.

The remainder of the world's deserts are found in the cold but dry high latitudes, such as Antarctica.

▼ CHARACTERISTICS OF DESERTS

To people who live in humid regions, deserts may seem stark and inhospitable. Instead of a landscape of rolling hills and gentle slopes with an almost continuous vegetation cover, deserts are dry, have little vegetation, and consist of nearly continuous rock exposures or sand dunes. And yet despite the great contrast between deserts and more humid areas, the same geologic processes are at work, only operating under different climatic conditions.

Temperature, Precipitation, and Vegetation

The heat and dryness of deserts are well known. Many of the deserts of the low latitudes have average summer temperatures ranging between 32° and 38°C. It is not uncommon for some low-elevation inland deserts to record daytime highs of 46° to 50°C for weeks at a time. The highest temperature ever recorded was 58°C in El Azizia, Libya, on September 13, 1922.

During the winter months when the angle of the Sun is lower and there are fewer daylight hours, daytime temperatures average between 10° and 18°C. Winter nighttime lows can be quite cold, however, with frost and freezing temperatures common in the more poleward deserts. Winter daily temperature fluctuations in low-latitude deserts are among the greatest in the world, ranging between 18° and 35°C. Temperatures have been known to fluctuate from below 0°C to more than 38°C in a single day!

The dryness of the low-latitude deserts results primarily from the year-round dominance of the subtropical high-pressure belt, while the dryness of the mid-latitude deserts is due to their isolation from moist marine

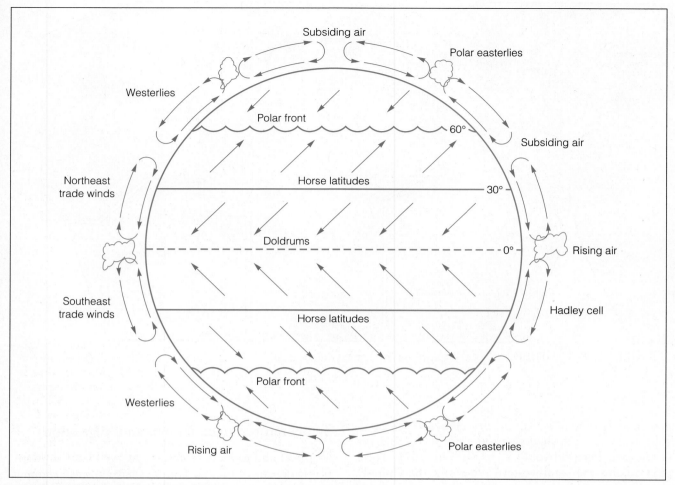

◆ FIGURE 15.19 The general circulation of the Earth's atmosphere.

winds and the rainshadow effect created by mountain ranges. The dryness of both is further accentuated by their high temperatures.

Although deserts are defined as regions receiving, on average, less than 25 cm of rain per year, the amount of rain that falls each year is very unpredictable and unreliable. It is not uncommon for an area to receive more than an entire year's average rainfall in one cloudburst and then to receive very little rain for several years. Thus, yearly rainfall averages can be quite misleading.

Deserts display a wide variety of vegetation (◆ Fig. 15.22, page 398). While the driest deserts, or those with large areas of shifting sand, are almost devoid of vegetation, most deserts support at least a sparse plant cover. Compared to humid areas, desert vegetation may appear monotonous. A closer examination, however, reveals an amazing diversity of plants that have evolved the ability to live in the near-absence of water.

Desert plants are widely spaced, typically small, and have low growth rates. Their stems and leaves are usually hard and waxy to minimize water loss by evaporation and protect the plant from sand erosion. Most plants have a widespread shallow root system to absorb the dew that forms each morning in all but the driest deserts and to help anchor the plant in what little soil there may be. In extreme cases, many plants lie dormant during particularly dry years and spring to life after the first rain shower with a beautiful profusion of flowers.

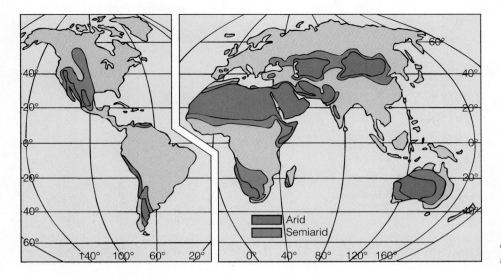

◆ FIGURE 15.20 The distribution of the Earth's arid and semiarid regions.

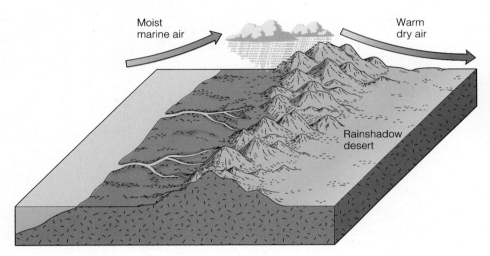

◆ FIGURE 15.21 Many deserts in the middle and high latitudes are rainshadow deserts, so named because they form on the leeward sides of mountain ranges. When moist marine air moving inland meets a mountain range, it is forced upward where it cools and forms clouds that produce rain. This rain falls on the windward side of the mountain. The air descending on the leeward side is much warmer and drier, producing a rainshadow desert.

Weathering and Soils

Mechanical weathering is dominant in desert regions. Daily temperature fluctuations and frost wedging are the primary forms of mechanical weathering (see Chapter 5). The breakdown of rocks by roots and from salt crystal growth are of minor importance. Some chemical weathering does occur, but its rate is greatly reduced by aridity and the scarcity of organic acids produced by the sparse vegetation. Most chemical weathering occurs during the winter months when there is more precipitation, particularly in the mid-latitude deserts.

An interesting feature seen in many deserts is a thin, red, brown, or black shiny coating on the surface of many rocks. This coating, called **rock varnish,** is composed of iron and manganese oxides (◆ Fig. 15.23, page 398). Because many of the varnished rocks contain little or no iron and manganese oxides, the varnish is thought to result from either windblown iron and manganese dust that settles on the ground or from the precipitated waste of microorganisms.

Desert soils, if developed, are usually thin and patchy because the limited rainfall and the resultant scarcity of vegetation reduce the efficiency of chemical weathering and hence soil formation. Furthermore, the

Perspective 15.2

DEATH VALLEY NATIONAL MONUMENT

Death Valley National Monument was established in 1933 and encompasses 7,700 km² of southeastern California and part of western Nevada (◆ Fig. 1). The hottest, driest, and lowest of the U.S. National Monuments and Parks, it receives less than 5 cm of rain per year and features normal daytime summer temperatures above 42°C. The highest temperature ever recorded was 57°C in the shade! The topographic relief in Death Valley is impressive. Telescope Peak near the southwestern border is 3,368 m high, while the lowest point in the Western Hemisphere—86 m below sea level—is less than 32 km to the east at Badwater.

Within Death Valley and its bordering mountains are excellent examples of a wide variety of desert landforms and economically valuable evaporite deposits. In addition,

numerous folds, faults, landslides, and considerable evidence of volcanic activity can be seen.

The geologic history of Death Valley is complex and still being worked out, but rocks from every geologic era can be found in the valley or the surrounding mountains. Although the geologic history of the region reaches back to the Precambrian, Death Valley itself formed less than 4 million years ago.

Death Valley formed during the Pliocene Epoch, when the Earth's crust was stretched and rifted, forming various horsts and grabens. Death Valley continues to subside along normal faults and is sinking most rapidly along its western side. This movement has been so great that more than 3,000 m of sediments are buried beneath the present valley floor.

◆ FIGURE 1 Death Valley National Monument, California, encompasses 7,700 km² of southeastern California and part of western Nevada.

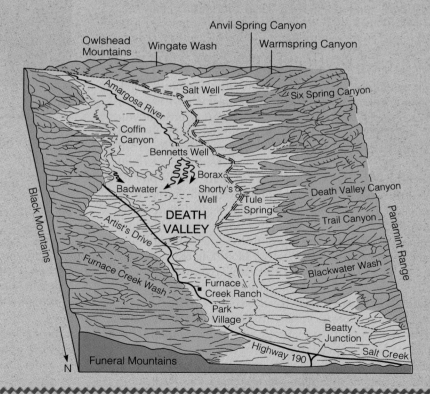

During the Pleistocene Epoch, when the climate of this region was more humid than it is today, numerous lakes spread over the valley (see Chapter 24). Lake Manly, the largest of these lakes (145 km long and 178 m deep), dried up about 10,000 years ago, when the climate became arid.

Volcanic activity has also been occurring during the last several thousand years. The most famous volcanic feature in Death Valley is Ubehebe Crater, an explosion crater that formed approximately 2,000 years ago (◆ Fig. 2).

In addition to the usual desert features, Death Valley also includes some unusual ones such as the Devil's Golf Course, a bed of solid rock salt displaying polygonal ridges and pinnacles that are almost impossible to traverse (◆ Fig. 3). The Harmony Borax Works was home to the famous 20-mule teams that hauled out countless wagons of borax (◆ Fig. 4). The borax, used for ceramic glazes, fertilizers, glass, solder, and pharmaceuticals, was leached from volcanic ash by hot groundwater and then accumulated in layers of lake sediment.

Besides the numerous geologic features that have made Death Valley famous, it is also home to more than 600 species of plants as well as numerous animal species.

◆ FIGURE 3 Devil's Golf Course consists of a layer of solid rock salt that has formed a network of polygonal ridges and pinnacles making it very difficult to traverse.

◆ FIGURE 2 Ubehebe Crater, an explosion crater, last erupted approximately 2,000 years ago.

◆ FIGURE 4 Twenty-mule teams carried borax out of Death Valley.

◆ FIGURE 15.22 Tucson, Arizona. Desert vegetation is typically sparse, widely spaced, and characterized by slow growth rates. Cacti are an excellent example of the type of vegetation that has adapted to the harsh desert environment. (Photo courtesy of B. M. C. Pape.)

◆ FIGURE 15.23 The shiny black coating on this rock exposed at Castle Valley, Utah, is rock varnish. It is composed of iron and manganese oxides.

sparseness of the vegetative cover enhances wind and water erosion of what little soil actually forms.

Mass Wasting, Streams, and Groundwater

When traveling through a desert, most people are impressed by such wind formed features as moving sand, sand dunes, and sand and dust storms. They may also notice the dry washes and dry stream beds. Because of the lack of running water, most people would

conclude that wind is the most important erosional geologic agent in deserts. They would be wrong. Running water, even though it occurs infrequently, causes most of the erosion in deserts. The dry conditions and sparse vegetation characteristic of deserts enhance water erosion. If you look closely, you will see the evidence of erosion and transportation by running water nearly everywhere except in areas covered by sand dunes.

Recall that most of a desert's annual rainfall of 25 cm or less comes in brief, heavy, localized cloudbursts. During these times, considerable erosion occurs because the ground cannot absorb all of the rainwater. With so little vegetation to hinder its flow, runoff is rapid, especially on moderately to steeply sloping surfaces, resulting in flash floods and sheetflows. Dry stream channels quickly fill with raging torrents of muddy water and mudflows, which carve out steep-sided gullies and overflow their banks. During these times, a tremendous amount of sediment is rapidly transported and deposited far downstream.

While water is the major erosive agent in deserts today, recall that it was even more important during the Pleistocene Epoch when these regions were more humid (see Chapter 24). During that time, many of the major topographic features of deserts were forming. Today that topography is being modified by wind and infrequently flowing streams.

Most desert streams are poorly integrated and flow only intermittently. Many of them never reach the sea because the water table is usually far deeper than the channels of most streams, so they cannot draw upon groundwater to replace water lost to evaporation and absorption into the ground. This type of drainage in which a stream's load is deposited within the desert is called **internal drainage** and is common in most arid regions.

While the majority of deserts have internal drainage, some deserts have permanent through-flowing streams such as the Nile and Niger rivers in Africa, the Rio Grande and Colorado rivers in the southwestern United States, and the Indus River in Asia. These streams are able to flow through the desert region because their headwaters are well outside the desert and water is plentiful enough to offset losses resulting from evaporation and infiltration. However, demands for greater amounts of water for agriculture and domestic use from the Colorado River are leading to increased salt concentrations in its lower reaches and causing political problems between the United States and Mexico.

The water table in most desert regions is below the stream channels and is only recharged for a short time after a rainfall. In deserts with through-flowing streams, the water table slopes away from the streams. The through-flowing streams help to recharge the groundwater supply and can support vegetation along their banks. Trees, which have high moisture requirements, are rare in deserts, but may occasionally occur along the banks of both ephemeral and permanent streams, where their roots can reach the higher water table.

Wind

Although running water does most of the erosional work in deserts, wind can also be an effective geologic agent capable of producing a variety of distinctive erosional and depositional features. It is very effective in transporting and depositing unconsolidated sand, silt, and dust-sized particles. Contrary to popular belief, however, most deserts are not sand-covered wastelands, but rather consist of vast areas of rock exposures. Sand-covered regions, or sandy deserts, constitute less than 25% of the world's deserts. The sand in these areas has accumulated primarily by the action of wind.

⯆ DESERT LANDFORMS

Because of differences in temperature, precipitation, and wind, as well as the underlying rocks and recent tectonic events, the landforms in arid regions vary considerably. Although wind is an important geologic

agent in deserts, many distinctive landforms are produced and modified by running water.

After an infrequent and particularly intense rainstorm, excess water that is not absorbed by the ground may accumulate in low areas and form **playa lakes** (◆ Fig. 15.24a). Such lakes are very temporary, lasting from a few hours to several months. Most of them are very shallow and have rapidly shifting boundaries as water flows in or leaves by evaporation and seepage into the ground. Furthermore, the water in playa lakes is often very saline.

When a playa lake evaporates, the dry lake bed is called a **playa** or *salt pan* and is characterized by mudcracks and precipitated salt crystals (◆ Fig. 15.24b).

◆ FIGURE 15.24 (*a*) Playa lake formed after a rainstorm filled Croneis Dry Lake, Mojave Desert, California. (*b*) Racetrack Playa, Death Valley, California. The Inyo Mountains can be seen in the background.

(a)

(b)

Salts in some playas are thick enough to be mined commercially. For example, borates have been mined in Death Valley, California, for more than a hundred years (see Perspective 15.2).

Other common features of deserts, particularly in the Basin and Range region of the United States, are *alluvial fans* and *bajadas*. **Alluvial fans** form after a cloudburst, when sediment-laden streams flowing out from the generally straight, steep mountain fronts deposit their load on the relatively flat desert floor. Because there are no valley walls to contain it, the sediment spreads out laterally, forming a gently sloping and poorly sorted fan-shaped sedimentary deposit (◆ Fig. 15.25). Alluvial fans are similar in origin and

◆ FIGURE 15.25 Aerial view of an alluvial fan, Death Valley, California.

◆ FIGURE 15.26 Coalescing alluvial fans forming a bajada at the base of the Black Mountains, Death Valley, California.

shape to deltas (see Chapter 12) but are formed entirely on land. Alluvial fans may coalesce to form a **bajada.** This broad alluvial apron typically has an undulating surface resulting from the overlap of adjacent fans (◆ Fig. 15.26).

Large alluvial fans and bajadas are frequently important sources of groundwater for domestic and agricultural use. Their outer portions are typically composed of fine-grained sediments suitable for cultivation, and their gentle slopes allow good drainage of water. Many alluvial fans and bajadas are also the sites of large towns and cities, such as San Bernardino, California, Salt Lake City, Utah, and Teheran, Iran.

Most mountains in desert regions, including those of the Basin and Range region, rise abruptly from gently sloping surfaces called pediments. **Pediments** are erosional bedrock surfaces of low relief that slope gently away from mountain bases (◆ Fig. 15.27). Most pediments are covered by a thin layer of debris or by alluvial fans or bajadas.

The origin of pediments has been the subject of much controversy. Most geologists agree that they are erosional features developed on bedrock in association with the erosion and retreat of a mountain front (Fig. 15.27a). The disagreement concerns how the erosion occurred. While not all geologists would agree, it appears

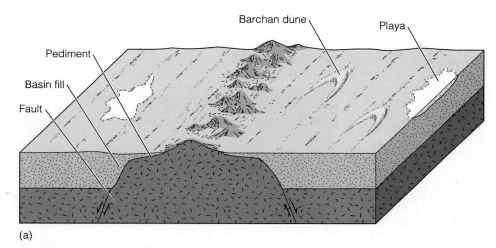

(a)

◆ FIGURE 15.27 (a) Pediments are erosional bedrock surfaces formed by erosion along a mountain front. (b) Pediment north of Mesquite, Nevada.

(b)

that pediments are produced by the combined erosional activities of lateral erosion by streams, sheet flooding, and various weathering processes along the retreating mountain front. Thus, pediments grow at the expense of the mountain, and they will continue to expand as the mountain is eroded away or partially buried.

Rising conspicuously above the flat plains of many deserts are isolated steep-sided erosional remnants called **inselbergs,** a German word meaning "island mountain" (◆ Fig. 15.28). Inselbergs have survived for a longer period of time than other mountains because of their greater resistance to weathering.

Other easily recognized erosional remnants common to arid and semiarid regions are mesas and buttes (◆ Fig. 15.29). A **mesa** is a broad, flat-topped erosional remnant bounded on all sides by steep slopes. Continued weathering and stream erosion will form isolated pillarlike structures known as **buttes.** Buttes and mesas consist of relatively easily weathered sedimentary rocks capped by nearly horizontal, resistant rocks such as sandstone, limestone, or basalt. They form when the resistant rock layer is breached, allowing rapid erosion of the less resistant underlying sediment. One of the best-known areas of mesas and buttes in the United States is Monument Valley on the Arizona-Utah border.

◆ FIGURE 15.28 Ayers Rock and the Olgas in central Australia are good examples of inselbergs. (*a*) Ayers Rock at sunset. (*b*) Aerial view of the Olgas with Ayers Rock in the background.

◆ FIGURE 15.29 (*a*) Mesas southeast of Zuni Pueblo, New Mexico. (*b*) Butte in Monument Valley, Arizona.

(a)

(a)

(b)

(b)

Chapter Summary

1. Wind can transport sediment in suspension or by saltation and surface creep as part of the bed load.
2. Wind erodes material either by abrasion or deflation. Abrasion is a near-surface effect caused by the impact of saltating sand grains. Ventifacts are common wind-abraded features.
3. Deflation is the removal of loose surface material by the wind. Deflation hollows resulting from differential erosion of surface material are common features of many deserts, as is desert pavement, which effectively protects the underlying surface from additional deflation.
4. The two major wind deposits are dunes and loess. Dunes are mounds or ridges of wind-deposited sand, whereas loess is wind-deposited silt and clay.
5. The four major dune types are barchan, longitudinal, transverse, and parabolic. The amount of sand available, the prevailing wind direction, the wind velocity, and the amount of vegetation present determine which type will form.
6. Loess is derived from deserts, Pleistocene glacial outwash deposits, and river floodplains in semiarid regions. Loess covers approximately 10% of the Earth's land surface and weathers to a rich and productive soil.
7. Deserts are very dry (less than 25 cm rain/year), have poorly developed soils, and are mostly or completely devoid of vegetation.
8. The winds of the major east-west–oriented air pressure belts resulting from rising and cooling air are deflected by the Coriolis effect. These belts help control the world's climate.
9. Dry climates are located in the low and middle latitudes where the potential loss of water by evaporation exceeds the yearly precipitation. Dry climates cover 30% of the Earth's surface and are subdivided into semiarid and arid regions.
10. The majority of the world's deserts are in the low-latitude dry climate zone between 20° and 30° north and south latitudes. Their dry climate results from a high-pressure belt of descending dry air. The remaining deserts are in the middle latitudes where their distribution is related to the rainshadow effect and in the dry polar regions.
11. Deserts are characterized by lack of precipitation and high evaporation rates. Furthermore, rainfall is unpredictable and, when it does occur, tends to be very intense and of short duration. As a consequence of such aridity, desert vegetation and animals are scarce.
12. Mechanical weathering is the dominant form of weathering in deserts. The sparse precipitation and slow rates of chemical weathering result in poorly developed soils.
13. Running water is the dominant agent of erosion in deserts and was even more important during the Pleistocene Epoch when wetter climates resulted in humid conditions.
14. Wind is an erosional agent in deserts and is very effective in transporting and depositing unconsolidated fine-grained sediments.
15. Important desert landforms include playas, which are dry lake beds; when temporarily filled with water, they form playa lakes. Alluvial fans are poorly sorted, fan-shaped sedimentary deposits that may coalesce to form bajadas.
16. Pediments are erosional bedrock surfaces of low relief gently sloping away from mountain bases. The origin of pediments is controversial, although most geologists think that they form by the combined activities of lateral erosion by streams, sheet flooding, and various weathering processes.
17. Inselbergs are isolated steep-sided erosional remnants that rise above the surrounding desert plains. Buttes and mesas are, respectively, pinnacle-like and flat-topped erosional remnants with steep sides.

Important Terms

abrasion
alluvial fan
bajada
barchan dune
barchanoid dune
butte
Coriolis effect
deflation
deflation hollow
desert

desert pavement
desertification
dry climate
dune
inselberg
internal drainage
loess
longitudinal dune
mesa

parabolic dune
pediment
playa
playa lake
rainshadow desert
rock varnish
transverse dune
ventifact
yardang

Review Questions

1. Deserts:
   a. ___ can be found in the low, middle, and high latitudes; b. ___ receive more than 25 cm of rain per year; c. ___ are mostly or completely devoid of vegetation; d. ___ answers (a) and (c); e. ___ answers (b) and (c).
2. Between what latitudes in both hemispheres do the driest deserts in the world occur?
   a. ___ 10° and 20°; b. ___ 20° and 30°; c. ___ 30° and 40°; d. ___ 40° and 60°; e. ___ 60° and 80°.
3. The Coriolis effect causes wind to be deflected:
   a. ___ to the right in the Northern Hemisphere and the left in the Southern Hemisphere; b. ___ to the left in the Northern Hemisphere and the right in the Southern Hemisphere; c. ___ only to the left for both hemispheres; d. ___ only to the right for both hemispheres; e. ___ not at all.
4. The primary process by which bed load is transported is:
   a. ___ suspension; b. ___ abrasion; c. ___ saltation; d. ___ precipitation; e. ___ answers (a) and (c).
5. Which particle size constitutes most of a wind's suspended load?
   a. ___ sand; b. ___ silt; c. ___ clay; d. ___ answers (a) and (b); e. ___ answers (b) and (c).
6. Which of the following is a feature produced by wind erosion?
   a. ___ playa; b. ___ loess; c. ___ dune; d. ___ yardang; e. ___ none of these.
7. What is the approximate angle of repose for dry sand?
   a. ___ 15°; b. ___ 25°; c. ___ 35°; d. ___ 45°; e. ___ 55°.
8. Which of the following is a crescent-shaped dune whose tips point downwind?
   a. ___ barchan; b. ___ longitudinal; c. ___ parabolic; d. ___ transverse; e. ___ barchanoid.
9. Which of the following dunes form long ridges of sand aligned roughly parallel to the direction of the prevailing wind?
   a. ___ barchan; b. ___ longitudinal; c. ___ parabolic; d. ___ transverse; e. ___ barchanoid.
10. Where are the thickest and most extensive loess deposits in the world?
    a. ___ United States; b. ___ Pampas of Argentina; c. ___ Belgium; d. ___ Ukraine; e. ___ northeast China.
11. What is the primary cause of the dryness of low-latitude deserts?
    a. ___ rainshadow effect; b. ___ isolation from moist marine winds; c. ___ dominance of the subtropical high-pressure belt; d. ___ Coriolis effect; e. ___ all of these.
12. The dominant form of weathering in deserts is _____ , desert vegetation is _____ , and soils are _____ :
    a. ___ mechanical, limited, thick; b. ___ mechanical, diverse, thin; c. ___ mechanical, limited, thin; d. ___ chemical, diverse, thick; e. ___ chemical, diverse, thin.
13. The major agent of erosion in deserts today is _____ , while during the Pleistocene Epoch it was _____ :
    a. ___ wind, running water; b. ___ wind, wind; c. ___ running water, wind; d. ___ running water, running water; e. ___ wind, glaciers.
14. The dry lake beds in many deserts are:
    a. ___ playas; b. ___ bajadas; c. ___ inselbergs; d. ___ pediments; e. ___ mesas.
15. An important source of groundwater for domestic and agricultural use is:
    a. ___ alluvial fans; b. ___ playa lakes; c. ___ bajadas; d. ___ answers (a) and (b); e. ___ answers (a) and (c).
16. What are some of the problems associated with desertification?
17. Describe how the global distribution of air pressure belts and winds operates.
18. What are the two ways that sediments are transported by wind?
19. Describe the two ways that wind erodes. How effective an erosional agent is wind?
20. What is the difference between a ventifact and yardang and how do they both form?
21. Why is desert pavement important in a desert environment?
22. How do sand dunes migrate?
23. Describe the four major dune types and the conditions necessary for their formation.
24. What is loess and why is it important?
25. What is meant by arid and semiarid climate?
26. Why are most of the world's deserts located in the low latitudes?
27. How are temperature, precipitation, and vegetation interrelated in desert environments?
28. What is the dominant form of weathering in desert regions, and why is it so effective?
29. Why are deserts characterized by internal drainage? What role does groundwater play in this type of drainage?
30. Explain the difference between a butte and a mesa, and describe how they form.

Additional Readings

Agnew, C., and A. Warren. 1990. Sand trap. *The Sciences* March/April:14–19.

Brookfield, M. E., and T. S. Ahlbrandt. 1983. *Eolian sediments and processes.* New York: Elsevier.

Dorn, R. I. 1991. Rock Varnish. *American Scientist* 79, no. 6:542–53.

Ellis, W. S. 1987. Africa's Sahel: The stricken land. *National Geographic* 172, no. 2:140–79.

Greeley, R., and J. Iversen. 1985. *Wind as a geologic process.* Cambridge, Mass.: Cambridge Univ. Press.

Hunt, C. B. 1975. *Death Valley: Geology, ecology, archaeology.* Berkeley, Calif.: Univ. of California Press.

Sheridan, D. 1981. *Desertification of the United States.* Washington, D.C.: Council on Environmental Quality.

Thomas, D. S. G., ed. 1989. *Arid zone geomorphology.* New York: Halsted.

Walker, A. S. 1982. Deserts of China. *American Scientist* 70, no. 4:366–76.

Wells, S. G., and D. R. Haragan. 1983. *Origin and evolution of deserts.* Albuquerque, N. Mex.: Univ. of New Mexico Press.

Whitney, M. A. 1985. Yardangs. *Journal of Geological Education* 33, no. 2:93–96.

CHAPTER
16

Chapter Outline

View of the Pacific shoreline at Point Reyes National Seashore, California.

Shorelines and Shoreline Processes

Prologue

In 1900, Galveston, Texas, was a busy port city of 38,000 on Galveston Island, a long, narrow barrier island a short distance from the mainland. On September 8, a hurricane swept in from the Caribbean, destroying much of the city and killing between 6,000 and 8,000 of Galveston's residents in the greatest natural disaster in U.S. history. When the hurricane struck, storm waves surged inland, eventually covering the entire island. Buildings and other structures near the shoreline were battered to pieces, and "great beams and railway ties were lifted by the [waves] and driven like battering rams into dwellings and business houses"* farther inland. Finally, after the first four shoreline blocks were destroyed, the debris piled up high enough to form a protective barrier for the rest of the city.

At about 10:00 P.M., the wind suddenly died down and soon thereafter the water began to subside. The next morning was calm and clear, but the city was in utter ruins; property damage was estimated at more than $20 million, and at least 15% of the city's population had been killed. Hurricanes had swept through Galveston before, some of them causing damage and deaths, and the residents were aware of how vulnerable the city was. The highest part of the island was only 2.7 m above mean low tide; thus storm waves could sweep across the entire island.

In order to protect the city from future hurricanes, a colossal two-part project was begun in 1902. First, a seawall 5.6 km long was constructed along the side of

. . . .
*L. W. Bates, Jr., "Galveston—A City Built upon Sand," *Scientific American* 95(1906): 64.

the city facing the shore (the south); with government assistance, the seawall was eventually extended to 16 km (◆ Fig. 16.1). The wall is 4.8 m wide at its base and 1.5 m wide at its top and has a concave face so that waves are deflected upward. Its top is just over 5 m above mean low tide, about 0.4 m above the highest water level recorded during the 1900 hurricane. A wide apron of granite riprap (a layer of large stones to prevent erosion) protects the wall on its seaward side.

The seawall was constructed to protect the city from the direct impact of waves, and it has been successful in doing so. However, the seawall alone would not prevent the city from flooding during

◆ FIGURE 16.1 Construction of this seawall to protect Galveston, Texas from storm waves began in 1902.

storms. To protect against this hazard, the second part of the project had to be completed. This entailed filling the area behind the seawall with sand and raising parts of the city to the level of the top of the wall. Filling such an area would have been a rather simple task had it not been for the streetcar lines, sewers, power lines, roadways, sidewalks, and nearly 3,000 buildings that lay in the area to be filled.

Before filling could begin, jacks were placed beneath the buildings so that they could be raised to the appropriate height and supported on stilts until fill was pumped beneath them (◆ Fig. 16.2). To raise a church estimated to weigh more than 2,700 metric tons required 700 jacks. In short, most of the city was raised anywhere from a few centimeters to as much as 3.6 m above its former level!

The last of the more than 8.5 million m³ of fill was in place on August 9, 1910. Seven years and more than $3.5 million had been invested, and subsequent events indicate that the time and expense were justified. During 1961, hurricane Carla hit the city, and although some flooding occurred and some buildings were damaged by wind, the flooding was not serious and no deaths occurred. At the west end of the seawall, where the island is unprotected, the shoreline has been eroded back about 45 m. Had the seawall not been constructed, the shoreline along Galveston would no doubt have been eroded as well.

◆ FIGURE 16.2 Some of the nearly 3,000 buildings in Galveston, Texas that were raised and supported on stilts until sand fill was pumped beneath them.

▼ INTRODUCTION

Shorelines are the areas between low tide and the highest level on land affected by storm waves. In this chapter, we are concerned mostly with ocean shorelines where processes such as waves, nearshore currents, and tides continually modify existing shoreline features. However, waves and nearshore currents are also effective geologic agents in large lakes, the shorelines of which exhibit many of the same features present along seashores. The most notable differences are that waves and nearshore currents are more energetic on seashores, and even the largest lakes lack appreciable tides.

The continents possess more than 400,000 km of shorelines. They vary from rocky, steep shorelines, such as those in Maine and much of the western United States and Canada, to those with broad sandy beaches as in eastern North America from New Jersey southward. Whatever their type, on all shorelines there is a continual interplay between the energy levels of shoreline processes and the shoreline materials.

Scientists from several disciplines have contributed to our understanding of shorelines as dynamic systems. Elected officials and city planners of coastal communities must become familiar with shoreline processes so they can develop policies that serve the public as well as protect the fragile shoreline environment. In short, the study of shorelines is not only interesting, but has many practical applications.

▼ WAVE DYNAMICS

Waves are oscillations of a water surface. They occur on all bodies of water, but are most significant in large lakes and the oceans where they serve as agents of erosion, transport, and deposition. Many of the erosional and depositional features of the world's shorelines form and are modified by the energy of incoming waves.

◆ Figure 16.3 shows a typical series of waves in deep water and the terminology applied to them. The highest part of a wave is its **crest,** whereas the low point between crests is the **trough. Wave length** is the distance between successive wave crests (or troughs), and **wave height** is the vertical distance from trough to crest. The speed at which a wave advances, generally called celerity (C), can be calculated if one knows the wave length (L) and the **wave period** (T), which is the

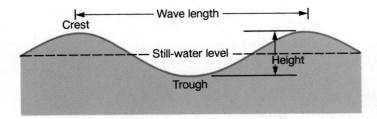

◆ FIGURE 16.3 Wave terminology.

time required for two successive wave crests (or troughs) to pass a given point:

$$C = L/T$$

The speed of wave advance (*C*) is actually a measure of the velocity of the wave form rather than a measure of the speed of the molecules of water. In fact, water waves are somewhat similar to the waves moving across a grass-covered field; the grass moves forward and back as a wave passes but has no net forward movement. Likewise, as waves move across a water surface, the water "particles" rotate in circular orbits, with little or no net movement in the direction of wave travel (◆ Fig. 16.4). They do, however, transfer energy in the direction of wave advance.

The diameters of the orbits followed by water particles in waves diminish rapidly with depth, and at a depth of about one-half wave length (*L*/2), called **wave base,** they are essentially zero (Fig. 16.4). Thus, at depths exceeding wave base, the water and sea floor, or lake floor, are unaffected by surface waves. The significance of wave base will be explored more fully in later sections.

Wave Generation

Waves can be generated by several processes including displacement of water by landslides, displacement of the sea floor by faulting, and volcanic explosions. However, most of the geologic work done on shorelines occurs from wind-generated waves. When wind blows over water, some of its energy is transferred to the water, causing the water surface to oscillate. The mechanism whereby energy is transferred from wind to water is related to the frictional drag resulting from one fluid (air) moving over another (water).

As one would expect, the harder and longer the wind blows, the larger are the waves generated. Wind velocity and duration, however, are not the only factors

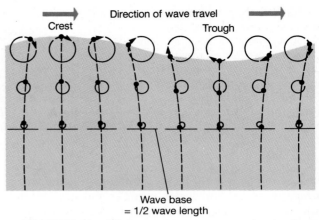

◆ FIGURE 16.4 The water in waves moves in circular orbits. The diameters of these orbits are equal to wave height at the surface, but they decrease in magnitude with depth. Wave base is the depth at which the diameters of these orbits are essentially zero.

controlling the size of waves. For example, high-velocity wind blowing over a small pond will never generate large waves regardless of how long it blows. In fact, waves occur on ponds and most lakes only while the wind is blowing; once the wind stops, the water quickly smooths out. In contrast, the surface of the ocean is always in motion, and during storms, waves with heights of 20 to more than 30 m have been recorded; the highest ever reliably measured had a height of 34 m.

The reason for the disparity between the wave sizes on ponds and lakes and on the oceans is the **fetch,** which is the distance the wind blows over a continuous water surface. The greater the fetch, the greater the size of the waves. Fetch is limited by the available water surface, so on ponds and lakes it corresponds to their length or width, depending on wind direction. A wind blowing the length of Lake Superior, for example, can generate large waves, and even larger ones develop in the oceans. To produce waves of greater length and height, more energy must be transferred from wind to water; hence large waves form beneath large storms at sea.

Shallow-Water Waves and Breakers

Waves moving out from the area of generation form swells and lose only a small amount of energy as they

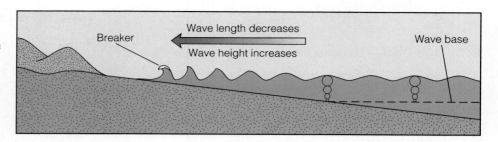

◆ FIGURE 16.5 As deep-water waves move toward shore, the orbital motion of water within them is disrupted when they reach the point at which wave base intersects the sea floor. Wave length decreases while wave height increases, causing the waves to oversteepen and eventually break.

travel across the ocean. In deep-water swells, the water surface oscillates and water particles move in orbital paths, with little net displacement of water occurring in the direction of wave advance (◆ Fig. 16.5). When these waves enter progressively shallower water, the water is displaced in the direction of wave advance (Fig. 16.5).

When deep-water waves enter shallow water, they are transformed from broad, undulating swells into sharp-crested waves. This transformation begins at a water depth of wave base; that is, it begins where wave base intersects the sea floor. At this point, the waves "feel" the bottom, and the orbital motions of water particles within waves are disrupted (Fig. 16.5). As they move further shoreward, the speed of wave advance and wave length decrease, and wave height increases. In effect, as waves enter shallower water, they become oversteepened; the wave crest advances faster than the wave form, until eventually the crest plunges forward as a **breaker** (Fig. 16.5). Breakers are commonly several times higher than deep-water waves, and when they plunge forward, their kinetic energy is expended on the shoreline. Exceptionally large waves generated during storms or by faulting, volcanic explosions, and rockfalls can cause serious flooding in coastal regions.

⯆ NEARSHORE CURRENTS

It is convenient to identify the *nearshore zone* as the area extending seaward from the shoreline to just beyond the area where waves break. It includes a *breaker zone* and a *surf zone,* which is where breaking waves rush forward onto the shore followed by seaward movement of the water as backwash. The width of the nearshore zone varies depending on the wave length of the approaching waves, because long waves break at a greater depth, and thus farther offshore, than do short waves. Two types of currents are important in the nearshore zone, *longshore currents* and *rip currents.*

Wave Refraction and Longshore Currents

Deep-water waves are characterized by long, continuous crests, but rarely are their crests parallel with the shoreline (◆ Fig. 16.6). In other words, they seldom approach a shoreline head on. Thus, one part of a wave enters shallow water where it encounters wave base and begins breaking before other parts of the same wave. As a wave begins breaking, its velocity diminishes, but the part of the wave still in deep water races ahead until it too encounters wave base. The net effect of this oblique approach is that the waves bend so that they more nearly parallel the shoreline (Fig. 16.6). Such a phenomenon is called **wave refraction.**

Even though waves are refracted, they still usually strike the shoreline at some angle, causing the water between the breaker zone and the beach to flow parallel to the shoreline. These **longshore currents,** as they are called, are long and narrow and flow in the

◆ FIGURE 16.6 Wave refraction. These oblique waves are refracted and more nearly parallel the shoreline as they enter progressively shallower water.

same general direction as the approaching waves. These currents are particularly important agents of transport and deposition in the nearshore zone.

Rip Currents

Waves carry water into the nearshore zone, so there must be a mechanism for mass transfer of water back out to sea. One way in which water moves seaward from the nearshore zone is in **rip currents,** which are narrow surface currents that flow out to sea through the breaker zone (◆ Fig. 16.7). Surfers commonly take advantage of rip currents for an easy ride out beyond the breaker zone, but such currents pose a danger to inexperienced swimmers. Some rip currents flow at several kilometers per hour, so if a swimmer is caught in one, it is useless to try to swim directly back to shore. Instead, because rip currents are narrow and usually nearly perpendicular to the shore, one can swim parallel to the shoreline for a short distance and then turn shoreward with no difficulty.

The configuration of the sea floor plays an important role in determining the location of rip currents. Rip currents commonly develop where wave heights are lower than in adjacent areas. Such differences in wave height are commonly controlled by variations in water depth. For example, if waves move over a depression, the height of the waves over the depression tends to be less than in adjacent areas.

◆ FIGURE 16.7 Suspended sediment, indicated by discolored water, being carried seaward by a rip current.

▼ SHORELINE DEPOSITION

Depositional features of shorelines include *beaches, spits, baymouth bars,* and *barrier islands.* The characteristics of beaches are determined by wave energy, and they are continually modified by waves and longshore currents. Spits and baymouth bars both result from deposition by longshore currents, but the origin of barrier islands is controversial. Rip currents play only a minor role in the configuration of shorelines, but they do transport fine-grained sediment seaward through the breaker zone.

Beaches

Beaches are the most familiar of all coastal landforms, attracting millions of visitors each year and providing the economic base for many communities. Depending on shoreline configuration and wave intensity, beaches may be discontinuous, existing only as *pocket beaches* in protected areas such as embayments, or they may be continuous for long distances (◆ Fig. 16.8).

By definition a **beach** is a deposit of unconsolidated sediment extending landward from low tide to a change in topography such as a line of sand dunes, a sea cliff, or the point where permanent vegetation begins (◆ Fig. 16.9). Typically, a beach has several component parts (Fig. 16.9) including a **backshore** that is usually dry, being covered by water only during storm waves or exceptionally high tides. The backshore consists of one or more **berms,** platforms composed of sediment deposited by waves; the berms are nearly horizontal or slope gently in a landward direction. The sloping area below the berm that is exposed to wave swash is called the **beach face.** The beach face is part of the **foreshore,** an area covered by water during high tide but exposed during low tide.

Some of the sediment on beaches is derived from weathering and wave erosion of the shoreline, but most of it is transported to the coast by streams and redistributed along the shoreline by longshore currents. **Longshore drift** is the phenomenon by which sand is transported along a shoreline by longshore currents (◆ Fig. 16.10a). As previously noted, waves usually strike beaches at some angle, causing the sand grains to move up the beach face at a similar angle; as the sand grains are carried seaward in the backwash, however, they move perpendicular to the long axis of the beach. Thus, individual sand grains move in a zigzag pattern in the direction of longshore currents. This movement is

(a)

(b)

◆ FIGURE 16.8 (*a*) A pocket beach in California.
(*b*) The Grand Strand of South Carolina, shown here at
Myrtle Beach, is 100 km of nearly continuous beach.

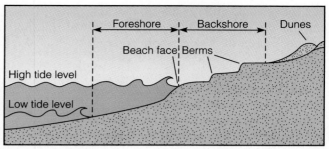

◆ FIGURE 16.9 Cross section of a typical beach showing
its component parts.

not restricted to the beach; it extends seaward to the
outer edge of the breaker zone (Fig. 16.10a).

In an attempt to widen a beach or prevent erosion,
shoreline residents often build *groins,* structures that
project seaward at right angles from the shoreline
(◆ Fig. 16.10b). They interrupt the flow of longshore
currents, causing sand to be deposited on their upcurrent side and widening the beach at that location.
However, erosion inevitably occurs on the downcurrent
side of a groin.

Although quartz is the most common mineral in
most beach sands, there are some notable exceptions.
For example, the black sand beaches of Hawaii are
composed of sand-sized basalt rock fragments, and
some Florida beaches are composed of the fragmented
calcium carbonate shells of marine organisms. In short,
beaches are composed of whatever material is available;
quartz is most abundant simply because it is available
in most areas and is the most durable and stable of the
common rock-forming minerals.

Seasonal Changes in Beaches

A beach is an area where wave energy is dissipated, so
the loose grains composing the beach are constantly
affected by wave motion. But the overall configuration
of a beach remains unchanged as long as equilibrium
conditions persist. The beach profile consisting of a
berm or berms and a beach face shown in Figure 16.9
can be thought of as a profile of equilibrium; that is, all
parts of the beach are adjusted to the prevailing
conditions of wave intensity and nearshore currents.

Tides and longshore currents affect the configuration of beaches to some degree, but by far the most
important agent modifying their equilibrium profile is
storm waves. In many areas, beach profiles change with

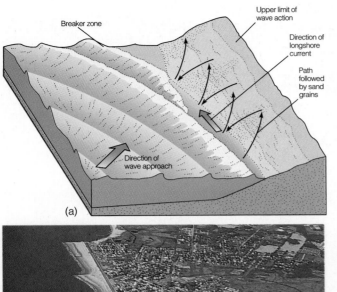

(a)

(b)

◆ FIGURE 16.10 (*a*) Longshore currents transport sediment along the shoreline between the breaker zone and the upper limit of wave action. Such sediment transport is longshore drift. (*b*) These groins at Cape May, New Jersey interrupt the flow of longshore currents so sand is trapped on their upcurrent side. On the downcurrent side of the groins sand is eroded because of continuing longshore drift.

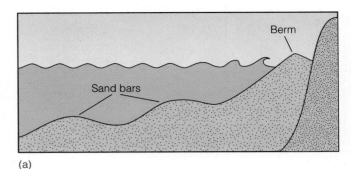

(a)

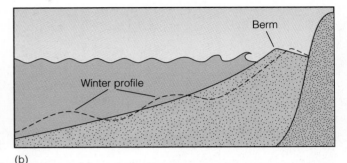

(b)

◆ FIGURE 16.11 Seasonal changes in beach profiles. (*a*) A winter beach showing offshore sand bars. (*b*) A summer beach with its wider berm and more gently sloping beach face.

the seasons; thus, we recognize *summer beaches* and *winter beaches,* each of which is adjusted to the conditions prevailing at these times (◆ Fig. 16.11). Summer beaches are generally covered with sand and are characterized by a wide berm, a gently sloping beach face, and a smooth offshore profile. Winter beaches, on the other hand, tend to be coarser grained and steeper; they have a small berm or none at all, and their offshore profiles reveal sand bars paralleling the shoreline (Fig. 16.11).

Seasonal changes in beach profiles are related to changing wave intensity. During the winter, energetic storm waves erode the sand from the beach and transport it offshore where it is stored in sand bars (Fig. 16.11). The same sand that was eroded from the beach during the winter returns the next summer when it is driven onshore by the more gentle swells that occur during that season. Thus, the volume of sand in the system remains more or less constant; it simply moves farther offshore or onshore depending on the energy of waves.

Spits and Baymouth Bars

Other than the beach itself, some of the most common depositional landforms on shorelines are *spits* and *baymouth bars,* both of which are variations of the same feature. A **spit** is simply a continuation of a beach forming a point, or "free end," that projects into a body of water, commonly a bay. A **baymouth bar** is a spit that has grown until it completely closes off a bay from the open sea (◆ Fig. 16.12).

Both spits and baymouth bars form and grow as a result of longshore drift. Where currents are weak, as in the deeper water at the opening to a bay, longshore

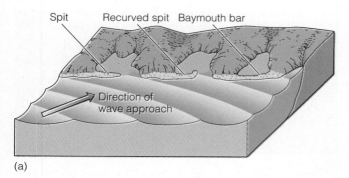

(b)

◆ FIGURE 16.12 (*a*) Spits form where longshore currents deposit sand in deeper water as at the entrance to a bay. A baymouth bar is simply a spit that has grown until it extends across the mouth of a bay. (*b*) A spit at the mouth of the Klamath River in California.

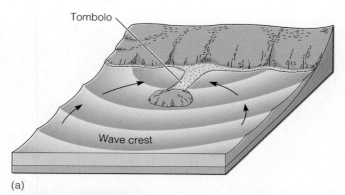

(b)

◆ FIGURE 16.13 (*a*) Origin of a tombolo. Wave refraction around an island causes longshore currents to converge and deposit a sand bar that joins the island with the mainland. (*b*) This small island is connected to the Oregon shoreline by a tombolo.

current velocity diminishes, and sediment is deposited, forming a sand bar. The free ends of many spits are curved by wave refraction or waves approaching from a different direction. Such spits are called *hooks* or *recurved spits* (Fig. 16.12).

A rarer type of spit, called a **tombolo,** extends out into the sea and connects an island to the mainland. Tombolos develop on the shoreward sides of islands as shown in ◆ Figure 16.13. Wave refraction around an island causes converging currents that turn seaward and deposit a sand bar connecting the shore with the island.

Although spits, baymouth bars, and tombolos are most commonly found on irregular seacoasts, many examples of the same features occur in large lakes. Whether along seacoasts or lakeshores, these sand

deposits present a continuing problem where bays must be kept open for pleasure boating or commercial shipping. The entrances to such bays must either be regularly dredged or protected. The most common way to protect entrances to bays is to build *jetties,* which are structures extending seaward (or lakeward) that protect the bay from deposition by longshore currents.

Barrier Islands

Long, narrow islands composed of sand and separated from the mainland by a lagoon are **barrier islands** (◆ Fig. 16.14). On their seaward margins, barrier islands are smoothed by waves, but their lagoon sides are irregular. During large storms, waves completely overtop these islands and deposit lobes of sand in the

◆ FIGURE 16.14 This chain of barrier islands comprises the Outer Banks of North Carolina. Cape Hatteras, near the center of the photo, juts the furthest out into the Atlantic.

lagoon. Once deposited, these lobes are modified only slightly because they are protected from further wave action. Windblown sand dunes are common on barrier islands and are generally the highest part of these islands.

The origin of barrier islands has been long debated and is still not completely resolved. It is known that they form on gently sloping continental shelves with abundant sand in areas where both tidal fluctuations and wave energy levels are low. Although barrier islands occur in many areas, most of them are along the east coast of the United States from New York to Florida and along the U.S. Gulf Coast. According to one model, barrier islands formed as spits that became detached from the land, while another model proposes that they formed as beach ridges on coasts that subsequently subsided.

Because sea level is currently rising, most barrier islands are migrating in a landward direction. Such migration is a natural consequence of the evolution of these islands, but it is a problem for the island residents and communities. Barrier islands generally migrate rather slowly, but the rates for many are rapid enough to cause shoreline problems (see Perspective 16.1).

The Nearshore Sediment Budget

We can think of the gains and losses of sediment in the nearshore zone in terms of a budget. If a nearshore system has a balanced budget, sediment is supplied to it as fast as it is removed, and the volume of sediment remains more or less constant, although sand may shift offshore and onshore with the changing seasons (Fig. 16.11). A positive budget means gains exceed losses, whereas a negative budget results when losses exceed gains. If a negative budget prevails long enough, a nearshore system is depleted and beaches may completely disappear.

Erosion of sea cliffs provides some sediment to beaches, but in most areas probably no more than 5 to 10% of the total sediment supply is derived from this source. There are exceptions, however. For example, almost all the sediment on the beaches of Maine is derived from the erosion of shoreline rocks. Most of the sediment on typical beaches is transported to the shoreline by streams and then redistributed along the shoreline by longshore drift. Thus, longshore drift also plays a role in the nearshore sediment budget because it continually moves sediment into and away from beach systems.

The primary ways in which a nearshore system loses sediment include offshore transport, wind, and deposition in submarine canyons. Offshore transport mostly involves fine-grained sediment that is carried seaward where it eventually settles in deeper water. Wind is an important process because it removes sand from beaches and blows it inland where it commonly piles up as sand dunes.

If the heads of submarine canyons are nearshore, huge quantities of sand are funneled into them and deposited in deeper water. La Jolla and Scripps submarine canyons off the coast of southern California funnel off an estimated 2 million m^3 of sand each year. In most areas, however, submarine canyons are too far offshore to interrupt the flow of sand in the nearshore zone.

It should be apparent from the preceding discussion that if a nearshore system is in equilibrium, its incoming supply of sediment exactly offsets its losses. Such a delicate balance tends to continue unless the system is somehow disrupted. One common way in which this balance is affected is the construction of dams across the streams supplying sand. The sediment contribution from a stream is proportional to its drainage area. Once dams have been built, all sediment from the upper reaches of the drainage systems is trapped in reservoirs and thus cannot reach the shoreline.

Perspective 16.1

RISING SEA LEVEL AND COASTAL MANAGEMENT

•••••••••• ▨ ••••••••••

Shorelines in the United States are eroding as sea level rises. According to one study, 54% of U.S. shorelines are eroding at rates ranging from millimeters per year to more than 10 m in a few areas (◆Fig. 1). Many other areas of the world are experiencing shoreline problems as well.

During the last century, sea level rose about 12 cm worldwide, and all indications are that it will continue to rise. The absolute rate of sea level rise in a particular shoreline region depends on two factors. The first is the volume of water in the ocean basins, which is increasing as

a result of the melting of glacial ice and the thermal expansion of near-surface seawater. Many scientists think that sea level will continue to rise as a consequence of global warming resulting from concentrations of greenhouse gases in the atmosphere.

The second factor controlling sea level is the rate of uplift or subsidence of a coastal area. In some areas, uplift is occurring so fast that sea level is actually falling with respect to the land. In other areas, however, sea level is rising while the coastal region is simultaneously subsiding, resulting in a net change in sea level of as much as 30 cm

◆ FIGURE 1 Shoreline erosion in the United States. No data are available for shoreline areas left uncolored.

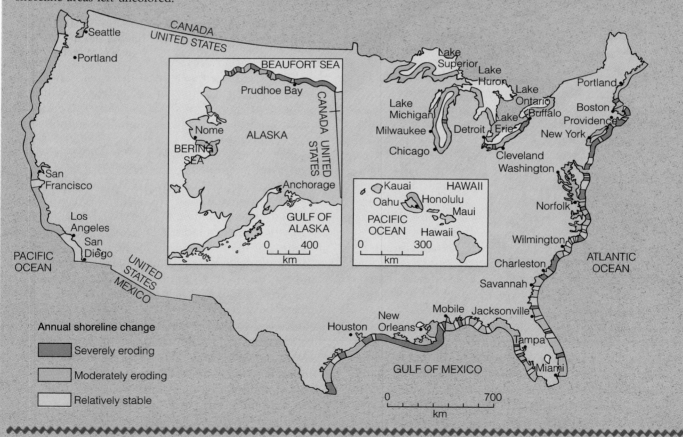

per century. Perhaps such a "slow" rate of sea level change seems insignificant; after all it amounts to only a few millimeters per year. However, in gently sloping coastal areas, as in the eastern United States from New Jersey southward, even a slight rise in sea level would eventually have widespread effects.

Many of the nearly 300 barrier islands along the east and Gulf coasts of the United States are migrating landward as a natural consequence of rising sea level. During storms, their beaches are eroded by large waves that carry the beach sand over the islands and into their lagoons. Thus, sand is removed from the seaward sides of barrier islands and deposited on their landward sides, resulting in a gradual landward shift of the entire island

complex (◆ Fig. 2). During the last 120 years, Hatteras Island, North Carolina, has migrated nearly 500 m landward so that Cape Hatteras lighthouse, which was 460 m from the shoreline when it was built in 1870, now stands on a promontory in the Atlantic Ocean.

Landward migration of barrier islands would pose few problems if it were not for the numerous communities, resorts, and vacation homes located on them. Moreover, barrier islands are not the only threatened areas. For example, Louisiana's coastal wetlands, an important wildlife habitat and seafood-producing area, are currently being lost at a rate of about 90 km² per year. Much of this loss results from sediment compaction, but sea level rise exacerbates the problem.

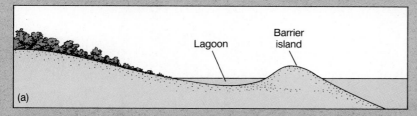

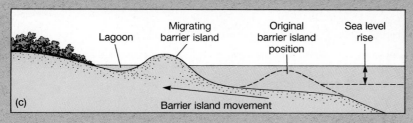

◆ FIGURE 2 Rising sea level and the landward migration of barrier islands. (*a*) Barrier island before landward migration in response to rising sea level. (*b*) Landward movement occurs when storm waves wash sand from the seaward side of the islands and deposit it in the lagoon. (*c*) Over time, the entire complex migrates landward.

continued on next page

Rising sea level also directly threatens many beaches upon which communities depend for revenue. The beach at Miami Beach, Florida, for example, was disappearing at an alarming rate until the Army Corps of Engineers began replacing the eroded beach sand (◆ Fig. 3). The problem is even more serious in other countries. A rise in sea level of only 2 m would inundate large areas of the east and Gulf coasts, but would cover 20% of the entire country of Bangladesh. Other problems associated with sea level rise include increased coastal flooding during storms and saltwater incursions that may threaten groundwater supplies (see Chapter 13).

Because nothing can be done to prevent sea level from rising, engineers and scientists must examine what can be done to prevent or minimize the effects of shoreline erosion. At present, only a few viable options exist. One is to put strict controls on coastal development. North Carolina, for example, permits large structures to be sited no closer to the shoreline than 60 times the annual erosion rate. Although a growing awareness of shoreline processes has resulted in similar legislation elsewhere, some states have virtually no restrictions on coastal development.

Regulating coastal development is commendable, but it has no impact on existing structures and coastal communities. A general retreat from the shoreline may be possible, but expensive, for individual dwellings and small communities, but it is impractical for large population centers. Such communities as Atlantic City, New Jersey, Miami Beach, Florida, and Galveston, Texas, have adopted one of two strategies to combat coastal erosion. One is to build protective barriers such as seawalls. Seawalls, such as the one at Galveston, Texas (see the Prologue), can be effective, but they are tremendously expensive to construct and maintain. For example, more than $50 million has been spent to replenish the beach and build a seawall at Ocean City, Maryland in just the last 5 years. Furthermore, they retard erosion only in the area directly behind the seawall; recall that Galveston Island west of the seawall has been eroded back about 45 m.

Another option, adopted by both Atlantic City, New Jersey, and Miami Beach, Florida, is to pump sand onto the beaches to replace that lost to erosion (Fig. 3). This, too, is expensive as the sand must be replenished periodically because erosion is a continuing process. In

◆ FIGURE 3 The beach at Miami Beach, Florida, before (right) and after the U.S. Army Corps of Engineers' beach nourishment project.

many areas, groins are constructed to preserve beaches, but unless additional sand is artificially supplied to the beaches, longshore currents invariably erode sand from the downcurrent sides of the groins.

Rising sea level has already had a significant economic impact, and all options for dealing with this phenomenon are expensive. Fortifying the shoreline with seawalls, groins, and other structures is initially expensive, requires constant maintenance, and in the long run will be ineffective if sea level continues to rise. A general retreat from the shoreline is simply impractical for most coastal communities. Perhaps the best option is to replace sand lost to erosion by pumping it from elsewhere, usually farther offshore. In some areas, however, little can be done to offset the effects of rising sea level.

In addition to rising sea level, many coastal communities are threatened by storm waves, especially from hurricanes, vast storms with winds that may exceed 300 km/hr. When these storms sweep across coastal areas, the intense wind can cause considerable damage, but most of the damages and about 90% of all hurricane fatalities are caused by coastal flooding. In fact, of the nearly $2 billion paid out by the federal government's National Flood Insurance Program since 1974, most has gone to owners of beachfront homes.

Flooding during hurricanes is caused by large storm-generated waves being driven onshore and by intense rainfall, in some cases more than 60 cm in 24 hours. In addition, as the storm moves over the ocean, low atmospheric pressure beneath the eye of the storm causes the ocean surface to bulge upward as much as 0.5m. When the eye of the storm reaches the shoreline, the bulge, coupled with wind-driven waves, piles up in a storm surge that can rise several meters above normal high tide and inundate areas several kilometers inland.

Several coastal areas in the United States have been devastated by storm surges, including Galveston, Texas, in 1900 (see the Prologue) and Charleston, South Carolina, in 1989 (◆ Fig. 4). One of the greatest natural disasters of the twentieth century occurred in 1970 when a storm surge estimated at 8 to 10 m high flooded the low-lying coastal areas of Bangladesh, drowning 300,000 people. Since 1970, coastal Bangladesh has been flooded several more times, the most recent and most tragic being on April 30, 1991, when more than 100,000 people were drowned.

◆ FIGURE 4 Charleston, South Carolina was flooded by a storm surge produced by Hurricane Hugo on September 22, 1989.

⩒ SHORELINE EROSION

Along seacoasts where erosion rather than deposition predominates, beaches are lacking or poorly developed and a sea cliff commonly develops (◆ Fig. 16.15). Sea cliffs are erosional features frequently pounded by waves, especially during storms: the cliff face retreats landward as a consequence of *corrosion, hydraulic action,* and *abrasion,* the same processes that account for erosion by running water (see Chapter 12). *Corrosion* is an erosional process involving the wearing away of rock by chemical processes, especially the solvent action of seawater. The force of the water itself, called *hydraulic action,* is a particularly effective erosional process. Waves exert tremendous pressure on shorelines by direct impact, but are most effective on sea cliffs composed of unconsolidated sediment or rocks that are highly fractured. *Abrasion* is an erosional process involving the grinding action of rocks and sand carried by waves.

Wave-Cut Platforms and Associated Landforms

The rate at which sea cliffs are eroded and retreat in a landward direction depends on wave intensity and the resistance of the coastal rocks or sediments. Most sea cliff retreat occurs during storms and, as one would expect, occurs most rapidly in sea cliffs composed of

unconsolidated sediment. For example, a sea cliff composed of unconsolidated glacial drift on Cape Cod, Massachusetts, retreats as much as 30 m per century, and some parts of the White Cliffs of Dover in Great Britain are retreating at a rate of more than 100 m per century. By comparison, sea cliffs consisting of dense igneous or metamorphic rocks may retreat at negligible rates.

Sea cliffs retreat mostly as a consequence of hydraulic action and abrasion at their bases. As a sea cliff is undercut by such erosion, the upper part is left unsupported and susceptible to mass wasting processes. Thus, sea cliffs retreat little by little, and as they do, they leave a beveled surface called a **wave-cut platform** that slopes gently in a seaward direction (◆ Fig. 16.16). Broad wave-cut platforms exist in many areas, but invariably the water over them is shallow because the abrasive planing action of waves is only effective to a depth of about 10 m.

Wave-cut platforms are surfaces of sediment transport. The sediment eroded from sea cliffs is transported seaward until it reaches deeper water at the edge of the wave-cut platform. There it is deposited and forms a *wave-built platform,* which is a seaward extension of the wave-cut platform (Fig. 16.16).

Sea cliffs do not retreat uniformly, however, because some of the materials of which they are composed are more resistant to erosion than others. **Headlands** are seaward-projecting parts of the shoreline that are eroded on both sides due to wave refraction (◆ Fig. 16.17). *Sea caves* may form on opposite sides of a headland, and if these join, they form a *sea arch* (◆ Fig. 16.18a and b). Continued erosion generally causes the span of an arch to collapse, yielding isolated *sea stacks* on wave-cut

◆ FIGURE 16.15 On shorelines where erosion rather than deposition predominates, a sea cliff develops. Wave erosion of sea cliffs causes them to migrate landward and leave a beveled surface—a wave-cut platform.

◆ FIGURE 16.16 Wave erosion of a sea cliff produces a gently sloping surface called a wave-cut platform. Deposition at the seaward margin of the wave-cut platform forms a wave-built platform.

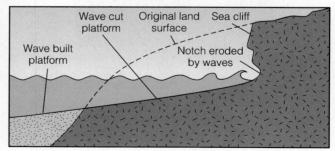

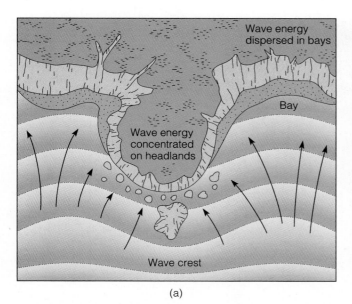

(a)

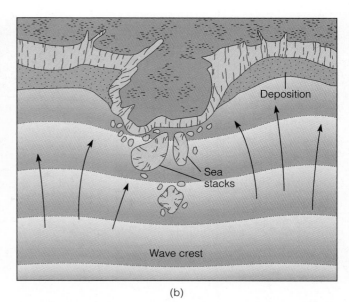

(b)

◆ FIGURE 16.17 (*a*) Wave refraction acts to straighten shorelines by concentrating wave energy on headlands. (*b*) The same shoreline after extensive erosion of the headlands and deposition in the bays.

platforms (◆ Fig. 16.18c). In the long run, shoreline processes tend to straighten an initially irregular shoreline. They do so because wave refraction causes more wave energy to be expended on headlands and less on embayments. Thus, headlands become eroded, and some of the sediment yielded by erosion is deposited in the embayments. The net effect of these processes is to straighten the shoreline.

⩛ TYPES OF COASTS

Coasts can be classified in different ways, but none of them are completely satisfactory because of variations in the factors controlling coastal development and variations in the composition and configuration of coasts. Rather than attempt to categorize all coasts, we shall simply note that two types of coasts have already been discussed, those dominated by deposition and those dominated by erosion, and shall look further at the changing relationships between coasts and sea level.

Depositional coasts, such as the U.S. Gulf Coast, are characterized by an abundance of detrital sediment and the presence of such depositional landforms as deltas and barrier islands. Erosional coasts are generally steep and irregular and typically lack well-developed beaches except in protected areas. They are further character-

ized by erosional features such as sea cliffs, wave-cut platforms, and sea stacks. Many of the beaches along the west coast of North America fall into this category.

The following section examines coasts in terms of their changing relationships to sea level. But note that while some coasts, such as those in southern California, are described as emergent (uplifted), these same coasts may be erosional as well. In other words, coasts commonly possess features allowing them to be classified in several ways.

Submergent and Emergent Coasts

If sea level rises with respect to the land or the land subsides, coastal regions are flooded and said to be **submergent** or *drowned* (◆ Fig. 16.19). Much of the east coast of North America from Maine southward through South Carolina was flooded during the post-Pleistocene rise in sea level, so that it is now an extremely irregular coast. Recall that during the expansion of glaciers during the Pleistocene, sea level was as much as 130 m lower than at present, and that streams eroded their valleys more deeply as they adjusted to a lower base level. When sea level rose, the lower ends of these valleys were drowned, forming *estuaries* such as Delaware and Chesapeake bays (Fig. 16.19). Estuaries

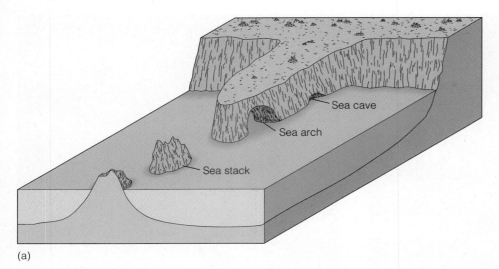

(a)

(b)

(c)

◆ FIGURE 16.18 (*a*) Erosion of a headland and the origin of sea caves, sea arches, and sea stacks. (*b*) This sea stack in Australia has an arch developed in it. (*c*) Sea stacks south of La Push, Washington.

are the seaward ends of river valleys where seawater and freshwater mix. The divides between adjacent drainage systems on submergent coasts project seaward as broad headlands or a line of islands.

Submerged coasts also occur at higher latitudes where Pleistocene glaciers flowed into the sea. When sea level rose, the lower ends of the glacial troughs were drowned, forming fiords (Fig. 14.16).

Emergent coasts are found where the land has risen with respect to sea level (◆ Fig. 16.20). Emergence can occur when water is withdrawn from the oceans, as occurred during the Pleistocene expansion of glaciers. At present, however, coasts are emerging as a consequence of isostasy or tectonism. In northeastern

Canada and the Scandinavian countries, for example, the coasts are irregular because isostatic rebound is elevating formerly glaciated terrain from beneath the sea.

Coasts rising in response to tectonism, on the other hand, tend to be straight because the sea-floor topography being exposed as uplift proceeds is smooth. The west coasts of North and South America are rising as a consequence of plate tectonics. Distinctive features of such coasts are **marine terraces** (◆ Fig. 16.21), which are old wave-cut platforms now elevated above sea level. Uplift in such areas appears to be episodic rather than continuous, as indicated by the multiple levels of terraces in some areas. In southern California, for

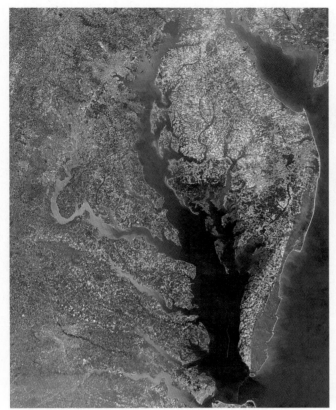

◆ FIGURE 16.19 Chesapeake Bay is a large estuary. It formed when the east coast of the United States was flooded as sea level rose following the Pleistocene Epoch.

◆ FIGURE 16.20 An emergent coast in California. Such coasts are characterized by a sea cliff, and they tend to be straighter than submergent coasts. (Photo courtesy of Jerry Westby.)

◆ FIGURE 16.21 This gently sloping surface in Ireland is a marine terrace.

example, several terrace levels are present, each of which probably represents a period of tectonic stability followed by uplift. The highest of these terraces is now about 425 m above sea level.

⩔ TIDES

On seacoasts the surface of the ocean rises and falls once or twice daily in response to the gravitational attraction of the Moon and Sun. Such regular fluctuations in the sea's surface are **tides.** Two high tides and two low tides occur daily in most areas as sea level rises and falls anywhere from a few centimeters to more than 15 m (◆ Fig. 16.22). During rising or *flood tide,* more and more of the nearshore area is flooded until high tide is reached. *Ebb tide* occurs when currents flow seaward during a decrease in the height of the tide.

Both the Moon and the Sun have sufficient gravitational attraction to exert tide-generating forces strong enough to deform the solid body of the Earth, but they have a much greater influence on the oceans. The Sun is 27 million times more massive than the Moon, but it is 390 times as far from the Earth, and its tide-generating force is only 46% as strong as that of the Moon. Accordingly, the tides are dominated by the Moon, but the Sun does play an important role in generating tides as well.

If we consider only the Moon acting on a spherical, water-covered Earth, the tide-generating forces produce two bulges on the ocean surface (◆ Fig. 16.23a). One bulge extends toward the Moon because it is on

(a)

(b)

◆ FIGURE 16.22 (a) Low tide and (b) high tide.

(b) Spring tides

(c) Neap tides

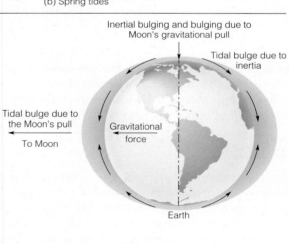
(a) Tidal forces

◆ FIGURE 16.23 (a) The tides are caused by the gravitational pull of the Moon and, to a lesser degree, the Sun. The Earth-Moon-Sun alignments at the times of the (b) spring and (c) neap tides are shown.

the side of the Earth where the Moon's gravitational attraction is greatest. The other bulge occurs on the opposite side of the Earth, where the Moon's gravitational attraction is least. These two bulges always point toward and away from the Moon, (Fig. 16.23a), so as the Earth rotates and the Moon's position changes, an observer at a particular shoreline location experiences the rhythmic rise and fall of tides twice daily. The heights of two successive high tides may vary depending on the Moon's inclination with respect to the equator.

The Moon revolves around the Earth every 28 days, so its position with respect to any latitude changes slightly each day. That is, as the Moon moves in its orbit and the Earth rotates on its axis, it takes the Moon 50 minutes longer each day to return to the same position it was in the previous day. Thus, an observer would experience a high tide at 1:00 P.M. on one day, for example, and at 1:50 P.M. on the following day.

Tides are also complicated by the combined effects of the Moon and the Sun. Even though the Sun's tide-generating force is weaker than the Moon's, when the Moon and Sun are aligned every two weeks, their forces are added together and generate *spring tides,* which are about 20% higher than average tides (◆ Fig. 16.23b). When the Moon and Sun are at right angles to one another, also at two-week intervals, the Sun's tide-generating force cancels some of that of the Moon, and *neap tides* about 20% lower than average occur (◆ Fig. 16.23c).

Tidal ranges are also affected by shoreline configuration. Broad, gently sloping continental shelves as in the Gulf of Mexico have low tidal ranges, whereas steep, irregular shorelines experience a much greater rise and fall of tides. Tidal ranges are greatest in some narrow, funnel-shaped bays and inlets. For example, in the Bay of Fundy in Nova Scotia a tidal range of 16.5 m has been recorded, and ranges greater than 10 m occur in several other areas.

Tides have an important impact on shorelines because the area of wave attack constantly shifts onshore and offshore as the tides rise and fall. Tidal currents themselves, however, have little modifying effect on shorelines, except in narrow passages where tidal current velocity is great enough to erode and transport sediment. Indeed, if it were not for strong tidal currents, some passageways would be blocked by sediments deposited by longshore currents.

Chapter Summary

1. Shorelines are continually modified by the energy of waves and longshore currents and, to a lesser degree, by tidal currents.
2. Waves are oscillations on water surfaces that transmit energy in the direction of wave movement. Surface waves affect the water and sea floor only to wave base, which is equal to one-half the wave length.
3. Little or no net forward motion of water occurs in waves in the open sea. When waves enter shallow water, they are transformed into waves in which water moves in the direction of wave advance.
4. Wind-generated waves, especially storm waves, are responsible for most geologic work on shorelines, but waves can also be generated by faulting, volcanic explosions, and rockfalls.
5. Breakers form where waves enter shallow water and the orbital motion of water particles is disrupted. The waves become oversteepened and plunge forward onto the shoreline, thus expending their kinetic energy.
6. Waves approaching a shoreline at an angle generate a longshore current. Such currents are capable of considerable erosion, transport, and deposition.
7. Rip currents are narrow surface currents that carry water from the nearshore zone seaward through the breaker zone.
8. Beaches are the most common shoreline depositional features. They are continually modified by nearshore processes, and their profiles generally exhibit seasonal changes.
9. Spits, baymouth bars, and tombolos all form and grow as a consequence of longshore current transport and deposition.
10. Barrier islands are nearshore sediment deposits of uncertain origin. They parallel the mainland but are separated from it by a lagoon.
11. The volume of sediment in a nearshore system remains rather constant unless the system is somehow disrupted as when dams are built across the streams supplying sand to the system.
12. Many shorelines are characterized by erosion rather than deposition. Such shorelines have sea cliffs and wave-cut platforms. Other features commonly present include sea caves, sea arches, and sea stacks.
13. Submergent and emergent coasts are defined on the basis of their relationships to changes in sea level.
14. The gravitational attraction of the Moon and Sun causes the ocean surface to rise and fall as tides twice daily in most shoreline areas. Most tidal currents have little effect on shorelines.

Important Terms

backshore
barrier island
baymouth bar
beach
beach face
berm
breaker
crest (wave)
emergent coast
fetch

foreshore
headland
longshore current
longshore drift
marine terrace
rip current
shoreline
spit
submergent coast

tide
tombolo
trough (wave)
wave base
wave-cut platform
wave height
wave length
wave period
wave refraction

Review Questions

1. Which of the following is not a depositional landform?
   a. ___ spit; b. ___ tombolo; c. ___ baymouth bar;
   d. ___ beach; e. ___ sea stack.
2. The speed at which a wave form advances over a water surface is:
   a. ___ celerity; b. ___ wave length; c. ___ refraction;
   d. ___ wave base; e. ___ fetch.
3. Wave base is:
   a. ___ the distance offshore that waves break; b. ___ the width of a longshore current; c. ___ the depth at which the orbital motion in surface waves dies out; d. ___ the distance wind blows over a water surface; e. ___ the height of storm waves.
4. Waves approaching a shoreline obliquely generate:
   a. ___ flood tides; b. ___ longshore currents; c. ___ tidal currents; d. ___ berms; e. ___ marine terraces.
5. Most beach sand is composed of what mineral?
   a. ___ basalt; b. ___ calcite; c. ___ gravel; d. ___ quartz;
   e. ___ feldspar.
6. Erosion of a sea cliff produces a gently sloping surface called a(an):
   a. ___ submergent coast; b. ___ wave-cut platform;
   c. ___ beach; d. ___ backshore; e. ___ emergent coast.
7. Islands composed of sand and separated from the mainland by a lagoon are:
   a. ___ barrier islands; b. ___ atolls; c. ___ baymouth bars;
   d. ___ sea stacks; e. ___ sea arches.
8. The force of waves impacting on shorelines is:
   a. ___ corrosion; b. ___ wave oscillation; c. ___ hydraulic action; d. ___ terracing; e. ___ translation.
9. The distance the wind blows over a water surface is the:
   a. ___ fetch; b. ___ berm; c. ___ spit; d. ___ wave period;
   e. ___ wave trough.
10. In deep-water waves, the water moves in orbital paths but with little net movement in the direction of wave advance. Such waves are:

    a. ___ breakers; b. ___ refracted waves; c. ___ swells;
    d. ___ longshore drift waves;e. ___ rip currents.
11. The bending of waves so that they more nearly parallel the shoreline is wave ___.
    a. ___ translation; b. ___ oscillation; c. ___ deflection;
    d. ___ refraction; e. ___ reflection.
12. The excess water in the nearshore zone returns to the open sea by:
    a. ___ tombolos; b. ___ longshore currents; c. ___ wave refraction; d. ___ emergence; e. ___ rip currents.
13. A sand deposit extending into the mouth of a bay is a:
    a. ___ headland; b. ___ beach; c. ___ spit; d. ___ wave-cut platform; e. ___ sea stack.
14. Although there are exceptions, most beaches receive most of their sediment from:
    a. ___ wave erosion of sea cliffs; b. ___ erosion of offshore reefs; c. ___ streams; d. ___ breakers; e. ___ coastal submergence.
15. Erosional remnants of a shoreline now rising above a wave-cut platform are:
    a. ___ barrier islands; b. ___ sea stacks; c. ___ beaches;
    d. ___ marine terraces; e. ___ spits.
16. Which of the following is a distinctive feature of emergent coasts?
    a. ___ marine terraces; b. ___ estuaries; c. ___ drowned river valleys; d. ___ very high tidal range; e. ___ fiords.
17. How do deep- and shallow-water waves differ?
18. What is wave base and how does it affect waves as they enter shallow water?
19. Explain how a longshore current is generated.
20. What is longshore drift?
21. Sketch a north-south shoreline along which several groins have been constructed. Assume that waves approach from the northwest.
22. Explain why quartz is the most common mineral composing beach sands.

23. Sketch the profiles of a summer beach and a winter beach, and explain why they differ.
24. How does a tombolo form?
25. Explain the concept of a nearshore sediment budget.
26. How does a wave-cut platform develop?

27. Explain how an initially irregular shoreline is straightened. A sketch may be helpful.
28. Why does an observer at a shoreline experience two high and two low tides each day?
29. What are the characteristics of a submergent coast?

Additional Readings

Bird, E. C. F. 1984. *Coasts: An introduction to coastal geomorphology.* New York: Blackwell.

Bird, E. C. F., and M. L. Schwartz. 1985. *The world's coastline.* New York: Van Nostrand Reinhold.

Flanagan, R. 1993. Beaches on the brink. *Earth* 2, no. 6:24–33.

Fox, W. T. 1983. *At the sea's edge.* Englewood Cliffs, N.J.: Prentice-Hall.

Garrett, C. and L. R. M. Maas. 1993. Tides and their effects. *Oceanus* 36, no. 1:27–37.

Hecht, J. 1988. America in peril from the sea. *New Scientist* 118:54–59.

Komar, P. D. 1976. *Beach processes and sedimentation.* Englewood Cliffs, N.J.: Prentice-Hall.

_____. 1983. *CRC handbook of coastal processes and erosion.* Boca Raton, Fla.: CRC Press.

Pethick, J. 1984. *An introduction to coastal geomorphology.* London: Edward Arnold.

Schneider, S. H. 1990. *Global warming: Are we entering the greenhouse century?* San Francisco: Sierra Club Books.

Snead, R. 1982. *Coastal landforms and surface features.* Stroudsburg, Pa.: Hutchinson Ross.

Walden, D. 1990. Raising Galveston. *American Heritage of Invention & Technology* 5:8–18.

Williams, S. J., K. Dodd, and K. K. Gohn. 1990. Coasts in crisis. *U.S. Geological Survey Circular 1075.*

CHAPTER
17

Chapter Outline

Massive cross-bedded sandstones of the Jurassic-aged Navajo formation as viewed from Emerald Pool Trail, Zion National Park, Utah.

Geologic Time: Concepts and Principles

Prologue

What is time? We seem obsessed with it, and organize our lives around it with the help of clocks, calendars, and appointment books. Yet most of us feel we don't have enough of it—we are always running "behind" or "out of time." According to biologists and psychologists, children less than two years old and animals exist in a "timeless present," where there is no past or future. They have no conscious concept of time. Some scientists believe that our early ancestors may also have lived in a state of timelessness with little or no perception of a past or future. According to Buddhist, Taoist, and Mayan beliefs, time is circular, and like a circle, all things are destined to return to where they once were. Thus, in these belief systems, there is no beginning or end, but rather a cyclicity to everything.

In some respects, time is defined by the methods used to measure it. Many prehistoric monuments are oriented to detect the summer solstice. Sundials were used to divide the day into measurable units. As civilization advanced, mechanical devices were invented to measure time, the earliest being the water clock, first used by the ancient Egyptians and further developed by the Greeks and Romans. The pendulum clock was invented in the seventeenth century and provided the most accurate timekeeping for the next two and a half centuries.

Today the quartz watch is the most popular timepiece. Powered by a battery, a quartz crystal vibrates approximately 100,000 times per second. An integrated circuit counts these vibrations and converts them into a digital or dial reading on your watch face. An inexpensive quartz watch today is more accurate than the best mechanical watch, and precision-manufactured quartz clocks are accurate to within one second per 10 years.

Precise timekeeping is important in our technological world. Ships and aircraft plot their locations by satellite, relying on a time signal that has an accuracy of a millionth of a second. Deep-space probes such as the *Voyagers* (see Chapter 19) require radio commands timed to billionths of a second, while physicists exploring the motion inside the nucleus of an atom deal in trillionths of a second as easily as we talk about minutes.

To achieve such accuracy, scientists use atomic clocks. First developed in the 1940s, these clocks rely on an atom's oscillating electrons, a rhythm so regular that they are accurate to within a few thousandths of a second per day. Recently, an atomic clock known to be accurate to within one second per three million years was installed at the National Institute of Standards and Technology (NIST). Named the NIST-7, this clock is a refinement over previous atomic clocks because it uses lasers instead of a magnetic field to stimulate cesium 133 atoms. Cesium 133 is used in atomic clocks because its electrons oscillate at a very predictable rate (9,192,631,770 vibrations per second). In fact, it was a cesium clock that was used to prove Einstein's prediction that a clock will slow down as its speed increases.

While physicists deal with incredibly short intervals of time, astronomers and geologists deal with "deep time," millions or billions of years. When astronomers look at a distant galaxy, they are seeing what it looked like billions of years ago. Geologists looking into the Grand Canyon are viewing nearly two billion years of Earth history preserved in the rocks of the canyon walls. Geologists can measure decay rates of such radioactive elements as uranium, thorium, and rubidium to determine how long ago an igneous rock formed. Furthermore, geologists know that the Earth's rotational velocity has been slowing down a few thousandths of a second per century as a result of the frictional effects of tides, ocean currents, and varying thicknesses of polar ice. Five hundred million years ago a day was only 20 hours long, and at the current rate of slowing, 200 million years from now a day will be 25 hours long.

Time is a fascinating topic that has been the subject of numerous essays and books. And while we can comprehend concepts like milliseconds and understand how a quartz watch works, deep time, or geologic time, is still very difficult for most people to comprehend.

▼ INTRODUCTION

Time is what sets geology apart from most of the other sciences, and an appreciation of the immensity of geologic time is fundamental to understanding both the physical and biological history of our planet (◆ Fig. 17.1). Most people have difficulty comprehending geologic time because they tend to view time from the perspective of their own existence. Ancient history is what occurred hundreds or even thousands of years ago, but when geologists talk in terms of ancient geologic history, they are referring to events that happened millions or even billions of years ago!

Geologists use two different frames of reference when speaking of geologic time. **Relative dating** involves placing geologic events in a sequential order as determined from their position in the rock record. Relative dating will not tell us how long ago a particular event occurred, only that one event preceded another. The various principles used to determine relative dating were discovered hundreds of years ago and since then have been used to construct the *relative geologic time scale*. These principles are still widely used today.

Absolute dating results in specific dates for rock units or events expressed in years before the present. Radiometric dating is the most common method of obtaining absolute age dates. Such dates are calculated from the natural rates of decay of various radioactive elements occurring in trace amounts in some rocks. It was not until the discovery of radioactivity near the end of the nineteenth century that absolute ages could be accurately applied to the relative geologic time scale. Today the geologic time scale is really a dual scale: a

relative scale based on rock sequences with radiometric dates expressed as years before the present added to it (◆ Fig. 17.2).

▼ EARLY CONCEPTS OF GEOLOGIC TIME AND THE AGE OF THE EARTH

The concept of geologic time and its measurement have changed through human history. For example, many early Christian scholars and clerics tried to establish the date of creation by analyzing historical records and the genealogies found in Scripture. Based on their analyses, they generally believed that the Earth and all of its features were no more than about 6,000 years old. The idea of a very young Earth provided the basis for most western chronologies of Earth history prior to the eighteenth century.

During the eighteenth and nineteenth centuries, several attempts were made to determine the age of the Earth on the basis of scientific evidence rather than revelation. For example, the French zoologist Georges Louis de Buffon (1707–1788) assumed that the Earth gradually cooled to its present condition from a molten beginning. To simulate this history, he melted iron balls of various diameters and allowed them to cool to the surrounding temperature. By extrapolating their cooling rate to a ball the size of the Earth, he determined that the Earth was at least 75,000 years old. While this age was much older than that derived from Scripture, it was still vastly younger than we now know the Earth to be.

Other scholars were equally ingenious in attempting to calculate the Earth's age. For example, if deposition rates could be determined for various sediments, geologists reasoned that they could calculate how long it would take to deposit any rock layer. Furthermore, they could then extrapolate how old the Earth was from the total thickness of sedimentary rock in the Earth's crust. Rates of deposition vary, however, even for the same type of rock. Furthermore, it is impossible to estimate how much rock has been removed by erosion, or how much a rock sequence has been reduced by compaction. As a result of these variables, estimates ranged from less than a million years to more than a billion years.

Another attempt to determine the Earth's age involved ocean salinity. Scholars assumed that the Earth's ocean waters were originally fresh and that their present salinity was the result of dissolved salt being carried into the ocean basins by streams. Knowing the volume of ocean water and its salinity, John Joly, a nineteenth-century Irish geologist, measured the amount

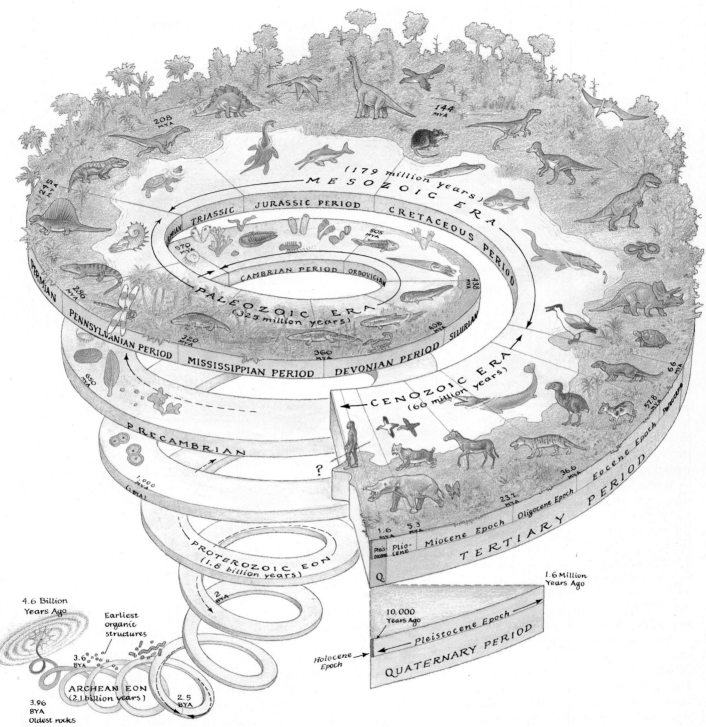

◆ FIGURE 17.1 Geologic time is depicted in this spiral history of the Earth from the time of its formation 4.6 billion years ago to the present. (B.Y.=billion years; M.Y.=million years.)

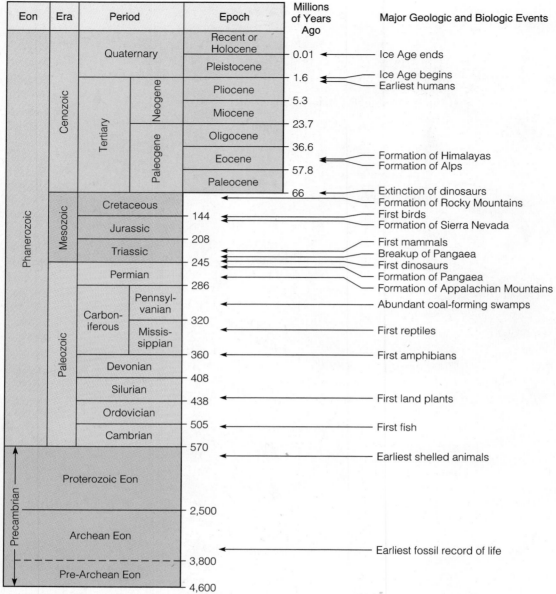

◆ FIGURE 17.2 The geologic time scale. Some of the major biological and geological events are indicated along the right-hand margin.

of salt currently in the world's streams. He then calculated that it would have taken at least 90 million years for the oceans to reach their present salinity level. This was still much younger than the now accepted age of 4.6 billion years for the Earth, mainly because Joly had no way of calculating how much salt had been recycled or the amount of salt stored in continental salt deposits and sea-floor clay deposits.

In addition to these efforts, the naturalists of the eighteenth and nineteenth centuries were also formulating some of the fundamental geologic principles that are used in deciphering the history of the Earth. From the evidence preserved in the rock and fossil record, it was clear to them that the Earth is very old and that geologic processes have operated over long periods of time.

JAMES HUTTON AND THE RECOGNITION OF GEOLOGIC TIME

The Scottish geologist James Hutton (1726–1797) is considered by many to be the father of modern geology. His detailed studies and observations of rock exposures and present-day geological processes were instrumental in establishing the concept that geological processes had operated over vast amounts of time. Because Hutton relied on known processes to account for Earth history, he concluded that the Earth must be very old and wrote that "we find no vestige of a beginning, and no prospect of an end."

Observing the processes of wave action, erosion by running water, and sediment transport, Hutton concluded that given enough time these processes could account for the geologic features of his native Scotland. He believed that "the past history of our globe must be explained by what can be seen to be happening now." This assumption that present-day processes have operated throughout geologic time was the basis for the **principle of uniformitarianism** (see Chapter 1).

Unfortunately, Hutton was not a particularly good writer, so his ideas were not widely disseminated or accepted. In 1830, however, Charles Lyell published a landmark book, *Principles of Geology,* in which he championed Hutton's concept of uniformitarianism. Instead of relying on catastrophic events to explain various features of the Earth, Lyell recognized that imperceptible changes brought about by present-day processes could, over long periods of time, have tremendous cumulative effects. Through his writings, Lyell firmly established uniformitarianism as the guiding philosophy of geology. Furthermore, the recognition of virtually limitless time was also necessary for, and instrumental in, the acceptance of Darwin's 1859 theory of evolution.

After finally establishing that present-day processes have operated over vast periods of time, geologists were nevertheless nearly forced to accept a very young age for the Earth when a highly respected English physicist, Lord Kelvin (1824–1907), claimed in 1866 to have destroyed the uniformitarian foundation of Huttonian-Lyellian geology. Starting with the generally accepted belief that the Earth was originally molten, Kelvin assumed that the Earth has gradually been losing heat and that, by measuring this heat loss, he could determine the age of the Earth.

Kelvin knew from deep mines that the Earth's temperature increases with depth, and he reasoned that the Earth is therefore losing heat from its interior. By knowing the melting temperatures of the Earth's rocks, the size of the Earth, and the rate of heat loss, Kelvin was able to calculate the age at which the Earth was entirely molten. From these calculations, he concluded that the Earth could not be older than 100 million years or younger than 20 million years. This wide range in age reflected uncertainties over average temperature increases with depth and the various melting points of the Earth's constituent materials.

After finally establishing that the Earth was very old, and showing how present-day processes operating over long periods of time can explain geological features, geologists were in a quandary. They either had to accept Kelvin's dates and squeeze events into a shorter time frame or reject his calculations.

While Kelvin's reasoning and calculations were sound, his basic premises were false, thereby invalidating his conclusions. Kelvin was unaware that the Earth has an internal heat source, radioactivity, that has allowed it to maintain a fairly constant temperature through time.* His 40-year campaign for a young Earth ended with the discovery of radioactivity near the end of the nineteenth century. His calculations were therefore no longer valid, and his proof for a geologically young Earth collapsed. Moreover, while the discovery of radioactivity destroyed Kelvin's arguments, it provided geologists with a clock that could measure the Earth's age and validate what geologists had been saying all along, namely, that the Earth was indeed very old!

RELATIVE DATING METHODS

Before the development of radiometric dating techniques, geologists had no acceptable means of absolute age dating and thus depended on relative dating methods. These methods only allow events to be placed in sequential order and do not tell us how long ago an event took place. While the principles of relative dating may now seem self-evident, their discovery was an important scientific achievement because they provided geologists with a means to interpret geologic history and develop a relative geologic time scale.

Six fundamental geologic principles are used in relative dating: *superposition, original horizontality, lateral*

. . . .

*Actually, the Earth's temperature has decreased through time because the original amount of radioactive materials has been decreasing and thus is not supplying as much heat. However, the temperature is decreasing at a rate considerably slower than would be required to lend any credence to Kelvin's calculations.

continuity, cross-cutting relationships, inclusions, and *fossil succession.*

Fundamental Principles of Relative Dating

The seventeenth century was an important time in the development of geology as a science because of the widely circulated writings of the Danish anatomist, Nicolas Steno (1638–1686). Steno observed that during flooding, streams spread out across their floodplains and deposit layers of sediment that bury organisms dwelling on the floodplain. Subsequent flooding events produce new layers of sediments that are deposited or superposed over previous deposits. When lithified, these layers of sediment become sedimentary rock. Thus, in an undisturbed succession of sedimentary rock layers, the oldest layer is at the bottom and the youngest layer is at the top. This **principle of superposition** is the basis for relative age determinations of strata and their contained fossils (◆ Fig. 17.3).

Steno also observed that because sedimentary particles settle from water under the influence of gravity, sediment is deposited in essentially horizontal layers, thus illustrating the **principle of original horizontality** (Fig. 17.3). Therefore, a sequence of sedimentary rock layers that is steeply inclined from the horizontal must have been tilted after deposition and lithification.

Steno's third principle, the **principle of lateral continuity**, states that sediment extends laterally in all directions until it thins and pinches out or terminates against the edge of the depositional basin (Fig. 17.3).

James Hutton is credited with discovering the **principle of cross-cutting relationships**. Based on his detailed studies and observations of rock exposures in Scotland, Hutton recognized that an igneous intrusion or fault must be younger than the rocks it intrudes or displaces (◆ Fig. 17.4).

◆ FIGURE 17.4 The principle of cross-cutting relationships. (*a*) A dark-colored dike has been intruded into older light-colored granite, north shore of Lake Superior, Ontario, Canada. (*b*) A small fault displacing horizontal beds in central Texas. (Photo (*b*) courtesy of David J. Matty.)

(a)

(b)

◆ FIGURE 17.3 The Grand Canyon of Arizona illustrates three of the six fundamental principles of relative dating. The sedimentary rocks of the Grand Canyon were originally deposited horizontally in a variety of marine and continental environments (principle of original horizontality). The oldest rocks are therefore at the bottom of the canyon, and the youngest rocks are at the top, forming the rim (principle of superposition). The exposed rock layers extend laterally for some distance (principle of lateral continuity).

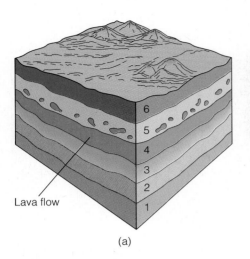

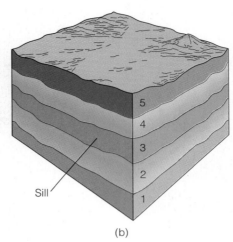

(a)

(b)

◆ FIGURE 17.5 Relative ages of lava flows, sills, and associated sedimentary rocks may be difficult to determine. (*a*) A buried lava flow (4) baked the underlying bed, and bed 5 contains inclusions of the lava flow. The lava flow is younger than bed 3 and older than beds 5 and 6. (*b*) The rock units above and below the sill (3) have been baked, indicating that the sill is younger than beds 2 and 4, but its age relative to bed 5 cannot be determined.

While this principle illustrates that an intrusive igneous structure is younger than the rocks it intrudes, the association of sedimentary and igneous rocks may cause problems in relative dating. Buried lava flows and intrusive igneous bodies such as sills look very similar in a sequence of strata (◆ Fig. 17.5). A buried lava flow, however, is older than the rocks above it (principle of superposition), while a sill is younger than all the beds below it and younger than the bed immediately above it.

To resolve such relative age problems as these, geologists look to see if the sedimentary rocks in contact with the igneous rocks show signs of baking or alteration by heat (see Chapter 6, Contact Metamorphism). A sedimentary rock showing such effects must be older than the igneous rock with which it is in contact. In Figure 17.5, for example, a sill produces a zone of baking immediately above and below it because it intruded into previously existing sedimentary rocks. A lava flow, on the other hand, bakes only those rocks below it.

Another way to determine relative ages is by using the **principle of inclusions**. This principle holds that inclusions, or fragments of one rock contained within a layer of another, are older than the rock layer itself. For example, the batholith shown in ◆ Figure 17.6a contains sandstone inclusions, and the sandstone unit shows the effects of baking. Accordingly, we conclude that the sandstone is older than the batholith. In ◆ Figure 17.6b, however, the sandstone contains granite rock fragments, indicating that the batholith was the source rock for the inclusions and is therefore older than the sandstone.

Fossils have been known for centuries (see Chapter 18), yet their utility in relative dating and geologic

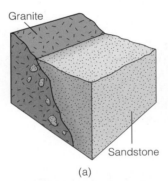

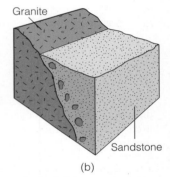

(a)

(b)

◆ FIGURE 17.6 (*a*) The granite batholith is younger than the sandstone because the sandstone has been baked at its contact with the granite and the granite contains sandstone inclusions. (*b*) Granite inclusions in the sandstone indicate that the batholith was a source of the sandstone and therefore is older.

mapping was not fully appreciated until the early nineteenth century. William Smith (1769–1839), an English civil engineer involved in surveying and building canals in southern England, independently recognized the principle of superposition by reasoning that the fossils at the bottom of a sequence of strata are older than those at the top of the sequence. This recognition served as the basis for the **principle of fossil succession** or the *principle of faunal and floral succession* as it is sometimes called (◆ Fig. 17.7).

According to this principle, fossil assemblages succeed one another through time in a regular and predictable order. The validity and successful use of this

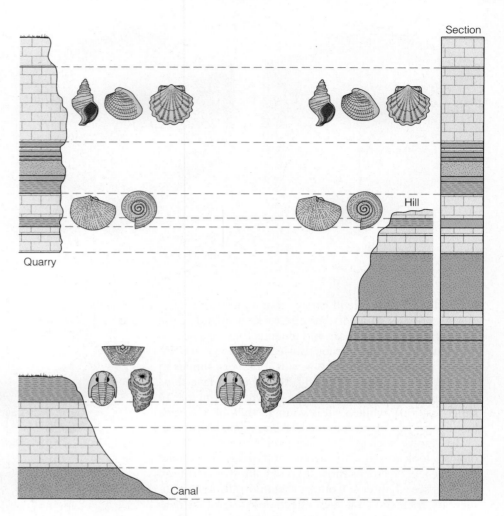

◆ FIGURE 17.7 This generalized diagram shows how William Smith used fossils to identify strata of the same age in different areas (principle of fossil succession). The composite section on the right shows the relative ages of all strata in this area.

principle depends on three points: (1) life has varied through time, (2) fossil assemblages are recognizably different from one another, and (3) the relative ages of the fossil assemblages can be determined. Observations of fossils in older versus younger strata clearly demonstrate that life-forms have changed. Because this is true, fossil assemblages (point 2) are recognizably different. Furthermore, superposition can also be used to demonstrate the relative ages of the fossil assemblages.

Unconformities

Our discussion thus far has been concerned with conformable sequences of strata, sequences in which no depositional breaks of any consequence occur. A sharp

bedding plane (see Fig. 5.27) separating strata may represent a depositional break of minutes, hours, years, or even tens of years, but it is inconsequential when considered in the context of geologic time.

Surfaces of discontinuity representing significant amounts of geologic time are **unconformities**, and any interval of geologic time not represented by strata in a particular area is a **hiatus** (◆ Fig. 17.8). Thus, an unconformity is a surface of nondeposition or erosion that separates younger strata from older rocks. As such, it represents a break in our record of geologic time. The famous 12-minute gap in the Watergate tapes of Richard Nixon's presidency is somewhat analogous. Just as we have no record of the conversations that were occurring during this period of time,

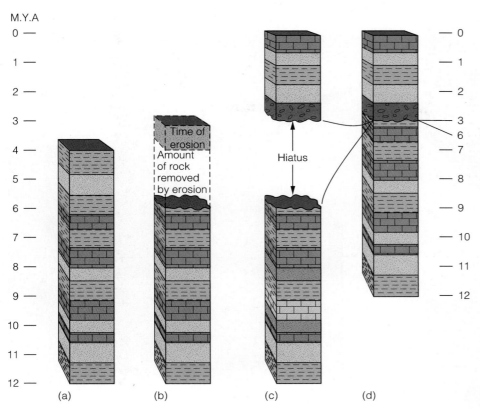

◆ FIGURE 17.8 Simplified diagram showing the development of an unconformity and a hiatus. (a) Deposition began 12 million years ago (M.Y.A.) and continued more or less uninterrupted until 4 M.Y.A. (b) A 1-million-year episode of erosion occurred, and during that time strata representing 2 million years of geologic time were eroded. (c) A hiatus of 3 million years exists between the older strata and the strata that formed during a renewed episode of deposition that began 3 M.Y.A. (d) The actual stratigraphic record. The unconformity is the surface separating the strata and represents a major break in our record of geologic time.

we have no record of the events that occurred during a hiatus.

Three types of unconformities are recognized. A **disconformity** is a surface of erosion or nondeposition between younger and older beds that are parallel with one another (◆ Fig. 17.9). Unless a well-defined erosional surface separates the older from the younger parallel beds, the disconformity frequently resembles an ordinary bedding plane. Accordingly, many disconformities are difficult to recognize and must be identified on the basis of fossil assemblages.

An **angular unconformity** is an erosional surface on tilted or folded strata over which younger strata have been deposited (◆ Fig. 17.10). Both younger and older strata may dip, but if their dip angles are different (generally the older strata dip more steeply), an angular unconformity is present.

The angular unconformity illustrated in Figure 17.10b is probably the most famous in the world. It was here at Siccar Point, Scotland, that James Hutton realized that severe upheavals had tilted the lower rocks

and formed mountains that were then worn away and covered by younger, flat-lying rocks. The erosional surface between the older tilted rocks and the younger flat-lying strata meant that there was a significant gap in the rock record. Although Hutton did not use the term unconformity, he was the first to understand and explain the significance of such discontinuities in the rock record.

The third type of unconformity is a **nonconformity.** Here an erosion surface cut into metamorphic or igneous rocks is covered by sedimentary rocks (◆ Fig. 17.11, page 440). This type of unconformity closely resembles an intrusive igneous contact with sedimentary rocks. The principle of inclusions is helpful in determining whether the relationship between the underlying igneous rocks and the overlying sedimentary rocks is the result of an intrusion or erosion. In the case of an intrusion, the igneous rocks are younger, but in the case of erosion, the sedimentary rocks are younger. Being able to distinguish between a nonconformity and an intrusive contact is very important because they represent different sequences of events.

Geologic Time: Concepts and Principles   437

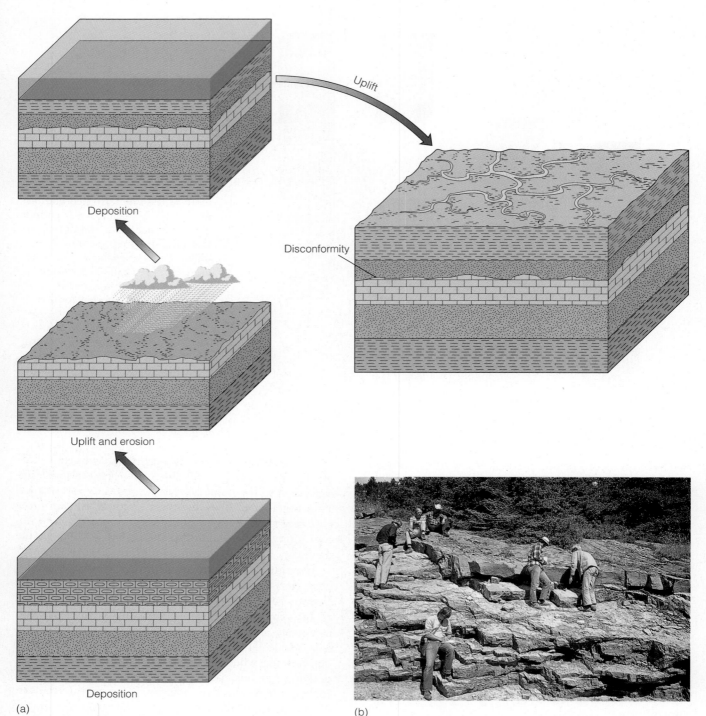

Deposition

Uplift

Disconformity

Uplift and erosion

Deposition

(a)

(b)

◆ FIGURE 17.9 (*a*) Formation of a disconformity. (*b*) Disconformity between Mississippian and Jurassic strata in Montana. The geologist at the upper left is sitting on Jurassic strata, and his right foot is resting upon Mississippian rocks.

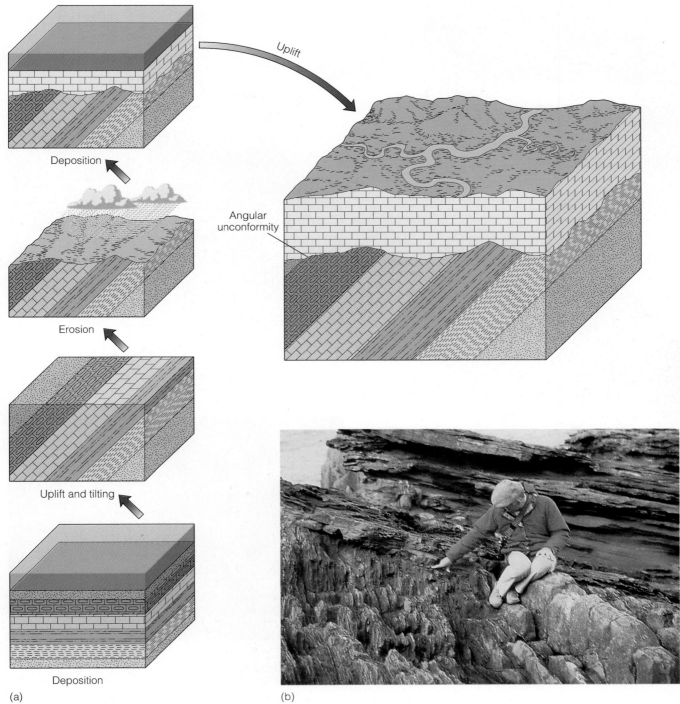

(a)   (b)

◆ FIGURE 17.10 (*a*) Formation of an angular unconformity. (*b*) Angular
unconformity at Siccar Point, Scotland. (Photo courtesy of Dorothy L. Stout.)

Labels within figure (a): Deposition, Uplift, Erosion, Uplift and tilting, Deposition, Angular unconformity

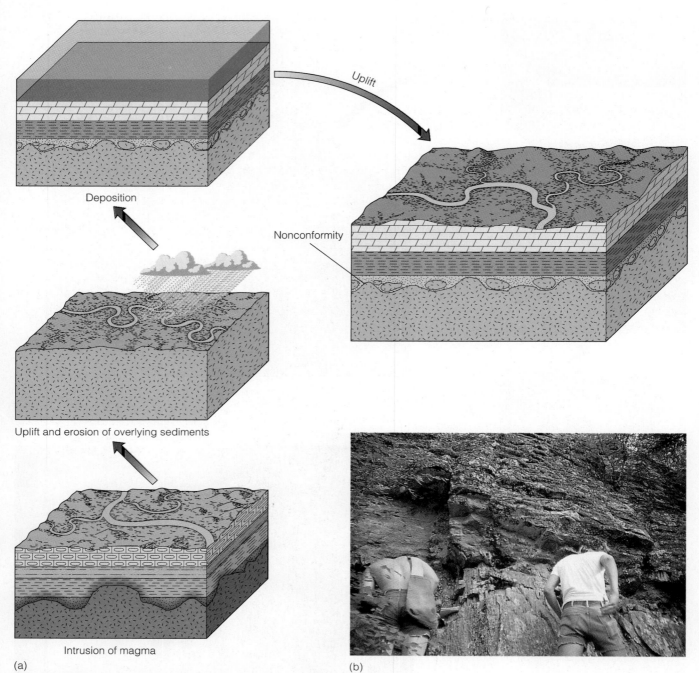

Deposition

Uplift

Nonconformity

Uplift and erosion of overlying sediments

Intrusion of magma

(a)

(b)

◆ FIGURE 17.11 (*a*) Formation of a nonconformity. (*b*) Nonconformity between Precambrian granite and the overlying Cambrian-age Deadwood Formation, Wyoming.

Applying the Principles of Relative Dating to Determine the Geologic History of an Area

We can decipher the geologic history of the area represented by the block diagram in ◆ Figure 17.12 by applying the various relative dating principles just discussed. The methods and logic used in this example are the same as those applied by nineteenth-century geologists in constructing the geologic time scale.

According to the principles of superposition and original horizontality, beds A, B, C, D, E, F, and G were deposited horizontally; then they were either tilted, faulted (H), and eroded, or after deposition, they were faulted (H), tilted, and then eroded (◆ Fig. 17.13a, b, and c). Because the fault cuts beds A–G, it must be younger than the beds according to the principle of cross-cutting relationships.

Beds J, K, and L were then deposited horizontally over this erosional surface producing an angular unconformity (I) (◆ Fig. 17.13d). Following deposition of these three beds, the entire sequence was intruded by a dike (M), which, according to the principle of cross-cutting relationships, must be younger than all the rocks it intrudes (◆ Fig. 17.13e).

The entire area was then uplifted and eroded; next beds P and Q were deposited, producing a disconformity (N) between beds L and P and a nonconformity (O) between the igneous intrusion M and the sedimentary bed P (◆ Fig. 17.13f and g). We know that the relationship between igneous intrusion M and the overlying sedimentary bed P is a nonconformity because of the presence of inclusions of M in P (principle of inclusions).

At this point, there are several possibilities for the geologic history of this area. According to the principle of cross-cutting relationships, dike R must be younger than bed Q because it intrudes into it. It can have intruded anytime *after* bed Q was deposited; however, we cannot determine whether R was formed right after Q, right after S, or after T was formed. For purposes of this history, we will say that it intruded after the deposition of bed Q (◆ Fig. 17.13g and h).

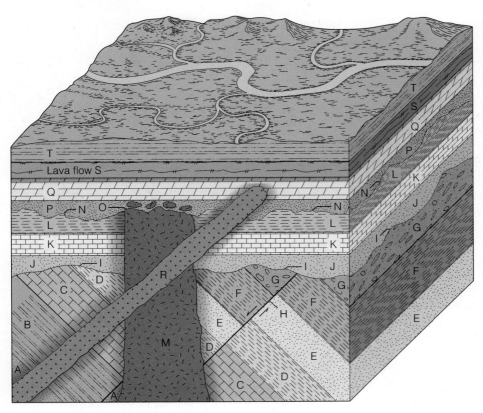

◆ FIGURE 17.12 A block diagram of a hypothetical area in which the various relative dating principles can be applied to reconstruct the geologic history.

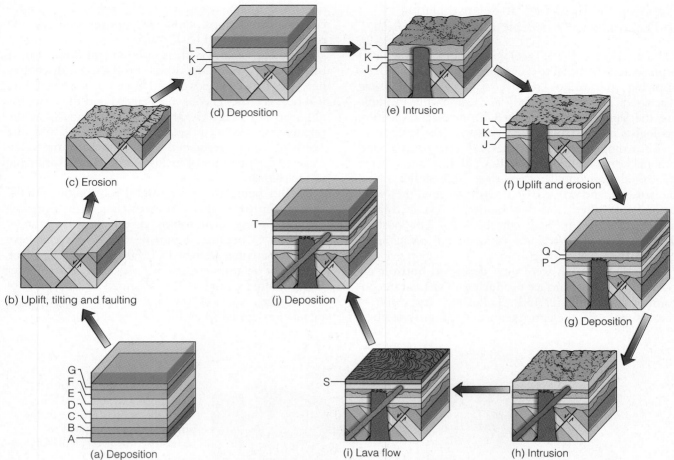

◆ FIGURE 17.13 (*a*) Beds A, B, C, D, E, F, and G are deposited. (*b*) The preceding beds are tilted and faulted. (*c*) Erosion. (*d*) Beds J, K, and L are deposited, producing an angular unconformity. (*e*) The entire sequence is intruded by a dike. (*f*) The entire sequence is uplifted and eroded. (*g*) Beds P and Q are deposited, producing a disconformity (N) and a nonconformity (O). (*h*) Dike R intrudes. (*i*) Lava (S) flows over bed Q, baking it. (*j*) Bed T is deposited.

Following the intrusion of dike R, lava S flowed over bed Q, followed by the deposition of bed T (◆ Fig. 17.13i and j). Although the lava flow (S) is not a sedimentary unit, the principle of superposition still applies because it flowed on the Earth's surface, just as sediments are deposited on the Earth's surface.

Thus, we have established a relative chronology for the rocks and events of this area by using the principles of relative dating. Remember, however, that we have no way of knowing how many years ago these events occurred unless we can obtain radiometric dates for the igneous rocks. With these dates we can establish the range of absolute ages between which the different

sedimentary units were deposited and also determine how much time is represented by the unconformities.

▼ CORRELATION

If geologists are to interpret Earth history, they must demonstrate the time equivalency of rock units in different areas. This process is known as **correlation.**

If exposures are adequate, units may simply be traced laterally (principle of lateral continuity), even if occasional gaps exist (◆ Fig. 17.14). Other criteria used to correlate units are similarity of rock type, position in a sequence, and **key beds**. Key beds are units, such as

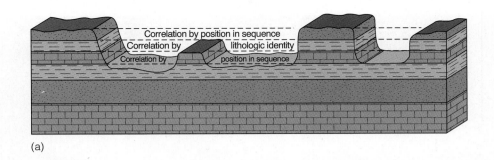

(a)

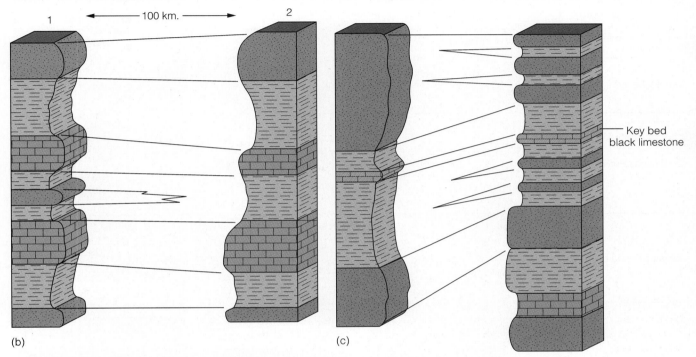

1 ← 100 km. → 2

Key bed
black limestone

(b)   (c)

◆ FIGURE 17.14 Correlation of rock units. (*a*) In areas of adequate exposures, rock units can be traced laterally even if occasional gaps exist. (*b*) Correlation by similarities in rock type and position in a sequence. The sandstone in section 1 is assumed to intertongue or grade laterally into the shale at section 2. (*c*) Correlation using a key bed, a distinctive black limestone.

coal beds or volcanic ash layers, that are sufficiently distinctive to allow identification of the same unit in different areas (Fig. 17.14). In addition to surface correlation, geologists frequently use well logs, cores, or cuttings to correlate subsurface rock units when exploring for minerals, coal, and petroleum.

Generally, no single location in a region has a geologic record of all the events that occurred during its history; therefore geologists must correlate from one area to another in order to determine the complete geologic history of the region. An excellent example is provided by the history of the Colorado Plateau (◆ Fig. 17.15). Within this region a record of approximately 2.0 billion years of Earth history is present. Because of the forces of erosion, however, the entire record is not preserved at any single location. Within the walls of the Grand Canyon are rocks of the Precambrian and Paleozoic Era, while Paleozoic and Mesozoic Era rocks are found in Zion National Park, and Mesozoic and Cenozoic Era rocks are exposed in Bryce Canyon (Fig.

Grand Canyon
National Park
Arizona

Zion
National Park
Utah

Bryce Canyon
National Park
Utah

UTAH
Bryce
Canyon
Zion
Grand
Canyon

ARIZONA

Tertiary Period			Wasatch Fm
Cretaceous Period			Kaiparowits Fm
			Wahweap Ss
			Straight Cliffs Ss
			Tropic Shale
			Dakota Ss
Jurassic Period			Winsor Fm
			Curtis Fm
			Entrada Ss
		Carmel Fm	Carmel Fm
		Navajo Ss	Navajo Ss
		Kayenta Fm	Older rocks not exposed
Triassic Period		Wingate Ss	
		Chinle Fm	
	Moenkopi Fm	Moenkopi Fm	
Permian Period	Kaibab Ls	Kaibab Ls	
	Toroweap Fm	Older rocks not exposed	
	Coconino Ss		
	Hermit Shale		
Pennsylvanian Period	Supai Fm		
Mississippian Period	Redwall Ls		
Devonian Period*	Temple Butte Ls		
Cambrian Period	Mauv Fm		
	Bright Angel Shale		
	Tapeats Ss		
Precambrian Eon	Vishnu Schist		

Colorado
River

Fm = Formation   Ss = Sandstone   Ls = Limestone   *Rocks of Ordovician and Silurian age are not present in the Grand Canyon.

◆ FIGURE 17.15 (left) Correlation of rocks within the Colorado Plateau. By correlating the rocks from various locations, the history of the entire region can be deciphered.

17.15). By correlating the uppermost rocks at one location with the lowermost equivalent rocks of another area, the history of the entire region can be deciphered.

Although geologists can match up rocks on the basis of similar rock type and stratigraphic position, correlation of this type can only be done in a limited area where beds can be traced from one site to another. In order to correlate rock units over a large area or to correlate age-equivalent units of different composition, fossils and the principle of fossil succession must be used.

Fossils are useful as time indicators because they are the remains of organisms that lived for a certain length of time during the geologic past. Fossils that are easily identified, are geographically widespread, and existed for a rather short geologic time are particularly useful. Such fossils are called **guide fossils** or *index fossils* (◆ Fig. 17.16). For example, the trilobite *Isotelus* and the clam *Inoceramus* meet all of these criteria and are therefore good guide fossils. In contrast, the brachiopod *Lingula* is easily identified and widespread, but its geologic range of Ordovician to Recent makes it of little use in correlation.

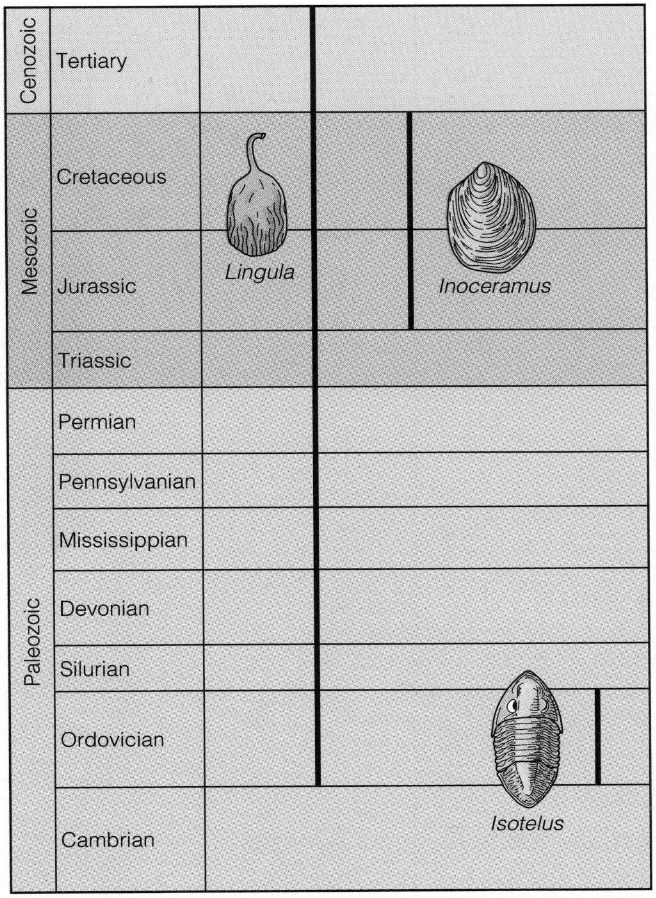

◆ FIGURE 17.16 The geologic ranges of three marine invertebrates. The brachiopod *Lingula* is of little use in correlation because of its long geologic range. The trilobite *Isotelus* and the bivalve *Inoceramus* are good guide fossils because they are geographically widespread, are easily identified, and have short geologic ranges.

Because most fossils have fairly long geologic ranges, geologists construct **assemblage range zones** to determine the age of the sedimentary rocks containing the fossils. Assemblage range zones are established by plotting the overlapping geologic ranges of different species of fossils. The first and last occurrences of two species are used to establish an assemblage zone's boundaries (◆ Fig. 17.17). Correlation of assemblage zones generally yields correlation lines that are considered time equivalent. In other words, the strata encompassed by the correlation lines are thought to be the same age.

ABSOLUTE DATING METHODS

Thus far, our discussion has largely concerned the concept of geologic time and the formulation of principles used to determine relative ages. It is somewhat ironic that radioactivity, the very process that invalidated Lord Kelvin's calculations, now serves as the basis for determining absolute dates.

Although most of the isotopes of the 91 naturally occurring elements are stable, some are radioactive and spontaneously decay to other more stable isotopes of elements, releasing energy in the process. The discovery, in 1903 by Pierre and Marie Curie, that radioactive decay produces heat as a by-product meant that geologists finally had a mechanism for explaining the internal heat of the Earth that did not rely on residual cooling from a molten origin. Furthermore, geologists and paleontologists had a powerful tool to date geologic events accurately, and thus verify the long time periods postulated by Hutton, Lyell, and Darwin.

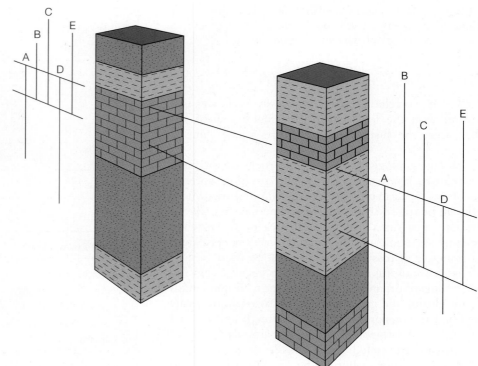

◆ FIGURE 17.17 Correlation of two sections by using assemblage range zones. These zones are established by the overlapping ranges of fossils A through E.

Atoms, Elements, and Isotopes

As we discussed in Chapter 2, all matter is made up of chemical elements, each of which is composed of extremely small particles called *atoms*. The nucleus of an atom is composed of *protons* and *neutrons* with *electrons* encircling it (see Fig. 2.2). The number of protons defines an element's *atomic number* and helps determine its properties and characteristics. The combined number of protons and neutrons in an atom is its *atomic mass number*. However, not all atoms of the same element have the same number of neutrons in their nuclei. These variable forms of the same element are called *isotopes*. Most isotopes are stable, but some are unstable and spontaneously decay to a more stable form. It is the decay rate of unstable isotopes that geologists measure to determine the absolute age of rocks.

Radioactive Decay and Half-Lives

Radioactive decay is the process whereby an unstable atomic nucleus is spontaneously transformed into an atomic nucleus of a different element. Three types of radioactive decay are recognized, all of which result in a change of atomic structure (◆ Fig. 17.18). In **alpha decay**, two protons and two neutrons are emitted from the nucleus, resulting in a loss of two atomic numbers and four atomic mass numbers. In **Beta decay**, a fast-moving electron is emitted from a neutron in the nucleus, changing that neutron to a proton and consequently increasing the atomic number by one, with no resultant atomic mass number change. **Electron capture** results when a proton captures an electron from an electron shell and thereby converts to a neutron, resulting in a loss of one atomic number and no change in the atomic mass number.

Some elements undergo only one decay step in the conversion from an unstable form to a stable form. For example, rubidium 87 decays to strontium 87 by a single beta emission, and potassium 40 decays to argon 40 by a single electron capture. Other radioactive elements undergo several decay steps (see Perspective 17.1). Uranium 235 decays to lead 207 by seven alpha and six beta steps, while uranium 238 decays to lead 206 by eight alpha and six beta steps (◆ Fig. 17.19, page 450).

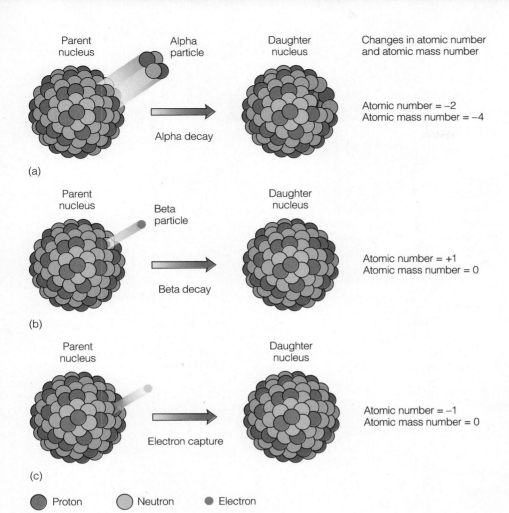

(a)

Parent nucleus — Alpha particle — Daughter nucleus — Changes in atomic number and atomic mass number

Alpha decay

Atomic number = –2
Atomic mass number = –4

(b)

Parent nucleus — Beta particle — Daughter nucleus

Beta decay

Atomic number = +1
Atomic mass number = 0

(c)

Parent nucleus — Daughter nucleus

Electron capture

Atomic number = –1
Atomic mass number = 0

● Proton   ● Neutron   ● Electron

◆ FIGURE 17.18 Three types of radioactive decay. (*a*) Alpha decay, in which an unstable parent nucleus emits two protons and two neutrons. (*b*) Beta decay, in which an electron is emitted from the nucleus. (*c*) Electron capture, in which a proton captures an electron and is thereby converted to a neutron.

When discussing decay rates, it is convenient to refer to them in terms of half-lives. The **half-life** of a radioactive element is the time it takes for one-half of the atoms of the original unstable **parent element** to decay to atoms of a new, more stable **daughter element**. The half-life of a given radioactive element is always constant and can be precisely measured in the laboratory. Half-lives of various radioactive elements range from less than a billionth of a second to 49 billion years.

Radioactive decay occurs at a geometric rate rather than a linear rate. Therefore, a graph of the decay rate produces a curve rather than a straight line (◆ Fig. 17.20, page 451). For example, an element with *1,000,000* parent atoms will have *500,000* parent atoms and 500,000 daughter atoms after one half-life. After two half-lives, it will have *250,000* parent atoms (one-half of the previous parent atoms, which is equivalent to one-fourth of the original parent atoms) and 750,000 daughter atoms. After three half-lives, it will have *125,000* parent atoms (one-half of the previous parent atoms or one-eighth of the original parent atoms) and 875,000 daughter atoms, and so on until the number of parent atoms remaining is so few that they cannot be accurately measured by present-day instruments.

By measuring the parent-daughter ratio and knowing the half-life of the parent (determined in the laboratory), geologists can calculate the age of a sample containing the radioactive element. The parent-daughter ratio is usually determined by a *mass spectrometer,* an instrument that measures the proportions of elements of different masses.

RADON: THE SILENT KILLER

According to the U.S. National Research Council, approximately 20,000 people die prematurely each year from cancers induced by exposure to indoor radon. In fact, radon is the second leading cause of lung cancer in the United States.

Your chances of being adversely affected by radon depend on numerous interrelated factors such as geographic location, the geology of the area, the climate, how the building is constructed, and how much time you spend in the building. While there are, as yet, no federal standards defining unacceptable indoor radon levels, the Environmental Protection Agency (EPA) recommends radon levels not exceed 4 picocuries per liter (pCi/L) of air (a curie is the standard measure of radiation, and a picocurie is one-trillionth of a curie).

Radon is a colorless, odorless, naturally occurring radioactive gas that has a three-day half-life and is part of the uranium 238–lead 206 radioactive decay series (Fig. 17.19). It occurs in any rock or soil that contains uranium 238. Outdoors, radon escapes into the atmosphere where it is diluted and dissipates to harmless levels (0.2 pCi/L is the average ambient outdoor level of radon). In an enclosed area such as a home, however, radon can accumulate to unhealthy levels (>4 pCi/L). Continued exposure to these elevated levels over several years can greatly increase the risk of lung cancer.

As one of the natural decay products of uranium 238, radon itself decays into other radioactive elements called radon daughters (Fig. 17.19). Every time you breathe, these daughter elements become trapped in your lungs and eventually break down, releasing high-energy alpha and beta decay particles (Fig. 17.18) that damage lung tissue and can cause lung cancer.

Concern about the health risks posed by radon first arose during the 1960s when the news media revealed that some homes in the West had been built with uranium mine tailings. Since then, geologists have found that high indoor radon levels can be caused by natural uranium in minerals of the rock and soil on which buildings are constructed. In response to the high cost of energy during the 1970s and 1980s, old buildings were insulated, and new buildings were constructed to be as energy efficient and airtight as possible. Ironically, these energy-saving measures also sealed in radon.

Radon enters buildings through dirt floors, cracks in the floor or walls, joints between floors and walls, floor drains, sumps, and utility pipes as well as any cracks or pores in hollow-block walls (◆ Fig. 1). Radon can also be released into a building whenever the water is turned on if the water comes from a private well. Municipal water is generally safe because it has usually been aerated before it gets to your home.

To find out if your home has a radon problem, you must test for it with commercially available, relatively inexpensive, simple home testing devices. If radon readings are above the recommended EPA levels of 4 pCi/L, several remedial measures can be taken to reduce your risk. These include sealing up all cracks in the foundation, pouring a concrete slab over a dirt floor, increasing the circulation of air throughout the house, especially in the basement and crawl space, providing filters for drains and other utility openings, and limiting the time spent in areas with higher concentrations of radon.

◆ FIGURE 1 Some of the common entry points where radon can enter a house.

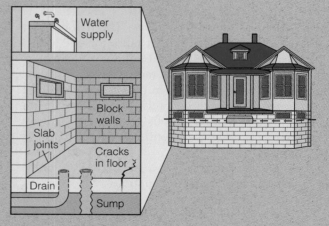

It is important to remember that although the radon hazard covers most of the country, some areas are more likely to have higher natural concentrations of radon than others (◆ Fig. 2). For example, such rocks as uranium-bearing granites, metamorphic rocks of granitic composition, and black shales (high carbon content) are quite likely to cause indoor radon problems. Other rocks such as marine quartz sandstone, noncarbonaceous shales and siltstones, most volcanic rocks, and igneous and metamorphic rocks rich in iron and magnesium typically do not cause radon problems. The permeability of the soil overlying the rock can also affect the indoor levels of radon gas. Some soils are more permeable than others and allow more radon to escape into the overlying structures.

The climate and type of construction affect not only how much radon gets into a structure, but how much escapes. Concentrations of radon are highest during the winter in northern climates because houses are sealed as tightly as possible. Homes with basements are more likely to have higher radon levels than those built on concrete slabs. While research continues into the sources of indoor radon and ways of controlling it, the most important thing people can do is to test their home, school, or business for radon.

◆ FIGURE 2 Areas in the United States where granite, phosphate-bearing rocks, carbonaceous shales, and uranium occur. These rocks are all potential sources of radon gas.

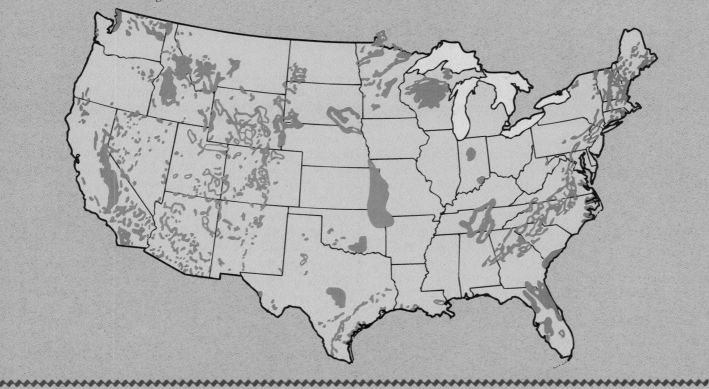

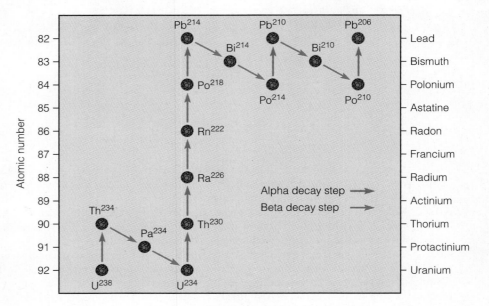

◆ FIGURE 17.19 Radioactive decay series for uranium 238 to lead 206. Radioactive uranium 238 decays to its stable end product, lead 206, by eight alpha and six beta decay steps. A number of different isotopes are produced as intermediate steps in the decay series.

Sources of Uncertainty

The most accurate radiometric dates are obtained from igneous rocks. As a magma cools and begins to crystallize, radioactive parent atoms are separated from previously formed daughter atoms. Because they are the right size, some radioactive parent atoms are incorporated into the crystal structure of certain minerals. The stable daughter atoms, however, are a different size than the radioactive parent atoms and consequently cannot fit into the crystal structure of the same mineral as the parent atoms. Therefore when the magma begins to crystallize, the mineral will contain radioactive parent atoms but no stable daughter atoms (◆ Fig. 17.21). Thus, the time that is being measured is the time of crystallization of the mineral containing the radioactive atoms, and not the time of formation of the radioactive atoms.

Except in unusual circumstances, sedimentary rocks cannot be radiometrically dated, because one would be measuring the age of a particular mineral rather than the time that it was deposited as a sedimentary particle. One of the few instances in which radiometric dates can be obtained on sedimentary rocks is when the mineral glauconite is present. Glauconite is a greenish mineral containing radioactive potassium 40, which decays to argon 40 (● Table 17.1). It forms in certain marine environments as a result of chemical reactions with clay minerals during the conversion from sedi-

ments to sedimentary rock. Thus, it forms when the sedimentary rock forms, and a radiometric date indicates the time of the sedimentary rock's origin. However, because the daughter product argon is a gas, it can easily escape from a mineral. Therefore, any date obtained from glauconite, or any other mineral containing the potassium 40–argon 40 pair, must be considered a minimum age.

To obtain accurate radiometric dates, geologists must be sure that they are dealing with a closed system, meaning that neither parent nor daughter atoms have been added or removed from the system since crystallization and that the ratio between them results only from radioactive decay. Otherwise, an inaccurate date will result. If daughter atoms have leaked out of the mineral being analyzed, the calculated age will be too young; if parent atoms have been removed, the calculated age will be too great.

Leakage may occur if the rock is heated as occurs during metamorphism. If this happens, some of the parent or daughter atoms may be driven from the mineral being analyzed, resulting in an inaccurate age determination. If the daughter product was completely removed, then one would be measuring the time since metamorphism (a useful measurement itself), and not the time since crystallization of the mineral (◆ Fig. 17.22). Because heat affects the parent-daughter ratio, metamorphic rocks are difficult to age date accurately. Remember that while the parent-daughter ratio may be

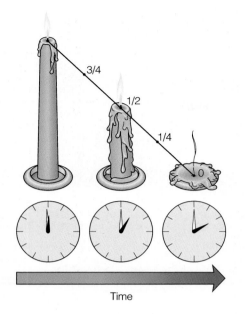

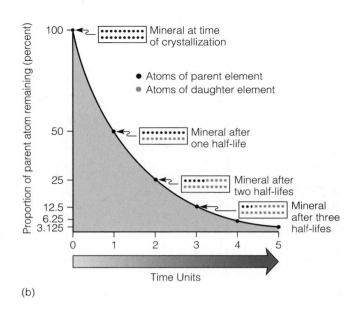

(b)

(a)

◆ FIGURE 17.20 (*a*) Uniform, linear depletion is characteristic of many familiar processes. (*b*) Geometric radioactive decay curve, in which each time unit represents one half-life, and each half-life is the time it takes for one-half of the parent element to decay to the daughter element.

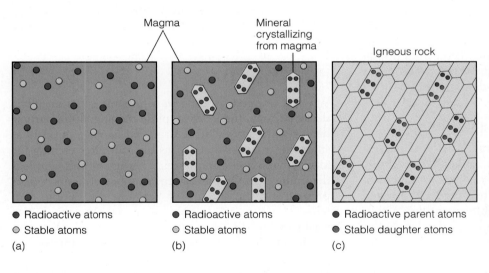

◆ FIGURE 17.21 (*a*) A magma contains both radioactive and stable atoms. (*b*) As the magma cools and begins to crystallize, some radioactive atoms are incorporated into certain minerals because they are the right size and can fit into the crystal structure. Therefore, at the time of crystallization, the mineral will contain 100% radioactive parent atoms and 0% stable daughter atoms. (*c*) After one half-life, 50% of the radioactive parent atoms will have decayed to stable daughter atoms.

affected by heat, the decay rate of the parent element remains constant, regardless of any physical or chemical changes.

To ensure that an accurate radiometric date is obtained, geologists must make sure that the sample is fresh and unweathered and that it has not been subjected to high temperatures or intense pressures after crystallization. Furthermore, it is sometimes possible to cross-check the date obtained by measuring the parent-daughter ratio of two different radioactive elements in

● **TABLE 17.1** Five of the Principal Long-Lived Radioactive Isotope Pairs Used in Radiometric Dating

Isotopes		Half-Life of Parent (Years)	Effective Dating Range (Years)	Minerals and Rocks That Can Be Dated
Parent	Daughter			
Uranium 238	Lead 206	4.5 billion	10 million to 4.6 billion	Zircon Uraninite
Uranium 235	Lead 207	704 million		
Thorium 232	Lead 208	14 billion		
Rubidium 87	Strontium 87	48.8 billion	10 million to 4.6 billion	Muscovite Biotite Potassium-feldspar Whole metamorphic or igneous rock
Potassium 40	Argon 40	1.3 billion	100,000 to 4.6 billion	Glauconite Hornblende Muscovite Whole volcanic rock Biotite

the same mineral. For example, naturally occurring uranium consists of both uranium 235 and uranium 238 isotopes. Through various decay steps, uranium 235 decays to lead 207, whereas uranium 238 decays to lead 206 (Fig. 17.19). If the minerals containing both uranium isotopes have remained closed systems, the ages obtained from each parent-daughter ratio should be in close agreement and therefore should indicate the time of crystallization of the magma. If the ages do not closely agree, other samples must be taken and ratios measured to see which, if either, date is correct.

Long-Lived Radioactive Isotope Pairs

Table 17.1 shows the five common, long-lived parent-daughter isotope pairs used in radiometric dating. Long-lived pairs have half-lives of millions or billions of years. All of these were present when the Earth formed and are still present in measurable quantities. Other shorter-lived radioactive isotope pairs have decayed to the point that only small quantities near the limit of detection remain.

The most commonly used isotope pairs are the uranium-lead and thorium-lead series, which are used principally to date ancient igneous intrusives, lunar samples, and some meteorites. The rubidium-strontium pair is also used for very old samples and has been effective in dating the oldest rocks on Earth as well as meteorites. The potassium-argon method is typically used for dating fine-grained volcanic rocks from which

individual crystals cannot be separated; hence the whole rock is analyzed. However, argon is a gas, so great care must be taken to assure that the sample has not been subjected to heat, which would allow argon to escape; such a sample would yield an age that is too young. Other long-lived radioactive isotope pairs exist, but they are rather rare and are used only in special situations.

Fission Track Dating

When a uranium isotope in a mineral emits an alpha decay particle, the heavy, rapidly moving alpha particle damages the crystal structure. The damage appears as small linear tracks that are visible only after etching the mineral with hydrofluoric acid. The age of the sample is determined on the basis of the number of fission tracks present and the amount of uranium the sample contains. The older the sample, the greater the number of tracks (◆ Fig. 17.23).

Fission track dating is of particular interest to geologists because the technique can be used to date samples ranging from only a few hundred to hundreds of millions of years in age. It is most useful for dating samples between about 40,000 and one million years ago, a period for which other dating techniques are not particularly suitable. One of the problems in fission track dating occurs when the rocks have been subjected to high temperatures. If this happens, the damaged crystal structures are repaired by annealing, and conse-

452   Chapter 17

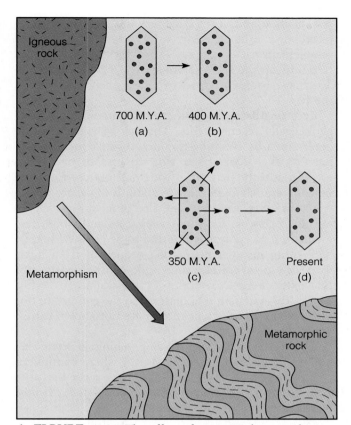

◆ FIGURE 17.22 The effect of metamorphism in driving out daughter atoms from a mineral that crystallized 700 million years ago (M.Y.A.). The mineral is shown immediately after crystallization (*a*), then at 400 million years (*b*), when some of the parent atoms had decayed to daughter atoms. Metamorphism at 350 M.Y.A. (*c*) drives the daughter atoms out of the mineral into the surrounding rock. (*d*) Assuming the rock has remained a closed chemical system throughout its history, dating the mineral today yields the time of metamorphism, while dating the whole rock provides the time of its crystallization, 700 M.Y.A.

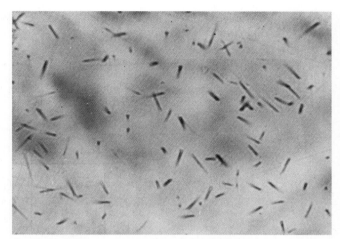

◆ FIGURE 17.23 Each fission track (about 16 μm in length) in this apatite crystal is the result of the radioactive decay of a uranium atom. The crystal, which has been etched with hydrofluoric acid to make the fission tracks visible, comes from one of the dikes at Shiprock, New Mexico, and has a calculated age of 27 million years. (Photo courtesy of Charles W. Naeser, U.S. Geological Survey.)

quently the tracks disappear. In such instances, the calculated age will be younger than the actual age.

Radiocarbon and Tree-Ring Dating Methods

Carbon is an important element in nature and is one of the basic elements found in all forms of life. It has three isotopes; two of these, carbon 12 and 13, are stable, whereas carbon 14 is radioactive. Carbon 14 has a half-life of 5,730 years plus or minus 30 years. The **carbon 14 dating technique** is based on the ratio of carbon 14 to carbon 12 and is generally used to date once-living material.

The short half-life of carbon 14 makes this dating technique practical only for specimens younger than about 70,000 years. Consequently, the carbon 14 dating method is especially useful in archeology and has greatly aided in unraveling the events of the latter portion of the Pleistocene Epoch.

Carbon 14 is constantly formed in the upper atmosphere by the bombardment of cosmic rays, which are high-energy particles (mostly protons). These high-energy particles strike the atoms of upper-atmospheric gases, splitting their nuclei into protons and neutrons. When a neutron strikes the nucleus of a nitrogen atom (atomic number 7, atomic mass number 14), it may be absorbed into the nucleus and a proton emitted. Thus, the atomic number of the atom decreases by one, while the atomic mass number stays the same. Because the atomic number has changed, a new element, carbon 14 (atomic number 6, atomic mass number 14), is formed. The newly formed carbon 14 is rapidly assimilated into the carbon cycle and, along with carbon 12 and 13, is absorbed in a nearly constant ratio by all living organisms (◆ Fig. 17.24). When an organism dies, however, carbon 14 is not replenished, and the ratio of carbon 14

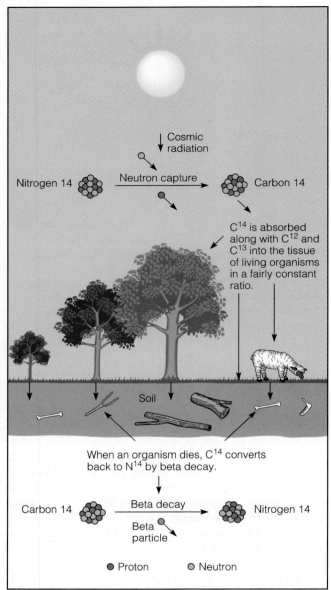

Cosmic radiation

Nitrogen 14   Neutron capture   Carbon 14

C^{14} is absorbed along with C^{12} and C^{13} into the tissue of living organisms in a fairly constant ratio.

Soil

When an organism dies, C^{14} converts back to N^{14} by beta decay.

Carbon 14   Beta decay   Nitrogen 14

Beta particle

● Proton   ○ Neutron

◆ FIGURE 17.24 The carbon cycle showing the formation, dispersal, and decay of carbon 14.

to carbon 12 decreases as carbon 14 decays back to nitrogen by a single beta decay step (Fig. 17.24).

Currently, the ratio of carbon 14 to carbon 12 is remarkably constant in both the atmosphere and living organisms. There is good evidence, however, that the production of carbon 14, and thus the ratio of carbon 14 to carbon 12, has varied somewhat over the past

several thousand years. This was determined by comparing ages established by carbon 14 dating of wood samples against those established by counting annual tree-rings in the same samples. As a result, carbon 14 ages have been corrected to reflect such variations in the past.

Tree-ring dating is another useful method for dating geologically recent events. The age of a tree can be determined by counting the growth-rings in the lower part of the stem. Each ring represents one year's growth, and the pattern of wide and narrow rings can be compared among trees to establish the exact year in which the rings were formed. The procedure of matching ring patterns from numerous trees and wood fragments in a given area is referred to as cross-dating. By correlating distinctive tree-ring sequences from living to nearby dead trees, a time scale can be constructed that extends back to about 14,000 years ago (◆ Fig. 17.25). By matching ring patterns to the composite ring scale, wood samples whose ages are not known can be accurately dated.

The applicability of tree-ring dating is somewhat limited because it can only be used where continuous tree records are found. It is therefore most useful in arid regions, particularly the southwestern United States.

▼ THE DEVELOPMENT OF THE GEOLOGIC TIME SCALE

The geologic time scale is a hierarchical scale in which the 4.6-billion-year history of the Earth is divided into time units of varying duration (Fig. 17.2). The geologic time scale was not developed by any one individual, but rather evolved, primarily during the nineteenth century, through the efforts of many people. By applying relative dating methods to rock outcrops, geologists in England and western Europe defined the major geologic time units without the benefit of radiometric dating techniques. Using the principles of superposition and fossil succession, they were able to correlate various rock exposures and piece together a composite geologic section. This composite section is, in effect, a relative time scale because the rocks are arranged in their correct sequential order.

By the beginning of the twentieth century, geologists had developed a relative geologic time scale, but did not yet have any absolute dates for the various time unit boundaries. Following the discovery of radioactivity

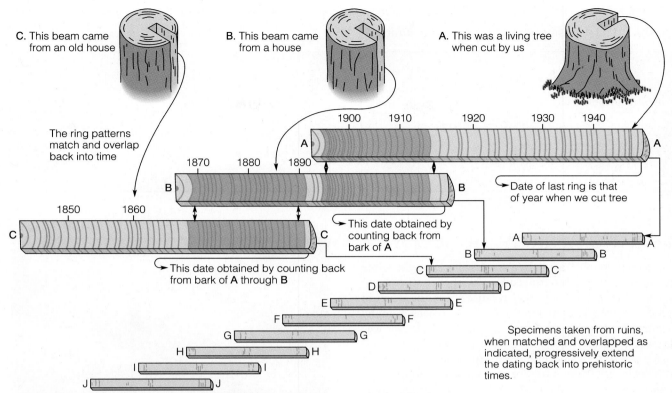

◆ FIGURE 17.25 In the cross-dating method, tree-ring patterns from different woods are matched against each other to establish a ring-width chronology backward in time.

near the end of the last century, radiometric dates were added to the relative geologic time scale.

Because sedimentary rocks, with rare exceptions, cannot be radiometrically dated, geologists have had to rely on interbedded volcanic rocks and igneous intrusions to apply absolute dates to the boundaries of the various subdivisions of the geologic time scale (◆ Fig. 17.26). An ash fall or lava flow provides an excellent marker bed that is a time-equivalent surface, supplying a minimum age for the sedimentary rocks below and a maximum age for the rocks above. Ash falls are particularly useful because they may fall over both marine and nonmarine sedimentary environments and can provide a connection between these different environments.

Thousands of absolute ages are now known for sedimentary rocks of known relative ages, and these absolute dates have been added to the relative time scale. In this way, geologists have been able to determine the absolute ages of the various geologic periods and to determine their durations (Fig. 17.2).

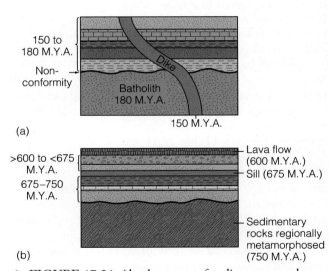

◆ FIGURE 17.26 Absolute ages of sedimentary rocks can be determined by dating associated igneous rocks. In (*a*) and (*b*), sedimentary rocks are bracketed by rock bodies for which absolute ages have been determined.

● **TABLE 17.2** Classification of Stratigraphic Units

Units Defined by Content		Units Expressing or Related to Geologic Time	
Lithostratigraphic Units	Biostratigraphic Units	Time-Stratigraphic Units	Time Units

Supergroup
  Group
    Formation
      Member
        Bed
	Biozones	Eonothem..........................Eon
		  Erathem ....Era
		    System..........................Period
		      Series..........................Epoch
		        Stage ....Age

▼ STRATIGRAPHIC TERMINOLOGY

The recognition of a relative time scale brought some order to stratigraphy (the study of the composition, origin, and areal and age relationships of layered rocks). Problems remained, however, because many rock units are time transgressive, that is, a rock unit was deposited during one period in a particular area and during another period elsewhere. In order to deal with both rocks and time, modern stratigraphic terminology includes two fundamentally different kinds of units: those defined by their content and those related to geologic time (● Table 17.2).

Units defined by their content include **lithostratigraphic** and **biostratigraphic units**. Lithostratigraphic (*lith-* and *litho-* are prefixes meaning stone or stonelike) units are defined by physical attributes of the rocks, such as rock type, with no consideration of time of origin. The basic lithostratigraphic unit is the **formation**, which is a mappable rock unit with distinctive upper and lower boundaries (◆ Fig. 17.27). Formations may be lumped together into larger units called *groups* or *supergroups* or divided into smaller units called *members* and *beds* (Table 17.2).

Biostratigraphic units are bodies of strata containing recognizably distinct fossils. Fossil content is the only criterion used to define them. Biostratigraphic unit boundaries do not necessarily correspond to lithostratigraphic boundaries (◆ Fig. 17.28). The fundamental biostratigraphic unit is the **biozone**. Several types of biozones are recognized, one of which, the *assemblage zone,* was discussed in the section on correlation.

The units related to geologic time include **time-stratigraphic units** (also know as chronstratigraphic units) and **time units** (Table 17.2). Time-stratigraphic units are units of rock that were deposited during a

◆ FIGURE 17.27 The Madison Group in Montana consists of two formations, the Lodgepole Formation and the overlying Mission Canyon Formation. The Mission Canyon Formation is the rock unit exposed on the skyline. The underlying Lodgepole Formation is the rock covered on the slopes below.

specific interval of time. The **system** is the fundamental time-stratigraphic unit. It is based on rocks in a particular area, the stratotype, and is recognized beyond the stratotype area primarily on the basis of fossil content.

Time units are simply units designating specific intervals of geologic time. The basic time unit is the **period**, but smaller units including epoch and age are also recognized. The time units period, epoch, and age correspond to the time-stratigraphic units system, series, and stage, respectively (Table 17.2). For example, the Cambrian Period is defined as the time during which strata of the Cambrian System were deposited. Time units of higher rank than period also exist. Eras include several periods, while eons include two or more eras.

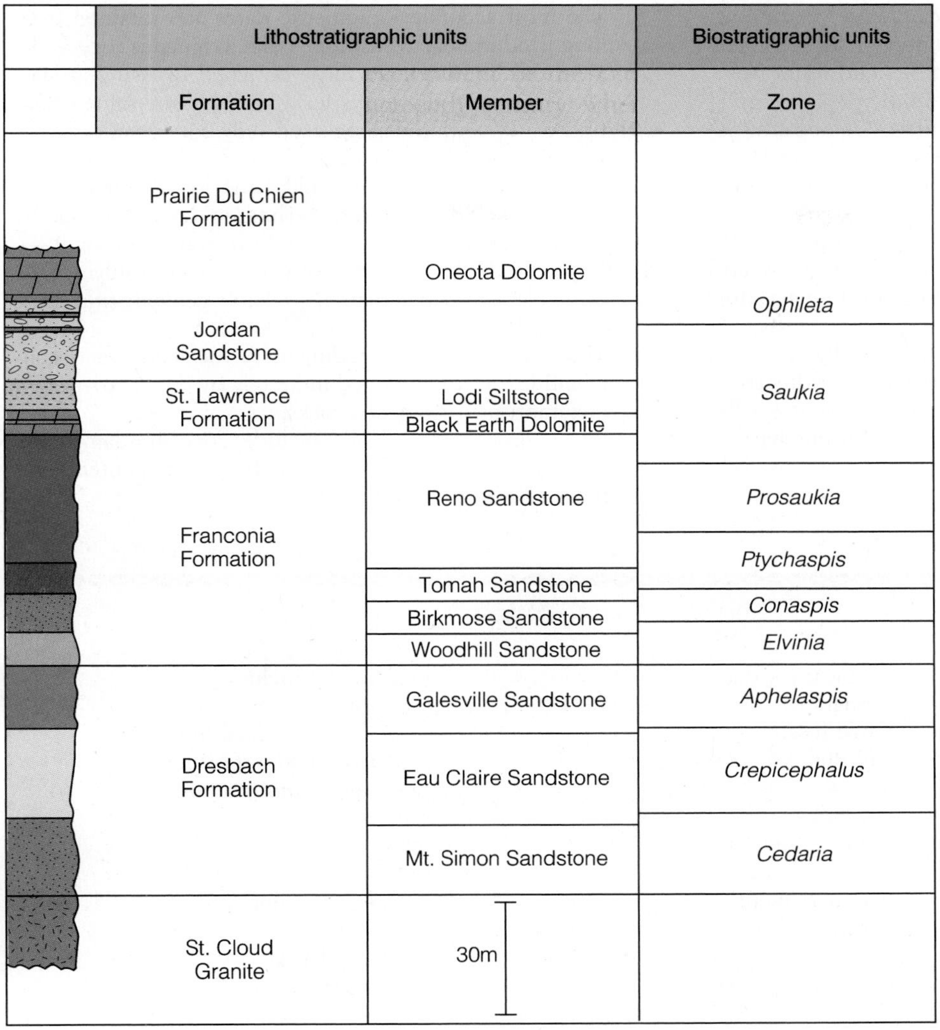

Lithostratigraphic units		Biostratigraphic units
Formation	Member	Zone
Prairie Du Chien Formation		
	Oneota Dolomite	*Ophileta*
Jordan Sandstone		*Saukia*
St. Lawrence Formation	Lodi Siltstone	
	Black Earth Dolomite	
Franconia Formation	Reno Sandstone	*Prosaukia*
		Ptychaspis
	Tomah Sandstone	*Conaspis*
	Birkmose Sandstone	*Elvinia*
	Woodhill Sandstone	
Dresbach Formation	Galesville Sandstone	*Aphelaspis*
	Eau Claire Sandstone	*Crepicephalus*
	Mt. Simon Sandstone	*Cedaria*
St. Cloud Granite	30m	

◆ FIGURE 17.28 Relationships of biostratigraphic units to lithostratigraphic units in southeastern Minnesota. Notice that the biozone boundaries do not necessarily correspond to lithostratigraphic boundaries.

Chapter Summary

1. Relative dating involves placing geologic events in a sequential order as determined from their position in the rock record. Absolute dating results in specific dates for events, expressed in years before the present.

2. During the eighteenth and nineteenth centuries, attempts were made to determine the age of the Earth based on scientific evidence rather than revelation. While some attempts were quite ingenious, they yielded a variety of ages that now are known to be much too young.

3. James Hutton believed that present-day processes operating over long periods of time could explain all the geologic features of his native Scotland. His observations were instrumental in establishing the basis for the principle of uniformitarianism.

4. Uniformitarianism, as articulated by Charles Lyell, soon became the guiding principle of geology. It holds that the laws of nature have been constant through time and that the same processes operating today have operated in the past, although not necessarily at the same rates.

5. In addition to uniformitarianism, the principles of superposition, original horizontality, lateral continuity, crosscutting relationships, inclusions, and fossil succession are basic for determining relative geologic ages and for interpreting the geologic history of the Earth.

6. Surfaces of discontinuity that encompass significant amounts of geologic time are common in the geologic record. Such surfaces are unconformities and result from times of nondeposition, erosion, or both.

7. Correlation is the stratigraphic practice of demonstrating equivalency of units in different areas. Time equivalence is most commonly demonstrated by correlating strata containing similar fossils.

8. Radioactivity was discovered during the late nineteenth century, and soon thereafter radiometric dating techniques allowed geologists to determine absolute ages for geologic events.

9. Absolute age dates for rock samples are usually obtained by determining how many half-lives of a radioactive parent element have elapsed since the sample originally crystallized. A half-life is the time it takes for one-half of the radioactive parent element to decay to a stable daughter element.

10. The most accurate radiometric dates are obtained from long-lived radioactive isotope pairs in igneous rocks. The most reliable dates are those obtained by using at least two different radioactive decay series in the same rock.

11. Carbon 14 dating can be used only for organic matter such as wood, bones, and shells and is effective back to about 70,000 years ago. Unlike the long-lived isotopic pairs, the carbon 14 dating technique determines age by the ratio of radioactive carbon 14 to stable carbon 12.

12. Through the efforts of many geologists applying the principles of relative dating, a relative geologic time scale was established.

13. Most absolute ages of sedimentary rocks and their contained fossils are obtained indirectly by dating associated metamorphic or igneous rocks.

14. Stratigraphic terminology includes two fundamentally different kinds of units: those based on content and those related to geologic time.

Important Terms

absolute dating
alpha decay
angular unconformity
assemblage range zone
beta decay
biostratigraphic unit
biozone
carbon 14 dating technique
correlation
daughter element
disconformity
electron capture

fission track dating
formation
guide fossil
half-life
hiatus
key bed
lithostratigraphic unit
nonconformity
parent element
period
principle of cross-cutting relationships
principle of fossil succession

principle of inclusions
principle of lateral continuity
principle of original horizontality
principle of superposition
principle of uniformitarianism
radioactive decay
relative dating
system
time-stratigraphic unit
time unit
tree-ring dating
unconformity

Review Questions

1. In which type of unconformity are the beds parallel to each other?
   a. ___ nonconformity; b. ___ angular unconformity;
   c. ___ disconformity; d. ___ hiatus; e. ___ none of these.

2. Placing geologic events in sequential order as determined by their position in the rock record is called:
   a. ___ absolute dating; b. ___ uniformitarianism;
   c. ___ relative dating; d. ___ correlation; e. ___ historical dating.

3. If a rock is heated during metamorphism and the daughter atoms migrate out of a mineral that is subsequently radiometrically dated, an inaccurate date will be obtained. This date will be _____the actual date.
   a. ___ younger than; b. ___ older than; c. ___ the same as; d. ___ it cannot be determined; e. ___ none of these.

4. Which of the following methods can be used to demonstrate age equivalency of rock units?
   a. ___ lateral tracing; b. ___ radiometric dating;
   c. ___ guide fossils; d. ___ position in a sequence;
   e. ___ all of these.

5. Which fundamental geologic principle states that the oldest layer is on the bottom of a vertical succession of sedimentary rocks and the youngest is on top?
   a. ___ lateral continuity; b. ___ fossil succession;
   c. ___ original horizontality; d. ___ superposition;
   e. ___ cross-cutting relationships.

6. In which type of radioactive decay are two protons and two neutrons emitted from the nucleus?
   a. ___ alpha; b. ___ beta; c. ___ electron capture;
   d. ___ fission track; e. ___ radiocarbon.

7. The author of *Principles of Geology* and the principal advocate and interpreter of uniformitarianism was:
a. ___ Hutton; b. ___ Steno; c. ___ Lyell; d. ___ Smith;
e. ___ Playfair.
8. The era younger than the Mesozoic is the:
a. ___ Proterozoic; b. ___ Archean; c. ___ Paleozoic;
d. ___ Phanerozoic; e. ___ Cenozoic.
9. Which of the following is not a long-lived radioactive isotope pair?
a. ___ uranium-lead; b. ___ thorium-lead;
c. ___ potassium-argon; d. ___ carbon-nitrogen;
e. ___ none of these.
10. What is being measured in radiometric dating?
a. ___ the time when the radioactive isotope formed;
b. ___ the time of crystallization of a mineral containing an isotope; c. ___ the amount of the parent isotope only;
d. ___ when the dated mineral became part of a sedimentary rock; e. ___ when the stable daughter isotope was formed.
11. If a radioactive element has a half-life of 4 million years, the amount of parent material remaining after 12 million years of decay will be what fraction of the original amount?
a. ___ $\frac{1}{32}$; b. ___ $\frac{1}{16}$; c. ___ $\frac{1}{8}$; d. ___ $\frac{1}{4}$; e. ___ $\frac{1}{2}$.
12. In carbon 14 dating, which ratio is being measured?
a. ___ the parent to daughter isotope; b. ___ C^{14}/N^{14};
c. ___ C^{12}/C^{13}; d. ___ C^{12}/N^{14}; e. ___ C^{12}/C^{14}.
13. How many half-lives are required to yield a mineral with 625 atoms of U^{238} and 19,375 atoms of Pb^{206}?
a. ___ 4; b. ___ 5; c. ___ 6; d. ___ 8; e. ___ 10.
14. What is the difference between relative and absolute dating of geologic events?
15. What are the six fundamental principles used in relative age dating? Why are they so important in deciphering Earth history?
16. How do lithostratigraphic units differ from time-stratigraphic units?
17. Define the three types of unconformities. Why are unconformities important in relative age dating?
18. Explain how a geologist would determine the relative ages of a granite batholith and an overlying sandstone formation.
19. Why is the principle of uniformitarianism important to geologists?

20. Are volcanic eruptions, earthquakes, and storm deposits geologic events encompassed by uniformitarianism?
21. What is radon, and why is it so dangerous to humans?
22. What are assemblage range zones? How can such zones be used to demonstrate time equivalency of strata in widely separated areas?
23. If you wanted to calculate the absolute age of an intrusive body, what information would you need?
24. Assume a hypothetical radioactive isotope with an atomic number of 150 and an atomic mass number of 300 emits five alpha decay particles and three beta decay particles and undergoes two electron capture steps. What are the atomic number and atomic mass number of the resulting stable daughter product?
25. What are some of the potential sources of error in radiometric dating?
26. How can geologists be sure that the absolute age dates they obtain from igneous rocks are accurate?
27. Why is it difficult to date sedimentary and metamorphic rocks radiometrically?
28. How does the carbon 14 dating technique differ from uranium-lead dating methods?
29. How did the geologic time scale evolve?
30. Using the principles of relative dating, give the geologic history for the diagram below.

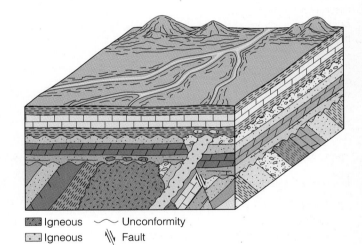

Igneous   Unconformity
Igneous   Fault

Additional Readings

Albritton, C. C., Jr. 1980. *The abyss of time.* San Francisco: Freeman, Cooper.
_____ . 1984. Geologic time. *Journal of Geological Education* 32, no. 1:29–37.
Berry, W. B. N. 1987. *Growth of a prehistoric time scale.* 2d ed. Palo Alto, Calif.: Blackwell Scientific.

Boslough, J. 1990. The enigma of time. *National Geographic* 177, no. 3:109–32.
Geyh, M. A., and H. Schleicher. 1990. *Absolute age determination.* New York: Springer-Verlag.
Gould, S. J. 1987. *Time's arrow, time's cycle.* Cambridge, Mass.: Harvard Univ. Press.

Harland, W. B., R. L. Armstrong, A. V. Cox, L. E. Craig, A. G. Smith, and D. G. Smith. 1990. *A geologic time scale 1989.* New York: Cambridge Univ. Press.

Itano, W. M. and N. F. Ramsey. 1993. Accurate measurement of time. *Scientific American* 269, no. 1:56–57.

Ramsey, N. F. 1988. Precise measurement of time. *American Scientist* 76, no. 1:42–49.

Wetherill, G. W. 1982. Dating very old objects. *Natural History* 91, no. 9:14–20.

CHAPTER
18

Chapter Outline

The French botanist-geologist Jean Baptiste Pierre Antoine de Monet de Lamarck (1744–1829) was the first scientist to propose a formal, widely accepted theory of evolution.

Fossils and Evolution

Prologue

On December 27, 1831, Charles Robert Darwin departed from Devonport, England, aboard the H.M.S. *Beagle* as an unpaid naturalist. Nearly five years later the *Beagle* returned to England, and Darwin never ventured far from home again. Nevertheless, the 64,360-kilometer voyage was the most important experience of his life, an experience that changed his view of nature and ultimately revolutionized all of science.

When Darwin sailed aboard the *Beagle,* he was a little-known recent graduate in theology from Christ's College. In fact, his father, Dr. Robert Darwin, had sent him to Christ's College as a last resort; Charles showed little aptitude for academics, except perhaps for science, and had already withdrawn from medical studies at Edinburgh. Although he completed his theological studies, he was rather indifferent to religion. He nevertheless fully accepted the biblical account of creation as historical fact. Darwin's belief in biblical creation initially endeared him to the *Beagle*'s captain, Robert Fitzroy, but his views changed during the voyage, and his relationship with Captain Fitzroy became strained.

When the voyage began, Darwin was nominally a clergyman with interests in the sciences, especially botany, zoology, and geology. He suffered from prolonged attacks of seasickness, but nevertheless kept detailed notes on his observations; he collected, cataloged, and dissected specimens and eventually became a professional naturalist. The *Beagle* made several lengthy stops in South America where Darwin explored rain forests, experienced an earthquake, and collected fossils. The fossils he collected were clearly related to the living sloth, armadillo, and llama and implied that living species descended from fossil species. Such evidence caused him to question the concept of fixity of species, a concept held by those who accepted biblical creation.

Darwin was particularly fascinated by the plants and animals of oceanic islands. The Cape Verde Islands and the Galapagos Islands are comparable distances west of Africa and South America, respectively. Each island group has its own unique plants and animals, yet these plants and animals most closely resemble those of the nearby continent. The Galapagos, for example, are populated by 13 species of finches (◆ Fig. 18.1), but only one species exists in South America. These finches are adapted to the different habitats occupied on the mainland by various species such as parrots, flycatchers, and toucans.

Darwin reasoned that the Cape Verdes and Galapagos had received colonists from the nearby continents and that "such colonists would be liable to modification—the principle of inheritance still betraying their original birthplace."* The significance of this statement is that it revealed a change in Darwin's view of nature; he no longer accepted the idea of fixity of species. For example, he proposed that an ancestral species of finch had somehow reached the Galapagos Islands from South America and differentiated into the various types he observed (Fig. 18.1). Furthermore, he had read Charles Lyell's *Principles of Geology* during his voyage and accepted

. . . .

*C. Darwin, *The Origin of Species* (New York: New American Library, 1958), p. 377.

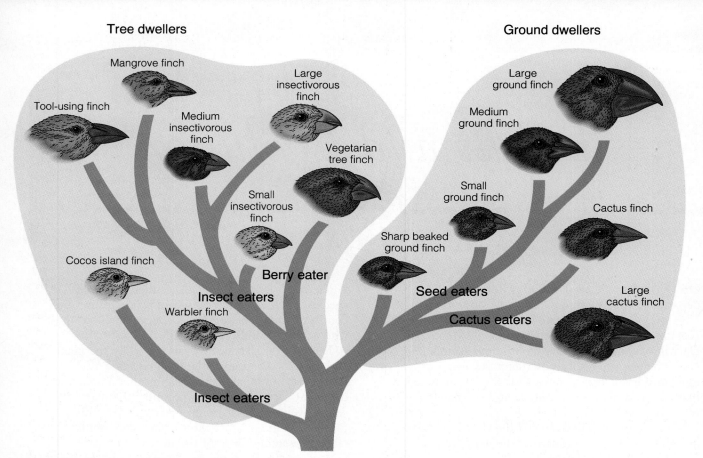

Tree dwellers

Ground dwellers

Mangrove finch

Tool-using finch

Medium insectivorous finch

Large insectivorous finch

Vegetarian tree finch

Small insectivorous finch

Cocos island finch

Berry eater

Insect eaters

Warbler finch

Insect eaters

Large ground finch

Medium ground finch

Small ground finch

Sharp beaked ground finch

Cactus finch

Seed eaters

Cactus eaters

Large cactus finch

◆ FIGURE 18.1 Darwin's finches arranged to show evolutionary relationships. The six species on the right are ground-dwellers, and the others are adapted for life in trees. Notice that the shapes of the beaks of finches in both groups vary depending on diet.

uniformitarianism and the great age of the Earth. In short, he had come to view nature as dynamic rather than static.

⇞ INTRODUCTION

The term *evolution* comes from the Latin, meaning unrolled. Roman books were written on parchment and rolled on wooden rods, so they were unrolled, or evolved, as they were read. In modern usage, evolution usually refers to change through time. If evolution is defined simply as change through time, it is a pervasive phenomenon; stars evolve, the Earth has evolved and continues to do so, languages and social systems evolve, and organisms evolve.

The biological concept of evolution, most appropriately called *organic evolution,* is concerned with changes in organisms that are inheritable. Evolution so defined is observable, at least on a small scale. For example, the proportion of hereditary determinants in populations of moths in Great Britain changed in such a way that their color changed from light to dark in several generations. Large-scale evolution is not directly observable but can be inferred from fossil evidence and from studies of comparative anatomy, embryology, and biochemistry of living organisms.

As we discussed in Chapter 1, theories are scientific explanations of natural phenomena that have a large body of supporting evidence. The **theory of evolution** is such an explanation. The central claim of this theory is that all living organisms are the evolutionary descendants of life-forms that existed during the past. As a scientific theory, it meets the usual criteria for theories: it is a naturalistic explanation, and it can be tested so that its validity can be assessed.

Fossils are the remains or traces of prehistoric organisms that have been preserved in rocks of the Earth's crust. In addition to their use in determining relative ages of strata (see Chapter 17), fossils are important in determining environments of deposition (see Chapter 5), and they provide some of the evidence for evolution.

During most of historic time, most people did not recognize that fossils were the remains of organisms. Such perceptive observers as Leonardo da Vinci in 1508, Robert Hooke in 1665, and Nicolas Steno in 1667 recognized the true nature of fossils, but their views were largely ignored. By the late eighteenth century, however, the evidence had become overwhelming that fossils had indeed once been living organisms. By the nineteenth century, it was also apparent that many fossils represented types of organisms now extinct.

⩔ THE EVOLUTION OF AN IDEA

Evolution as a biological concept was seriously considered by some naturalists even before Charles Darwin was born. Indeed, as early as 600 B.C., the ancient Greeks speculated on possible interrelationships among organisms; but overall neither the ancient world nor the medieval gave much thought to evolution. In fact, the prevailing belief well into the eighteenth century was that all important knowledge was contained in the works of Aristotle and the Bible. One did not have to observe nature to understand it, since all the answers were in these two sources, particularly in the first two chapters of Genesis. Literally interpreted, Genesis was taken as the final word on the origin and diversity of life and much of Earth history. According to this view, all species of plants and animals were perfectly fashioned by God during the days of creation and have remained fixed and immutable ever since. To question divine creation in any way was regarded as heresy.

In eighteenth-century Europe, the social and intellectual climate changed, the absolute authority of the Church in all matters declined, and ironically, the very naturalists who were determined to find physical evidence supporting Genesis as a factual, historical account were responsible for finding more and more evidence that could not be explained by a literal reading of Scripture. This was particularly true of the evidence for a worldwide catastrophic flood. For example, the sedimentary deposits that traditionally had been attributed to the biblical deluge showed clear indications of a noncatastrophic origin. Many naturalists reasoned that this evidence truly reflected the conditions that prevailed when the strata were deposited, for to infer otherwise implied a deception on the part of the Creator, a thesis they could not accept.

Most eighteenth-century naturalists did not abandon the Judeo-Christian tradition, however. Rather, they came to accept Genesis as a symbolic account containing important spiritual messages, not a factual account of creation, a view shared today by most theologians and biblical scholars.

These changing attitudes can be attributed in large part to the Enlightenment, an eighteenth-century philosophical movement that relied on rationalism. In science this meant that one could discover natural laws to explain observations of natural systems; all did not depend on divine revelation. The sciences, particularly biology and chemistry, began to acquire a modern look.

In the changed intellectual atmosphere of the eighteenth century, the concepts of uniformitarianism and the great age of the Earth were gradually accepted. The French zoologist Georges Cuvier clearly demonstrated that many types of plants and animals had become extinct, and that fossils differed through time. In view of the fossil evidence, as well as observations and studies of living organisms, many naturalists became convinced that species were not immutable. Change from one species to another—that is, evolution—became an acceptable idea. What was lacking, however, was an overall theoretical framework to explain evolution.

Lamarck and the Giraffe's Neck

Erasmus Darwin, a physician, poet, and naturalist and Charles Darwin's grandfather, was the first to propose a process by which gradual evolution could occur. But it was Jean Baptiste de Lamarck (1744–1829), a French botanist-geologist, who was first taken seriously by his colleagues. Pointing out that organisms are adapted to their environments, he proposed that evolution is an adaptive process whereby organisms acquire traits or characteristics during their lifetimes and pass these acquired traits on to their descendants.

According to Lamarck's theory of *inheritance of acquired characteristics* ancestral short-necked giraffes, for example, stretched their necks to browse on leaves of trees (♦ Fig. 18.2). The environmentally imposed characteristic of neck-stretching was thus expressed in future generations as a longer neck. In other words, the capacity for giraffes to produce offspring with longer necks was acquired through the habit of neck-stretching.

Considering the data available in the early 1800s, Lamarck formulated a reasonable theory. Little was known about how traits are passed from parent to offspring, so Lamarck's inheritance mechanism seemed logical and was accepted by many naturalists as a viable mechanism for evolution. In fact, Lamarck's concept of inheritance was not completely refuted until more modern concepts of inheritance were developed. Based on the work of Gregor Mendel in the 1860s, and developments in the early 1900s, we now know that all organisms possess hereditary determinants called *genes* that cannot be modified by any effort of the organism during its lifetime. Numerous attempts have been made to validate Lamarckian inheritance, including one during this century in the former Soviet Union, but all have failed.

Charles Darwin's Contribution

In 1859 Charles Robert Darwin published *The Origin of Species* in which his ideas on evolution were outlined. Most naturalists almost immediately recognized that evolution provides a unifying theory that explains an otherwise encyclopedic collection of biologic facts. Most, but not all, churches and theologians eventually accepted the idea of evolution as a naturalistic explanation that does not conflict with the purpose and intent of Scripture.

While 1859 may mark the beginning of modern evolutionary biology, Darwin had actually formulated his ideas more than 20 years earlier, but aware of the furor his ideas would generate, had been reluctant to publish. As a matter of fact, when the furor did erupt, Darwin, a semi-invalid most of his life, never defended his theories in public; he left their defense to others.

As discussed in the Prologue, Darwin concluded during the voyage of the *Beagle* that species are not fixed. In fact, by the time Darwin returned home in 1836, he had what he thought was clear evidence that species had indeed changed through time, but he had no idea what might bring about change. By 1838, however, his observations of the selection practiced by animal and plant breeders, and a chance reading of Thomas Malthus's essay on population, gave him the elements necessary to formulate his theories.

Animal and plant breeders selected organisms for desirable traits, bred organisms with those traits, and thereby induced a great amount of change. This practice of **artificial selection** yielded a fantastic variety of

♦ FIGURE 18.2 According to Lamarck's theory of inheritance of acquired characteristics, ancestral short-necked giraffes stretched their necks to reach leaves higher up on trees, and their descendants were born with longer necks.

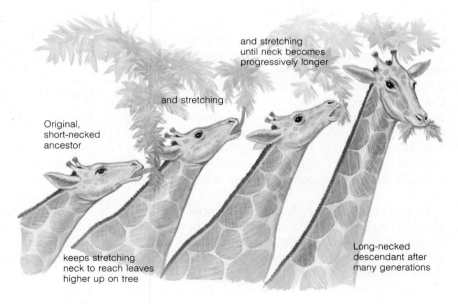

and stretching until neck becomes progressively longer

and stretching

Original, short-necked ancestor

keeps stretching neck to reach leaves higher up on tree

Long-necked descendant after many generations

domestic pigeons, dogs, and plants. If this much change could be produced artificially, was a process that selected among variant types perhaps also acting on natural populations?

Darwin came to appreciate fully the power of selection after reading Malthus's book on population. Malthus argued that far more animals are born than reach maturity, yet the numbers of adult animals of a species remain rather constant. He reasoned that competition for resources resulted in a high infant mortality rate, thus limiting the size of the population. Darwin proposed that a natural process was selecting only a few animals for survival, a process he called **natural selection.**

Natural Selection

For 20 years Darwin kept his ideas on evolution and natural selection largely to himself. But in 1858 he received a letter from Alfred Russell Wallace, a young naturalist working in southern Asia. Wallace had also read Malthus's essay on population and had come to precisely the same conclusion as Darwin. Friends convinced Darwin that his paper and Wallace's should be read at the same session of the Linnaean Society of London. Later some of Darwin's correspondence was published, establishing his scientific priority, and indeed Wallace even insisted that Darwin be given credit

in recognition of his earlier discovery and more thorough documentation of the theory.

The Darwin-Wallace theory of natural selection can be summarized as follows:

1. All populations contain heritable variations—size, speed, agility, coloration, and so forth.
2. Some variations are more favorable than others; that is, some variant types have an edge in the competition for resources and avoidance of predators.
3. Not all young survive to reproductive maturity.
4. Those with favorable variations are more likely to survive and pass on their favorable variations.

Evolution by natural selection is then largely a matter of reproductive success, for only those that reproduce can pass on favorable variations. Of course, favorable variations do not guarantee survival for an individual, but within a population, those with favorable variations are more likely to survive and reproduce.

Suppose, for example, that in an ancestral population of giraffes some individuals had longer necks than most of the others (◆ Fig. 18.3). These long-necked animals could obviously browse higher on trees. The important point here is that some simply have longer necks and therefore enjoy a selective advantage in that they can acquire resources more effectively. Consequently, they are more likely to survive, reproduce, and

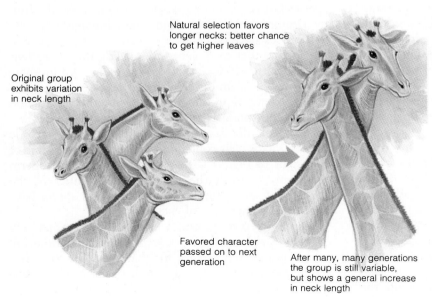

Natural selection favors longer necks: better chance to get higher leaves

Original group exhibits variation in neck length

Favored character passed on to next generation

After many, many generations the group is still variable, but shows a general increase in neck length

◆ FIGURE 18.3 According to the Darwin-Wallace theory of natural selection, the giraffe's long neck evolved as animals with favorable variations were selected for survival.

pass on their favorable variations. Even in the second generation, a few more giraffes might have longer necks, which would give a competitive edge. And so it would go, generation by generation, thereby increasing the proportion of giraffes with longer necks.

▼ MENDEL AND THE BIRTH OF GENETICS

Critics of natural selection were quick to point out that Darwin and Wallace could not account for the origin of variation or explain how it was maintained in populations. They reasoned that should a variant trait arise, it would simply blend with other traits in the population and be lost. At the time, this was a valid criticism, and neither Darwin nor Wallace could effectively answer it. Actually, information that could have given them the answer existed, but it remained in obscurity until 1900.

A monastery in what is now Czechoslovakia may seem to be an unlikely setting for the discovery of the rules of inheritance, and an Austrian monk may seem an unlikely candidate for the "father of genetics." Nevertheless, during the 1860s Gregor Mendel, who later became abbot of the monastery, was doing research that answered some of the inheritance problems that plagued Darwin and Wallace. Unfortunately, Mendel's work was published in an obscure journal and went largely unnoticed. Mendel even sent Darwin a copy of his manuscript, but it apparently lay unread on his bookshelf.

Mendel performed a series of controlled experiments with true-breeding strains of garden peas (strains that when self-fertilized always display the same trait, such as a particular flower color). In one experiment, he transferred pollen from white-flowered plants to red-flowered plants, which produced a second generation of all red-flowered plants. But when left to self-fertilize, these plants yielded a third generation with a ratio of red-flowered plants to white-flowered plants of slightly over 3 to 1 (◆ Fig. 18.4).

Mendel concluded from his experiments that traits such as flower color are controlled by a pair of factors, or what we now call **genes.** He also concluded that genes controlling the same trait occur in alternate forms, or **alleles;** that one allele may be dominant over another; and that offspring receive one allele of each pair from each parent. When an organism produces sex cells—pollen and ovules in plants, and sperm and eggs in animals—only one allele for a trait is present in each sex cell (Fig. 18.4). For example, if R represents the allele for red flower color, and r represents white, the

offspring may inherit the combinations of alleles symbolized as RR, Rr, or rr. And since R is dominant over r, only those offspring with the rr combination will have white flowers.

The most important aspects of Mendel's work can be summarized as follows: the factors (genes) controlling traits do not blend during inheritance, but are transmitted as discrete entities; and even though particular traits may not be expressed in each generation, they are not lost. Therefore, some variation in populations is accounted for by alternate expression of genes (alleles), and variation can be maintained.

Even though Mendelian genetics explains much about heredity, we now know the situation is much more complex. For example, our discussion has relied upon a single gene controlling a trait, but, in fact, most traits are controlled by many genes. Nevertheless, Mendel discovered the answers Darwin and Wallace needed, though they went unnoticed until three independent researchers rediscovered them in 1900.

Genes and Chromosomes

The cells of all organisms contain threadlike structures called **chromosomes.** Chromosomes are complex, double-stranded, helical molecules of **deoxyribonucleic acid (DNA)** (◆ Fig. 18.5). Specific segments, or regions, of the DNA molecule are the basic hereditary units, the genes.

The number of chromosomes is specific for a single species but varies among species. For example, the fruit fly *Drosophila* has 8 chromosomes, humans have 46, and horses have 64. However, chromosomes occur in pairs, pairs that carry genes controlling the same characteristics. Remember that the genes on chromosome pairs may occur in different forms, alleles.

In sexually reproducing organisms, the production of eggs and sperm results when parent cells undergo a type of cell division known as *meiosis.* The meiotic process yields cells containing only one chromosome of each pair (◆ Fig. 18.6a). Accordingly, eggs and sperm have only one-half the chromosome number of the parent cell; for example, human eggs and sperm have 23 chromosomes, one of each pair.

During reproduction, a sperm fertilizes an egg (◆ Fig. 18.6b), producing a fertilized egg with the full chromosome complement for that species—46 in humans. As Mendel deduced from his garden pea experiments, one-half of the genetic makeup of the fertilized egg comes from each parent. The fertilized egg, how-

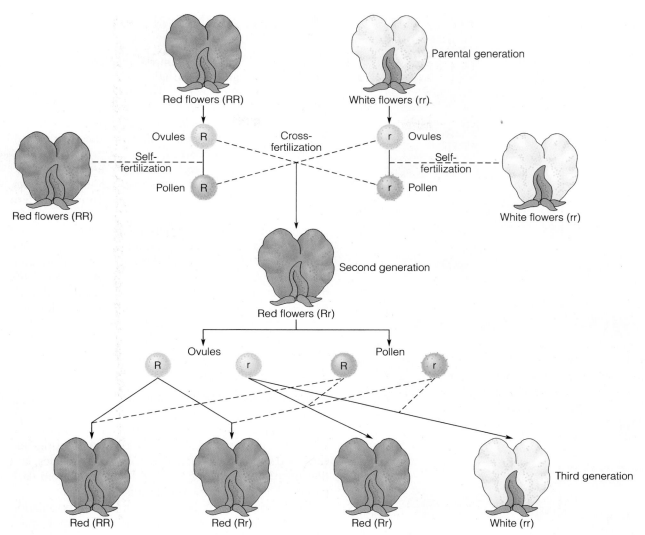

◆ FIGURE 18.4 In his experiments with flower color in garden peas, Mendel used true-breeding strains. Such plants, shown here as the parental generation, when self-fertilized always yield offspring with the same trait as the parent. However, if the parental generation is cross-fertilized, all plants in the second generation will receive the combination of alleles indicated by Rr; these plants will have red flowers because R is dominant over r. The second generation of plants produces ovules and pollen with the alleles shown and, when left to self-fertilize, produces a third generation with a ratio of three plants with red flowers to one plant with white flowers.

ever, develops and grows by a cell division process called *mitosis* that does not reduce the chromosome number (◆ Fig. 18.6c).

According to the chromosomal theory of inheritance, chromosomes carrying the genetic determinants,

the genes, are passed from one generation to the next. However, changes called **mutations** may occur in genes, and any such change occurring in sex cells is inheritable. We shall have more to say about mutations in the section on "Sources of Variation."

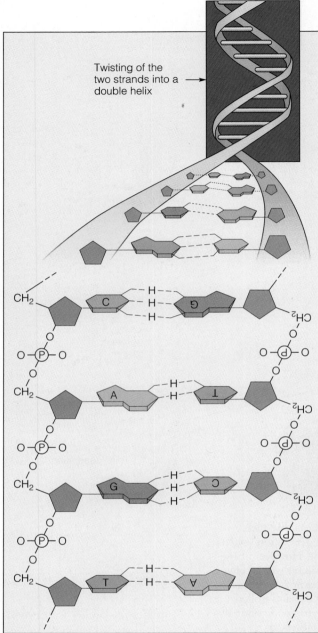

Twisting of the two strands into a double helix

CH₂

C — H, H, H — G

O—P—O   O—P—O

CH₂

A — H, H — T

O—P—O   O—P—O

CH₂

G — H, H, H — C

O—P—O   O—P—O

CH₂

T — H, H — A

◆ FIGURE 18.5 Chromosomes are double-stranded, helical molecules of deoxyribonucleic acid (DNA) shown here diagrammatically. Specific segments of chromosomes are genes. The two strands of the molecule are joined by hydrogen bonds (H).

THE MODERN SYNTHESIS

As noted previously, the Darwin-Wallace concept of evolution by natural selection was challenged because it could not explain how variation arose nor how it was maintained in populations. The developing science of genetics partly answered the inheritance problems, but early in this century, geneticists believed that mutation rather than natural selection was the mechanism whereby evolution occurred.

During the 1930s and 1940s, the ideas developed by geneticists, paleontologists, population biologists, and others were brought together to form a **modern synthesis** or neo-Darwinian view of evolution. The chromosome theory of inheritance was incorporated into evolutionary thinking; mutation was seen as one source of variation in populations; Lamarckian concepts of inheritance were completely rejected; and the importance of natural selection was reaffirmed. The modern synthesis also emphasized that evolution is a gradual process, a point that has been challenged by some recent investigators, as will be discussed later.

Sources of Variation

The raw material for evolution by natural selection is variation within populations. Most variations can be accounted for by sexual reproduction and the reshuffling of alleles from generation to generation. Considering that each of thousands of genes may have several alleles, and that offspring receive one-half of their genes from each parent, the potential for variation is enormous.

New variation may be introduced into a population by a change in genes, a mutation. To understand mutations, we must explore the function of chromosomes. One function of chromosomes is to direct the synthesis of protein molecules. During protein synthesis, information in the chromosome structure directs the formation of a protein by selecting the appropriate amino acids in a cell and arranging these amino acids into a sequence. The proteins produced determine the characteristics of an organism. Any change in the information directing protein synthesis is a mutation.

Two additional aspects of mutations are important. First, only mutations that occur in sex cells are inheritable. And second, mutations are random with respect to fitness. That is, there is no evidence of a predetermination of a mutation's effect; it may be harmful (many are), neutral, or beneficial. However, the attributes of harmful versus beneficial can be considered only with respect to the environment.

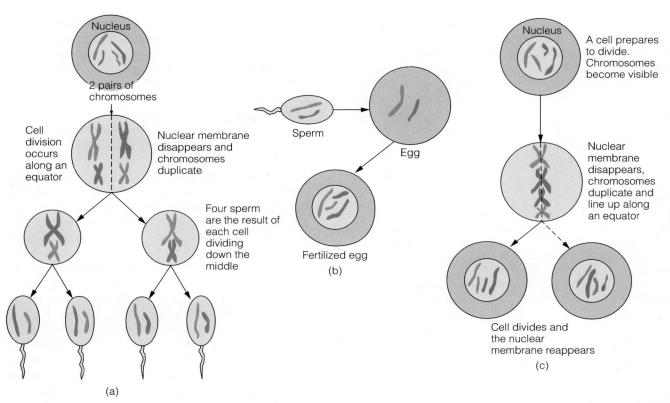

◆ FIGURE 18.6 (*a*) Meiosis is a type of cell division in which sex cells are formed; each cell contains one member of each chromosome pair. The formation of sperm cells is shown; eggs form in a similar manner, but only one of the four final cells is a functional egg. (*b*) When a sperm fertilizes an egg, the full complement of chromosomes is present in the fertilized egg. (*c*) Mitosis results in the complete duplication of a cell. In this simplified example, a cell with four chromosomes (two pairs) produces two cells, each with four chromosomes. Mitosis occurs in all body cells except sex cells. Once an egg is fertilized, the developing embryo grows by mitosis.

If a species were well adapted to its environment, most mutations would not be particularly useful and perhaps would be harmful. But if the environment changed, what was once a harmful mutation could become beneficial. For example, some plants have developed a tolerance for contaminated soils around mines. Plants of the same species from the normal environment do poorly or die in these contaminated soils, while contaminant-resistant plants do poorly in the normal environment. The mutations for contaminant resistance probably occurred repeatedly in the population, but their adaptive significance was not realized until contaminated soils were present.

Speciation and the Rate of Evolution

In classifying organisms, the term **species** is used for populations of similar individuals that in nature can interbreed and produce fertile offspring. The process of speciation involves a change in the genetic makeup of the population, that is, a change in the frequency of alleles in its **gene pool**. A gene pool is simply all of the genes available in the genetic makeup of a population. There may also be a marked change in form and structure (*morphology*), but morphologic change is not necessary in the origin of a new species. In fact, a descendant species may look very much like its ancestral species.

According to the concept of **allopatric speciation**, most species arise when a small part of a population is geographically isolated from its parent population by some kind of barrier. A rising mountain range, a stream, or an invasion of part of a continent by the sea may effectively isolate parts of a once-interbreeding population (◆ Fig. 18.7). Isolation may also occur when a few individuals of a species are accidentally introduced to a remote area.

◆ FIGURE 18.7 Allopatric speciation. (*a*) Reduction of the area occupied by a species may leave small, isolated populations, *peripheral isolates*, at the periphery of the once more extensive range. In this example, members of both peripheral isolates have evolved into new species. (*b*) Barriers have formed across parts of a central population's range, thereby isolating small populations. (*c*) Out-migration and the origin of a peripheral isolate.

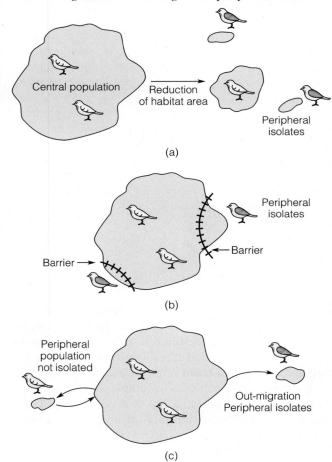

Once a small population becomes isolated, it no longer shares in the parent population's gene pool, and it is subjected to different selection pressures. Under these conditions, the isolated population evolves independently and eventually may give rise to a reproductively isolated species. Numerous examples of allopatric speciation have been well documented, especially in extremely remote areas such as oceanic islands; the finches of the Galapagos Islands are a good example (Fig. 18.1).

Most investigators agree that allopatric speciation is a common phenomenon, but they disagree about how rapidly new species evolve. Darwin proposed that the origin of a new species is generally a gradual process. That is, the gradual accumulation of minor changes brings about a transition from one species to another, a process commonly called **phyletic gradualism** (◆ Fig. 18.8a). According to this concept, an ancestral species changes gradually and continuously and grades imperceptibly into a descendant species.

Another view, called **punctuated equilibrium**, holds that a species changes little or not at all during most of its history and then evolves rapidly to give rise to a new species. In other words, long periods of equilibrium are occasionally punctuated by short periods of rapid evolution. Accordingly, a new species arises rapidly, perhaps in a few thousand years, but once it has evolved, it remains much the same for the rest of its existence (◆ Fig. 18.8b).

Proponents of punctuated equilibrium claim that the fossil record supports their hypothesis. They argue that few examples of gradual transitions from one species to another are found in the fossil record, and that many species appear to have existed for millions of years without noticeable change. According to this view, transitional forms connecting ancestral and descendant species are unlikely to be preserved in the fossil record because species arise rapidly in small, geographically isolated populations.

Critics, however, point out that neither Darwin nor those who formulated the modern synthesis insisted that all evolutionary change was gradual and continuous, a view shared by many present-day biologists and paleontologists. Indeed, Darwin allowed for times during which evolutionary change in small populations could be quite rapid. Furthermore, deposition of sediments in most environments is not continuous; thus, the lack of gradual transitions in many cases is simply an artifact of the fossil record. And finally, despite the incomplete nature of the fossil record, there are a

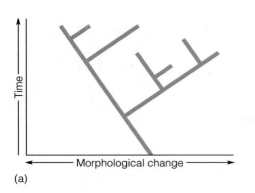

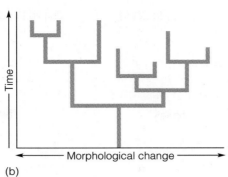

(a)

(b)

◆ FIGURE 18.8 Comparison of two models for differentiation of organisms from a common ancestor. (*a*) In the phyletic gradualism model, slow, continuous change occurs as one species evolves into another. (*b*) According to the punctuated equilibrium model, change occurs rapidly, and new species evolve rapidly. However, little or no change occurs in a species during most of its existence.

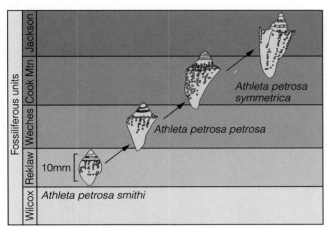

◆ FIGURE 18.9 Gradual evolution of the small snail *Athleta* showing changes in size, shape, and the development of spines on the shell.

number of examples of gradual transitions from ancestral to descendant species (◆ Fig. 18.9).

Divergent, Convergent, and Parallel Evolution

When an interbreeding population gives rise to diverse descendant types of organisms, the process is referred to as **divergent evolution** (◆ Fig. 18.10). Divergence into numerous related types involves an **adaptive radiation**, which occurs when species of related ancestry exploit different aspects of the environment. At some time in the past, a population of finches, apparently from South America, reached the Galapagos Islands. This ancestral population diversified into the adaptive types of species of finches that Darwin observed (Fig. 18.1). The fossil record provides many examples of

divergence and adaptive radiation. The diversification and adaptive radiation of placental mammals from a shrewlike common ancestor is an excellent example (◆ Fig. 18.11).

While divergent evolution leads to organisms that differ markedly from their ancestors, **convergent evolution** and **parallel evolution** are processes whereby similar adaptations arise in different groups (Fig. 18.10). Convergent evolution can be defined as the development of similiar characteristics in distantly related organisms, and parallel evolution as the development of similar characteristics in closely related organisms. The distinction between these two types of evolution depends on the degree of relatedness between the organisms in question, which is not always easy to determine. However, both convergent and parallel evolution occur as a result of different organisms adapting to similar ways of life.

Convergent evolution is well illustrated by the similar adaptations of Australian marsupial (pouched) mammals and placental mammals of the other continents. Convergence accounts for the fact that many Australian marsupials superficially resemble placental mammals elsewhere. Another impressive example of convergent evolution is shown by the Cenozoic mammals of North and South America (◆ Fig. 18.12). During most of the Cenozoic, South America was an island continent, and its mammalian fauna, consisting of both marsupials and placentals, evolved independently. A number of mammals on each continent adapted to similar environments and developed many features in common.

Parallel evolution is also a common phenomenon, but, as previously noted, it is not always easy to distinguish from convergent evolution. For example, jerboas and kangaroo rats are closely related, but each independently developed similar structures (◆ Fig. 18.13).

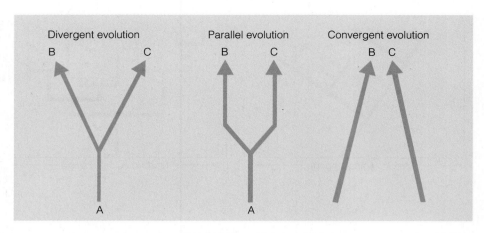

◆ FIGURE 18.10 In this example, divergent evolution results as species A diverges and gives rise to two descendant species, B and C. Parallel evolution also involves divergence from a common ancestor, A, but descendant species, B and C, then develop in a similar way as they adapt to a similar life-style. Convergent evolution is much like parallel evolution except that species B and C are distantly related but resemble one another in some aspects because of similar adaptations.

Divergent evolution   Parallel evolution   Convergent evolution

◆ FIGURE 18.11 Divergent evolution of the placental mammals. An ancestor, probably a small, shrewlike animal, gave rise to diverse descendant groups, each of which adapted to different life-styles.

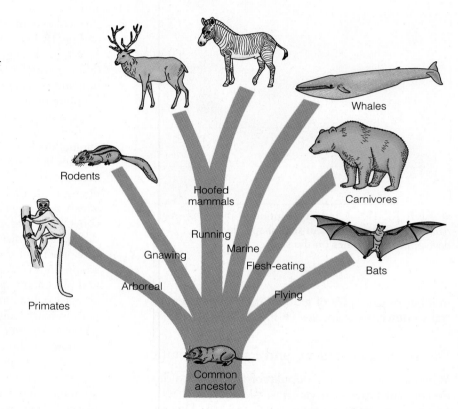

Evolutionary Trends

The evolutionary history, or *phylogeny,* of a group can be worked out in some detail if sufficient fossil material is known. Furthermore, constructing such phylogenies reveals the *evolutionary trends* that characterize a group of organisms (see Perspective 18.1). The phylogeny of ammonites, an extinct group related to the living squid and octopus, is well known; one evolutionary trend in this group was an increasingly complex shell structure (◆ Fig. 18.14, page 478).

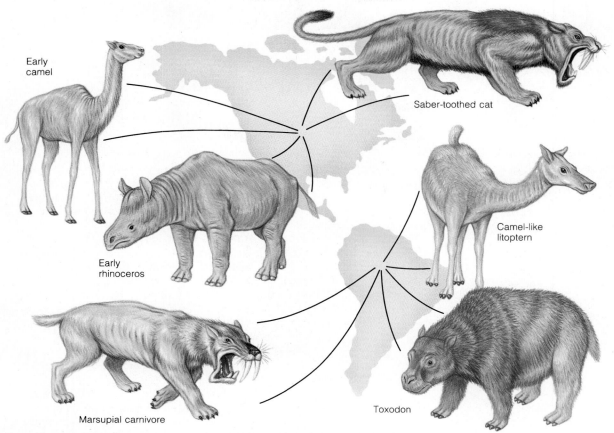

◆ FIGURE 18.12 Before the origin of the Isthmus of Panama a few million years ago, mammals of North and South America evolved independently. Similar adaptations of animals on each continent account for convergence.

◆ FIGURE 18.13 Parallel evolution. The kangaroo rat and jerboa are closely related but have independently developed similar features.

CLADISTICS AND CLADOGRAMS

Traditionally, the evolutionary history of many groups of organisms has been shown by *phylogenetic trees* (◆Fig. 1). Such trees are constructed with the horizontal axis representing anatomical differences between organisms and the vertical axis representing time. These trees show patterns of ancestry and descent based on a wide range of characteristics, although the characteristics used are rarely specified.

Cladistics is derived from the word *clade,* which refers to a group of organisms whose members are more closely related among themselves than they are to other groups. A *cladogram* is a diagram showing these relationships. Cladograms are constructed on the basis of derived characteristics as opposed to primitive characteristics. For example, land-dwelling vertebrate animals (amphibians, reptiles, birds, and mammals) possess bone and paired limbs for locomotion on land. Such characteristics are considered primitive because they are shared by all land-dwelling vertebrates and are therefore of little use in establishing relationships.

Derived characteristics, on the other hand, are specializations that evolved after the primitive characteristics became established. For example, the amniote egg is a primitive characteristic because it is shared by all reptiles, birds, and mammals (◆Fig. 2). Hair and mammary glands, however, are derived characteristics because they serve to differenti-

◆ FIGURE 1 A phylogenetic tree showing the inferred relationships among reptiles, birds, and mammals.

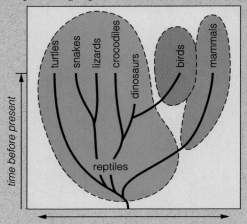

◆ FIGURE 2 Cladogram showing the relationships among living reptiles, birds, and mammals. Some of the characteristics used to differentiate the subgroups are indicated.

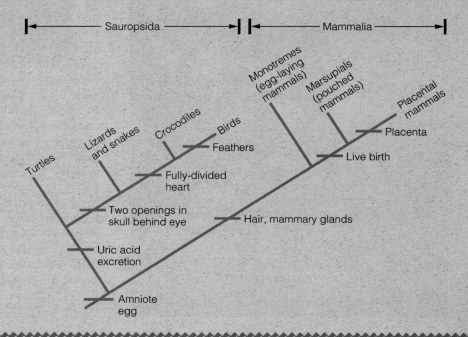

ate the mammals from those animals collectively called the Sauropsida (Fig. 2). Live birth is derived and indicates that monotremes (egg-laying mammals) differ from the marsupials (pouched mammals) and placental mammals.

Cladograms usually do not have an absolute time scale associated with them, but they nevertheless show the relative times of appearance of the groups possessing derived characteristics. In other words, they show evolutionary relationships. Figure 2 indicates that all mammals are more closely related to one another than they are to any of the sauropsids, and that crocodiles are more closely related to birds than either is to turtles, snakes, and lizards.

Cladistics and cladograms work well for living organisms, but are somewhat limited when applied to the fossil record. The main problem is recognizing and differentiating primitive characteristics from derived characteristics,

especially in groups with poor fossil records. Even in the better known evolutionary lineages, the earliest diversification is usually the most poorly represented by fossils. Nevertheless, cladistics has been applied to a number of groups. The relationships among the Ceratopsidae, a family of horned dinosaurs, have been determined by cladistics (◆ Fig. 3). Notice in Figure 3 that in addition to the family Ceratopsidae, two subfamilies are also shown, the Centrosaurinae and the Chasmosaurinae, both of which were defined by cladistics. So, in addition to showing evolutionary relationships, cladograms can be used to classify organisms. Not all paleontologists and biologists accept cladistics because such analyses depend on traits that are assumed to have arisen solely by descent from a common ancestor. Parallel and convergent evolution are assumed to be of little importance, an assumption that may not be justified in many cases.

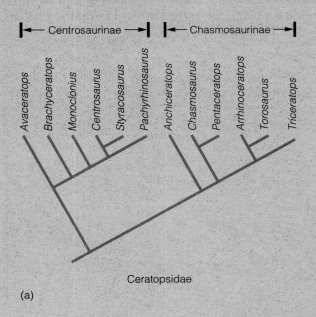

(a)

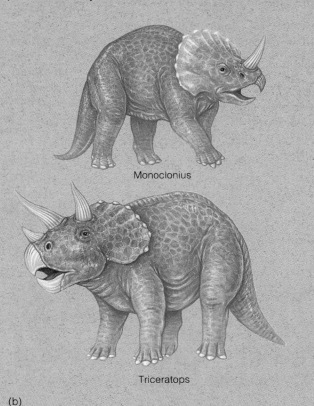

Monoclonius

Triceratops

◆ FIGURE 3 (a) Cladogram showing the relationships among the Ceratopsidae, a family of horned dinosaurs. (b) Examples of dinosaurs from each of the subfamilies of the Ceratopsidae.

(b)

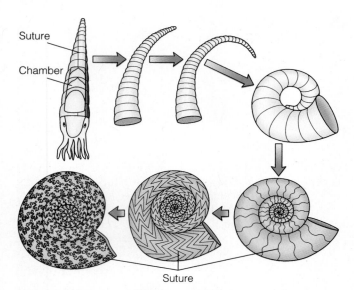

Suture

Chamber

Suture

◆ FIGURE 18.14 Evolutionary trends in the ammonites. Ammonites with coiled shells evolved from straight-shelled ancestors. Notice that an ammonite's shell is internally divided into chambers, and that the animal lives only in the outermost chamber. The sutures, the lines formed where chamber walls meet the wall of the outer shell, became increasingly more complex as ammonites evolved.

Titanotheres were mammals that lived in North America and eastern Asia from Early Eocene to Early Oligocene time. A good fossil record records their phylogeny and several evolutionary trends (◆ Fig. 18.15). One trend was simply an increase in size, but titanotheres also developed large horns, and the shape of the skull changed.

Increase in size is one of the most common evolutionary trends. However, trends are complex in that they do not always proceed at the same rate, they may be reversed, or several may occur in the same group. Horses evolved from fox-sized ancestors that lived during the Eocene. One trend was an increase in size, but it was not a steady, uniform increase, and, in fact, some horses actually show a reversal in this trend. Horse evolution is also characterized by a reduction in the number of toes, lengthening of the legs and feet, and changes in the teeth and skull. All of these trends are well documented by fossils, but they did not occur at the same rate (see Chapter 24).

One can view evolutionary trends as a series of adaptations to changing environments or adaptations that occur in response to exploitation of new habitats. The same trends occur repeatedly in the fossil record.

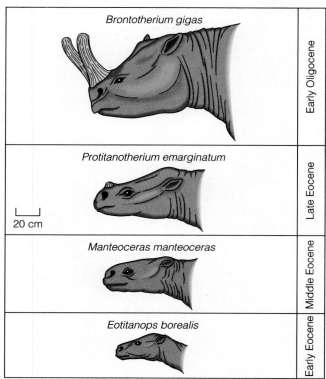

Brontotherium gigas — Early Oligocene

Protitanotherium emarginatum — Late Eocene

20 cm

Manteoceras manteoceras — Middle Eocene

Eotitanops borealis — Early Eocene

◆ FIGURE 18.15 Evolutionary trends in the titanotheres, an extinct group of mammals related to horses and rhinoceroses. Titanotheres evolved from small ancestors to giants about 2.4 m tall at the shoulder and developed large horns.

Increase in size is one of the most often repeated trends, but a number of others are known as well. Ammonites, for example, nearly became extinct at the end of the Paleozoic Era, but those that survived into the Mesozoic diversified and gave rise to many adaptive types very similar to those that had existed earlier. The evolutionary significance of this phenomenon is that very similar adaptations have occurred in many groups of organisms in different places throughout the history of life.

Extinctions

Extinctions are the rule in the history of life. In fact, extinction seems to be the ultimate fate of all species; probably more than 99% of all species that ever lived are now extinct. The term *extinct* seems so final, and, in a very real sense, it is. But we must distinguish between two types of extinction. In one case a species evolves into a new species that differs so much from its ancestral group that the parent species can be consid-

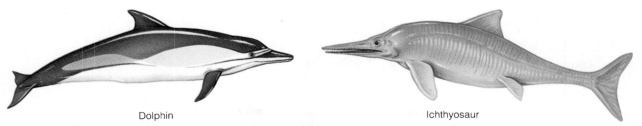

Dolphin   Ichthyosaur

◆ FIGURE 18.16 Ichthyosaurs were fish-eating, marine reptiles that became extinct at the end of the Mesozoic Era. Following their extinction, no air-breathing animal occupied their niche until the dolphins and porpoises appeared about 30 million years later. Notice that ichthyosaurs and dolphins resemble one another in several features. Both groups adapted to a similar life-style and provide an excellent example of convergent evolution.

ered extinct. Perhaps this is best called *pseudoextinction* since, strictly speaking, no extinction event occurred. The second case of extinction is terminal. That is, a species dies out without giving rise to anything else. Many examples of terminal extinction are known from the fossil record.

Extinction by one method or another appears to be a continual occurrence in the history of life. But so is the origin of new species that usually quickly exploit the opportunities created by the disappearance of other species. In some cases of extinction, an ecologically equivalent organism may not appear for some time. Ichthyosaurs were Mesozoic marine reptiles that lived like and superficially resembled present-day porpoises and dolphins (◆ Fig. 18.16). Yet, whereas ichthyosaurs are unknown after the Mesozoic, porpoises and dolphins did not occupy their niche until some 30 million years later.

At times extinction rates have been greatly accelerated in events called *mass extinctions.* The two greatest crises in the history of life were the mass extinctions at the end of the Paleozoic and Mesozoic eras. In both cases, the diversity of life was sharply reduced. These extinction events, and some of lesser magnitude, will be discussed in greater detail in the following chapters.

⩔ EVIDENCE FOR EVOLUTION

When Charles Darwin proposed his theory, he cited such supporting evidence as classification, embryology, comparative anatomy, the geographic distribution of organisms, and, to a limited extent, the fossil record. Darwin had little knowledge of the mechanism of inheritance or biochemistry, and molecular biology was completely unknown during his lifetime. Studies in these areas coupled with a better understanding of the

fossil record have added to the body of evidence. Today, most scientists accept that the theory of evolution is as well supported by evidence as any major theory.

Classification—A Nested Pattern of Similarities

The Swedish naturalist Carolus Linnaeus (1707–1778) proposed a formal classification of organisms. According to the Linnaean system, an organism is given a two-part name; the coyote, for example, is referred to by its genus (plural, *genera*)) name and its species name, *Canis latrans.*

● Table 18.1 shows the basic arrangement of the Linnaean classification although the scheme now used contains more categories. The arrangement is hierarchical. That is, as one proceeds up the list, the categories become more inclusive (◆ Fig. 18.17). The coyote

~~~~~~~~~~~~~~~~~~~~~~~~~~~~~~~~~~~~

● TABLE 18.1 The Classification Scheme Now in Use Showing the Hierarchical Arrangement of the Categories

The animal classified in this example is the coyote, *Canis latrans.*
Kingdom----------------------------------Animalia
  Phylum--------------------------------------Chordata
    Subphylum-------------------------------Vertebrata
      Class---Mammalia
        Order---Carnivora
          Family---Canidae
            Genus---*Canis*
              Species---*latrans*

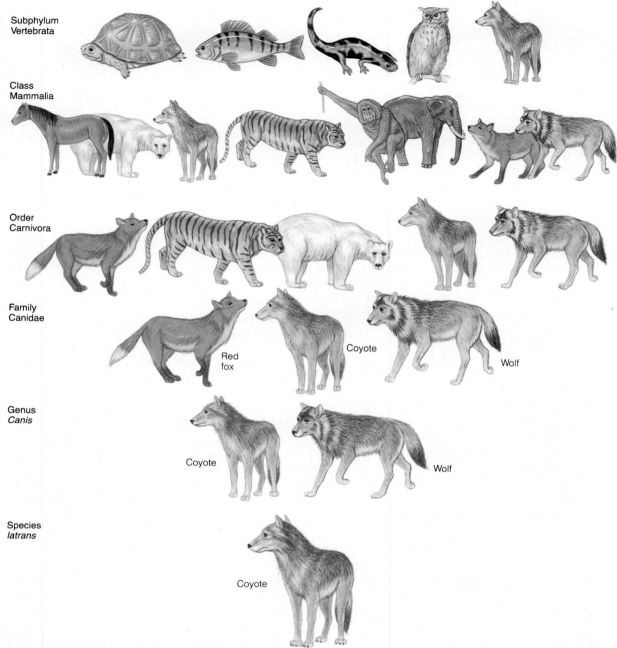

Subphylum
Vertebrata

Class
Mammalia

Order
Carnivora

Family
Canidae

Red
fox

Coyote

Wolf

Genus
Canis

Coyote

Wolf

Species
latrans

Coyote

◆ FIGURE 18.17 Traditional classification of organisms based on shared characteristics. All members of the subphylum Vertebrata, including fishes, amphibians, reptiles, birds, and mammals, have a segmented vertebral column. Among the vertebrates, however, only those warm-blooded animals having hair or fur and mammary glands are mammals. Eighteen orders of mammals are recognized, including the order Carnivora shown here. All members of this order have teeth specialized for a diet of meat. The family Canidae includes only the doglike carnivores, and the genus *Canis* includes closely related species. The coyote, *Canis latrans,* stands alone as a species.

(*Canis latrans*) and the wolf (*Canis lupus*) are members of different species, but they share a large number of characteristics, so both belong to the same genus. The red fox (*Vulpes fulva*) also shares some, but fewer, characteristics with coyotes and wolves, so it, along with other related doglike animals, make up the family Canidae. In turn, all canids share some characteristics with cats, bears, weasels, and raccoons, all of which belong to the order Carnivora. Likewise, all carnivores, rodents, bats, primates, and elephants have hair and mammary glands and constitute the class Mammalia (Fig. 18.17). And so it goes, up to kingdom, the most inclusive category in the classification scheme.

Following the publication of *The Origin of Species*, biologists soon realized that classification based on shared characteristics among organisms constituted a strong argument for evolution. In our example, coyotes and wolves share many characteristics because they evolved from a common ancestor in the not too distant past. They also share some characteristics with bears and cats because all members of the order Carnivora had a common ancestor in the more remote past.

Biological Evidence for Evolution

According to evolutionary theory, all life-forms are related, and all living organisms descended with modification from ancestors that lived during the past. If this statement is correct, then there should be evidence of fundamental similarities among all life-forms, and closely related organisms should be more similar to one another than to more distantly related organisms. As a matter of fact, all living things, from bacteria to whales, are composed of the same chemical elements—mostly carbon, nitrogen, hydrogen, and oxygen. Furthermore, the chromosomes of all organisms are composed of DNA, except bacteria, which have RNA; and in all cells, proteins are synthesized in essentially the same way.

Thousands of biochemical tests of numerous organisms have provided compelling evidence for evolutionary relationships. Blood proteins, for example, are similar among all mammals, but they also indicate that among the primates humans are most closely related to the great apes, followed, in order, by the Old World monkeys, the New World monkeys, and the lower primates such as lemurs. Biochemical tests indicate that all birds are related among themselves and are more closely related to turtles and crocodiles than to snakes and lizards, a finding corroborated by the fossil record.

Studies of developing embryos reveal that some organisms are similar to one another until very late in embryonic development, while others are less similar. Fish, chimpanzee, and human embryos are very similar in the earliest stages, but differences soon become apparent in fish embryos. Chimpanzees and humans, in contrast, retain remarkable embryonic similarities until very late in the developmental process. The inference is that chimpanzees and humans are more closely related to one another than they are to fish.

Evidence for evolution is also provided by comparing anatomical structures of organisms. The forelimbs of humans, whales, dogs, and birds (◆ Fig. 18.18) are superficially dissimilar, yet all are composed of the same bones and have basically the same arrangement of muscles, nerves, and blood vessels; all are similarly arranged with respect to other structures, and all have a similar pattern of embryonic development. Such structures are said to be **homologous**. Descent with modification from a common ancestor—that is, a common genetic heritage—is the best explanation for the existence of homologous organs.

Not all similarities among organisms are evidence of evolutionary relationships; in fact, some similarities are rather superficial. Insects are not closely related to bats and birds, but all three of these animals have wings. Such structures that serve the same function are **analogous organs** (◆ Fig. 18.19). The wings of bats and birds, but not of insects, are homologous, however, and provide evidence for a common ancestry. Insect wings, on the other hand, represent convergence—a similar solution to an adaptive problem, flight—but they differ from bat and bird wings in both structure and embryological development.

Probably every living species has **vestigial structures**—nonfunctional or partly functional remnants of organs that were functional in their ancestors. Vestigial structures are leftovers in the evolutionary sense, structures that are probably being lost. For example, one trend in human evolution has been shortening of the jaw, which leaves too little room for the third molars, the so-called wisdom teeth. Many mammals have dewclaws, which are small, functionless, vestiges of fingers and toes. Dewclaws are particularly obvious in dogs (◆ Fig. 18.20), but vestigial toes also occur in pigs and deer.

Another type of evidence for evolution is provided by observations of small-scale evolution in living organisms (◆ Fig. 18.21). We have already mentioned one example, the adaptations of plants to contaminated

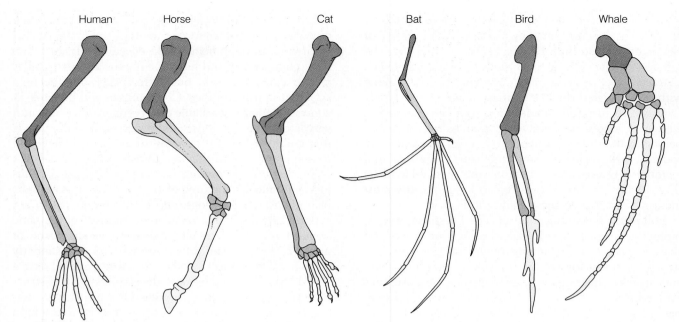

◆ FIGURE 18.18 Homologous organs such as the forelimbs of vertebrates may serve different functions but are composed of the same elements and undergo similar embryological development.

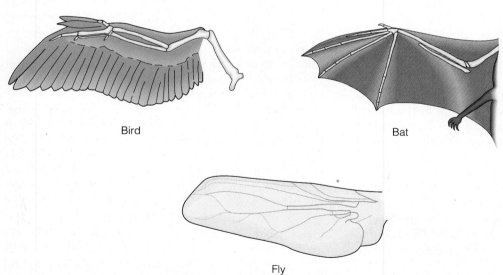

◆ FIGURE 18.19 The fly's wings serve the same function as wings of birds and bats but have a different structure and different embryological development. Organs that serve the same function are analogous, but the bird and bat wings are also homologous.

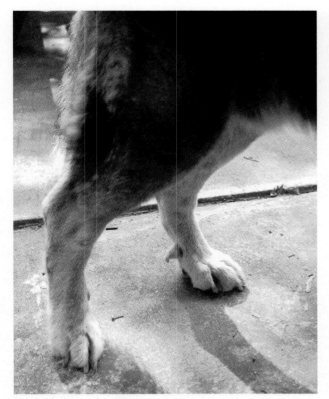

◆ FIGURE 18.20 Left hindfoot of a dog showing the dewclaw, which is a vestige of the big toe.

◆ FIGURE 18.21 Small-scale evolution of the peppered moths of Great Britain. In pre-industrial England most of these moths were gray and blended well with the lichens on trees. Black varieties were known, but rare, since they stood out against the light background and were easily spotted by birds. With industrialization and pollution, the trees became soot-covered and dark, and the frequency of dark-colored moths increased while that of light-colored moths decreased. However, in rural areas unaffected by pollution, the light-dark frequency did not change.

soils. New insecticides and pesticides must be developed continually as insects and rodents develop resistance to existing ones. Development of antibiotic-resistant strains of bacteria constitutes a continuing problem in medicine.

▼ FOSSILS AND FOSSILIZATION

Fossils provide some of the evidence for organic evolution. Before discussing that evidence, however, we shall consider how fossils are preserved.

Our definition of fossils (see Chapter 5) includes the phrase "remains or traces of ancient organisms." Remains, commonly called **body fossils** (◆ Fig. 18.22a), are usually the hard skeletal elements such as bones, teeth, and shells. In exceptional cases, preservation by freezing and mummification provides us with partial or nearly complete body fossils with flesh, hair, and internal organs. **Trace fossils** include tracks, trails, burrows, borings, nests, or any other indication of the activities

of organisms (◆ Fig. 18.22b and c). Fossilized feces, called **coprolites,** provide evidence of activity of organisms (◆ Fig. 18.22d) and may provide important information on the diet and size of the animal that produced them.

Fossils in general are quite common. The remains of various microorganisms are by far the most abundant and in many ways the most useful. Shells of marine animals are also common and are easily collected in many areas. Despite their abundance, however, fossils represent very few of the organisms that lived at any one time, since any potential fossil must escape the ravages of destructive processes such as waves or running water, scavengers, exposure to the atmosphere, and bacterial action. Hard skeletal elements are more resistant to destructive processes than soft parts, but even these must be buried in some protective medium such as mud or sand if they are to be preserved; and even if buried they may be dissolved by groundwater or destroyed by alteration of the host rock.

Considering all the ways in which potential fossils can be destroyed, it comes as no surprise that the fossil record is biased toward those organisms with hard parts and those organisms that lived in areas of active

(a)

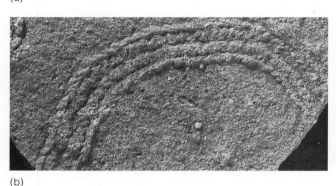

(b)

(c)

◆ FIGURE 18.22 Body fossils are the actual remains of organisms. (*a*) Miocene horse teeth from Montana. (*b*) and (*c*) Trace fossils. (*b*) Crawling trace in the Lower Pennsylvanian Fentress Formation, Tennessee. (Photo courtesy of Molly Fritz Miller, Vanderbilt University.) (*c*) Dinosaur tracks in Cretaceous rocks in the bed of the Paluxy River, Texas. (*d*) Fossilized feces (coprolite) of a carnivorous mammal. Specimen measures 5.5 cm long. (Photos (*a*) and (*d*) courtesy of Sue Monroe.)

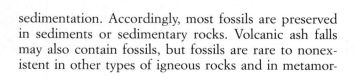

(d)

sedimentation. Accordingly, most fossils are preserved in sediments or sedimentary rocks. Volcanic ash falls may also contain fossils, but fossils are rare to nonexistent in other types of igneous rocks and in metamorphic rocks. Despite its biases, the fossil record is our only record of prehistoric life. Indeed, it is only through fossils that we have any knowledge of such extinct lifeforms as trilobites and dinosaurs.

Types of Preservation

Fossils may be preserved as **unaltered remains**, in which case they retain their original structure and composition. The pollen and spores of plants have a tough outer cuticle that is chemically resistant to change. Partial or complete insects may be trapped in the resin secreted by some plants, especially by coniferous trees. When this resin is buried, it hardens to become amber and preserves in exquisite detail the remains contained therein (◆ Fig. 18.23a).

◆ FIGURE 18.23 (*a*) Insects in amber. (*b*) Excavation of mammoth bones from tar pit 9 at Rancho La Brea in Los Angeles, California in 1914.

(a)

(b)

Unusual types of preservation of unaltered remains include preservation in tar, mummification, and freezing. The La Brea Tar Pits of southern California contain numerous tar-impregnated bones of Pleistocene animals that otherwise retain their original composition and structure (◆ Fig. 18.23b). Mummification involves air drying and shriveling of soft parts such as muscles and tendons before burial (◆ Fig. 18.24a). Only a few types of animals have been preserved by freezing, including mammoths, woolly rhinoceroses, musk oxen, and a few others, and all are from Late Pleistocene–age deposits. The frozen mammoths of Siberia are no doubt the best known example of this mode of preservation (◆ Fig. 18.24b).

Altered remains are fossils that have been changed structurally, chemically, or both. Quite commonly, mineral matter is added to the pores and cavities of bones, teeth, and shells after burial. This process, called *permineralization,* increases the preservation potential of fossils by increasing their durability.

The shells of many marine invertebrates are composed of an unstable form of calcium carbonate known as aragonite. When buried, these shells commonly recrystallize in the more stable form of calcium carbonate called calcite. A few plants and animals have shells or skeletal elements composed of opal, a variety of silicon dioxide (SiO_2). These also commonly recrystallize. Although recrystallization does not change the chemical composition or outward appearance, the microscopic crystal structure of the shell is altered.

We know that clams, snails, and many other invertebrates have skeletons of calcium carbonate, yet in some strata we find these shells composed of silicon dioxide (SiO_2) or pyrite (FeS_2). These shells have been altered by *replacement,* during which the original skeletal material is replaced by a compound of a different composition (◆ Fig. 18.25a).

Wood may be preserved by the complete replacement of woody tissue by silicon dioxide in the form of chert or opal and, more rarely, by calcite. Wood preserved in this manner is often called *petrified,* a term that means to become stone. The replacement may be so exact that even details of cell structure are finely preserved. In many cases the form of the woody tissue actually remains unaltered, and the pore spaces have simply been filled in by silicon dioxide.

The altered remains of plants may be preserved by *carbonization.* During carbonization, the volatile elements of organic material, such as a leaf, vaporize, leaving a thin carbon film (◆ Fig. 18.25b).

(a)

(b)

(c)

(a)

(b)

◆ FIGURE 18.24 (*a*) Close-up of the mummified remains of the dinosaur *Anatosaurus* showing detail of the surface pattern of the skin. (*b*) Frozen baby mammoth found in Siberia in 1977. The baby was six or seven months old, 1.15m long, 1.0 m tall, and when alive had a hairy coat. Most of the hair has fallen out except around the feet. The carcass is about 40,000 years old.

◆ FIGURE 18.25 Altered remains. (*a*) Replacement by pyrite. (*b*) Carbonization as seen in a leaf from the Pennsylvanian Mazon Creek flora in Illinois. (*c*) Mold (left) and cast. (Photos courtesy of Sue Monroe.)

In addition to unaltered and altered remains, fossils may be preserved as **molds** and **casts** (◆ Fig. 18.25c). Molds are formed when the remains of a buried organism are dissolved, thereby leaving a cavity with the external shape of the organism (an external mold).

If the mold is later filled with sediment or mineral matter, an external cast is formed. Internal molds are cavities showing the inner features of shells, and internal casts are formed when such cavities are filled with sediment or mineral matter.

FOSSILS AND EVOLUTION

The occurrence of fossil marine animals far from the sea, even high in mountains, led many early naturalists to conclude that they were deposited during the world-wide biblical flood. As early as 1508, Leonardo da Vinci realized that the fossil distribution was not what one would expect in a rising flood. Nevertheless, the flood explanation persisted, and John Woodward (1665–1728) proposed that the fossils had separated out of the floodwater according to their density. In other words, Woodward's hypothesis predicted that the fossils should be arranged in a vertical sequence, with the densest at the bottom followed upward by those of decreasing density. This hypothesis was quickly rejected because field observations clearly did not support the prediction; fossils of various densities are found throughout the fossil record.

The fossil record does show a sequence, but not one based on density, size, or shape of the fossils. Rather, it shows a sequence of appearances of different life-forms through time (◆ Fig. 18.26). The fact is that older and older fossiliferous strata contain organisms increasingly different from those living today. One-celled organisms appeared before multicelled organisms, plants before animals, and invertebrates before vertebrates. Among the vertebrates, fish appear first, followed in order by amphibians, reptiles, mammals, and birds (Fig. 18.26).

Fossils also provide evidence for the evolutionary origins of some of the groups in Figure 18.26 and evidence for evolution within groups. Some examples of such fossil evidence have already been discussed in other contexts, such as the evolution of ammonites and titanotheres.

The evolution of horses and their living relatives, the rhinoceroses and tapirs, is well documented by fossils (◆ Fig. 18.27). Horses, rhinoceroses, and tapirs may seem to be an odd assortment of animals, but they all share several characteristics that imply they are related by evolutionary descent. If so, one would predict that as each family is traced back in the fossil record, differentiating one from the other would become increasingly difficult. In fact, the earliest members of each family are differentiated from one another by minor differences in size and characteristics of their teeth.

Many people have the idea that the fossil record provides the main evidence for evolution. Fossils are unquestionably important in evolutionary studies, but one must not overlook the fact that fossils are only one of several lines of evidence that support the concept of evolution. Shared characteristics, biochemistry, comparative anatomy, and small-scale evolution of living organisms are equally important.

◆ FIGURE 18.26 The times of appearance of invertebrates and major groups of vertebrate animals.

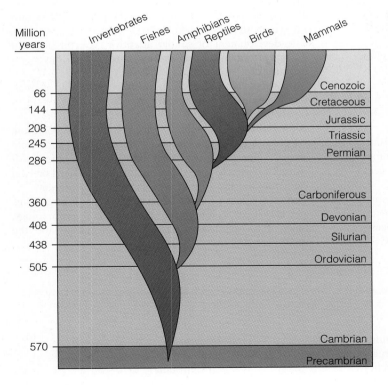

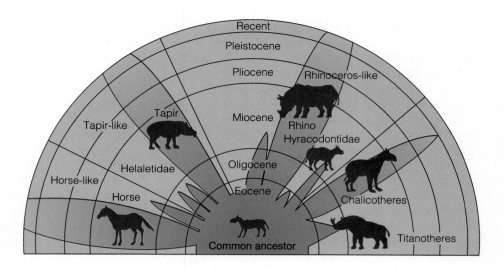

◆ FIGURE 18.27 The divergence and adaptive radiation of the odd-toed hoofed mammals (order Perissodactyla) which includes horses, rhinoceroses, and tapirs, and their extinct relatives, is well documented by fossils. The evolution of these animals is discussed in more detail in Chapter 24.

Chapter Summary

1. The first formal theory of evolution to be taken seriously was proposed by Jean Baptiste de Lamarck. Inheritance of acquired characteristics was his mechanism for evolution.

2. In 1859 Charles Darwin and Alfred Russell Wallace simultaneously published their views on evolution. They proposed natural selection as the mechanism for evolutionary change.

3. Darwin's observations of variation in natural populations and artificial selection, as well as his reading of Thomas Malthus's essay on population, helped him formulate the idea that natural processes select favorable variants for survival.

4. Gregor Mendel's breeding experiments with garden peas provided some of the answers regarding how variation is maintained and passed on. Mendel's work is the basis for modern genetics.

5. Genes are the hereditary determinants in all organisms. This genetic information is carried in the chromosomes of cells. Only the genes in the chromosomes of sex cells are inheritable.

6. Most variation is maintained in populations by sexual reproduction and mutations.

7. Evolution by natural selection is a two-step process. First, variation must be produced and maintained in interbreeding populations, and second, favorable variants must be selected for survival.

8. An important way in which new species evolve is by allopatric speciation. When a group is isolated from its parent population, gene flow is restricted or eliminated, and the isolated group is subjected to different selection pressures.

9. Divergent evolution involves an adaptive radiation during which an ancestral stock gives rise to diverse species. The development of similar adaptive types in different groups of organisms results from parallel and convergent evolution.

10. Most fossils are preserved in sedimentary rocks as unaltered remains, altered remains, molds and casts, or traces of organic activity.

11. Although fossils of some organisms are quite common, very few of the organisms that lived at any one time were actually preserved as fossils. Furthermore, the fossil record is biased toward those organisms with hard skeletal elements that lived in areas of active sedimentation.

12. The evidence for evolution includes the way organisms are classified, comparative anatomy and embryology, biochemical similarities, small-scale evolution, and the fossil record.

Important Terms

adaptive radiation	altered remains	body fossil
allele	analogous organ	cast
allopatric speciation	artificial selection	chromosome

convergent evolution
coprolite
divergent evolution
DNA (deoxyribonucleic acid)
gene
gene pool
homologous organ

modern synthesis
mold
mutation
natural selection
parallel evolution
phyletic gradualism

punctuated equilibrium
species
theory of evolution
trace fossil
unaltered remains
vestigial structure

Review Questions

1. Genes controlling the same trait occur in different forms known as:
   a. ___ homologous organs; b. ___ alleles;
   c. ___ species; d. ___ factors; e. ___ DNA.
2. The type of cell division that produces functional sex cells containing one-half of the chromosomes typical of a species is:
   a. ___ mosaic evolution; b. ___ punctuated equilibrium;
   c. ___ allopatric speciation; d. ___ meiosis; e. ___ adaptation.
3. The first person to propose a formal theory of evolution was:
   a. ___ Alfred Russell Wallace; b. ___ Jean Baptiste de Lamarck; c. ___ Erasmus Darwin; d. ___ Charles Lyell; e. ___ William Smith.
4. One of the main sources of variation in populations is:
   a. ___ spontaneous generation; b. ___ fixity of species;
   c. ___ sexual reproduction; d. ___ phyletic gradualism;
   e. ___ mitosis.
5. According to the concept of punctuated equilibrium:
   a. ___ species change gradually and continually over millions of years; b. ___ most species that ever existed are now extinct; c. ___ the most common evolutionary trend is increase in size; d. ___ new species arise rapidly, perhaps in a few thousands of years.
6. Clams and brachiopods are very distantly related but possess a number of similarities because of:
   a. ___ punctuated equilibrium; b. ___ convergent evolution; c. ___ adaptive divergence; d. ___ parallel adaptation; e. ___ artificial selection.
7. In the biological classification scheme now used, organisms are assigned to genera, families, and orders on the basis of:
   a. ___ type of fossilization; b. ___ shared characteristics;
   c. ___ time of existence; d. ___ inheritance of acquired characteristics; e. ___ vestigial structures.
8. All of the genes available to an interbreeding population constitute its:
   a. ___ source of mutations; b. ___ chromosome base;
   c. ___ gene pool; d. ___ phylogeny; e. ___ species resource.
9. Organs containing the same basic structural elements, such as a bird's wing and a dog's forelimb, are said to be:

a. ___ homologous; b. ___ artificial; c. ___ vestigial;
   d. ___ mutations; e. ___ mosaics.
10. Genes, the basic units of inheritance, are specific segments of:
    a. ___ sex cells; b. ___ analogous organs;
    c. ___ fossils; d. ___ species; e. ___ chromosomes.
11. Which of the following is not a fossil?
    a. ___ dinosaur footprint; b. ___ Paleozoic clam˙ shell;
    c. ___ Roman coin; d. ___ Pleistocene elephant tusk;
    e. ___ Mesozoic worm burrow.
12. Suppose that a clam shell is buried in sand, then dissolved leaving a cavity with the shape of a clam shell. This cavity is a:
    a. ___ formation; b. ___ coprolite; c. ___ cast; d. ___ system; e. ___ mold.
13. The diversification of a species into two or more descendant species is an example of:
    a. ___ the modern synthesis; b. ___ parallel evolution;
    c. ___ phyletic equilibrium; d. ___ mosaic evolution;
    e. ___ divergent evolution.
14. Chromosomes are complex molecules of:
    a. ___ drosophila; b. ___ deoxyribonucleic acid;
    c. ___ carbon dioxide; d. ___ potassium, aluminum, and silicon; e. ___ calcium carbonate.
15. Speciation resulting from isolation of a small part of an interbreeding population is referred to as _____ speciation.
    a. ___ adaptive radiation; b. ___ convergent;
    c. ___ gradual; d. ___ allopatric; e. ___ punctuated.
16. Compare evolution by inheritance of acquired characteristics with evolution by natural selection.
17. How does sexual reproduction maintain variation in interbreeding populations?
18. What observations led Darwin to conclude that natural selection is a mechanism for evolution?
19. What is a mutation? Explain what is meant by harmful, neutral, and beneficial mutations.
20. Explain the process whereby an organism accidentally introduced to a remote island may give rise to a new species.
21. Give examples of divergent and convergent evolution.
22. How does the way in which organisms are classified provide evidence for evolution?

23. Explain how the concept of punctuated equilibrium accounts for the origin of new species. Compare this with phyletic gradualism.
24. How does pseudoextinction differ from terminal extinction?
25. What are the most important aspects of Gregor Mendel's breeding experiments with garden peas?
26. Explain how the fossil record provides evidence for organic evolution.
27. How are adaptive radiation and divergent evolution related?
28. Compare the processes of permineralization, replacement, and recrystallization of fossils.

Additional Readings

Bajema, C. J. 1985. Charles Darwin and selection as a cause of adaptive evolution, 1837–1859. *The American Biology Teacher* 47:226–32.

Bowler, P. J. 1990. *Charles Darwin: The man and his influence.* Cambridge, Mass.: Blackwell.

Dobzhansky, T., F. J. Ayala, G. L. Stebbins, and J. W. Valentine. 1977. *Evolution.* San Francisco: W. H. Freeman.

Donovan, S. K., ed. 1991. *The processes of fossilization.* New York: Columbia Univ. Press.

Futuyma, D. J. 1986. *Evolutionary biology.* 2d ed. Sunderland, Mass.: Sinauer.

Grant, P. R. 1991. Natural selection and Darwin's finches. *Scientific American* 265, no. 4:82–87.

Hoffman, A. 1989. *Arguments on evolution: A paleontologist's perspective.* New York: Oxford Univ. Press.

Mayr, E. 1978. Evolution. *Scientific American* 239, no. 3: 47–55.

Moody, R. 1986. *Fossils.* New York: Macmillan.

Pinna, G. 1990. *Illustrated encyclopedia of fossils.* New York: Facts on File.

Simpson, G. G. 1983. *Fossils and the history of life.* New York: Scientific American Books.

CHAPTER
19

The Andromeda Galaxy, 2.2 million light years away, is the closest major galaxy to our Milky Way. Its full diameter is nearly 150,000 light years and it contains some 300,000,000,000 stars.

A History of the Universe, Solar System, and Planets

Prologue

On August 20 and September 5, 1977, *Voyagers 1* and *2* were launched on an ambitious mission to explore the outer planets. They both flew by Jupiter and Saturn, but *Voyager 1* took a course out of the solar system while *Voyager 2* went on to Uranus and Neptune. Twelve years and 7.13 billion km after it was launched, *Voyager 2* radioed back spectacular images of the blue planet Neptune (◆ Fig. 19.1) and

◆ FIGURE 19.1 A shimmering blue planet set against the black backdrop of space, Neptune reveals itself to *Voyager 2*'s instruments during its August 1989 flyby. Shown here is Neptune's turbulent atmosphere with its Great Dark Spot and various wispy clouds.

its pink and blue mottled moon Triton. Its primary mission completed, and with all but a few of its instruments turned off to conserve power, *Voyager 2*'s last act will be to measure the exotic fields and subatomic particles it passes through on its voyage to infinity.

The discoveries made by these two space probes were truly fantastic and in many cases totally unexpected by scientists. In addition to discovering three new moons of Jupiter, the *Voyagers* found dusty rings encircling the planet, thus demonstrating that rings are a common feature in the outer solar system. They showed that the Great Red Spot is an enormous, persistent eddy in Jupiter's atmosphere, and they detected lightning discharges that are 10,000 times more powerful than those on Earth. The *Voyagers* sent back images of one of Jupiter's moons, Io, spewing forth hot sulfurous gases 320 km into space. Another Jovian moon, Europa, was revealed to be encrusted with a thick shell of ice covering a liquid ocean several kilometers below its surface. This ice is crisscrossed with what appear to be cracks that occasionally open to erupt water and then refreeze.

As the *Voyagers* flew past Saturn, they revealed the spectacular complexity of its 70,000-km-wide ring system and sent back images of spiral bands of debris only 35 m thick. The *Voyagers* also discovered that the Saturnian moon Titan has an atmosphere rich in hydrocarbons and nitrogen.

As *Voyager 2* passed by Uranus on January 24, 1986, it found nine dark, compact rings encircling the planet, discovered 10 new moons, and revealed a corkscrew-shaped magnetic field that extends for millions of kilometers from the planet.

Voyager 2 reached its final target, Neptune, in August 1989 and sent back spectacular images and data that were, for the most part, completely unanticipated. Instead of a quiet, placid planet, Neptune turned out to be a dynamic world cloaked in a thin atmosphere composed predominantly of hydrogen and helium mixed with some methane. Winds up to 2,000 km/hour blow over the planet creating tremendous storms, the largest of which, the Great Dark Spot, is in the southern hemisphere. It is nearly as big as the Earth and is similar to the Great Red Spot on Jupiter. Indeed, one of the mysteries raised by *Voyager 2*'s discovery is where Neptune gets the energy to drive such a storm system.

Equally intriguing were the discoveries of six new Neptunian moons and three rings encircling the planet. The most astonishing discoveries, however, were found on Neptune's largest moon Triton, which has a diameter of 2,720 km, 700 km less than our own moon (◆ Fig. 19.2). Triton, with a mottled

surface of delicate pinks, reds, and blues, is turning out to be one of the most colorful objects in the solar system. Its surface consists primarily of water ice, with minor amounts of nitrogen and a methane frost. Geysers were also discovered to be erupting carbon-rich material and frozen nitrogen particles some 8 km above its surface. This makes Triton only the second place other than Earth undergoing active volcanism.

Some areas of Triton are smooth while others are very irregular, indicating numerous episodes of deformation. Heavily cratered areas bear witness to bombardment by meteorites or the collapse of its surface. Perhaps the most intriguing aspect of Triton is that it may have once been a planet—much like Pluto, which it resembles in size and possibly composition—that was captured by Neptune's gravitational field soon after the formation of the solar system. However, much still needs to be learned about Triton and Neptune's other moons before this hypothesis can be accepted.

⩔ INTRODUCTION

Of all the known planets and their moons only one, Earth, has life on it. This unique planet, revolving around the Sun every 365.25 days, is a dynamic and complex body. When viewed from the blackness of space, the Earth is a brilliant, shimmering, bluish planet, wrapped in a veil of swirling white clouds (see Chapter 1 opening photo). Beneath these clouds is a surface covered by oceans, seven continents, and numerous islands.

The Earth has not always looked the way it does today. Based on various lines of evidence, many scientists think that the Earth began as a homogeneous mass of rotating dust and gases that contracted, heated, and differentiated during its early history to form a medium-sized planet with a metallic core, a mantle composed of iron- and magnesium-rich rocks, and a thin crust. Overlying this crust is an atmosphere currently composed of 78% nitrogen and 21% oxygen.

As the third planet from the Sun, Earth seems to have formed at just the right distance from the Sun (149,600,000 km) so that it is neither too hot nor too cold to support life as we know it. Furthermore, its size is just right to hold an atmosphere. If the Earth were smaller, its gravity would be so weak that it could retain little, if any, atmosphere.

◆ FIGURE 19.2 Neptune's moon Triton is described by scientists as "a world unlike any other." In this composite of numerous high-resolution images taken by *Voyager 2* during its August 1989 flyby, various features can be seen. The large south polar ice cap at the bottom consists mostly of frozen nitrogen that was deposited during the previous Tritonian winter and is slowly evaporating. The dark plumes in the lower right may be the result of volcanic activity. Smooth plains and fissures in the upper half are evidence of geologic activity in which the surface has been cracked and flooded by slushy ice that refroze.

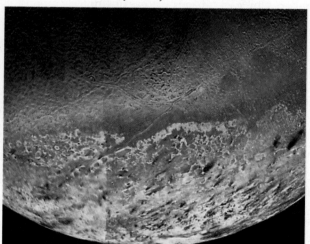

THE ORIGIN OF THE UNIVERSE

Most scientists think that the universe originated between 13 and 20 billion years ago in what is popularly called the "**Big Bang**." In a region infinitely smaller than an atom, both time and space were set at zero. As explained by Einstein's theory of relativity, space and time are unalterably linked to form a space-time continuum. In other words, without space there can be no time. Therefore, there is no "before the Big Bang," only what occurred after it.

Two fundamental phenomena indicate that the Big Bang occurred. The first is the expansion of the universe. When astronomers look beyond our own solar system, they observe that everywhere in the universe galaxies are apparently moving away from each other at tremendous speeds. By measuring this expansion rate, they can calculate how long ago the galaxies were all together at a single point. Secondly, a background radiation of 2.7° above absolute zero (absolute zero equals −273°C) permeates the entire universe. This background radiation is thought to be the faint afterglow of the Big Bang.

At the time of the Big Bang, matter as we know it did not exist, and the universe consisted of pure energy. Within the first second after the Big Bang, the four basic forces—gravity, electromagnetic force, strong nuclear force, and weak nuclear force (● Table 19.1)—had all separated, and the universe experienced enormous expansion. Matter and antimatter collided and annihilated each other. Fortunately, there was a slight excess of matter left over that would become the universe. When the universe was three minutes old, temperatures were cool enough for protons and neutrons to fuse together to form the nuclei of hydrogen and helium atoms. Approximately 100,000 years later, electrons joined with the previously formed nuclei to make complete atoms of hydrogen and helium. At the same time, photons (the energetic particles of light) separated from matter, and light burst forth for the first time.

As the universe continued expanding and cooling, stars and galaxies formed, and the chemical makeup of the universe changed. Early in its history, the universe was 100% hydrogen and helium, whereas today it is 98% hydrogen and helium by weight.

Over the course of their history, stars undergo many nuclear reactions whereby lighter elements are converted into heavier elements by nuclear fusion in which atomic nuclei combine to form more massive nuclei. Such reactions, which convert hydrogen to helium, occur in the cores of all stars. The subsequent conversion of helium to heavier elements, such as carbon, depends on the mass of the star. When a star dies, often explosively, the heavier elements that were formed in its core are returned to interstellar space and are available for inclusion in new stars. When new stars form, they will have a small amount of these heavier elements, which may be converted to still heavier elements. In this way the chemical composition of the galaxies has gradually changed.

● **TABLE 19.1** The Four Basic Forces of the Universe

Four forces appear to be responsible for all interactions of matter:

1. **Gravity** is the attraction of one body toward another.
2. The **electromagnetic force** combines electricity and magnetism into the same force and binds atoms into molecules. It also transmits radiation across the various spectra at wavelengths ranging from gamma rays (shortest) to radio waves (longest) through massless particles called *photons.*
3. The **strong nuclear force** binds protons and neutrons together in the nucleus of an atom.
4. The **weak nuclear force** is responsible for the breakdown of an atom's nucleus, producing radioactive decay.

THE ORIGIN AND EARLY DEVELOPMENT OF THE SOLAR SYSTEM

Having briefly examined the origin and history of the universe, we can now examine how our own solar system, which is part of the Milky Way Galaxy, formed. The solar system consists of the Sun, nine planets, 61 known moons, a tremendous number of asteroids—most of which orbit the Sun in a zone between Mars and Jupiter—and millions of comets, meteorites, and interplanetary dust and gases (● Table 19.2).

General Characteristics of the Solar System

Any theory that attempts to explain the origin and history of the solar system must take into account several general characteristics (● Table 19.3).

● **TABLE 19.2** Characteristics of the Sun, Planets, and Moon

Object	Mean Distance to Sun (km × 10⁶)	Orbital Period (days)	Rotational Period (days)	Tilt of Axis	Equatorial Diameter (km)	Mass(kg)	Mean Density (g/cm³)	Number of Satellites
Sun	—	—	25.5	—	1,391,400	1.99×10^{30}	1.41	—
Terrestrial planets								
Mercury	57.9	88.0	58.7	28°	4,880	3.33×10^{23}	5.43	0
Venus	108.2	224.7	243	3°	12,104	4.87×10^{27}	5.24	0
Earth	149.6	365.3	1	24°	12,760	5.97×10^{24}	5.52	1
Mars	227.9	687.0	1.03	24°	6,787	6.42×10^{23}	3.96	2
Jovian planets								
Jupiter	778.3	4,333	0.41	3°	142,796	1.90×10^{27}	1.33	16
Saturn	1,428.3	10,759	0.43	27°	120,660	5.69×10^{26}	0.69	18
Uranus	2,872.7	30,685	0.72	98°	51,200	8.69×10^{25}	1.27	15
Neptune	4,498.1	60,188	0.67	30°	49,500	1.03×10^{26}	1.76	8
Pluto	5,914.3	90,700	6.39	122°	2,300	1.20×10^{22}	2.03	1
Moon	0.38 (from Earth)	27.3	27.32	7°	3,476	7.35×10^{22}	3.34	—

● **TABLE 19.3** General Characteristics of the Solar System

> **1. Planetary orbits and rotation**
> - Planetary and satellite orbits lie in a common plane.
> - Nearly all of the planetary and satellite orbital and spin motions are in the same direction.
> - The rotation axes of nearly all the planets and satellites are roughly perpendicular to the plane of the ecliptic.
>
> **2. Chemical and physical properties of the planets**
> - The terrestrial planets are small, have a high density (4.0 to 5.5 g/cm³), and are composed of rock and metallic elements.
> - The Jovian planets are large, have a low density (0.7 to 1.7 g/cm³), and are composed of gases and frozen compounds.
>
> **3. The slow rotation of the Sun**
>
> **4. Interplanetary material**
> - The existence and location of the asteroid belt.
> - The distribution of interplanetary dust.

All of the planets revolve around the Sun in the same direction, in nearly circular orbits, and in approximately the same plane (called the *plane of the ecliptic*), except for Pluto whose orbit is both highly elliptical and tilted 17° to the orbital plane of the rest of the planets (◆ Fig. 19.3).

All of the planets, except Venus and Uranus, and nearly all the planetary moons rotate counterclockwise when viewed from a point high in space above the Earth's North Pole. Furthermore, the axes of rotation of the planets, except for those of Uranus and Pluto, are nearly perpendicular to the plane of the ecliptic (Fig. 19.3b).

The nine planets can be divided into two groups based on their chemical and physical properties. The four inner planets—Mercury, Venus, Earth, and Mars—are all small and have high mean densities (Table 19.2), indicating that they are composed of rock and metallic elements. They are known as the **terrestrial planets** because they are similar to *terra,* which is Latin for Earth.

The next four planets—Jupiter, Saturn, Uranus, and Neptune—are called the **Jovian planets** because they all resemble Jupiter. The Jovian planets are large and have low mean densities, indicating they are composed of lightweight gases such as hydrogen and helium, as well as frozen compounds such as ammonia and methane. The outermost planet, Pluto, is small and has a low mean density of slightly more than 2.0 g/cm³.

The slow rotation of the Sun is another feature that must be accounted for in any comprehensive theory of the origin of the solar system. If the solar system

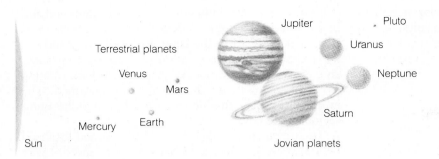

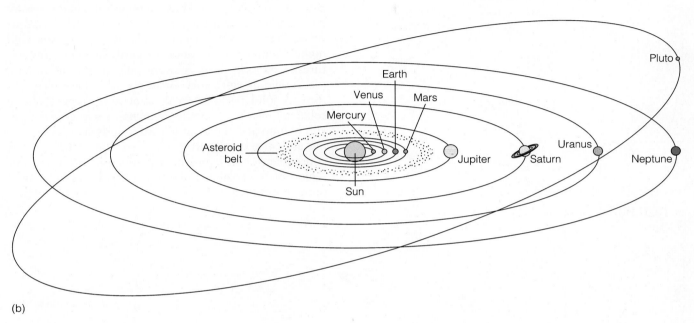

(b)

formed from the collapse of a rotating cloud of gas and dust as is currently accepted, the Sun, which was at the center of that cloud, should have a very rapid rate of rotation, instead of its leisurely 25-day rotation.

Finally, any theory of the origin of the solar system must accommodate the nature and distribution of the various interplanetary objects such as the asteroid belt, comets, and interplanetary gases and dust.

Current Theory of the Origin and Early History of the Solar System

Various scientific theories of the origin of the solar system have been proposed, modified, and discarded since the French scientist and philosopher René Des-

cartes first proposed in 1644 that the solar system formed from a gigantic whirlpool within a universal fluid. Most theories have involved an origin from a primordial rotating cloud of gas and dust. Through the forces of gravity and rotation, this cloud then shrank and collapsed into a rotating disk. Detached rings within the disk condensed into planets, and the Sun condensed in the center of the disk.

The problem with most of these theories is that they failed to explain the slow rotation of the Sun, which according to the laws of physics should be rotating rapidly. This problem was finally solved with the discovery of the solar wind, which is an outflow of ionized gases from the Sun that interact with its magnetic field and slow down its rotation through a magnetic braking process (◆ Fig. 19.4).

According to the currently accepted **solar nebula theory** (◆ Fig. 19.5), interstellar material in a spiral arm of the Milky Way Galaxy condensed and began collapsing about 4.6 billion years ago. As this cloud gradually collapsed under the influence of gravity, it flattened and began rotating counterclockwise, with about 90% of its mass concentrated in the central part of the cloud. As the rotation and concentration of material continued, an embryonic Sun, surrounded by a turbulent, rotating cloud of material called a solar nebula, formed. The inner portions of this nebula were hot and the outer regions were cold.

The turbulence in this solar nebula formed localized eddies where gas and solid particles condensed. Every element and compound has a temperature and pressure combination at which it condenses from the gaseous phase, just as frost forms from water vapor on a cold night. Elements that condense easily at high temperatures, such as iron, magnesium, silicon, and aluminum, are known as **refractory elements**, and these elements formed solid particles in the hot inner region of the solar nebula. The **volatile elements**, such as hydrogen and helium and the volatile compounds such as ammonia and methane condense at very low temperatures; consequently, they remained gaseous in the hot inner region of the solar nebula, but formed ices in its cold outer portion.

◆ FIGURE 19.4 The slow rotation of the Sun is the result of the interaction of its magnetic force lines with ionized gases of the solar nebula. Thus, the rotation is slowed by a magnetic braking process.

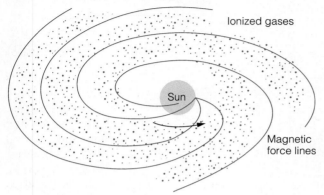

Ionized gases

Sun

Magnetic force lines

◆ FIGURE 19.5 The solar nebula theory for the origin of our solar system involves (*a*) a huge nebula condensing under its own gravitational attraction, then (*b*) contracting, rotating, and (*c*) flattening into a disk, with (*d*) the Sun forming in the center and eddies gathering up material to form planets. As the Sun contracts and begins to visibly shine, (*e*) intense solar radiation blows away unaccreted gas and dust until finally, (*f*) the Sun begins burning hydrogen and the planets complete their formation.

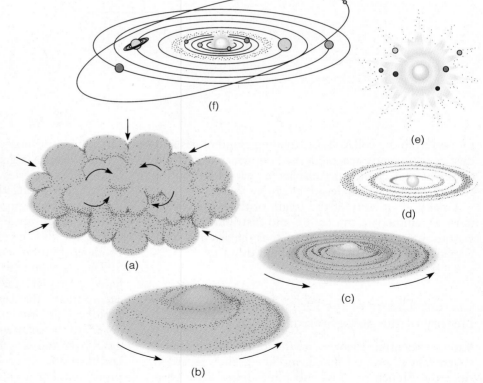

(f)

(e)

(d)

(c)

(a)

(b)

As condensation took place, gaseous, liquid, and solid particles began accreting into ever-larger masses called **planetesimals** that eventually became true planetary bodies. The composition and evolutionary history of the planets are indicated, in part, by their distance from the Sun. For example, the terrestrial planets are composed of rock and metallic elements that condensed at the high temperatures of the inner nebula. The Jovian planets, all of which have small central rocky cores compared to their overall size, are composed mostly of hydrogen, helium, ammonia, and methane, which condense at low temperatures. Thus, the farther away from the Sun that condensation occurred, the lower the temperature, and hence the higher the percentage of volatile elements relative to refractory elements.

While the planets were accreting, material that had been pulled into the center of the nebula also condensed, collapsed, and was heated to several million degrees by gravitational compression. The result was the birth of a star, our Sun.

During the early accretionary phase of the solar system's history, collisions between various bodies were common, as indicated by the craters on many planets and moons. An unusually large collision involving Venus could explain why it rotates clockwise rather than counterclockwise, and a collision could also explain why Uranus and Pluto do not rotate nearly perpendicular to the plane of the ecliptic.

It is thought that the *asteroids* probably formed as planetesimals in a localized eddy between what eventually became Mars and Jupiter in much the same way as other planetesimals formed the terrestrial planets. However, the tremendous gravitational field of Jupiter prevented this material from ever accreting into a planet.

The *comets,* which are interplanetary bodies composed of loosely bound rocky and icy material, are thought to have condensed near the orbits of Uranus and Neptune. Each time the comets pass by Jupiter and Saturn, however, the gravitational slingshot effect of those planets increases their speed, forcing them further out into the solar system.

The solar nebula theory of the formation of the solar system thus accounts for the similarities in orbits and rotation of the planets and their moons, the differences in composition between the terrestrial and Jovian planets, the slow rotation of the Sun, and the presence of the asteroid belt. Based on the available data, the solar nebula theory best explains the features of the solar system and provides a logical explanation for its evolutionary history.

▼ METEORITES

Meteorites are thought to be pieces of material that originated during the formation of the solar system 4.6 billion years ago. Early in the history of the solar system, a period of heavy meteorite bombardment occurred as the solar system cleared itself of the many pieces of material that had not yet accreted into planetary bodies or moons. Since then, meteorite activity has greatly diminished. Most of the meteorites that currently reach the Earth are probably fragments resulting from collisions between asteroids.

Meteorites are classified into three broad groups based on their proportions of metals and silicate minerals (♦ Fig. 19.6). About 93% of all meteorites are composed of iron and magnesium silicate minerals and thus are known as **stones**. Stony meteorites are not all the same and can be divided into three different types.

Ordinary chondrites are the most common type of stony meteorite and are composed of such high-temperature ferromagnesian silicate minerals as olivine and certain pyroxenes (Fig. 19.6b). Age dating reveals they are 4.6 billion years old and thus represent material that existed when the solar system was forming. Most chondrites contain **chondrules**, which are small mineral bodies that formed by rapid cooling when the first solid material of the meteorite was condensing.

The second type of stony meteorites is **carbonaceous chondrites**, which have the same general composition as ordinary chondrites but also contain about 5% organic compounds (Fig. 19.6c). These organic molecules, which include some amino acids, are not biogenically produced, but may represent chemical precursors of organisms. **Acondrites**, the third type of stony meteorite, do not contain chondrules (Fig. 19.6d). Their composition is similar to that of terrestrial basalts, and they are thought to be the result of collisions between larger asteroids.

Irons, the second meteorite group, account for about 6% of all meteorites. They are composed primarily of a combination of iron and nickel alloys (Fig. 19.6e). Their large crystal size and chemical composition indicate that they cooled very slowly in large objects such as asteroids where the hot iron-nickel interior could be insulated from the cold of space. Collisions between such slowly cooling asteroids produced the iron meteorites that we find today.

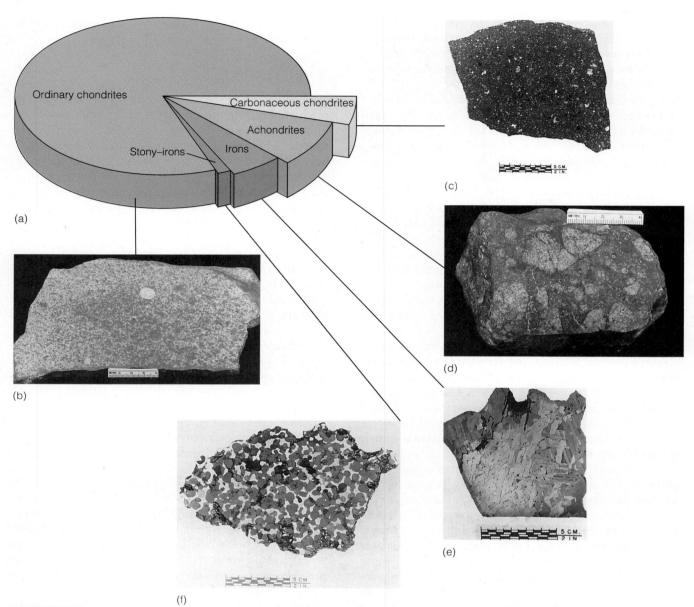

◆ FIGURE 19.6 (*a*) Relative proportions of the different types of meteorites. (*b*) Polished slab of an ordinary chondrite. (*c*) Polished slab of a carbonaceous chondrite from Allende, Mexico. (*d*) Acondrite. (*e*) Polished slab of an iron meteorite from Bogou, Upper Volta, Brazil. (*f*) Polished slab of a stony-iron from Thiel Mountain, Antarctica. The white minerals are iron-nickel and the grey minerals are olivine.

Stony-irons, the third group, are composed of nearly equal amounts of iron and nickel and silicate minerals; they make up less than 1% of all meteorites (Fig. 19.6f). Stony-irons are generally believed to represent fragments from the zone between the silicate and metallic portions of a large differentiated asteroid.

Astronomers have identified at least 40 asteroids larger than a kilometer in diameter whose orbits cross the Earth's and estimate that there may be as many as 1,000 such asteroids. A collision between a large asteroid and the Earth formed the famous Meteor Crater in Arizona (◆ Fig. 19.7). While asteroid-Earth collisions

◆ FIGURE 19.7 Meteor Crater, Arizona, is the result of an Earth-asteroid collision that occurred between 25,000 and 50,000 years ago. It produced a crater 1.2 km in diameter and 180 m deep.

◆ FIGURE 19.8 Artistic rendition of what the moment of impact would look like if the nucleus of a comet, 48 km in diameter, hit northern New Jersey. Everything visible in this picture, including the buildings of lower Manhattan in the foreground, would be vaporized, and a plume of fine material would be ejected into the atmosphere and circulated around the Earth.

are rare, they do happen and could have devastating results if they occurred in a populated area (◆ Fig. 19.8; see also Perspective 19.1). Many scientists think that a collision with a meteorite about 10 km in diameter led to the extinctions of dinosaurs and several other groups of animals 66 million years ago. Such a collision would have generated a tremendous amount of dust that would have blocked out the Sun, thereby lowering global temperatures and preventing photosynthesis, which, in turn, would have triggered a collapse of the ecosystem and massive extinctions. We know that the ash released into the atmosphere from volcanic eruptions has affected climates (see Perspective 4.1), and studies indicate that a collision with a large meteorite could produce enough dust to similarly affect global climate.

▼ THE PLANETS

A tremendous amount of information about each planet in the solar system has been derived from Earth-based observations and measurements as well as from the numerous space probes launched during the past 30 years. Such information as a planet's size, mass, density, composition, presence of a magnetic field, and atmospheric composition has allowed scientists to formulate hypotheses concerning the origin and history of the planets and their moons.

The Terrestrial Planets

It appears that all of the terrestrial planets had a similar early history during which volcanism and cratering from meteorite impacts were common. After accretion, each planet appears to have undergone differentiation as a result of heating by radioactive decay. The mass, density, and composition of the planets indicate that each formed a metallic core and a silicate mantle-crust during this phase. Images sent back by the various space probes also clearly show that volcanism and cratering by meteorites continued during the differentiation phase. Volcanic eruptions produced lava flows, and an atmosphere developed on each planet by **outgassing**, a process whereby light gases from the interior rise to the surface during volcanic eruptions (see Perspective 19.2, page 504).

Mercury

Mercury, the closest planet to the Sun, apparently has changed very little since it was heavily cratered during its early history (◆ Fig. 19.9a, page 506). Most of what we know about this small (4,880 km diameter) planet comes

Perspective 19.1

THE TUNGUSKA EVENT

On June 30, 1908, a bright object crossed the sky moving from southeast to northwest over central Siberia, and a few seconds later a huge explosion occurred in the Tunguska River basin (◆ Fig. 1). The noise from the explosion was heard up to 1,000 km away, a column of incandescent matter rose to a height of about 20 km, the shock wave from the explosion traveled around the world twice, and seismographs around the world registered an earthquake. Eyewitnesses reported that the concussion wave threw people to the ground as much as 60 km away from the blast site.

What the object was that caused this massive explosion remains uncertain. Part of the uncertainty is because the event occurred in an extremely remote area, and it was not until 1921 that an expedition was launched to investigate. Unfortunately, illness and exhaustion prevented this expedition from reaching the explosion site. Finally, in 1927, 19 years after the explosion, an expedition led by Leonid Kulik successfully reached the Tunguska basin. A vast peat bog called the Southern Swamp was identified as the site above which the explosion occurred; subsequent investigations and studies indicate

◆ FIGURE 1 The Tunguska explosion occurred in central Siberia.

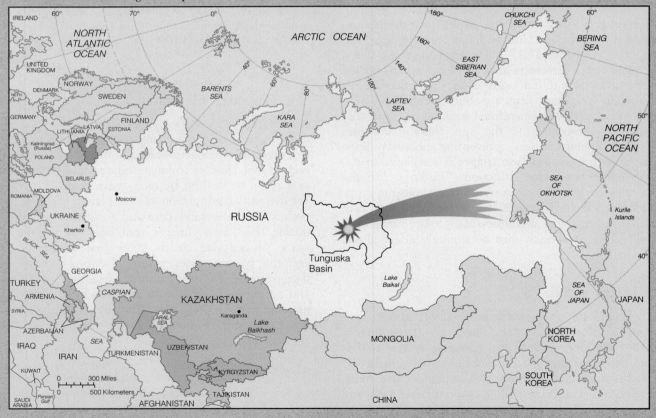

that the explosion occurred about 8 km above the surface, and it is estimated to have been about 12.5 megatons (equivalent to 12.5 million tons of TNT). More than 1,000 km² of forest were leveled by the explosion (◆ Fig. 2), and, according to earlier accounts, tens of thousands of animals perished. Fortunately, there were no human casualties.

The most widely accepted explanation for the Tunguska event is that it was caused by a small, icy comet that exploded in the atmosphere. According to this hypothesis, a comet, perhaps 50 m in diameter, entered the atmosphere and began heating up; as this heating occurred, frozen gases were instantaneously converted to the gaseous state, releasing a tremendous amount of energy and causing a large explosion.

This explanation has recently been challenged by a group of scientists from the National Aeronautics and Space Administration Ames Research Center, who suggest that the Tunguska explosion was caused by a stony meteorite. According to the scientists, computer simulations indicate that the nuclei of comets explode at too high an altitude to account for the pattern and magnitude of destruction found in the Tunguska River Basin. Instead, the researchers conclude that a stony meteorite about 30 m in diameter traveling at 15 km per second would explode at approximately the same altitude as the Tunguska object did, thus producing the same type of destruction.

The computer simulation doesn't completely rule out the possibility of the Tunguska object being either a very fast iron meteorite or a carbonaceous meteorite. The likelihood that the Tunguska projectile was a type of stony meteorite (the most common group of meteorites) has a greater probability, however, than it being a relatively rare low-density comet, as many people have hypothesized.

◆ FIGURE 2 Evidence of the Tunguska event is still apparent in this photograph taken 20 years later. The destruction was caused by some type of explosion in central Siberia in 1908.

Perspective 19.2

THE EVOLUTION OF CLIMATE
ON THE TERRESTRIAL PLANETS

The origins and early evolution of the terrestrial planets appear to have been similar, yet each of these planets has acquired a dramatically different climate. Why? All four planets were initially alike, with atmospheres high in carbon dioxide and water vapor derived by outgassing. Mercury, because of its small size and proximity to the Sun, lost its atmosphere by evaporation early in its history. Venus, Earth, and Mars, however, all were temperate enough during their early histories to have had fluid water on their surfaces, yet only Earth still has surface water and a climate capable of supporting life.

The reason that these three planets evolved such different climates is related to the recycling of carbon dioxide between the atmosphere and the crust (carbon-silicate geochemical cycle) as well as their distances from the Sun. Carbon dioxide recycling is an important regulator of climates because carbon dioxide, other gases, and water vapor allow sunlight to pass "through" them but trap the heat reflected back from the planet's surface. Heat is thus retained, and the temperature of the atmosphere and surface increases in what is known as the *greenhouse effect.*

Carbon dioxide combines with water in the atmosphere to form carbonic acid. When this slightly acidic rain falls, it decomposes rocks, releasing calcium and bicarbonate ions into streams and rivers and, ultimately, the oceans. In the oceans, marine organisms use some of these ions to construct shells of calcium carbonate. When the organisms die, their shells become part of the total carbonate sediments. During subduction these carbonate sediments are heated under pressure and release carbon dioxide gas that reenters the atmosphere primarily through volcanic eruptions (◆ Fig. 1).

The recyling of carbon dioxide has allowed the Earth to maintain a moderate climate throughout its history. For example, when the Earth's surface cools, less water vapor is present in the atmosphere and there is less rain. The amount of carbon dioxide leaving the atmosphere thus decreases and less decomposition of rocks occurs. This leads to a temporary increase in carbon dioxide in the atmosphere, greater greenhouse warming, and, thus, higher surface temperatures.

Just the opposite would happen if the surface temperature should increase. Oceanic evaporation would then increase, leading to greater rainfall and more rapid decomposition of rock; as a result, carbon dioxide would be removed from the atmosphere. Greenhouse warming would then decrease and surface temperatures would fall.

Venus today is almost completely waterless. Many scientists, however, think that during its early history when the Sun was dimmer, Venus perhaps had vast oceans. During this time, water vapor as well as carbon dioxide was being released into the atmosphere by volcanism. The water vapor condensed and formed oceans, while carbon dioxide cycled (by plate tectonics) just as it does on Earth. As the Sun's energy output increased, however, these oceans eventually evaporated. Once the oceans disappeared, there was no water to return carbon to the crust, and carbon dioxide began accumulating in the atmosphere, creating a greenhouse effect and raising temperatures.

Mars, like Venus and Earth, probably once had a moderate climate and surface water, as indicated by the crisscrossing network of valleys on its oldest terrain. Because Mars is smaller than the Earth, it had less internal heat when it formed and hence cooled rapidly. Eventually, the interior of Mars became so cold that it no longer released carbon dioxide. As a result, the amount of atmospheric carbon dioxide decreased to its current low level. The greenhouse effect was thus weakened, and the Martian atmosphere became thin and cooled to its present low temperature.

If Mars had been the size of Earth or Venus, it very likely would have had enough internal heat to continue recycling carbon dioxide, thus offsetting the effects of low sunlight levels caused by its distance from the Sun. In other words, Mars would still have enough carbon dioxide in its atmosphere so that it could maintain a "temperate climate."

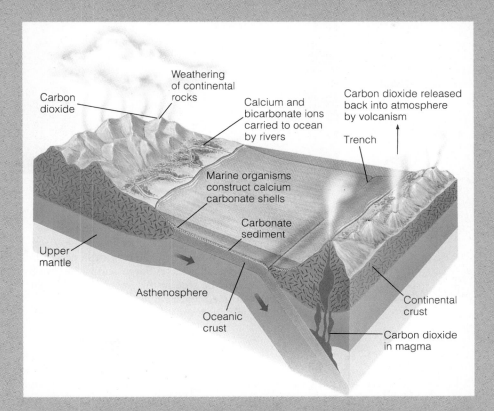

Carbon dioxide

Weathering of continental rocks

Calcium and bicarbonate ions carried to ocean by rivers

Carbon dioxide released back into atmosphere by volcanism

Trench

Marine organisms construct calcium carbonate shells

Carbonate sediment

Upper mantle

Asthenosphere

Oceanic crust

Continental crust

Carbon dioxide in magma

◆ FIGURE 1 The carbon-silicate geochemical cycle illustrates how carbon dioxide is recycled. Carbon dioxide is removed from the atmosphere by combining with water and forming slightly acidic rain that falls on the Earth's surface and decomposes rocks. This decomposition releases calcium and bicarbonate ions that ultimately reach the oceans. Marine organisms use these ions to construct shells of calcium carbonate. When they die, the shells become part of the carbonate sediments that are eventually subducted. As the sediments are subjected to heat and pressure, they release carbon dioxide gas back into the atmosphere primarily through volcanic eruptions.

(a)

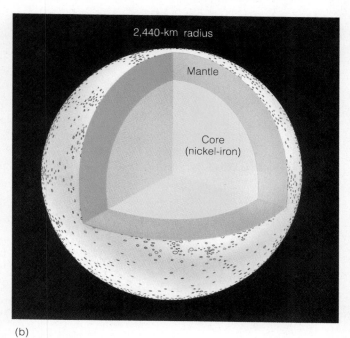

(b)

(c)

◆ FIGURE 19.9 (*a*) Mercury has a heavily cratered surface that has changed very little since its early history. (*b*) Internal structure of Mercury, showing its large solid core relative to its overall size. (*c*) Seven scarps (indicated by arrows) can clearly be seen in this image. It is thought that these scarps formed when Mercury cooled and contracted early in its history.

from measurements and observations made during the flybys of *Mariner 10* in 1974 and 1975 (Table 19.2). Its high overall density of 5.4 g/cm³ indicates that it has a large metallic core measuring 3,600 km in diameter; the core accounts for 80% of Mercury's mass (◆ Fig. 19.9b). Furthermore, Mercury has a weak magnetic field (about 1% as strong as the Earth's), indicating that the core is probably partially molten.

Images sent back by *Mariner 10* show a heavily cratered surface with the largest impact basins filled with what appear to be lava flows similar to the lava plains on the Moon. The lava plains are not deformed, however, indicating that there has been little or no tectonic activity. Another feature of Mercury's surface is a large number of scarps (◆ Fig. 19.9c). It is suggested that these scarps formed when Mercury cooled and contracted.

Because Mercury is so small, its gravitational attraction is insufficient to retain atmospheric gases; any atmosphere that it may have held when it formed probably escaped into space very quickly. Nevertheless, very small quantities of hydrogen and helium, thought to have originated from the solar winds that stream by Mercury, were detected by *Mariner 10*.

Venus

Of all the planets, Venus is the most similar in size and mass to the Earth (Table 19.2). It differs, however, in most other respects. Venus is searingly hot with a surface temperature of 475°C and an oppressively thick atmosphere composed of 96% carbon dioxide and 3.5% nitrogen with traces of sulfur dioxide and sulfuric and hydrochloric acid (◆ Fig. 19.10a). From information obtained by the various space probes that have passed by, orbited Venus, and descended to its surface, we know that three distinct cloud layers composed of droplets of sulfuric acid envelop the planet. Furthermore, winds up to 360 km/hour occur at the top of the clouds, whereas the planet's surface is calm.

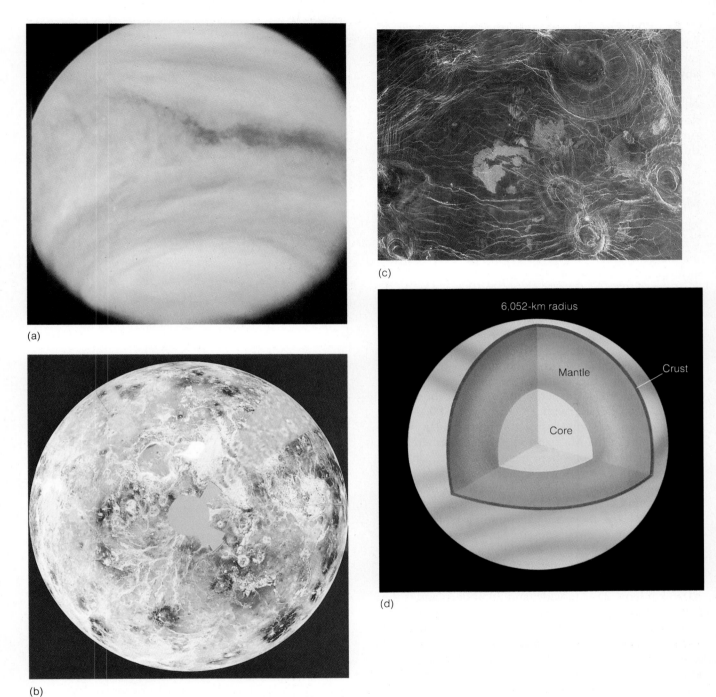

(a)

(b)

(c)

(d)

◆ FIGURE 19.10 (*a*) Venus has a searingly hot surface and is surrounded by an oppressively thick atmosphere composed largely of carbon dioxide. (*b*) A nearly complete map of the northern hemisphere of Venus based on radar images beamed back to Earth from the *Magellan* space probe. (*c*) This radar image of Venus made by the *Magellan* spacecraft reveals circular and oval-shaped volcanic features. A complex network of cracks and fractures extends outward from the volcanic features. Geologists think these features were created by blobs of magma rising from the interior of Venus with dikes filling some of the cracks. (*d*) The internal structure of Venus.

Radar images from orbiting spacecraft as well as from the Venusian surface indicate a wide variety of terrains (◆ Fig. 19.10b), some of which are unlike anything seen elsewhere in the solar system.

Even though no active volcanism has been observed on Venus, the presence of volcanoes, numerous lava flows, folded mountain ranges, and a network of fractures indicate internal and surface activity has occurred during the past (◆ Fig. 19.10c). There is, however, no evidence for active plate tectonics such as on Earth.

Mars

Mars, the red planet, is differentiated, as are all the terrestrial planets, into a metallic core and a silicate mantle and crust (◆ Fig. 19.11a). The thin Martian atmosphere consists of 95% carbon dioxide, 2.7% nitrogen, 1.7% argon, and traces of other gases. Mars has distinct seasons during which its polar ice caps of frozen carbon dioxide expand and recede.

Perhaps the most striking aspect of Mars is its surface, many features of which have not yet been satisfactorily explained. Like the surfaces of Mercury and the Moon, the southern hemisphere is heavily cratered, attesting to a period of meteorite bombardment. *Hellas*, a crater with a diameter of 2,000 km, is the largest known impact structure in the solar system and is found in the Martian southern hemisphere.

The northern hemisphere is much different, having large smooth plains, fewer craters, and evidence of extensive volcanism. The largest known volcano in the solar system, *Olympus Mons* has a basal diameter of 600 km, rises 27 km above the surrounding plains, and is topped by a huge circular crater 80 km in diameter.

The northern hemisphere is also marked by huge canyons that are essentially parallel to the Martian equator. One of these canyons, *Valles Marineris*, is at least 4,000 km long, 250 km wide, and 7 km deep and is the largest yet discovered in the solar system (◆ Fig. 19.11b). If it were present on Earth, it would stretch from San Francisco to New York! It is not yet known how these vast canyons formed, although geologists postulate that they may have started as large rift zones that were subsequently modified by running water and wind erosion. Such hypotheses are based on comparison to rift structures found on Earth and topographic features formed by geologic agents of erosion such as water and wind.

Tremendous wind storms have strongly influenced the surface of Mars and led to dramatic dune formations (see Perspective 15.1, Fig. 3). Even more stunning than the dunes, however, are the channels that appear

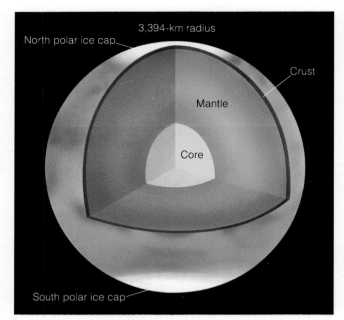

(a)

(b)

◆ FIGURE 19.11 (*a*) The internal structure of Mars. (*b*) A striking view of Mars is revealed in this mosaic of 102 *Viking* images. The largest canyon known in the solar system, *Valles Marineris*, can be clearly seen in the center of this image, while three of the planet's volcanoes are visible on the left side of the image.

to be the result of running water (see Fig. 12.1). Mars is currently too cold for surface water to exist, yet the channels strongly indicate that there was running water on the planet during the past.

The fresh-looking surfaces of its many volcanoes strongly suggest that Mars was a tectonically active planet during the past and may still be. There is, however, no evidence that plate movement, such as occurs on Earth, has ever occurred.

The Jovian Planets

The Jovian planets are completely unlike any of the terrestrial planets in size or chemical composition (Table 19.2) and have had completely different evolutionary histories. While they all apparently contain a small core in relation to their overall size, the bulk of a Jovian planet is composed of volatile elements and compounds that condense at low temperatures such as hydrogen, helium, methane, and ammonia.

Jupiter

Jupiter is the largest of the Jovian planets (Table 19.2; ◆ Fig. 19.12a). With its moons, rings, and radiation belts, it is the most complex and varied planet in the solar system. Jupiter's density is only one-fourth that of Earth, but because it is so large, it has 318 times the mass (Table 19.2). It is an unusual planet in that it emits almost 2.5 times more energy than it receives from the Sun. One explanation is that most of the excess energy is left over from the time of its formation. When Jupiter formed, it heated up because of gravitational contraction (as did all the planets) and is still cooling. Jupiter's massive size insulates its interior, and hence it has cooled very slowly.

Jupiter has a relatively small central core of solid rocky material formed by differentiation. Above this core is a thick zone of liquid metallic hydrogen followed by a thicker layer of liquid hydrogen; above that is a thin layer of clouds (◆ Fig. 19.12b). Surrounding Jupiter are a strong magnetic field and an intense radiation belt.

Jupiter has a dense atmosphere of hydrogen, helium, methane, and ammonia, which some think are the same gases that composed the Earth's first atmosphere. Jupiter's cloudy atmosphere is divided into a series of different colored bands as well as a variety of spots (the Great Red Spot) and other features, all interacting in incredibly complex motions.

Revolving around Jupiter are 16 moons varying greatly in geologic activity. Also surrounding Jupiter is a thin, faint ring, a feature shared by all the Jovian planets.

(a)

(b)

◆ FIGURE 19.12 (*a*) Jupiter, the largest planet in the solar system, is also the most complex with its moons, rings, and radiation belts. Its cloudy atmosphere displays a regular pattern of bands and spots, the largest of which is the Great Red Spot shown in the lower part of this photograph. (*b*) The internal structure of Jupiter.

Saturn

Saturn is slightly smaller than Jupiter, about one-third as massive, and about one-half as dense, but has a similar internal structure and atmosphere (Table 19.2). Saturn, like Jupiter, gives off more energy (2.2 times as much) than it gets from the Sun. Saturn's most conspicuous feature is its ring system, consisting of thousands of rippling, spiraling bands of countless particles (◆ Fig. 19.13a).

The composition of Saturn is similar to Jupiter's, but consists of slightly more hydrogen and less helium. Saturn's core is not as dense as Jupiter's, and as in the case of Jupiter, a layer of liquid metallic hydrogen overlies the core, followed by a zone of liquid hydrogen and helium, and, lastly, a layer of clouds (◆ Fig. 19.13b). Because liquid metallic hydrogen can exist only at very high pressures, and since Saturn is smaller than Jupiter, such high pressures are found at greater depths in Saturn. Therefore, there is less of this conducting material than on Jupiter, and as a consequence, Saturn has a weaker magnetic field.

Even though the atmospheres of Saturn and Jupiter are similar, Saturn's atmosphere contains little ammonia because it is farther from the Sun and therefore is colder. The cloud layer on Saturn is thicker than on

◆ FIGURE 19.13 Saturn and three of its moons. (a) This image of Saturn was taken by *Voyager 2* from several million kilometers and shows the ring system of the planet as well as its banded atmosphere. Saturn has an atmosphere similar to that of Jupiter, but has a thicker cloud cover and contains little ammonia. (b) The internal structure of Saturn.

(a)

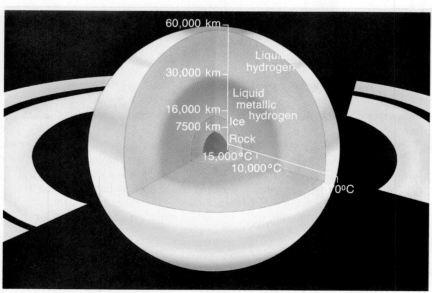

(b)

Jupiter, but it lacks the contrast between the different bands. Unlike Jupiter, Saturn has seasons because its axis tilts 27°.

Saturn has 18 known moons, most of which are small with low densities. Titan with its nitrogen and methane atmosphere is the most distinctive and perhaps the most interesting moon in the solar system.

Uranus

Uranus is much smaller than Jupiter, but their densities are about the same (Table 19.2). It is the only planet that lies on its side (◆ Fig. 19.14a and b); that is, its axis of rotation nearly parallels the plane of the ecliptic. Some scientists think that a collision with an Earth-sized

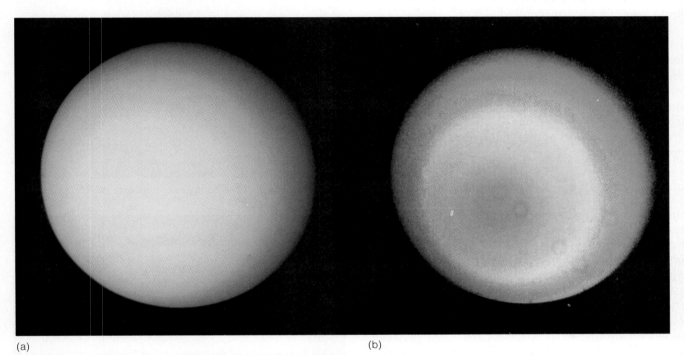

(a)

(b)

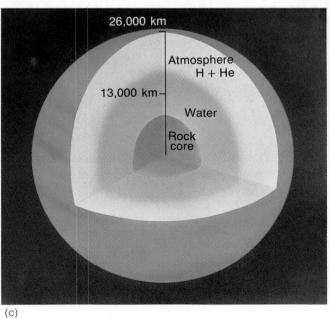

26,000 km

Atmosphere
H + He

13,000 km

Water

Rock core

(c)

◆ FIGURE 19.14 (*a*) Images of Uranus taken by *Voyager 2* under ordinary light show a featureless planet. (*b*) When color is enhanced by computer processing techniques, Uranus is seen to have zonal flow patterns in its atmosphere. (*c*) The internal structure of Uranus.

body early in its history may have knocked Uranus on its side.

Data gathered by the flyby of *Voyager 2* suggest that Uranus has a water zone beneath its cloud cover. Because the planet's density is greater than if it were composed entirely of hydrogen and helium, it is thought that Uranus must have a dense, rocky core, and this core may be surrounded by a deep global ocean of liquid water (◆ Fig. 19.14c).

The atmospheric composition of Uranus is similar to that of Jupiter and Saturn with hydrogen being the dominant gas, followed by helium and some methane. Uranus also has a banded atmosphere and a circulation pattern much like those of Jupiter and Saturn. Surrounding Uranus is a huge corkscrew-shaped magnetic field that stretches for millions of kilometers into space. Uranus has at least nine thin, faint rings and 15 small moons circling it.

Neptune and Pluto

The flyby of *Voyager 2* in August 1989 provided the first detailed look at Neptune and showed it to be a dynamic, stormy planet (Fig. 19.1). Its atmosphere is similar to those of the other Jovian planets, and it exhibits a pattern of zonal winds and giant storm systems comparable to those of Jupiter. The internal structure of Neptune is similar to that of Uranus (Table 19.2); it has a rocky core approximately 17,000 km in diameter surrounded by a semifrozen slush of water and liquid methane (◆ Fig. 19.15). Its atmosphere is composed of hydrogen and helium with some methane. Encircling Neptune are three faint rings and eight moons.

With a diameter of only 2,300 km, Pluto is the smallest planet and, strictly speaking, it is not one of the Jovian planets (Table 19.2). Little is known about Pluto, but recent studies indicate it has a rocky core overlain by a mixture of methane gas and ice (◆ Fig. 19.16). It has a thin, two-layer atmosphere with a clear upper layer overlying a more opaque lower layer. Pluto's mid-latitude areas are dark, while its north pole region is distinctly brighter. Furthermore, its south polar region is extremely bright, which is interpreted to indicate seasonality.

Pluto differs from all the other planets in that it has a highly elliptical orbit that is tilted with respect to the plane of the ecliptic. It has one known moon, Charon, that is nearly half its size with a surface that appears to differ markedly from Pluto's.

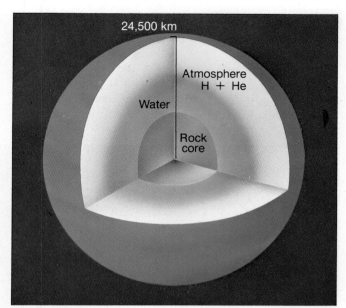

◆ FIGURE 19.15 The internal structure of Neptune.

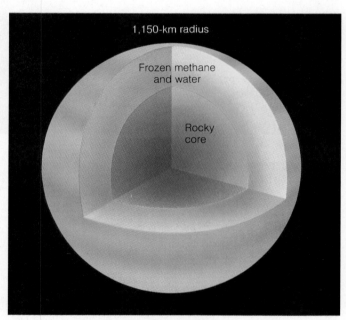

◆ FIGURE 19.16 The internal structure of Pluto.

❧ THE ORIGIN AND DIFFERENTIATION OF THE EARLY EARTH

As matter was accreting in the various turbulent eddies that swirled around the early Sun, enough material eventually gathered together in one eddy to form the planet Earth. The Earth consists of concentric layers of different composition and densities (◆ Fig. 19.17). This differentiation into concentric layers is a fundamental characteristic of all the terrestrial planets, and presumably this differentiation occurred very early in their history.

Geologists know that the Earth is 4.6 billion years old. The oldest known rocks, however, are 3.96-billion-year-old metamorphic rocks from Canada. Like the younger crustal rocks, these rocks are composed of relatively light silicate minerals. It appears that a crust, a heavier silicate mantle, and an iron-nickel core were already present 3.96 billion years ago, or 640 million years after the Earth formed. What did the early Earth look like, and how did it come to be differentiated into its three concentric layers?

The early Earth was probably of generally uniform composition and density throughout (◆ Fig. 19.18a). It was composed mostly of silicate compounds, iron and magnesium oxides, and smaller amounts of all the other chemical elements. Scientists believe this early Earth was rather cool, so the elements and nebular rock fragments accreting to it were solids rather than gases

or liquids. In order for the iron and nickel to concentrate in the core, the Earth must have heated up enough for them to melt and sink through the surrounding lighter silicate minerals.

This initial heating could have occurred in three ways. Some heat was no doubt generated by the impact of meteorites; most of this heat was radiated back into space, but some was probably retained by the accreting planet. Heat was also generated within the early Earth as it was reduced to a smaller volume by gravitational compression. Rock is a poor conductor of heat, so this heat accumulated within the Earth. The third cause of internal heating was the decay of radioactive elements such as uranium, thorium, and others. Even though these elements form only a very small portion of the Earth, the heat generated during radioactive decay was absorbed by the surrounding rock.

The combination of meteorite impacts, gravitational compression, and heat from radioactive decay increased the temperature of the early Earth enough to melt iron and nickel, which, being denser than silicate minerals, settled to the center of the Earth and formed

◆ FIGURE 19.17 Cross section of the Earth showing the various layers and their average density. The crust is divided into a continental and oceanic portion. Continental crust is 20 to 90 km thick; oceanic crust is 5 to 10 km thick.

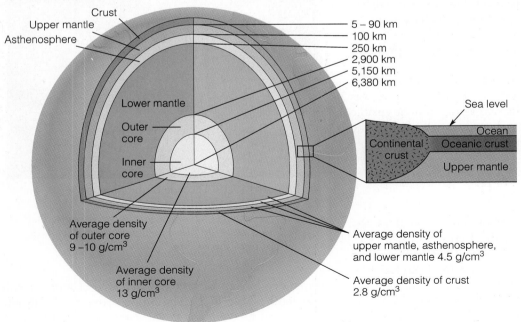

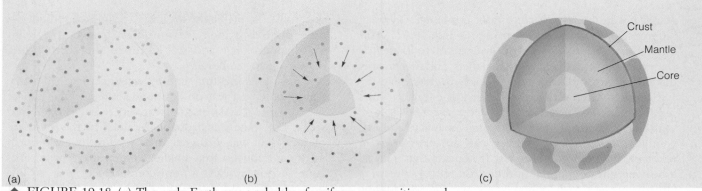

(a)   (b)   (c)

◆ FIGURE 19.18 (*a*) The early Earth was probably of uniform composition and density throughout. (*b*) Heating of the early Earth reached the melting point of iron and nickel, which, being denser than silicate minerals, settled to the Earth's center. At the same time, the lighter silicates flowed upward to form the mantle and the crust. (*c*) In this way, a differentiated Earth formed, consisting of a dense iron-nickel core, an iron-rich silicate mantle, and a silicate crust with continents and ocean basins.

the core (◆ Fig. 19.18b). Simultaneously, the lighter silicates slowly flowed upward, beginning the differentiation of the mantle from the core.

Calculations indicate that with a uniform distribution of elements in an early solid Earth, enough heat could be generated to begin melting iron and nickel at depths between 400 and 850 km. Melting would have had to begin at shallow depths because the temperature at which melting begins increases with pressure; therefore the melting point of any material increases toward the Earth's center.

The differentiation into a layered planet is probably the most significant event in the history of the Earth. Not only did it lead to the formation of a crust and eventually to continents (◆ Fig. 19.18c), but it was probably responsible for the outgassing of light volatile elements from the interior that eventually led to the formation of the oceans and atmosphere.

▼ THE ORIGIN OF THE EARTH-MOON SYSTEM

We probably know more about our Moon than any other celestial object except the Earth (◆ Fig. 19.19). Nevertheless, even though the Moon has been studied for centuries through telescopes and has been sampled directly, many questions remain unanswered.

The Moon is one-fourth the diameter of the Earth, has a low density (3.3 g/cm³) relative to the terrestrial planets, and exhibits an unusual chemistry in that it is

◆ FIGURE 19.19 The side of the Moon as seen from Earth. The light-colored areas are the lunar highlands which were heavily cratered by meteorite impacts. The dark-colored areas are maria, which formed when lava flowed out onto the surface.

bone-dry, having been largely depleted of most volatile elements (Table 19.2). The Moon orbits the Earth and rotates on its own axis at the same rate, so we always see the same side. Furthermore, the Earth-Moon system is unique among the terrestrial planets. Neither Mercury nor Venus has a moon, and the two small moons of Mars—Phobos and Deimos—are probably captured asteroids.

The surface of the Moon can be divided into two major parts: the low-lying dark-colored plains, called *maria,* and the light-colored *highlands* (Fig. 19.19). The highlands are the oldest parts of the Moon and are heavily cratered, providing striking evidence of the massive meteorite bombardment that occurred in the solar system more than four billion years ago.

Study of the several hundred kilograms of rocks returned by the *Apollo* missions indicates that three kinds of materials dominate the lunar surface: igneous rocks, breccias, and dust. Basalt, a common dark-colored igneous rock on Earth, is one of the several different types of igneous rocks on the Moon and makes up the greater part of the maria. The presence of igneous rocks that are essentially the same as those on Earth shows that magmas similar to those on Earth were generated on the Moon long ago.

The lunar surface is covered with a regolith that is estimated to be 3 to 4 m thick. This gray covering, which is composed of compacted aggregates of rock fragments called breccia, glass spherules, and small particles of dust, is thought to be the result of debris formed by meteorite impacts.

The interior structure of the Moon is quite different from that of the Earth, indicating a different evolutionary history (◆ Fig. 19.20). The highland crust is thick (65 to 100 km) and comprises about 12% of the Moon's volume. It was formed about 4.4 billion years ago, immediately following the Moon's accretion. The highlands are composed principally of the igneous rock anorthosite, which is made up of light-colored feldspar minerals that are responsible for their white appearance.

A thin covering (1 to 2 km thick) of basaltic lava fills the maria; lava covers about 17% of the lunar surface, mostly on the side facing the Earth. These maria lavas came from partial melting of a thick underlying mantle of silicate composition. Moonquakes occur at a depth of about 1,000 km, but below that depth seismic shear waves apparently are not transmitted. Because shear waves do not travel through liquid, their lack of transmission implies that the innermost mantle may be partially molten. There is increasing evidence that the Moon has a small (600 km to 1,000 km diameter) metallic core comprising 2 to 5% of its volume.

The origin and earliest history of the Moon are still unclear, but the basic stages in its subsequent develop-

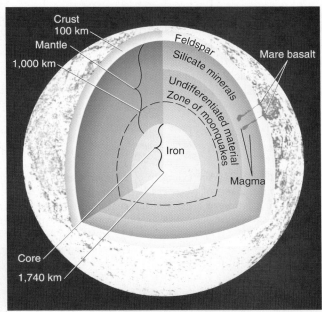

◆ FIGURE 19.20 The internal structure of the Moon is different from that of the Earth. The upper mantle is the source for the maria lavas. Moonquakes occur at a depth of 1,000 km. Because seismic shear waves are not transmitted below this depth, it is believed that the innermost mantle is liquid. Below this layer is a small metallic core.

ment are well understood. It formed some 4.6 billion years ago and shortly thereafter was partially or wholly melted, yielding a silicate melt that cooled and crystallized to form the mineral anorthite. Because of the low density of the anorthite crystals and the lack of water in the silicate melt, the thick anorthosite highland crust formed. The remaining silicate melt cooled and crystallized to produce the zoned mantle, while the heavier metallic elements formed the small metallic core.

The formation of the lunar mantle was completed by about 4.4 to 4.3 billion years ago. The maria basalts, derived from partial melting of the upper mantle, were extruded during great lava floods between 3.8 and 3.2 billion years ago.

Numerous models have been proposed for the origin of the Moon, including capture from an independent orbit, formation with the Earth as part of an integrated two-planet system, breaking off from the Earth during accretion, and formation resulting from a collision between the Earth and a large planetesimal. These various models are not mutually exclusive, and elements of some occur in others. At this time, scientists cannot agree on a single model, as each has some inherent problems. However, the model that seems to account best for the Moon's particular composition and structure involves an impact by a large planetesimal with a young Earth (◆ Fig. 19.21).

In this model, a giant planetesimal, the size of Mars or larger, crashed into the Earth about 4.6 to 4.4 billion years ago, causing the ejection of a large quantity of hot material that formed the Moon. The material that was ejected was mostly in the liquid and vapor phase and came primarily from the mantle of the colliding planetesimal. As it cooled, the various lunar layers crystallized out in the order we have discussed.

◆ FIGURE 19.21 According to one hypothesis for the origin of the Moon, a large planetesimal the size of Mars crashed into the Earth 4.6 to 4.4 billion years ago, causing the ejection of a mass of hot material that formed the Moon. This computer simulation shows the formation of the Moon as a result of an Earth-planetesimal collision.

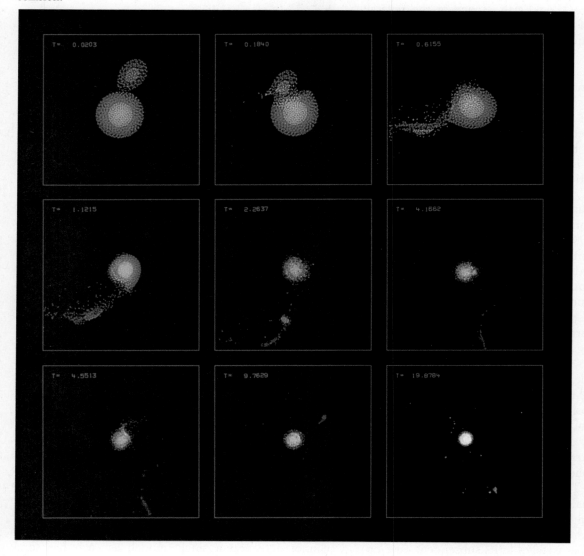

Chapter Summary

1. The universe began with a Big Bang approximately 13 to 20 billion years ago. Astronomers have deduced this age from the fact that celestial objects are moving away from each other in what appears to be an ever-expanding universe.
2. The universe has a background radiation of 2.7° above absolute zero, which is thought to be the faint afterglow of the Big Bang.
3. About 4.6 billion years ago, the solar system formed from a rotating cloud of interstellar matter. As this cloud condensed, it eventually collapsed under the influence of gravity and flattened into a counterclockwise rotating disk. Within this rotating disk, the Sun, planets, and moons formed from the turbulent eddies of nebular gases and solids.
4. Meteorites provide vital information about the age and composition of the solar system. The three major groups are stones, irons, and stony-irons. Each has a different composition, indicating a different origin.
5. Temperature as a function of distance from the Sun played a major role in the type of planets that evolved. The inner terrestrial planets are composed of rock and metallic elements that condense at high temperatures. The outer Jovian planets plus Pluto are composed mostly of hydrogen, helium, ammonia, and methane, all of which condense at lower temperatures.
6. All of the terrestrial planets are differentiated into a core, mantle, and crust, and all seem to have had a similar early history during which volcanism and cratering from meteorite impacts were common.
7. The Jovian planets differ from the terrestrial planets in size and chemical composition and followed completely different evolutionary histories. All of the Jovian planets have a small core compared to their overall size; they are mainly composed of volatile elements and compounds that condense at low temperatures, such as hydrogen, helium, methane, and ammonia.
8. The Earth formed from one of the swirling eddies of nebular material 4.6 billion years ago, and by at least 3.8 billion years ago it had differentiated into its present-day structure. It accreted as a solid body and then underwent differentiation during a period of internal heating.
9. The Moon probably formed as a result of a Mars-sized planetesimal crashing into Earth 4.6 to 4.4 billion years ago and causing it to eject a large quantity of hot material. As the material cooled, the various lunar layers crystallized, forming a zoned body.

Important Terms

acondrite
Big Bang
carbonaceous chondrite
chondrule
irons
Jovian planet

meteorite
ordinary chondrite
outgassing
planetesimal
refractory element

solar nebula theory
stones
stony-irons
terrestrial planet
volatile element

Review Questions

1. The most abundant meteorites are:
   a. ___ stones; b. ___ irons; c. ___ stony-irons;
   d. ___ acondrites; e. ___ peridotites.
2. The age of the universe is generally accepted by scientists as:
   a. ___ 570 million years; b. ___ 4.6 billion years;
   c. ___ 8 to 15 billion years; d. ___ 13 to 20 billion years;
   e. ___ greater than 50 billion years.
3. Which of the following is not one of the four basic forces?
   a. ___ gravity; b. ___ electromagnetic; c. ___ strong nuclear; d. ___ weak nuclear; e. ___ photon.
4. The composition of the universe has been changing since the Big Bang. Yet 98% of it by weight still consists of the elements:
   a. ___ hydrogen and carbon; b. ___ helium and carbon;
   c. ___ hydrogen and helium; d. ___ carbon and nitrogen;
   e. ___ hydrogen and nitrogen.
5. Which of the following is not a terrestrial planet?
   a. ___ Mercury; b. ___ Jupiter; c. ___ Earth;
   d. ___ Venus; e. ___ Mars.
6. The age of the solar system is generally accepted by scientists as ___ billion years.
   a. ___ 4.6; b. ___ 10; c. ___ 15.5; d. ___ 20; e. ___ 50.

7. The major problem that plagued most early theories of the origin of the solar system involved the:
a. ___ distribution of elements throughout the solar system; b. ___ rotation of the planets around their axes; c. ___ slow rotation of the Sun; d. ___ revolution of the planets around the Sun; e. ___ source of meteorites and asteroids.

8. The surface of the Moon is divided into light-colored highlands and low-lying, dark-colored plains called:
a. ___ anorthosites; b. ___ regolith; c. ___ cratons; d. ___ nebulas; e. ___ maria.

9. The most widely accepted theory regarding the origin of the Moon involves:
a. ___ capture from an independent orbit; b. ___ an independent origin from the Earth; c. ___ breaking off from the Earth during the Earth's accretion; d. ___ formation resulting from a collision between the Earth and a large planetesimal; e. ___ none of these.

10. Images radioed back by *Voyagers 1* and *2* revealed that:
a. ___ all of the Jovian planets have rings; b. ___ Neptune is a placid planet; c. ___ Uranus has a large spot like those of Jupiter and Neptune; d. ___ Pluto has an atmosphere similar to that of Mars; e. ___ all of these.

11. The planets can be separated into terrestrial and Jovian primarily on the basis of which property?
a. ___ size; b. ___ atmosphere; c. ___ density; d. ___ color; e. ___ none of these.

12. It is currently thought that the Tunguska explosion was caused by a(n):
a. ___ meteor; b. ___ asteroid; c. ___ nuclear explosion; d. ___ volcanic eruption; e. ___ comet.

13. Which of the following events did all of the terrestrial planets experience early in their history?
a. ___ accretion; b. ___ differentiation; c. ___ volcanism; d. ___ meteorite impacting; e. ___ all of these.

14. Which of the following is not characteristic of Mercury?
a. ___ a strong magnetic field; b. ___ heavy cratering of its surface; c. ___ scarps; d. ___ numerous lava flows; e. ___ small amounts of atmospheric hydrogen and helium.

15. The atmosphere of Venus is:
a. ___ thick and composed of carbon dioxide; b. ___ similar to Earth's; c. ___ nonexistent; d. ___ thin, like that of Mars; e. ___ none of these.

16. The surface of Mars possesses:
a. ___ huge valleys; b. ___ massive volcanoes; c. ___ large craters; d. ___ smooth plains; e. ___ all of these.

17. Which planets give off more energy than they receive?
a. ___ Jupiter; b. ___ Saturn; c. ___ Uranus; d. ___ answers (a) and (b); e. ___ answers (a) and (c).

18. Both Jupiter and Saturn have a relatively small rocky core overlain by a zone of:
a. ___ helium; b. ___ liquid metallic hydrogen; c. ___ frozen ammonia; d. ___ hydrogen; e. ___ carbon dioxide.

19. The only planet whose axis of rotation nearly parallels the plane of the ecliptic is:
a. ___ Venus; b. ___ Saturn; c. ___ Uranus; d. ___ Neptune; e. ___ Pluto.

20. What was the main source of heat for the Earth early in its history?
a. ___ meteor impact; b. ___ radioactivity; c. ___ gravitational compression; d. ___ an initial molten condition; e. ___ spontaneous combustion.

21. What two fundamental phenomena indicate that the Big Bang occurred?

22. How does the solar nebula theory account for the general characteristics of the solar system?

23. What are the three major groups of meteorites?

24. How do the terrestrial planets differ from the Jovian planets?

25. What are the similarities and differences in the origin and history of the four terrestrial planets?

26. Discuss why Venus, Earth, and Mars currently have quite different atmospheres.

27. What are the similarities and differences in the origin and history of the four Jovian planets?

28. Discuss the origin and differentiation of the Earth into three concentric layers.

29. Discuss the origin of the Earth-Moon system.

30. Discuss how the *Voyager* space probes have changed our ideas about the planets they have flown by.

Additional Readings

Benzel, R. 1990. Pluto. *Scientific American* 262, no. 6:50–59.

Fernie, J. D. 1993. The Tunguska Event. *American Scientist,* 81, no. 5:412–15.

Freedman, W. L. 1992. The expansion rate and size of the universe. *Scientific American* 267, no. 5:54–61.

Grieve, R. A. F. 1990. Impact cratering on the Earth. *Scientific American* 262, no. 4:66–73.

Horgan, J. 1990. Universal truths. *Scientific American* 263, no. 4:108–17.

Ingersoll, A. P. 1987. Uranus. *Scientific American* 256, no. 1: 38–45.

Kasting, J. F., O. B. Toon, and J. B. Pollack. 1988. How climate evolved on the terrestrial planets. *Scientific American* 258, no. 2:90–97.

Kinoshita, J. 1989. Neptune. *Scientific American* 261, no. 5: 82–91.

Kuhn, K. F. 1991. *In quest of the universe.* St. Paul, Minn.: West Publishing.

McSween, H. Y., Jr. 1989. Chondritic meteorites and the formation of planets. *American Scientist* 77, no. 2:146–53.

Osterbrock, D. E., J. A. Gwinn, and R. S. Brashear. 1993. Edwin Hubble and the expanding universe. *Scientific American* 269, no. 1:84–89.

Saunders, R. S. 1990. The surface of Venus. *Scientific American* 263, no. 6:60–65.

Taylor, S. R. 1987. The origin of the Moon. *American Scientist* 75, no. 5:468–77.

Van den Bergh, S., and J. E. Hesser. 1993. How the Milky Way formed. *Scientific American* 268, no. 1:72–81.

CHAPTER
20

Chapter Outline

The Late Proterozoic Jacobsville Sandstone exposed along the shore of Lake Superior at Marquette, Michigan.

Precambrian Earth and Life History

Prologue

Imagine a barren, lifeless, waterless, hot planet with a poisonous atmosphere. Volcanoes erupt nearly continuously, meteorites and comets flash through the atmosphere, and cosmic radiation is intense. The planet's crust, which is composed entirely of dark-colored igneous rock, is thin and unstable. Storms form in the turbulent atmosphere, and lightning discharges are common, but no rain falls. And because the atmosphere contains no oxygen, nothing burns. Rivers and pools of molten rock emit a continuous reddish glow.

This may sound like a science fiction novel, but it is probably a reasonably accurate description of the Earth shortly after it formed (◆ Fig. 20.1). We emphasize "probably" because no record exists for the earliest chapter of Earth history, the interval from 4.6 to 3.8 billion years ago, although one area of rocks 3.96 billion years old is now known in Canada.

We can only speculate about what the Earth was like during this time, based on our knowledge of how planets form and about other Earth-like planets.

When the Earth formed, it had a tremendous reservoir of primordial heat, heat generated by colliding particles as the Earth accreted, by gravitational compression, and by the decay of short-lived radioactive elements. Many geologists believe that the early Earth was so hot that it was partly or perhaps almost entirely molten. No one knows what the Earth's surface temperature was during its earliest history, but it was almost certainly too hot for liquid water to exist or for any known organism to survive. Volcanism must have been ubiquitous and nearly continuous. Molten rock later solidified to form a thin, discontinuous, dark-colored crust, only to be disrupted by upwelling magmas.

◆ FIGURE 20.1 The Earth as it is thought to have appeared about 4.6 billion years ago.

Assuming that visitors to the early Earth could tolerate the high temperatures, they would also have to contend with other factors. The atmosphere would be unbreathable by any of today's inhabitants. It probably contained considerable carbon dioxide and water vapor, but little or no oxygen. No ozone layer existed in the upper atmosphere, so our hypothetical visitors would receive a lethal dose of ultraviolet radiation, unless protected, and would be threatened constantly by comet and meteorite impacts. The view of the Moon would have been spectacular because it was much closer to the Earth. However, its gravitational attraction would have caused massive Earth tides. And finally, our visitors would experience a much shorter day because the Earth rotated on its axis in as little as 10 hours.

Eventually, much of the Earth's primordial heat was dissipated into space and its surface cooled. As the Earth cooled, water vapor began to condense, rain fell, and surface water began to accumulate. The bombardment by comets and meteorites slowed. By 3.8 billion years ago, a few small areas of continental crust existed. The atmosphere still lacked oxygen and an ozone layer, but by as much as 3.5 billion years ago, life appeared. Some inconclusive evidence indicates that the most primitive life-forms existed even earlier.

⮛ INTRODUCTION

Precambrian is often used informally to refer to both rocks and time. All crustal rocks lying beneath strata of the Cambrian System are called Precambrian. As a geochronologic term, Precambrian includes all geologic time from the Earth's origin 4.6 billion years ago to the beginning of the Phanerozoic Eon 570 million years ago. If all geologic time were represented by a 24-hour day, slightly more than 21 hours of it would be Precambrian. Unfortunately for geologists, not all of this vast interval of time is recorded; only one area of rocks older than 3.8 billion years is known on Earth.

Establishing formal, widely recognized subdivisions of the Precambrian is a difficult task. Precambrian rocks are exposed on all continents, but many have been complexly deformed and altered by metamorphism, and much of Precambrian Earth history is recorded by nonstratified rocks. Thus, the principle of superposition cannot be applied, particularly in older Precambrian rocks, making relative age determinations difficult. In addition, most correlations must be based

on radiometric age dates because these rocks contain few biostratigraphically useful fossils.

In 1982, in an effort to standardize Precambrian terminology, the North American Commission on Stratigraphic Nomenclature approved a proposal that recognizes two Precambrian eons, the **Archean** and **Proterozoic** (⬤ Table 20.1). This usage has gained wide acceptance in North America and is followed in this book.

The Precambrian subdivisions in Table 20.1 are geochronologic, based on radiometric age dates rather than time-stratigraphic ages. This departs from normal

⬤ **TABLE 20.1** Two Classification Schemes for Precambrian Rocks and Time in the United States

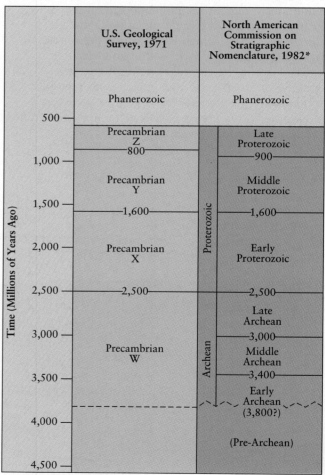

*This scheme, which was proposed by Harrison and Peterman, is followed in this book.

practice in which geologic systems based on stratotypes are the basic time-stratigraphic units. An example will help clarify this point. The Cambrian Period is a geochronologic term, but it corresponds to the Cambrian System, a time-stratigraphic unit based on a body of rock with a stratotype in Wales. By contrast, Precambrian terminology is strictly geochronologic. There are no stratotypes for the subdivisions of the Precambrian.

An alternative scheme for designations of Precambrian time was adopted in 1971 by the U.S. Geological Survey (USGS) (Table 20.1). Instead of names, this scheme uses simple letter designations, which are also based on age dates rather than stratotypes. Although this usage is not followed in this book, students will encounter it on recent USGS maps and in some books and articles.

The basic difference between the Archean and Proterozoic is the style of crustal evolution. Archean crust-forming processes generated greenstone belts and granite-gneiss complexes that were shaped into cratons; and although these rock assemblages continued to form during the Proterozoic, they did so at a considerably

reduced rate. Many Archean rocks have been metamorphosed, although the degree of metamorphism varies considerably, and some are completely unaltered. In contrast, many Proterozoic rocks are unmetamorphosed and little deformed, and in many areas Proterozoic rocks are separated from Archean rocks by a profound unconformity. Finally, the Proterozoic is characterized by widespread assemblages of sedimentary rocks that are rare in the Archean and by a plate tectonic style that is essentially the same as that of the present.

⩔ SHIELDS AND CRATONS

Recall from Chapter 10 that each continent is characterized by a **Precambrian shield** consisting of a vast area of exposed ancient rocks. Continuing outward from the shields are broad platforms of buried Precambrian rocks that underlie much of the continents. The shields and buried platforms are collectively called **cratons** (◆ Fig. 20.2). We can think of the cratons as the ancient nuclei of the continents. The cratons form the

◆ FIGURE 20.2 Precambrian cratons of the world. The areas of exposed Precambrian rocks are the shields, whereas the buried Precambrian rocks are the platforms. Shields and platforms collectively make up the cratons.

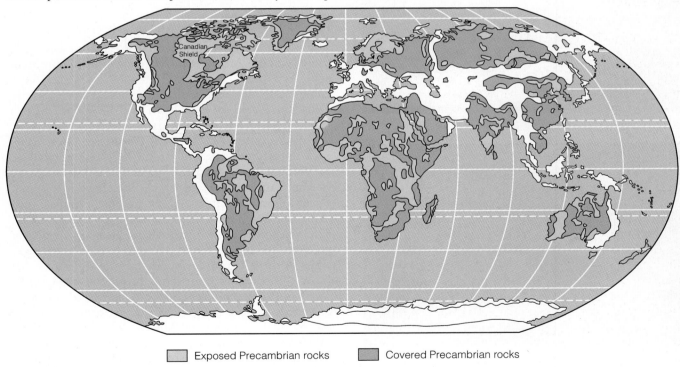

☐ Exposed Precambrian rocks   ☐ Covered Precambrian rocks

Precambrian Earth and Life History   523

foundations upon which Phanerozoic sediments were deposited. In addition the cratons, including their exposed Precambrian shields, have been extremely stable since the beginning of the Phanerozoic Eon. Their stability during that time contrasts sharply with their Precambrian history of orogenic activity.

In the North American **Canadian Shield** (Fig. 20.2) both Archean and Proterozoic rocks are present, including intrusives, lava flows, various sedimentary rocks, and metamorphic equivalents of all of these (◆ Fig. 20.3). Beyond the Canadian Shield, exposures of Precambrian rocks are limited to areas of uplift and erosion, as in the Appalachians, the southwestern United States, the Rocky Mountains, and the Black

◆ FIGURE 20.3 Rocks of the Canadian Shield.
(*a*) Outcrop of gneiss, Georgian Bay, Ontario, Canada.
(*b*) Basalt (dark) and granite (light) along the banks of the Chippewa River, Ontario, Canada.

(a)

(b)

Hills of South Dakota (Fig. 20.2). Geophysical evidence and deep drilling demonstrate that Precambrian rocks underlie most of North America.

The geologic history of the Canadian Shield is complex and not fully understood. Nevertheless, we can recognize several smaller cratons within the shield, each of which is delineated on the basis of radiometric ages and structural trends. These cratons and other buried cratons beyond the shield are the subunits that constitute the North American craton. Each of these subunits may have been independent minicontinents that were later assembled into the larger cratonic unit. The amalgamation of these small cratons occurred along deformation belts during the Early Proterozoic and will be considered more fully later in this chapter.

⯆ ARCHEAN ROCKS

Areas underlain by Archean rocks are characterized by two main types of rock bodies: **greenstone belts** and **granite-gneiss complexes.** By far the most abundant rocks are granites and gneisses. For example, the Rhodesian Province of southern Africa consists of about 83% gneiss and various granitic rocks; the remaining 17% is mostly greenstone belts.

Greenstone Belts

The oldest large, well-preserved greenstone belts are those of South Africa, which date from 3.6 billion years ago. In North America, greenstone belts are most common in the Superior and Slave cratons of the Canadian Shield (◆ Fig. 20.4), and most formed between 2.7 and 2.5 billion years ago.

An idealized greenstone belt consists of three major rock units; the lower and middle units are dominated by volcanic rocks, and the upper unit is sedimentary (◆ Fig. 20.5a). Most greenstone belts have a synclinal structure and are intruded by granitic magmas, and many are complexly folded and cut by thrust faults. The volcanic rocks of greenstone belts are typically greenish due to the abundance of the mineral chlorite, which formed during low-grade metamorphism.

As the common occurrence of pillow basalts indicates, much of the volcanism responsible for the igneous rocks of greenstone belts was subaqueous (◆ Fig. 20.5b). Shallow water and subaerial eruptions are indicated by pyroclastics, and in some areas large volcanic centers built up above sea level. Perhaps the most

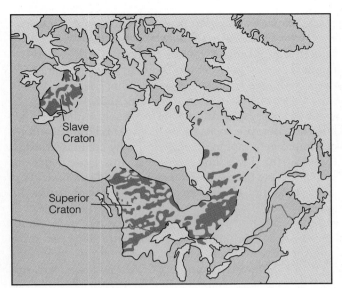

◆ FIGURE 20.4 Greenstone belts (shown in green) of the Canadian Shield are mostly in the Superior and Slave cratons.

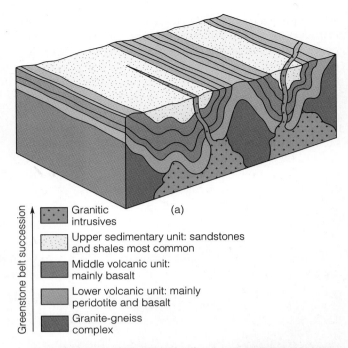

Greenstone belt succession

- Granitic intrusives
- Upper sedimentary unit: sandstones and shales most common
- Middle volcanic unit: mainly basalt
- Lower volcanic unit: mainly peridotite and basalt
- Granite-gneiss complex

(a)

(b)

◆ FIGURE 20.5 (a) Two adjacent greenstone belts showing their synclinal structure. Older greenstone belts—those more than 2.8 billion years old—have an ultramafic lower unit succeeded upward by a basaltic unit as shown here. In younger greenstone belts, the succession is a basaltic lower unit overlain by an andesite-rhyolite unit. The upper unit in both older and younger greenstone belts consists of sedimentary rocks. (b) Pillow structures in the Ispheming greenstone belt, Marquette, Michigan.

interesting igneous rocks in greenstone belts are ultramafic lava flows; such flows are rare in rocks younger than Archean (see Perspective 3.1).

Sedimentary rocks are a minor component in the lower parts of greenstone belts but become increasingly abundant toward the top (Fig. 20.5). The most common sedimentary rocks are successions of *graywacke* and *argillite*. Graywacke is a variety of sandstone containing abundant clay, and those in the greenstone belts are rich in volcanic rock fragments. Argillites are simply slightly metamorphosed mudrocks such as shale. Small-scale graded bedding and cross-bedding indicate that the graywacke-argillite successions were deposited by turbidity currents. Some of the sedimentary rocks, such as quartz sandstones and shales, in the upper units show clear evidence of shallow-water deposition in delta, tidal-flat, barrier-island, and shallow marine-shelf environments.

Most currently popular models for the development of greenstone belts rely on Archean plate movements. According to one model, greenstone belts develop in *back-arc basins* that subsequently close. Thus, there is an early stage of extension when the back-arc basin opens, accompanied by volcanism and sedimentation, followed by an episode of compression. During this compressional stage, the greenstone belt assumes its synclinal form (Fig. 20.5a) and is metamorphosed and intruded by granitic magmas. An alternate model proposes that some greenstone belts formed in *intracontinental rifts*.

⩔ DEVELOPMENT OF ARCHEAN CRATONS

Geologists know that several continental cratons had formed by the beginning of the Archean, because rocks this old are known from several areas, including South Africa, Minnesota, Greenland, and Canada. Moreover, many of these rocks are metamorphic, which means they formed from even older rocks. Recently, 3.96-billion-year-old rocks were discovered in Canada. Sedimentary rocks in Australia contain detrital minerals dated at 4.2 billion years, indicating that source rocks at least that old were present (● Table 20.2).

These earliest cratons were probably rather small, however, because rocks older than 3.0 billion years are of limited geographic extent, especially compared with those 3.0 to 2.5 billion years old (this latter interval seems to have been a time of rapid crustal evolution). A plate tectonic model has been proposed for the

● **TABLE 20.2** Chronologic Summary of Events Important in the Archean Development of Cratons (Ages in Thousands of Millions of Years)

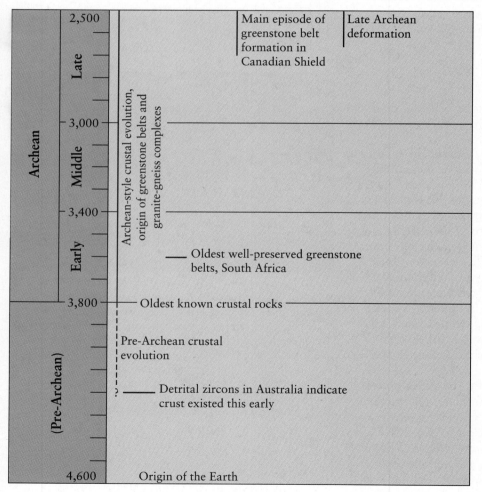

Archean crustal evolution of the southern Superior craton of Canada (◆ Fig. 20.6). This model incorporates sialic plutonism, greenstone belt formation, and collisions of microcontinents. It accounts for the origin of both greenstone belts and granite-gneiss complexes by the sequential accretion of island arcs (Fig. 20.6).

◆ FIGURE 20.6 Origin of the southern Superior craton. (a) Geologic map showing greenstone belts (green areas) and granite-gneiss subprovinces (tan areas). (b) Plate tectonic model for development of the southern Superior craton. The figure represents a north-south section, and the upper diagram is an earlier stage of the lower diagram.

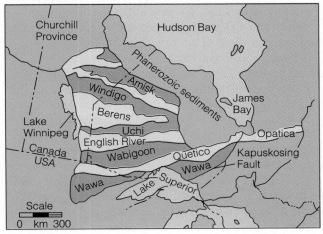

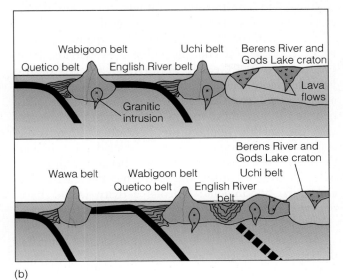

The events leading to the origin of the southern Superior craton are part of a more extensive orogenic episode that occurred near the end of the Archean. This episode of deformation was responsible for the formation of the Superior and Slave cratons as well as some Archean terranes within other parts of the Canadian Shield. It also affected Archean rocks in Wyoming, Montana, and the Minnesota River Valley.

Deformation during the Late Archean was the last major Archean event in North America. Several sizable cratons had formed, constituting the older parts of the Canadian Shield. These cratons, however, were independent units or microcontinents that were later assembled during the Early Proterozoic to form a larger craton.

◥ ARCHEAN PLATE TECTONICS

Undoubtedly, the present tectonic regime of opening and closing oceans has been a primary agent in Earth evolution for at least the last 2 billion years. In fact, this regime probably became established in the Early Proterozoic. Many geologists are becoming convinced that some sort of plate tectonics was operating in the Archean as well, but they disagree about the details.

Some Archean rocks appear to record plate movements as shown by deformation belts between presumed colliding cratons and island arcs. But ophiolite complexes, which mark younger convergent plate margins, are rare, although Late Archean ophiolites have recently been reported from several areas.

Apparently, the Earth's radiogenic heat production has diminished through time. Thus, during the Archean, when more heat was available, sea-floor spreading and plate motions probably occurred faster, and magma was generated more rapidly. Furthermore, Archean plates seem to have behaved differently from those in the Proterozoic. For example, sedimentary sequences typical of passive continental margins are uncommon in the Archean but quite common in the Proterozoic. Their near-absence in the Archean indicates that continents with adjacent shelves and slopes were either not present or only poorly developed.

Another factor that favors some kind of Archean plate tectonics is the episode of rapid crustal growth that occurred 3.0 to 2.5 billion years ago. Like continents today, Archean continents probably grew by accretion at convergent plate margins, although they probably grew more rapidly because plate motions were faster.

Today most geologists would probably agree that plate tectonics was operative in the Archean. Its details differed, however, from the present style of plate tectonics, which began during the Proterozoic when large, stable cratons were present.

▼ PROTEROZOIC CRUSTAL EVOLUTION

We noted earlier that the Archean cratons assembled through a series of island arc and minicontinent collisions. These provided the nuclei around which Proterozoic continental crust accreted, thereby forming much larger cratons. One large landmass, called **Laurentia,** consisted mostly of North America and Greenland, parts of northwestern Scotland, and perhaps parts of the Baltic Shield of Scandinavia. Here we will emphasize the geologic evolution of Laurentia. • Table 20.3 summarizes the crust-forming events discussed in the following sections.

The first major episode in the Proterozoic evolution of Laurentia occurred between 2.0 and 1.8 billion years ago, during the Early Proterozoic (Table 20.3). During this time, several major **orogens** developed; these are zones of deformed rocks, many of which have been metamorphosed and intruded by plutons. The Archean cratons were sutured along these deformed belts (◆ Fig. 20.7a). In other words, this was an episode of continental growth during which collisions between Archean cratons formed a larger craton, and new crust was accreted at the margins of the cratons so formed. By 1.8 billion years ago, much of what is now Greenland, central Canada, and the north-central United States formed a large craton (Fig. 20.7a).

One Early Proterozoic episode of colliding Archean cratons is recorded in rocks of the Thelon orogen in the northwestern Canadian Shield, where the Slave and Rae cratons collided (Fig. 20.7a). Rocks of the Trans-Hudson orogen in northern Saskatchewan record island arc collisions, granitic plutonism, intense deformation, and regional metamorphism. The events leading to the formation of the Trans-Hudson orogen resulted in the suturing of the Superior, Wyoming, and Hearne cratons.

Other notable events in the Early Proterozoic evolution of Laurentia include the development of the Wopmay and Penokean orogens (Fig. 20.7a). Rocks of the Wopmay orogen, adjacent to the Slave craton in northwestern Canada, record the oldest completely preserved Wilson cycle (see Perspective 9.1). Furthermore, some rocks in the Wopmay orogen form a **quartzite-carbonate-shale assemblage,** a suite of rocks that are characteristic of passive margins and first became abundant and widespread during the Proterozoic.

Following the initial episode of amalgamation of Archean cratons, considerable accretion occurred along the southern margin of Laurentia. Between 1.8 and 1.6 billion years ago, growth continued in what is now the southwestern and central United States as successively younger belts were sutured to Laurentia, forming the Central Plains, Yavapai, and Mazatzal orogens (◆ Fig. 20.7b). Continental accretion occurred as northward-migrating island arcs collided with Laurentia, resulting in deformation, metamorphism, and the emplacement of granitic batholiths. The net effect was the accretion of more than 1,000 km of continental crust along the southern margin of Laurentia (Fig. 20.7b).

No major episode in the growth of Laurentia occurred between 1.6 and 1.3 billion years ago. Nevertheless, during this interval extensive igneous activity occurred that was unrelated to orogenic activity (Table 20.3; ◆ Fig. 20.8). These rocks did not increase the area of Laurentia because they were intruded into or erupted upon already existing crust. These igneous rocks are buried beneath Phanerozoic strata in many areas, but they are exposed in eastern Canada, extend across southern Greenland, and also occur in the Baltic Shield of Scandinavia. Recent investigations show that they are mostly granitic plutons, calderas and their fill, and vast sheets of rhyolite and ash flows, although gabbros and anorthosites (plutonic rocks composed mostly of plagioclase) are known as well.

The origin of these Middle Proterozoic igneous rocks continues to be the subject of debate. Paul Hoffmann of the Geological Survey of Canada postulates that large-scale mantle upwelling, or what he calls a superswell, rose beneath a Proterozoic supercontinent to produce these rocks. According to this hypothesis, the mantle temperature beneath a Proterozoic supercontinent would have been considerably higher than beneath later supercontinents because radiogenic heat production within the Earth has decreased. Accordingly, nonorogenic igneous activity would have occurred following the amalgamation of the first supercontinent.

Another major episode in the evolution of Laurentia, the **Grenville orogeny** in the eastern United States and Canada, occurred between 1.3 and 1.0 billion years ago

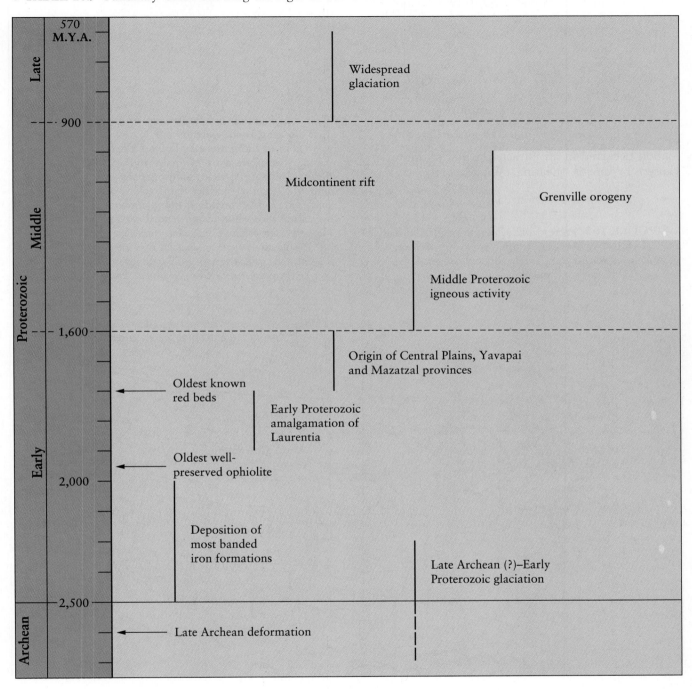

● **TABLE 20.3** Summary Chart Showing the Ages of Some of the Proterozoic Events Discussed in the Text

(◆ Fig. 20.7c; Table 20.3). Rocks of the Grenville orogen are extensively exposed in southeastern Canada where they abut the Archean Superior Province (Fig. 20.7c). This belt of deformed rocks extends northeast into Greenland and continues into Scandinavia. It also continues southwest through the area of the present-day Appalachian Mountains (Fig. 20.7c).

Many geologists think the Grenville orogen can be explained by the opening and then closing of an ocean basin. If so, the Grenville Supergroup may represent sediments deposited on a passive continental margin. However, some geologists think that Grenville deformation occurred in an intracontinental setting or was caused by major shearing. Whatever the cause, it

◆ FIGURE 20.7 Proterozoic evolution of the Laurentian craton. (*a*) During the Early Proterozoic, the Archean cratons were sutured along deformation belts called *orogens*. (*b*) Laurentia grew along its southern margin by accretion of the Central Plains, Yavapai, and Mazatsal orogens. (*c*) A final episode of Proterozoic accretion occurred during the Grenville orogeny.

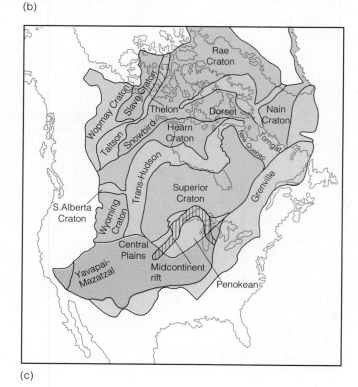

(b)

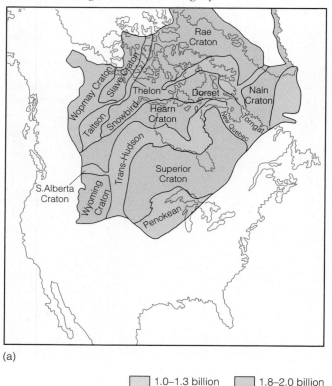

(a)

1.0–1.3 billion	1.8–2.0 billion
1.6–1.8 billion	>2.5 billion

(c)

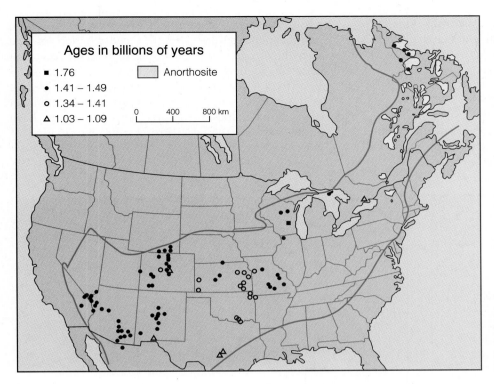

◆ FIGURE 20.8 Distribution of Middle Proterozoic igneous rock complexes unrelated to orogenic activity. Most of this activity occurred between 1.34 and 1.49 billion years ago.

represents the final episode of Proterozoic continental accretion of Laurentia.

Contemporaneous with Grenville deformation was an episode of continental rifting in Laurentia. The **Midcontinent rift** is a major feature extending from the Lake Superior basin southwest into Kansas, while a southeasterly trending branch extends through Michigan and into Ohio (◆ Fig. 20.9). It cuts across Archean and Early Proterozoic rocks and, in the east, terminates against the Grenville orogen.

Most of the rift is concealed by younger strata except in the Lake Superior region where rocks that filled the rift are well exposed. The central part of the rift contains thick accumulations of basaltic lava flows that form extensive lava plateaus. For example, the Portage Lake Volcanics consist of numerous overlapping lava flows forming a volcanic pile several kilometers thick (◆ Fig. 20.10a and d). For the entire rift, the volume of volcanic rocks has been estimated at 300,000 to 1,000,000 km³.

Intertonguing with and overlying the volcanics are thick sequences of sedimentary rocks. Near the rift margins coarse-grained rocks such as the Copper Har-

◆ FIGURE 20.9 Location of the Midcontinent rift. The volcanic and sedimentary rocks that fill the rift are exposed in the Lake Superior region, but elsewhere they are buried beneath younger rocks.

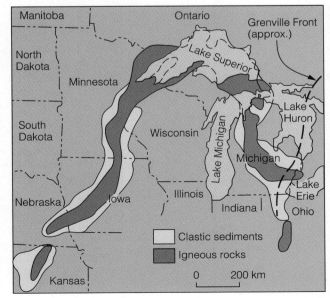

bor Conglomerate formed large alluvial fans that graded into sandstones and shales along the rift axis (◆ Fig. 20.10a and c).

▼ PROTEROZOIC SUPERCONTINENTS

Thus far this chapter has largely reviewed the geologic evolution of Laurentia, a large landmass composed mostly of North America and Greenland (Fig. 20.7). Paleomagnetic studies and the continuation of the Grenville orogen into Scandinavia indicate that the Baltic Shield may also have been part of or situated close to Laurentia during the Proterozoic (◆ Fig. 20.11). Indeed, some geologists suggest that a collision between the Baltic and Laurentian cratons may have caused some of the early Proterozoic deformation of Laurentia.

◆ FIGURE 20.10 Rocks of the Midcontinent rift exposed in the Lake Superior region. (*a*) Section showing vertical relationships. (*b*) Freda Sandstone. (Photo courtesy of Albert B. Dickas, University of Wisconsin, Superior.) (*c*) Copper Harbor Conglomerate. (*d*) Portage Lake Volcanics in foreground. (Photos *c* and *d* courtesy of Sue Monroe.)

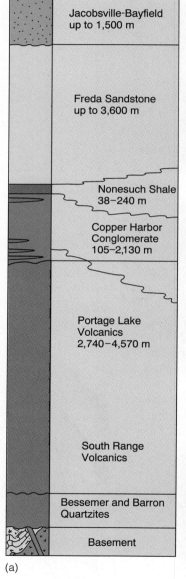

(a)

(b)

(c)

(d)

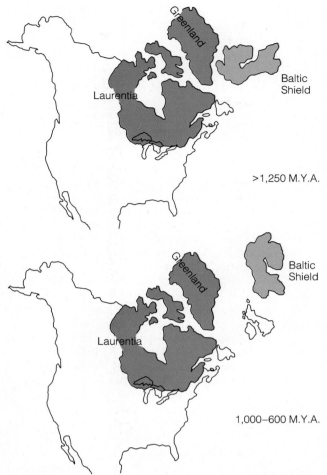

>1,250 M.Y.A.

1,000–600 M.Y.A.

◆ FIGURE 20.11 Reconstructions of the Baltic Shield relative to Laurentia for two times during the Proterozoic.

But what of Laurasia and Gondwana, the two major components of the Late Paleozoic supercontinent Pangaea? The components that ultimately formed Pangaea were evolving during the Precambrian, but few investigators agree on their size, shape, or associations. One group holds that a single Pangaea-like supercontinent persisted from the Late Archean to the end of the Proterozoic. Other investigators, pointing to uncertainties in the data, maintain that this reconstruction is not justified.

During the latest Proterozoic, all the Southern Hemisphere continents were strongly deformed. These events may mark the time during which the supercontinent Gondwana was assembled. In fact, some geolo-

gists think that Laurentia, Australia, and Antarctica were connected during the Late Proterozoic, forming a supercontinent (◆ Fig. 20.12, page 536).

⩔ PROTEROZOIC ROCKS

Some Proterozoic rocks and rock associations deserve special attention. Greenstone belts, for example, are more typical of the Archean, but they continued to form during the Proterozoic, though they differed in detail. Proterozoic greenstone belts occur on several continents, including North America, and at least one has been reported from the Cambrian of Australia. Quartzite-carbonate-shale assemblages also merit our attention because they closely resemble sedimentary rock associations of the Phanerozoic. Deposits attributed to two episodes of Proterozoic glaciation are recognized as well.

The onset of a present style of plate tectonics probably occurred in the Early Proterozoic, and the first well-preserved *ophiolites,* which mark convergent plate margins, are of this age (see Perspective 20.1). Banded iron formations and red beds, too, are important Proterozoic rock types, but these are more conveniently discussed in the section on the atmosphere.

Fully 60% of known Proterozoic rock successions are composed of quartzite-carbonate-shale assemblages. Widespread deposition of this assemblage occurred throughout the Proterozoic along rifted continental margins and in intracratonic basins. The most important aspect of these assemblages is that they represent suites of rocks similar to those of the Phanerozoic. Hence their appearance indicates the widespread distribution of stable cratons with depositional environments much like those of the present.

Early Proterozoic quartzite-carbonate-shale assemblages are widely distributed in the Great Lakes region (◆ Fig. 20.13, page 536). The quartzites were derived from mature quartz sandstones, and many show wave-formed ripple marks and cross-bedding (◆ Fig. 20.14a, page 537). Thick carbonates, mostly dolostone, containing abundant stromatolites, are also present (◆ Fig. 20.14b, page 537).

In the western United States and Canada, quartzite-carbonate-shale assemblages were deposited along the continental margin. These Middle to Late Proterozoic rocks are best preserved in three large basins (◆ Fig. 20.15, page 537). In the Belt Basin of the northwestern United States and adjacent parts of Canada, a thick sequence of sedimentary rocks was deposited between 1.45 billion and 850 million years ago (◆ Fig. 20.16,

THE OLDEST COMPLETE OPHIOLITE

According to plate tectonic theory, oceanic crust is produced along the axes of oceanic ridges, then moves away from these ridges by sea-floor spreading, and is eventually destroyed at subduction zones. This continuous destruction of oceanic crust explains why no oceanic crust older than 180 million years is known except in fragments that were tectonically emplaced on continents. Such fragments consisting of a sequence of upper mantle, oceanic crust, and deep-sea sedimentary rocks are known as *ophiolites*. Ophiolites are one of the primary features used to recognize ancient convergent plate margins.

Probably less than 0.001% of all oceanic crust that ever existed is now preserved as ophiolites. Nevertheless, these small fragments provide geologists with information on the processes by which oceanic crust is generated at oceanic ridges and on the locations of ancient plate collisions. Furthermore, their presence indicates a style of plate tectonics essentially the same as that operating at present. Thus, the oldest ophiolites mark the time when a plate tectonic style comparable to that of the present began.

Ophiolites are well known from Phanerozoic-aged rocks, and a probable 2.65-billion-year-old (Late Archean)

ophiolite has recently been reported from Wyoming. The oldest complete ophiolite so far recognized is the 1.96-billion-year-old (Early Proterozoic) Jormua mafic-ultramafic complex of Finland (◆ Fig. 1). Even though the Jormua complex is highly deformed, it is similar to younger, well-documented ophiolites. Clearly, the Jormua complex represents a sequence of rocks that can be interpreted as upper mantle, lower oceanic crust, submarine eruptions represented by pillow basalts (Fig. 1b), as well as metamorphosed deep-sea clay, oozes, and turbidites.

Asko Kontinen of the Geological Survey of Finland thinks that the Jormua complex formed along a divergent plate margin during an extensional stage in the breakup of an Archean craton. A major ocean basin formed in which deep-sea sediments and turbidites were deposited. A compressional stage occurred about 1.96 billion years ago during which most of the oceanic crust was destroyed by subduction. The Jormua complex, however, was tectonically emplaced on the craton and represents a small fragment of this ancient oceanic crust.

page 538). Like the deposits of the Great Lakes region, these rocks are predominantly quartzite, shale, and thick stromatolite-bearing carbonates.

Early Proterozoic glacial deposits are known from Canada, the United States, Australia, and South Africa.

An example is the Huronian Supergroup of Ontario where one glacial deposit may date from 2.7 billion years ago, making it Late Archean in age. Similar deposits thought to be the same age also occur in Michigan, the Medicine Bow Mountains of Wyoming,

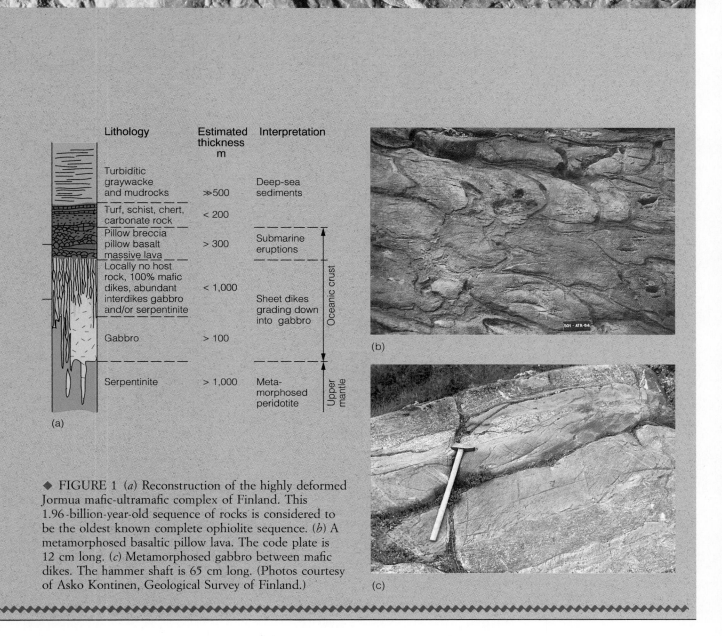

Lithology	Estimated thickness m	Interpretation	
Turbiditic graywacke and mudrocks	≫500	Deep-sea sediments	
Turf, schist, chert, carbonate rock	< 200		
Pillow breccia pillow basalt massive lava	> 300	Submarine eruptions	Oceanic crust
Locally no host rock, 100% mafic dikes, abundant interdikes gabbro and/or serpentinite	< 1,000	Sheet dikes grading down into gabbro	
Gabbro	> 100		
Serpentinite	> 1,000	Metamorphosed peridotite	Upper mantle

(a)

(b)

(c)

◆ FIGURE 1 (*a*) Reconstruction of the highly deformed Jormua mafic-ultramafic complex of Finland. This 1.96-billion-year-old sequence of rocks is considered to be the oldest known complete ophiolite sequence. (*b*) A metamorphosed basaltic pillow lava. The code plate is 12 cm long. (*c*) Metamorphosed gabbro between mafic dikes. The hammer shaft is 65 cm long. (Photos courtesy of Asko Kontinen, Geological Survey of Finland.)

and Quebec, Canada. This distribution indicates that North America may have had an extensive Early Proterozoic ice sheet centered southwest of Hudson Bay (◆ Fig. 20.17, page 538). Deposits of about the same age also occur on other continents; the dating of these deposits is not precise enough to determine whether there was a single, widespread glacial episode or a number of glacial events at different times in different places.

Widespread Late Proterozoic glaciation occurred between 900 and 600 million years ago (Table 20.3).

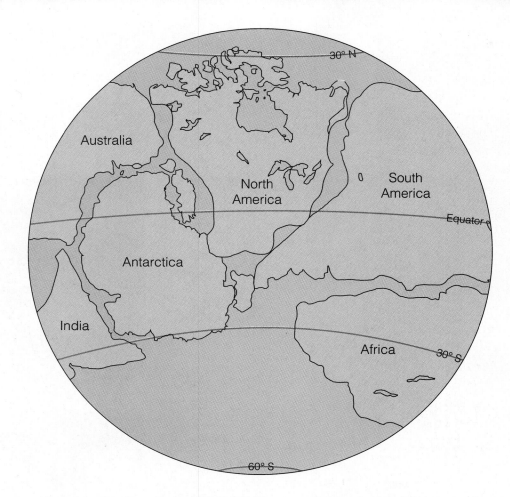

◆ FIGURE 20.12 Possible arrangement of the continents during the Late Proterozoic.

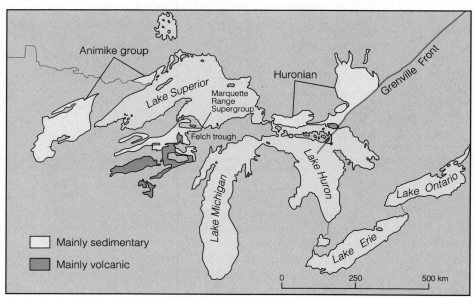

◆ FIGURE 20.13 The geographic distribution of Early Proterozoic rocks in the Great Lakes region.

(a)

(b)

◆ FIGURE 20.14 Early Proterozoic sedimentary rocks of the Great Lakes region. (*a*) Wave-formed ripple marks in the Mesnard Quartzite. The crests of the ripples point toward the observer. (*b*) Outcrop of the Kona Dolomite showing stromatolites.

Glacial deposits have been recognized on all continents except Antarctica. Glaciation was not continuous during this entire interval but was episodic with four major glacial periods recognized. ◆ Figure 20.18 shows the approximate distribution of these Late Proterozoic glaciers, but we should emphasize that they are approximate, because the geographic extent of glacial ice is unknown. In addition, the glaciers shown in Figure 20.18 were not all present at the same time. Nevertheless, the most extensive glaciation in Earth history was during· the latest Proterozoic when ice sheets seem to have been present even in near-equatorial areas.

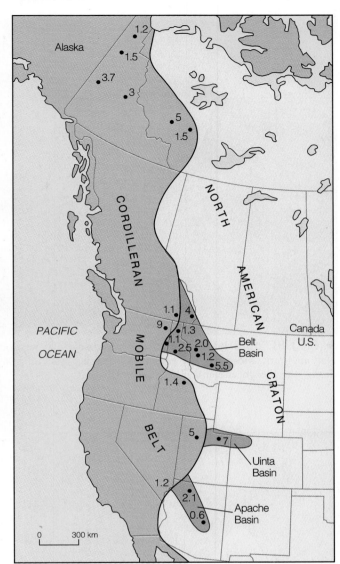

◆ FIGURE 20.15 Late Proterozoic basins of sedimentation in the western United States and Canada. The numbers indicate the approximate thickness of Late Proterozoic rocks in kilometers.

▼ THE ARCHEAN ATMOSPHERE AND OCEANS

If we could somehow go back and visit the early Earth, we would witness a barren, waterless surface and numerous meteorite impacts, be subjected to intense ultraviolet radiation, and be unable to breathe the atmosphere. Today our atmosphere is rich in nitrogen

◆ FIGURE 20.16 Rocks of the Late Proterozoic Belt Supergroup in Glacier National Park, Montana. The present topography resulted from erosion by glaciers during the Pleistocene and Holocene epochs.

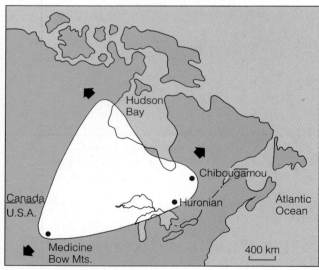

◆ FIGURE 20.17 Glacial deposits of about the same age in Ontario and Quebec, Canada, and in Michigan and Wyoming suggest that an Early Proterozoic ice sheet was centered southwest of Hudson Bay.

(78%) and oxygen (21%) and contains important trace amounts of carbon dioxide, water vapor, and other gases. In the upper atmosphere, ozone (O_3) blocks most of the Sun's ultraviolet radiation.

Before the Earth had a differentiated core, it lacked a magnetic field and *magnetosphere,* the area around the Earth within which the magnetic field is confined. The absence of a magnetosphere ensured that a strong solar wind, an outflow of ions from the Sun, would sweep away any gases that might otherwise have formed an atmosphere. Once the magnetosphere was established, internally derived volcanic gases began to accumulate. The process by which atmospheric gases are derived from within the Earth is known as **outgassing** (◆ Fig. 20.19). The gases formed an early atmosphere but one notably deficient in free oxygen; and without free oxygen, there could have been no ozone layer. Furthermore, the early atmosphere may have contained ammonia (NH_3) and methane (CH_4), both of which could have resulted from volcanic gases reacting chemically in the atmosphere.

An oxygen-deficient atmosphere appears to have persisted throughout the Archean as indicated by de-

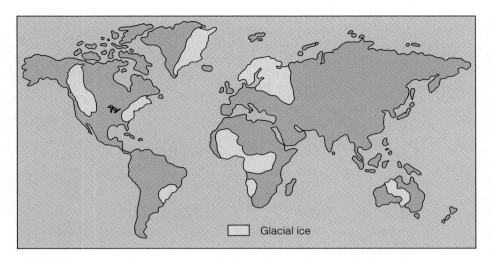

◆ FIGURE 20.18 Major Late Proterozoic glacial centers shown on a map with continents in their present positions. The extent of ice centers is hypothetical and approximate.

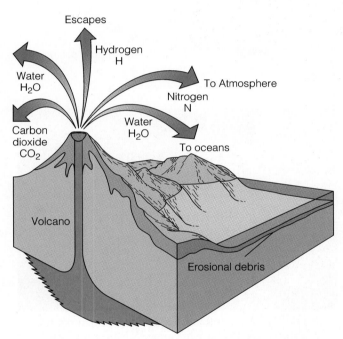

◆ FIGURE 20.19 Outgassing supplied gases to form an early atmosphere composed of the gases shown. Chemical reactions in the atmosphere also probably yielded methane (CH_4) and ammonia (NH_3).

Two processes, both of which began in the Archean, can account for the introduction of free oxygen into the atmosphere. The first was **photochemical dissociation** of water vapor, in which water molecules were broken up by ultraviolet radiation in the upper atmosphere (◆ Fig. 20.20). This process eventually may have supplied up to 2% of present-day oxygen levels. At 2% free oxygen, ozone (O_3) will form, creating a barrier against incoming ultraviolet radiation and thus limiting the formation of more free oxygen by photochemical dissociation. Even more important were the activities of photosynthesizing organisms. During **photosynthesis** carbon dioxide and water combine into organic molecules, and oxygen is released as a waste product (Fig. 20.20). Even so, the atmosphere at the end of the Archean may have contained no more than 1% of its present oxygen level.

The major gas emitted by volcanoes is water vapor, so the early atmosphere was also rich in this compound. Once the Earth cooled sufficiently, water vapor condensed and surface waters began to accumulate. Oceans existed during the Early Archean, although their volumes and extent are unknown. We can envision an early hot Earth with considerable volcanic activity and a rapid accumulation of surface waters.

▼ THE PROTEROZOIC ATMOSPHERE

We noted that many geologists are convinced that the Archean atmosphere contained little or no free oxygen. In other words, the atmosphere was not strongly oxidizing as it is now. The late Preston Cloud, a specialist

trital deposits containing pyrite (FeS_2) and uraninite (UO_2). Both of these minerals are quickly oxidized in the presence of free oxygen. Iron in the oxidized state became quite common in the Proterozoic, indicating that at least some free oxygen was present by that time.

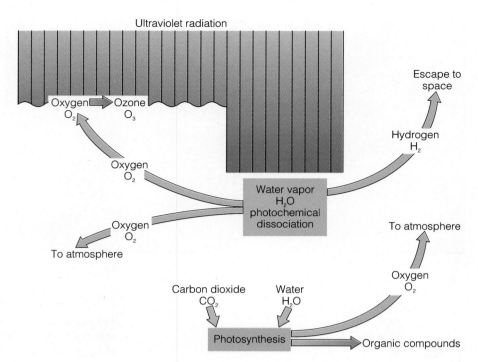

◆ FIGURE 20.20 The processes of photochemical dissociation and photosynthesis added free oxygen to the atmosphere. Once free oxygen was present, an ozone layer formed in the upper atmosphere and blocked most incoming ultraviolet radiation.

on the early history of life, estimated that the free oxygen content of the atmosphere increased from about 1% to 10% through the Proterozoic. It was not until about 400 million years ago that oxygen reached its present level. Most of the atmospheric oxygen was released as a waste product by photosynthesizing cyanobacteria (Fig. 20.20). As indicated by fossil stromatolites, cyanobacteria became common about 2.3 billion years ago.

Banded Iron Formations

Banded iron formations (BIFs) are sedimentary rocks consisting of alternating thin layers of silica (chert) and iron minerals (◆ Fig. 20.21). The iron in these formations is mostly iron oxide (hematite and magnetite), but iron silicate, iron carbonate, and iron sulfide also occur.

Archean BIFs are small lenticular (lens-shaped) bodies measuring a few kilometers across and a few meters thick; most appear to have been deposited in greenstone belts. Proterozoic BIFs are much more common; they are typically hundreds of meters thick and can be traced for hundreds of kilometers. Most were deposited in shallow marine waters.

BIFs are found throughout the geologic column, but the period from 2.5 to 2.0 billion years represents a

◆ FIGURE 20.21 Precambrian banded iron formations. Early Proterozoic BIF at Ishpeming, Michigan. At this location the rocks are brilliantly colored alternating layers of red chert and silver iron minerals.

unique time in Earth history, a time during which 92% of the Earth's BIFs formed. Also remarkable are the consistent layering and extent of BIFs and their chemical purity; BIFs are composed mostly of iron oxides and silicon dioxide as chert.

Iron is a highly reactive element. In the presence of oxygen, it combines to form rustlike oxides that are not readily soluble in water. In the absence of free oxygen, however, iron is easily taken into solution and can accumulate in large quantities in the world's oceans. And because the Archean atmosphere was deficient in free oxygen, it seems likely that little oxygen was dissolved in seawater. The increase in abundance of stromatolites about 2.3 billion years ago resulted in an increase in free oxygen in the oceans, because oxygen is a metabolic waste product of photosynthesizing cyanobacteria that form stromatolites. Apparently, this introduction of free oxygen into the world's oceans helped cause the precipitation of dissolved iron and silica and thus the formation of the BIFs.

Red Beds

Continental **red beds**—red sandstones and shales—first appeared about 1.8 billion years ago, following the deposition of the Proterozoic banded iron formations. The color of red beds is caused by the presence of ferric oxide, usually as the mineral hematite (Fe_2O_3), which forms under oxidizing conditions. These deposits become increasingly abundant through the Proterozoic and are quite common in the Phanerozoic.

Red beds from the Waterberg Group of South Africa are considered to be the oldest deposit of this type. A detailed study of these rocks shows that the red color is restricted to those deposits that accumulated in continental environments and particularly in stream systems, while those facies attributed to nearshore marine sedimentation are unpigmented. The presence of such deposits indicates that the atmosphere was oxidizing, although it may have contained as little as 1 to 2% free oxygen during the Early Proterozoic.

☞ THE ORIGIN OF LIFE

The fossil record reveals that life existed on Earth as much as 3.5 billion years ago. Compared to the present, however, the Archean seems to have been biologically impoverished. Today the Earth's biosphere consists of millions of species of animals, plants, and other organisms, all of which are thought to have evolved from one or a few primordial types. In Chapter 18 we considered the evolutionary processes whereby the diversification of life occurred, but here we are concerned with how life originated in the first place.

First, we must be very clear about what life is; that is, what is living and what is nonliving? Minimally, a living organism must reproduce and practice some kind of

metabolism. Reproduction ensures the long-term survival of a group of organisms as a species; metabolism ensures short-term survival of an individual organism as a chemical system.

All investigators agree that two requirements were necessary for the origin of life: (1) a source of the appropriate elements from which organic molecules could have been synthesized and (2) an energy source to promote chemical reactions that synthesized organic molecules. All organisms are composed mostly of carbon, hydrogen, nitrogen, and oxygen, all of which were present in the early atmosphere in the form of carbon dioxide (CO_2), water vapor (H_2O), and nitrogen (N_2) and possibly methane (CH_4) and ammonia (NH_3). It is postulated that the elements necessary for life (C, H, N, and O) combined to form simple organic molecules called **monomers.** The energy sources that promoted these reactions were probably ultraviolet radiation and lightning. Typical monomers characteristic of organisms are amino acids.

In the 1950s, Stanley Miller synthesized several amino acids by circulating gases approximating the composition of the early atmosphere in a closed glass vessel (◆ Fig. 20.22). This mixture of gases was subjected to an electric spark (simulating lightning), and in a few days the mixture became cloudy. Analysis showed that several amino acids typical of organisms had formed. In more recent experiments, all 20 amino acids common in organisms have been successfully synthesized.

Making monomers in a test tube is one thing, but the molecules of organisms are **polymers** such as proteins and nucleic acids, which consist of linked monomers. Therefore the next question is, how did the process of polymerization occur? This is more difficult to answer, especially if polymerization occurred in an aqueous solution, which usually results in depolymerization. However, Sidney Fox of the University of Southern Illinois has synthesized small molecules he calls *proteinoids,* some of which consist of more than 200 amino acid units (◆ Fig. 20.23). He dehydrated concentrated amino acids and found that when heated they spontaneously polymerized to form proteinoids.

At this stage we can refer to these molecules as *protobionts,* which are intermediate between inorganic chemical compounds and living organisms. These protobionts, however, would have been diluted and would have ceased to exist if some kind of outer membrane had not developed. In other words, they had to be self-contained as present-day cells are. Fox's experiments demonstrated that proteinoids will spontaneously aggregate into microspheres (Fig. 20.23), which

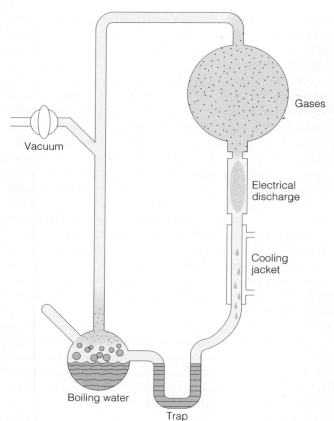

◆ FIGURE 20.22 Experimental apparatus used by Stanley Miller. Several amino acids characteristic of organisms were artificially synthesized during Miller's experiments.

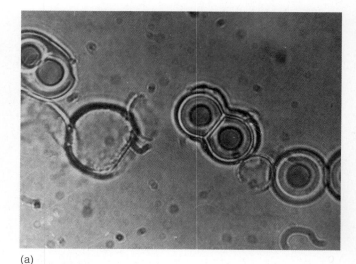

(a)

(b)

◆ FIGURE 20.23 (*a*) Bacterium-like proteinoid. (*b*) Proteinoid microspheres. (Photos courtesy of Sidney W. Fox, Coastal Research and Development Institute, University of South Alabama.)

are bounded by a cell-like membrane and grow and divide much as bacteria do.

Fox's experimental results are interesting, but how can they be related to what may have taken place in the early history of life? Monomers likely formed continuously and in great abundance, accumulated in the oceans, and formed what the British biochemist J. B. S. Haldane characterized as a hot, dilute soup. According to Fox, the amino acids in this hot dilute soup may have washed up onto a beach or perhaps cinder cones, where they were concentrated by evaporation and polymerized by heat. The polymers were then washed back into the sea, where they reacted further.

Not much is known about the next step in the origin of life—the development of a reproductive mechanism. Fox's microspheres divide and are considered by some experts to represent a protoliving system. However, in present-day organisms nucleic acids, either as RNA or DNA, are necessary for reproduction. The problem is that nucleic acids cannot replicate without enzymes, and enzymes cannot be made without nucleic acids. Or so it seemed until recently.

Recent experimental evidence has demonstrated that small RNA molecules can replicate themselves without the aid of protein enzymes. In view of this evidence, it seems that the first replicating system may have been an

RNA molecule. In fact, some researchers propose an early "RNA world" in which these molecules were intermediate between inorganic chemical compounds and the DNA-based molecules of organisms. Just how RNA molecules were naturally synthesized, however, remains a mystery, because they cannot easily be synthesized under the conditions that probably prevailed on the early Earth.

▼ THE EARLIEST ORGANISMS

Prior to the mid-1950s, we had very little knowledge of Precambrian life. Investigators had long assumed that the fossils so abundant in Cambrian strata must have had a long earlier history, but no such record was known. In the early 1900s, Charles Walcott described layered moundlike structures from the Early Proterozoic Gunflint Iron Formation of Ontario. These structures are now called **stromatolites.** Walcott proposed

that they represented reefs constructed by algae, but paleontologists did not demonstrate that stromatolites are the products of organic activity until 1954. Studies of present-day stromatolites show that these structures originate by the entrapment of sediment grains on sticky mats of photosynthesizing cyanobacteria or blue-green algae (◆ Fig. 20.24a through c). Morphologically distinct types of stromatolites may form (◆ Fig. 20.24d), all of which are known from Precambrian rocks.

Currently, the oldest known stromatolites are from the 3.3- to 3.5-billion-year-old Warrawoona Group near North Pole, Australia (◆ Fig. 20.25). Indirect evidence for even more ancient life comes from Archean rocks of western Greenland. These 3.8-billion-year-old rocks contain small carbon spheres that may be of biological origin, but the evidence is not conclusive.

The oldest known fossils are of photosynthesizing organisms, but photosynthesis is a complex metabolic process, and it seems reasonable that it was preceded

◆ FIGURE 20.24 (*a*) through (*c*). The entrapment of sediment by a mat of cyanobacteria; brown masses are sediment grains. (*a*) Uncovered mat at beginning of daylight period. (*b*) Sediment trapping during daylight. (*c*) Regrowth and sediment binding during darkness. This process is repeated many times and yields the layered structure of stromatolites. (*d*) Morphological types of stromatolites include irregular mats, columns, and columns linked by mats. (*e*) Recent stromatolites, Shark Bay, Australia. (Photo courtesy of Phillip E. Playford, Geological Survey of Western Australia.)

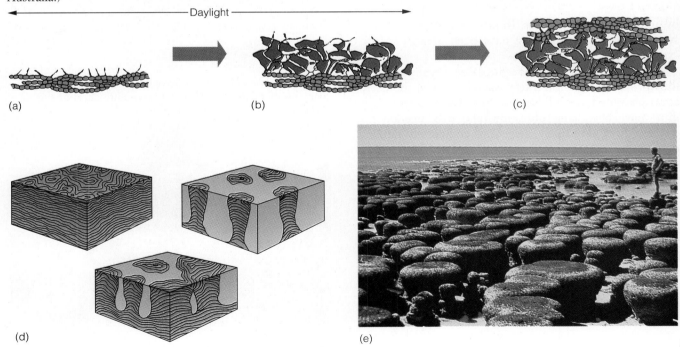

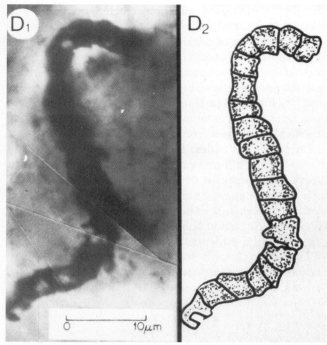

D₁   D₂

0   10μm

◆ FIGURE 20.25 Photomicrograph and schematic restoration of fossil prokaryote from the 3.3 to 3.5 billion-year-old Warrawoona Group, Western Australia. (Photo courtesy of J. William Schopf, University of California, Los Angeles.)

by an even simpler process. In other words, nonphotosynthesizing organisms must have been present before cyanobacteria appeared.

We have no fossils of these earliest organisms, but they probably resembled tiny bacteria. Since the early atmosphere contained little or no free oxygen, they must have been **anaerobic,** meaning they needed no oxygen. And very likely they were completely dependent on an external source of nutrients. We refer to such organisms as **heterotrophic** as opposed to **autotrophic** organisms, which make their own nutrients, as in photosynthesis. All these early life-forms were unicellular and lacked a cell nucleus. Cells of this type are called **prokaryotic cells.**

We can characterize these earliest organisms as anaerobic, heterotrophic prokaryotes. Their nutrient source was probably adenosine triphosphate (ATP), which was used to drive the energy-requiring reactions in cells. ATP can be synthesized from simple gases and phosphate, so it was probably available in the early

Earth environment. The earliest life-forms may have simply acquired their ATP from their surroundings. This situation could not have persisted for long, though, because as more and more cells competed for the same resources, the supply must have diminished. The first organisms to develop a more sophisticated metabolism probably used *fermentation* to meet their energy needs. Fermentation is an anaerobic process in which molecules such as sugars are split, releasing carbon dioxide, alcohol, and energy. In fact, most living prokaryotes ferment.

Other than the origin of life itself, the most significant biological event of the Archean was the development of the autotrophic process of photosynthesis as much as 3.5 billion years ago. These cells were still anaerobic prokaryotes (Fig. 20.25), but as autotrophs they were no longer completely dependent on an external source of preformed organic molecules as a source of nutrients.

▼ PROTEROZOIC LIFE

We noted that Archean fossils are not very common, and all are varieties of bacteria—unicellular prokaryotes. The Early Proterozoic is likewise characterized by these organisms. Even the well-known, 1.8- to 2.1-billion-year-old Gunflint Iron Formation of Ontario, Canada, with 12 species of microorganisms, contains only bacteria and cyanobacteria (◆ Fig. 20.26).

Before the appearance of cells capable of sexual reproduction, evolution was a comparatively slow pro-

◆ FIGURE 20.26 Photomicrographs of spheroidal and filamentous microfossils from stromatolitic chert of the Gunflint Iron Formation, Ontario.

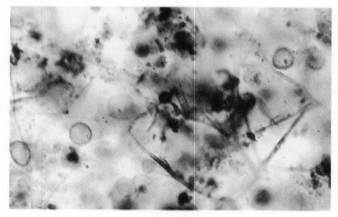

cess, accounting for the low organic diversity during the Archean and Early Proterozoic. This situation did not persist, however. By the Middle Proterozoic, cells appeared that reproduced sexually, and the tempo of evolution picked up markedly.

A New Type of Cell Appears

The appearance of **eukaryotic cells** marks one of the most important events in the history of life. Eukaryotic cells are considerably more organizationally complex than prokaryotic cells and have a membrane-bounded nucleus that contains the genetic material (◆ Fig. 20.27). Organisms composed of eukaryotic cells are

◆ FIGURE 20.27 Prokaryotic and eukaryotic cells. Note that eukaryotes have a cell nucleus containing the genetic material and several organelles such as mitochondria and plastids.

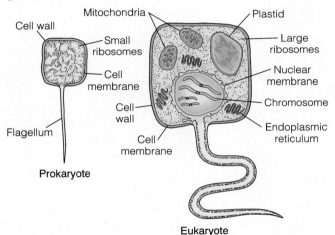

Prokaryote

Eukaryote

eukaryotes whereas those with prokaryotic cells are *prokaryotes.* Most eukaryotes are multicellular, and in marked contrast to prokaryotes, most reproduce sexually and most are aerobic; thus, eukaryotes could not have appeared until some free oxygen was present in the atmosphere.

A number of Proterozoic fossil localities have yielded unicellular eukaryotes, but the oldest one appears to be in the 1.2- to 1.4-billion-year-old Beck Springs Dolomite of southeastern California (◆ Fig. 20.28, ● Table 20.4, page 548) and the 1.4-billion-year-old Greyson shale of Montana and rocks of the same age in China. Fossils that appear to be unicellular algae from the 1.0-billion-year-old Bitter Springs Formation of Australia show evidence of both mitosis and meiosis, processes used only by eukaryotic cells.

The fossil evidence for the appearance of eukaryotic cells comes from size and relative complexity. Eukaryotic cells are larger, commonly much larger, than prokaryotic cells (Fig. 20.27). Cells larger than 60 microns appear in abundance about 1.4 billion years ago. As for relative complexity, prokaryotic cells are typically simple spherical or platelike structures. Proterozoic fossils of branched filaments, flask-shaped organisms, and some containing what appear to be internal, membrane-bounded structures all indicate a eukaryotic level of organization. Theories of how eukaryotic cells originated from prokaryotic cells are considered in Perspective 20.2.

Additional evidence for eukaryotes in the Proterozoic comes from a fossil group called *acritarchs.* These hollow fossils are probably the cysts of planktonic algae (◆ Fig. 20.29a and b, page 549), which became quite abundant during the Late Proterozoic but first appeared about 1.4 billion years ago. Numerous microfossils with vase-shaped skeletons have been recovered

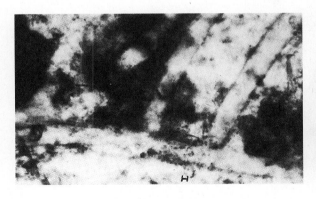

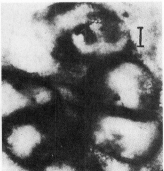

◆ FIGURE 20.28 Microfossils from the Beck Springs Dolomite, California. These fossils may be the oldest known eukaryotic cells. (Photos courtesy of Preston Cloud, University of California, Santa Barbara.)

THE ORIGIN OF EUKARYOTIC CELLS

Stratigraphic position and radiometric ages tell us that eukaryotic cells were present by the Middle Proterozoic. But although we know when eukaryotes appeared, the fossil record reveals nothing about how they evolved from prokaryotic ancestors. The study of living microorganisms gives us some idea of how this may have occurred.

A currently popular theory among evolutionary biologists holds that eukaryotic cells formed from several prokaryotic cells that had established a symbiotic relationship. Symbiosis, the living together of two or more dissimilar organisms, is quite common among organisms today. It may take the form of parasitism in which one organism lives at the expense of another, or the two may coexist with mutual benefit. For example, lichens, once considered to be plants, are actually symbiotic fungi and algae.

In a symbiotic relationship, each symbiont must be capable of metabolism and reproduction, but the degree of dependence in some symbiotic relationships is such that one symbiont cannot live independently. Many parasites, for example, cannot exist outside a host organism. This may have been the case with Proterozoic prokaryotes: two or more prokaryotes may have entered into a symbiotic relationship (◆Fig. 1), and the symbionts became increasingly interdependent until the unit could exist only as a whole.

Supporting evidence for the symbiosis theory comes from the study of living eukaryotic cells. For example, eukaryotic cells contain internal structures called organelles that have their own complements of genetic material. Although these organelles, such as plastids and mitochondia, cannot exist independently today, they apparently were once capable of reproduction as free-living organisms.

Another way of evaluating the symbiosis theory is to look at protein synthesis. Prokaryotic cells synthesize proteins, but can be thought of as a single system, whereas eukaryotes are a combination of protein-synthesizing systems; that is, some of the organelles within eukaryotes such as mitochondria and plastids are capable of protein synthesis. These organelles, with their own genetic material and protein-synthesizing capabilities, are thought to have been free-living bacteria that entered into a symbiotic relationship. With time, the interdependence of the various units grew until life was possible only as an integrated whole (Fig. 1).

More recently, another theory for the origin of eukaryotic cells has been proposed. According to this theory, the sequence of events leading to the origin of eukaryotes began when early prokaryotes lost the ability to synthesize an essential component of cell walls. One solution to this problem was to develop an internal skeleton consisting of microtubules and microfilaments

◆ FIGURE 1 Symbiosis theory for the origin of eukaryotic cells. An aerobic bacterium and a larger host of the kingdom Monera united to form a mitochondria-containing amoeboid. An amoeboflagellate was formed by a union of the amoeboid and a bacterium of the spirochete group; this amoeboflagellate was the direct ancestor of two kingdoms—Fungi and Animalia. Another kingdom, Plantae, was founded when this amoeboflagellate formed a union with blue-green algae (cyanobacteria) that became plastids.

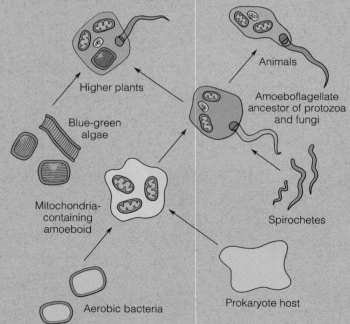

(◆Fig. 2a). This cytoskeleton, as it is called, allowed for an outer fluid cell membrane that infolded so that material coming into the cell could be enveloped. Such infolding is thought to have resulted in the origin of such intracellular structures as the cell nucleus (◆Fig. 2b). According to this theory, the first eukaryotic cell was anaerobic; later it acquired a free-living aerobic organism by symbiosis, thus accounting for the origin of aerobic eukaryotes.

Although the fossil record does not record the acquisition of organelles or symbiosis, living eukaryotes can give some idea of what the first eukaryotes may have been like. The present-day giant amoeba *Pelomyxa,* which lives in the mud of ponds, lacks mitochondria. Two types andhundreds of individual bacteria, however, have a symbiotic relationship with *Pelomyxa* and perform the same function as mitochondria.

Pelomyxa provides evidence for the symbiotic theory for the origin of eukaryotes. However, recent studies of *Giardia,* a single-celled eukaryote, seem to indicate that eukaryotes may have acquired their internal membrane-bounded organelles by infolding of the cell wall. Although *Giardia* is a eukaryote, it shares many characteristics with prokaryotes and is capable of acquiring materials from outside by infolding of the cell wall.

◆ FIGURE 2 (*a*) Diagrammatic view of a cell showing the cytoskeleton consisting of microtubules and microfilaments. (*b*) According to one theory for the origin of eukaryotic cells, cells acquired such structures as a nucleus, mitochondria, and plastids by infolding andenveloping materials.

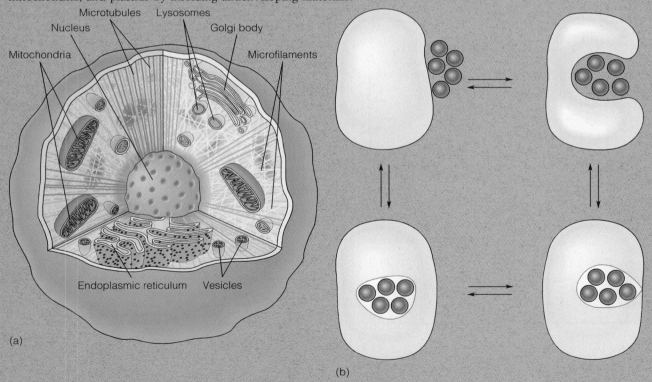

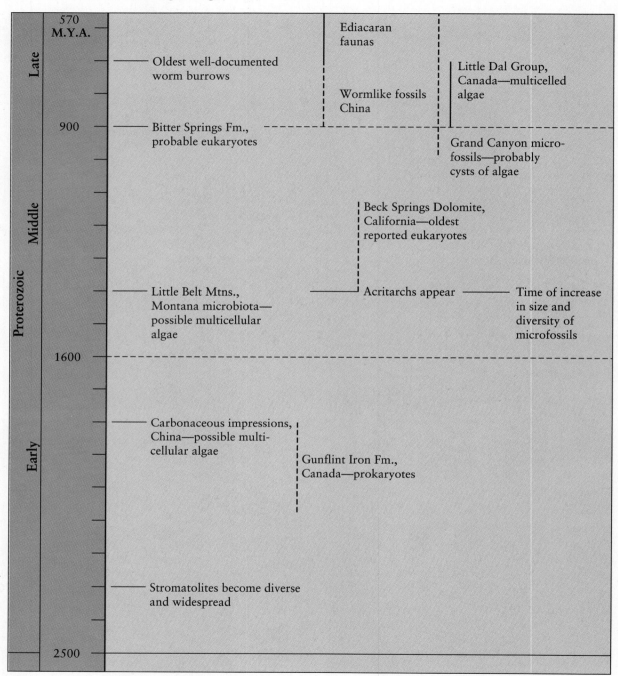

from Late Proterozoic rocks of the Grand Canyon (◆ Fig. 20.29c). These have been tentatively identified as cysts of some kind of algae.

Multicellular Organisms

Multicellular organisms are not only composed of many cells, commonly billions, but also have cells specialized to perform specific functions such as reproduction and respiration. We know from fossils that multicellular organisms appeared in the Late Proterozoic, but we have no fossil evidence of the transition from their unicellular ancestors.

The study of present-day organisms gives some clues as to how this transition may have occurred. Perhaps some unicellular organism divided and formed a group of cells that did not disperse but remained together as a colony. The cells in some colonies may have become somewhat specialized, similar to the situation in the living *colonial organisms* (◆ Fig. 20.30). Further specialization might have led to simple multicellular organisms such as sponges that consist of cells specialized for reproduction, respiration, and food gathering. In other words, specialization of cells might have led to the development of organs with specific functions.

Multicellular Algae

Well-preserved carbonaceous impressions of multicellular algae are known from rocks 800 million years old. And probable algae are preserved in 1,000- to 700-million-year-old rocks in Spitzbergen, China, India, and Canada (Table 20.4). The size, composition, and general shape of these impressions indicate photosynthesizing eukaryotes, probably planktonic algae. In fact, some organic sheets from northwestern Canada closely resemble the living green alga known as sea lettuce.

Even older rocks contain carbonaceous impressions that may be multicellular algae, but this is uncertain. For example, the 1.4-billion-year-old microbiota from the Little Belt Mountains of Montana contains filaments and spherical forms of uncertain affinities (◆ Fig. 20.31; Table 20.4). Carbonaceous macroscopic filaments from 1.8-billion-year-old rocks of China are also suggestive of multicellular algae.

The Ediacaran Fauna

In 1947, the Australian geologist R. C. Sprigg discovered impressions of soft-bodied animals in rocks of the Ediacara Hills of South Australia. Additional discoveries by geologists and amateur paleontologists turned up

(a)

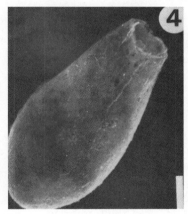

(b)

(c)

◆ FIGURE 20.29 These common Late Proterozoic microfossils are thought to represent eukaryotic organisms. (*a*) and (*b*) Acritarchs are probably the cysts of algae. (Photos courtesy of Andrew H. Knoll, Harvard University.) (*c*) Vase-shaped microfossil, probably a cyst of some kind of algae. (Photo courtesy of Bonnie Bloeser.)

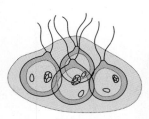
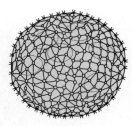

Gonium

Volvox

◆ FIGURE 20.30 Unicellular versus multicellular organisms. Although *Gonium* consists of as few as four cells, all cells are alike and can produce a new colony. *Volvox* has some cells specialized to perform different functions and has thus crossed the threshold that separates unicellular from multicellular organisms.

◆ FIGURE 20.31 Carbonaceous impressions in Proterozoic rocks in the Little Belt Mountains, Montana. These may be impressions of multicellular algae, but this is uncertain. (Photo courtesy of Robert Horodyski, Tulane University.)

what appeared to be impressions of algae and various animals, many of which bear no resemblance to living organisms. These discoveries have provided some of the evidence that has partly resolved one of the great mysteries in the history of life—the apparent absence of animal fossils in strata older than the Cambrian.

The rock unit in which the Ediacara Hills fossils were discovered, the Pound Quartzite, was initially thought to be Cambrian. However, a joint investigation by the South Australian Museum and the University of Adelaide demonstrated that the fossil-bearing strata lie more than 150 m below the oldest recognized Cam-

brian strata. Eventually, it became clear that what had been discovered was a unique assemblage of soft-bodied animals preserved as molds and casts on the undersides of sandstone layers (◆ Fig. 20.32). Martin Glaessner of the University of Adelaide believes that these Ediacaran animals lived in a nearshore, shallow marine environment (Fig. 20.32f).

Some investigators, including Glaessner, are of the opinion that at least three present-day invertebrate phyla are represented: jellyfish and sea pens (phylum Cnidaria), segmented worms (phylum Annelida), and primitive members of the phylum Arthropoda, the phylum that includes insects, spiders, and crabs. One wormlike Ediacaran fossil, *Spriggina,* has been cited as a possible ancestor of trilobites, and another, *Tribachidium,* may be a primitive echinoderm (Fig. 20.32b and d).

Researchers disagree, however, on exactly what these Ediacaran animals were and how they should be classified. Adolph Seilacher of Tübingen, Germany, for example, thinks that they represent an early evolutionary radiation quite distinct from the ancestry of the present-day invertebrate phyla. In fact, Seilacher thinks they represent an evolutionary "dead end." He also believes that these animals fed by passive absorption because there is no conclusive evidence of a mouth in the preserved specimens.

Ediacara-type faunas are now known on all continents except Antarctica. Collectively, these **Ediacaran faunas,** as they are commonly called, existed from 670 to 570 million years ago. These animals were widespread during this time, but their fossils are rare. Their scarcity should come as no surprise, however, since all lacked durable skeletons.

The discovery of pre-Paleozoic fossil-bearing strata has prompted Preston Cloud, formerly of the University of California at Santa Barbara, and Glaessner to propose a new geologic period and system, the *Ediacarian.* According to this proposal, the Ediacarian Period constitutes the first period of the Paleozoic Era. Thus, strata of the Ediacarian System were deposited during the Ediacarian Period, which began 670 million years ago when multicelled organisms appeared in the fossil record and ended 570 million years ago when shelly faunas of the Cambrian Period first appeared.

Cloud and Glaessner's proposal is a reasonable one and has been accepted by many geologists. Many other geologists, however, prefer the more traditional time scale, and in this book we consider the Ediacaran fauna to be latest Proterozoic in age.

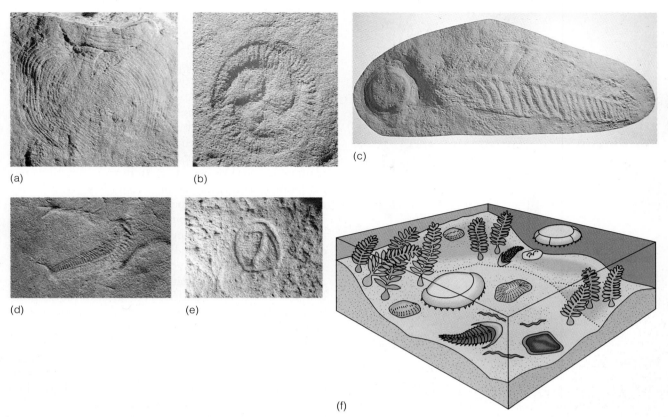

◆ FIGURE 20.32 The Ediacaran fauna of Australia. Impressions of multicelled animals: (*a*) *Ovatoscutum concentricum;* (*b*) *Tribrachidium heraldicum,* a possible primitive echinoderm; (*c*) *Charniodiscus arboreus;* (*d*) *Spriggina floundersi,* a possible ancestor of trilobites; and (*e*) *Parvancorina minchami.* (*f*) Reconstruction of the Ediacaran environment. (Photos courtesy of Neville Pledge, South Australian Museum.)

Other Proterozoic Animal Fossils

Although scarce, there is some evidence for pre-Ediacaran animals. For example, a jellyfish-like impression is known from rocks 2,000 m below the Pound Quartzite. And in many areas burrows, presumably made by worms, occur in rocks at least 700 million years old.

Wormlike fossils associated with fossil algae were recently reported from 700- to 900-million-year-old strata in China (◆ Fig. 20.33; Table 20.4). Perhaps these are worms, but both their biological affinities and their age have been questioned. For the present, we can consider these wormlike fossils as persuasive but not conclusive evidence for pre-Ediacaran animals.

All known Proterozoic animals were soft-bodied; that is, they lacked the durable exoskeletons that characterize many Phanerozoic invertebrates. There is some evidence, however, that the earliest stages of skeletonization occurred during the Late Proterozoic. For example, some Ediacaran animals may have had chitin, possibly a chitinous carapace, and others may have possessed some calcareous skeletal elements.

By the latest Proterozoic, several skeletonized animals probably existed. Evidence for this conclusion comes from minute scraps of shell-like material and denticles from larger animals, and spicules, presumably from sponges. Durable skeletons of chitin (a complex organic substance), silica, and calcium carbonate, however, began appearing in abundance at the beginning of the Phanerozoic Eon 570 million years ago.

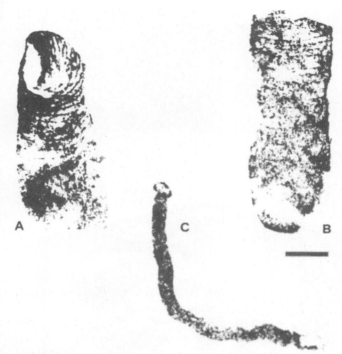

◆ FIGURE 20.33 Wormlike body fossils from the Late Proterozoic of China. (Photos courtesy of Sun Weiguo, Nanjing Institute of Geology and Palaeontology, Academia Sinica, Nanjing, People's Republic of China.)

▼ ARCHEAN MINERAL DEPOSITS

Although a variety of mineral resources occur in Archean rocks, the mineral most commonly associated with them is gold. Archean and Proterozoic rocks near Johannesburg, South Africa, have yielded more than 50% of the world's gold since 1886, much of it from conglomerate layers. A number of gold mining areas also occur within the Superior craton in Ontario, Canada, and the second largest gold mine in the United States is in the Archean Homestake Formation at Lead, South Dakota.

A number of Archean-aged deposits of massive sulfides of zinc, copper, and nickel are known from several areas, including Western Australia, Zimbabwe, and the Abitibi greenstone belt of Ontario, Canada. Many of these deposits probably formed as the result of hydrothermal activity associated with greenstone belt volcanism. Indeed, similar deposits are currently form-ing adjacent to black smokers, which are types of hydrothermal vents at or near spreading ridges.

About one-fourth of the world's chrome reserves are in Archean rocks, especially those in Zimbabwe. The deposits occur in greenstone belt rocks and appear to have formed when crystals settled and became concen-trated in the lower parts of mafic and ultramafic sills and other intrusive bodies.

One area of chrome deposits in the United States is in the Stillwater Complex in Montana. Low-grade ores from this area were mined during both World Wars as well as the Korean War, but they were simply stockpiled; that is, they were not refined for their chrome content. The Stillwater Complex is also a potential source of platinum, as are some other Archean rocks, but most platinum mined today comes from Proterozoic mafic rocks of the Bushveld Complex of South Africa.

Although Archean banded iron formations are mined in some areas, they are neither as thick nor as extensive as those of the Proterozoic Eon.

Pegmatites are very coarsely crystalline igneous rocks, commonly associated with plutons. Most are composed of quartz and feldspars and thus correspond to the composition of granite. Archean pegmatites are mostly granitic and of little economic value. Some, such as those of the Herb Lake district in Manitoba, Canada, and the Rhodesian Province in Africa, contain valuable minerals. In addition to minerals of gem quality, Archean pegmatites contain minerals mined for their lithium, beryllium, and rubidium.

▼ PROTEROZOIC MINERAL DEPOSITS

The most notable mineral deposits of Proterozoic age are banded iron formations (BIFs) (Fig. 20.21). BIFs are present in all Precambrian cratons; as noted earlier, 92% of all BIFs were deposited during the Late Proterozoic, although a few Archean and Phanerozoic examples are known. These deposits constitute the world's major iron ores. The largest producers of iron ores include Brazil, Australia, China, the Ukraine, Sweden, South Africa, Canada, and the United States. Even though the United States is a major producer, it must still import about 30% of the iron ore used, mostly from Canada and Venezuela.

In North America, most of the large iron mines are in the Lake Superior region. Huge deposits of BIFs in Ontario, Canada, and adjacent states have been mined extensively for decades. In 1988, Minnesota was the leading producer of iron ore in the United States, accounting for 72% of the total production, while Michigan produced 25% of the total.

The richest ores—those containing up to 70% iron—in the Lake Superior region were depleted by the time of World War II. Continued mining has been pos-

◆ FIGURE 20.34 Iron pellets from a mine in northern Michigan.

◆ FIGURE 20.35 Tourmaline from the Dunton mine in Maine.

sible because a method was developed to separate the iron ore from unusable rock of lower-grade ores and then shape the iron into pellets (◆ Fig. 20.34). These pellets contain about 65% iron and are easily shipped via the Great Lakes to the steel-producing centers.

The Sudbury mining district in Ontario, Canada, is an important area of nickel and platinum production. Nickel is essential in the production of nickel alloys such as stainless steel and Monel metal (nickel plus copper), which are valued for their strength and resistance to corrosion and heat. The United States must import more than 50% of all nickel used; most of these imports come from the Sudbury mining district.

Some platinum for jewelry, surgical instruments, and chemical and electrical equipment is also exported to the United States from Canada, but the major exporter is South Africa. The Bushveld Complex of South Africa is a layered complex of igneous rocks from which both platinum and chromite, the only ore of chromium, are mined. Much of the chromium used in the United

States is imported from South Africa; it is used mostly in the manufacture of stainless steel.

Economically recoverable oil and gas have been discovered in Proterozoic rocks in China and Siberia, arousing some interest in the Midcontinent rift as a potential source of hydrocarbons. So far, considerable land has been leased for exploration, and numerous geophysical studies have been done. However, even though some rock units within the rift are known to contain petroleum, no producing oil or gas wells are currently operating.

A number of Proterozoic pegmatites are important economically. The Dunton pegmatite in Maine, whose age is generally considered to be Late Proterozoic, has yielded magnificent gem-quality specimens of tourmaline and other minerals (◆ Fig. 20.35). Other pegmatites are mined for gemstones as well as for tin; industrial minerals, such as feldspars, micas, and quartz; and minerals containing such elements as cesium, rubidium, lithium, and beryllium. In addition, some of the world's largest known mineral crystals have been discovered in these pegmatites.

Chapter Summary

1. Precambrian time is divided into two eons, the Archean and Proterozoic.

2. Each continent has an ancient, stable craton. The exposed part of a craton is a Precambrian shield. The Canadian Shield of North America is made up of several cratons, which are delineated by age dates and structural trends.

3. Archean rocks are predominantly greenstone belts and granite-gneiss complexes. Greenstone belts have a syncli-

nal structure and occur as linear bodies within much more extensive areas of granite and gneiss.

4. Typical greenstone belts can be divided into three major rock sequences. The two lower units are dominated by volcanic rocks, some of which are ultramafic. The upper unit consists mostly of sedimentary rocks, particularly graywacke-argillite assemblages deposited by turbidity currents.

5. Archean cratons served as nuclei about which Proterozoic crust accreted. One large landmass that formed by this process is called Laurentia. It consisted mostly of North America and Greenland.

6. The major events in the Proterozoic evolution of Laurentia were an Early Proterozoic episode of amalgamation of cratons, Middle Proterozoic igneous activity, and the Middle Proterozoic Grenville orogeny and Midcontinent rift.

7. Plate tectonics similar to that of the present did not begin until the Early Proterozoic, but many geologists think some type of Archean plate tectonics occurred. Archean plates may have moved more rapidly, however, because the Earth possessed more radiogenic heat.

8. Quartzite-carbonate-shale assemblages are known from the Late Archean but become common in Proterozoic rocks. These rock assemblages were deposited on passive continental margins and in intracratonic basins.

9. Widespread glaciation occurred during the Early and Late Proterozoic.

10. The atmosphere and surface waters were derived from internally generated volcanic gases by a process known as outgassing. The atmosphere so formed was deficient in free oxygen. The atmosphere became progressively richer in free oxygen during the Proterozoic.

11. During the period from 2.5 to 2.0 billion years ago, most of the world's iron ores were deposited as banded iron formations, and the first continental red beds were deposited about 1.8 billion years ago. The widespread occurrence of oxidized iron in sedimentary rocks indicates an oxidizing atmosphere.

12. Most models for the origin of life require a nonoxidizing atmosphere. Atmospheric gases contained the elements necessary for simple organic molecules. Ultraviolet radiation and lightning probably provided the energy for synthesizing organic molecules. Some investigators think RNA molecules may have been the first molecules capable of self-replication.

13. The Archean fossil record is very poor. A few localities contain unicellular, prokaryotic bacteria. Stromatolites formed by photosynthesizing bacteria may date from 3.5 billion years ago.

14. Eukaryotic cells first appeared during the Middle Proterozoic.

15. The oldest fossils of multicellular organisms are carbonaceous impressions, probably of algae, in rocks between 1 billion and 700 million years old.

16. The Late Proterozoic Ediacaran faunas include the oldest well-documented animal fossils other than burrows. Animals were widespread at this time, but all were soft-bodied, so fossils are not common.

Important Terms

anaerobic
Archean Eon
autotrophic
banded iron formation (BIF)
Canadian Shield
craton
Ediacaran faunas
eukaryotic cell
granite-gneiss complex
greenstone belt

Grenville orogeny
heterotrophic
Laurentia
Midcontinent rift
monomer
multicellular organism
orogen
outgassing
photochemical dissociation

photosynthesis
polymer
Precambrian
Precambrian shield
prokaryotic cell
Proterozoic Eon
quartzite-carbonate-shale assemblage
red reds
stromatolite

Review Questions

1. The lower units in a typical greenstone belt are composed mostly of:
   a. ___ granite; b. ___ volcanic rocks; c. ___ limestone and shale; d. ___ volcanic ash; e. ___ graywacke-argillite.

2. Amino acids are examples of:
   a. ___ photosynthesis; b. ___ heterotrophism; c. ___ cratons; d. ___ stromatolites; e. ___ monomers.

3. Which of the following was probably responsible for adding some free oxygen to the Earth's early atmosphere?
   a. ___ fermentation; b. ___ outgassing; c. ___ carbonization; d. ___ abiotic synthesis; e. ___ photochemical dissociation.

4. The largest exposed area of the North American craton is the:
   a. ___ Wyoming craton; b. ___ American platform; c. ___ Canadian Shield; d. ___ Appalachian Mountains; e. ___ Grand Canyon.

5. The presence of _____ indicates that much of the greenstone belt volcanism was subaqueous.
   a. ___ pyroclastics; b. ___ argillites; c. ___ andesite;
   d. ___ pillow basalts; e. ___ ultramafic magma.

6. The process whereby gases are derived from within the Earth and released into the atmosphere by volcanism is:
   a. ___ photochemical dissociation; b. ___ oxygen respiration; c. ___ outgassing; d. ___ photosynthesis; e. ___ polymerization.

7. The only known Archean organisms were:
   a. ___ prokaryotic cells; b. ___ sponges; c. ___ jellyfish;
   d. ___ microspheres; e. ___ monomers.

8. Some scientists think that the first self-replicating system may have been a(n):
   a. ___ stromatolite; b. ___ RNA molecule; c. ___ bacterium; d. ___ ATP cell; e. ___ proteinoid.

9. The large landmass consisting mostly of North America and Greenland that formed during the Proterozoic was:
   a. ___ Gondwana; b. ___ Hudsonia; c. ___ Laurentia;
   d. ___ the Trans-Hudson orogen; e. ___ the Wyoming craton.

10. Widespread Middle Proterozoic igneous rocks consist mostly of:
    a. ___ ultramafic lava flows and andesite; b. ___ ash flow deposits and plutons; c. ___ colliding island arcs;
    d. ___ greenstone belts; e. ___ sedimentary and metamorphic rocks.

11. Thick accumulations of basaltic lava flows forming extensive lava plateaus occur in the:
    a. ___ Gowganda Formation; b. ___ Midcontinent rift;
    c. ___ Penokean orogen; d. ___ Grenville orogen;
    e. ___ Wyoming craton.

12. By far the most common associations of Proterozoic rocks are:
    a. ___ basalt-andesite-ash fall; b. ___ sandstone-granite-basalt; c. ___ granite-andesite-tillite; d. ___ banded iron formation-tillite-andesite; e. ___ quartzite-carbonate-shale.

13. Most banded iron formations (BIFs) were deposited during the:
    a. ___ Late Proterozoic; b. ___ Middle Archean;
    c. ___ Early Proterozoic; d. ___ Trans-Hudson orogen;
    e. ___ Midcontinent rift.

14. The presence of red beds during the Proterozoic indicates:
    a. ___ widespread glaciation; b. ___ that the atmosphere contained some free oxygen; c. ___ that animals had appeared; d. ___ a chemically reducing atmosphere; e. ___ that carbonate deposition was becoming increasingly common.

15. Probably the most important event in the evolution of life during the Proterozoic was the appearance of:
    a. ___ photosynthesizing organisms; b. ___ algae;
    c. ___ the first autotroph; d. ___ eukaryotic cells;
    e. ___ ophiolites.

16. What are Precambrian shields and cratons?

17. Describe the vertical succession of rock types in a typical greenstone belt.

18. What sequence of events led to the origin of the southern Superior craton of the Canadian Shield?

19. How does a greater amount of radiogenic heat account for a different style of plate tectonics during the Archean?

20. Explain how photosynthesis and photochemical dissociation of water vapor supplied oxygen to the Archean atmosphere.

21. Summarize the experimental evidence indicating that monomers could have formed by natural processes during the Archean.

22. Explain why ultramafic lava flows are so rare in rocks younger than Archean.

23. Summarize the major difference between the Archean and Proterozoic.

24. What is the Midcontinent rift, and what kinds of rocks does it contain?

25. Discuss the significance of ophiolites and quartzite-carbonate-shale assemblages in establishing that the Proterozoic was characterized by a modern style of plate tectonics.

26. How are BIFs thought to have been deposited?

27. What evidence indicates that eukaryotic cells appeared between 1.4 and 1.0 billion years ago?

28. Explain how a symbiotic relationship among Proterozoic prokaryotes may have given rise to eukaryotes.

29. Briefly review the evidence for the presence of animals in the Late Proterozoic.

Additional Readings

Condie, K. C. 1981. *Archean greenstone belts.* New York: Elsevier Scientific Publishing.

Condie, K. C. 1989. *Plate tectonics and crustal evolution.* 3d ed. New York: Pergamon Press.

Fox, S. W. 1991. Synthesis of life in the lab? Defining a proto-living system. *The Quarterly Review of Biology* 66:181–85.

Glaessner, M. F. 1984. *The dawn of animal life.* New York: Cambridge University Press.

Hambrey, M. 1992. Secrets of a tropical ice age. *New Scientist* 133, no. 1806:42–49.

Hoffmann, P. F. 1988. United plates of America, the birth of a craton: Early Proterozoic assembly and growth of

Laurentia. *Annual Review of Planetary Sciences* 16:543–603.

Horgan, J. 1991. In the beginning. *Scientific American* 264, no. 2:116–25.

Kabnick, K. S., and D. A. Peattie. 1991. *Giardia:* A missing link between prokaryotes and eukaryotes. *American Scientist* 79, no. 1:34–43.

Knoll, A. H. 1991. End of the Proterozoic Eon. *Scientific American* 265, no. 4:64–73.

McCall, G. J. H., ed. 1977. *The Archean.* Stroudsburg, Penn.: Dowden, Hutchinson & Ross.

Margulis, L. 1982. *Early life.* New York: Van Nostrand Reinhold.

Margulis, L., and L. Olendzenski, eds. 1992. *Environmental evolution: Effects of the origin and evolution of life on planet Earth.* Cambridge, Mass.: MIT Press.

Nisbet, E. G. 1987. *The young Earth: An introduction to Archean geology.* Boston: Allen & Unwin.

Schopf, J. W., ed. 1992. *Major events in the history of life.* Boston: James and Bartlett.

Schofp, J. W., and C. Klein, eds., 1992. *The Proterozoic biosphere.* New York: Cambridge University Press.

Windley, B. F. 1984. *The evolving continents.* New York: John Wiley & Sons.

York, D. 1993. The earliest history of the Earth. *Scientific American* 268, no. 1:90–96.

CHAPTER
21

Chapter Outline

Major John Wesley Powell, who led the first geologic expedition down the Grand Canyon in Arizona.

Paleozoic Earth History

Prologue

"The Grand Canyon is the one great sight which every American should see," declared President Theodore Roosevelt. "We must do nothing to mar its grandeur." And so, in 1908, he named the Grand Canyon a national monument to protect it from exploitation. In 1919 the Grand Canyon National Monument was upgraded to a national park primarily because both its scenery and the geology exposed in the canyon are unparalleled.

When people visit the Grand Canyon, many are astonished by the seemingly limitless time represented by the rocks exposed in the walls. For most people, staring down 1.5 km at the rocks in the canyon is their only exposure to the concept of geologic time.

Major John Wesley Powell was the first geologist to explore the Grand Canyon region. Major Powell, a Civil War veteran who lost his right arm in the battle of Shiloh, led a group of hardy explorers down the uncharted Colorado River through the Grand Canyon in 1869. Without any maps or other information, Powell and his group ran the many rapids of the Colorado River in fragile wooden boats, hastily recording what they saw. Powell wrote in his diary that "all about me are interesting geologic records. The book is open and I read as I run."

From this initial reconnaissance, Powell led a second expedition down the Colorado River in 1871. This second trip included a photographer, a surveyor, and three topographers. This expedition made detailed topographic and geologic maps of the Grand Canyon area as well as the first photographic record of the region.

Probably no one has contributed as much to the understanding of the Grand Canyon as Major Powell. In recognition of his contributions, the Powell Memorial was erected on the South Rim of the Grand Canyon in 1969 to commemorate the hundredth anniversary of his first expedition.

When we stand on the rim and look down into the Grand Canyon, we are really looking far back in time, all the way back to the early history of our planet. More than one billion years of history are recorded in the rocks of the Grand Canyon, ranging from mountain-building episodes to periods of transgressions and regressions of shallow seas.

The oldest rocks exposed in the Grand Canyon record two major mountain-building episodes during the Proterozoic Eon. The first episode, represented by the Vishnu and Brahma schists, records a time of uplift, deformation, and metamorphism. This mountain range was eroded to a rather subdued landscape and was followed by deposition of approximately 4,000 m of sediments and lava flows of the Grand Canyon Supergroup. These rocks and the underlying Vishnu and Brahma schists were uplifted and formed a second Proterozoic mountain range. This mountain range was also eroded to a nearly flat surface by the end of the Proterozoic Eon.

The first sea of the Paleozoic Era transgressed over the region during the Cambrian Period, depositing sandstones, siltstones, and limestones. A major unconformity separates the Cambrian rocks from the Mississippian limestones exposed as the cliff-forming Redwall Limestone. Another unconformity separates the Redwall Limestone from the overlying Permian Kaibab Limestone, which forms the rim of the Grand Canyon.

The Grand Canyon in all its grandeur is a most appropriate place to start our discussion of the Paleozoic history of North America.

INTRODUCTION

Having reviewed the geologic history of the Archean and Proterozoic eons, we now turn our attention to the Phanerozoic Eon, comprising the remaining 12% of geologic time. At the beginning of the Phanerozoic, there were six major continental landmasses, four of which straddled the paleoequator. Plate movements during the Phanerozoic created a changing panorama of continents and ocean basins whose positions affected atmospheric and oceanic circulation patterns and created new environments for habitation by the rapidly evolving biota.

The Paleozoic history of most continents involves major mountain-building activity along the continental borders and numerous shallow-water marine transgressions and regressions over their interiors. These transgressions and regressions were caused by global changes in sea level probably related to plate activity and glaciation.

The following chapters present the geologic history of North America in terms of those major transgressions and regressions rather than a period-by-period chronology. While we will focus on North American geologic history, we will endeavor to place those events in a global context.

CONTINENTAL ARCHITECTURE: CRATONS AND MOBILE BELTS

During the Precambrian, continental accretion and orogenic activity led to the formation of sizable continents. At least three large continents existed during the Late Proterozoic, and some geologists believe that these landmasses later collided to form a single Pangaea-like supercontinent (see Fig. 20.12). This supercontinent began breaking apart sometime during the latest Proterozoic. By the beginning of the Paleozoic Era, six major continents were present. Each continent can be divided into two major components: a craton and one or more mobile belts.

Recall that cratons are the relatively stable and immobile parts of continents and form the foundation upon which Phanerozoic sediments were deposited (◆ Fig. 21.1). Cratons typically consist of two parts: a shield and a platform.

Shields are the exposed portion of the crystalline basement rocks of a continent and are composed of Precambrian metamorphic and igneous rocks (see Figs. 6.4 and 20.2) that reveal a history of extensive orogenic activity during the Precambrian. During the Phanerozoic, however, shields were extremely stable and formed the foundation of the continents.

Extending outward from the shields are buried Precambrian rocks that constitute a platform, another part of the craton. Overlying the platform are flat-lying or gently dipping Phanerozoic sedimentary rocks. Phanerozoic rocks deposited on the platform include detrital and chemical sedimentary rocks that were deposited in widespread shallow seas that transgressed and regressed over the craton. These seas, called **epeiric seas,** were a common feature of most Paleozoic cratonic histories. Changes in sea-level caused primarily by continental glaciation as well as by plate movement were responsible for the advance and retreat of the seas.

While most of the Paleozoic platform rocks are still essentially flat-lying, in some places they were gently folded into regional arches, domes, and basins (Fig. 21.1). In many cases some of these structures stood out as low islands during the Paleozoic Era and supplied sediments to the surrounding epeiric seas.

Mobile belts are elongated areas of mountain-building activity. They occur along the margins of continents where sediments are deposited in the relatively shallow waters of the continental shelf and the deeper waters at the base of the continental slope. During plate convergence along these margins, the sediments are deformed and intruded by magma, creating mountain ranges.

Four mobile belts formed around the margin of the North American craton during the Paleozoic; these were the **Franklin, Cordilleran, Ouachita,** and **Appalachian mobile belts** (Fig. 21.1). Each was the site of mountain building in response to compressional forces along a convergent plate boundary and formed such mountain ranges as the Appalachians and Ouachitas.

PALEOZOIC PALEOGEOGRAPHY

One of the major lessons plate tectonic theory teaches us is that the Earth's geography is constantly changing. The present-day configuration of the continents and ocean basins is merely a snapshot in time. As the plates move about the Earth, the location of continents and ocean basins is constantly changing and being modified. One of the goals of historical geology is to provide paleogeographic reconstructions of the world for the geologic past. By synthesizing all of the pertinent paleoclimatic, paleomagnetic, paleontologic, sedimentologic, stratigraphic, and tectonic data available, geolo-

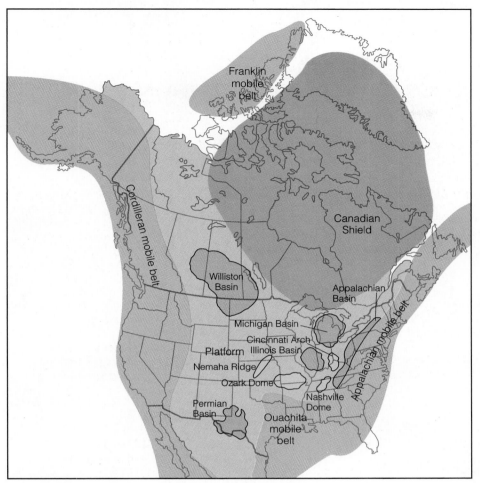

◆ FIGURE 21.1 The mobile belts and major cratonic structures of North America that formed during the Paleozoic Era.

gists can prepare paleogeographic maps of what the world looked like at a particular time in the geologic past (see Perspective 21.1).

Recall that by the beginning of the Paleozoic, six major continents were present. In addition to these large landmasses, geologists have also identified numerous small microcontinents and island arcs associated with various microplates that were present during the Paleozoic. We will be primarily concerned, however, with the history of the six major continents and their relationship to each other. The six major Paleozoic continents are **Baltica** (Russia west of the Ural Mountains and the major part of northern Europe), **China** (a complex area consisting of at least three Paleozoic continents that were not widely separated and are here considered to include China, Indochina, and the Malay

Peninsula), **Gondwana** (Africa, Antarctica, Australia, Florida, India, Madagascar, and parts of the Middle East and southern Europe), **Kazakhstania** (a triangular continent centered on Kazakhstan, but considered by some to be an extension of the Paleozoic Siberian continent), **Laurentia** (most of present North America, Greenland, northwestern Ireland, Scotland, and part of eastern Russia), and **Siberia** (Russia east of the Ural Mountains and Asia north of Kazakhstan and south of Mongolia). The paleogeographic reconstructions that follow (Figs. 21.2, 21.3, and 21.4) are based on the methods used to determine and interpret the location, geographic features, and environmental conditions on the paleocontinents (see Perspective 21.1).

In contrast to today's global geography, the Cambrian world consisted of six major continents dispersed

Perspective 21.1

PALEOGEOGRAPHIC RECONSTRUCTIONS
AND MAPS

The key to any reconstruction of world paleogeography is the correct positioning of the continents in terms of latitude and longitude as well as orientation of a paleocontinent relative to the paleonorth pole. The main criteria used for paleogeographic reconstructions are paleomagnetism, biogeography, tectonic patterns, and climatology.

Paleomagnetism provides the only source of quantitative data on the orientations of the continents. For the Paleozoic Era, however, the paleomagnetic data are often inconsistent and contradictory due to secondary magnetizations acquired through the effects of metamorphism or weathering.

The distribution of faunas and floras provides a useful check on the latitudes determined by paleomagnetism and can provide additional limits on longitudinal separation of continents. As is well known, the distribution of plants and animals is controlled by both climatic and geographic barriers. Such information can be used to position continents and ocean basins in a way that accounts for the biogeographic patterns indicated by fossil evidence.

Tectonic activity is indicated by deformed sediments associated with andesitic volcanics and ophiolites. Such features allow geologists to recognize ancient mountain ranges and zones of subduction. These mountain ranges may subsequently have been separated by plate movement, so the identification of large, continuous mountain ranges provides important information about continental positions in the geologic past.

Climate-sensitive sedimentary rocks are used to interpret past climatic conditions. Desert dunes are typically well sorted and cross-bedded on a large scale and associated with other deposits, indicating an arid environment. Coals form in freshwater swamps where climatic conditions promote abundant plant growth. Evaporites result when evaporation exceeds precipitation, such as in desert regions or along hot, dry shorelines. Tillites result from glacial activity and indicate cold, wet environments.

Paleogeographic features can be determined by associations of sedimentary rocks and sedimentary structures. For example, large-scale cross-beds may indicate aeolian or windblown conditions such as in deserts. Delta complexes and deep-sea fans have characteristic internal features and three-dimensional forms that can be recognized in the rock record, just as coal and associated deposits usually follow a particular sequence. These features can be used to interpret such geographic features as lakes, streams, swamps, and shallow and deep marine areas.

Former mountain ranges can be recognized by folded and faulted sedimentary rocks associated with metamorphic and igneous rocks. We have already mentioned the association of andesites and ophiolites as evidence of former mountain building.

By combining all relevant geologic, paleontologic, and climatologic information, geologists can construct paleogeographic maps. Such maps are simply interpretations of the geography of an area for a particular time in the geologic past. The majority of paleogeographic maps show the distribution of land and sea, probable climatic regimes, and such geographic features as mountain ranges, swamps, and glaciers.

around the globe at low tropical latitudes (◆ Fig. 21.2a). Water circulated freely among ocean basins, and the polar regions were apparently ice-free. By the Late Cambrian, epeiric seas had covered large areas of Laurentia, Baltica, Siberia, Kazakhstania, and China, while major highlands were present in northeastern Gondwana, eastern Siberia, and central Kazakhstania.

During the Ordovician and Silurian periods, plate movement played a major role in the changing global geography (◆ Fig. 21.2b and c). Gondwana moved southward during the Ordovician and began to cross the South Pole as indicated by Upper Ordovician tillites found today in the Sahara Desert. In contrast to the passive continental margin Laurentia exhibited during the Cambrian, an active convergent plate boundary formed along its eastern margin during the Ordovician as indicated by the Late Ordovician *Taconic orogeny* that occurred in New England. During the Silurian, Baltica moved northwestward relative to Laurentia and collided with it to form the larger continent of **Laurasia.** This collision, which closed the northern Iapetus Ocean, is marked by the *Caledonian orogeny.* Following this orogeny, the southern part of the Iapetus Ocean still remained open between Laurentia and Gondwana (Fig. 21.2c). Siberia and Kazakhstania moved from a southern equatorial position during the Cambrian to north temperate latitudes by the end of the Silurian Period.

During the Devonian, as the southern Iapetus Ocean narrowed between Laurasia and Gondwana, mountain building continued along the eastern margin of Laurasia with the *Acadian orogeny* (◆ Fig. 21.3a, page 566). The erosion of the resulting highlands provided vast amounts of reddish fluvial sediments that covered large areas of northern Europe (Old Red Sandstone) and eastern North America (the Catskill Delta). Other Devonian tectonic events, probably related to the collision of Laurentia and Baltica, include the Cordilleran *Antler orogeny,* the *Ellesmere orogeny* along the northern margin of Laurentia (which may reflect the collision of Laurentia with Siberia), and the change from a passive continental margin to an active convergent plate boundary in the Uralian mobile belt of eastern Baltica. The distribution of reefs, evaporites, and red beds, as well as the existence of similar floras throughout the world, suggests a rather uniform global climate during the Devonian Period.

During the Carboniferous Period, southern Gondwana moved over the South Pole, resulting in extensive continental glaciation (◆ Figs. 21.3b and ◆ 21.4a, page 567). The advance and retreat of these glaciers produced global changes in sea level that affected sedimentation patterns on the cratons. As Gondwana continued moving northward, it first collided with Laurasia during the Early Carboniferous and continued suturing with it throughout the rest of the Carboniferous (Figs. 21.3b and 21.4a). Because Gondwana rotated clockwise relative to Laurasia, deformation generally progressed in a northeast-to-southwest direction along the Hercynian, Appalachian, and Ouachita mobile belts of the two continents. The final phase of collision between Gondwana and Laurasia is indicated by the Ouachita Mountains of Oklahoma, which were formed by thrusting during the Late Carboniferous and Early Permian.

Elsewhere, Siberia collided with Kazakhstania and moved toward the Uralian margin of Laurasia (Baltica), colliding with it during the Early Permian. It has recently been suggested that the northwestern margin of China collided with the southwestern margin of Siberia during the Late Carboniferous. Thus, by the end of the Carboniferous, the various continental landmasses were fairly close together as Pangaea began taking shape.

The Carboniferous coal basins of eastern North America, western Europe, and the Donets Basin of Ukraine all lay in the equatorial zone, where rainfall was high and temperatures were consistently warm. The absence of strong seasonal growth-rings in fossil plants from these coal basins is indicative of such a climate. The fossil plants found in the coals of Siberia and China, however, show well-developed growth-rings, signifying seasonal growth with abundant rainfall and distinct seasons such as occur in the temperate zones (latitudes 40 degrees to 60 degrees north).

Glacial conditions and the movement of large continental ice sheets in the high southern latitudes are indicated by widespread tillites and glacial striations in southern Gondwana (see Figs. 9.6 and 9.8). These ice sheets spread toward the equator and, at their maximum growth, extended well into the middle temperate latitudes.

The assembly of Pangaea was essentially concluded during the Permian with the completion of many of the continental collisions that began during the Carboniferous (◆ Fig. 21.4b, page 567). Although geologists generally agree on the configuration and location of the western half of the supercontinent, there is no consensus on the number or configuration of the various terranes and continental blocks that composed the eastern half of Pangaea. Regardless of the exact configuration of the eastern portion, geologists know that the supercontinent was surrounded by various subduction zones and moved steadily northward during the Permian. Furthermore, an enormous single ocean, **Panthalassa,** surrounded Pangaea and spanned the Earth from pole to pole. Waters of this ocean probably circulated more freely than at present, resulting in more equable water temperatures.

The formation of a single large landmass had climatic consequences for the terrestrial environment as

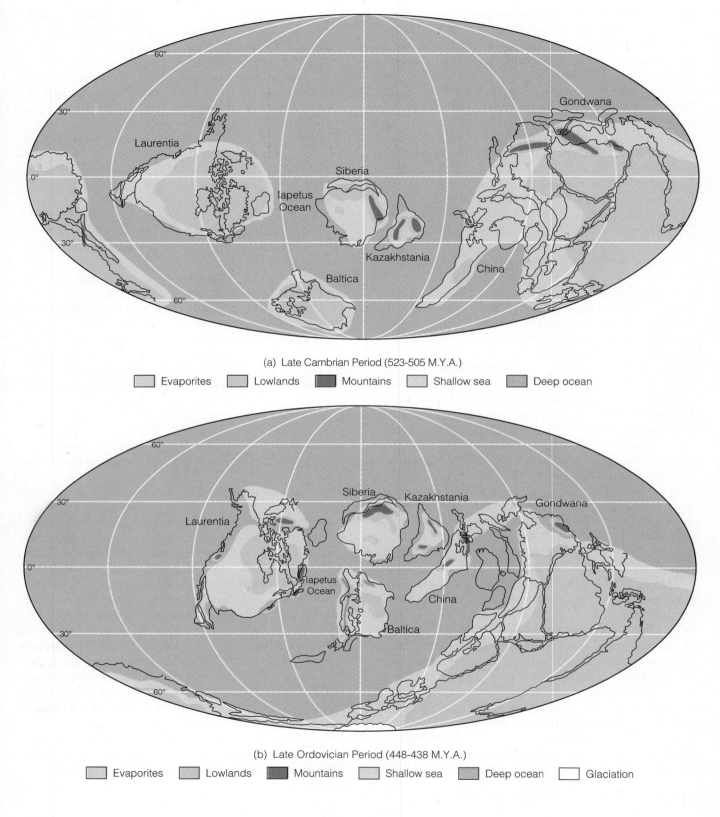

(a) Late Cambrian Period (523-505 M.Y.A.)

Evaporites   Lowlands   Mountains   Shallow sea   Deep ocean

(b) Late Ordovician Period (448-438 M.Y.A.)

Evaporites   Lowlands   Mountains   Shallow sea   Deep ocean   Glaciation

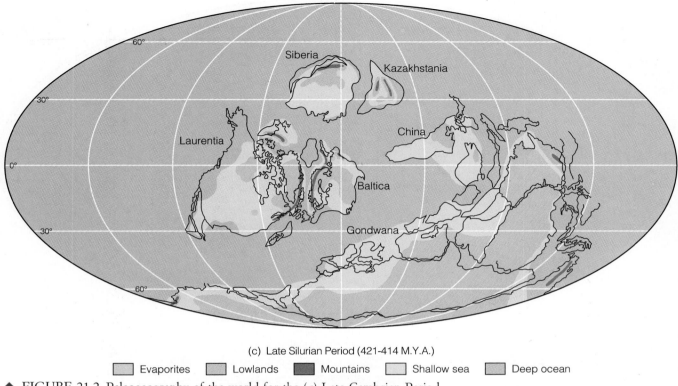

(c) Late Silurian Period (421-414 M.Y.A.)

Evaporites | Lowlands | Mountains | Shallow sea | Deep ocean

◆ FIGURE 21.2 Paleogeography of the world for the (*a*) Late Cambrian Period, (*b*) Late Ordovician Period, and (*c*) Late Silurian Period.

well. Terrestrial Permian sediments indicate that arid and semiarid conditions were widespread over Pangaea. The mountain ranges produced by the *Hercynian, Alleghenian,* and *Ouachita orogenies* were high enough to create rain shadows that blocked the moist, subtropical, easterly winds—much as the southern Andes Mountains do in western South America today. This produced very dry conditions in North America and Europe, as is evident from the extensive Permian evaporites found in western North America, central Europe, and parts of Russia. Permian coals, indicative of abundant rainfall, were mostly limited to the northern temperate belts (latitude 40 degrees to 60 degrees north), while the last remnants of the Carboniferous ice sheets retreated to the mountainous regions of eastern Australia.

▼ PALEOZOIC EVOLUTION OF NORTH AMERICA

It is convenient to divide the history of the North American craton into two parts, the first dealing with the relatively stable continental interior over which epeiric seas transgressed and regressed, and the second with the mobile belts where mountain building occurred.

In 1963 the American geologist Laurence L. Sloss proposed that the sedimentary-rock record of North America could be subdivided into six cratonic sequences. A **cratonic sequence** is a large-scale (greater than supergroup) lithostratigraphic unit representing a major transgressive-regressive cycle bounded by cratonwide unconformities (◆ Fig. 21.5, page 568). The transgressive phase, which is usually covered by younger sediments, commonly is well preserved, while the regressive phase of each sequence is marked by an unconformity. Where rocks of the appropriate age are preserved, each of the six unconformities can be shown to extend across the various sedimentary basins of the North American craton and into the mobile belts along the cratonic margin.

Geologists have also recognized major unconformity bounded sequences in cratonic areas outside North America. Such global transgressive and regressive

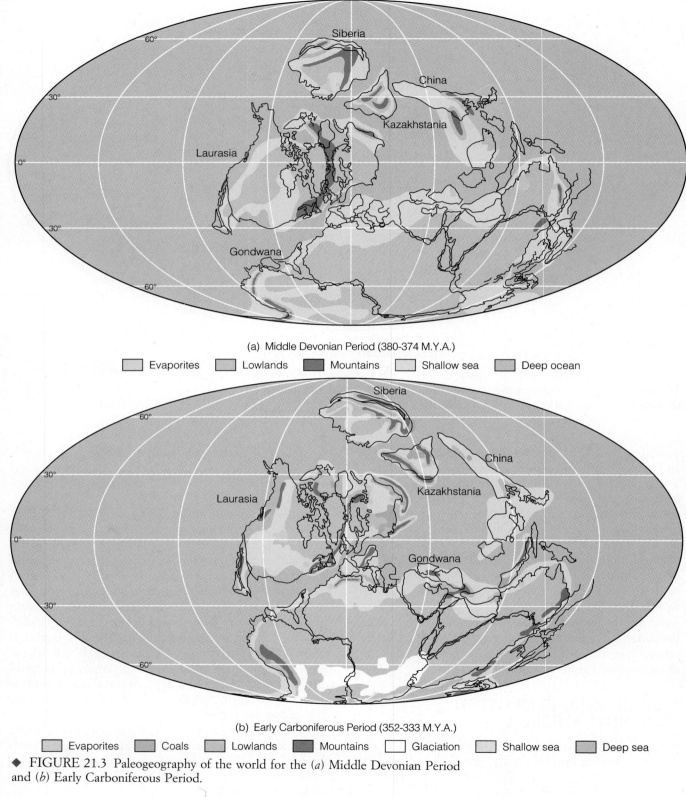

(a) Middle Devonian Period (380-374 M.Y.A.)

Evaporites　　Lowlands　　Mountains　　Shallow sea　　Deep ocean

(b) Early Carboniferous Period (352-333 M.Y.A.)

Evaporites　　Coals　　Lowlands　　Mountains　　Glaciation　　Shallow sea　　Deep sea

◆ FIGURE 21.3 Paleogeography of the world for the (*a*) Middle Devonian Period and (*b*) Early Carboniferous Period.

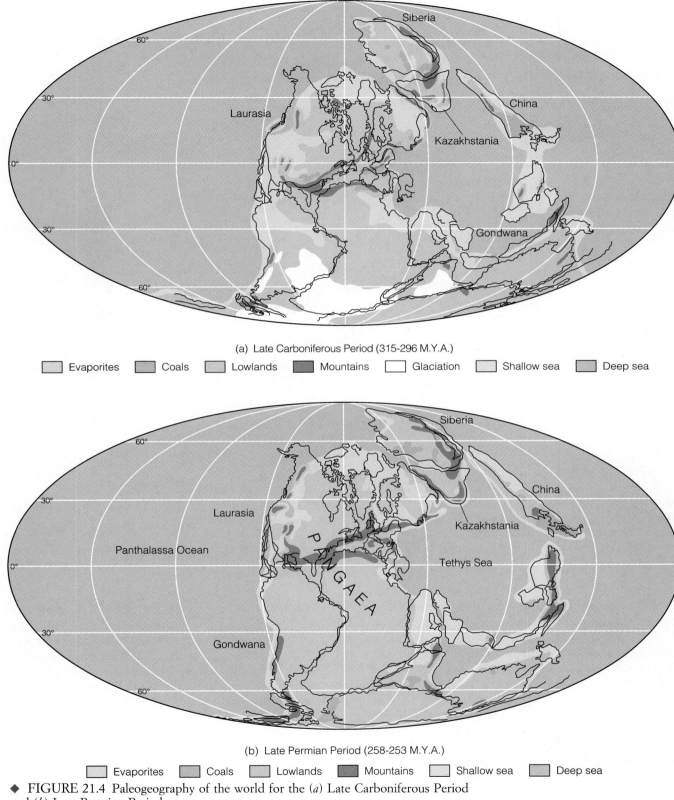

(a) Late Carboniferous Period (315-296 M.Y.A.)

Evaporites   Coals   Lowlands   Mountains   Glaciation   Shallow sea   Deep sea

(b) Late Permian Period (258-253 M.Y.A.)

Evaporites   Coals   Lowlands   Mountains   Shallow sea   Deep sea

◆ FIGURE 21.4 Paleogeography of the world for the (a) Late Carboniferous Period and (b) Late Permian Period.

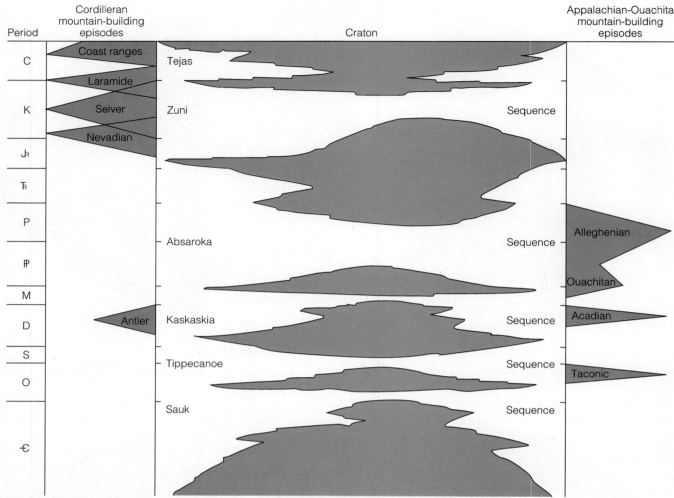

◆ FIGURE 21.5 Cratonic sequences of North America. The white areas represent sequences of rocks that are separated by large scale unconformities shown as brown areas. The major Cordilleran orogenies are shown on the left side of the figure, and the major Appalachian orogenies are shown on the right side.

cycles are caused by sea level changes and are thought to result from major tectonic and glacial events.

The realization that rock units can be divided into cratonic sequences and that these sequences can be further subdivided and correlated provides the foundation for an important new concept in geology that allows high-resolution analysis of time and facies relationships within sedimentary rocks. **Sequence stratigraphy** is the study of rock relationships within a time-stratigraphic framework of related facies bounded by erosional or nondepositional surfaces. The basic unit of sequence

stratigraphy is the *sequence,* which is a succession of rocks bounded by unconformities and their equivalent conformable strata. Sequence boundaries form as a result of the relative drop in sea level. Sequence stratigraphy is becoming an important tool in geology because it allows geologists to subdivide sedimentary rocks into related units that are bounded by time-stratigraphically significant boundaries. Geologists are using sequence stratigraphy for high-resolution correlation and mapping as well as interpreting and predicting depositional environments.

THE SAUK SEQUENCE

Rocks of the **Sauk sequence** record the first major transgression onto the North American craton (Fig. 21.5). During the Late Proterozoic and Early Cambrian, deposition of marine sediments was limited to the passive shelf areas of the Appalachian and Cordilleran borders of the craton. The craton itself was above sea level and experiencing extensive weathering and erosion. Because North America was located in a tropical climate at this time and there is no evidence of any terrestrial vegetation, weathering and erosion of the exposed Precambrian basement rocks must have proceeded at a very rapid rate. During the Middle Cambrian, the transgressive phase of the Sauk began with epeiric seas encroaching over the craton. By the Late Cambrian, the Sauk Sea had covered most of North America, leaving only a portion of the Canadian Shield and a few large islands above sea level (◆ Fig. 21.6). These islands, collectively referred to as the **Transcontinental Arch,** extended from New Mexico to Minnesota and the Lake Superior region.

The sediments deposited on both the craton and along the shelf area of the craton margin show abundant evidence of shallow-water deposition. The only difference between the shelf and craton deposits is that the shelf deposits are thicker. In both areas, the sands are generally clean and well sorted and commonly contain ripple marks and small-scale cross-bedding. Many of the carbonates are bioclastic (composed of fragments of organic remains), contain stromatolites, or have oolitic (small, spherical calcium carbonate grains) textures. Such sedimentary structures and textures are evidence of shallow-water deposition.

THE TIPPECANOE SEQUENCE

As the Sauk Sea regressed from the craton during the Early Ordovician, it revealed a landscape of low relief. The rocks exposed were predominantly limestones and dolostones that experienced deep (in some places up to 50 m) and extensive erosion because North America was still located in a tropical environment (◆ Fig. 21.7). The resulting cratonwide unconformity marks the boundary between the Sauk and Tippecanoe sequences.

Like the Sauk sequence, deposition of the **Tippecanoe sequence** began with a major transgression onto the craton. This transgressing sea deposited clean quartz sands over most of the craton. The best known of the Tippecanoe basal sandstones is the St. Peter Sandstone, an almost pure quartz sandstone used in manufacturing glass. It occurs throughout much of the midcontinent and resulted from numerous cycles of weathering and erosion of Proterozoic and Cambrian sandstones deposited during the Sauk transgression (◆ Fig. 21.8, page 572).

The Tippecanoe basal sandstones were followed by widespread carbonate deposition (Fig. 21.7). The limestones were generally the result of deposition by calcium carbonate–secreting organisms such as corals, brachiopods, stromatoporoids, and bryozoans. In addition to the limestones, there were also many dolostones. Most of the dolostones formed as magnesium was substituted for some of the calcium in calcite and, in the process, converted the limestones into dolostones.

In the eastern portion of the craton, the carbonates grade laterally into shales. These shales mark the farthest extent of detrital sediments derived from weathering and erosion of the highlands formed during the Taconic orogeny, a tectonic event we will discuss later.

Tippecanoe Reefs and Evaporites

Organic reefs are limestone structures constructed by living organisms, some of which contribute skeletal materials to the reef framework (◆ Fig. 21.9, page 573). Today corals and calcareous algae are the most prominent reefbuilders, but in the geologic past, other organisms played a major role. Regardless of the organisms dominating reef communities, reefs appear to have occupied the same ecological niche in the geologic past that they do today. Because of the ecological requirements of reef-building organisms, reefs today are confined to a narrow latitudinal belt between 30 degrees north and south of the equator. Corals, the major reef-building organisms today, require warm, clear, shallow water of normal salinity for optimal growth.

The size and shape of a reef are largely the result of the interaction between the reef-building organisms, the bottom topography, wind and wave action, and subsidence of the sea floor. Reefs also alter the area around them by forming barriers to water circulation or wave action.

Reefs have been common features since the Cambrian and have been built by a variety of organisms. The first skeletal builders of reeflike structures were archaeocyathids. These conical-shaped organisms lived during the Cambrian and had double, perforated, calcareous shell walls. Archaeocyathids built small mounds that have been found on all continents except South America (see Fig. 22.7). Beginning in the Middle

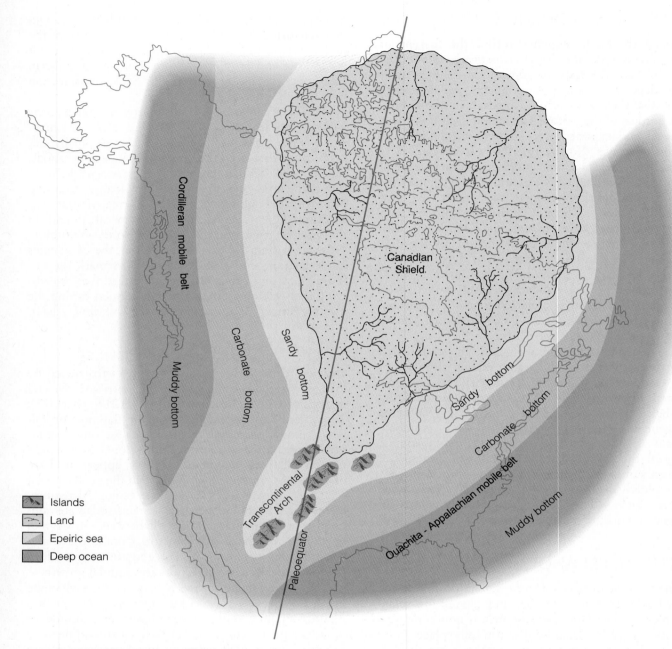

Islands
Land
Epeiric sea
Deep ocean

Cordilleran mobile belt

Muddy bottom

Carbonate bottom

Sandy bottom

Canadian
Shield

Sandy bottom

Carbonate bottom

Transcontinental
Arch

Paleoequator

Ouachita - Appalachian mobile belt

Muddy bottom

◆ FIGURE 21.6 Paleogeography of North America during the Cambrian Period.
Note the position of the Cambrian paleoequator. During this time North America
straddled the equator as indicated in Figure 21.2a.

Ordovician, stromatoporoid-coral reefs became common in the low latitudes, and similar reefs remained so throughout the rest of the Phanerozoic Eon. The burst of reef building seen in the Late Ordovician through Devonian probably occurred in response to evolutionary changes triggered by the appearance of extensive carbonate seafloors and platforms beyond the influence of detrital sediments.

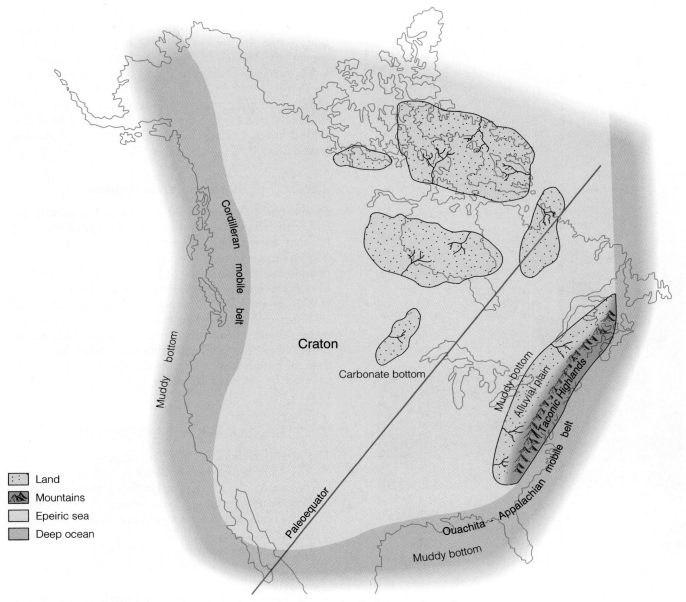

Land

Mountains

Epeiric sea

Deep ocean

Cordilleran mobile belt

Muddy bottom

Craton

Carbonate bottom

Muddy bottom

Alluvial Plain

Taconic Highlands

Appalachian mobile belt

Ouachita

Muddy bottom

Paleoequator

◆ FIGURE 21.7 Paleogeography of North America during the Ordovician Period. Note that the position of the equator has changed, indicating North America was rotating in a counterclockwise direction.

The Middle Silurian rocks (Tippecanoe sequence) of the present-day Great Lakes region are world famous for their reef and evaporite deposits and have been extensively studied (◆ Fig. 21.10, page 574). The most famous structure in the region is the Michigan Basin. It is a broad, circular basin surrounded by large barrier reefs. No doubt these reefs contributed to increasingly restricted circulation and the precipitation of Upper Silurian evaporites within the basin (◆ Fig. 21.11, page 575).

Within the rapidly subsiding interior of the basin, other types of reefs are found. *Pinnacle reefs* are tall, spindly structures up to 100 m high. They reflect the

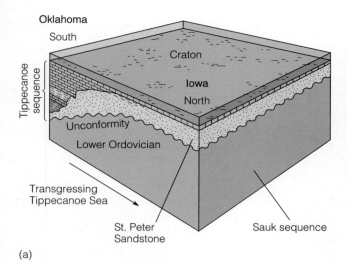

(a)

(b)

◆ FIGURE 21.8 (a) The transgression of the Tippecanoe Sea resulted in the deposition of the St. Peter Sandstone (Middle Ordovician) over a large area of the craton. (b) Outcrop of St. Peter Sandstone in Governor Dodge State Park, Wisconsin.

rapid upward growth needed to maintain themselves near sea level during subsidence of the basin (Fig. 21.11). In addition to the pinnacle reefs, bedded carbonates and thick sequences of salt and anhydrite are also found in the Michigan Basin.

As the Tippecanoe Sea gradually regressed from the craton during the Late Silurian, precipitation of evaporite minerals occurred in the Appalachian, Ohio, and Michigan basins. In the Michigan Basin alone, approximately 1,500 m of sediments were deposited, nearly

half of which are halite and anhydrite. How did such thick sequences of evaporites accumulate? One possibility is that a drop in sea level occurred so that the tops of the barrier reefs were as high as or above sea level, thus preventing the influx of new seawater into the basin. Evaporation of the basinal seawater would result in the precipitation of salts. A second possibility is that the reefs grew upward so close to sea level that they formed a sill or barrier that eliminated interior circulation (◆ Fig. 21.12, page 576).

The End of the Tippecanoe Sequence

By the Early Devonian, the regressing Tippecanoe Sea had retreated to the craton margin exposing an extensive lowland topography. During this regression, marine deposition was initially restricted to a few interconnected cratonic basins and, finally by the end of the Tippecanoe, to only the mobile belts surrounding the craton.

During the Early Devonian as the Tippecanoe Sea regressed, the craton experienced mild deformation resulting in the formation of many domes, arches, and basins. These structures were mostly eroded during the time the craton was exposed so that they were eventually covered by deposits from the encroaching Kaskaskia Sea.

▼ THE KASKASKIA SEQUENCE

The boundary between the Tippecanoe sequence and the overlying **Kaskaskia sequence** is marked by a major unconformity. As the Kaskaskia Sea transgressed over the low relief landscape of the craton the majority of the basal beds deposited consisted of clean, well-sorted, quartz sandstones.

The source areas for the basal Kaskaskia sandstones were primarily the eroding highlands of the Appalachian mobile belt area (◆ Fig. 21.13, page 577), exhumed Cambrian and Ordovician sandstones cropping out along the flanks of the Ozark Dome, and exposures of the Canadian Shield in the Wisconsin area. The lack of similar sands in the Silurian carbonate beds below the Tippecanoe-Kaskaskia unconformity indicates that the source areas of the basal Kaskaskia detrital rocks were submerged when the Tippecanoe sequence was deposited. Stratigraphic studies indicate that these source areas were uplifted and the Tippecanoe carbonates removed by erosion prior to the Kaskaskia transgression. Kaskaskian basal rocks elsewhere on the craton consist of carbonates that are frequently difficult to

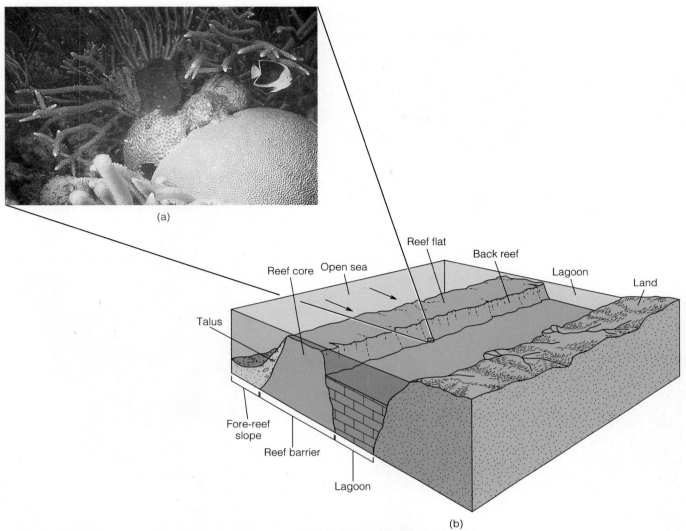

(a)

(b)

◆ FIGURE 21.9 (a) Present-day reef community showing the various reef-building organisms. (b) Diagrammatic cross section of a reef showing the various environments within the reef complex. (Photo courtesy of L. J. Lipke, Amoco Production Company.)

differentiate from the underlying Tippecanoe carbonates unless they are fossiliferous.

Except for widespread Late Devonian and Early Mississippian black shales, the majority of Kaskaskian rocks are carbonates, including reefs, and associated evaporite deposits. In many other parts of the world, such as southern England, Belgium, central Europe, Australia, and Russia, the Middle and early Late Devonian epochs were times of major reef building (see Perspective 21.2, page 578)

Reef Development in Western Canada

The Middle and Late Devonian reefs of western Canada contain large reserves of petroleum and have therefore been widely studied from outcrops and in the subsurface (◆ Fig. 21.14, page 579). These reefs began forming as the Kaskaskia Sea transgressed southward into western Canada. By the end of the Middle Devonian, they had coalesced into a large barrier-reef system that restricted the flow of oceanic water into the

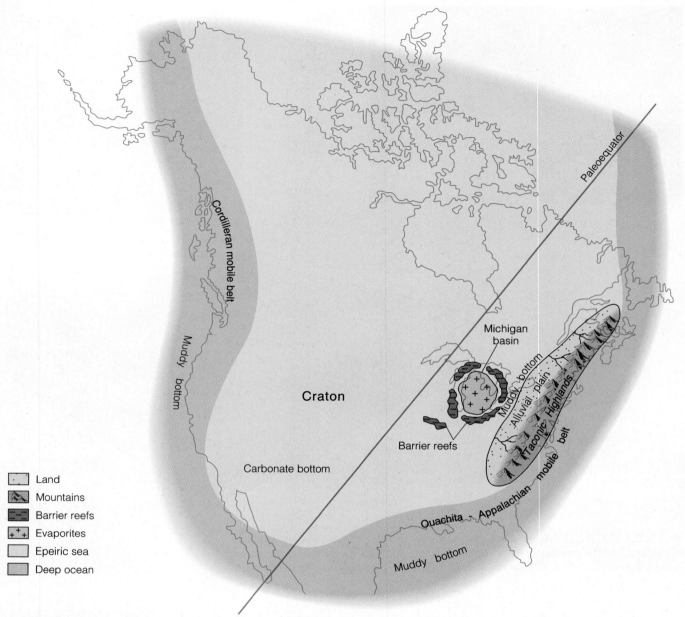

Land

Mountains

Barrier reefs

Evaporites

Epeiric sea

Deep ocean

◆ FIGURE 21.10 Paleogeography of North America during the Silurian Period. Note the development of reefs in the Michigan, Ohio, and Indiana-Illinois-Kentucky areas.

back-reef platform, thus creating conditions for evaporite precipitation. In the back reef, up to 300 m of evaporites were precipitated in much the same way as in the Michigan Basin during the Silurian (see Fig. 21.11). More than half of the world's potash, which is used in fertilizers, comes from these Devonian evapor-

Meters
0
100

(a)

Laminar
stromatoporoid
Barrier
reef
Anhydrite
Halite
Evaporite
Carbonate
Pinnacle
reef

Stromatoporoid
Barrier reef
Stromatolites
Algal
Coral algal
Crinoidal
Laminar
stromatoporoid
Clinton Fm.
Niagara Fm.

(b)

(c)

(d)

◆ FIGURE 21.11 (*a*) Generalized cross section of the northern Michigan Basin during the Silurian Period. (*b*) Stromatoporoid barrier-reef facies. (*c*) Evaporite facies. (*d*) Carbonate facies. (Photos courtesy of Sue Monroe.)

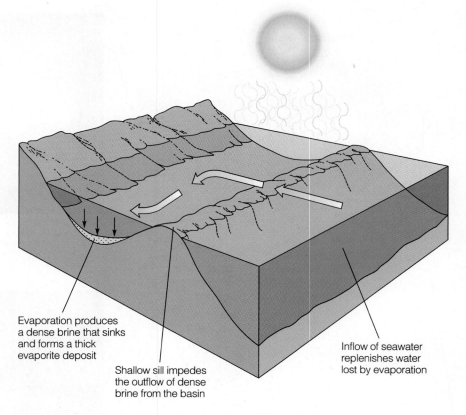

◆ FIGURE 21.12 Silled basin model for evaporite sedimentation by direct precipitation from seawater. Vertical scale is greatly exaggerated.

Evaporation produces a dense brine that sinks and forms a thick evaporite deposit

Shallow sill impedes the outflow of dense brine from the basin

Inflow of seawater replenishes water lost by evaporation

ites. By the middle of the Late Devonian, reef growth stopped in the western Canada region, although non-reef carbonate deposition continued.

Black Shales

In North America, many areas of carbonate-evaporite deposition gave way to a greater proportion of shales and coarser detrital rocks beginning in the Middle Devonian and continuing into the Late Devonian. This change to detrital deposition resulted from the formation of new source areas brought on by the mountain-building activity associated with the Acadian orogeny in North America (Fig. 21.13).

As the Devonian Period ended, a conspicuous change in sedimentation occurred over the craton with the appearance of widespread black shales. In the eastern United States, these black shales are commonly called the Chattanooga Shale, but are known by a variety of local names elsewhere (for example, New Albany Shale and Antrim Shale). Although these black shales are best developed from the cratonic margins along the Appalachian mobile belt to the Mississippi Valley, correlatives can also be found in many western states and in western Canada (◆ Fig. 21.15).

The Late Devonian–Early Mississippian black shales of North America are typically noncalcareous, thinly bedded, and usually less than 10 m thick. Fossils are usually rare, but some Upper Devonian black shales do contain rich conodont faunas with large numbers of individuals. Because most black shales lack body fossils, they are difficult to date and to correlate. In places where they can be dated, usually by conodonts (pelagic animals), acritarchs (pelagic algae), or plant spores, the lower beds are Late Devonian, and the upper beds are Early Mississippian in age.

Although the origin of these extensive black shales is still being debated, the essential features required to produce them include undisturbed anaerobic bottom water, a reduced supply of coarser detrital sediment, and high organic productivity in the overlying oxygenated waters. High productivity in the surface waters leads to a shower of organic material, which decom-

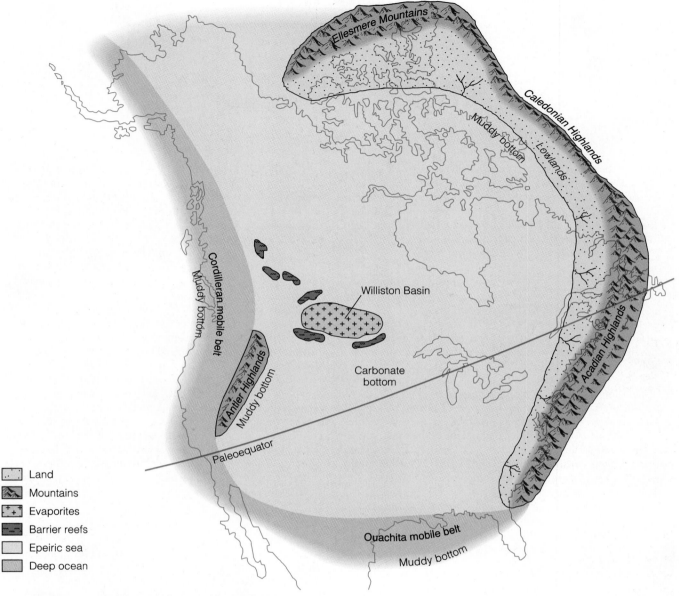

Legend:
- Land
- Mountains
- Evaporites
- Barrier reefs
- Epeiric sea
- Deep ocean

◆ FIGURE 21.13 Paleogeography of North America during the Devonian Period.

poses on the undisturbed substrate and depletes the dissolved oxygen at the sediment-water interface.

The wide extent of such apparently shallow-water black shales in North America remains puzzling. Nonetheless, these shales are rich in uranium and are an important source rock of oil and gas in the Appalachian region.

The Late Kaskaskia—A Return to Extensive Carbonate Deposition

Following deposition of the widespread Late Devonian—Early Mississippian black shales, carbonate sedimentation on the craton dominated the remainder of the Mississippian Period (◆ Fig. 21.16, page 580).

THE CANNING BASIN, AUSTRALIA—
A DEVONIAN GREAT BARRIER REEF

One of the largest and most spectacularly exposed fossil-reef complexes in the world is the Great Barrier Reef of the Canning Basin, Western Australia (◆ Fig. 1). This barrier-reef complex developed during the Middle and Late Devonian Period when the Canning Basin was covered by a tropical epeiric sea (Fig. 21.3a). The reefs are now exposed as limestone ridges that extend for some 350 km along the northern margin of the Canning Basin, but they probably continued around the present northern coastal region to join with similar reefs exposed in the Bonaparte Basin to the east.

The limestone reefs rise 50 to 100 m above the surrounding plains looking much the same as they did when the area was covered by the Devonian epeiric sea. The shales and other soft sediments deposited on the open ocean side of the reef complex (Fig. 21.9) have been eroded away, leaving the resistant limestone reefs standing as ridges.

The reefs themselves were constructed primarily by calcareous algae, stromatoporoids, and tabulate and rugose corals, which also were the main components of the other major reef complexes in the world at that time. An interesting feature of these Canning Basin reefs is the contribution of column-shaped stromatolites which are found in the reef, back-reef, and marginal-slope areas of the reef complex. Stromatolites are an unusual component because they ceased to be abundant by the end of the Proterozoic Eon. Throughout the Phanerozoic Eon, stromatolites typically formed only in areas generally inhospitable to other marine organisms.

The outcrop along Windjana Gorge beautifully reveals the various features and facies of the Devonian Great Barrier Reef complex (◆ Fig. 2). The reef core consists of unbedded limestones composed predominantly of calcareous algae, stromatoporoids, and corals. The back-reef facies (Fig. 21.9) is bedded and makes up the major part of the total reef complex environment. A diverse and abundant fauna of calcareous algae, stromatoporoids, various corals, some bivalves, gastropods, cephalopods, brachiopods, and crinoids lived in this lagoonal area behind the reef core.

In front of the reef core was the steep fore-reef slope (Fig. 21.9) that supported some organisms, including algae, sponges, and stromatoporoids. This facies contains considerable reef talus, an accumulation of debris eroded by waves from the reef front. The ocean-basin deposits contain the fossils of mainly nektonic and planktonic organisms such as fish, radiolarians, cephalopods, and conodonts.

Near the end of the Late Devonian, nearly all the reef-building organisms as well as much of the associated fauna of the Canning Basin Great Barrier Reef became extinct. As we will discuss in Chapter 22, few massive tabulate-rugose-stromatoporoid reefs are known from latest Devonian or younger rocks.

◆ FIGURE 2 Outcrop of the Devonian Great Barrier Reef along Windjana Gorge. The talus of the fore-reef area can be seen on the left side of the picture sloping away from the reef core, which is unbedded. To the right of the reef core is the back-reef facies, which is horizontally bedded. (Photos courtesy of Geoffrey Playford, The University of Queensland, Brisbane, Australia.)

◆ FIGURE 1 Aerial View of Windjana Gorge showing the Devonian Great Barrier Reef exposed as a limestone ridge.

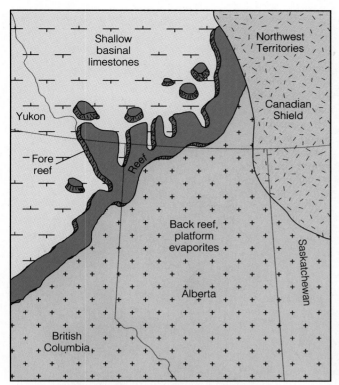

◆ FIGURE 21.14 Reconstruction of the extensive Devonian Reef complex of western Canada. These extensive reefs controlled the regional facies of the Devonian epeiric seas.

During this time, a variety of carbonate sediments were deposited in the epeiric sea as indicated by the extensive deposits of crinoidal limestones (rich in crinoid fragments), oolitic limestones, and various other limestones and dolostones (◆ Fig. 21.17). These Mississippian carbonates display cross-bedding, ripple marks, and well-sorted fossil fragments, all of which are indicative of a shallow-water environment. In addition, numerous small organic reefs occurred throughout the craton during the Mississippian. These were all much smaller than the large barrier-reef complexes that dominated the earlier Paleozoic seas.

During the Late Mississippian regression of the Kaskaskia Sea from the craton, carbonate deposition was replaced by vast quantities of detrital sediments. The resulting sandstones, particularly in the Illinois Basin, have been studied in great detail because they are excellent petroleum reservoirs. Prior to the end of the Mississippian, the Kaskaskia Sea had retreated to

◆ FIGURE 21.15 (*a*) The extent of the Upper Devonian to Lower Mississippian Chattanooga Shale and its equivalent units in North America. (*b*) Upper Devonian New Albany Shale, Button Mold Knob Quarry, Kentucky.

the craton margin, once again exposing the craton to widespread weathering and erosion that resulted in a cratonwide unconformity when the Absaroka Sea began transgressing back over the craton.

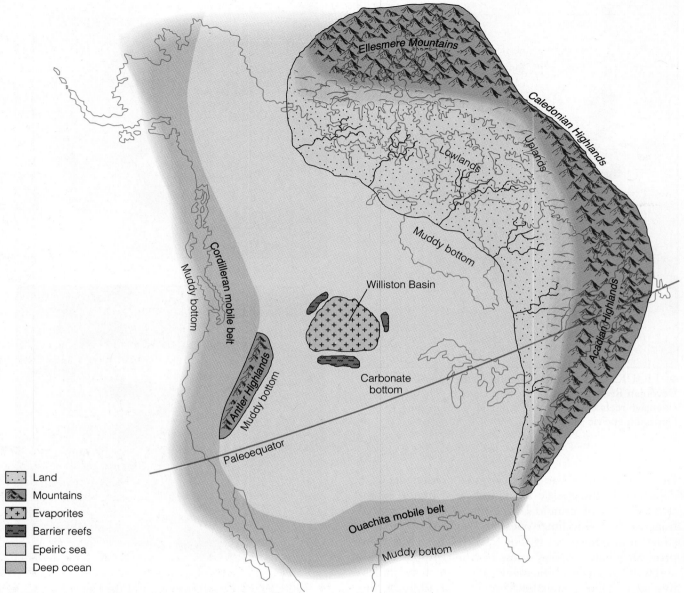

◆ FIGURE 21.16 Paleogeography of North America during the Mississippian Period.

⩢ THE ABSAROKA SEQUENCE

The **Absaroka sequence** includes uppermost Mississippian through Lower Jurassic rocks. In this chapter, however, we will only be concerned with the Paleozoic rocks of the Absaroka sequence. The extensive uncon-

formity separating the Kaskaskia and Absaroka sequences essentially divides the strata into the North American Mississippian and Pennsylvanian systems. These two systems are equivalent to the European Lower and Upper Carboniferous systems, respectively. The rocks of the Absaroka sequence are not only

◆ FIGURE 21.17 Mississippian limestones exposed near Bowling Green, Kentucky.

different from those of the Kaskaskia sequence, but they are also the result of quite different tectonic regimes affecting the North American craton.

The lowermost sediments of the Absaroka sequence are confined to the margins of the craton. These deposits are generally thickest in the east and southeast, near the emerging highlands of the Appalachian and Ouachita mobile belts, and thin westward onto the craton. The lithologies also reveal lateral changes from nonmarine detrital rocks and coals in the east, through transitional marine-nonmarine beds, to largely marine detrital rocks and limestones farther west (◆ Fig. 21.18).

Cyclothems

One of the characteristic features of Pennsylvanian rocks is their cyclical pattern of alternating marine and nonmarine strata. Such rhythmically repetitive sedimentary sequences are known as **cyclothems.** They result from repeated alternations of marine and nonmarine environments, usually in areas of low relief. Though seemingly simple, cyclothems reflect a delicate interplay between nonmarine deltaic and shallow-marine interdeltaic and shelf environments.

For purposes of illustration, we can look at a typical coal-bearing cyclothem from the Illinois Basin (◆ Fig. 21.19). Such a cyclothem contains nonmarine units, capped by a coal and overlain by marine units. Figure 21.19 shows the depositional environments that produced the cyclothem. The initial units represent deltaic and fluvial deposits. Above them is an underclay that frequently contains root casts from the plants and trees that comprise the overlying coal. The coal bed results from accumulations of plant material and is overlain by marine units of alternating limestones and shales, usually with an abundant marine invertebrate fauna. The marine cycle ends with an erosion surface. A new cyclothem begins with a nonmarine deltaic sandstone. It should be noted, however, that all of the beds illustrated in the idealized cyclothem are not always preserved due to abrupt changes from marine to nonmarine conditions or to removal of some units by erosion.

Cyclothems represent transgressive and regressive sequences with an erosional surface separating one cyclothem from another. Thus, an idealized cyclothem passes upward from fluvial-deltaic deposits, through coals, to detrital shallow-water marine sediments, and finally to limestones typical of an open marine environment.

Such regularity and cyclicity in sedimentation over a large area requires an explanation. The hypothesis currently favored by most geologists is a rise and fall of sea level related to advances and retreats of Gondwanan continental glaciers. When the Gondwanan ice sheets advanced, sea level dropped, and when they melted, sea level rose. Late Paleozoic cyclothem activity on all of the cratons closely corresponds to Gondwanan glacial-interglacial cycles.

Cratonic Uplift—The Ancestral Rockies

Recall that cratons are stable areas, and when they do experience deformation, it is usually only mild and results in structures of great dimensions. The Pennsylvanian Period, however, was a time of unusually severe deformation of the craton and resulted in uplifts of sufficient magnitude to expose Precambrian basement rocks. In addition to newly formed highlands and basins, many previously formed arches and domes, such as the Cincinnati Arch, Nashville Dome, and Ozark Dome, were also reactivated (Fig. 21.1).

During the Pennsylvanian, the area of greatest deformation occurred in the southwestern part of the North American craton where a series of fault-bounded uplifted blocks formed the **Ancestral Rockies** (◆ Fig. 21.20, page 584). These mountain ranges had diverse geologic histories and were not all elevated at the same time. Uplift of these mountains, some of which were elevated more than 2 km along near-vertical faults, resulted in the erosion of the overlying Paleozoic sediments and exposure of the Precambrian igneous and metamorphic basement rocks. As the mountains eroded, tremendous quantities of coarse, red arkosic sand and conglomerate

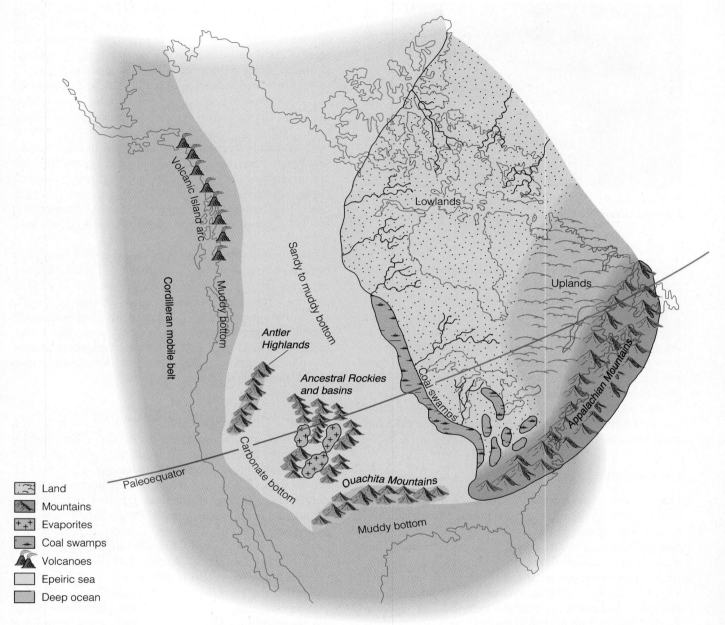

Legend:
- Land
- Mountains
- Evaporites
- Coal swamps
- Volcanoes
- Epeiric sea
- Deep ocean

Map labels: Volcanic Island arc, Cordilleran mobile belt, Muddy Bottom, Sandy to muddy bottom, Antler Highlands, Ancestral Rockies and basins, Carbonate bottom, Paleoequator, Ouachita Mountains, Muddy bottom, Lowlands, Uplands, Coal swamps, Appalachian Mountains

◆ FIGURE 21.18 Paleogeography of North America during the Pennsylvanian Period.

were deposited in the surrounding basins. Currently these sediments are preserved in many areas, including the Garden of the Gods near Colorado Springs and at the Red Rocks Amphitheatre near Morrison, Colorado.

Intracratonic mountain ranges are unusual, and their cause has long been debated. It is now thought that the collision of Gondwana with Laurasia (Fig. 21.4a) produced great stresses in the southwestern region of the North American craton. These crustal stresses were relieved by faulting which resulted in uplift of cratonic blocks and downwarp of adjacent basins, forming a series of ranges and basins.

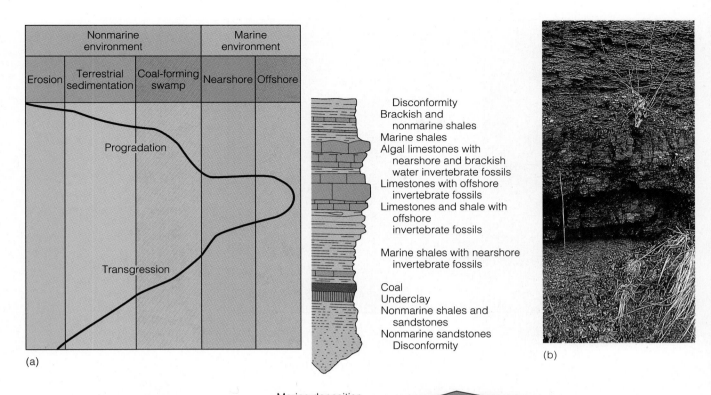

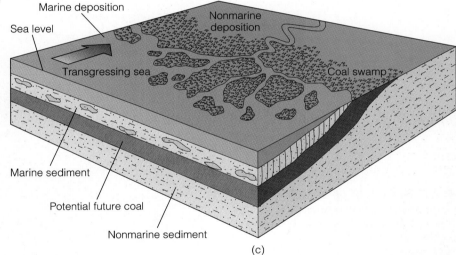

◆ FIGURE 21.19 (*a*) Columnar section of a complete cyclothem. (*b*) Pennsylvanian coal bed, West Virginia. (Photo courtesy of Wayne E. Moore.) (*c*) Reconstruction of the environment of a Pennsylvanian coal-forming swamp.

The Late Absaroka—More Evaporite Deposits and Reefs

While the various intracratonic basins were filling with sediment during the Late Pennsylvanian, the Absaroka Sea slowly began retreating from the craton. During the Early Permian, the Absaroka Sea occupied a narrow region from Nebraska through West Texas (◆ Fig. 21.21). By the Middle Permian, the sea had retreated to West Texas and southern New Mexico. The thick evaporite deposits in Kansas and Oklahoma provide evidence of the restricted nature of the Absaroka Sea during the Early and Middle Permian and its southwestward retreat from the central craton.

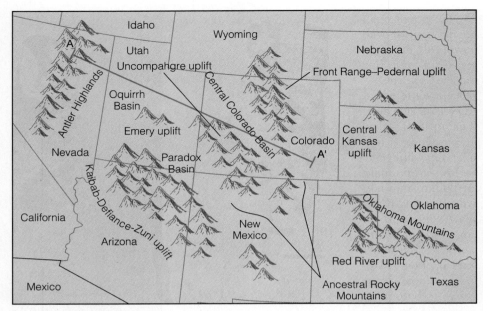

(a)

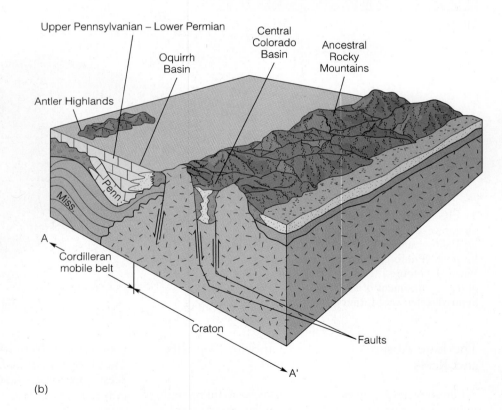

◆ FIGURE 21.20 (*a*) Location of the principal Pennsylvanian highland areas and basins of the southwestern part of the craton. (*b*) Cross section of the Ancestral Rockies, which were elevated by faulting during the Pennsylvanian Period. Erosion of these mountains produced coarse, red-colored sediments that were deposited in the adjacent basins.

(b)

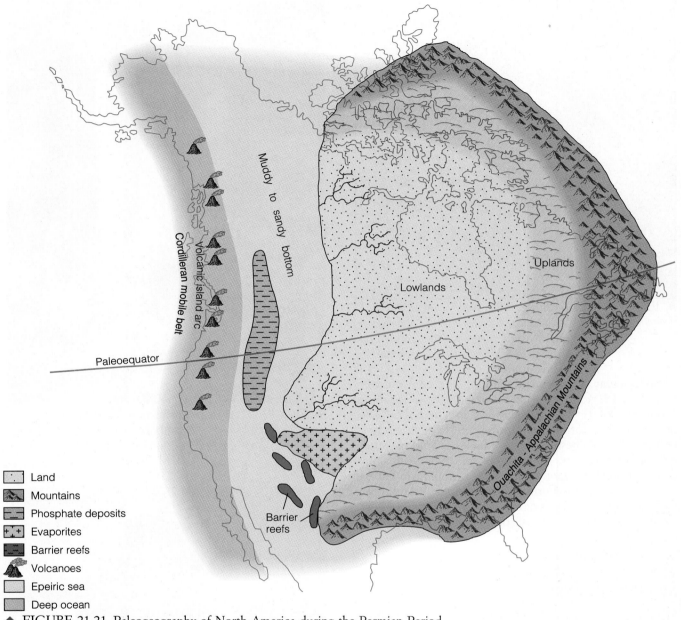

Land

Mountains

Phosphate deposits

Evaporites

Barrier reefs

Volcanoes

Epeiric sea

Deep ocean

◆ FIGURE 21.21 Paleogeography of North America during the Permian Period.

During the Middle and Late Permian, the Absaroka Sea was restricted to West Texas and southern New Mexico, forming an interrelated complex of lagoonal, reef, and open-shelf environments (◆ Fig. 21.22). Three basins separated by two submerged platforms formed in this area during the Permian. Massive reefs grew around the basin margins (◆ Fig. 21.23), while limestones, evaporites, and red beds were deposited in the lagoonal areas behind the reefs. As the barrier reefs grew and the passageways between the basins became more restricted, Late Permian evaporites gradually filled the individual basins with deposits up to 600 m thick.

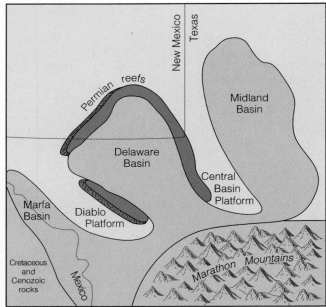

◆ FIGURE 21.22 Location of the West Texas Permian basins and surrounding reefs.

◆ FIGURE 21.23 A reconstruction of the Middle Permian Capitan Limestone reef environment. Shown are brachiopods, corals, bryozoans, and large glass sponges.

Spectacular deposits representing the geologic history of this region can be seen today in the Guadalupe Mountains of Texas and New Mexico where the Capitan Limestone forms the caprock of these mountains (◆ Fig. 21.24). These reefs have been extensively studied because of the tremendous oil production that comes from this region.

By the end of the Permian Period, the Absaroka Sea had retreated from the craton, and continental red beds were deposited over most of the southwestern and eastern region.

▼ HISTORY OF THE PALEOZOIC MOBILE BELTS

Having examined the Paleozoic history of the craton, we now turn out attention to the orogenic activity in the mobile belts. The mountain building that occurred during this time had a profound influence on the climate and sedimentary history of the craton. In addition, it was part of the global tectonic regime that sutured the continents together, forming Pangaea by the end of the Paleozoic Era. We will first examine the Appalachian mobile belt, where the first Phanerozoic orogeny began during the Middle Ordovician.

◆ FIGURE 21.24 The prominent light-colored Capitan Limestone forms the caprock of the Guadalupe Mountains. The Capitan Limestone is rich in fossil corals and associated reef organisms. (Photo courtesy of Bill Cornell, The University of Texas at El Paso.)

Appalachian Mobile Belt

Throughout Sauk time, the Appalachian region was a broad, passive, continental margin. Sedimentation was closely balanced by subsidence as thick, shallow marine sands were succeeded by extensive carbonate deposits. During this time, the **Iapetus Ocean** was widening as a result of movement along a divergent plate boundary (◆ Fig. 21.25a).

Taconic Orogeny

Beginning with the subduction of the Iapetus plate beneath Laurentia (an oceanic-continental convergent plate boundary), the Appalachian mobile belt was born (◆ Fig. 21.25b). The resulting **Taconic orogeny,** named after the present-day Taconic Mountains of eastern New York, central Massachusetts, and Vermont, was the first of several orogenies to affect the Appalachian region.

The Appalachian mobile belt can be divided into two depositional environments. The first is the extensive, shallow-water carbonate platform that formed the broad eastern continental shelf and stretched from Newfoundland to Alabama (Fig. 21.25a). It formed during the Sauk Sea transgression onto the craton when carbonates were deposited in a large, shallow sea. The shallow-water depth on the platform is indicated by

stromatolites, desiccation cracks, and other sedimentary structures.

Carbonate deposition ceased along the east coast during the Middle Ordovician and was replaced by deepwater deposits characterized by thinly bedded black shales, graded beds, coarse sandstones, graywackes, and associated volcanics. This suite of sediments marks the onset of mountain building, in this case, the Taconic orogeny. The subduction of the Iapetus plate beneath Laurentia resulted in volcanism and downwarping of the carbonate platform, forming an area where sediments accumulated (Fig. 21.25b).

The final piece of evidence for the Taconic orogeny is the development of a large **clastic wedge,** an extensive accumulation of mostly detrital sediments that are deposited adjacent to an uplifted area. These deposits are thickest and coarsest nearest the highland area and become thinner and finer grained away from the source area, eventually grading into the carbonate cratonic facies (◆ Fig. 21.26). The clastic wedge resulting from the erosion of the Taconic Highlands is referred to as the **Queenston Delta.** Careful mapping and correlation of these deposits indicate that more than 600,000 km^3 of rock were eroded from the Taconic Highlands. Based on this figure, geologists estimate the Taconic Highlands were at least 4,000 m high.

◆ FIGURE 21.25 Evolution of the Appalachian mobile belt from the Late Proterozoic to the Late Ordovician. (*a*) During the Late Proterozoic to the Early Ordovician, the Iapetus Ocean was opening along a divergent plate boundary. Both the east coast of Laurentia and the west coast of Baltica were passive continental margins where large carbonate platforms existed. (*b*) Beginning in the Middle Ordovician, the passive margins of Laurentia and Baltica became oceanic-continental plate boundaries resulting in orogenic activity.

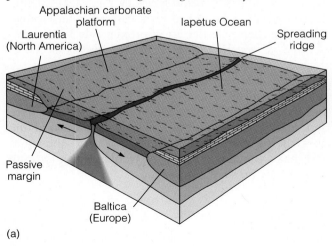

(a)

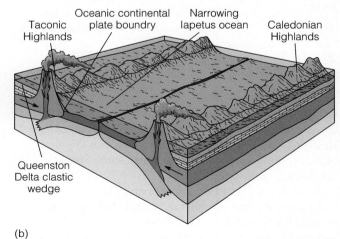

(b)

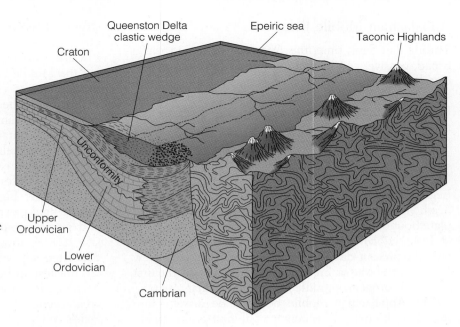

◆ FIGURE 21.26 Reconstruction of the Taconic Highlands and Queenston Delta clastic wedge. The clastic wedge consists of thick coarse-grained detrital sediments nearest the highlands and thins laterally into finer-grained sediments on the craton.

Caledonian Orogeny

The Caledonian mobile belt extends along the western border of Baltica and includes the present-day countries of Scotland, Ireland, and Norway (Fig. 21.2c). During the Middle Ordovician, subduction along the boundary between the Iapetus plate and Baltica (Europe) began, forming a mirror image of the convergent plate boundary off the east coast of Laurentia (North America).

The culmination of the **Caledonian orogeny** occurred during the Late Silurian and Early Devonian with the formation of a mountain range along the margin of Baltica. Red-colored sediments deposited along the front of the Caledonian highlands formed a large clastic wedge. These deposits are known as the Old Red Sandstone.

Acadian Orogeny

The third Paleozoic orogeny to affect Laurentia and Baltica began during the Late Silurian and concluded at the end of the Devonian Period. The **Acadian orogeny** affected the Appalachian mobile belt from Newfoundland to Pennsylvania as sedimentary rocks were folded and thrust against the craton.

As with the preceding Taconic and Caledonian orogenies, the Acadian orogeny occurred along an oceanic-continental convergent plate boundary. As the northern Iapetus Ocean continued to close during the Devonian, the plate carrying Baltica finally collided with Lauren-

tia, forming a continental-continental convergent plate boundary along the zone of collision (Fig. 21.3a).

Weathering and erosion of the Acadian Highlands produced a thick clastic wedge called the **Catskill Delta,** named for the Catskill Mountains in northern New York where it is well exposed. The Catskill Delta, composed of red, coarse conglomerates, sandstones, and shales, contains nearly three times as much sediment as the Queenston Delta.

The Devonian rocks of New York are among the best studied on the continent. A cross section of the Devonian strata clearly reflects an eastern source (Acadian Highlands) for the Catskill facies (◆ Fig. 21.27). These clastic rocks can be traced from eastern Pennsylvania, where the coarse clastics are approximately 3 km thick, to Ohio, where the deltaic facies are only about 100 m thick and consist of cratonic shales and carbonates.

The red beds of the Catskill Delta derive their color from the hematite found in the sediments. Plant fossils and oxidation of the hematite indicate the beds were deposited in a continental environment. Toward the west, the red beds grade laterally into gray sandstones and shales containing fossil tree trunks, which indicate a swamp or marsh environment.

The Old Red Sandstone

The red beds of the Catskill Delta have a European counterpart in the Devonian Old Red Sandstone of the

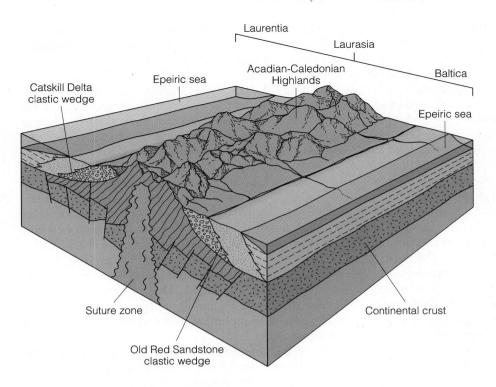

Laurentia

Laurasia

Acadian-Caledonian
Highlands

Baltica

Epeiric sea

Catskill Delta
clastic wedge

Epeiric sea

Suture zone

Continental crust

Old Red Sandstone
clastic wedge

◆ FIGURE 21.27 Cross section showing the area of collision between Laurentia and Baltica. Note the bilateral symmetry of the Catskill Delta clastic wedge and the Old Red Sandstone and their relationship to the Acadian and Caledonian Highlands.

British Isles (Fig. 21.27). The Old Red Sandstone was a Devonian clastic wedge that grew eastward from the Caledonian Highlands onto the Baltica craton. The Old Red Sandstone, just like its North American Catskill counterpart, contains numerous fossils of freshwater fish, early amphibians, and land plants.

By the end of the Devonian Period, Baltica and Laurentia were sutured together, forming Laurasia (Fig. 21.27). The red beds of the Catskill Delta can be traced north, through Canada and Greenland, to the Old Red Sandstone of the British Isles and into northern Europe. These beds were deposited in similar environments along the flanks of developing mountain chains formed by tectonic forces at convergent plate boundaries.

Geologists now think that the Taconic, Caledonian, and Acadian orogenies were all part of the same major orogenic event related to the closing of the Iapetus Ocean (Figs. 21.25 and 21.27). This event began with paired oceanic-continental convergent plate boundaries during the Taconic and Caledonian orogenies and culminated with a continental-continental convergent plate boundary during the Acadian orogeny as Laurentia and Baltica became sutured. Following this, the Hercynian-Alleghenian orogeny began, followed by orogenic activity in the Ouachita mobile belt.

Hercynian-Alleghenian Orogeny

The Hercynian mobile belt of southern Europe and the Appalachian and Ouachita mobile belts of North America mark the zone along which Europe (part of Laurasia) collided with Gondwana (Fig. 21.3). While Gondwana and southern Laurasia collided during the Pennsylvanian and Permian in the area of the Ouachita mobile belt, eastern Laurasia (Europe and southeastern North America) joined together with Gondwana (Africa) as part of the *Hercynian-Alleghenian orogeny* (Fig. 21.4).

Initial contact along the Hercynian mobile belt between eastern Laurasia and Gondwana began during the Mississippian Period. The greatest deformation occurred during the Pennsylvanian and Permian periods. This event is referred to as the **Hercynian orogeny.** The central and southern parts of the Appalachian mobile belt (from New York to Alabama) were folded and thrust toward the craton as eastern Laurasia and Gondwana were sutured. This event in North America is referred to as the **Alleghenian orogeny.**

Cordilleran Mobile Belt

During the Late Proterozoic and Early Paleozoic, the Cordilleran area was a passive continental margin along which extensive continental shelf sediments were deposited. Thick sections of marine sediments graded laterally into thin cratonic units as the Sauk Sea transgressed onto the craton. Beginning in the Middle Paleozoic, an island arc (called the *Klamath Arc*) formed on the western margin of the craton. A collision between this eastward-moving island arc and the western border of the craton occurred during the Late Devonian and Early Mississippian resulting in a highland area.

This orogenic event, called the **Antler orogeny,** was caused by subduction and resulted in the closing of the narrow ocean basin that separated the Klamath island arc system from the craton (◆ Fig. 21.28). Erosion of the Antler Highlands produced large quantities of sediment that were deposited to the east in the epeiric sea covering the craton and to the west in the deep sea. The Antler orogeny was the first in a series of orogenic events to affect the Cordilleran mobile belt. During the Mesozoic and Cenozoic, this area was the site of major tectonic activity by both oceanic-continental convergence and accretion of various terranes.

Ouachita Mobile Belt

The Ouachita mobile belt extends for approximately 2,100 km from the subsurface of Mississippi to the Marathon region of Texas. Approximately 80% of the former mobile belt is buried beneath a Mesozoic and Cenozoic sedimentary cover. The two major exposed areas in this region are the Ouachita Mountains of Oklahoma and Arkansas and the Marathon Mountains

◆ FIGURE 21.28 Reconstruction of the Cordilleran mobile belt during the Early Mississippian, showing the effects of the Antler orogeny.

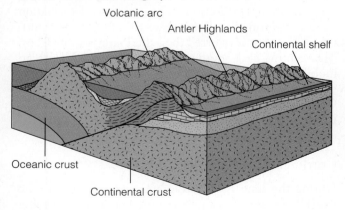

of Texas. Based on extensive study of the subsurface geology and the Ouachita and Marathon mountains, geologists have learned that this region had a very complex geologic history (◆ Fig. 21.29).

During the Late Proterozoic to Early Mississippian, shallow-water detrital and carbonate sediments were slowly deposited on a broad continental shelf, while in the deeper-water portion of the adjoining mobile belt, bedded cherts and shales were also slowly accumulating (Fig. 21.29a). Beginning in the Mississippian Period, the rate of sedimentation increased dramatically as the region changed from a passive continental margin to an active convergent plate boundary (Fig. 21.29b). Rapid deposition of sediments continued into the Pennsylvanian with the formation of a clastic wedge that thickened to the south. As much as 16,000 m of Mississippian- and Pennsylvanian-aged rocks crop out in the Ouachita Mountains. The formation of a clastic wedge marks the beginning of uplift of the area and formation of a mountain range during the **Ouachita orogeny.**

Thrusting of sediments continued throughout the Pennsylvanian and Early Permian as a result of the compressive forces generated along the zone of subduction as Gondwana collided with Laurasia (Fig. 21.29c). The collision of Gondwana and Laurasia is marked by the formation of a large mountain range, most of which was eroded during the Mesozoic Era. Only the rejuvenated Ouachita and Marathon mountains remain of this once lofty mountain range.

The Ouachita deformation was part of the general worldwide tectonic activity that occurred when Gondwana united with Laurasia. The Hercynian, Appalachian, and Ouachita mobile belts were continuous and marked the southern boundary of Laurasia (Fig. 21.4). The tectonic activity that resulted in the uplift in the Ouachita mobile belt was very complex and involved not only the collision of Laurasia and Gondwana, but several microplates between the continents that eventually became part of Central America. The compressive forces impinging on the Ouachita mobile belt also affected the craton by causing broad uplift of the southwestern part of North America.

▼ PALEOZOIC MICROPLATES AND THE FORMATION OF PANGAEA

We have generally presented the geologic history of the mobile belts surrounding the Paleozoic continents in terms of subduction along convergent plate bound-

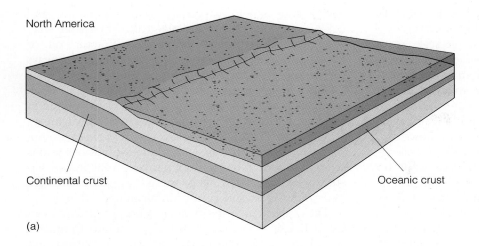

(a)

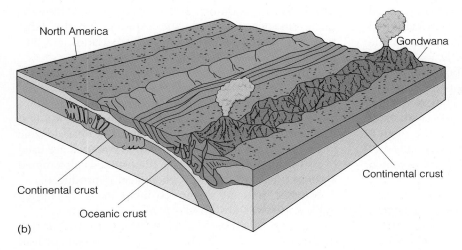

(b)

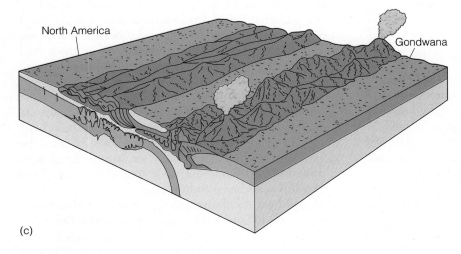

(c)

◆ FIGURE 21.29 Plate tectonic model for deformation of the Ouachita mobile belt. (*a*) Depositional environment prior to the beginning of orogenic activity. (*b*) Incipient continental collision between North America and Gondwana began during the Mississippian to Pennsylvanian. (*c*) Continental collision continued during the Pennsylvanian Period.

aries. It is becoming increasingly clear, however, that accretion along the continental margins is more complicated than the somewhat simple, large-scale plate interactions that we have described. Geologists now recognize that numerous terranes or microplates existed during the Paleozoic and were involved in the orogenic events that occurred during that time.

A careful examination of the Paleozoic global paleogeographic maps (Figs. 21.2, 21.3, and 21.4) shows that there were numerous microplates, and their location and role during the formation of Pangaea must be taken into account. For example, the small continent of Avalonia is composed of some coastal parts of New England, southern New Brunswick, much of Nova Scotia, the Avalon Peninsula of eastern Newfoundland, southeastern Ireland, Wales, England, and parts of Belgium and northern France. This microplate existed as a separate continent during the Ordovician and collided with Baltica during the Silurian and Laurentia during the Devonian (Figs. 21.2 and 21.3).

Florida and parts of the eastern seaboard of North America make up the Piedmont microplate that was part of the larger Gondwana continent. This microplate became sutured to Laurasia during the Pennsylvanian Period. Numerous microplates occupied the region between Gondwana and Laurasia that eventually became part of Central America during the Pennsylvanian collision between these continents.

Thus, while the basic history of the formation of Pangaea during the Paleozoic remains the same, geologists now realize that microplates also played an important role. Furthermore, the recognition of terranes within mobile belts helps explain some previously anomalous geologic situations.

▼ PALEOZOIC MINERAL RESOURCES

Paleozoic-aged rocks contain a variety of important mineral resources, including energy resources and metallic and nonmetallic mineral deposits. Important sources of industrial or silica sand are the Upper Cambrian Jordan Sandstone of Minnesota and Wisconsin, the Middle Ordovician St. Peter Sandstone, the Lower Silurian Tuscarora Sandstone in Pennsylvania and Virginia, the Devonian Ridgeley Formation in West Virginia, Maryland, and Pennsylvania, and the Devonian Sylvania Sandstone in Michigan.

Silica sand has a variety of uses, including the manufacture of glass, refractory bricks for blast fur-

naces, and molds for casting iron, aluminum, and copper alloys. Some silica sands, called hydraulic fracturing sands, are pumped into wells to fracture oil- or gas-bearing rocks and provide permeable passageways for the oil or gas to migrate to the well.

Thick deposits of Silurian evaporites, mostly rock salt (NaCl) and rock gypsum ($CaSO_4 \cdot H_2O$) altered to rock anhydrite ($CaSO_4$), underlie parts of Michigan, Ohio, New York, and adjacent areas in Ontario, Canada. These rocks are important sources of various salts. In addition, barrier and pinnacle reefs in carbonate rocks associated with these evaporites are the reservoirs for oil and gas in Michigan and Ohio.

The Zechstein evaporites of Europe extend from Great Britain across the North Sea and into Denmark, the Netherlands, Germany, and eastern Poland and Lithuania. In addition to the evaporites themselves, Zechstein deposits form the caprock for the large reservoirs of the gas fields of the Netherlands and part of the North Sea region.

Other important evaporite mineral resources include those of the Permian Delaware Basin of West Texas and New Mexico, and Devonian evaporites in the Elk Point basin of Canada. In Michigan, gypsum is mined and used in the construction of wallboard. Late Paleozoic limestones from many areas in North America are used in the manufacture of cement. Limestone is also mined and used in blast furnaces when steel is produced.

Metallic mineral resources including tin, copper, gold, and silver are known from Late Paleozoic-aged rocks, especially those that have been deformed during mountain building. The host rocks for deposits of lead and zinc in southeast Missouri are Cambrian dolostones, although some Ordovician rocks contain these metals as well. These deposits have been mined since 1720 but have been largely depleted. Now most lead and zinc mined in Missouri come from Mississippian-aged sedimentary rocks.

The Silurian Clinton Formation crops out from Alabama north to New York, and equivalent rocks are found in Newfoundland. This formation has been mined for iron in many places. In the United States, the richest ores and most extensive mining occurred near Birmingham, Alabama, but only a small amount of ore is currently produced in that area.

Petroleum and natural gas are recovered in commercial quantities from rocks ranging from the Devonian through Permian. For example, Devonian-aged rocks in the Michigan Basin, Illinois Basin, and the Williston

Basin of Montana, South Dakota, and adjacent parts of Alberta, Canada, have yielded considerable amounts of hydrocarbons. Permian reefs and other strata in the western United States, particularly Texas, have also been productive rocks.

Although Permian-aged coal beds are known from several areas including Asia, Africa, and Australia, much of the coal in North America and Europe is Pennsylvanian (Late Carboniferous). Large areas in the Appalachian region and the midwestern United States are underlain by vast coal deposits (◆ Fig. 21.30). These coal deposits formed from the lush vegetation such as the nonflowering seed-bearing plants called *lycopsids* that were the dominant plants in Pennsylvanian coal swamps (see Chapter 22).

Much of this coal is characterized as bituminous coal, which contains about 80% carbon. It is a dense,

black coal that has been so thoroughly altered that plant remains can be seen only rarely. Bituminous coal is used to make *coke,* a hard, gray substance made up of the fused ash of bituminous coal. Coke is used to fire blast furnances during the production of steel.

Some of the Pennsylvanian coal from North America is called *anthracite,* a metamorphic type of coal containing up to 98% carbon. Most anthracite is in the Appalachian region (Fig. 21.30). It is an especially desirable type of coal because it burns with a smokeless flame and it yields more heat per unit volume than other types of coal. Unfortunately, it is the least common type of coal, so much of the coal used in the United States is bituminous.

◆ FIGURE 21.30 Distribution of coal deposits in the United States. The age of the coals in the midwestern states and the Appalachian region are mostly Pennsylvanian, whereas those in the west are mostly Cretaceous and Tertiary.

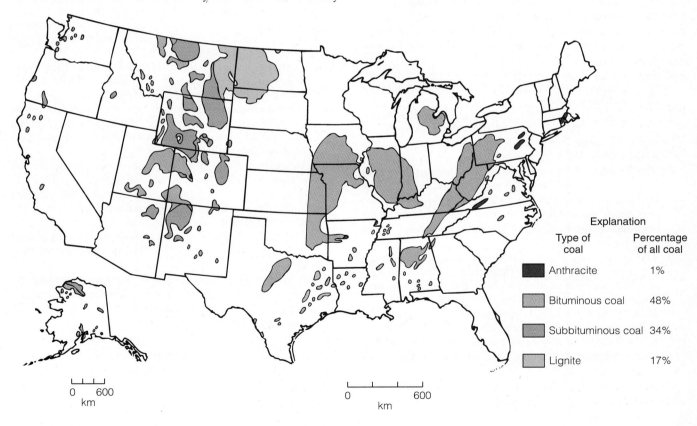

Type of coal	Percentage of all coal
Anthracite	1%
Bituminous coal	48%
Subbituminous coal	34%
Lignite	17%

Chapter Summary

● Tables 21.1 and 21.2 provide a summary of the geologic history of the North American craton and mobile belts as well as global events and sea level changes during the Paleozoic Era.

1. Six major continents existed at the beginning of the Paleozoic Era; four of them were located near the paleoequator.

2. During the Early Paleozoic (Cambrian-Silurian), Laurentia was moving northward and Gondwana moved to a south polar location, as indicated by tillite deposits.

3. During the Late Paleozoic, Baltica and Laurentia collided, forming Laurasia. Siberia and Kazakhstania collided and finally were sutured to Laurasia. Gondwana moved over the South Pole and experienced several glacial-interglacial periods, resulting in global changes in sea level and transgressions and regressions along the low-lying craton margins.

4. Laurasia and Gondwana underwent a series of collisions beginning in the Carboniferous. During the Permian, the formation of Pangaea was completed. Surrounding the supercontinent was a global ocean, Panthalassa.

5. Most continents consist of two major components: a relatively stable craton over which epeiric seas transgressed and regressed, surrounded by mobile belts in which mountain building took place.

6. The geologic history of North America can be divided into cratonic sequences that reflect cratonwide transgressions and regressions.

7. The Sauk Sea was the first major transgression onto the craton. At its peak it covered the craton except for a series of large, northeast-southwest trending islands called the Transcontinental Arch.

8. The Tippecanoe sequence began with deposition of an extensive sandstone over the exposed and eroded Sauk landscape. During Tippecanoe time, extensive carbonate deposition occurred. In addition, large barrier reefs enclosed basins, resulting in evaporite deposition within these basins.

9. The basal beds of the Kaskaskia sequence that were deposited on the exposed Tippecanoe surface consisted of either sandstones, derived from the eroding Taconic Highlands, or carbonate rocks.

10. Most of the Kaskaskia sequence was dominated by carbonates and associated evaporites. The Devonian Period was a time of major reef building in western Canada, southern England, Belgium, Australia, and Russia.

11. A persistent and widespread black shale, the Chattanooga Shale and its equivalents, was deposited over a large area of the craton during the Late Devonian and Early Mississippian.

12. The Mississippian Period was dominated for the most part by carbonate deposition.

13. Transgressions and regressions over the low-lying craton resulted in cyclothems and the formation of coals during the Pennsylvanian Period.

14. Cratonic mountain building occurred during the Pennsylvanian Period, and thick nonmarine detrital rocks and evaporites were deposited in the intervening basins.

15. By the Early Permian, the Absaroka Sea occupied a narrow zone of the south-central craton. Here, several large reefs and associated evaporites developed. By the end of the Permian Period, the Absaroka Sea had retreated from the craton.

16. The eastern edge of North America was a stable carbonate platform during Sauk time. During Tippecanoe time an oceanic-continental convergent plate boundary formed, resulting in the Taconic orogeny, the first of several orogenies to affect the Appalachian mobile belt. The newly formed Taconic Highlands shed sediments into the western epeiric sea, producing a clastic wedge called the Queenston Delta.

17. The Caledonian, Acadian, Hercynian, and Alleghenian orogenies were all part of the global tectonic activity resulting from the assembly of Pangaea.

18. The Cordilleran mobile belt was the site of a minor Devonian orogeny called the Antler orogeny during which deep-water sediments were thrust eastward over shallow-water sediments.

19. Mountain building occurred in the Ouachita mobile belt during the Pennsylvanian and Early Permian. This tectonic activity was partly responsible for the cratonic uplift that occurred in the southwest, producing the Ancestral Rockies.

20. During the Paleozoic Era, numerous microplates existed and played an important role in the formation of Pangaea.

21. Paleozoic age rocks contain a variety of mineral resources including building stone, limestone for cement, silica sand, evaporites, petroleum, coal, iron ore, lead, zinc, and other metallic deposits.

Important Terms

Absaroka sequence
Acadian orogeny
Alleghenian orogeny
Ancestral Rockies
Antler orogeny
Appalachian mobile belt
Baltica
Caledonian orogeny
Catskill Delta
China
clastic wedge
Cordilleran mobile belt

cratonic sequence
cyclothem
epeiric sea
Franklin mobile belt
Gondwana
Hercynian orogeny
Iapetus Ocean
Kaskaskia sequence
Kazakhstania
Laurasia
Laurentia
mobile belt

organic reef
Ouachita mobile belt
Ouachita orogeny
Panthalassa Ocean
Queenston Delta
Sauk sequence
sequence stratigraphy
Siberia
Taconic orogeny
Tippecanoe sequence
Transcontinental Arch

●—————————————————————————————————————●

Review Questions

1. An elongated area marking the site of former mountain building is a:
   a. ___ craton; b. ___ platform; c. ___ shield;
   d. ___ epeiric sea; e. ___ mobile belt.
2. Which of the following was not a Paleozoic continent?
   a. ___ Gondwana; b. ___ Baltica; c. ___ Kazakhstania;
   d. ___ Eurasia; e. ___ Laurentia.
3. A major transgressive-regressive cycle bounded by cratonwide unconformities is (a)n:
   a. ___ cratonic sequence; b. ___ epeiric sea; c. ___ orogeny; d. ___ biostratigraphic unit; e. ___ cyclothem.
4. During deposition of the Sauk sequence, the only area above sea level besides a portion of the Canadian Shield was the:
   a. ___ Appalachian mobile belt; b. ___ Transcontinental Arch; c. ___ Taconic Highlands; d. ___ Queenston Delta; e. ___ cratonic margin.
5. The predominant cratonic lithologies of the Tippecanoe sequence are:
   a. ___ coals; b. ___ sandstones and shales;
   c. ___ graywackes and cherts; d. ___ volcanics;
   e. ___ evaporites and reef carbonates.
6. The Paleozoic ocean separating Laurentia from Siberia and Baltica was the:
   a. ___ Panthalassa; b. ___ Tethys; c. ___ Caledonian;
   d. ___ Iapetus; e. ___ Kaskaskia.
7. Which was the first major transgressive sequence onto the North American craton?
   a. ___ Sauk; b. ___ Tippecanoe; c. ___ Kaskaskia;
   d. ___ Absaroka; e. ___ Zuni.
8. Extensive continental glaciation of the Gondwana continent occurred during which period?
   a. ___ Devonian; b. ___ Silurian; c. ___ Carboniferous;
   d. ___ Cambrian; e. ___ Permian.

9. Extensive cratonic black shales were deposited during which two periods?
   a. ___ Late Silurian–Early Devonian; b. ___ Late Devonian–Early Mississippian; c. ___ Late Mississippian–Early Pennsylvanian; d. ___ Late Pennsylvanian–Early Permian; e. ___ Late Permian–Early Triassic.
10. Rhythmically repetitive sedimentary sequences are:
    a. ___ tillites; b. ___ cyclothems; c. ___ orogenies;
    d. ___ reefs; e. ___ evaporites.
11. Uplift in the southwestern part of the craton during the Late Absaroka resulted in which mountainous region?
    a. ___ Ancestral Rockies; b. ___ Antler Highlands;
    c. ___ Appalachians; d. ___ Marathon; e. ___ none of these.
12. The Taconic orogeny resulted from what type of plate movement?
    a. ___ oceanic-oceanic convergent; b. ___ oceanic-continental convergent; c. ___ continental-continental convergent; d. ___ divergent; e. ___ transform.
13. Which orogeny took place along the western margin of Baltica during the Silurian Period?
    a. ___ Caledonian; b. ___ Acadian; c. ___ Taconic;
    d. ___ Antler; e. ___ Alleghenian.
14. During which period did the Ouachita mobile belt change from a passive plate margin to an active plate margin?
    a. ___ Silurian; b. ___ Devonian; c. ___ Mississippian;
    d. ___ Pennsylvanian; e. ___ Permian.
15. Weathering of which highlands or mountains produced the Catskill Delta?
    a. ___ Acadian; b. ___ Alleghenian; c. ___ Antler;
    d. ___ Appalachian; e. ___ Caledonian.

(Continued on page 600)

● **TABLE 21.1** Summary of Early Paleozoic Geologic Events

Age (Millions of Years)	Geologic Period	Sequence	Relative Changes in Sea Level		Cordilleran Mobile Belt
			Rising	Falling	
408 —	Silurian	Tippecanoe			
438 —	Ordovician	Tippecanoe			
505 —	Cambrian	Sauk		Present sea level	
570					

Craton	Ouachita Mobile Belt	Appalachian Mobile Belt	Major Events Outside North America
Extensive barrier reefs and evaporites common.			Caledonian orogeny
Queenston Delta clastic wedge. Transgression of Tippecanoe Sea. Regression exposing large areas to erosion.		Taconic orogeny	Continental glaciation in Southern Hemisphere.
Canadian Shield and Transcontinental Arch only areas above sea level. Transgression of Sauk Sea.			

● **TABLE 21.2** Summary of Late Paleozoic Geologic Events

Age (Millions of Years)	Geologic Period		Sequence	Relative Changes in Sea Level		Cordilleran Mobile Belt
				Rising	Falling	
245	Permian		Absaroka			
286	Carboniferous	Pennsylvanian	Absaroka			
320	Carboniferous	Mississippian	Kaskaskia			
360	Devonian		Kaskaskia			Antler orogeny
	Devonian		Tippecanoe	←Present sea level		Klamath Arc
408			Tippecanoe			

● **TABLE 21.2** Summary of Late Paleozoic Geologic Events (continued)

Craton	Ouachita Mobile Belt	Appalachian Mobile Belt	Major Events Outside North America
Deserts, evaporites, and continental red beds in southwestern United States. Extensive reefs in Texas area.		Formation of Pangaea	
		Allegheny orogeny	Hercynian orogeny
Coal swamps common.	Ouachita orogeny		
Formation of Ancestral Rockies.			Continental glaciation in Southern Hemisphere.
Transgression of Absaroka Sea.			
Widespread black shales and limestones.			
Widespread black shales. Catskill Delta clastic wedge.			Old Red Sandstone clastic wedge in British Isles.
Extensive barrier-reef formation in Western Canada. Transgression of Kaskaskia Sea.		Acadian orogeny	
			Caledonian orogeny

16. The European counterpart to the Devonian Catskill Delta is the:
a. ___ Phosphoria Formation; b. ___ Oriskany Sandstone; c. ___ Capitan Limestone; d. ___ Old Red Sandstone; e. ___ none of these.
17. What type of plate boundary resulted from the Alleghenian orogeny?
a. ___ divergent; b. ___ oceanic-oceanic convergent;
c. ___ oceanic-continental convergent;
d. ___ continental-continental convergent; e. ___ transform.
18. Which orogeny was not part of the closing of the Iapetus Ocean?
a. ___ Acadian; b. ___ Alleghenian; c. ___ Antler;
d. ___ Caledonian; e. ___ Taconic.
19. The first Paleozoic orogeny to occur in the Cordilleran mobile belt was the:
a. ___ Acadian; b. ___ Alleghenian; c. ___ Antler;
d. ___ Caledonian; e. ___ Ellesmere.
20. What is the main economic deposit of a cyclothem?
a. ___ evaporites; b. ___ carbonates; c. ___ gravels;
d. ___ ores; e. ___ coals.

21. Draw a diagrammatic cross section of a present-day reef, labeling and defining the various environments within the reef complex.
22. Briefly discuss the Paleozoic geologic history of the world.
23. Why are cratonic sequences a convenient way to study the geologic history of the Paleozoic Era?
24. Discuss how evaporites of the Michigan Basin may have formed during the Silurian Period.
25. What are cyclothems? How do they form? Why are they economically important?
26. What is the evidence for glaciation on Gondwana during the Late Paleozoic?
27. How did the sedimentary environments of the Kaskaskia sequence differ from those of the Absaroka sequence in North America?
28. Compare the Taconic, Caledonian, and Acadian orogenies in terms of the tectonic forces that caused them and the sedimentary features that resulted.
29. How did the formation of Pangaea and Panthalassa affect the world's climate at the end of the Paleozoic Era?
30. Briefly discuss the various natural resources that are found in Paleozoic-aged rocks.

Additional Readings

Bally, A. W., and A. R. Palmer, eds. 1989. *The geology of North America: An overview.* The Geology of North America. vol. A. Boulder, Colo.: Geological Society of America.

Bambach, R. K., C. R. Scotese, and A. M. Ziegler. 1980. Before Pangaea: The geographies of the Paleozoic world. *American Scientist* 68, no. 1:26–38.

Catacosinos, P. A., and P. A. Daniels, Jr., eds. 1991. *Early sedimentary evolution of the Michigan Basin.* Geological Society of America Special Paper 256. Boulder, Colo.: Geological Society of America.

Dallmeyer, R. D., ed. 1989. *Terranes in the Circum-Atlantic Paleozoic orogens.* Geological Society of America Special Paper 230. Boulder, Colo.: Geological Society of America.

Dineley, D. L. 1984. *Aspects of a stratigraphic system—The Devonian.* New York: John Wiley & Sons.

Fouch, T. D., and E. R. Magathan, eds. 1980. *Paleozoic paleogeography of the west-central United States.* Rocky Mountain Paleogeography Symposium 1. Denver, Colo.: Rocky Mountain Section, Society of Economic Paleontologists and Mineralogists.

Harwood, D. S., and M. M. Miller, eds. 1990. *Paleozoic and Early Mesozoic paleogeographic relations: Sierra Nevada, Klamath Mountains, and related terranes.* Geological

Society of America Special Paper 255. Boulder, Colo.: Geological Society of America.

Hatcher, R. D., Jr., W. A. Thomas, and G. W. Viele, eds. 1989. *The Appalachian-Ouachita orogen in the United States.* The Geology of North America. vol. F-2. Boulder, Colo.: Geological Society of America.

McKerrow, W. S., and C. R. Scotese, eds. 1990. *Palaeozoic palaeogeography and biogeography.* The Geological Society Memoir No. 12. London: Geological Society of London.

Moullade, M., and A. E. M. Nairn, eds. 1991. *The Phanerozoic geology of the world I: The Palaeozoic.* New York: Elsevier Science Publishing.

Sloss, L. L., ed. 1988. *Sedimentary cover—North American Craton: U.S.* The Geology of North America. vol. D-2. Boulder, Colo.: Geological Society of America.

Stewart, J. H., C. H. Stevens, and A. E. Fritsche, eds. 1977. *Paleozoic paleogeography of the western United States.* Pacific Coast Paleogeography Symposium 1. Tulsa, Okla.: Society of Economic Paleontologists and Mineralogists.

Wilgus, C. K., B. S. Hastings, C. G. St. C. Kendall, H. Posamentier, C. A. Ross, and J. Van Wagoner, eds. 1988. *Sea-level changes: An integrated approach.* Society of Economic Paleontologists and Mineralogists Special

Publication No. 42. Tulsa, Okla.: Society of Economic Paleontologists and Mineralogists.

Woodrow, D. L., and W. D. Sevan. 1985. The Catskill Delta. *Geological Society of America Special Paper* 201: 1–246.

Ziegler, P. A. 1989. *Evolution of Laurussia: A study in Late Palaeozoic plate tectonics.* Boston: Kluwer Academic Publishers.

CHAPTER
22

Chapter Outline

Diorama of the environment and biota of the Phyllopod bed of the Burgess Shale, British Columbia, Canada. In the background is the vertical wall of a submarine escarpment with algae growing on it. The large cylindrical ribbed organisms on the muddy bottom in the foreground are sponges.

Paleozoic Life History

Prologue

On August 30 and 31, 1909, near the end of the summer field season, Charles D. Walcott, geologist and head of the Smithsonian Institution, was searching for fossils along a trail on Burgess Ridge between Mount Field and Mount Wapta, near Field, British Columbia, Canada. On the west slope of this ridge, he discovered the first soft-bodied fossils from the Burgess Shale, a discovery of immense importance in deciphering the early history of life. During the following week, Walcott and his collecting party split open numerous blocks of shale, many of which yielded the impressions of a number of soft-bodied organisms beautifully preserved on bedding planes. Walcott returned to the site the following summer and located the shale stratum that was the source of his fossil-bearing rocks in the steep slope above the trail. He quarried the site and shipped back thousands of fossil specimens to the United States National Museum of Natural History, where he later cataloged and studied them.

The importance of Walcott's discovery is not that it was another collection of well-preserved Cambrian fossils, but rather that it allowed geologists a rare glimpse into a world previously almost unknown—that of the soft-bodied animals that lived some 530 million years ago. The beautifully preserved fossils from the Burgess Shale present a much more complete picture of a Middle Cambrian community than deposits containing only fossils of the hard parts of organisms. Specifically, the Burgess Shale contains species of trilobites, sponges, brachiopods, mollusks, and echinoderms, all of which have hard parts and are characteristic of Cambrian faunas throughout the world. But in addition to the diverse skeletonized fauna, a large and varied fossil assemblage of soft-bodied animals is also present. In fact, 60% of the total fossil assemblage is composed of soft-bodied animals, which usually are not preserved. In all, more than 100 genera of animals, at least 60 of which were soft-bodied and preserved as impressions, have been recovered from the Burgess Shale. This proportion of soft-bodied animals to those with hard parts is comparable to present-day marine communities.

What conditions led to the remarkable preservation of the Burgess Shale fauna? When it was deposited, the Burgess Shale was located at the base of a steep submarine escarpment. The animals whose exquisitely preserved fossil remains are found in the Burgess Shale lived in and on mud banks that formed along the top of this escarpment. Periodically, this unstable area would slump and slide down the escarpment as a turbidity flow. At the base, the mud and animals carried with it were deposited in a deep-water anaerobic environment devoid of life. In such an environment, bacterial degradation did not destroy the buried animals, and they were compressed by the weight of the overlying sediments, eventually resulting in their preservation as carbonaceous impressions.

⯆ INTRODUCTION

In this chapter, we examine the history of Paleozoic life as a series of interconnected biologic and geologic events in which the underlying principles and processes of evolution and plate tectonics played a major role.

The opening and closing of ocean basins, transgressions and regressions of epeiric seas, and the changing positions of the continents had a profound effect on the evolution of the marine and terrestrial communities.

A time of tremendous biologic change began with the appearance of skeletonized animals near the Precambrian-Cambrian boundary. Following this event, marine invertebrates began a period of adaptive radiation and evolution during which the Paleozoic marine invertebrate community greatly diversified. Indeed, the history of the Paleozoic marine invertebrate community was one of diversifications and extinctions.

Vertebrates also evolved during the Paleozoic. The earliest fossil records of vertebrates are of fish, which evolved during the Cambrian. One group of fish was ancestral to the first land animals, the amphibians, which evolved during the Devonian. Reptiles evolved from a group of amphibians during the Pennsylvanian Period and were the dominant vertebrate animals on land by the end of the Paleozoic Era.

Plants preceded animals onto the land. Both plants and animals had to solve the same basic problems in making the transition from water to land. The method of reproduction proved to be the major barrier to expansion into new environments for both groups. With the evolution of the seed in plants and the amniote egg in animals, this limitation was removed, and both groups were able to move into all terrestrial environments.

The end of the Paleozoic Era was a time of major extinctions. The marine invertebrate community was greatly decimated, and many amphibians and reptiles on land also became extinct.

▼ THE CAMBRIAN PERIOD AND THE EMERGENCE OF A SHELLY FAUNA

Most scientists recognize that the variety and complexity of Cambrian life suggest that multicelled organisms must have had a Precambrian history during which they lacked hard parts and thus did not leave a fossil record. Impressions of the first unequivocally multicelled animals belong to the widely distributed Ediacaran fauna and are found in rocks between 570 to 670 million years old (see Fig. 20.32). Associated with Ediacaran faunas in Namibia and southern China are the first shelled fossils (◆ Fig. 22.1). These are small calcium carbonate tubes, presumably housing wormlike suspension-feeding organisms. In addition, small organic tube-shaped fossils, also presumably housing

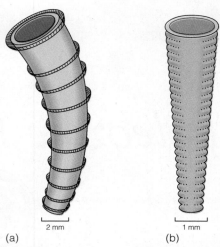

◆ FIGURE 22.1 The first shelled fossils. These small calcium carbonate tubes presumably housed wormlike suspension-feeding organisms. They are found associated with Ediacaran faunas from Namibia and China. (a) Cloudina. (b) Sinotubulites.

wormlike suspension-feeding animals occur with the calcareous tubes. By the latest part of the Proterozoic, several skeletonized animals had made their appearance, yet durable skeletons of chitin, silica, and calcium carbonate did not begin to appear in abundance until the beginning of the Phanerozoic Eon 570 million years ago.

The Early Cambrian was characterized by a low-diversity shelly fauna consisting of animals that used both calcium carbonate and calcium phosphate to construct their skeletons. They included small worm tubes, mollusks and echinoderms, archaeocyathids, and brachiopods (◆ Fig. 22.2). It is likely that this fauna was yet another "experiment," like the Ediacaran fauna of the Proterozoic Eon, but this experiment was very successful. By the Middle Cambrian, a large number of the major groups of invertebrate animals had evolved. Many, such as the brachiopods, are still around today, while others, including the archaeocyathids and trilobites, are extinct.

The Cambrian Period was also a time during which new body plans evolved and animals moved into new niches. As might be expected, the Cambrian witnessed a higher percentage of such experiments than any other period of geologic history.

The emergence of so many different organisms at this juncture in the Earth's history resulted in the

(a)

(b)

(c)

◆ FIGURE 22.2 Three small Lower Cambrian shelly fossils. (*a*) A conical sclerite (a piece of the armor covering) of *Lapworthella* from Australia. (*b*) *Archaeooides,* an enigmatic spherical fossil from the Mackenzie Mountains, Northwest Territories, Canada. (*c*) The tube of an anabaritid from the Mackenzie Mountains, Northwest Territories, Canada. (Photos courtesy of Simon Conway Morris and Stefan Bengtson, University of Cambridge, England.)

Cambrian			Ordovician	
Early	Middle	Late	Early	Middle
Arthropoda (Trilobites) →→→				
Brachiopoda →→→				
Echinodermata →→→				
Mollusca (Gastropoda) →→→				
Porifera →→→				
	Mollusca (Bivalvia) →→→			
	Mollusca (Cephalopoda) →→→			
		Protozoa →→→		
			Bryozoa →→→	
			Cnidaria (Tabulate corals) →→→	
				Cnidaria (Rugose corals) →→→

◆ FIGURE 22.3 The first recorded occurrence of selected members of the major marine invertebrate phyla.

development of new relationships and community structures as organisms underwent tremendous evolutionary change and filled previously unoccupied niches.

The Acquisition and Significance of Hard Parts

A striking aspect of the Early Cambrian fauna is that many animals already had fully developed features; that is, their anatomies indicate an extended period of evolution before the evolution of hard parts. Furthermore, the major skeletonized animal groups did not all evolve at once, but rather evolved throughout the Cambrian and Ordovician periods (◆ Fig. 22.3). A preskeletonized period of evolution occurred during which members of a phylum evolved for an unknown period of time as soft-bodied organisms.

But why did invertebrates initially acquire skeletons and why did such acquisitions occur over an extended period of time? Various explanations have been proposed. One that is particularly appealing to many paleontologists is that mineralized skeletons evolved as a response to invertebrates' need to eliminate mineral matter from their metabolic systems. One way to do this is to secrete a buildup of excess ions as a solid. In this way a mineralized skeleton may have evolved as a means of eliminating high levels of calcium and phosphate ions. The evolution of a skeleton would then prove to be advantageous to the organism in a variety of ways.

Among the advantages an exoskeleton confers on an organism are: (1) It provides protection against ultraviolet radiation, allowing animals to move more easily into shallower waters. (2) It helps prevent drying out in an intertidal environment. (3) It provides protection against predators. (4) A supporting skeleton, whether an exo- or endoskeleton, allows animals to increase their size. (5) It also provides attachment sites for development of strong muscles, thus increasing locomotor efficiency in mobile animals.

Currently, there is no clear answer as to why marine invertebrates acquired mineralized skeletons over an extended period of time. Hard parts undoubtedly originated because of a variety of biologic and environmental factors rather than a single one. Whatever the reason, the acquisition of a mineralized skeleton was a major evolutionary innovation allowing invertebrates to successfully occupy a wide variety of marine habitats.

▼ PALEOZOIC INVERTEBRATE MARINE LIFE

Rather than focusing on the history of each invertebrate phylum (● Table 22.1), we will survey the evolution of the Paleozoic marine invertebrate communities through time, concentrating on the major features and the changes that occurred. To do that, we need to first examine the nature and structure of living marine communities.

The Present Marine Ecosystem

In analyzing the present-day marine ecosystem, we must look at where organisms live and how they get around, as well as how they feed (◆ Fig. 22.4). Organisms that live in the water column above the sea floor are called *pelagic.* They can be divided into two main groups: the floaters, or **plankton,** and the swimmers, or **nekton.**

Plankton are mostly passive and go where the current carries them. Plant plankton such as diatoms, dinoflagellates, and various algae, are called *phytoplankton* and are mostly microscopic. Animal plankton are called *zooplankton* and are also mostly microscopic. Examples of zooplankton include foraminifera, radiolarians, and jellyfish. The nekton are swimmers and are mainly vertebrates such as fish; the invertebrate nekton include cephalopods.

Organisms that live on or in the sea floor make up the **benthos.** They can be characterized as *epifauna* (animals) or *epiflora* (plants), for those that live on the sea floor, or as *infauna,* which are animals living in and moving through the sediments. The benthos can be further divided into those organisms that stay in one place, called *sessile,* and those that move around on or in the sea floor, called *mobile.*

The feeding strategies of organisms are also important in terms of their relationships with other organisms in the marine ecosystem. There are basically four feeding groups: **suspension-feeding** animals remove or consume microscopic plants and animals as well as dissolved nutrients from the water; **herbivores** are plant eaters; **carnivore-scavengers** are meat eaters; and **sediment-deposit feeders** ingest sediment and extract the nutrients from it.

We can define an organism's place in the marine ecosystem by where it lives and how it eats. For example, an articulate brachiopod is a benthonic, epi-

---

● **TABLE 22.1** The Major Invertebrate Groups and Their Stratigraphic Ranges

Phylum Protozoa	Cambrian-Recent	**Phylum Mollusca**	Cambrian-Recent
Class Sarcodina	Cambrian-Recent	Class Monoplacophora	Cambrian-Recent
Order Foraminifera	Cambrian-Recent	Class Gastropoda	Cambrian-Recent
Order Radiolaria	Cambrian-Recent	Class Bivalvia	Cambrian-Recent
		Class Cephalopoda	Cambrian-Recent
Phylum Porifera	Cambrian-Recent		
Class Demospongea	Cambrian-Recent	**Phylum Annelida**	Precambrian-Recent
Order Stromatoporoida	Cambrian-Oligocene		
		Phylum Arthropoda	Cambrian-Recent
Phylum Archaeocyatha	Cambrian	Class Trilobita	Cambrian-Permian
		Class Crustacea	Cambrian-Recent
Phylum Cnidaria	Cambrian-Recent	Class Insecta	Silurian-Recent
Class Anthozoa	Ordovician-Recent		
Order Tabulata	Ordovician-Permian	**Phylum Echinodermata**	Cambrian-Recent
Order Rugosa	Ordovician-Permian	Class Blastoidea	Ordovician-Permian
Order Scleractinia	Triassic-Recent	Class Crinoidea	Cambrian-Recent
		Class Echinoidea	Ordovician-Recent
Phylum Bryozoa	Ordovician-Recent	Class Asteroidea	Ordovician-Recent
Phylum Brachiopoda	Cambrian-Recent	**Phylum Hemichordata**	Cambrian-Recent
Class Inarticulata	Cambrian-Recent	Class Graptolithina	Cambrian-Mississippian
Class Articulata	Cambrian-Recent		

◆ FIGURE 22.4 Where and how animals and plants live in the marine ecosystem.
Plankton: (*a*) jellyfish. Nekton: (*b*) fish and (*c*) cephalopod. Benthos: (*d*) through (*k*).
Sessile epiflora: (*d*) seaweed. Sessile epifauna: (*g*) bivalve, (*i*) coral, and (*j*) crinoid.
Mobile epifauna: (*k*) starfish and (*h*) gastropod. Infauna: (*e*) worm and (*f*) bivalve.
Suspension feeders: (*g*) bivalve, (*i*) coral, and (*j*) crinoid. Herbivores: (*h*) gastropod.
Carnivores-scavengers: (*k*) starfish. Sediment-deposit feeders: (*e*) worm.

faunal suspension feeder, whereas a cephalopod is a nektonic carnivore.

An ecosystem includes several **trophic levels** which are tiers of food production and consumption within a feeding hierarchy. The feeding hierarchy and hence energy flow in an ecosystem comprise a food web of complex interrelationships among the producers, consumers, and decomposers (◆ Fig. 22.5). The **primary producers,** or *autotrophs,* are those organisms that manufacture their own food. Virtually all marine primary producers are phytoplankton. Feeding on the primary producers are the primary consumers, which are mostly suspension feeders. Secondary consumers feed on the primary consumers, and thus are predators,

while tertiary consumers, which are also predators, feed on the secondary consumers. In addition to the producers and consumers, there are also transformers and decomposers. These are bacteria that break down the dead organisms that have not been consumed into organic compounds that are then recycled.

When we look at the marine realm today, we see a complex organization of organisms interrelated by trophic interactions and affected by changes in the physical environment. When one part of the system changes, the whole structure changes, sometimes almost insignificantly, other times catastrophically.

As we examine the evolution of the Paleozoic marine ecosystem, keep in mind how geologic and evolutionary

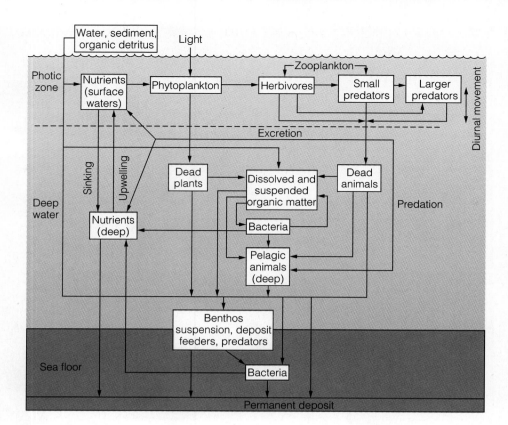

◆ FIGURE 22.5 Marine food web showing the relationships among the producers, consumers, and decomposers.

changes can have a significant impact on the composition and structure of the ecosystem. For example, the major transgressions onto the craton opened up vast areas of shallow seas that could be inhabited. The movement of continents affected oceanic circulation patterns as well as causing environmental changes as the continents and their epeiric seas moved through different climatic zones.

Cambrian Marine Community

Although almost all the major invertebrate phyla appeared in the fossil record during the Cambrian Period, many were represented by only a few species. While trace fossils are common, and echinoderms diverse, trilobites, brachiopods, and archaeocyathids comprised the majority of Cambrian skeletonized life (◆ Fig. 22.6).

Trilobites were by far the most conspicuous element of the Cambrian marine invertebrate community and made up about half of the total fauna. Trilobites were benthonic mobile sediment-deposit feeders that crawled or swam along the seafloor. Trilobite faunas

◆ FIGURE 22.6 Reconstruction of a Cambrian marine community. Floating jellyfish, swimming arthropods, benthonic sponges, and scavenging trilobites are shown.

can be grouped into distinct faunal realms that characterize the paleogeography of the time. Furthermore, trilobites are excellent guide fossils.

Cambrian **brachiopods** were mostly primitive types known as *inarticulates*. They secreted a shell composed of the organic compound chitin combined with calcium phosphate. Inarticulate brachiopods also lacked a tooth-and-socket-arrangement along the hinge line. The *articulate* brachiopods, which have a tooth-and-socket arrangement, were also present but did not become abundant until the Ordovician Period.

The third major group of Cambrian organisms were **archaeocyathids** (◆ Fig. 22.7). These organisms were benthonic sessile suspension feeders that constructed reeflike structures. The rest of the Cambrian fauna consisted of representatives of the other major phyla, including many organisms that were short-lived evolutionary experiments.

The Burgess Shale Biota

No discussion of Cambrian life would be complete without mentioning one of the best examples of a preserved soft-bodied fauna and flora, the Burgess Shale biota. As the Sauk Sea transgressed from the Cordilleran shelf onto the western edge of the craton, Early Cambrian sands were covered by a Middle Cambrian black mud that allowed a diverse soft-bodied benthonic community to be preserved. As we discussed in the Prologue, these fossils were discovered in 1909 by Charles D. Walcott near Field, British Columbia. They represent one of the most significant fossil finds of the century because they consist of impressions of soft-bodied animals and plants (◆ Fig. 22.8), which are very rarely preserved in the fossil record.

In recent years, the reconstruction, classification, and interpretation of many of the Burgess Shale fossils have undergone a major change that has led to new theories and explanations of the Cambrian explosion of life. Recall that during the Late Proterozoic multicellu-

◆ FIGURE 22.7 Restoration of a Cambrian reeflike structure built by archaeocyathids.

lar organisms evolved, and shortly thereafter animals with hard parts made their first appearance. These were followed by an explosion of invertebrate phyla during the Cambrian, some of which are now extinct. These Cambrian phyla represent the root stock and basic body plans from which all present-day invertebrates evolved. The question that paleontologists are now hotly debating is how many phyla arose during the Cambrian, and at the center of that debate are the Burgess Shale fossils. For years, most paleontologists placed the bulk of the Burgess Shale organisms into existing phyla, with only a few assigned to phyla that are now extinct. Thus, the phyla of the Cambrian world were viewed as being essentially the same in number as the phyla of the present-day world, but with fewer species in each phylum. According to this view, the history of life has been simply a gradual increase in the diversity of species within each phylum through time. The number of basic body plans has therefore remained more-or-less constant since the initial radiation of multicellular organisms.

This view has recently been challenged by Harvard paleobiologist Stephen Jay Gould in his 1989 book *Wonderful Life*. According to Gould, the initial explosion of varied life-forms in the Cambrian was promptly followed by a short period of experimentation and then extinction of many phyla. The richness and diversity of modern life-forms are the result of repeated variations of the basic body plans that survived the Cambrian extinctions. In other words, life was much more diverse in terms of phyla during the Cambrian than it is today. The reason members of the Burgess Shale biota look so strange to us is that no living organisms possess their basic body plan, and therefore many of them have been placed into new phyla.

Recent discoveries of age-equivalent fossils at other localities has led some paleontologists to reassign some of the Burgess Shale specimens back into extant phyla, thereby reducing the diversity of the Burgess Shale fauna. If these reassignments to known phyla prove to be correct, then no massive extinction event followed the Cambrian explosion, and life has gradually increased in diversity through time. Currently, there is no clear answer to this debate, and the outcome will probably be decided as more fossil discoveries are made.

Ordovician Marine Community

A major transgression that began during the Middle Ordovician (Tippecanoe sequence) resulted in the most

(a)

(b)

(c)

◆ FIGURE 22.8 Some of the fossil animals preserved in the Burgess Shale.
(*a*) *Burgessia bella,* an arthropod. (*b*) *Waptia fieldensis,* another arthropod.
(*c*) *Burgessochaeta setigera,* an annelid worm. (*d*) *Marrella splendens,* an arthropod
that is the most abundant fossil animal in the Phyllopod bed of the Burgess Shale.

(d)

widespread inundation of the craton. This vast epeiric sea, which experienced a uniformly warm climate during this time, opened numerous new marine habitats that were soon filled by a variety of organisms.

Not only did sedimentation patterns change dramatically from the Cambrian to the Ordovician, but the fauna underwent equally striking changes in composition. Whereas the Cambrian invertebrate community was dominated by three groups—trilobites, brachiopods, and archaeocyathids—the Ordovician was characterized by the adaptive radiation of many other animal phyla, (such as bryozoans and corals), with a consequent dramatic increase in the diversity of the total shelly fauna (◆ Fig. 22.9).

During the Cambrian, archaeocyathids were the main builders of reeflike structures, but bryozoans, stromatoporoids, and tabulate and rugose corals assumed that role beginning in the Middle Ordovician. Many of these reefs were small patch reefs similar in

◆ FIGURE 22.9 Recreation of a Middle Ordovician sea floor fauna. Cephalopods, crinoids, colonial corals, graptolites, trilobites, and brachiopods are shown. (From the Field Museum, Chicago, # Geo80820c.)

size to those of the Cambrian but of a different composition, whereas others were quite large. As with present-day reefs, Ordovician reefs exhibited a high diversity of organisms and were dominated by suspension feeders.

The end of the Ordovician was a time of mass extinctions in the marine realm. More than 100 families of marine invertebrates did not survive into the Silurian, and in North America alone, approximately one-half of the brachiopods and bryozoans died out. What caused such an event? Many geologists think that these extinctions were the result of the extensive glaciation that occurred in Gondwana at the end of the Ordovician Period (see Chapter 21).

Mass extinctions, those geologically rapid events in which an unusually high percentage of the fauna and/or flora becomes extinct, have occurred throughout geologic time (at or near the end of the Ordovician, Devonian, Permian, and Cretaceous periods) and currently are the focus of much research and debate (see Perspective 22.1).

Silurian and Devonian Marine Communities

The mass extinction at the end of the Ordovician was followed by rediversification and recovery of many of the decimated groups. Brachiopods, bryozoans, gastropods, bivalves, corals, crinoids, and graptolites were just some of the groups that rediversified again beginning during the Silurian.

As we discussed in Chapter 21, the Silurian and Devonian were times of major reef building. While most of the Silurian radiations of invertebrates represented repopulating of niches, organic reef-builders diversified in new ways, building massive reefs larger than any produced during the Cambrian or Ordovician. This repopulation was probably due in part to renewed transgressions over the craton, and although a major drop in sea level occurred at the end of the Silurian, the Middle Paleozoic sea level was generally high (see Table 21.1).

The Silurian and Devonian reefs were dominated by tabulate and colonial rugose corals and stromatoporoids (◆ Fig. 22.10). While the fauna of these Silurian and Devonian reefs was somewhat different from that of earlier reefs and reeflike structures, the general composition and structure are the same as in present-day reefs.

The Silurian and Devonian periods were also the time when *eurypterids* (arthropods with scorpionlike bodies and impressive pincers) were abundant, especially in brackish and freshwater habitats (◆ Fig. 22.11). *Ammonoids,* a subclass of the cephalopods, evolved from nautiloids during the Early Devonian and rapidly diversified. With their distinctive suture patterns, short stratigraphic ranges, and widespread distribution, ammonoids are excellent guide fossils for the Devonian through Cretaceous periods (Fig. 22.10).

Another mass extinction occurred near the end of the Devonian and resulted in a worldwide near-total collapse of the massive reef communities. On land, however, the seedless vascular plants were seemingly unaffected, although the diversity of freshwater fish was greatly reduced.

The demise of the Middle Paleozoic reef communities serves to highlight the geographic aspects of the Late Devonian mass extinction event. The tropical groups were most severely affected; in contrast, the polar communities were seemingly little affected. Apparently, an episode of global cooling was largely responsible for the extinctions near the end of the Devonian. During such a cooling, the disappearance of tropical conditions would have had a severe effect on reef and other warm-water organisms. Cool-water species, on the other hand, could have simply migrated toward the equator. While cooling temperatures certainly played an important role in the Late Devonian extinctions, the closing of the Iapetus Ocean and the orogenic events of this time undoubtedly also played a role by reducing the area of shallow shelf environments where many marine invertebrates lived.

Carboniferous and Permian Marine Communities

The Carboniferous invertebrate marine community responded to the Late Devonian extinctions in much the same way the Silurian invertebrate marine community responded to the Late Ordovician extinctions—that is, by renewed adaptive radiation and rediversification. The brachiopods and ammonoids quickly recovered and again assumed important ecologic roles, while other groups, such as the lacy bryozoans and crinoids, reached their greatest diversity during the Carboniferous. With the decline of the stromatoporoids and the tabulate and rugose corals, large organic reefs like those existing earlier in the Paleozoic virtually disappeared and were replaced by small patch reefs. These reefs were dominated by crinoids, blastoids, lacy bryozoans, brachiopods, and calcareous algae and flourished

EXTINCTIONS: CYCLICAL OR RANDOM?

Throughout geologic history, various plant and animal species have become extinct. In fact, extinction is a common feature of the fossil record, and the rate of extinction through time has fluctuated only slightly. Just as new species evolve, others become extinct. There have, however, been brief intervals in the past during which mass extinctions have eliminated large numbers of species. Extinctions of this magnitude could only occur due to radical changes in the environment on a regional or global scale.

When we look at the different mass extinction events that have occurred throughout the geologic past, several common themes stand out. The first is that mass extinctions have affected life both in the sea and on land. Second, tropical organisms, particularly in the marine realm, apparently are more affected than organisms from the temperate and high latitude regions. Third, some animal groups repeatedly experience mass extinctions. During the first mass extinction event, such groups are severely affected but not wiped out. Following the initial crisis, the survivors diversify, only to have their numbers reduced further by another mass extinction. Three marine invertebrate groups in particular display this characteristic: the trilobites, graptolites, and ammonoids. Each of these groups experienced high rates of extinction, followed by high rates of speciation. The fourth theme of mass extinctions is the apparent periodicity displayed during the Phanerozoic Eon. It has been proposed that mass extinctions have occurred approximately every 26 million years.

When we look at the mass extinctions for the last 570 million years, we see that several extinction events occurred during the Cambrian, and these affected only marine invertebrates, particularly trilobites. Three other marine mass extinctions took place during the Paleozoic Era: one at the end of the Ordovician, involving many invertebrates; one near the end of the Devonian, affecting the major barrier reef–building organisms as well as the primitive armored fish; and the most severe at the end of the Permian, when between 70 to 90% of the marine species became extinct. On land, a group of reptiles called pelycosaurs also became extinct at the end of the Permian.

The Mesozoic Era experienced several mass extinctions, the most devastating occurring at the end of the Cretaceous, when all large animals, including dinosaurs and seagoing animals such as plesiosaurs and ichthyosaurs, became extinct. Many scientists think that the terminal Cretaceous extinction event was caused by a meteorite impact, although that theory is still being vigorously debated.

Several mass extinction events occurred during the Cenozoic Era. The most severe was near the end of the Eocene Epoch.

Though many scientists think of the marine mass extinctions as sudden events from a geological perspective, they were rather gradual from a human perspective, occurring over hundreds of thousands or even millions of years. Furthermore, many geologists think that climatic changes, rather than some catastrophe, were primarily responsible for the extinctions, particularly in the marine realm. Evidence of glacial episodes or other signs of climatic change have been correlated with the extinction events recorded in the fossil record.

One of the more controversial issues to emerge from the mass extinction debate is whether they occur periodically. David Raup and John Sepkoski of the University of Chicago have suggested a periodicity of about 26 million years for mass extinction events since the end of the Paleozoic Era. According to them, the most recent mass extinction occurred nearly 11 million years ago. They based their conclusions on the geologic ranges of marine invertebrate families. By graphing all family-level extinctions of marine invertebrates since the Middle Permian, and treating each family extinction as if it occurred at the end of a geologic age, they found that their extinction peaks best fit a 26-million-year periodicity. Furthermore, some of the peaks, such as the one representing the end of the Cretaceous, coincided with recognized mass extinctions.

What Raup and Sepkoski's data show is not that the major extinction events occur precisely every 26 million years, but rather that the occurrence is closer to 26-million-year intervals than would be expected for a strictly random pattern. It has been suggested that some type of recurring extraterrestrial event, such as a meteorite impact, can account for the apparent periodicity (see Chapter 23).

◆ FIGURE 22.10 Reconstruction of a Middle Devonian reef from the Great Lakes area. Shown are corals, ammonoids, trilobites, and brachiopods. (From the Field Museum, Chicago # Geo80821c.)

◆ FIGURE 22.12 Marine life during the Mississippian based on an Upper Mississippian fossil site at Crawfordville, Indiana. Invertebrate animals shown include crinoids, blastoids, lacy bryozoans, and small corals.

◆ FIGURE 22.11 Restoration of a Silurian brackish-marine bottom scene near Buffalo, New York. Shown are algae, eurypterids, worms, and shrimp. (From the Field Museum, Chicago, # Geo80819c.)

◆ FIGURE 22.13 A Permian patch-reef community from the Glass Mountains of West Texas. Shown are algae, productid brachiopods, cephalopods, and corals.

during the Late Paleozoic (◆ Fig. 22.12). In addition, bryozoans, crinoids, and fusulinids (spindle-shaped foraminifera) contributed large amounts of skeletal debris to the formation of the vast bedded limestones that constitute the majority of Mississippian sedimentary rocks.

The Permian invertebrate marine faunas resembled those of the Carboniferous. Due to the restricted size of the shallow seas on the cratons and the reduced shelf space along the continental margins, however, they were more restricted in their distribution (see Fig.

21.22). The spiny and odd-shaped productids dominated the brachiopod assemblage and constituted an important part of the reef complexes that formed in the Texas region during the Permian (◆ Fig. 22.13). The fusulinids, which first evolved during the Late Mississippian and greatly diversified during the Pennsylvanian, experienced a further diversification during the Permian; more than 5,000 species are known from the Permian Period alone. Because of their abundance, diversity, and worldwide occurrence, fusulinids are important guide fossils for the Pennsylvanian and Permian. Bryozoans, sponges, and some types of calcareous algae also were common elements of the Permian invertebrate fauna.

The Permian Marine Invertebrate Extinction Event

The greatest recorded mass extinction event to affect the marine invertebrate community occurred at the end of the Permian Period (◆ Fig. 22.14). Before the Permian ended, roughly one-half of all marine invertebrate families and perhaps 90% of all marine invertebrate species became extinct. Fusulinids, rugose and tabulate corals, two bryozoan orders, and two brachiopod orders as well as trilobites and blastoids did not survive

◆ FIGURE 22.14 Phanerozoic diversity for marine invertebrate and vertebrate families. Note the three episodes of Paleozoic mass extinctions, with the greatest occurring at the end of the Permian Period.

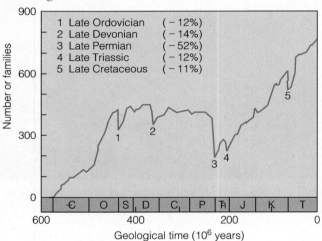

the end of the Permian. All of these groups had been very successful during the Paleozoic Era, although the trilobites had decreased in abundance during the Early Paleozoic.

What caused such a crisis for the marine invertebrates? Many hypotheses have been proposed, but no completely satisfactory answer has been found. Two currently discussed hypotheses are (1) a reduction of living area related to widespread regression of the seas and suturing of the continents when Pangaea formed and (2) decreased ocean salinity due to widespread arid climates. During the Paleozoic, there was a trend toward continental convergence that culminated in the formation of Pangaea by the end of the Permian. Such continental convergence resulted in regression of the epeiric seas from the cratons and reduction of the shallow-water shelf area surrounding each continent. Decreased ocean salinity would affect those organisms with narrow salinity tolerances, which includes most marine invertebrates. Extensive marginal marine evaporite deposits were formed during the Permian Period, and it is hypothesized that removal of the salts from the ocean to form these deposits lowered ocean salinity to lethal levels for many invertebrates. Other calculations, however, show that not enough salts could have been removed by this method to have significantly lowered the salinity level of the entire Panthalassa Ocean.

The Permian mass extinctions were probably caused by a combination of many interrelated geologic and biologic factors. In any case, the surviving marine invertebrate faunas of the Early Triassic were of very low diversity and were widely distributed around the world. This wide distribution indicates that either the groups were tolerant of a wide temperature range, or the latitudinal temperature gradient was very gentle, enabling species of normal temperature tolerance to spread far and wide.

◆ FIGURE 22.15 The structure of the lancelet *Amphioxus* illustrates the three characteristics of a chordate: a notochord, a dorsal hollow nerve cord, and gill slits.

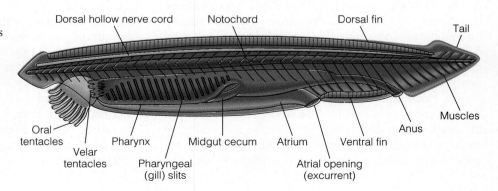

VERTEBRATE EVOLUTION

A **chordate** is an animal that has, at least during part of its life cycle, a notochord, a dorsal hollow nerve cord, and gill slits (◆ Fig. 22.15). **Vertebrates,** which are animals with backbones, are simply a subphylum of chordates.

The ancestors and early members of the phylum Chordata were soft-bodied organisms that left few fossils. Consequently, we know very little about the early evolutionary history of the chordates or vertebrates. Surprisingly, a very close relationship exists between echinoderms and chordates. They may even have shared a common ancestor, because the development of the embryo is the same in both groups and differs completely from other invertebrates (◆ Fig. 22.16). Furthermore, the biochemistry of muscle activity, blood proteins, and the larval stages are very similar in both echinoderms and chordates.

The evolutionary pathway to vertebrates may have begun with a sessile suspension-feeding animal with exposed cilia on its arms (◆ Fig. 22.17). Subsequently, these organisms evolved into nektonic gilled animals, perhaps looking somewhat like *Amphioxus* (Fig. 22.15). With the modification of the notochord to vertebrae, the first true vertebrates evolved.

FISH

The most primitive vertebrates are the fishes, and the oldest fish remains are found in the Upper Cambrian Deadwood Formation in northeastern Wyoming. Here phosphatic scales and plates of *Anatolepis,* a primitive member of the class Agnatha (jawless fish) have been recovered from marine sediments. All known Cambrian and Ordovician fossil fish have been found in shallow, nearshore marine deposits, while the earliest nonmarine

◆ FIGURE 22.17 A diagrammatic family tree suggesting the possible mode of evolution of vertebrates. The echinoderms may have arisen from forms somewhat similar to small pterobranchs; the acorn worm may have evolved from pterobranch descendants that had evolved a gill-feeding system but were somewhat more advanced in other regards. Tunicates represent a stage in which, in the adult, the gill apparatus has become highly evolved. The important point, though, is the development in some tunicates of a free-swimming larva with such advanced features as a notochord, nerve cord, and free-swimming habit. In further progress to the lancelet *Amphioxus* and the vertebrates, the old sessile adult stage has been abandoned, and it is the larval type that has initiated the advance.

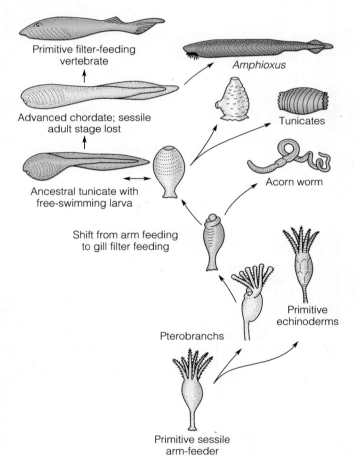

Primitive filter-feeding vertebrate

Amphioxus

Advanced chordate; sessile adult stage lost

Tunicates

Ancestral tunicate with free-swimming larva

Acorn worm

Shift from arm feeding to gill filter feeding

Primitive echinoderms

Pterobranchs

Primitive sessile arm-feeder

◆ FIGURE 22.16 (*a*) Arrangement of cells resulting from spiral cleavage. In this arrangement, cells in successive rows are nested between each other. Spiral cleavage is characteristic of all invertebrates except the echinoderms. (*b*) Arrangement of cells resulting from radial cleavage is characteristic of chordates and echinoderms. In this configuration, cells are directly above each other.

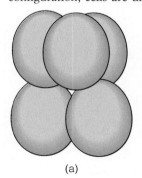

(a)

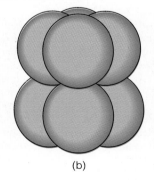

(b)

fish remains have been found in Silurian strata. This does not prove that fish originated in the oceans, but it does lend strong support to the idea.

As a group, fish range from the Late Cambrian to the present (◆ Fig. 22.18). The oldest and most primi-tive of the class Agnatha are the **ostracoderms,** whose name means "bony skin" (● Table 22.2). These are armored jawless fish that first evolved during the Late Cambrian, reached their zenith during the Silurian and Devonian, and then became extinct.

◆ FIGURE 22.18 Geologic ranges of the major fish groups.

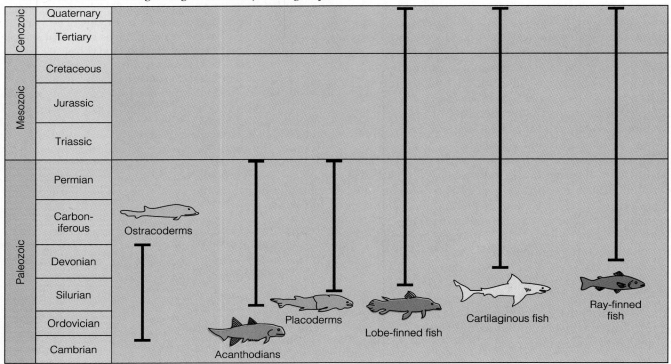

● **TABLE 22.2** Brief Classification of Fish Showing the Groups Referrred to in the Text

Classification	Geologic Range	Living Example
Class Agnatha (jawless fish)	Late Cambrian–Recent	Lamprey, hagfish
Early members of the class are called ostracoderms		No living ostracoderms
Class Acanthodii (the first fish with jaws)	Early Silurian–Permian	None
Class Placodermi (armored jawed fish)	Late Silurian–Permian	None
Class Chondrichthyes (cartilagenous fish)	Devonian-Recent	Sharks, rays, skates
Class Osteichthyes (bony fish)	Devonian-Recent	Tuna, perch, bass, pike, catfish, trout, salmon, lungfish, *Latimeria*
Subclass Actinopterygii (ray-finned fish)	Devonian-Recent	Tuna, perch, bass, pike, catfish, trout, salmon
Subclass Sarcopterygii (lobe-finned fish)	Devonian-Recent	Lungfish, *Latimeria*
Order Dipnoi	Devonian-Recent	Lungfish
Order Crossopterygii	Devonian-Recent	*Latimeria*
Suborder Rhipidistia	Devonian-Permian	None

The majority of ostracoderms lived on the sea bottom. *Hemicyclaspis* is a good example of a bottom-dwelling ostracoderm (◆ Fig. 22.19a). Vertical scales allowed *Hemicyclaspis* to wiggle sideways, propelling itself along the seafloor, while the eyes on the top of its head allowed it to see such predators as cephalopods and jawed fish approaching from above. While moving along the sea bottom, it probably sucked up small bits of food and sediments through its jawless mouth.

The evolution of jaws was a major evolutionary advance among primitive vertebrates. While their jawless ancestors could only feed on detritus, jawed fish could eat plants and also become active predators, thus opening many new ecological niches. The evolution of the vertebrate jaw is an excellent example of evolutionary opportunism. Various studies suggest that the jaw originally evolved from the first three gill arches of jawless fish. Because the gills are soft, they are supported by gill arches composed of bone or cartilage. The evolution of the jaw may thus have been related to respiration rather than feeding (◆ Fig. 22.20). By evolving joints in the forward gill arches, jawless fish could open their mouths wider. Every time a fish opened and closed its mouth, it would pump more water past the gills, thereby increasing the oxygen intake. The modification from rigid to hinged forward gill arches enabled fish to increase both their food consumption and oxygen intake, and the evolution of the jaw as a feeding structure rapidly followed.

The remains of the first jawed fish, called **acanthodians** (◆ Fig. 22.19c, Table 22.2), are found in Lower Silurian nonmarine rocks. Acanthodians are an enigmatic group of fish characterized by large spines, scales covering much of the body, jaws, teeth, and reduced bony armor. Their relationship to other fish has not been well established. The acanthodians were most abundant during the Devonian, declined in importance

◆ FIGURE 22.19 Recreation of a Devonian sea floor showing (*a*) an ostracoderm (*Hemicyclaspis*), (*b*) a placoderm (*Bothriolepis*), (*c*) an acanthodian (*Parexus*) and (*d*) a ray- finned fish (*Cheirolepis*).

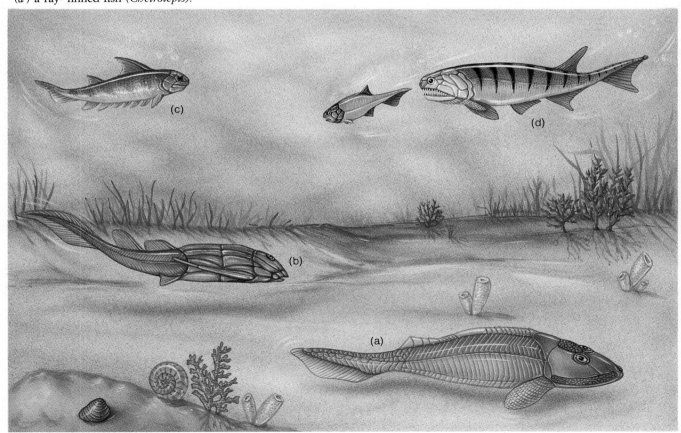

through the Carboniferous, and became extinct during the Permian.

While we do not know how the acanthodians were related to other, more complex fish, we do know that during the Devonian a major adaptive radiation of jawed fish occurred. **Placoderms** (Table 22.2), whose name means "plate-skinned," evolved during the Late Silurian and, like acanthodians, reached their peak of abundance and diversity during the Devonian. Placo-

derms were heavily armored jawed fish that lived in both fresh water and the ocean. The placoderms exhibited considerable variety, including small bottom-dwellers called *antiarchs* (◆ Fig. 22.19b) as well as *arthrodires,* which were the major predators of the Devonian seas. Arthrodires are best represented by *Dunkleosteus,* a Late Devonian fish that lived in the mid-continental North American epeiric seas (◆ Fig. 22.21a). It was by far the largest fish of the time,

◆ FIGURE 22.20 The evolution of the vertebrate jaw is thought to have occurred from the modification of the first two or three anterior gill arches. This theory is based on the comparative anatomy of living vertebrates.

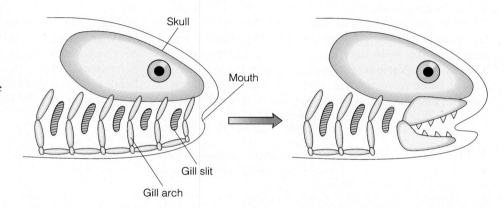

◆ FIGURE 22.21 A Late Devonian marine scene from the midcontinent of North America. (*a*) The giant placoderm *Dunkleosteus* (length more than 12 m) is pursuing (*b*) the shark *Cladoselache* (length up to 1.2 m). Also shown are (*c*) the bottom-dwelling placoderm *Bothriolepis* and (*d*) the swimming ray-finned fish *Cheirolepis,* both of which attained a length of 40–50 cm.

attaining a length of more than 12 m. It had a heavily armored head and shoulder region, a huge jaw lined with razor-sharp bony teeth, and a flexible tail, all features consistent with its status as a ferocious predator.

In addition to the abundant acanthodians, placoderms, and ostracoderms, other fish groups, such as the cartilaginous and bony fish, also evolved during the Devonian Period. It is small wonder, then, that the Devonian is informally called the "Age of Fish," since all major fish groups were present during this time period.

The **cartilaginous fish,** class Chrondrichthyes (Table 22.2), represented today by sharks, rays, and skates, first evolved during the Middle Devonian, and by the Late Devonian, primitive marine sharks such as *Cladoselache* were quite abundant (◆ Fig. 22.21b). Cartilaginous fishes have never been as numerous nor as diverse as their cousins, the bony fishes, but they were, and still are, important members of the marine vertebrate fauna.

Along with the cartilaginous fish, the **bony fish,** class Osteichthyes (Table 22.2), also first evolved during the Devonian. Because bony fish are the most varied and numerous of all the fishes, and because the amphibians evolved from them, their evolutionary history is particularly important. There are two groups of bony fish: the common **ray-finned fish** (◆ Figure 22.21d) and the less familiar **lobe-finned fish** (Table 22.2).

The term *ray-finned* refers to the way the fins are supported by thin bones that spread away from the body (◆ Fig. 22.22a). From a modest freshwater beginning during the Devonian, the ray-finned fish, which include most of the familiar fish such as trout, bass, perch, salmon, and tuna, rapidly diversified to dominate the Mesozoic and Cenozoic seas.

Present-day lobe-finned fish are characterized by muscular fins. The fins do not have radiating bones, but rather articulating bones with the fin attached to the body by a fleshy shaft (◆ Fig. 22.22b). Two major groups of lobe-finned fish are recognized: the *lung fish* and *crossopterygians* (Table 22.2). The lung fish were fairly abundant during the Devonian, but today only three freshwater genera exist, one each in South America, Africa, and Australia. Their present-day distribution presumably reflects the Mesozoic breakup of Gondwana.

The **crossopterygians** are a second group of lobe-finned fish and a most important group because it was from them that the amphibians evolved. During the Devonian, two separate branches of crossopterygians evolved. One led to the amphibians, while the other

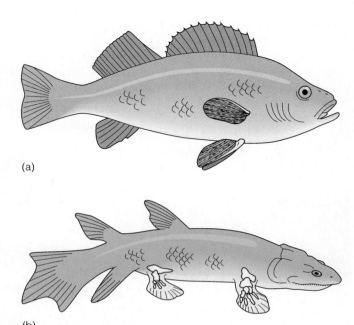

(a)

(b)

◆ FIGURE 22.22 Arrangement of fin bones for (*a*) a typical ray-finned fish and (*b*) a lobe-finned fish. The muscles extend into the fin of the lobe-finned fish, allowing greater flexibility of movement than for the ray-finned fish.

invaded the sea. This latter group, the *coelacanths,* were thought to have become extinct at the end of the Cretaceous. In 1938, however, fisherman caught a coelacanth in the deep waters off Madagascar, and since then several dozen more have been caught.

The group of crossopterygians that is ancestral to amphibians are *rhipidistians* (Table 22.2). These fish, attaining lengths of over 2 m, were the dominant freshwater predators of the Late Paleozoic. *Eusthenopteron,* a good example of a rhipidistian crossopterygian, had an elongate body that enabled it to move swiftly in the water, as well as paired muscular fins that could be used for locomotion on land (◆ Fig. 22.23). The structural similarity between crossopterygian fish and the earliest amphibians is striking and one of the better documented transitions from one major group to another (◆ Fig. 22.24).

▼ AMPHIBIANS—VERTEBRATES INVADE THE LAND

Although amphibians were the first vertebrates to live on land, they were not the first land-living organisms.

◆ FIGURE 22.23 *Eusthenopteron,* a member of the rhipidistian crossopterygians. The crossopterygians are the group from which the amphibians are thought to have evolved. *Eusthenopteron* had an elongate body and paired fins that could be used for moving about on land.

Land plants, which probably evolved from green algae, first evolved during the Ordovician. Furthermore, insects, millipedes, spiders, and even snails invaded the land before amphibians (see Perspective 22.2).

The transition from water to land required that several barriers be surmounted. The most critical for animals were desiccation, reproduction, the effects of gravity, and the extraction of oxygen from the atmosphere by lungs rather than from water by gills. These problems were partly solved by the crossopterygians; they already had a backbone and limbs that could be used for walking and lungs that could extract oxygen (Fig. 22.24).

The earliest amphibian fossils are found in the Upper Devonian Old Red Sandstone of eastern Greenland. These amphibians had streamlined bodies, long tails, and fins. In addition, they had four legs, a strong backbone, a rib cage, and pelvic and pectoral girdles, all of which were structural adaptations for walking on land (◆ Fig. 22.25). The earliest amphibians appear to have had many characteristics that were inherited from the crossopterygians with little modification.

The Late Paleozoic amphibians did not at all resemble the familiar frogs, toads, newts, and salamanders that make up the modern amphibian fauna. Rather they displayed a broad spectrum of sizes,

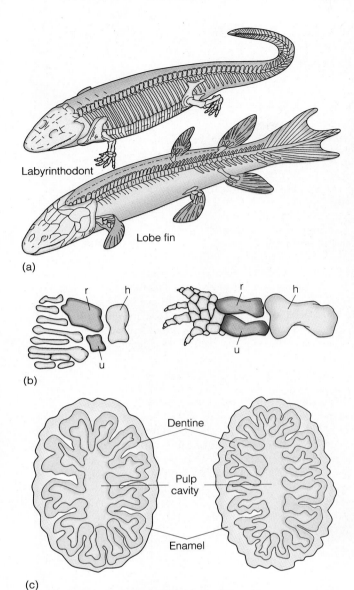

Labyrinthodont

Lobe fin

(a)

(b)

Dentine

Pulp cavity

Enamel

(c)

◆ FIGURE 22.24 Similarities between the crossopterygian lobe-finned fish and the labyrinthodont amphibians. (*a*) Skeletal similarity. (*b*) Comparison of the limb bones of a crossopterygian (left) and amphibian (right); colors identifies the bones (u = ulna, shown in blue, r = radius, mauve, h = humerus, gold) that the two groups have in common. (*c*) Comparison of tooth cross sections shows the complex and distinctive structure found in both the crossopterygians (left) and amphibians (right).

shapes, and modes of life (◆ Fig. 22.26). One group of amphibians were the **labyrinthodonts,** so named for the labyrinthine wrinkling and folding of the chewing surface of their teeth (Fig. 22.24). Most labyrinthodonts

◆ FIGURE 22.25 A Late Devonian landscape in the eastern part of Greenland. Shown is *Ichthyostega,* an amphibian that grew to a length of about 1 m. The flora of the time was diverse, consisting of a variety of small and large seedless vascular plants.

were large animals, as much as 2 m in length. These typically sluggish creatures lived in swamps and streams, eating fish, vegetation, insects, and other small amphibians (Fig. 22.26).

Labyrinthodonts were very abundant during the Carboniferous when swampy conditions were widespread, but soon declined in abundance during the Permian, perhaps in response to changing climatic conditions. Only a few species survived into the Triassic.

▼ EVOLUTION OF THE REPTILES— THE LAND IS CONQUERED

Amphibians were limited in colonizing the land because they had to return to water to lay their gelatinous eggs. The evolution of the **amniote egg** (◆ Fig. 22.27) freed reptiles from this constraint. In such an egg, the developing embryo is surrounded by a liquid-filled sac called the *amnion* and provided with both a yolk, or food sac, and an allantois, or waste sac. In this way the emerging reptile is, in essence, a miniature adult, bypassing the need for a larval stage in the water. The evolution of the amniote egg allowed vertebrates to

A LATE ORDOVICIAN INVASION OF THE LAND

Recently, the question of when the land was first colonized has been much discussed. Excellent evidence indicates that the land was colonized by vascular plants during the Silurian Period. Fragments of plant remains have been found in Middle Silurian rocks indicating primitive plants had invaded the land by that time. These early plants, like their Devonian descendants, had to live near bodies of water due to their reproductive requirements.

While Silurian plant fossil evidence is represented by a handful of geographically scattered fossils, the spore evidence for a Silurian land colonization is much more abundant and diverse. This may be due in part to the greater chance for preservation of spores, which have a resistant organic covering. In the North Atlantic region, Silurian vascular plant megafossil evidence is based primarily on specimens of *Cooksonia* (Fig. 22.31). Yet at least 14 spore genera are described for the Silurian Period.

Discoveries of probable vascular plant megafossils and characteristic spores indicate to many paleontologists that the evolution of vascular plants occurred well before the Middle Silurian. Sheets of cuticlelike cells—that is, the cells that cover the surface of modern land plants—and tetrahedral clusters that closely resemble the spore tetrahedrals of primitive land plants have been reported from Middle to Upper Ordovician rocks from western Libya by Jane Gray and her co-workers at the University of Oregon (◆Fig. 1).

The interpretation of these spores has been controversial and not completely accepted by all paleontologists. If they are in fact from land plants, this means that land plants had a long pre-Silurian record. And if land plants existed during the Ordovician, when did animals invade the land?

The first vertebrate animals made the transition from water to land during the Late Devonian. Arthropods, however, including scorpions and flightless insects, evolved at least by the Early Devonian.

The discovery of fossil burrows within a buried soil of the Upper Ordovician Juanita Formation in central Pennsylvania represents the oldest reported nonmarine trace fossils. While no hard parts or traces of organisms are associated with the burrows, their size, shape, and arrangement are consistent with a bilaterally symmetrical burrower that was resistant to desiccation. It is hypothesized that these burrows may have been made by millipedes.

The existence of sizable burrowing organisms on dry land during the Late Ordovician implies that some type of terrestrial vegetation existed for them to feed on. Burrowing animals could conceivably have fed on soil algae, now thought to have been in existence since the Late Precambrian. The discovery of possible primitive land plant spores from Middle to Upper Ordovician rocks means that these plants could have supported large populations of burrowing herbaceous arthropods as well as litter organisms.

◆ FIGURE 1 Fossils that closely resemble the spore tetrahedrals of primitive land plants. The sheet of cuticlelike cells (center) is from the Upper Ordovician Melez Chograne Formation of Libya. The others are from the Upper Ordovician Djeffara Formation of Libya. (Photos courtesy of Jane Gray, University of Oregon.)

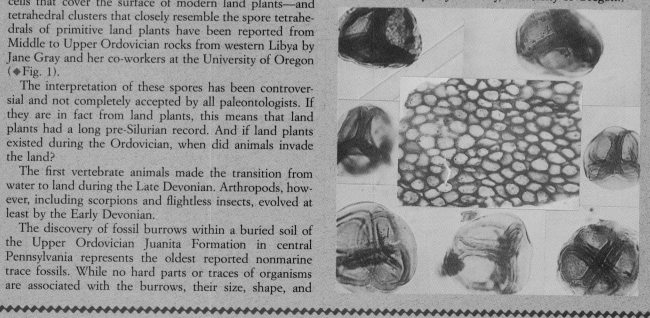

◆ FIGURE 22.26 Reconstruction of a Carboniferous coal swamp. The varied amphibian fauna of the time is shown, including the large labyrinthodont amphibian *Eryops* (foreground), the larval *Branchiosaurus* (center), and the serpent-like *Dolichosoma* (background).

colonize all parts of the land because they no longer had to return to the water as part of their reproductive cycle.

Many of the differences between amphibians and reptiles are physiological and are not preserved in the fossil record. Nevertheless, amphibians and reptiles differ sufficiently in skull structure, jawbones, ear location, and limb and vertebral construction to suggest that reptiles evolved from labyrinthodont ancestors by at least the early Pennsylvanian.

The oldest known reptiles are from the Lower Pennsylvanian Joggins Formation in Nova Scotia, Canada. Here, remains of *Hylonomus* are found in the sediments filling in tree trunks. These earliest reptiles were small and agile and fed largely on grubs and insects. They belonged to the group of reptiles known as **captorhinomorphs,** the group from which all other reptiles evolved (◆ Fig. 22.28). During the Permian Period, reptiles diversified and began displacing many amphibians. The success of the reptiles is due partly to their advanced method of reproduction and their more advanced jaws and teeth as well as to their ability to move rapidly on land.

The **pelycosaurs,** or finback reptiles, evolved from the captorhinomorphs during the Pennsylvanian and

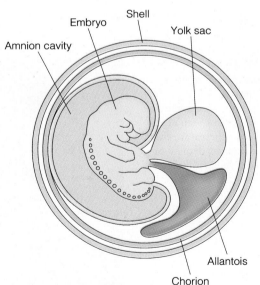

◆ FIGURE 22.27 The amniote egg. In an amniote egg, the embryo is surrounded by a liquid sac (amnion cavity) and provided with a food source (yolk sac) and waste sac (allantois). The evolution of the amniote egg freed reptiles from having to return to the water for reproduction and allowed them to inhabit all parts of the land.

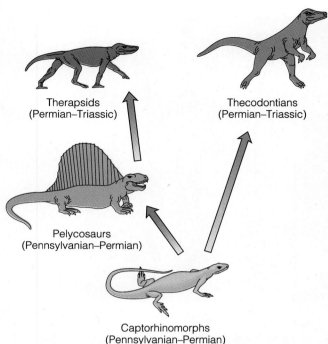

FIGURE 22.28 Evolutionary relationships among the earliest reptiles.

Therapsids
(Permian–Triassic)

Thecodontians
(Permian–Triassic)

Pelycosaurs
(Pennsylvanian–Permian)

Captorhinomorphs
(Pennsylvanian–Permian)

were the dominant reptile group by the Early Permian. They evolved into a diverse assemblage of herbivores, exemplified by *Edaphosaurus,* and carnivores such as *Dimetrodon* (◆ Fig. 22.29). An interesting feature of the pelycosaurs is their sail. It was formed by vertebral spines that, in life, were covered with skin. The sail has been variously explained as a type of sexual display, a means of protection, and a display to look more ferocious, but current consensus seems to be that the sail served as some type of thermoregulatory device, raising the reptile's temperature by catching the sun's rays or cooling it by facing the wind. Because pelycosaurs are considered to be the group from which therapsids evolved, it is interesting that they may have had some sort of body-temperature control.

The pelycosaurs became extinct during the Permian and were succeeded by the **therapsids,** mammal-like reptiles that evolved from the carnivorous pelycosaur lineage and rapidly diversified into herbivorous and carnivorous lineages (◆ Fig. 22.30). Therapsids were small to medium-sized animals displaying the beginnings of many mammalian features; fewer bones in the skull due to fusion of many of the small skull bones; enlargement of the lower jawbone; differentiation of the teeth for various functions such as nipping, tearing, and chewing food; and a more vertical position of the legs for greater flexibility, as opposed to the way the legs sprawled out to the side in primitive reptiles.

◆ **FIGURE 22.29** Most pelycosaurs, or finback reptiles have a characteristic sail on their back. One hypothesis explains the sail as a type of thermoregulatory device. Other hypotheses are that it was a type of sexual display or a device to make the reptile look more intimidating. Shown here are (*a*) the carnivore *Dimetrodon* and (*b*) the herbivore *Edaphosaurus.*

◆ FIGURE 22.30 A Late Permian scene in southern Africa showing various therapsids including *Dicynodon* (left foreground) and *Moschops* (right). Many paleontologists think therapsids were endothermic and may have had a covering of fur as shown here.

Furthermore, some paleontologists think therapsids were *endothermic,* or warm-blooded, enabling them to maintain a constant internal body temperature. This characteristic would have allowed them to expand into a variety of habitats, and indeed the Permian rocks do indicate a wide latitudinal distribution.

As the Paleozoic Era came to an end, the therapsids constituted about 90% of the known reptile genera and occupied a wide range of ecological niches. The mass extinctions that decimated the marine fauna at the close of the Paleozoic had an even greater effect on the terrestrial population. By the end of the Permian, about 50% of the invertebrate marine families were extinct compared with 75% of the amphibians and 80% of the reptile families. Plants, on the other hand, apparently did not experience as great a turnover as animals.

▽ PLANT EVOLUTION

When plants made the transition from water to land, they had to solve the same problems that animals did: desiccation, support, and the effects of gravity. Plants did so by evolving a variety of structural adaptations that were fundamental to the subsequent radiations and diversification that occurred during the Silurian, Devonian, and later periods (● Table 22.3). Most experts agree that the ancestors of land plants first evolved in a marine environment, then moved into a freshwater environment and finally onto land. In this way, the differences in osmotic pressures between salt and fresh water were overcome while the plant was still in the water.

The higher land plants are composed of two major groups, the nonvascular and vascular plants. Most land plants are **vascular,** meaning they have a tissue system of specialized cells for the movement of water and nutrients. The **nonvascular** plants, such as bryophytes (liverworts, hornworts, and mosses) and fungi, do not have these specialized cells and are typically small and usually live in low, moist areas.

The earliest land plants from the Middle to Late Ordovician were probably small and bryophyte-like in their overall organization (but not necessarily related to bryophytes). The evolution of vascular tissue in plants was an important step as it allowed for the transport of food and water. The ancestor of terrestrial vascular plants was probably some type of green alga. While no fossil record of the transition from green algae to terrestrial vascular plants exists, comparison of their physiology reveals a strong link. Primitive *seedless vascular plants* such as ferns resemble green algae in their pigmentation, important metabolic enzymes, and type of reproductive cycle. Furthermore, the green algae are one of the few plant groups to have made the transition from marine to fresh water. The evolution of terrestrial vascular plants from an aquatic, probably green algal ancestry was accompanied by various modifications that allowed them to occupy this new and harsh environment.

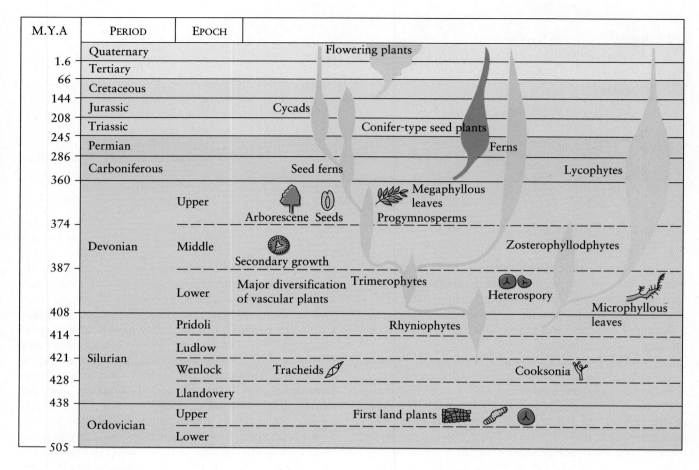

● **TABLE 22.3** Major Events in the Evolution of Land Plants. The Devonian Period was a time of rapid evolution for the land plants. Major events were the appearance of leaves, heterospory, secondary growth, and the emergence of seeds.

M.Y.A	PERIOD	EPOCH	
1.6	Quaternary		Flowering plants
66	Tertiary		
144	Cretaceous		
208	Jurassic		Cycads
245	Triassic		Conifer-type seed plants
286	Permian		Ferns
360	Carboniferous		Seed ferns · · · Lycophytes
	Devonian	Upper	Arborescene · Seeds · Megaphyllous leaves · Progymnosperms
374		Middle	Secondary growth · · · Zosterophyllodphytes
387		Lower	Major diversification of vascular plants · Trimerophytes · Heterospory · Microphyllous leaves
408	Silurian	Pridoli	Rhyniophytes
414		Ludlow	
421		Wenlock	Tracheids · Cooksonia
428		Llandovery	
438	Ordovician	Upper	First land plants
		Lower	
505			

Silurian and Devonian Floras

The earliest known vascular land plants are small Y-shaped stems assigned to the genus *Cooksonia* from the Middle Silurian of Wales and Ireland. Together with Upper Silurian and Lower Devonian species from Scotland, New York State, former Czechoslovakia, and the Commonwealth of Independent States, these earliest plants were small, simple, leafless stalks with a spore-producing structure at the tip (◆ Fig. 22.31); they are known as **seedless vascular plants** because they did not produce a seed. They also did not have a true root

system. A *rhizome,* the underground part of the stem, transferred water from the soil to the plant and anchored the plant to the ground. The sedimentary rocks in which these plant fossils are found indicate that they lived in low, wet, marshy, freshwater environments.

An interesting parallel can be seen between seedless vascular plants and amphibians. When they made the transition from water to land, they had to overcome the problems such a transition involved. Both groups, while very successful, nevertheless required a source of water in order to reproduce. In the case of amphibians, their gelatinous egg had to remain moist, while the

◆ FIGURE 22.31 The earliest known fertile land plant was *Cooksonia*, seen in this fossil from the Upper Silurian of South Wales. *Cooksonia* consisted of upright, branched stems terminating in sporangia (spore-producing structures). It also had a resistant cuticle and produced spores typical of a vascular plant. These plants probably lived in moist environments such as mud flats. This specimen is 1.49 cm long. (Photo courtesy of Dianne Edwards, University College, England.)

seedless vascular plants required water for the sperm to travel through to reach the egg.

From this simple beginning, the seedless vascular plants evolved many of the major structural features characteristic of present-day plants such as leaves, roots, and secondary growth. These features did not all evolve simultaneously but rather at different times, a pattern known as *mosaic evolution*. This diversification and adaptive radiation took place during the Late Silurian and Early Devonian and resulted in a tremendous increase in diversity. From the end of the Early Devonian to the end of the Devonian, the number of plant genera remained about the same, yet the composition of the flora changed. Whereas the Early Devonian landscape was dominated by relatively small, low-growing, bog-dwelling types of plants, the Late Devonian witnessed forests of large tree-sized plants up to 10 m tall.

In addition to the diverse seedless vascular plant flora of the Late Devonian, another significant floral event took place. The evolution of the seed at this time liberated land plants from their dependence on moist conditions and allowed them to spread over all parts of the land.

Seedless vascular plants require moisture for successful fertilization because the sperm must travel to the egg on the surface of the gamete-bearing plant (gametophyte) to produce a successful spore-generating plant (sporophyte). Without moisture, the sperm would dry out before reaching the egg (◆ Fig. 22.32a). In the seed method of reproduction, the spores are not released to the environment as they are in the seedless vascular plants, but are retained on the spore-bearing plant, where they grow into the male and female forms of the gamete-bearing generation. In the case of the **gymnosperms,** or flowerless seed plants, these are male and female cones (◆ Fig. 22.32b). The male cone produces pollen, which contains the sperm and has a waxy coating to prevent desiccation, while the egg, or embryonic seed, is contained in the female cone. After fertilization, the seed then develops into a mature, cone-bearing plant. In this way the need for a moist environment for the gametophyte generation is solved. The significance of this development is that seed plants, like reptiles, were no longer restricted to wet areas, but were free to migrate into previously unoccupied dry environments. While the seedless vascular plants dominated the flora of the Carboniferous coal-forming swamps, the gymnosperms made up an important element of the Late Paleozoic flora, particularly in the nonswampy areas.

Late Carboniferous and Permian Floras

As discussed earlier, the rocks of the Pennsylvanian Period (Late Carboniferous) are the major source of the world's coal. Coal results from the alteration of plants living in low, swampy areas. The geologic and geographic conditions of the Pennsylvanian were ideal for the growth of seedless vascular plants, and consequently these coal swamps had a very diverse flora (◆ Fig. 22.33).

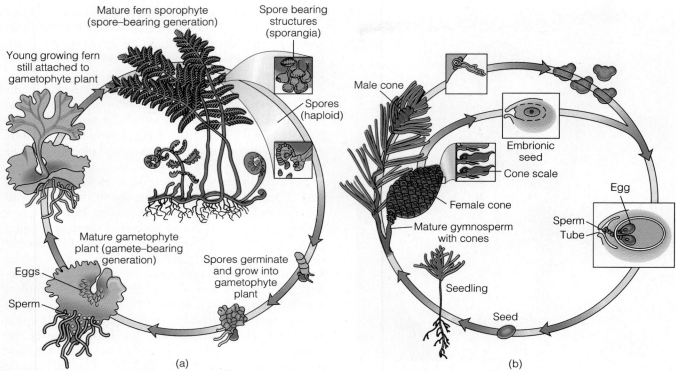

(a)

(b)

◆ FIGURE 22.32 (*a*) Generalized life history of a seedless vascular plant. The mature sporophyte plant produces spores, which upon germination grow into small gametophyte plants that produce sperm and eggs. The fertilized eggs grow into the spore-producing mature plant, and the sporophyte-gametophyte life cycle begins again. (*b*) Generalized life history of a gymnosperm plant. The mature plant bears both male cones that produce sperm-bearing pollen grains and female cones that contain embryonic seeds. Pollen grains are transported to the female cones by the wind. Fertilization occurs when the sperm moves through a moist tube growing from the pollen grain and unites with the embryonic seed which then grows into a cone-bearing mature plant.

It is evident from the fossil record that while the Early Carboniferous flora was similar to its Late Devonian counterpart, a great deal of evolutionary experimentation was occurring that would lead to the highly successful Late Paleozoic flora of the coal swamps and adjacent habitats. Among the seedless vascular plants, the *lycopsids* and *sphenopsids* were the most important coal-forming groups of the Pennsylvanian Period.

The lycopsids were the dominant element of the coal swamps, achieving heights up to 30 m in such genera as *Lepidodendron* and *Sigillaria.* The Pennsylvanian lycopsid trees are interesting because they lacked branches except at their top. The leaves were elongate and similar to the individual palm leaf of today. As the trees grew, the leaves were replaced from the top, leaving prominent and characteristic rows or spirals of scars on the trunk. Today, the lycopsids are represented by small temperate-forest ground pines.

The sphenopsids, the other important coal-forming plant group, are characterized by being jointed and having horizontal underground stem-bearing roots. Many of these plants, such as *Calamites,* averaged 5 to 6 m tall. Living sphenopsids include the horsetail *(Equisetum)* and scouring rushes (◆ Fig. 22.34). Small seedless vascular plants and seed ferns formed a thick

◆ FIGURE 22.33 Reconstruction of a Pennsylvanian coal swamp with its characteristic vegetation. The amphibian is *Eogyrinus.*

undergrowth or ground cover beneath these treelike plants.

Not all plants were restricted to the coal-forming swamps. Among those plants occupying higher and drier ground were some of the *cordaites,* a group of tall gymnosperm trees that grew up to 50 m and probably formed vast forests (◆ Fig. 22.35). Another important non-swamp dweller was *Glossopteris,* the famous plant so abundant in Gondwana (see Fig. 9.3), whose distribution is cited as critical evidence that the continents have moved through time.

The floras that were abundant during the Pennsylvanian persisted into the Permian, but due to climatic and geologic changes resulting from tectonic events they declined in abundance and importance. By the end of the Permian, the cordaites became extinct, while the lycopsids and sphenopsids were reduced to mostly small, creeping forms. The gymnosperms, whose lifestyle was more suited to the warmer and drier Permian climates, diversified and came to dominate the Permian, Triassic, and Jurassic landscapes.

◆ FIGURE 22.34 Living sphenopsids include the horsetail *Equisetum*.

◆ FIGURE 22.35 A cordaite forest from the Late Carboniferous. Cordaites were a group of gymnosperm trees that grew up to 50 m tall.

Chapter Summary

• Table 22.4 summarizes the major evolutionary and geologic events of the Paleozoic Era and shows their relationships to each other.

1. Soft-bodied multicelled organisms presumably had a long Precambrian history during which they lacked hard parts. Hard parts evolved in different invertebrate groups during the Cambrian and Ordovician periods and provided such advantages as protection against predators and support for muscles, enabling organisms to grow large and increase locomotor efficiency.

2. Marine organisms are classified as plankton if they are floaters, nekton if they swim, and benthos if they live on or in the sea floor.

3. Marine organisms can be divided into four basic feeding groups: suspension feeders, which remove or consume microscopic plants and animals as well as dissolved nutrients from water; herbivores, which are plant eaters; carnivore-scavengers, which are meat eaters; and sediment-deposit feeders, which ingest sediment and extract nutrients from it.

4. The marine ecosystem consists of various trophic levels of food production and consumption. At the base are primary producers, upon which all other organisms are dependent. Feeding on the primary producers are the primary consumers, which in turn can be fed upon by higher levels of consumers. The decomposers are bacteria that break down the complex organic compounds of dead organisms and recycle them within the ecosystem.

5. The Cambrian invertebrate community was dominated by three major groups—the trilobites, brachiopods, and archaeocyathids. Little specialization existed among the invertebrates, and most phyla were represented by only a few species.

6. The Ordovician marine invertebrate community marked the beginning of the dominance by the shelly fauna and the start of large-scale reef building. The end of the Ordovician Period was a time of major extinctions for many of the invertebrate phyla.

7. The Silurian and Devonian periods were times of diverse faunas dominated by reef-building animals, while the Carboniferous and Permian periods saw a great decline in invertebrate diversity.

8. Chordates are characterized by a notochord, dorsal hollow nerve cord, and gill slits. The earliest chordates were soft-bodied organisms that were rarely fossilized. Vertebrates are a subphylum of the chordates.

9. The fish are the earliest known vertebrates. Their first fossil occurrence is in Upper Cambrian rocks. They have had a long and varied history, including jawless and jawed armored forms (ostracoderms and placoderms), cartilaginous forms, and bony forms. The crossopterygians, a group of lobe-finned fish, gave rise to the amphibians.

10. The link between the crossopterygians and the earliest amphibians is very convincing and includes a close similiarity of bone and tooth structures. The transition from fish to amphibians occurred during the Devonian. During the Carboniferous, the labyrinthodont amphibians were the dominant terrestrial vertebrate animals.

11. The earliest fossil record of reptiles is from the Late Carboniferous. The evolution of an amniote egg was the critical factor in the reptiles' ability to colonize all parts of the land.

12. The pelycosaurs were the dominant reptile group during the Early Permian, while the therapsids dominated the landscape for the rest of the Permian Period.

13. Plants had to overcome the same basic problems as the animals, namely desiccation, reproduction, and gravity in making the transition from water to land.

14. The earliest fossil record of land plants is from Middle to Upper Ordovician rocks. These plants were probably small and bryophyte-like in their overall organization.

15. The evolution of vascular tissue was an important event in plant evolution as it allowed food and water to be transported throughout the plant and provided the plant with additional support.

16. The ancestor of terrestrial vascular plants was probably some type of green alga based on such similarities as pigmentation, metabolic enzymes, and the same type of reproductive cycle.

17. The earliest seedless vascular plants were small, leafless stalks with spore-producing structures on their tips. From this simple beginning, plants evolved many of the major structural features characteristic of today's plants.

18. By the end of the Devonian Period, forests with tree-sized plants up to 10 m had evolved. The Late Devonian also witnessed the evolution of the flowerless seed plants (gymnosperms) whose reproductive style freed them from having to stay near water.

19. The Carboniferous Period was a time of vast coal swamps, where conditions were ideal for the seedless vascular plants. With the onset of more arid conditions during the Permian, the gymnosperms became the dominant element of the world's flora.

20. A major extinction occurred at the end of the Paleozoic Era, affecting the invertebrates as well as the vertebrates.

● TABLE 22.4 Major Evolutionary and Geologic Events of the Paleozoic Era

Age (Millions of Years)	Geologic Period		Invertebrates	Vertebrates
245				
	Permian		Largest mass extinction event to affect the invertebrates.	Acanthodians, placoderms, and pelycosaurs become extinct. Therapsids and pelycosaurs the most abundant reptiles.
286	Carboniferous	Pennsylvanian	Fusulinids diversify.	Reptiles evolve. Amphibians abundant and diverse.
320		Mississippian	Crinoids, lacy bryozans, blastoids become abundant. Renewed adaptive radiation following extinctions of many reef-builders.	
360	Devonian		Extinctions of many reef-building invertebrates near end of Devonian. Reef building continues. Eurypterids abundant.	Amphibians evolve. All major groups of fish present—Age of Fish.
408	Silurian		Major reef building. Diversity of invertebrates remains high.	Ostracoderms common. Acanthodians, the first jawed fish, evolve.
438	Ordovician		Extinctions of a variety of marine invertebrates near end of Ordovician. Major adaptive radiation of all invertebrate groups. Suspension feeders dominant.	Ostracoderms diversify.
505	Cambrian		Many trilobites become extinct near end of Cambrian. Trilobites, brachiopods, and archaeocyathids are most abundant.	Earliest vertebrates—jawless fish called ostracoderms.
570				

Plants	Major Geologic Events
Gymnosperms diverse and abundant.	Formation of Pangaea. Alleghenian orogeny. Hercynian orogeny.
Coal swamps with flora of seedless vascular plants and gymnosperms.	Coal-forming swamps common. Formation of Ancestral Rockies. Continental glaciation in Gondwana.
Gymnosperms appear (may have evolved during Late Devonian).	Ouachita orogeny.
First seeds evolve. Seedless vascular plants diversify.	Widespread deposition of black shale. Antler orogeny. Acadian orogeny.
Early land plants—seedless vascular plants.	Caledonian orogeny. Extensive barrier reefs and evaporites.
Plants move to land?	Continental glaciation in Gondwana. Taconic orogeny.
	First Phanerozoic transgression (Sauk) onto North American craton.

Important Terms

acanthodian
amniote egg
archaeocyathid
benthos
bony fish
brachiopod
captorhinomorph
carnivore-scavenger
cartilaginous fish
chordate
crossopterygian

gymnosperm
herbivore
labyrinthodont
lobe-finned fish
nekton
nonvascular
ostracoderm
pelycosaur
placoderm
plankton

primary producer
ray-finned fish
sediment-deposit feeder
seedless vascular plant
suspension feeder
therapsid
trilobite
trophic level
vascualr
vertebrate

Review Questions

1. The first shelled fossils were:
   a. ___ trilobites; b. ___ small calcium carbonate tubes;
   c. ___ archaeocyathids; d. ___ brachiopods; e. ___ mollusks.
2. Organisms living on or in the sea floor are:
   a. ___ epifauna; b. ___ epiflora; c. ___ infauna;
   d. ___ benthos; e. ___ all of these.
3. An exoskeleton is advantageous because it:
   a. ___ provides protection against predators; b. ___ prevents drying out in an intertidal environment; c. ___ provides attachment sites for development of strong muscles; d. ___ provides protection against ultraviolet radiation; e. ___ all of these.
4. The greatest recorded mass extinction event to affect the marine invertebrate community occurred at the end of which period?
   a. ___ Cambrian; b. ___ Ordovician; c. ___ Silurian;
   d. ___ Devonian; e. ___ Permian.
5. Which two periods experienced the greatest reef-building activity during the Paleozoic?
   a. ___ Cambrian-Ordovician; b. ___ Ordovician-Silurian;
   c. ___ Silurian-Devonian; d. ___ Devonian-Mississippian;
   e. ___ Mississippian-Permian.
6. What type of invertebrates dominated the Ordovician invertebrate community?
   a. ___ epifaunal benthonic sessile suspension feeders;
   b. ___ infaunal benthonic sessile suspension feeders;
   c. ___ epifaunal benthonic mobile suspension feeders;
   d. ___ infaunal nektonic carnivores; e. ___ epifloral planktonic primary producers.
7. The Early Cambrian was characterized by a:
   a. ___ high-diversity soft-bodied fauna; b. ___ high-diversity shelly fauna; c. ___ low-diversity shelly fauna;
   d. ___ low-diversity soft-bodied fauna; e. ___ none of these.

8. A _____ is an example of an epifaunal benthonic suspension feeder.
   a. ___ trilobite; b. ___ cephalopod; c. ___ graptolite;
   d. ___ articulate brachiopod; e. ___ gastropod.
9. Which of the following must an organism possess during at least part of its life cycle to be classified as a chordate?
   a. ___ vertebrae, dorsal hollow nerve cord, gill slits;
   b. ___ notochord, ventral solid nerve cord, lungs;
   c. ___ notochord, dorsal hollow nerve cord, gill slits;
   d. ___ vertebrae, dorsal hollow nerve cord, lungs;
   e. ___ notochord, dorsal solid nerve cord, lungs.
10. Based on similarity of embryo development, which invertebrate phylum is most closely allied with the chordates?
    a. ___ Arthropoda; b. ___ Annelida; c. ___ Porifera;
    d. ___ Echinodermata; e. ___ Mollusca.
11. Jawless armored fish are:
    a. ___ ostracoderms; b. ___ placoderms; c. ___ cartilaginous; d. ___ bony; e. ___ none of these.
12. The "Age of Fish" refers to which period?
    a. ___ Cambrian; b. ___ Ordovician; c. ___ Silurian;
    d. ___ Devonian; e. ___ Mississippian.
13. The first jawed fish were:
    a. ___ ostracoderms; b. ___acanthodians; c. ___ placoderms; d. ___ cartilaginous; e. ___ lobe-finned.
14. The ancestors of amphibians belong to which group?
    a. ___ placoderms; b. ___ ostracoderms; c. ___ acanthodians; d. ___ crossopterygians; e. ___ antiarchs.
15. Labyrinthodonts are:
    a. ___ reptiles; b. ___ fish; c. ___ amphibians;
    d. ___ plants; e. ___ none of these.
16. The most significant evolutionary change that allowed reptiles to colonize all parts of the land was:
    a. ___ endothermy; b. ___ origin of limbs capable of supporting the animals on land; c. ___ evolution of the

amniote egg; d. ___ evolution of a watertight skin;
e. ___ evolution of tear ducts.

17. The finback reptiles *Dimetrodon* and *Edaphosaurus* were:
a. ___ therapsids; b. ___ pelycosaurs; c. ___ labyrinth-odonts; d. ___ acanthodians; e. ___ captorhinomorphs.

18. Which problems did plants have to contend with to make the successful transition from water to land?
a. ___ desiccation; b. ___ photosynthesis;
c. ___ gravity; d. ___ answers (a) and (b);
e. ___ answers (a) and (c).

19. Which algal group was the probable ancestor of vascular plants?
a. ___ red; b. ___ blue-green; c. ___ green;
d. ___ brown; e. ___ yellow.

20. The first plant group that did not require a wet area for part of its life cycle was the:
a. ___ seedless vascular plants; b. ___ gymnosperms;
c. ___ angiosperms; d. ___ flowering plants; e. ___ labyrinthodonts.

21. Discuss the significance of the appearance of the first shelled animals.

22. Draw a marine food web that shows the relationships among the producers, consumers, and decomposers.

23. Discuss the major differences between the Cambrian marine community and the Ordovician marine community.

24. Discuss the Paleozoic evolutionary history of the amphibians.

25. Discuss the evidence for an Ordovician faunal and floral colonization of the land.

26. Outline the evolutionary history of fish.

27. Describe the problems that had to be overcome before organisms could inhabit the land.

28. Why were the reptiles so much more successful at extending their habitat than the amphibians?

29. In what ways are therapsids more mammal-like than pelycosaurs?

30. What are the major differences between seedless vascular plants and gymnosperms? What is the significance of these differences in terms of exploiting the terrestrial environment?

Additional Readings

Carroll, R. L. 1988. *Vertebrate paleontology and evolution.* New York: W. H. Freeman.

Carroll, R. L. 1992. The primary radiation of terrestrial vertebrates. *Annual Review of Earth and Planetary Sciences* 20:45–84.

Clarkson, E. N. K. 1993. *Invertebrate palaeontology and evolution.* 3d ed. New York: Chapman & Hall.

Colbert, E. H., and M. Morales. 1991. *Evolution of the vertebrates.* 4th ed. New York: John Wiley & Sons.

Cowen, R. 1989. *History of life.* Palo Alto, Calif.: Blackwell Scientific Publications.

Donovan, S. K., ed. 1989. *Mass extinctions.* New York: Columbia University Press.

Gensel, P. G., and H. N. Andrews. 1987. The evolution of early land plants. *American Scientist* 75, no. 5:478–89.

Gore, R. 1993. The Cambrian Period explosion of life. *National Geographic* 184, no. 4:120–136.

Gorr, T., and Kleinschmidt, T. 1993. Evolutionary relationships of the coelacanth. *American Scientist* 81, no.1:72–82.

Gould, S. J. 1989. *Wonderful life.* New York: W. W. Norton.

Gray, J., and W. Shear. 1992. Early life on land. *American Scientist* 80, no. 5:444–56.

Lane, N. G. 1992. *Life of the past.* 3d ed. Columbus, Ohio: Charles E. Merrill Publishing.

Levinton, J.S. 1992. The big bang of animal evolution. *Scientific American* 267, no. 5:84–93.

McMenamin, M. A. S. 1987. The emergence of animals. *Scientific American* 256, no. 4:94–102.

McMenamin, M. A. S., and D. L. Schulte McMenamin. 1990. *The emergence of animals: The Cambrian breakthrough.* New York: Columbia University Press.

McNamara, K., and P. Selden. 1993. Strangers on the shore. *New Scientist* 139, no. 1885: 23–27.

Reid, M. 1992. Ghosts of the Burgess Shale. *Earth* 1, no. 5:38–45.

Raup, D. M., and J. J. Sepkoski. 1984. Periodicity of extinctions in the geologic past. *Proceedings of the National Academy of Sciences,* 81:801–5.

Schopf, J. W., ed. 1992. *Major events in the history of life.* Boston: Jones and Bartlett.

Stanley, S. M. 1987. *Extinction.* New York: Scientific American Books.

Stearn, C. W., and R. L. Carroll. 1989. *Paleontology: The record of life.* New York: John Wiley & Sons.

Stewart, W. N., and G. W. Rothwell. 1993. *Paleobotany and the evolution of plants.* 2d ed. New York: Cambridge University Press.

Thomas, B., and R. Spicer. 1987. *Evolution and palaeobiology of land plants.* Portland, Oregon: Dioscorides Press.

Thomson, K. W. 1991. Where did tetrapods come from? *American Scientist* 79, no. 6:488–90.

White, M. E. 1986. *The greening of Gondwana.* Frenchs Forest, NSW, Australia: Reed Books Limited.

Whittington, H. B. 1985. *The Burgess Shale.* New Haven, Conn.: Yale University Press.

CHAPTER
23

Chapter Outline

Ownes Valley and eastern scarp along the Sierra Nevada, California. (Photo © Peter Kresan.)

Mesozoic Earth and Life History

Prologue

In 1909, Earl Douglass of the Carnegie Museum discovered "Dinosaur Ledge," a sandstone unit in the Upper Jurassic Morrison Formation that contained numerous dinosaur bones. After 13 years of excavating the layer, the Carnegie Museum had removed parts of 300 dinosaur specimens, two dozen of which were sufficiently complete to be reassembled. Ten different dinosaur species were represented as well as many other reptiles. It was by far the finest collection of dinosaur remains in the world. In the years that followed, the Smithsonian Institution and the University of Utah also worked this rich quarry and recovered more fossil specimens. Even more bones remained buried in an untouched part of the dipping sandstone ledge, and all that was needed to reveal them was the removal of the overlying layers of shale and siltstone.

Recognizing the scientific importance of this unit, the dinosaur quarry and 80 acres surrounding it were designated a national monument by President Wilson on October 4, 1915. Less than a year later, it was included in the newly created National Park System, and in 1938, Dinosaur National Monument was further expanded to 200,000 acres.

As far back as 1915, Earl Douglass envisioned an exhibit in which the dinosaur bones would be exposed in relief in the tilted sandstone bed exactly where they came to rest 140 million years ago. In a letter to Dr. Charles Walcott, secretary of the Smithsonian Institution, Douglass wrote: "I hope that the Government, for the benefit of science and the people, will uncover a large area, leave the bones and skeletons in relief, and house them in. It would make one of the most astounding and instructive sights

imaginable." Not until 1953, however, did work begin on a truly unique museum constructed around the still-buried sandstone ledge. The sediment overlying the remaining tilted sandstone ledge was removed, and the dinosaur bones were carefully exposed in bas relief. This quarry wall now forms the north wall of the visitors' center at Dinosaur National Monument (◆ Fig. 23.1). The structure was completed and opened to the public in 1958.

What was the landscape like during the Jurassic when dinosaurs roamed the area that is now Dinosaur National Monument? The land for miles around the present-day quarry was a low-lying desert during the Early Jurassic. This is indicated by the large cross-bedded dune sands of the Navajo Sandstone. Following deposition of the Navajo, a shallow sea transgressed from the west, depositing sandstones, siltstones, and limestones. During the Late Jurassic, the area from Mexico to Canada and from central Utah to the Mississippi River was above sea level. To the west was a mountain chain. Streams flowing from these mountains deposited sand and silt in the adjacent plains. Small lakes were numerous and some swamps were present. These stream, lake, and swamp deposits of the Jurassic coastal plain comprise the Morrison Formation.

Semitropical conditions prevailed during the Late Jurassic in the area, and forests of ginkos, cycads, and tree-ferns covered the land. In this setting pterosaurs glided through the air while dinosaurs roamed the landscape below. Crocodiles, turtles, and small mammals were also present. Most of the bones of the dinosaurs and other animals were deposited in the stream beds. During Cretaceous through Eocene time,

(a)

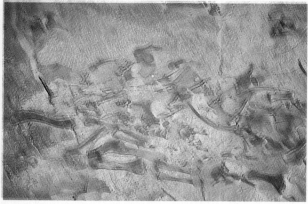

(b)

◆ FIGURE 23.1 (a) Visitors' center, Dinosaur National Monument. (b) North wall of visitors' center showing dinosaur bones in bas relief, just as they were deposited 140 million years ago.

the area was uplifted during the Laramide orogeny, and later the Green and Yampa rivers cut deep canyons in the area, exposing the rocks that make up Dinosaur National Monument.

⬇ INTRODUCTION

The Mesozoic Era (245 to 66 million years ago) is divided into three periods beginning with the Triassic, followed by the Jurassic, and finally the Cretaceous. The stratotypes for the systems from which these periods derive their names are in the Hercynian Mountains of Germany (Triassic), the Jura Mountains of Switzerland (Jurassic), and the Paris Basin of France (Cretaceous).

The Mesozoic Era is popularly known as the Age of Reptiles, a phrase emphasizing the fact that reptiles, and particularly dinosaurs, were the dominant land-dwelling vertebrate animals. It also was the time during which birds, mammals, and angiosperms (flowering plants) first evolved and diversified.

The dawn of the Mesozoic ushered in a new era in Earth history. The major geologic event was the breakup of Pangaea, which affected oceanic and climatic circulation patterns and influenced the evolution of the terrestrial biotas. Because most of the Mesozoic geologic history of North America involves the continental margins, we focus our attention on the eastern, gulf, and western coastal regions. The transgressions and regressions of the final two epeiric seas and their depositional sequences (Absaroka and Zuni) will be incorporated into the geologic history of each of these regions.

⬇ THE BREAKUP OF PANGAEA

Just as the formation of Pangaea influenced geologic and biologic events during the Paleozoic, the breakup of this supercontinent profoundly affected geologic and biologic events during the Mesozoic. The movement of continents affected the global climatic and oceanic regimes as well as the climates of the individual continents. Populations became isolated or were brought into contact with other populations, leading to evolutionary changes in the biota. So great was the effect of this breakup on the world, that it forms the central theme of this chapter.

Geologic, paleontologic, and paleomagnetic data indicate that the breakup of Pangaea occurred in four general stages. The first stage involved rifting between Laurasia and Gondwana during the Late Triassic. By the end of the Triassic, the expanding Atlantic Ocean separated North America from Africa (◆ Fig. 23.2a). This was followed by the rifting of North America from South America sometime during the Late Triassic and Early Jurassic.

Separation of the continents allowed water from the Tethys Sea to flow into the expanding central Atlantic Ocean, while Pacific Ocean waters flowed into the newly formed Gulf of Mexico, which at that time was little more than a restricted bay (◆ Fig. 23.3). During that time, these areas were located in the low tropical latitudes where high temperatures and high rates of evaporation were ideal for the formation of thick evaporite deposits.

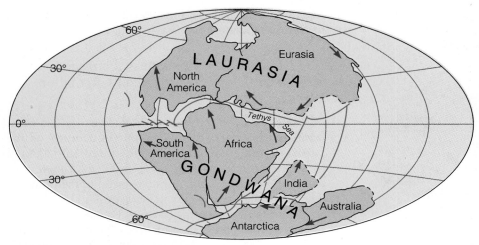

(a) Triassic Period (245–208 M.Y.A)

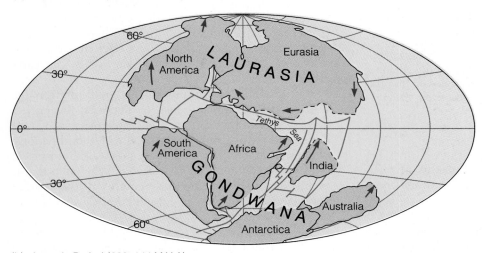

(b) Jurassic Period (208–144 M.Y.A)

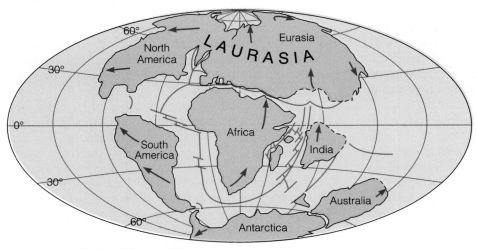

(c) Cretaceous Period (144–66 M.Y.A)

◆ FIGURE 23.2 Paleogeography of the world during the Mesozoic. Blue arrows show the direction of movement for the continents. (*a*) The Triassic Period, (*b*) the Jurassic Period, and (*c*) the Cretaceous Period.

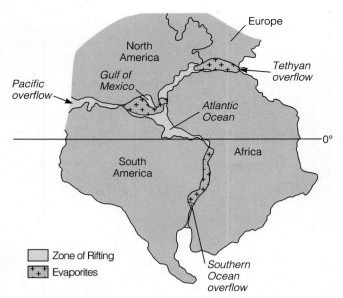

North America

Europe

Gulf of
Mexico

Tethyan
overflow

Pacific
overflow

Atlantic
Ocean

0°

South
America

Africa

Zone of Rifting

Evaporites

Southern
Ocean
overflow

◆ FIGURE 23.3 Evaporites accumulated in shallow basins as Pangaea broke apart during the Early Mesozoic. Water from the Tethys Sea flowed into the central Atlantic Ocean, while water from the Pacific Ocean flowed into the newly formed Gulf of Mexico. Marine water from the south flowed into the southern Atlantic Ocean.

The second stage in Pangaea's breakup involved rifting and movement of the various Gondwana continents during the Late Triassic and Jurassic periods. As early as the Late Triassic, Antarctica and Australia, which remained sutured together, separated from South America and Africa, while India split away from all four Gondwana continents and began moving northward (Figs. 23.2a and b).

The third stage of breakup began during the Late Jurassic, when South America and Africa began separating (◆ Fig. 23.2b). During this stage, the eastern end of the Tethys Sea began closing as a result of the clockwise rotation of Laurasia and the northward movement of Africa. This narrow Late Jurassic and Cretaceous seaway between Africa and Europe was the forerunner of the present Mediterranean Sea.

By the end of the Cretaceous, Australia and Antarctica had separated, India had nearly reached the equator, South America and Africa were widely separated, and the eastern side of what is now Greenland had begun separating from Europe (◆ Fig. 23.2c).

The final stage in the breakup of Pangaea occurred during the Cenozoic. During this stage, Australia continued moving northward, and Greenland completely

separated from Europe and rifted from North America to form a separate landmass.

The Effects of the Breakup of Pangaea on Global Climates and Ocean Circulation Patterns

By the end of the Permian Period, Pangaea extended from pole to pole, covered about one-fourth of the Earth's surface, and was surrounded by Panthalassa, a global ocean that encompassed about 300 degrees of longitude. Such a configuration exerted tremendous influence on the world's climate and resulted in generally arid conditions over large parts of Pangaea's interior.

The world's climates result from the complex interaction between wind and ocean currents and the location and topography of the continents. In general, dry climates occur on large landmasses in areas remote from sources of moisture and where barriers to moist air, such as mountain ranges, exist. Wet climates occur near large bodies of water or where winds can carry moist air over land.

Past climatic conditions can be inferred from the distribution of climate-sensitive deposits. Evaporites occur where evaporation exceeds precipitation. Desert dunes and red beds may form locally in humid regions, but are characteristic of arid regions. Coal forms in both warm and cool humid climates. Vegetation that is eventually converted into coal requires at least a good seasonal water supply; thus, coal deposits are indicative of humid conditions.

Widespread Triassic evaporites, red beds, and desert dunes in the low and middle latitudes of North and South America, Europe, and Africa indicate dry climates in those regions. Coal deposits formed mainly in the high latitudes, indicating humid conditions. These high-latitude coals are analogous to today's Scottish peat bogs or Canadian muskeg. The lands bordering the Tethys Sea were probably dominated by seasonal monsoon rains resulting from the warm moist winds and warm oceanic currents impinging against the east-facing coast of Pangaea.

The temperature gradient between the tropics and the poles also affects oceanic and atmospheric circulation. The greater the temperature difference between the tropics and the poles, the steeper the temperature gradient, and the faster the circulation of the oceans and atmosphere. Oceans absorb about 90% of the solar radiation they receive, while continents absorb only about 50%, even less if they are snow covered. The rest of the solar radiation is reflected back into space. Therefore, areas dominated by seas are warmer than those dominated by continents. By knowing the distribution of continents and ocean basins, geologists can calculate the average annual temperature for any region on Earth. From this information a temperature gradient can be determined. For example, the temperature gradient of the Northern Hemisphere is currently 41°C, but is calculated to have been 20°C during the Triassic. This means that the present-day circulation of the atmosphere and oceans is faster than it was during the Triassic.

The breakup of Pangaea during the Late Triassic caused the global temperature gradient to increase because the Northern Hemisphere continents moved further northward, displacing higher-latitude ocean waters. Due to the steeper global temperature gradient caused by a decrease in temperature in the high latitudes and the changing positions of the continents, oceanic and atmospheric circulation patterns greatly accelerated during the Mesozoic (◆ Fig. 23.4). Though the temperature gradient and seasonality on land were increasing during the Jurassic and Cretaceous, the middle- and higher-latitude oceans were still quite warm, because warm waters from the Tethys Sea were circulating to the higher latitudes. This resulted in a

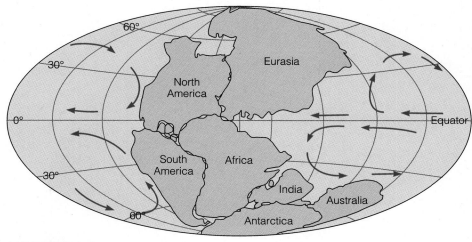

(a) Triassic Period

◆ FIGURE 23.4 Oceanic circulation evolved from (*a*) a simple pattern in a single ocean (Panthalassa) with a single continent (Pangaea) to (*b*) a more complex pattern in the newly formed oceans of the Cretaceous Period.

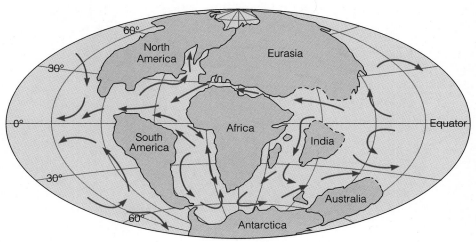

(b) Cretaceous Period

relatively equable worldwide climate to the end of the Cretaceous.

◥ THE MESOZOIC HISTORY OF NORTH AMERICA

In terms of tectonism and sedimentation, the beginning of the Mesozoic Era was essentially the same as the preceding Permian Period in North America. Terrestrial sedimentation continued over much of the craton, while block faulting and igneous activity began in the Appalachian region as North America and Africa began separating. The newly forming Gulf of Mexico experienced extensive evaporite deposition during the Late Triassic and Jurassic as North America separated from South America (Fig. 23.3).

◆ FIGURE 23.5 Paleogeography of North America during the Triassic Period.

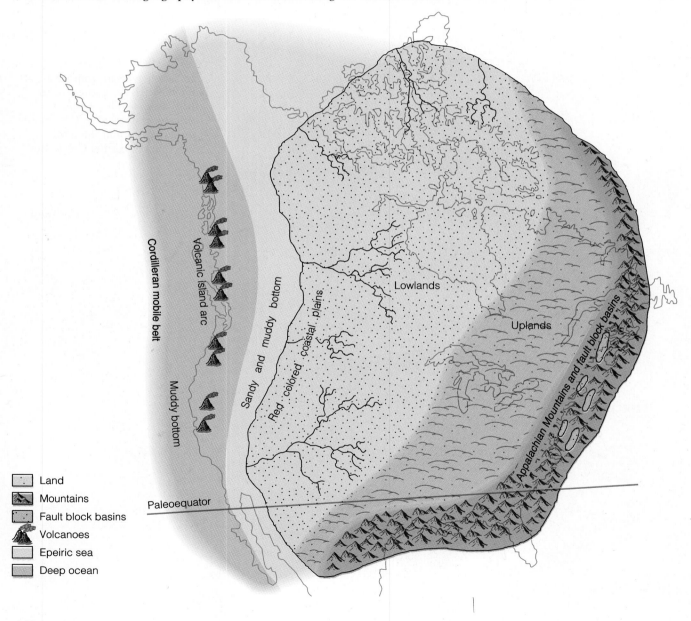

Land

Mountains

Fault block basins

Volcanoes

Epeiric sea

Deep ocean

A global rise in sea level during the Cretaceous resulted in worldwide transgressions onto the continents (Figs. ◆ 23.5, ◆ 23.6, and ◆ 23.7). These transgressions were caused by higher heat flow along the oceanic ridges due to increased rifting and the consequent expansion of oceanic crust. By the Middle Cretaceous, sea level probably was as high as at any time since the Ordovician, and approximately one-third of the present land area was inundated by epeiric seas.

Marine deposition was continuous over much of the North American Cordillera. A volcanic island arc system that formed off the western edge of the craton during the Permian was sutured to North America sometime later during the Permian or Triassic. This event is referred to as the *Sonoma orogeny* and will be discussed later in the chapter. During the Jurassic, the entire Cordilleran area was involved in a series of major mountain-building episodes that resulted in the forma-

◆ FIGURE 23.6 Paleogeography of North America during the Jurassic Period.

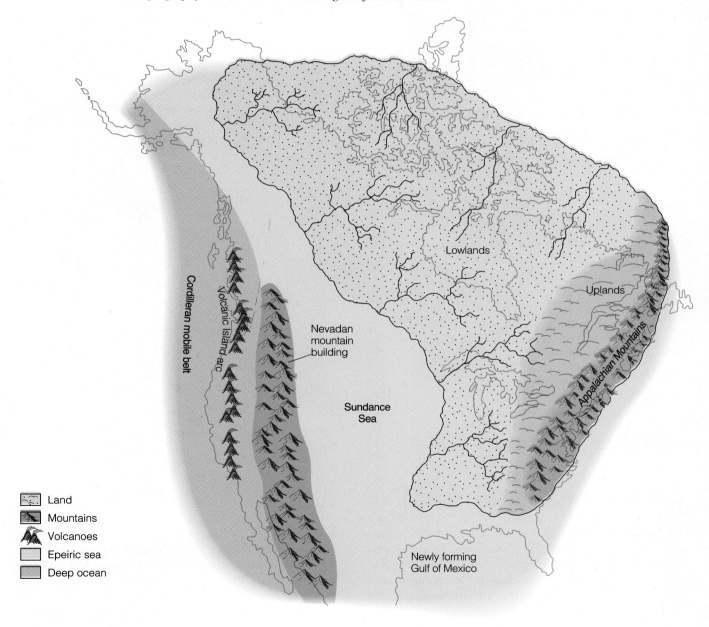

Land

Mountains

Volcanoes

Epeiric sea

Deep ocean

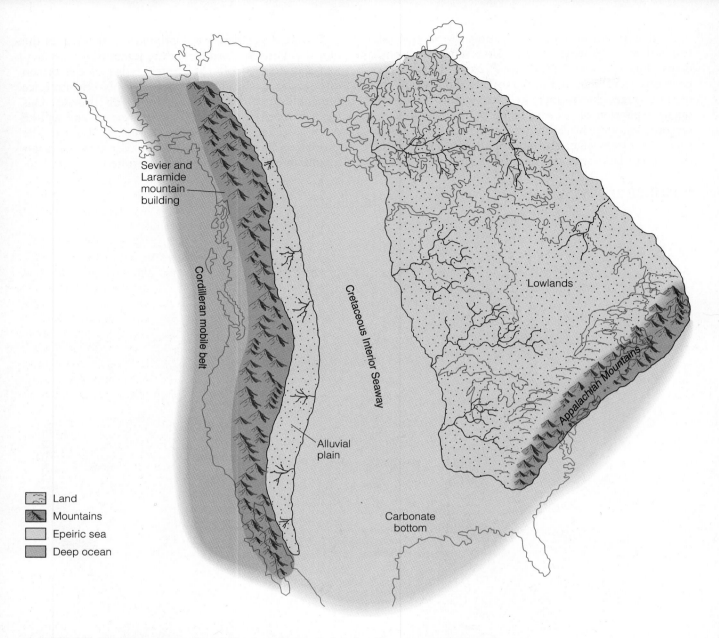

◆ FIGURE 23.7 Paleogeography of North America during the Cretaceous Period.

tion of the Sierra Nevada, the Rocky Mountains, and other lesser mountain ranges. While each orogenic episode has its own name, the entire mountain-building event is simply called the *Cordilleran orogeny* (also discussed later in this chapter). With this simplified overview of the Mesozoic history of North America in mind, we will now examine the specific regions of the continent.

Continental Interior

Recall that the history of the North American craton can be divided into unconformity-bound sequences reflecting advances and retreats of epeiric seas over the craton. While these transgressions and regressions played a major role in the Paleozoic geologic history of the continent, they were not as important during the

Mesozoic. During the Mesozoic, most of the continental interior was well above sea level and did not experience epeiric sea inundation. Consequently, the two Mesozoic cratonic sequences, the Absaroka sequence (Late Mississippian to Early Jurassic) and **Zuni sequence** (Early Jurassic to Early Paleocene) (see Fig. 21.5) are incorporated as part of the history of the three continental margin regions of North America.

Eastern Coastal Region

During the Early and Middle Triassic, coarse detrital sediments derived from the erosion of the recently uplifted Appalachians (Alleghenian orogeny) filled the various intermontane basins and spread over the surrounding areas. As erosion continued during the Mesozoic, this once lofty mountain system was reduced to a low-lying plain. During the Late Triassic, the first stage in the breakup of Pangaea began with North America separating from Africa. Fault-block basins developed in response to upwelling magma beneath Pangaea in a zone stretching from present-day Nova Scotia to North Carolina (◆ Fig. 23.8). Erosion of the adjacent fault-block mountains filled these basins with great quanti-

ties (up to 6,000 m) of poorly sorted red nonmarine detrital sediments known as the *Newark Group*. Reptiles roamed along the margins of the various lakes and streams that formed in these basins, leaving their footprints and trackways in the soft sediments. The Newark Group is mostly Late Triassic, but in some areas deposition began in the Early Jurassic.

Concurrent with sedimentation in the fault-block basins were extensive lava flows that blanketed the basin floors as well as intrusions of numerous dikes and sills. The most famous intrusion is the prominent 200-million-year-old Palisades sill along the Hudson River in the New York–New Jersey area.

As the Atlantic Ocean grew, rifting ceased along the eastern margin of North America, and this once active plate margin became a passive, trailing continental margin. The fault-block mountains that were produced by this rifting continued to erode during the Jurassic and Early Cretaceous until only a broad, low-lying erosional surface remained. The sediments resulting from erosion contributed to the growing eastern continental shelf. During the Cretaceous Period, the Appalachian region was reelevated and once again shed sediments onto the continental shelf, forming a gently dipping, seaward-

◆ FIGURE 23.8 (*a*) Areas where Triassic fault-block basin deposits crop out in eastern North America. (*b*) After the Appalachians were eroded to a low-lying plain by the Middle Triassic, fault-block basins formed as a result of Late Triassic rifting between North America and Africa. (*c*) These valleys accumulated tremendous thicknesses of sediments and were themselves broken by a complex of normal faults during rifting.

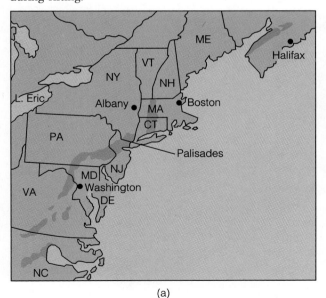

(a)

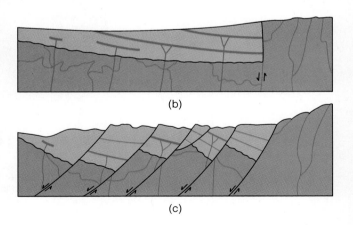

(b)

(c)

thickening wedge of rocks up to 3,000 m thick. These rocks are currently exposed in a belt extending from Long Island, New York, to Georgia.

Gulf Coastal Region

The Gulf Coastal region was above sea level until the Late Triassic (Fig. 23.5). As North America separated from South America during the Late Triassic, however, the Gulf of Mexico began to form (Fig. 23.6). With oceanic waters flowing into this newly formed, shallow, restricted basin, conditions were ideal for evaporite formation. More than 1,000 m of evaporites were precipitated at this time, and most geologists think that these Jurassic evaporites are the source for the Tertiary salt domes found today in the Gulf of Mexico and southern Louisiana.

By the Late Jurassic, circulation in the Gulf of Mexico was less restricted, and evaporite deposition ended. Normal marine conditions returned to the area with alternating transgressing and regressing seas, resulting in the deposition of sandstones, shales, and limestones. These sedimentary rocks were later covered and deeply buried by great thicknesses of Cretaceous and Cenozoic sediments.

During the Cretaceous, the Gulf Coastal region, like the rest of the continental margin, was inundated by northward-transgressing seas (Fig. 23.7). As a result, nearshore sandstones are overlain by finer sediments characteristic of deeper waters. Following an extensive regression at the end of the Early Cretaceous, a major transgression began during which a wide seaway extended from the Arctic Ocean to the Gulf of Mexico (Fig. 23.7). Cretaceous sediments that were deposited in the Gulf Coastal region formed a seaward-thickening wedge.

Reefs were widespread in the Gulf Coastal region during the Cretaceous. Bivalves called *rudists* were the main constituent of many of these reefs. Because of their high porosity and permeability, rudistoid reefs make excellent petroleum reservoirs.

Western Region

Mesozoic Tectonics

With the exception of the Late Devonian–Early Mississippian Antler orogeny (see Fig. 21.28), the Cordilleran region of North America experienced little tectonism during the Paleozoic. During the Permian, however, an island arc and ocean basin formed off the western North American craton (Fig. 23.5). This was followed by subduction of an oceanic plate beneath the island arc and the thrusting of oceanic and island arc rocks eastward against the craton margin (◆ Fig. 23.9). This

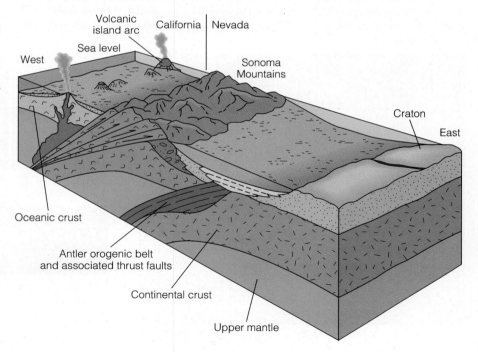

◆ FIGURE 23.9 Tectonic activity that culminated in the Permian-Triassic Sonoma orogeny in western North America. The Sonoma orogeny was the result of a collision between the southwestern margin of North America and an island arc system.

event initiated the **Sonoma orogeny** at or near the Permian-Triassic boundary.

Following the Late Paleozoic-Early Mesozoic destruction of the volcanic island arc during the Sonoma orogeny, the western margin of North America became an oceanic-continental convergent plate boundary. During the Late Triassic, a steeply dipping subduction zone developed along the western margin of North America in response to the westward movement of North America over the Pacific plate. This newly created oceanic-continental plate boundary controlled Cordilleran tectonics for the rest of the Mesozoic Era and for most of the Cenozoic Era; this subduction zone marks the beginning of the circum-Pacific orogenic system.

Two subduction zones dipping in opposite directions formed off the west coast of North America during the Middle and early Late Jurassic (◆ Fig. 23.10). The more westerly subduction zone was eliminated by the westward-moving North American plate, which overrode the oceanic Pacific plate.

The Franciscan Group, which is up to 7,000 m thick, is an unusual rock unit consisting of a chaotic mixture of rocks that accumulated during the Late Jurassic and Cretaceous. The various rock types such as graywacke, volcanic breccia, siltstone, black shale, chert, pillow basalt, and blueshist metamorphic rocks indicate that continental shelf, slope, and deep-sea environments were brought together in a submarine trench when North America overrode the Pacific plate (Fig. 23.10).

East of the Franciscan Group and currently separated from it by a major thrust fault is the Great Valley Group. It consists of more than 16,000 m of Cretaceous conglomerates, sandstones, siltstones, and shales. These sediments were deposited on the continental shelf and slope in a fore-arc basin setting at the same time the Franciscan deposits were accumulating in the submarine trench (see Fig. 6.22).

The general term **Cordilleran orogeny** is applied to the mountain-building activity that began during the Jurassic and continued into the Cenozoic (◆ Fig. 23.11). The Cordilleran orogeny consisted of a series of individual mountain-building events that occurred in different regions at different times. Most of this Cordilleran orogenic activity is related to the continued westward movement of the North American plate.

The first phase of the Cordilleran orogeny, the **Nevadan orogeny** (Fig. 23.11), began during the Late Jurassic and continued into the Cretaceous as large volumes of granitic magma were generated at depth beneath the western edge of North America. These granitic masses ascended as huge batholiths that are

now recognized as the Sierra Nevada, Southern California, Idaho, and Coast Range batholiths (◆ Fig. 23.12).

By the Late Cretaceous, most of the volcanic and plutonic activity had migrated eastward into Nevada and Idaho. This migration was probably caused by a change from high-angle to low-angle subduction, which resulted in the subducting oceanic plate reaching its melting depth farther east (◆ Fig. 23.13, page 650). Thrusting occurred progressively further east so that by the Late Cretaceous, it extended all the way to the Idaho-Washington border.

The second phase of the Cordilleran orogeny, the **Sevier orogeny,** was mostly a Cretaceous event (Fig. 23.11). As subduction of the Pacific plate beneath the North American plate continued, compressive forces generated along this convergent plate boundary were transmitted eastward, resulting in numerous overlapping, low-angle thrust faults (◆ Fig. 23.14, page 650). This thrusting produced generally north-south-trending mountain ranges consisting of blocks of Paleozoic shelf and slope strata. Though the term *Sevier* is usually applied to orogenic events occurring from southern California to Utah, the same deformational style was also occurring within a tectonic belt stretching from Montana northward into Canada.

During the Late Cretaceous to Early Cenozoic, the final pulse of the Cordilleran orogeny occurred (Fig. 23.11). The **Laramide orogeny** developed east of the Sevier orogenic belt in the present-day Rocky Mountain areas of New Mexico, Colorado, and Wyoming. Most of the features of the present-day Rocky Mountains resulted from the Cenozoic phase of the Laramide orogeny, and for that reason, it will be discussed in Chapter 24.

Mesozoic Sedimentation

Concurrent with the tectonism occurring in the Cordilleran mobile belt, Early Triassic sedimentation on the western continental shelf consisted of shallow-water marine sandstones, shales, and limestones. During the Middle and Late Triassic, the western shallow seas regressed further west, exposing large areas of former sea floor to erosion. Marginal marine and nonmarine Triassic rocks, particularly red beds, contribute to the spectacular and colorful scenery of the region (◆ Fig. 23.15a, page 652).

These rocks represent a variety of depositional environments, including fluvial, deltaic, floodplain, fresh and brackish water ponds, and desert dunes. The Upper Triassic Chinle Formation, for example, is widely exposed over the Colorado Plateau and is probably most famous for its petrified wood in Petri-

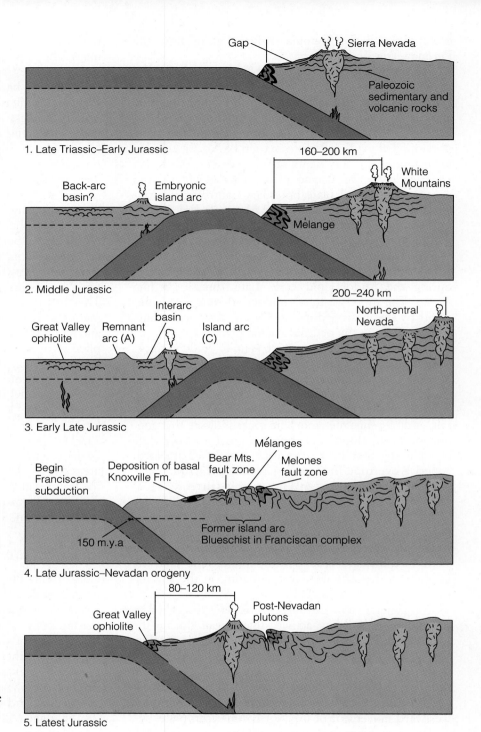

1. Late Triassic–Early Jurassic

Gap — 🌲🌲 Sierra Nevada

Paleozoic sedimentary and volcanic rocks

2. Middle Jurassic

160–200 km

Back-arc basin?   Embryonic island arc   White Mountains

Mélange

3. Early Late Jurassic

200–240 km

Interarc basin

Great Valley ophiolite   Remnant arc (A)   Island arc (C)   North-central Nevada

4. Late Jurassic–Nevadan orogeny

Mélanges

Begin Franciscan subduction   Deposition of basal Knoxville Fm.   Bear Mts. fault zone   Melones fault zone

150 m.y.a

Former island arc
Blueschist in Franciscan complex

5. Latest Jurassic

80–120 km

Great Valley ophiolite   Post-Nevadan plutons

◆ FIGURE 23.10 Interpretation of the tectonic evolution of the Sierra Nevada during the Mesozoic Era.

fied Forest National Park, Arizona (see Perspective 23.1). This formation, as well as other Triassic formations in the Southwest, also contain the fossilized remains and tracks of amphibians and reptiles.

The Early Jurassic deposits in a large part of the western region consist mostly of clean, cross-bedded sandstones indicative of windblown deposits. The thickest and most prominent of these is the Navajo

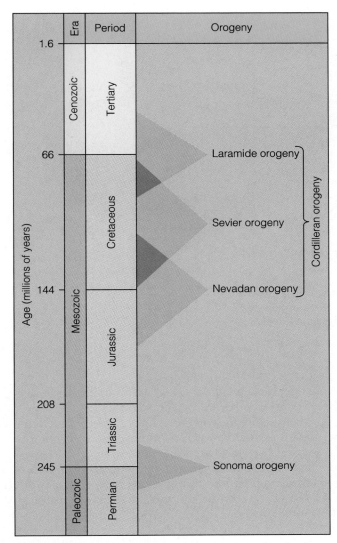

◆ FIGURE 23.11 Mesozoic orogenies occurring in the Cordilleran mobile belt.

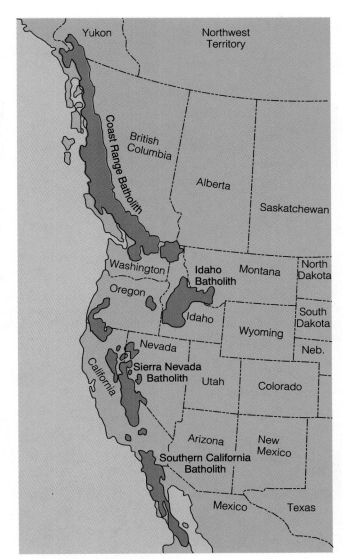

◆ FIGURE 23.12 Location of Jurassic and Cretaceous batholiths in western North America.

Sandstone, a widespread cross-bedded sandstone that accumulated in a coastal dune environment along the southwestern margin of the craton. The sandstone's most distinguishing feature is its large-scale cross-beds, some of which are more than 25 m high (◆ Fig. 23.15b, page 652).

Marine conditions returned to the area during the Middle Jurassic when a wide seaway called the **Sundance Sea** twice flooded the interior of western North America (Fig. 23.6). The resulting deposits, called the Sundance Formation, were largely derived from tec-

tonic highlands to the west that paralleled the shoreline and were the result of intrusive igneous activity and associated volcanism that began during the Triassic.

During the Late Jurassic, the folding and thrust faulting that began as part of the Nevadan orogeny in Nevada, Utah, and Idaho formed a large mountain chain paralleling the coastline (◆ Fig. 23.16a, page 652). As the mountain chain grew and shed sediments eastward, the Sundance Sea retreated northward. A large part of the area formerly occupied by the Sundance Sea was then covered by multicolored

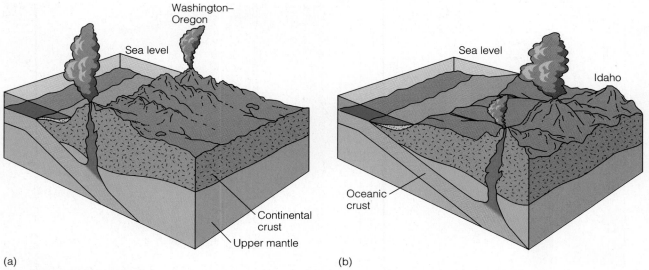

◆ FIGURE 23.13 A possible cause for the eastward migration of igneous activity in the Cordilleran region during the Cretaceous Period was a change from (a) high-angle to (b) low-angle subduction. As the subducting plate moved downward at a lower angle, the depth of melting moved farther to the east.

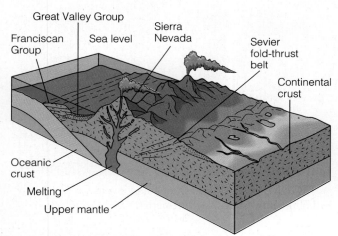

◆ FIGURE 23.14 Restoration showing the associated tectonic features of the Late Cretaceous Sevier orogeny due to subduction of the Pacific plate under the North American plate.

sandstones, mudstones, shales, and occasional lenses of conglomerates that comprise the world-famous Morrison Formation (◆ Fig. 23.16b). The Morrison Formation contains the world's richest assemblage of Jurassic dinosaur remains. Although most of the dinosaur skel-

etons are broken up, as many as 50 individuals have been found together in a small area.

Shortly before the end of the Early Cretaceous, Arctic waters spread southward over the craton, forming a large inland sea in the Cordilleran foreland basin area. By the beginning of the Late Cretaceous, this incursion joined the northward-transgressing waters from the Gulf area to create an enormous **Cretaceous Interior Seaway** that occupied the area east of the Sevier orogenic belt. Extending from the Gulf of Mexico to the Arctic Ocean, and more than 1,500 km wide at its maximum extent, this seaway effectively divided North America into two large landmasses until just before the end of the Late Cretaceous (Fig. 23.7). Mid-Cretaceous transgressions also occurred on other continents, and all were part of the global mid-Cretaceous rise in sea level that resulted from accelerated sea-floor spreading as Pangaea continued to fragment.

Cretaceous deposits less than 100 m thick indicate that the eastern margin of the Cretaceous Interior Seaway subsided slowly and received little sediment from the emergent, low relief craton to the east. The western shoreline, however, shifted back and forth, primarily in response to fluctuations in the supply of sediment from the Cordilleran Sevier orogenic belt to the West. The facies relationships show lateral changes

PETRIFIED FOREST NATIONAL PARK

Petrified Forest National Park is located in eastern Arizona about 42 km east of Holbrook. The park consists of two sections: the Painted Desert, which is north of Interstate 40, and the Petrified Forest, which is south of the Interstate.

The Painted Desert is a brilliantly colored landscape where colors and hues change constantly throughout the day. The multicolored rocks of the Triassic Chinle Formation have been weathered and eroded to form a badlands topography of numerous gullies, valleys, ridges, mounds, and mesas. The Chinle Formation is composed predominantly of various-colored shale beds. These shales and associated volcanic ash layers are easily weathered and eroded. Interbedded locally with the shales are lenses of conglomerates, sandstones, and limestones, which are more resistant to weathering and erosion than the shales and form resistant ledges.

The Petrified Forest was originally set aside as a national monument to protect the large number of petrified logs that lay exposed in what is now the southern part of the park (◆ Fig. 1). When the transcontinental railroad constructed a coaling and watering stop in Adamana, Arizona, passengers were encouraged to take excursions to "Chalcedony Park," as the area was then called, to see the petrified forests. In a short time, collectors and souvenir hunters hauled off tons of petrified wood, quartz crystals, and Indian relics. It was not until a huge rock crusher was built to crush the logs for the manufacture of abrasives that the area was declared a national monument and the petrified forests preserved and protected.

During the Triassic Period, the climate of the area was much wetter than today, with many rivers, streams, and lakes. About 40 different fossil plant species have been identified from the Chinle Formation. These include numerous seedless vascular plants such as rushes and ferns as well as gymnosperms such as cycads and conifers. Such plants thrive in floodplains and marshes. Most of the logs are conifers and belong to the genus *Araucarioxylon*. Some of these trees were more than 60 m tall and up to 4 m in diameter. Apparently, most of the conifers grew on higher ground or riverbanks. Although many trees were buried in place, most appear to have been uprooted and transported by raging streams during times of flooding. Burial of the logs was rapid, and ground water saturated with silica from the ash of nearby volcanic eruptions quickly permineralized the trees.

Deposition continued in the Colorado Plateau region during the Jurassic and Cretaceous, further burying the Chinle Formation. During the Laramide orogeny, the Colorado Plateau area was uplifted and eroded, exposing the Chinle Formation. Since the Chinle is mostly shales, it was easily eroded, leaving the more resistant petrified logs and log fragments exposed on the surface—much as we see them today.

◆ FIGURE 1 Petrified Forest National Park, Arizona. All of the logs here are *Araucarioxylon,* which is the most abundant tree in the park. The petrified logs have been weathered from the Chinle Formation and are mostly in the position in which they were buried some 200 million years ago.

(a)

(b)

◆ FIGURE 23.15 (*a*) Triassic red beds near Jamez Pueblo, New Mexico. (*b*) Large cross-beds of the Jurassic Navajo Sandstone in Zion National Park, Utah.

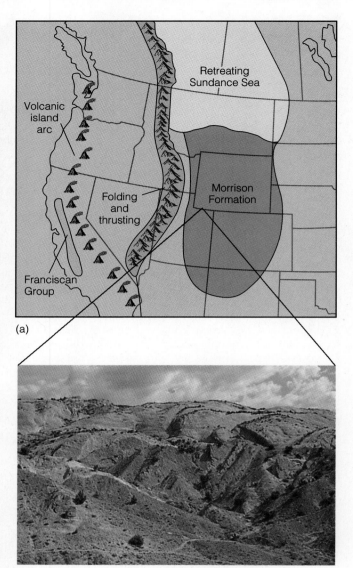

(a)

(b)

◆ FIGURE 23.16 (*a*) Paleogeography of western North America during the Late Jurassic. As the Sundance Sea withdrew from the western interior, the nonmarine Morrison Formation accumulated in part of the area formerly occupied by the Sundance Sea. (*b*) Panoramic view of the Jurassic Morrison Formation as seen from the visitors' center at Dinosaur National Monument, Utah.

from conglomerate and coarse sandstone adjacent to the mountain belt through finer sandstones, siltstones, shales, and even limestones and chalks in the east (◆ Fig. 23.17). During times of particularly active thrusting and uplift, these coarse clastic wedges of gravel and sand prograded even further east.

As the Mesozoic Era ended, the Cretaceous Interior Seaway withdrew from the craton. During this regression, marine waters retreated to the north and south, and marginal marine and continental deposition formed widespread coal-bearing deposits on the coastal plain.

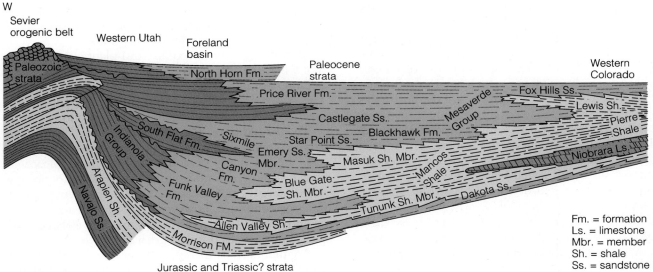

W
Sevier
orogenic belt

Western Utah

Foreland basin

Paleocene strata

Western Colorado

Paleozoic strata

North Horn Fm.

Price River Fm.

Fox Hills Ss.

Lewis Sh.

Castlegate Ss.

Mesaverde Group

Blackhawk Fm.

Pierre Shale

South Flat Fm.

Sixmile

Star Point Ss.

Indianola Group

Emery Ss. Mbr.

Masuk Sh. Mbr.

Niobrara Ls.

Canyon Fm.

Blue Gate Sh. Mbr.

Mancos Shale

Arapien Sh.

Funk Valley Fm.

Tununk Sh. Mbr.

Dakota Ss.

Navajo Ss.

Allen Valley Sh.

Morrison FM.

Jurassic and Triassic? strata

Fm. = formation
Ls. = limestone
Mbr. = member
Sh. = shale
Ss. = sandstone

◆ FIGURE 23.17 This restored west-east cross section of Cretaceous facies of the western Cretaceous Interior Seaway shows their relationship to the Sevier orogenic belt.

▼ MESOZOIC MINERAL RESOURCES

Although much of the coal in North America is Pennsylvanian or Tertiary in age, important Mesozoic coals occur in the Rocky Mountains states. These are mostly lignite and bituminous coals, but some local anthracites occur as well. Particularly widespread in western North America are coals of Cretaceous age. Mesozoic coals are also known from Alberta and British Columbia, Canada, as well as from Australia, Russia, and China.

Large concentrations of petroleum occur in many areas of the world, but more than 50% of all proven reserves are in the Persian Gulf region. During the Mesozoic, what is now the Gulf region was a broad passive continental margin extending eastward from Africa. This continental margin lay near the equator where countless microorganisms lived in the surface waters, particularly during the Cretaceous Period when most of the petroleum formed. The remains of these organisms accumulated with the bottom sediments and were buried, beginning the complex processes of oil generation and formation of source beds. Several transgressions and regressions occurred during which some of the reservoir rocks formed as extensive, thick regressive sandstones, oolitic limestones, algal reef limestones, and reefs composed of the shells of clams.

Overlying the reservoir rocks are cap rocks that include widespread shale and evaporite units.

Similar conditions existed in what is now the Gulf Coast region of the United States and Central America. Here petroleum and natural gas also formed on a broad shelf over which transgressions and regressions occurred. In this region, the hydrocarbons are largely in reservoir rocks that were deposited as distributary channels on deltas and as barrier-island and beach sands. Some of these hydrocarbons are associated with structures formed adjacent to rising salt domes. The salt, called the Louann Salt, initially formed in a long, narrow sea when North America separated from Europe and North Africa during the fragmentation of Pangaea (Fig. 23.3).

The richest uranium ores in the United States are widespread in the Colorado Plateau area of Colorado and adjoining parts of Wyoming, Utah, Arizona, and New Mexico. These ores, consisting of fairly pure masses of a complex potassium-, uranium-, vanadium-bearing mineral called *carnotite,* are associated with plant remains in sandstones that were deposited in ancient stream channels. Some petrified trees also contain large quantities of uranium.

As noted in Chapter 20, Proterozoic banded iron formations are the main sources of iron ores. There are, however, important exceptions. For example, the Jurassic-aged "Minette" iron ores of Western Europe, composed of oolitic limeonite and hematite, are important ores in France, Germany, Belgium, and Luxembourg. In Great Britain, low-grade iron ores of Jurassic age consist of oolitic siderite, which is an iron carbonate. And in Spain, Cretaceous rocks are the host rocks for iron minerals.

South Africa, the world's leading producer of gem-quality diamonds and among the leaders in industrial diamond production, mines these minerals from conical igneous intrusions called kimberlite pipes. Kimberlite pipes are composed of dark gray or blue igneous rock known as kimberlite. Diamonds, which form at great depth where pressure and temperature are high, are brought to the surface during the explosive volcanism that forms kimberlite pipes. Although kimberlite pipes have formed throughout geologic time, the most intense episode of such activity in South Africa and adjacent countries was during the Cretaceous Period. Emplacement of Triassic and Jurassic diamond-bearing kimberlites also occurred in Siberia.

The mother lode or source for the placer deposits mined during the California gold rush is in Jurassic-aged intrusive rocks of the Sierra Nevada. Gold placers are also known in Cretaceous-aged conglomerates of the Klamath Mountains of California and Oregon.

Porphyry copper was originally named for copper deposits in the western United States mined from porphyritic granodiorite, but the term now applies to large, low-grade copper deposits disseminated in a variety of rocks. These prophyry copper deposits are an excellent example of the relationship between convergent plate boundaries and the distribution, concentration, and exploitation of valuable metallic ores. Magma generated by partial melting of a subducting plate rises toward the surface, and as it cools, it precipitates and concentrates various metallic ores. The world's largest copper deposits were formed during the Mesozoic and Tertiary in a belt along the western margins of North and South America (see Fig. 9.30).

⩔ MESOZOIC LIFE

The Mesozoic Era is commonly referred to as the "Age of Reptiles," alluding to the fact that reptiles were the most diverse and abundant land-dwelling vertebrate animals. The Mesozoic diversification of reptiles was an important evolutionary event, but other equally important events also occurred. For example, mammals evolved from the mammal-like reptiles during the Triassic, while birds evolved from reptiles, probably small carnivorous dinosaurs, during the Jurassic.

Vast changes occurred in land-plant communities too, when the first flowering plants evolved and soon became the most diverse and abundant land plants. The Mesozoic was also a time of resurgence of marine invertebrates following the Permian extinction event.

The breakup of Pangaea that began during the Triassic continued throughout the Mesozoic. Nevertheless, the proximity of continents and mild Mesozoic climates allowed land animals and plants to occupy extensive geographic areas. As the fragmentation of Pangaea continued, however, some continents, Australia and South America especially, became isolated, and their faunas evolved independently.

Another mass extinction event occurred at the end of the Mesozoic. Once again organic diversity declined markedly as the dinosaurs and several other reptiles and some marine invertebrates died out. Because dinosaurs were victims of this extinction event, it has received more publicity than any other event although the Permian extinctions were more severe (see Chapter 22).

Marine Invertebrates

Following the wave of extinctions at the end of the Paleozoic, the Mesozoic was a time when the marine invertebrates repopulated the seas. The Early Triassic invertebrate marine fauna was not very diverse but by the Late Triassic the seas were once again richly populated with invertebrates. The mollusks became increasingly diverse and abundant throughout the Mesozoic. The brachiopods, however, never completely recovered from their near extinction at the end of the Paleozoic. In areas of warm, relatively clear, shallow marine waters, corals again proliferated. These corals were of a new and familiar type, the *scleractinians* (◆ Fig. 23.18). Echinoids, which were rare during the Paleozoic, greatly diversified during the Mesozoic.

One of the major differences between Paleozoic and Mesozoic marine invertebrate communities was the increased abundance and diversity of burrowing organisms. With few exceptions, Paleozoic burrowers were soft-bodied animals such as worms. The bivalves and echinoids, which were epifaunal elements during the Paleozoic, evolved various means of entering the infaunal habitats. This trend toward an infaunal existence may reflect an adaptive response to increasing predation from the rapidly evolving fish and cephalopods.

◆ FIGURE 23.18 Scleractinian corals evolved during the Triassic and proliferated in the warm, clear, shallow marine waters of the Mesozoic Era. Most living corals are scleractinians, represented here by the so-called staghorn coral. (Photo courtesy of Sue Monroe.)

Beginning with the primary producers, we will now examine some of the major Mesozoic marine plant and invertebrate groups. The *coccolithophores* are an important group of living phytoplankton (◆ Fig. 23.19a). They first evolved during the Jurassic and diversified tremendously during the Cretaceous. *Diatoms* evolved during the Cretaceous, but were more important as primary producers during the Cenozoic. Diatoms construct their shells out of silica, and are presently most abundant in cooler waters (◆ Fig. 23.19b). *Dinoflagellates* were common during the Mesozoic and today are the major primary producers in warm waters. (◆ Fig. 23.19c).

The mollusks, the major invertebrate phylum of the Mesozoic, include six classes, only three of which—the gastropods, bivalves, and cephalopods—are significant members of the marine invertebrate fauna. The gastropods increased in abundance and diversity during the Mesozoic, becoming most abundant during the Cretaceous, when carnivorous forms appeared. It was the bivalves and cephalopods, however, that dominated the invertebrate community of the Mesozoic.

Mesozoic bivalves diversified to inhabit many epifaunal and infaunal niches. Oysters and other clams became particularly diverse and abundant epifaunal suspension feeders and, continued to be important throughout the Cenozoic to the present (◆ Fig. 23.20). The reef-forming rudists were a significant group of Mesozoic bivalves. Bivalves also expanded into the infaunal niche during the Mesozoic. By burrowing into

◆ FIGURE 23.19 (*a*) *Calcidiscus macintyrei,* a Miocene coccolith from the Gulf of Mexico (left); *Discoaster variabilis,* a Pliocene-Miocene coccolith from the Gulf of Mexico (right). (*b*) *Actinoptychus senarius,* an Upper Miocene centric diatom from Java (left); *Cocconeis pellucida,* an Upper Miocene pinnate diatom from Java (right). (*c*) *Deflandrea spinulosa,* an Eocene dinoflagellate from Alabama (left); *Spiniferites mirabilis,* a Neogene dinoflagellate from the Gulf of Mexico (right). (Scanning electron photomicrographs courtesy of *a,* Merton E. Hill; *b,* John Barron, United States Geological Survey; *c,* John H. Wrenn, Louisiana State University.)

the sediment, they escaped predation from cephalopods and fish.

Cephalopods were one of the most important Mesozoic invertebrate groups. Their rapid evolution and nektonic life-style make them excellent guide fossils (◆ Fig. 23.21). The Ammonoidea, cephalopods with wrinkled sutures, are divided into three groups: the goniatites, ceratites, and ammonites. The ammonites, which are characterized by extremely complex suture patterns, were present during all three Mesozoic periods but were most prolific during the Jurassic and Cretaceous. While most ammonites were coiled, some attaining diameters of 2 m, others were uncoiled and led a near benthonic existence.

◆ FIGURE 23.20 Bivalves, represented here by two Cretaceous forms, were particularly diverse and abundant during the Mesozoic. (Photo courtesy of Sue Monroe.)

◆ FIGURE 23.21 Cephalopods were an important Mesozoic invertebrate group. The ammonites, which are characterized by extremely complex suture patterns, were particularly abundant and diverse during the Jurassic and Cretaceous and are excellent guide fossils for those periods. This specimen, *Scaphites preventricosus,* is from the Upper Cretaceous Colorado Formation, Toole Country, Montana, and shows the complex suture pattern characteristic of ammonites.

Although the ammonites became extinct at or near the end of the Cretaceous, two other groups of cephalopods survived into the Cenozoic—the nautiloids and the belemnoids, a group of squidlike cephalopods that was highly successful during the Jurassic and Cretaceous. The stalked echinoderms were minor members of the Mesozoic marine invertebrate community. However, the echinoids, which were exclusively epifaunal during the Paleozoic, branched out into the infaunal habitat and became very diverse and abundant. Bryozoans, although rare in Triassic strata, diversified and expanded during the Jurassic and Cretaceous.

As is true today, where shallow marine waters were warm and clear, coral reefs proliferated. Mesozoic corals belong to the order Scleractinia (Fig. 23.18). Whether scleractinian corals evolved from the rugose order or from an as yet unknown soft-bodied group that left no fossil record is still unresolved.

Lastly, the foraminifera underwent an explosive radiation during the Jurassic and Cretaceous that continued to the present. The planktonic forms in particular underwent rapid diversification, but most genera became extinct at the end of the Cretaceous.

We can think of the Mesozoic as a time of increasing complexity of the marine invertebrate community. At the beginning of the Triassic diversity was low and food chains were short. Near the end of the Cretaceous the marine invertebrate community was highly complex with interrelated food chains.

Primary Producers on Land—Plants

Triassic and Jurassic land-plant communities consisted of seedless vascular plants and various gymnosperms. Among the gymnosperms, however, the large seed ferns became extinct by the end of the Triassic, the *ginkgos* remained abundant throughout the Mesozoic Era, and *conifers* continued to diversify. A new type of gymnosperm, the *cycads,* evolved during the Triassic. Cycads superficially resemble palm trees, and several varieties still exist in tropical and subtropical areas.

The long dominance of seedless plants and gymnosperms ended during the Cretaceous as many were replaced by **angiosperms,** or flowering plants. The earliest angiosperms probably evolved from a specialized group of seed ferns. In any case, since the angiosperms evolved, they have adapted to nearly every terrestrial habitat. Several factors account for the phenomenal success of flowering plants, but chief among them is their method of reproduction (◆ Fig. 23.22). Two developments were particularly important: the evolution of flowers, which attract animal pollinators, especially insects; and the evolution of enclosed seeds.

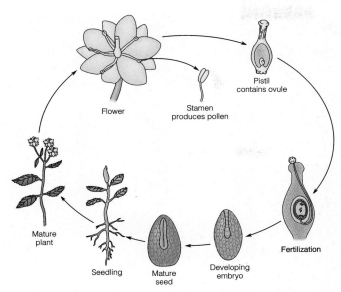

◆ FIGURE 23.22 The reproductive cycle in angiosperms.

Reptiles

Reptile diversification began during Pennsylvanian time with the evolution of the captorhinomorphs, apparently the first animals to lay amniote eggs (see Chapter 22). From this basic stock of *stem reptiles* all other reptiles evolved. Birds and mammals, too, have their ancestors among the reptiles, so they are a part of this major evolutionary diversification.

Thecodontians and the Ancestry of Dinosaurs

Thecodontian is an informal term for a variety of Late Permian and Triassic reptiles characterized by teeth set into individual sockets. Thecodontians are important because they included the ancestors of dinosaurs, flying reptiles, birds, and crocodiles. Some thecodontians were small, lightly built carnivores (◆ Fig. 23.23). These small predators had well-developed forelimbs and moved primarily on all four limbs; that is, they were **quadrupedal.** When running, however, they rose onto their hind limbs and moved in a **bipedal** fashion. The largest thecodontians were fully quadrupedal and covered with bony armor.

Dinosaurs evolved from thecodontians during the Late Triassic, but their specific thecodontian ancestor is uncertain. The traditional interpretation is that two distinct orders of dinosaurs were established when they first appeared, each of which may have had an independent origin from thecodontians similar to the one in Figure 23.23. Pelvic structure is the basis for recognizing the orders **Saurischia** and **Ornithischia** (Fig. 23.23). Saurischian dinosaurs had a lizard-like pelvis and are therefore referred to as lizard-hipped dinosaurs. Ornithischians had a bird-like pelvis; hence they are called bird-hipped dinosaurs.

Dinosaurs

A common but erroneous perception of dinosaurs is that they were poorly adapted animals that had trouble surviving. True, they became extinct, but to consider this a failure is to ignore the fact that for more than 140 million years they were the dominant land vertebrates. During their existence, they diversified into numerous types, adapted to a wide variety of environments, and some may have been warm-blooded (see Perspective 23.2). Eventually, the dinosaurs did die out, an event that then enabled mammals to become the dominant land vertebrates.

Two groups of saurischian dinosaurs are recognized: theropods and sauropods (Fig. 23.23). Theropods were carnivorous bipeds that ranged in size from tiny *Compsognathus*, which was the size of a chicken, to *Tyrannosaurus* (◆ Fig. 23.24, page 660), the largest terrestrial carnivore known. Also among the theropods is *Eoraptor* (dawn plunderer), recently discovered in earliest Late Triassic rocks in Argentina. Although *Eoraptor* is too specialized to be the common ancestor of dinosaurs, its size (about 1 m long) and structure indicate that it is close to that ancestry.

Included among the sauropods were the giant, quadrupedal herbivores such as *Apatosaurus, Diplodocus,* and *Brachiosaurus* (Fig. 23.23), the largest known land animals of any kind. Evidence from fossil trackways indicates that sauropods moved in herds. They depended on their size and herding behavior rather than speed as their primary protection from predators.

The great diversity of ornithischians is manifested by the fact that five distinct groups are recognized: ornithopods, pachycephalosaurs, ankylosaurs, stegosaurs, and ceratopsians (Fig. 23.23 and 23.24). Ornithopods include the duck-billed dinosaurs, which had flattened, bill-like mouths (Fig. 23.23). These dinosaurs were particularly varied and abundant during the Cretaceous, and some species were characterized by head crests that may have functioned as resonating chambers

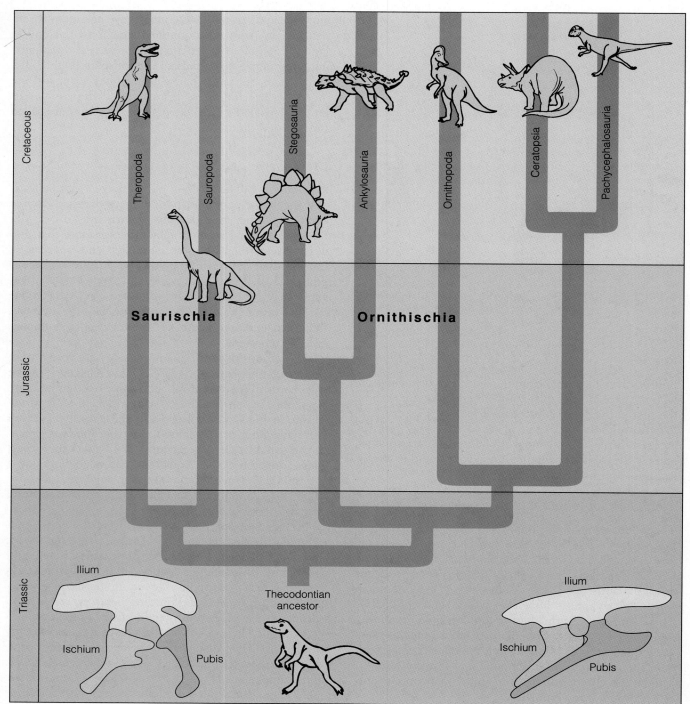

◆ FIGURE 23.23 Origin of and inferred relationships among dinosaurs. Both ornithischian and saurischian dinosaurs evolved from thecodontians, but each may have had an independent origin from that group. The pelvis of each order of dinosaurs is shown.

WARM-BLOODED DINOSAURS?

· ·

All living reptiles are *ectotherms,* that is, cold-blooded animals whose body temperature varies in response to the outside temperature. *Endotherms,* warm-blooded animals such as birds and mammals, are capable of maintaining a rather constant body temperature regardless of the outside temperature. Some investigators think that dinosaurs, or at least some dinosaurs, were endotherms.

Proponents of dinosaur endothermy note that dinosaur bones are penetrated by numerous passageways that, when the animals were living, contained blood vessels. Bones of endotherms typically have this structure, but considerably fewer of these passageways are found in bones of ectotherms. Living crocodiles and turtles have this so-called endothermic bone structure, yet they are ectotherms. And in some small mammals the bone structure is more typical of ectotherms, yet we know that they are capable of maintaining a constant body temperature. It may be that bone structure is more related to body size and growth patterns than to endothermy.

Because endotherms have high metabolic rates, they must eat more than ectotherms of comparable size. Consequently, endothermic predators require large prey populations. They would therefore constitute a much smaller proportion of the total animal population than their prey. In contrast, the proportion of ectothermic predators to their prey population is much greater. Where data are sufficient to allow an estimate, dinosaur predators appear to have made up 3 to 5 percent of the total population. These figures are comparable to present-day mammalian populations. However, a number of uncertainties about the composition of fossil communities make this argument for endothermy unconvincing to many paleontologists.

Living endotherms have a large brain in relation to body size. A relatively large brain is not necessary for endothermy, but endothermy does seem to be a prerequisite for having a large brain because a complex nervous system requires a rather constant body temperature. Some dinosaurs, particularly the small carnivores, did have a large brain in relation to their body, but many did not. That the small carnivorous ones had a large brain seems to be a good argument for endothermy, but there is an even more compelling argument. The relationship of birds to small carnivorous dinosaurs implies that these dinosaurs were endothermic or at least trending in that direction.

The large sauropods were probably not endothermic, but nevertheless may have been able to maintain their body temperatures within narrow limits as endotherms do. A large animal heats up and cools down slowly because it has a small surface area compared to its volume. With proportionately less surface area to allow heat loss, sauropods probably retained body heat more efficiently than smaller dinosaurs.

One further point on endothermy in dinosaurs is that the flying reptiles, the *pterosaurs,* evolved from thecodontians as did the dinosaurs. At least one species of pterosaur had hair or hairlike feathers. This is interesting because an insulating covering of hair or feathers is known only in endotherms. Furthermore, the physiology of active flight requires endothermy. Such evidence indicates that perhaps both thecodontians and dinosaurs were endothermic.

Obviously, considerable disagreement exists on dinosaur endothermy. In general, a fairly good case can be made for endothermic, small, carnivorous dinosaurs and pterosaurs, but for the others the question is still open.

to amplify bellowing. Some duck-billed dinosaurs practiced colonial nesting and care of the young. All ornithopods were herbivores and primarily bipedal, but their well-developed forelimbs allowed them to walk in a quadrupedal fashion, too.

The pachycephalosaurs constitute a most peculiar group of ornithischian dinosaurs. The most distinctive feature of these bipedal herbivores is the dome-shaped skull that resulted from thickening of the bones (Fig. 23.23). According to one hypothesis, these domed

◆ FIGURE 23.24 Scene from the Late Cretaceous showing the ankylosaur *Euoplocephalus* (foreground), the large theropod *Tyrannosaurus,* and the ceratopsian *Triceratops* (right).

skulls were used in intraspecific butting contests for dominance and mates.

Ankylosaurs were heavily armored, quadrupedal herbivores, and some were quite large (Figs. 23.23 and 23.24). Bony armor protected the back, flanks, and top of the head, and the tail ended in a large, bony clublike growth. No doubt a blow delivered by the powerful tail could seriously injure an attacking predator.

The stegosaurs, represented by the familiar genus *Stegosaurus* (Fig. 23.23), were quadrupedal herbivores with bony spikes on the tail, which were undoubtedly used for defense, and body plates on the back. The exact arrangement of these plates is debated but many paleontologists think they functioned as a device to absorb and dissipate heat.

A rather good fossil record indicates that large, Late Cretaceous ceratopsians such as *Triceratops* evolved from small, Early Cretaceous ancestors (Figs. 23.23 and 23.24). The later ceratopsians were characterized by huge heads, a large bony frill over the top of the neck, and a large horn or horns on the skull. Fossil trackways indicate that these large, quadrupedal herbivores moved in herds.

Flying Reptiles

Pterosaurs, the first vertebrate animals to fly, evolved from thecodontians during the Triassic and were abundant until their extinction at the end of the Mesozoic (◆ Fig. 23.25). Pterosaur flight adaptations include a wing membrane supported by an elongate fourth finger, light hollow bones, and development of those parts of the brain associated with muscular coordination and sight. Size varied considerably. Some early species ranged from sparrow to robin size, while one Cretaceous pterosaur from Texas had a wingspan of at least 12 m. At least one pterosaur had a coat of hair or hairlike feathers. That this pterosaur had a hair- or feather-covered body and was a flier strongly suggests that it, and perhaps all pterosaurs, were endotherms.

Marine Reptiles

Ichthyosaurs are probably the most familiar of the Mesozoic marine reptiles (◆ Fig. 23.26a). These animals were about 3 m long and were completely aquatic. Aquatic adaptations included a streamlined, somewhat fishlike body, a powerful tail for propulsion, and flipperlike forelimbs for maneuvering. The numerous sharp teeth indicate that ichthyosaurs were fish eaters. Some fossils with young ichthyosaurs within the body cavity support the interpretation that female ichthyosaurs retained the eggs in their bodies and gave birth to live young.

A second group of Mesozoic marine reptiles, the plesiosaurs, occurred in two varieties: short necked and long necked (◆ Fig. 23.26b). Most plesiosaurs were between 3.6 and 6 m long, but one species from Antarctica measures 15 m. Long-necked plesiosaurs were heavy-bodied animals with mouthfuls of sharp teeth and limbs specialized into oarlike paddles. They

(a)

(b)

◆ FIGURE 23.25 (*a*) Long-tailed and (*b*) short-tailed pterosaurs from the Jurassic of Europe. *Pteranodon* (*b*) was a Cretaceous pterosaur with a wingspan of more than 6 m.

probably rowed themselves through the water and may have used their long necks in snakelike fashion to capture fish. Plesiosaurs probably came ashore to lay their eggs.

Birds

Fossils from the Jurassic Solnhofen Limestone of Germany show the first features we associate with birds.

(a)

(b)

◆ FIGURE 23.26 Mesozoic marine reptiles.
(*a*) Ichthyosaurs and (*b*) a long-necked plesiosaur.

Several fossil specimens showing feather impressions have been discovered, but in almost every other known physical feature, these fossils are more similar to small carnivorous dinosaurs. The birdlike creature known as *Archaeopteryx* retained dinosaurlike teeth, tail, hind limb structure, and brain size but also possessed feathers and a wishbone, characteristics typical of birds. (◆ Fig. 23.27a).

Until recently, *Archaeopteryx* was the only known pre-Cretaceous bird, but the discovery of fossils of two crow-sized individuals called *Protoavis* has perhaps changed that situation. These fossils are from Triassic rocks, so they predate *Archaeopteryx,* and some investigators think they were more birdlike than *Archaeopteryx. Protoavis* had hollow bones and the breastbone structure of birds, but because no impressions of feathers were found on these specimens, some investigators believe they may simply have been small carnivorous dinosaurs. If *Protoavis* proves to be a bird rather than a reptile, it would imply that birds evolved earlier than originally thought.

Two more Mesozoic birds have recently been discovered. One specimen, from China, is slightly younger than *Archaeopteryx* and possesses both primitive and advanced characteristics. For example, it retains abdominal ribs similar to those of *Archaeopteryx* and the

theropod dinosaurs, but it has a reduced tail typical of present-day birds (◆ Fig. 23.27b). Another specimen, this one from Spain, is 20 to 30 million years younger than *Archaeopteryx;* it, too, is a mix of primitive and advanced characteristics, but it does appear to lack abdominal ribs.

Even more recently, another fossil bearing on the origin of birds has been reported. This fossil, a fragmentary skull from the Late Cretaceous of Mongolia, possesses both bird and theropod dinosaur characteristics. Its discoverers claim it is the closest dinosaur relative of *Archaeopteryx.*

From Reptile to Mammal

Therapsids, or the advanced mammal-like reptiles, were briefly described in Chapter 22. These reptiles diversified into numerous species of herbivores and carnivores, and during the Permian they were the dominant terrestrial vertebrates. One particular group of carnivorous therapsids called **cynodonts** was the most mammal-like of all and by the Late Triassic gave rise to the class Mammalia.

The transition from cynodonts to mammals is well documented by fossils and is so gradational that classi-

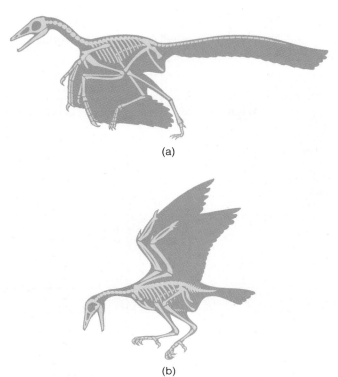

(a)

(b)

◆ FIGURE 23.27 (a) *Archaeopteryx,* a Jurassic age bird from Germany, has feathers and a wishbone and is therefore classified as a bird. In almost all other anatomical features, however, it more closely resembles small theropod dinosaurs. (b) This sparrow-sized bird, which was recently found in China, has a shortened tail, a characteristic more typical of birds.

fication of some fossils as either reptile or mammal is difficult. We can easily recognize living mammals as those warm-blooded animals with hair or fur that have mammary glands and, except for the platypus and spiny anteater, give birth to live young.

Obviously, most of these criteria for recognizing living mammals are inadequate for classifying fossils. For them, distinctions between mammals and mammal-like reptiles are based largely on details of the middle ear, the lower jaw, and the teeth (● Table 23.1). Reptiles have only one small bone in the middle ear—the stapes—while mammals have three—the incus, the malleus, and the stapes. Also, the lower jaw of a mammal is composed of a single bone called the *dentary,* but a reptile's jaw is composed of several bones (◆ Fig. 23.28). In addition, a reptile's jaw is hinged to the skull at a contact between the articular and quad-

rate bones, while in mammals the dentary contacts the squamosal bone of the skull (Fig. 23.28).

During the transition from cynodonts to mammals, the quadrate and articular bones that had formed the joint between the jaw and skull in reptiles were modified into the incus and malleus of the mammalian middle ear (Fig. 23.28). Fossils clearly document the progressive enlargement of the dentary until it became the only element in the mammalian jaw. Likewise, a progressive change from the reptile to mammal jaw joint is documented by fossil evidence. In fact, some of the most advanced cynodonts were truly transitional because they had a compound jaw joint consisting of (1) the articular and quadrate bones typical of reptiles and (2) the dentary and squamosal bones as in mammals (Fig. 23.28).

In Chapter 18 we noted that the study of embryos provides evidence for evolution. Opossum embryos clearly show that the middle ear bones of mammals were originally part of the jaw. In fact, even when opossums are born, the middle ear elements are still attached to the dentary (Fig. 23.28), but as they develop further, these elements migrate to the middle ear, and a typical mammal jaw joint develops.

Several other aspects of cynodonts also indicate they were ancestral to mammals. Their teeth were somewhat differentiated into distinct types in order to perform specific functions. In mammals the teeth are fully differentiated into incisors, canines, and chewing teeth, but typical reptiles do not have differentiated teeth. Another mammalian feature, the secondary palate, was partially developed in advanced cynodonts (◆ Fig. 23.29). This secondary palate is a bony shelf above the mouth that separates the nasal passages from the mouth cavity. It is an adaptation for eating and breathing at the same time, a necessary requirement for endotherms with their high demands for oxygen.

Mesozoic Mammals

Even though mammals appeared during the Late Triassic, their diversity remained low during the rest of the Mesozoic. The first mammals retained several reptilian characteristics, but had mammalian features as well. For example, the Triassic triconodonts had the fully differentiated teeth typical of mammals, but they had both the reptile and mammal types of jaw joints. In short, some mammalian features evolved more rapidly than others, thereby accounting for animals that possessed characteristics of both reptiles and mammals.

● **TABLE 23.1** Summary Chart Showing Some Characteristics and How They Changed during the Transition from Reptiles to Mammals

	Typical Reptile	Cynodont	Mammal
Lower jaw	Dentary and several other bones.	Dentary enlarged, other bones reduced.	Dentary bone only, except in earliest mammals.
Jaw-skull joint	Articular-quadrate	Articular-quadrate; some advanced cynodonts had both the reptile jaw-skull joint and the mammalian jaw-skull joint.	Dentary-squamosal
Middle ear bones	Stapes	Stapes	Stapes, incus, malleus
Secondary palate	Absent	Partially developed	Well developed
Teeth	No differentiation	Some differentiation	Fully differentiated
Cold-versus warm-blooded	Cold-blooded	Probably warm-blooded	Warm-blooded

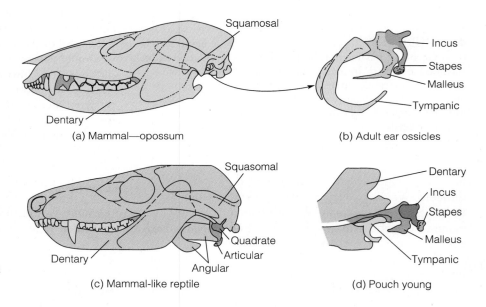

◆ FIGURE 23.28 (*a*) The skull of an opossum showing the typical mammalian dentary-squamosal jaw joint. (*b*) The skull of a cynodont shows the articular-quadrate jaw joint of reptiles. (*c*) Enlarged view of an adult opossum's middle ear bones. (*d*) View of the inside of a young opossum's jaw showing that the elements of the middle ear are attached to the dentary during early development. This is the same arrangement of bones that is found in the adults of the ancestral mammals.

The early mammals diverged into two distinct branches. One branch includes the triconodonts and their probable evolutionary descendants, the **monotremes,** or egg-laying mammals. Living monotremes are the platypus and spiny anteater of the Australian region. The second evolutionary branch included the **marsupial** (pouched) **mammals,** and the **placental mammals** and their ancestors, the **eupantotheres.** All living mammals except monotremes have ancestries that can be traced back through this branch.

Eupantotheres were shrew-sized animals with a poor fossil record, but details of their teeth indicate they were ancestral to both marsupial and placental mammals. The divergence of marsupials and placentals from a common ancestor probably occurred during the Early Cretaceous, but undoubtedly both were present by the Late Creta-

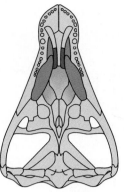

(a) Eutheriodont

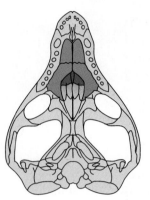

(b) Thrinaxodon

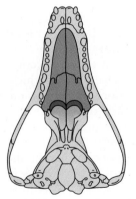

(c) Morganucodon

◆ FIGURE 23.29 Views of the bottoms of skulls of (*a*) an early therapsid (*b*) a cynodont, and (*c*) an early mammal showing the progressive development of the bony secondary palate (brown).

ceous. The earliest known placental mammals were members of the order *Insectivora* (◆ Fig. 23.30), an order represented today by shrews, moles, and hedgehogs.

⩔ MESOZOIC CLIMATES AND PALEOGEOGRAPHY

The present continental positions and climatic patterns largely restrict the distribution of organisms. Land plants and animals have little opportunity to colonize distant areas because of physical barriers (especially the ocean basins) and climatic barriers. Mesozoic barriers to migration were apparently not as effective as they are now, because some Mesozoic organisms are known from areas that are now widely separated.

Fragmentation of the supercontinent Pangaea began by the Late Triassic and continues to the present, but during much of the Mesozoic, close connections existed between the various landmasses. The proximity of these landmasses, however, is not sufficient to explain Mesozoic biogeographic distributions, because climates are also effective barriers to wide dispersal. During much of the Mesozoic, though, climates were more equable and lacked the strong north and south zonation characteristic of the present. In short, Mesozoic plants and animals had greater opportunities to occupy much more extensive geographic ranges.

Pangaea persisted as a single unit through most of the Triassic (Fig. 23.2a). The Triassic climate was warm-temperate to tropical, although some areas, such as the present southwestern United States, were arid. Mild temperatures extended 50 degrees north and south of the equator, and even the polar regions may have been temperate. The fauna was truly worldwide in its distribution. Some dinosaurs had continuous ranges across Laurasia and Gondwana, the peculiar gliding lizards were in New Jersey and England, and thecodontians known as phytosaurs lived in North America, Europe, and Madagascar.

By the Late Jurassic, Laurasia had become partly fragmented by the opening North Atlantic, but a connection still existed (Fig. 23.2b). The South Atlantic had begun to open so that a long, narrow sea separated the southern parts of Africa and South America. Otherwise the southern continents were still close together.

The mild Triassic climate persisted into the Jurassic. Ferns, whose living relatives are now restricted to the tropics of southeast Asia, are known from areas as far as 63° south latitude and 75° north latitude. Dinosaurs roamed widely across Laurasia and Gondwana. For example, the giant dinosaur *Brachiosaurus* is known from western North America and eastern Africa. Stegosaurs and some families of carnivorous dinosaurs lived throughout Laurasia and in Africa.

◆ FIGURE 23.30 The oldest known placental mammals were members of the order Insectivora such as those in this scene from the Late Cretaceous. These animals probably fed on insects, worms, and grubs.

By the Late Cretaceous, the North Atlantic had opened further, and Africa and South America were completely separated (Fig. 23.2c). South America remained an island continent until late in the Cenozoic. Its fauna, evolving in isolation, became increasingly different from faunas of the other continents. Marsupial mammals reached Australia from South America via Antarctica, but the South American connection was eventually severed. Placentals, other than bats and a few rodents, never reached Australia. This explains why the marsupials continue to dominate the continent's fauna even today.

Cretaceous climates were more strongly zoned by latitude, but they remained warm and equable until the close of that period. Climates then became more seasonal and cooler, a trend that persisted into the Cenozoic. Dinosaur and mammal fossils demonstrate that interchange was still possible, especially between the various components of Laurasia.

▼ MASS EXTINCTIONS—A CRISIS IN THE HISTORY OF LIFE

The mass extinction event at the close of the Mesozoic was second in magnitude only to the extinctions at the end of the Paleozoic (see Chapter 22). Casualties of the Mesozoic extinction event included dinosaurs, flying reptiles, marine reptiles, and several kinds of marine invertebrates. Among the latter were the ammonites, which had been so abundant through the Mesozoic, the rudistid clams, and some planktonic organisms.

Numerous ideas have been proposed to explain Mesozoic extinctions, but most have been dismissed as improbable or untestable. A new proposal was made recently based on a discovery at the Cretaceous-Tertiary boundary in Italy—a clay layer 2.5 cm thick, with an abnormally high concentration of the platinum group element iridium. Since this discovery, high iridium concentrations have been identified at many other Cretaceous-Tertiary boundary sites (◆ Fig. 23.31). The significance of this discovery lies in the fact that iridium is rare in crustal rocks but occurs in much higher concentrations in some meteorites. Several investigators proposed a meteorite impact to explain this iridium anomaly and further postulated that the impact of a large meteorite, perhaps 10 km in diameter, set in motion a chain of events that led to extinctions. Some Cretaceous-Tertiary boundary sites also contain soot and shock-metamorphosed quartz grains, both of which are cited as further evidence of an impact event.

The meteorite-impact scenario goes something like this. Upon impact, about 60 times the mass of the

(a)

(b)

◆ FIGURE 23.31 (a) View of the Cretaceous-Tertiary boundary in the Raton Basin, Colorado. The boundary, the thin white clay layer, is at the level of the knee of R. Farley Fleming of the U.S. Geological Survey. (b) Close-up of the clay layer. (Photos courtesy of D. J. Nichols, U.S. Geological Survey.)

meteorite was blasted from the Earth's crust high into the atmosphere, and the heat generated at impact started raging fires that added more particulate matter to the atmosphere. Sunlight was blocked for several months, causing a temporary cessation of photosynthesis; food chains collapsed, and extinctions followed. In addition, with sunlight greatly diminished, the Earth's surface temperatures were drastically reduced and could have added to the biologic stress.

Some investigators now claim that they have found the probable impact site centered on the town of Chicxulub on the Yucatán Peninsula of Mexico. The structure is about 180 km in diameter and lies beneath layers of sedimentary rock, so it has been detected only in drill holes and by geophysical work. Evidence supporting the conclusion that the Chicxulub structure is an impact crater includes shocked quartz, what appear to be the deposits of huge waves, and tektites, which are small pieces of rock that were melted during the proposed impact and hurled into the atmosphere. Although an impact origin for this structure is gaining acceptance, some geologists think it is some kind of volcanic feature.

Even if a meteorite did hit the Earth, did it lead to these extinctions? Some paleontologists think that dinosaurs, some marine invertebrates, and many plants were already on the decline and headed for extinction before the end of the Cretaceous. A meteorite impact, if one actually occurred, may have simply hastened the process. There is even some evidence indicating that dinosaurs survived into the Early Cenozoic, several tens of thousands of years after the proposed impact.

Investigators at the University of Chicago have proposed that the Mesozoic extinction event was only one of several to occur at 26-million-year intervals during the last 250 million years. One possible cause of these cyclic extinctions is periodic meteorite showers. Some investigators have suggested that a companion star to the Sun with a highly eccentric orbit could provide a mechanism for periodic meteor showers. In this scenario, when this star is close to the Sun, it perturbs cometary orbits, thereby causing terrestrial impacts and mass extinctions. The critics of cyclic mass extinctions think that some minor extinction events have been overemphasized. Furthermore, they point out that if mass extinctions are caused by periodic meteorite impacts, iridium anomalies corresponding with extinction events should be present, but few such anomalies have been identified.

In the final analysis, we have no widely accepted explanation for Mesozoic extinctions. We do know that vast shallow seas occupied large parts of the continents during the Cretaceous, and that by the latest Cretaceous they had largely withdrawn. We also know that the mild, equable climates of the Mesozoic became harsher and more seasonal by the end of the Mesozoic. Changes such as these seem adequate to many paleontologists to explain Mesozoic mass extinctions. But the fact remains that this extinction event was very selective, and no explanation accounts for all aspects of this crisis in the history of life.

Chapter Summary

● Tables 23.2 and 23.3 provide summaries of Mesozoic geologic and biologic events, respectively.

1. The breakup of Pangaea can be divided into four stages.
   a. The first stage involved the separation of North America from Africa during the Late Triassic, followed by the separation of North America from South America.
   b. The second stage involved the separation of Antarctica, India, and Australia from South America and Africa during the Jurassic. During this stage, India broke away from the still-united Antarctica and Australia landmass.
   c. During the third stage, South America separated from Africa, while Europe and Africa began to converge.
   d. In the last stage, Greenland separated from North America and Europe.
2. The breakup of Pangaea influenced global climatic and atmospheric circulation patterns. While the temperature gradient from the tropics to the poles gradually increased during the Mesozoic, overall global temperatures remained equable.
3. Except for incursions along the continental margin and two major transgressions (the Sundance Sea and the Cretaceous Interior Seaway), the North American craton was above sea level during the Mesozoic Era.
4. The Eastern Coastal Plain was the initial site of the separation of North America from Africa that began during the Late Triassic. During the Cretaceous Period, it was inundated by marine transgressions.
5. The Gulf Coastal region was the site of major evaporite accumulation during the Jurassic as North America rifted from South America. During the Cretaceous, it was inundated by a transgressing sea, which, at its maximum, connected with a sea transgressing from the north to create the Cretaceous Interior Seaway.
6. Mesozoic rocks of the western region of North America were deposited in a variety of continental and marine environments. One of the major controls of sediment distribution patterns was tectonism.
7. Western North America was affected by four interrelated orogenies: the Sonoma, Nevadan, Sevier, and Laramide. Each involved batholithic intrusions as well as eastward thrust faulting and folding.
8. The cause of the Sonoma, Nevadan, Sevier, and Laramide orogenies was the changing angle of subduction of the oceanic Pacific plate under the continental North American plate. The timing, rate, and to some degree the direction of plate movement was related to sea-floor spreading and the opening of the Atlantic Ocean.

9. Among the marine invertebrates, survivors of the Permian extinction event diversified and gave rise to increasingly complex Mesozoic marine invertebrate communities.
10. Triassic and Jurassic land-plant communities were composed of seedless plants and gymnosperms. Angiosperms, or flowering plants, appeared during the Early Cretaceous, diversified rapidly, and soon became the dominant land plants.
11. Dinosaurs evolved from thecodontians during the Late Triassic, but were most abundant and diverse during the Jurassic and Cretaceous. Based on pelvic structure, two distinct orders of dinosaurs are recognized—Saurischia (lizard-hipped) and Ornithischia (bird-hipped).
12. Pterosaurs were the first flying vertebrate animals. Small pterosaurs were probably active, wing-flapping fliers, while large ones may have depended more on thermal updrafts and soaring to stay aloft. At least one pterosaur species had hair or feathers, so it was very likely endothermic.
13. The fish-eating, porpoise-like ichthyosaurs were thoroughly adapted to an aquatic life. Female ichthyosaurs probably retained eggs within their bodies and gave birth to live young. Plesiosaurs were heavy-bodied marine reptiles that probably came ashore to lay eggs.
14. Birds probably evolved from small carnivorous dinosaurs. The oldest known bird, *Archaeopteryx,* appeared during the Jurassic, but few other Mesozoic birds are known. Recent finds of *Protoavis* in Triassic rocks may represent a bird older than *Archaeopteryx.*
15. The earliest mammals evolved during the Late Triassic, but they are difficult to distinguish from advanced cynodonts. Details of the teeth, the middle ear, and lower jaw are used to distinguish the two.
16. Several types of Mesozoic mammals existed, but all were small, and their diversity was low. A group of Mesozoic mammals called eupantotheres gave rise to both marsupials and placentals during the Cretaceous.
17. Because the continents were close together during much of the Mesozoic and climates were mild even at high latitudes, animals and plants dispersed very widely.
18. Mesozoic mass extinctions account for the disappearance of dinosaurs, several other groups of reptiles, and a number of marine invertebrates. One hypothesis holds that the extinctions were caused by the impact of a large meteorite with the Earth. Many paleontologists reject the meteorite proposal and claim that withdrawal of epeiric seas and climatic changes can account for this extinction event.

Important Terms

angiosperm	marsupial mammal	Sevier orogeny
bipedal	monotreme	Sonoma orogeny
Cordilleran orogeny	Nevadan orogeny	Sundance Sea
Cretaceous Interior Seaway	ornithischian	thecodontian
cynodont	placental mammal	therapsid
eupantothere	quadrupedal	Zuni sequence
Laramide orogeny	saurischian	

◈◈

Review Questions

1. Evidence for the breakup of Pangaea includes:
   a. ___ rift valleys; b. ___ dikes; c. ___ sills; d. ___ great quantities of poorly sorted nonmarine detrital sediments; e. ___ all of these.

2. Which cratonic sequence was deposited during Early Jurassic to Early Paleocene?
   a. ___ Tippecanoe; b. ___ Absaroka; c. ___ Zuni; d. ___ Sauk; e. ___ Kaskaskia.

3. The time of greatest post-Paleozoic inundation of the craton occurred during which geologic period?
   a. ___ Triassic; b. ___ Jurassic; c. ___ Cretaceous; d. ___ Paleogene; e. ___ Neogene.

4. The first Mesozoic orogeny in the Cordilleran region was the:
   a. ___ Antler; b. ___ Laramide; c. ___ Nevadan; d. ___ Sevier; e. ___ Sonoma.

5. What type of climates dominated the Triassic low and middle latitudes?
   a. ___ hot and humid; b. ___ cool and humid; c. ___ cold and dry; d. ___ warm and dry; e. ___ glacial.

6. The breakup of Pangaea began with initial Triassic rifting between which two continental landmasses?
   a. ___ South America and Africa; b. ___ Laurasia and Gondwana; c. ___ North America and Eurasia; d. ___ Antarctica and India; e. ___ India and Australia.

7. A possible cause for the eastward migration of igneous activity in the Cordilleran region during the Cretaceous was a change from:
   a. ___ oceanic-oceanic convergence to oceanic-continental convergence; b. ___ high-angle to low-angle subduction; c. ___ divergent to convergent plate margin activity; d. ___ divergent plate margin activity to subduction; e. ___ subduction to divergent plate margin activity.

8. The age of the thick evaporite deposits of the Gulf Coastal region that form the Tertiary salt domes is:
   a. ___ Permian; b. ___ Triassic; c. ___ Jurassic; d. ___ Cretaceous; e. ___ Tertiary.

9. The Sierra Nevada, Southern California, Idaho, and Coast Range batholiths formed as a result of which orogeny?
   a. ___ Sonoma; b. ___ Nevadan; c. ___ Sevier; d. ___ Laramide; e. ___ none of these.

10. The mammal-like reptiles that were most like mammals were the:
    a. ___ cynodonts; b. ___ monotremes; c. ___ symmetrodonts; d. ___ thecodontians; e. ___ bipeds.

11. Sauropod dinosaurs were:
    a. ___ carnivorous; b. ___ fast-running bipeds; c. ___ the largest dinosaurs; d. ___ descendants of therapsids; e. ___ particularly varied and abundant during the Early Triassic.

12. An important Mesozoic event in the history of land plants was the:
    a. ___ extinction of cycads; b. ___ origin of ferns; c. ___ first appearance of angiosperms; d. ___ dominance of ginkgos; e. ___ all of these.

13. All carnivorous dinosaurs are classified as:
    a. ___ phytosaurs; b. ___ theropods; c. ___ *Euparkeria;* d. ___ marsupials; e. ___ pachycephalosaurs.

14. A typical reptile's jaw-skull joint is between which of these pairs of bones?
    a. ___ squamosal-secondary palate; b. ___ articular-quadrate; c. ___ ethmoid-occipital; d. ___ ilium-dentary; e. ___ stapes-incus.

15. Because of their rapid evolution and nektonic life-style, the _____ are excellent Mesozoic guide fossils.
    a. ___ bivalves; b. ___ cephalopods; c. ___ cynodonts; d. ___ angiosperms; e. ___ trilobites.

16. The probable ancestors of dinosaurs and flying reptiles were:
    a. ___ gymnosperms; b. ___ phytosaurs; c. ___ ammonoids; d. ___ thecodontians; e. ___ ceratopsians.

17. Discuss the depositional environments, tectonic setting, and depositional processes of the Triassic Newark Group in the Eastern Coastal region.

(Continued on page 674)

Age (Millions of Years)	Geologic Period	Sequence	Relative Changes in Sea Level		Cordilleran Mobile Belt	
			Rising	Falling		
66					Jurassic and Cretaceous tectonism controlled by eastward subduction of the Pacific plate beneath North America and accretion of microplates.	Cordilleran orogeny
	Cretaceous	Zuni				Laramide orogeny
						Sevier orogeny
144				Present sea level		Nevadan orogeny
	Jurassic					
208		Absaroka				
	Triassic				Subduction zone develops as a result of westward movement of North America.	
245					Sonoma orogeny	

North American Interior	Gulf Coastal Region	Eastern Coastal Region	Global Plate Tectonic Events
	Major Late Cretaceous transgression. Reefs particularly abundant.	Appalachian region uplifted.	South America and Africa are widely separated. Greenland begins separating from Europe.
	Regression at end of Early Cretaceous.		
	Early Cretaceous transgression and marine sedimentation.	Erosion of fault-block mountains formed during the Late Triassic to Early Jurassic.	
	Sandstones, shales, and limestones are deposited in transgressing and regressing seas.		South America and Africa begin separating in the Late Jurassic.
	Thick evaporites are deposited in newly formed Gulf of Mexico.		
		Fault-block mountains and basins develop in eastern North America.	
		Deposition of Newark Group; lava flows, sills, and dikes.	Breakup of Pangaea begins with rifting between Laurasia and Gondwana.
	Gulf of Mexico begins forming during Late Triassic.		Supercontinent Pangaea still in existence.

● **TABLE 23.3** Summary of Biological Events for the Mesozoic Era

Age (Millions of Years)	Geologic Period	Invertebrates		Vertebrates
66				
	Cretaceous	Continued diversification of ammonites and belemnoids. Rudist become major reef-builders. Extinction of ammonites, rudists, and most planktonic foraminifera at end of Creataceous.	GREATEST DIVERSITY OF DINOSAURS	Extinctions of dinosaurs, flying reptiles, and marine reptiles. Placental and marsupial mammals diverge.
144	Jurassic	Ammonites and belemnoid cephalopods increase in diversity. Scleractinian coral reefs common. Appearance of rudist bivalves.		First birds (may have evolved in Late Triassic). Time of giant sauropod dinosaurs.
208	Triassic	The seas are repopulated by invertebrates that survived the Permian extinction event. Bivalves and echinoids expand into the infaunal niche.		Mammals evolve from cynodonts. Cynodonts become extinct. Thecodontians give rise to dinosaurs. Flying reptiles and marine reptiles evolved.
245				

Plants	Climate	Plate Tectonics
Angiosperms evolve and diversify rapidly. Seedless plants and gymnosperms still common but less varied and abundant.	North-south zonation of climates more marked, but remains equable. Climate becomes more seasonal and cooler at end of Cretaceous.	Further fragmentation of Pangaea. South America and Africa have separated. Australia separated from South America but remains connected to Antarctica. North Atlantic continues to open.
Seedless vascular plants and gymnosperms only.	Much like Triassic. Ferns with living relatives restricted to tropics live at high latitudes, indicating mild climates.	Fragmentation of Pangaea continues, but close connections exist among all continents.
Land flora of seedless vascular plants and gymnosperms as in Late Paleozoic.	Warm-temperate to tropical. Mild temperatures extend to high latitudes; polar regions may have been temperate. Local areas of aridity.	Fragmentation of Pangaea begins in Late Triassic.

18. Provide a general global history of the breakup of Pangaea.
19. Compare the tectonics of the Sonoma and Antler orogenies.
20. How did the breakup of Pangaea affect oceanic and climatic circulation patterns?
21. Discuss the tectonics of the Cordilleran mobile belt during the Mesozoic Era.
22. Compare the tectonic setting and depositional environment of the Gulf of Mexico evaporites with the evaporite sequences of the Paleozoic Era.
23. Briefly outline the major changes in the composition of Mesozoic marine invertebrate communities.

24. Discuss the changing aspects of land-plant communities during the Mesozoic.
25. What are the names of the two dinosaur orders, and how are they differentiated from one another?
26. Briefly discuss the adaptations for flight seen in pterosaurs. Is there any evidence for endothermy in pterosaurs? If so, what?
27. In what ways was *Archaeopteryx* similar to and different from small carnivorous dinosaurs?
28. What skeletal features of cynodonts indicate that they were endotherms?
29. Briefly summarize the evidence for and against the proposal that a meteorite impact caused Mesozoic mass extinctions.

Additional Readings

Bakker, R. T. 1986. *The dinosaur heresies.* New York: William Morrow.

Bakker, R. 1993 (a). Jurassic sea monsters. *Discover* 14, no. 9:78–85.

———. 1993 (b). Bakker's field guide to Jurassic dinosaurs. *Earth* 2, no. 5:33–43.

Bally, A. W., and A. R. Palmer, eds. 1989. *The geology of North America: An overview.* The Geology of North America. vol. A. Boulder, Colo.: Geological Society of America.

Bonatti, E. 1987. The rifting of continents. *Scientific American* 256, no. 3:96–103.

Donovan, S. K., ed. 1989. *Mass extinctions.* New York: Columbia University Press.

Gore, R. 1993. Dinosaurs. *National Geographic* 183, no. 1:2–53.

Hopson, J. A. 1987. The mammal-like reptiles: A study of transitional fossils. *The American Biology Teacher* 49, no. 1:16–26.

Jones, D. L., A. Cox, P. Coney, and M. Beck. 1982. The growth of western North America. *Scientific American* 247, no. 5:70–128.

Lucas, S. G. 1994. *Dinosaurs: The textbook.* Dubuque, Iowa: Wm. C. Brown.

McGowan, C. 1991. *Dinosaurs, spitfires, and sea dragons.* Cambridge, Mass.: Harvard University Press.

Nations, J. D., and J. G. Eaton, eds. 1991. *Stratigraphy, depositional environments, and sedimentary tectonics of the western margin, Cretaceous Western Interior Seaway.* Geological Society of America Special Paper 260. Boulder, Colo.: Geological Society of America.

Russell, D. A. 1988. *An odyssey in time: The dinosaurs of North America.* Toronto: Ont.: University of Toronto Press.

Salvador, A., ed. 1991. *The Gulf of Mexico Basin.* The Geology of North America. vol. J. Boulder, Colo.: Geological Society of America.

Savage, R. J. G. 1986. *Mammalian evolution: An illustrated guide.* New York: Facts on File Publications.

Vickers-Rich, P. And T. H. Rich. 1993. Australia's polar dinosaurs. *Scientific American* 269, no. 1:42–49.

Weishampel, D. B. P. Dodston, and H. Osmóska, eds. 1990. *The Dinosauria.* Berkeley, Calif.: University of California Press.

Wellenhofer, P. 1990. *Archaeopteryx. Scientific American* 262, no. 5:70–77.

Wellenhofer, P. 1991. *The illustrated encyclopedia of pterosaurs.* New York: Crescent Books.

CHAPTER
24

Chapter Outline

Shiprock is a volcanic neck 550 m high in northwest New Mexico. It formed about 27 million years ago during the Oligocene Epoch. (Photo courtesy of Sue Monroe.)

Cenozoic Earth and Life History

Prologue

According to Navajo legend, a young man named Nayenezgani asked his grandmother where the mythical birdlike creatures known as Tse'na'hale lived. She replied, "They dwell at Tsae-bidahi," which means Winged Rock, or Rock with Wings. We know Winged Rock as Shiprock, New Mexico, a volcanic neck rising nearly 550 m above the surrounding plain (see chapter-opening photo). Radiating outward from this conical volcanic neck are three dikes. One Navajo legend holds that Winged Rock represents a giant bird that brought the Navajo people from the north, and the dikes are snakes that have turned to stone.

Shiprock is the most impressive of many volcanic necks exposed in the Four Corners region of the Southwest. (Four Corners is a designation for the point at which the state boundaries of Colorado, Utah, Arizona, and New Mexico converge.) Shiprock is visible from more than 160 km, and for many years was a favorite with rock climbers, until the Navajos put a stop to all climbing on the reservation. The country rock penetrated by this volcanic neck includes Precambrian crystalline rocks and about 1,000 m of overlying Cambrian to Cretaceous sedimentary rocks. The rock unit exposed at the surface is the Upper Cretaceous Mancos Shale, a formation that was deposited in the Cretaceous Interior Seaway (see Fig. 23.17). Shiprock was emplaced during the Oligocene Epoch; fission track dating of one of the dikes indicates an age of 27 million years.

Shiprock is one of several breccia-filled volcanic necks in the Navajo volcanic field that formed as a result of explosive volcanic eruptions. During these eruptions, volcanic materials along with large pieces of country rock that were torn from the vent walls were hurled high into the air and fell randomly around the area. The material composing Shiprock is characterized as a tuff-breccia containing inclusions of red sandstone, shale, and some granite and gneiss. Because Shiprock now stands about 550 m above the surrounding plain, at least that much erosion must have occurred to expose it in its present form. We can only speculate as to how much higher and larger it was when it was an active volcano.

The dikes radiating from Shiprock formed when magma ascended rather quietly and was emplaced in the country rocks. However, the fractures along which this magma rose may have resulted from the explosive emplacement of the vent material. The dike on the northeast side of Shiprock extends more than 2,900 m outward from the vent and averages 2.3 m thick. The dike rock, like the material composing the volcanic neck, is more resistant to erosion than the adjacent Mancos Shale, and thus the dikes stand as near vertical walls above the surrounding plain.

⇞ INTRODUCTION

Traditionally, the Cenozoic has been divided into two periods, the **Tertiary** (66 to 1.6 million years ago) and the **Quaternary** (1.6 million years ago to the present). Both periods are further divided into epochs (◆ Fig. 24.1). The terms Tertiary and Quaternary are widely used among geologists, but a different scheme for designations of Cenozoic time is becoming increasingly popular. This scheme also recognizes two periods, the

Era	Period		Epoch	Duration, millions of years (approx.)	Millions of years ago (approx.)
Cenozoic Era	Tertiary Period	Quarternary Period	Pleistocene Epoch	1.99	
					1.6
		Neogene Period	Pliocene Epoch	3.3	
					5.3
			Miocene Epoch	18.7	
					24
		Paleogene Period	Oligocene Epoch	13	
					37
			Eocene Epoch	21	
					58
			Paleocene Epoch	8	
					66

◆ FIGURE 24.1 The geologic time scale for the Cenozoic Era.

Paleogene and *Neogene,* but they do not correspond directly to the Tertiary and Quaternary periods (Fig. 24.1). In this book we will follow the traditional usage.

Many of the features of the Earth have long histories, but the present distribution of land and sea and the topographic expression of continents and their landforms are all the end products of Cenozoic processes. The Appalachian Mountain region, for example, began its evolution during the Precambrian, but its present expression is largely the product of Cenozoic uplift and erosion. In short, the present distinctive aspect of the Earth developed very recently in the context of geologic time.

▼ CENOZOIC PLATE TECTONICS AND OROGENY

The Late Triassic fragmentation of the supercontinent Pangaea (see Fig. 23.2) began an episode of plate motions that continues even now. As a consequence of these plate motions, Cenozoic orogenic activity was largely concentrated in two major zones or belts, the *Alpine-Himalayan belt* and the *circum-Pacific belt* (see Fig. 10.22). The Apline-Himalayan belt includes the mountainous regions of southern Europe and north Africa and extends eastward through the Middle East and India and into southeast Asia, whereas the circum-Pacific belt, as its name implies, nearly encircles the Pacific Ocean basin.

The Alpine and Himalayan orogens developed in the Alpine-Himalayan orogenic belt. The *Alpine orogeny* began during the Mesozoic, but major deformation also occurred from the Eocene to Late Miocene as the African and Arabian plates moved northward against Eurasia. Deformation resulting from plate convergence formed the Pyrenees Mountains between Spain and France, the Alps of mainland Europe, the Apennines of Italy, and the Atlas Mountains of North Africa (see Fig. 10.22). Active volcanoes in Italy and seismic activity in much of southern Europe and the Middle East indicate that this orogenic belt remains geologically active.

Farther east in the Alpine-Himalayan orogenic belt, the Himalayan orogen resulted from the collision of India with Asia (see Fig. 10.25). The exact time of this collision is uncertain, but sometime during the Eocene India's northward drift rate decreased abruptly, indicating the probable time of collision. In any event, a *collision orogen* resulted as two continental plates became sutured, accounting for the location of the present-day Himalayas far inland rather than at a continental margin.

Plate subduction in the circum-Pacific orogenic belt occurred throughout the Cenozoic, giving rise to orogens in the Aleutians, the Philippines, Japan, and along the west coasts of North, Central, and South America. For example, the Andes Mountains in western South America formed as a result of convergence of the Nazca and South American plates (see Fig. 10.24). Spreading at the East Pacific Rise and subduction of the Cocos and Nazca plates beneath Central and South America, respectively, accounts for continuing orogenic activity in these regions.

▼ THE NORTH AMERICAN CORDILLERA

The **North American Cordillera** is a complex segment of the circum-Pacific orogenic belt extending from Alaska into Mexico (◆ Fig. 24.2). Beginning in the Late Jurassic and continuing into the Early Tertiary, it was more or less continuously deformed as the Nevadan, Sevier, and Laramide orogenies progressively affected areas from west to east (see Fig. 23.11). The **Laramide**

◆ FIGURE 24.2 The North American Cordillera and the major provinces of North America discussed in the text.

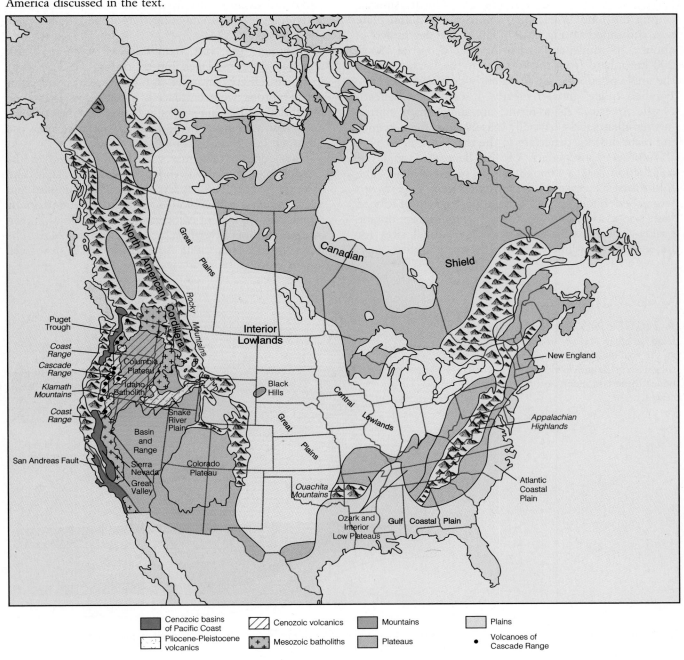

Legend:
- Cenozoic basins of Pacific Coast
- Pliocene-Pleistocene volcanics
- Cenozoic volcanics
- Mesozoic batholiths
- Mountains
- Plateaus
- Plains
- • Volcanoes of Cascade Range

orogeny, a Late Cretaceous to Eocene event, differs from the previous orogenies in that deformation occurred much farther inland than is typical of arc orogenies, and deformation was not accompanied by significant batholithic intrusions and volcanism.

To account for these observations, geologists have modified the classic model for arc orogens (◆ Fig. 24.3). According to this view, a subduction zone existed along the entire west coast of North America during the Late Cretaceous, where the **Farallon plate** was consumed (◆ Fig. 24.4). This plate descended at about a 50° angle, and arc magmatism occurred 150 to 200 km inland from the trench. By the Early Tertiary, the angle of subduction apparently decreased, and the Farallon plate moved nearly horizontally beneath North America. As a result, arc magmatism shifted farther inland and eventually ceased, since the descending plate did not penetrate to the mantle (Fig. 24.3).

Another consequence of the decreased angle of subduction was a change in the tectonic style. The fold-thrust tectonism of the Sevier orogeny gave way to large-scale buckling and fracturing, which produced fault-bounded uplifts with intervening intermontane basins. These basins were the depositional sites for Early Tertiary sediments eroded from the uplifts.

The Laramide orogen is centered in the middle and southern Rockies in Wyoming and Colorado, but deformation also occurred far to the north and south. In Montana and Alberta, Canada, large slabs of strata were transported to the east along overthrust faults (◆ Fig. 24.5). A major fold-thrust belt also developed in northern Mexico, and Laramide structural trends can be traced south into the area of Mexico City. Apparently, the cessation of Laramide deformation coincided with an increasing angle of descent of the Farallon plate (Fig. 24.3).

Although the vast batholiths of western North America were emplaced mostly during the Mesozoic, intrusive activity continued into the Tertiary. Numerous small plutons were emplaced, including copper- and molybdenum-bearing stocks in Utah, Nevada, Arizona, and New Mexico, and the igneous body we now call Shiprock (see the Prologue).

Considerable Cenozoic volcanism occurred in the Cordillera, although it varied in eruptive style and location (◆ Fig. 24.6). A large area in Washington and adjacent parts of Oregon and Idaho is underlain by about 200,000 km³ of Miocene basalt lava flows known as the Columbia River basalts (Figs. 24.2 and 24.6). The flows issued from long fissures and are now well

◆ FIGURE 24.3 Arc orogens resulting from (*a*) steep and (*b*) shallow subduction. In the shallow-subduction model, the subducted slab moves nearly horizontally beneath the continent, and arc volcanism ceases.

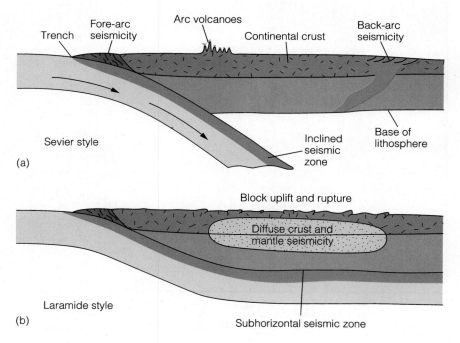

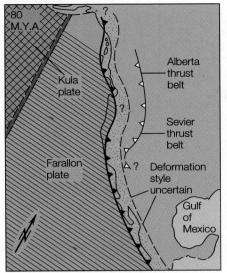

Late Cretaceous

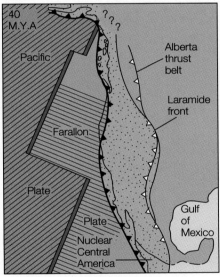

Eocene

◆ FIGURE 24.4 Schematic maps showing the Laramide style of deformation. The Laramide orogeny was caused by subduction of the Farallon plate beneath North America.

(a)

(b)

◆ FIGURE 24.5 (*a*) The Lewis overthrust in Glacier National Park, Montana. Late Proterozoic strata of the Belt Supergroup rest upon Cretaceous strata. The trace of the fault is the nearly horizontal light line on the side of the mountain. (*b*) Chief Mountain, Montana, is an erosional remnant of the Lewis overthrust.

exposed in the deep gorges cut by the Snake and Columbia rivers. The Snake River Plain in Idaho is actually a depression that has been filled mostly by Pliocene and younger basalts. Bordering the Snake River Plain on the northeast is the Yellowstone Plateau of Wyoming, an area of Late Pliocene and Quaternary rhyolitic and some basaltic volcanism. The volcanoes of the Cascade Range were built by andesitic volcanism during the Pliocene, Pleistocene, and Recent epochs (see Chapter 4).

A large area in the Cordillera known as the **Basin and Range Province** (Fig. 24.2) has been subjected to crustal extension and the development of north-south oriented normal faults since the Late Miocene. Differential movement on these faults has produced what is called a basin-and-range structure, consisting of mountain ranges separated by broad valleys (◆ Fig. 24.7).

(a)

(b)

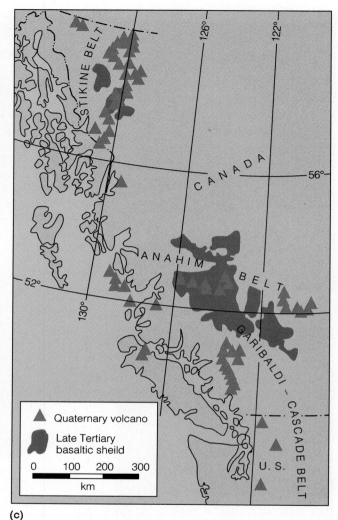

(c)

◆ FIGURE 24.6 Cenozoic volcanism in the North American Cordillera. (*a*) Sunset Crater, a cinder cone near Flagstaff, Arizona, formed mostly during the winter of A.D. 1064–1065. (*b*) The Columbia River basalts in Washington. (*c*) Distribution of Quaternary volcanoes in western Canada.

◆ FIGURE 24.7 Cross section of part of the Basin and Range Province in Nevada. The ranges and valleys are bounded by normal faults.

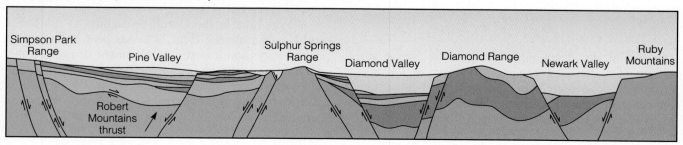

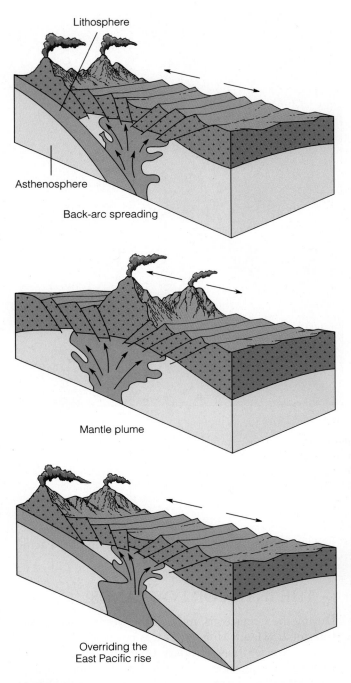

Back-arc spreading

Mantle plume

Overriding the
East Pacific rise

◆ FIGURE 24.8 Geologists agree that tensional forces caused by crustal extension are responsible for basin-and-range structure. There is no agreement, however, on what caused the crustal extension. These illustrations show three models that have been proposed to account for crustal extension and the resulting basin-and-range structure.

Several models have been proposed to account for this structure (◆ Fig. 24.8).

The Colorado Plateau (Fig. 24.2) is a vast area of deep canyons, broad mesas, volcanic mountains, and brilliantly colored rocks (◆ Fig. 24.9). During the Early Tertiary this area, which was formerly near sea level, was uplifted and deformed into broad anticlines and arches and basins. A number of large normal faults also cut the area, but overall deformation was far less intense than it was elsewhere in the Cordillera. Late Tertiary uplift elevated the region to the 1,200 to 1,800 m elevations seen today. As uplift proceeded, deposition ceased and erosion of the deep canyons began.

The present plate tectonic elements of the Pacific Coast developed as a consequence of the westward drift of the North American plate, the partial consumption of the Farallon plate, and the collision of North America with the Pacific-Farallon ridge (◆ Fig. 24.10). Most of the Farallon plate was consumed at a subduction zone along the west coast of North America, and now only two small remnants exist—the Juan de Fuca

◆ FIGURE 24.9 Rocks of the Colorado Plateau. The Grand Canyon, Arizona.

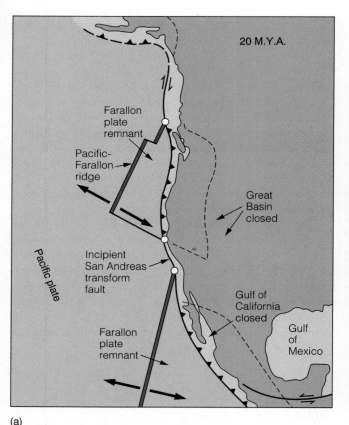

(a)

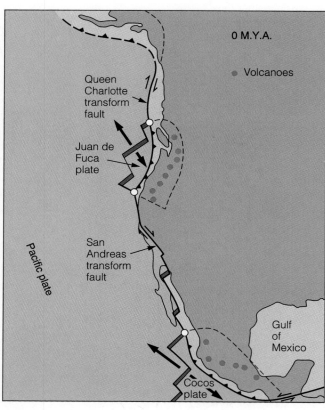

(b)

◆ FIGURE 24.10 (*a*) and (*b*) The westward drift of North America caused it to collide with the Pacific-Farallon ridge. As North America overrode the ridge, the plate margin became bounded by transform faults rather than a subduction zone.

and Cocos plates (Fig. 24.10). Westward drift of the North American plate also resulted in its collision with the Pacific-Farallon ridge and the origin of the Queen Charlotte and San Andreas transform faults along the west coasts of British Columbia, Canada, and California, respectively (Fig. 24.10).

Much of the Cordilleran deformation and volcanism occurred during the Tertiary Period, but these activities continued through the Quaternary to the present. For example, many of California's coastal oil- and gas-producing fold structures formed during the Pleistocene. Pleistocene tectonism occurred in the Sierra Nevada in California, the Grand Tetons of Wyoming, and parts of the central and northern Rocky Mountains. Continued subduction of the remnants of the Farallon and Cocos plates accounts for Pleistocene and

Recent seismic activity and volcanism in the Pacific Northwest and Mexico, respectively.

▼ THE INTERIOR LOWLANDS

During the Cretaceous, most of the western part of the Interior Lowlands (Fig. 24.2) was covered by the Zuni epeiric sea. By Early Tertiary time, the Zuni Sea had withdrawn from most of North America, but a sizable remnant arm of the sea was still present in North Dakota during the Paleocene Epoch. Sediments derived from Laramide highlands to the west and southwest were transported to this remnant sea where they were deposited in marginal marine and marine environments.

Elsewhere within the western Interior Lowlands, sediment was also transported eastward from the Cordillera, deposited in terrestrial environments, and formed large, eastward-thinning wedges that now underlie the Great Plains (♦ Fig. 24.11).

Igneous activity was not widespread in the Interior Lowlands, but in some local areas it was significant. In northeastern New Mexico, for example, Late Tertiary extrusive volcanism produced volcanoes and numerous lava flows. A number of small intrusive bodies were emplaced in Colorado, Montana, South Dakota, and Wyoming.

Eastward, beyond the Great Plains section of the Interior Lowlands, Cenozoic deposits, other than those of Pleistocene glaciers, are uncommon. Much of the Interior Lowlands was subjected to erosion during the Tertiary. Of course, the eroded material had to be deposited somewhere, and that was on the Gulf Coastal Plain (Fig. 24.2).

♦ THE GULF COASTAL PLAIN

Following the final withdrawal of the Cretaceous Zuni Sea, the Cenozoic **Tejas epeiric sea** made a brief appearance on the continent. But even at its maximum extent, it was largely restricted to the Atlantic and Gulf Coastal plain. In fact, its greatest incursion onto North America was in the area of the Mississippi Valley, where it extended as far north as southern Illinois.

Sedimentary facies development on the Gulf Coastal Plain was controlled largely by a regression of the Cenozoic Tejas epeiric sea. Its regression, however, was periodically reversed by minor transgressions; eight transgressive-regressive episodes are recorded in Gulf Coastal Plain sedimentary rocks.

The Gulf Coast sedimentation pattern was established during the Jurassic and persisted through the Cenozoic. Sediments were derived from the eastern Cordillera, western Appalachians, and Interior Lowlands and were transported toward the Gulf of Mexico. In general, the sediments form seaward-thickening wedges that grade from terrestrial facies in the north to progressively more offshore marine facies in the south (♦ Fig. 24.12).

Much of the Gulf Coastal Plain was dominated by detrital sediment deposition during the Cenozoic. In the Florida section of the coastal plain and the Gulf Coast of Mexico, however, significant carbonate deposition occurred. A carbonate platform was established in Florida during the Cretaceous, and shallow-water carbonate deposition continued through the Early Tertiary. Carbonate deposition continues in Florida at the present, but now it occurs only in Florida and the Florida Keys.

Southeast of Florida, across the 80-km-wide Florida Strait, lies the Great Bahama Bank. This area has been a carbonate bank from the Cretaceous to the present. Thick, shallow-water carbonates accumulated there, and the region has been used for a long time as a modern-day laboratory to help us understand the conditions under which limestone is deposited (♦ Fig. 24.13).

♦ FIGURE 24.11 Cenozoic sedimentary rocks of the Great Plains at Scott's Bluff, Nebraska.

♦ FIGURE 24.12 Cenozoic deposition on the Gulf Coastal Plain. This cross section of the Eocene Claiborne Group shows facies changes and the seaward thickening of the deposits.

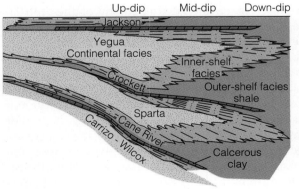

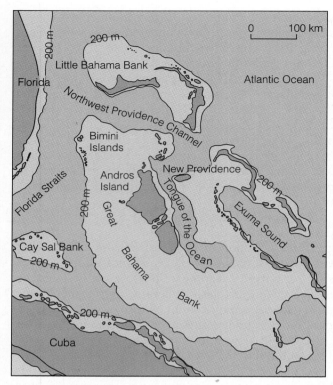

◆ FIGURE 24.13 Location of the Great Bahama Bank. This area is underlain by thick, shallow-water carbonate rocks. Carbonate deposition continues at the present.

◆ EASTERN NORTH AMERICA

The eastern seaboard has been a passive continental margin since Late Triassic rifting separated North America from North Africa and Europe. Some seismic activity still occurs there, but overall the region lacks the geologic activity characteristic of active, convergent margins.

The present distinctive topography of the Appalachian Mountains is the product of Cenozoic uplift and erosion. By the end of the Mesozoic, the Appalachian Mountains had been eroded to a plain (◆ Fig. 24.14). Cenozoic uplift rejuvenated the streams, which responded by renewed downcutting. As the streams eroded downward, they were superposed on resistant strata and cut large canyons across these strata. For example, the distinctive topography of the Valley and Ridge Province is the product of Cenozoic erosion and preexisting geologic structures. It consists of northeast-southwest trending ridges of resistant upturned strata

and intervening valleys eroded into less resistant strata (Fig. 24.14).

The Atlantic continental margin includes the Atlantic Coastal Plain (Fig. 24.2) and extends seaward across the continental shelf, slope, and rise (◆ Fig. 24.15). It possesses a number of Mesozoic and Cenozoic sedimentary basins that formed as a consequence of rifting. Deposition in these basins began during the Jurassic, and even though sediments of this age are known only from a few deep wells, they are presumed to underlie the entire margin of the continent. The distribution of Cretaceous and Cenozoic sediments is better known because both crop out in the Atlantic Coastal Plain, and both have been penetrated by wells on the continental shelf.

In general, the sedimentary rocks of the Atlantic Coastal Plain are part of a seaward-thickening wedge that dips gently seaward. In some places, such as off the coast of New Jersey, these deposits are up to 14 km thick. The best-studied seaward-thickening wedge of sedimentary rocks is in the Baltimore Canyon Trough, an area that also exhibits the structures typical of a passive continental margin (Fig. 24.15).

▼ PLEISTOCENE GLACIATION

In 1837 the Swiss naturalist Louis Agassiz argued that large displaced boulders (called *erratics*), coarse-grained sedimentary deposits, polished and striated bedrock, and U-shaped valleys found throughout parts of Europe were the result of huge ice masses moving over the land. Based on Agassiz's observations and arguments, geologists soon accepted that an Ice Age had indeed occurred during the recent geologic past.

We know today that the last Ice Age began about 1.6 million years ago and consisted of several intervals of glacial expansion separated by warmer interglacial periods. It appears that the present interglacial period began about 10,000 years ago, but geologists do not know whether we are still in an interglacial period or are entering another colder glacial interval.

The onset of glacial conditions really began about 40 million years ago when surface ocean waters at high southern latitudes rapidly cooled, and the water in the deep-ocean basins soon cooled to about 10°C colder than it had been previously. By the Middle Miocene an Antarctic ice cap had formed, accelerating the formation of very cold waters. Following a brief warming trend during the Pliocene, ice sheets began forming in the Northern Hemisphere about 1.6 million years ago, and the Pleistocene Ice Age was underway.

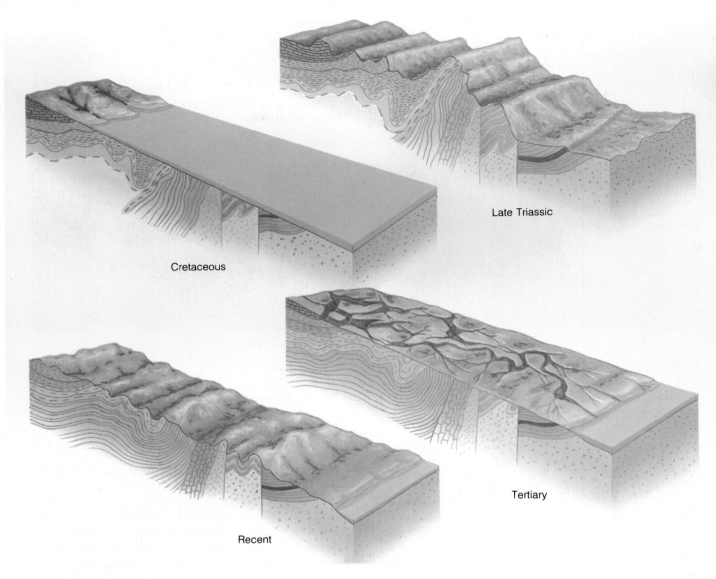

Late Triassic

Cretaceous

Tertiary

Recent

◆ FIGURE 24.14 The origin of the present topography of the Appalachian Mountains. Erosion in response to Cenozoic uplift accounts for this topography.

The climatic effects responsible for Pleistocene glaciation were, as one would expect, worldwide. Nevertheless, the world was not as frigid as it is commonly portrayed in cartoons and movies, nor was the onset of glacial conditions as rapid as many people believe. In fact, evidence from various lines of research indicates that the world's climate gradually cooled from the beginning of the Eocene through the Pleistocene. And oxygen isotope data (the ratio of O^{18} to O^{16}) from deep-sea cores reveal that during the past 2 million years there were at least 20 major warm-cold cycles.

From such glacial features as the distribution of moraines, erratic boulders, and drumlins (◆ Fig. 24.16), it seems at their greatest extent, Pleistocene glaciers covered about three times as much of the Earth's surface as they do now and were up to 3 km thick (◆ Fig. 24.17). Geologists have determined that at least four major **glacial stages,** each followed by an **intergla-**

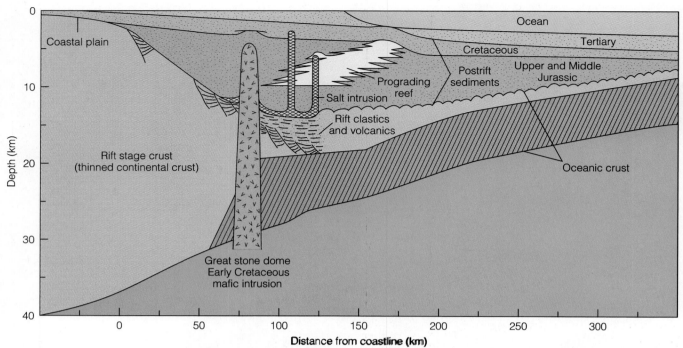

◆ FIGURE 24.15 The continental margin in eastern North America. The coastal plain and the continental shelf are covered mostly by Cenozoic sandstones and shales. Beneath these sediments are Cretaceous-aged and probably Jurassic-aged sedimentary rocks.

cial stage, occurred during the Pleistocene of North America. These four stages, the *Wisconsin, Illinoian, Kansan,* and *Nebraskan,* are named for the states in which the most southerly glacial deposits are well exposed. The three interglacial stages, the *Sangamon, Yarmouth,* and *Aftonian,* are named for localities of well-exposed interglacial soils and other deposits (◆ Fig. 24.18). In Europe, six or seven major glacial advances and retreats are recognized.

Recent studies indicate that there were an as yet undetermined number of pre-Illinoian glacial events, and that the history of glacial advances and retreats in North America is more complex than previously believed. In view of these data, the traditional four-part subdivision of the Pleistocene of North America must be modified.

▼ THE EFFECTS OF GLACIATION

Glaciation has had many direct and indirect effects. Movement of glaciers over the Earth's surface has produced distinctive landscapes in much of Canada, the northern tier of states, and in the mountains of the West (see Chapter 14). Sea level has risen and fallen with the formation and melting of glaciers, and these changes in turn have affected the margins of continents. Glaciers have also altered the world's climate, causing cooler and wetter conditions in some areas that are arid to semiarid today. In addition to the usual evidence of glacial activity (Fig. 24.16), one of the largest floods in history was caused by the collapse of an ice dam in eastern Washington (see Perspective 14.1).

More than 70 million km^3 of snow and ice covered the continents during the maximum glacial coverage of the Pleistocene. The storage of ocean waters in glaciers lowered sea level 130 m and exposed large areas of the present-day continental shelves, which were soon blanketed by vegetation.

Lowering of sea level also affected the base level of most major streams. When sea level dropped, streams eroded downward as they sought to adjust to a new lower base level. Stream channels in coastal areas were

(a)

(b)

◆ FIGURE 24.16 Features characteristic of glaciated areas. (*a*) Glacial till is the unsorted sediment deposited by a glacier. (*b*) Erratics are ice-transported boulders that have been carried far from their original source by glaciers, such as this one at South Hammond, New York. (Photo b courtesy of R. V. Dietrich.)

extended and deepened along the emergent continental shelves. When sea level rose with the melting of the glaciers, the lower ends of stream valleys along the east coast of North America were flooded and are now important harbors, while just off the west coast, they form impressive submarine canyons. Great amounts of sediment eroded by the glaciers were transported by streams to the sea and thus contributed to the growth of submarine fans along the base of the continental slope.

We noted in Chapter 14 that as the Pleistocene ice sheets formed and increased in size, the weight of the

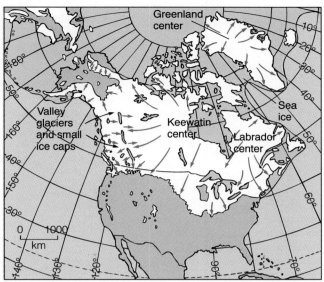

◆ FIGURE 24.17 Centers of ice accumulation and maximum extent of Pleistocene glaciation in North America.

ice caused the crust to slowly subside deeper into the mantle. In some places, the Earth's surface was depressed as much as 300 m below the preglacial elevations. As the ice sheets retreated by melting, the downwarped areas gradually rebounded to their former positions. (see Fig. 10.30).

During the Wisconsin glacial stage many large lakes existed in what are now dry basins in the southwestern United States. These lakes formed as a result of greater precipitation and overall cooler temperatures (especially during the summer), which lowered the evaporation rate. At the same time, increased precipitation and runoff helped maintain high water levels. Lakes that formed during those times are *pluvial lakes,* and they correspond to the expansion of glaciers elsewhere. The largest of these lakes was Lake Bonneville, which attained a maximum size of 50,000 sq km^2 and a depth of at least 335 m (◆ Fig. 24.19). The vast salt deposits of the Bonneville Salt Flats west of Salt Lake City, Utah, formed as parts of this ancient lake dried up; Great Salt Lake is simply the remnant of this once great lake. Another large pluvial lake (Lake Manly) existed in Death Valley, California, which is now the hottest, driest place in North America.

In contrast to pluvial lakes, which form far from glaciers, *proglacial lakes* are formed by the meltwater accumulating along the margins of glaciers (see Perspec-

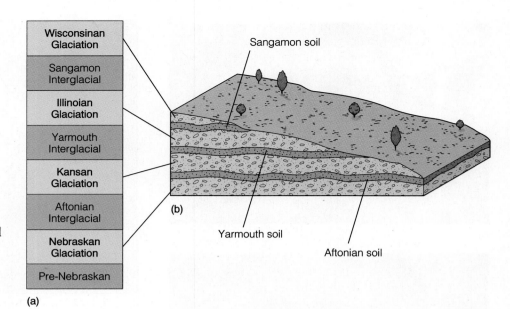

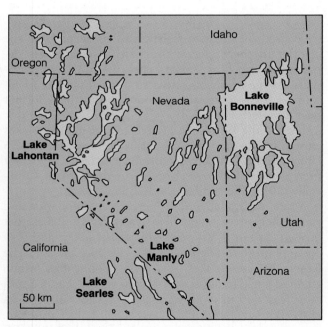

◆ FIGURE 24.18 (*a*) Standard terminology for Pleistocene glacial and interglacial stages in North America. (*b*) An idealized succession of deposits and soils developed during the glacial and interglacial stages.

◆ FIGURE 24.19 Pleistocene pluvial lakes in the western United States.

tive 24.1). In fact, in many proglacial lakes one shoreline is the ice front itself, while the other shorelines consist of moraines. Lake Agassiz was a large proglacial lake covering about 250,000 km² of North Dakota, Manitoba, Saskatchewan, and Ontario. It persisted until the glacial ice along its northern margin melted, at which time the lake was able to drain northward into Hudson Bay.

⩔ CENOZOIC MINERAL RESOURCES

The United States is the second largest producer of petroleum, accounting for more than 18% of the world's total. Much of this production comes from Tertiary reservoirs of the Gulf Coastal Plain and the adjacent continental shelf. On the Gulf Coastal Plain, much of the petroleum is in structural traps related to salt domes and other structures. Several Tertiary basins in southern California are also important areas of petroleum production. The Green River Formation of Wyoming, Utah, and Colorado has huge reserves of oil shale and evaporites (see the Prologue to Chapter 5).

In 1989, more than half of Florida's total mineral resources came from Upper Miocene phosphate-bearing rock in the Central Florida Phosphate District. Phosphorus from phosphate rock has a variety of uses in metallurgy, preserved foods, ceramics, and matches.

Diatomite is a sedimentary rock composed of the microscopic shells of diatoms, single-celled marine and freshwater plants that secrete skeletons of silica (SiO_2). This rock, also called diatomaceous earth, is used chiefly in gas purification and to filter a number of liquids such as molasses, fruit juices, water, and sewage. The United States is the leader in diatomite production, mostly from mines in Tertiary deposits in California, Oregon, and Washington.

Huge deposits of low-grade lignite and subbituminous coals in the Northern Great Plains are becoming increasingly important resources. These coal deposits are Late Cretaceous to Early Tertiary and are most extensive in the Williston and Powder River basins of Montana, Wyoming, and North and South Dakota. In addition to having a low sulfur content, some of these coals occur in beds 30 to 60 m thick!

Gold production from the Pacific Coast, particularly California, comes mostly from Tertiary and Quaternary gravels. The gold occurs as placer deposits, which formed as concentrations of minerals separated from weathered debris by fluvial processes.

A variety of other Tertiary mineral deposits are important. For example, the United States must import almost all manganese used in the manufacture of steel. The largest manganese deposits are in Lower Tertiary rocks in Russia. One Tertiary molybdenum deposit in Colorado accounts for much of the world production of this element. Tertiary sand and gravel, as well as evaporites, building stone, and clay deposits, are quarried from areas around the world.

Sand and gravel deposits resulting from glacial activity are a valuable Quaternary resource in many formerly glaciated areas. Most Pleistocene sand and gravel deposits originated as floodplain or terrace gravels, outwash sediment, or esker deposits. The bulk of the sand and gravel in the United States and Canada is used in construction and as roadbase and fill for highway and railway construction.

The periodic evaporation of pluvial lakes in the Death Valley region of California during the Pleistocene led to the concentration of many evaporite minerals such as borax. During the 1880s, borax was transported from Death Valley by the famous 20-mule team wagon trains (see Perspective 15.2).

Another Quaternary resource is peat, a vast potential energy resource that has been developed in Canada and Ireland. Peatlands formed from plant assemblages as the result of particular climate conditions.

⩔ CENOZOIC LIFE

The world's flora and fauna continued to change during the Cenozoic Era as more familiar types of plants and animals appeared. In this chapter we are concerned mainly with the adaptive radiation of mammals, especially some of the more familiar types such as carnivores, elephants, hoofed mammals, and primates, particularly the hominids—humans and their extinct ancestors.

Although we emphasize mammalian evolution in this chapter, one should be aware of other important events. The flowering plants continued to dominate land-plant communities, the present-day groups of birds appeared early in the Tertiary, and some marine invertebrates continued to diversify.

Marine Invertebrates and Phytoplankton

The Cenozoic marine ecosystem was populated mostly by those plants, animals, and single-celled organisms that survived the terminal Mesozoic extinction event. Cenozoic invertebrate groups that were especially prolific were the foraminifera, radiolarians, corals, bryozoans, mollusks, and echinoids. The marine invertebrate community in general became more provincial during the Cenozoic because of changing ocean currents and latitudinal temperature gradients.

Only a few species in each major group of phytoplankton survived into the Tertiary. The coccolithophores, diatoms, and dinoflagellates all recovered from their Late Cretaceous reduction in numbers. The diatoms were particularly abundant during the Miocene, probably because of increased volcanism during this time. Volcanic ash provided increased dissolved silica in seawater and was used by the diatoms to construct their skeletons.

The foraminifera were a major component of the Cenozoic marine invertebrate community. Though dominated by relatively small forms (◆ Fig. 24.20, page 694), it included some exceptionally large forms that lived in the warm waters of the Cenozoic Tethys Sea. Shells of these larger forms accumulated to form thick limestones, some of which were used by the ancient Egyptians to construct the Sphinx and the Pyramids of Gizeh.

The corals, having relinquished their reef-building role to the rudists during the mid-Cretaceous, again became the dominant reef-builders during the Cenozoic. Other suspension feeders such as the bryozoans and crinoids were also abundant and successful during the Cenozoic. Perhaps the least important of the Cenozoic marine invertebrates were the brachiopods, with fewer than 60 genera surviving today.

A BRIEF HISTORY OF THE GREAT LAKES

Before the Pleistocene, no large lakes existed in the Great Lakes region, which was then an area of generally flat lowlands with broad stream valleys draining to the north. As the glaciers advanced southward, they eroded the stream valleys more deeply, forming what were to become the basins of the Great Lakes. During these glacial advances, the ice front moved forward as a series of lobes, some of which flowed into the preexisting lowlands where the ice became thicker and moved more rapidly. As a consequence, the lowlands were deeply eroded—four of the five Great Lakes basins were eroded below sea level.

At their greatest extent, the glaciers covered the entire Great Lakes region and extended far to the south. As the ice sheet retreated northward during the late Pleistocene, the ice front periodically stabilized, and numerous recessional moraines were deposited. By about 14,000 years ago, parts of the Lake Michigan and Lake Erie basins were ice-free, and glacial meltwater began forming proglacial lakes (◆ Fig. 1). As the retreat of the ice sheet continued—although periodically interrupted by minor readvances of the ice front—the Great Lakes basins were uncovered, and the lakes expanded until they eventually reached their present size and configuration (Fig. 1). Currently, the Great Lakes contain nearly 23,000 km³ of water, about 18% of the water in all fresh water lakes.

Although the history of the Great Lakes just presented is generally correct, it is oversimplified. For instance, the areas and depths of the evolving Great Lakes fluctuated widely in response to minor readvances of the ice front. Furthermore, as the lakes filled, they spilled over the lowest parts of their margins, thus cutting outlets that partly drained them. And finally, as the glaciers retreated northward, isostatic rebound raised the southern parts of the Great Lakes region, greatly altering their drainage systems.

The present-day Great Lakes and their St. Lawrence River drainage constitute one of the great commercial waterways of the world. Oceangoing vessels can sail into the interior of North America as far west as Duluth, Minnesota and western Ontario. To do so, however, they must bypass Niagara Falls between Lake Erie and Lake Ontario via a system of locks. Niagara Falls plunges 51 m over the Niagaran escarpment, which consists of resistant dolostone that was exposed during the last glacial retreat. Erosion of softer shale at the base of the falls is causing Niagara Falls to retreat upstream at a rate of about a meter per year.

It is estimated that in about 25,000 years the escarpment between the lakes will have been eliminated entirely! As a consequence, Lake Erie will become a small lake adjacent to vast swampy areas, and the upper Great Lakes will be considerably smaller than at present.

Just as during the Mesozoic, bivalves and gastropods were two of the major groups of marine invertebrates, and they had a markedly modern appearance. Following the extinction of the ammonites and belemnites at the end of the Cretaceous, the Cenozoic cephalopod fauna consisted of nautiloids and shell-less cephalopods such as squids and octopuses.

The echinoids continued their expansion in the infaunal habitat and were particularly prolific during the Tertiary. New forms such as sand dollars evolved during this time from biscuit-shaped ancestors (◆ Fig. 24.21).

Diversification of Mammals

For more than 100 million years, mammals coexisted with dinosaurs; yet, their fossil record indicates that during this entire time they were neither diverse nor

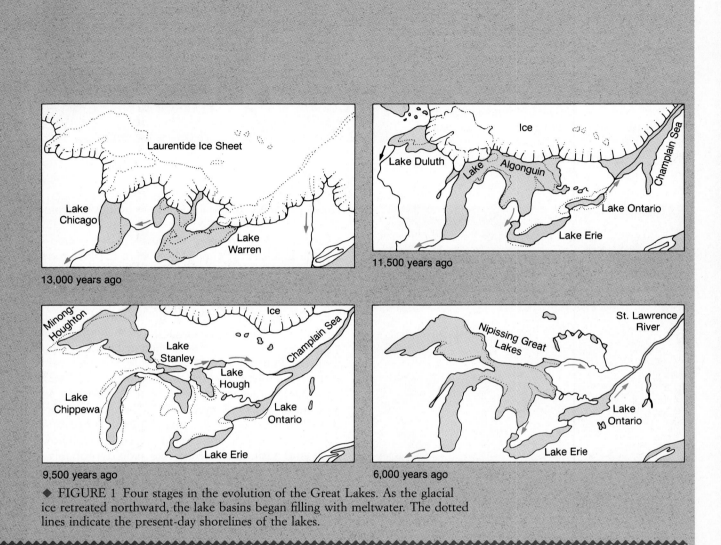

◆ FIGURE 1 Four stages in the evolution of the Great Lakes. As the glacial ice retreated northward, the lake basins began filling with meltwater. The dotted lines indicate the present-day shorelines of the lakes.

abundant. Mesozoic extinctions eliminated the dinosaurs and many of their relatives, thereby creating numerous adaptive opportunities that were quickly exploited by mammals. The Age of Mammals had begun.

Among living mammals only **monotremes** lay eggs, whereas marsupials and placentals give birth to live young. **Marsupials** are born in a very immature, almost embryonic condition, and then undergo further devel-

opment in the mother's pouch. **Placentals,** on the other hand, have developed a different reproductive method. In these animals, the amnion of the amniote egg (see Fig. 22.27) has fused with the walls of the uterus, forming a *placenta.* Nutrients and oxygen are carried from mother to embryo through the placenta, permitting the young to develop much more fully before birth. The phenomenal success of placental mammals is related in part to their reproductive method. A measure

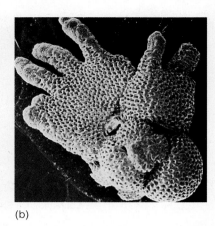

◆ FIGURE 24.20 Foraminifera of the Cenozoic Era. *(a) Cibicides americanus,* a benthonic form from the Early Miocene of California. *(b)* A planktonic form, *Globigerinoides fistulosus,* Pleistocene, South Pacific Ocean. (Photos courtesy of B. A. Masters.)

(a)   (b)

◆ FIGURE 24.21 Echinoids were particularly abundant during the Tertiary, and new infaunal forms such as this sand dollar evolved from their Mesozoic biscuit-shaped ancestors. (Photo courtesy of Sue Monroe.)

of this success is that more than 90% of all mammals, fossil and living, are placentals.

A major adaptive radiation of mammals began during the Paleocene and continued through the Cenozoic Era. Paleocene mammalian faunas are considered archaic because they were composed of primitive mammals, some of which—including marsupials, insectivores, and the rodentlike multituberculates—were holdovers from the Mesozoic Era (◆ Fig. 24.22).

Thirteen new orders of mammals first appeared during the Paleocene. Among these were the first rodents, rabbits, primates, and carnivores, but many of the other new orders soon became extinct. Most of these Paleocene mammals, even those assigned to living orders, had not yet become clearly differentiated from their ancestors, and the differences between herbivores and carnivores were slight. Large mammals did not evolve until the Late Paleocene, and the first giant terrestrial mammals appeared during the Eocene (◆ Fig. 24.23).

Diversification continued during the Eocene, when nine more orders evolved; all but one of these orders still exist. Most of the existing mammalian orders were present by Eocene time; yet if we could somehow go back and visit the Eocene, we would probably not recognize many of these animals. Some would be at least vaguely familiar to us, but the horses, camels, rhinoceroses, and elephants, for example, would bear little resemblance to their living descendants.

By Oligocene time, all of the living orders of mammals had evolved, while a number of archaic mammals became extinct during the Late Eocene and Oligocene. Diversification continued during the Oligocene, but it was within the existing orders as more familiar families and genera appeared. Miocene and Pliocene mammals were mostly animals that we could easily identify, and most of those of the Pleistocene would be quite familiar.

Cenozoic Mammals

Mammals evolved from cynodonts during the Late Triassic, but they were small and not very diverse

◆ FIGURE 24.22 The archaic mammalian fauna of the Paleocene Epoch included such animals as the multituburculate *Ptilodus* (right foreground), insectivores (right background), *Protictis,* an early carnivore, (left background), and the pantodont *Pantolambda* that stood about 1 m tall.

through the rest of the Mesozoic. Following the Mesozoic extinctions, however, mammals diversified and soon became the most abundant land-dwelling vertebrate animals. Throughout the Cenozoic, more and more familiar types of mammals appeared, eventually giving rise to the present-day mammalian fauna.

The Hoofed Mammals

Ungulate is an informal term referring to several types of mammals, especially the orders Artiodactyla and Perissodactyla. **Artiodactlys,** the even-toed hoofed mammals, are the most diverse living ungulates, and are represented today by 171 species. In contrast, the **perissodactyls,** or odd-toed hoofed mammals, consist

of only 16 living species of horses, rhinoceroses, and tapirs (see Perspective 24.2).

Some ungulates are **grazers,** meaning that they feed on grasses. However, grasses contain tiny particles of silicon dioxide (SiO_2) and are very abrasive, so they wear down teeth. Accordingly, the grazing ungulates have developed high-crowned chewing teeth that are more resistant to abrasion (◆ Fig. 24.24a). In contrast, those ungulates characterized as **browsers,** which eat the tender shoots, twigs, and leaves of trees and shrubs, did not develop high-crowned teeth.

Many ungulates live in open grassland habitats and depend on speed to escape predators. Adaptations for running include elongation of the bones of the palm and sole, which increases the length of the limbs (◆ Fig.

◆ FIGURE 24.23 The uintatheres were Eocene rhinoceros-sized mammals with three pairs of bony protuberances on the skull and saberlike upper canine teeth. (From the Field Museum, Chicago, Neg. # CK46T.)

◆ FIGURE 24.24 (a) Comparison of low-crowned and high-crowned teeth. Grazing ungulates developed high-crowned chewing teeth as an adaptation for eating abrasive grasses. The cusps of a high-crowned tooth are elevated into tall, slender pillars, and the entire tooth is covered by enamel and cement, both of which are hard substances. (b) Modifications in the limbs of ungulates. *Oxydactylus,* a Miocene camel, shows the evolutionary trend of limb elongation. The bones between the wrist and toes and between the ankle and toes became longer, thereby increasing the length of ungulate limbs.

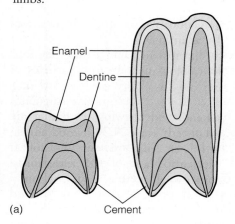

Enamel
Dentine
Cement
(a)

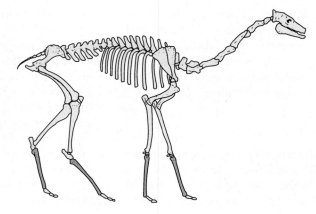

(b)

Perspective 24.2

A MIOCENE CATASTROPHE IN NEBRASKA

Ten million years ago, in what is now northeastern Nebraska, a vast grassland was inhabited by short-legged, aquatic rhinoceroses, camels, horses, saber-toothed deer, land turtles, and many other animals. This was a temperate savanna habitat with life as varied and abundant as it is now on the savannas of East Africa. But many of these animals perished when a vast cloud of volcanic ash rolled in, probably from the southwest.

Michael Voorhies of the University of Nebraska State Museum and his crews have recovered the remains of hundreds of victims of this catastrophe (◆Fig. 1). The magnitude of this event is difficult to imagine, since the source of the ash cloud may have been in New Mexico, more than 1,000 km away. We know from historic eruptions that an ash cloud can travel this far. For example, the April 11, 1815, eruption of Tambora in Indonesia was the deadliest volcanic eruption in recorded history. About 92,000 people perished, an ash layer 22 cm thick accumulated 400 km to the west, and some ash fell more than 1,500 km from the volcano.

The Nebraska ash fall was probably 10 to 20 cm thick, although it was redistributed by the wind and collected to 1 m deep in a depression where the fossils were recovered. This depression was the site of a water hole where animals congregated. Many animals, including three-toed horses, camels, deer, turtles, and birds, perished here when the ash fell. Their skeletons show signs of partial decomposition, scavenging, and trampling before they were completely buried.

Very soon after the initial ash fall, the depression was visited by herds of rhinoceroses and a few horses and camels. These, too, were suffocated by ash, but in this case the ash may simply have been blown into the depression by the wind. In any case, these animals were quickly buried, as indicated by the large number of complete skeletons.

One of the most remarkable things about these fossils is the preserved detail. According to Voorhies,

> Rarely found parts such as tongue bones, cartilages, tendons, and tiny bones in the middle ear all survive in exquisite detail and in their correct positions.*

Additionally, the association of babies, juveniles, young adults, and mature adults gives us a better understanding of herd structure. The rhinoceros herds, for example, were probably made up of a single bull and several cows with their calves.

Only rarely are paleontologists fortunate enough to find and recover so many well-preserved, associated vertebrate animals. A Late Miocene catastrophe turns out to be our good fortune since it provides us with a unique glimpse of what life was like in Nebraska 10 million years ago.

◆ FIGURE 1 Paleontologists excavating rhinoceros (foreground) and horse (background) skeletons from volcanic ash near Orchard, Nebraska.

*"Ancient Ashfall Creates a Pompeii of Prehistoric Animals," *National Geographic* 159, no. 1: (1981): 69.

24.24b). Also associated with running is the trend to reduce the number of bony elements in the limbs, especially toes. Thus, the limbs of running ungulates are long and slender.

Rabbit-sized ancestral artiodactlys appeared during the Early Eocene, but at this early stage they differed little from their ancestors. From these tiny ancestors, artiodactyls rapidly diversified into numerous families, many of which are now extinct (◆ Fig. 24.25). For example, small four-toed ancestral camels appeared early in the diversification of artiodactyls. Camels were common in North America from Eocene to Pleistocene times, and during the Pliocene they migrated to South America and Asia where their descendants still survive.

The *bovids* (Fig. 24.25), the most diverse artiodactlys, include cattle, bison, sheep, goats, and antelopes. Bovids evolved during the Miocene and diversified during the Pliocene. They evolved mostly on the Old World northern continents, but have since migrated to southern Asia and Africa, where they are most common today.

The living perissodactyls are the horses, rhinoceroses, and tapirs. These animals and the extinct titanotheres and chalicotheres are united by several shared characteristics, and the fossil record indicates that they share a common ancestor (◆ Fig. 24.26). Perissodactyls appeared during the Eocene and increased in diversity through the Oligocene, but have declined markedly since then.

Among the perissodactyls the horse family has a particularly good fossil record which shows evolution from a tiny Eocene ancestor known as *Hyracotherium* to present-day *Equus* (◆ Fig. 24.27). Several evolutionary trends are recognized, including an increase in size,

◆ FIGURE 24.25 Simplified family tree of the artiodactyls. Early in their history, artiodactyls split into three major groups: the suids include the pigs, hippopotamuses, and extinct giant hogs; the tylopoda are represented by the camels; and the ruminants consist of the cud-chewing animals.

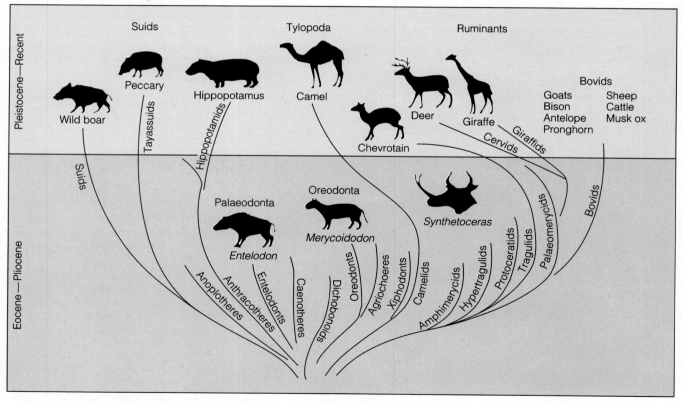

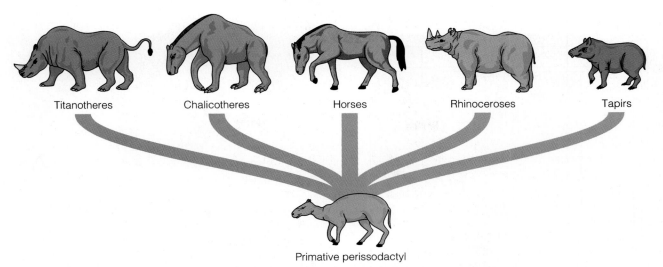

Titanotheres   Chalicotheres   Horses   Rhinoceroses   Tapirs

Primative perissodactyl

◆ FIGURE 24.26 Evolution of perissodactyls. Adaptive radiation and divergence from a common ancestor explain why perissodactyls share several characteristics yet differ markedly in appearance.

lengthening of the limbs, reduction of toes, and development of high-crowned chewing teeth as horses adapted to a grass diet and became speedy runners (Fig. 24.27). Horse evolution occurred mostly in North America, where they existed until the end of the Pleistocene.

Other Mammals

During the Paleocene, the *miacids,* small carnivorous mammals with short heavy limbs, made their appearance. They were not very fast runners, but then neither were their prey. Miacids were ancestral to all later members of the order Carnivora, which includes such animals as dogs, weasels, bears, seals, and cats (◆ Fig. 24.28). All carnivores have well developed canine teeth for slashing and tearing, but the most remarkable development of these teeth occurred in the saber-toothed cats.

The elephants (order Proboscidea) are the largest living land animals. They evolved from pig-sized ancestors in the Eocene, and by Oligocene time they clearly showed the trend to large size and the development of a long proboscis and large tusks (◆ Fig. 24.29, page 702). Mastodons with teeth adapted for browsing had evolved by the Miocene. The last major evolutionary event was the divergence of the existing elephant and mammoth lines during the Pliocene and Pleistocene (Fig. 24.29). During most of the Cenozoic elephants were widespread, especially on the northern continents, but they are now restricted to Africa and southeast Asia.

Whales first appeared during the Early Eocene, and by the Late Eocene had become diverse and widespread. Some were quite large. Eocene whales still possessed vestigial rear limbs and teeth resembling those of their land-dwelling ancestors, and were proportioned differently than living whales. By the Oligocene both groups of living whales, the toothed whales and the baleen whales, had evolved.

Many of the more familiar mammals are sizable animals, but most mammals are rather small. For example, more than 40% of all mammals are species of rodents, most of which are quite small. And bats (21%) and insectivores (10%) constitute another large part of the present-day mammalian fauna. These small animals have long evolutionary histories, and they have successfully adapted to a wide range of microhabitats unavailable to larger animals. Primates, too, appeared during the early adaptive radiation of mammals. They are discussed more fully in a later section of this chapter.

⬇ PLEISTOCENE FAUNAS

One of the most remarkable aspects of the Pleistocene mammalian fauna is that so many very large species existed. In North America, for example, there were mastodons and mammoths, giant bison, huge ground

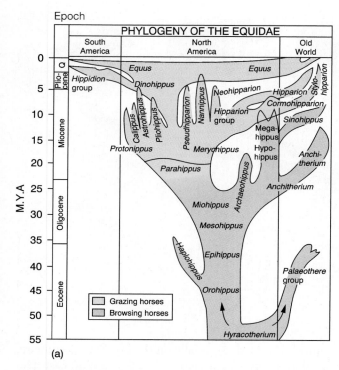

PHYLOGENY OF THE EQUIDAE

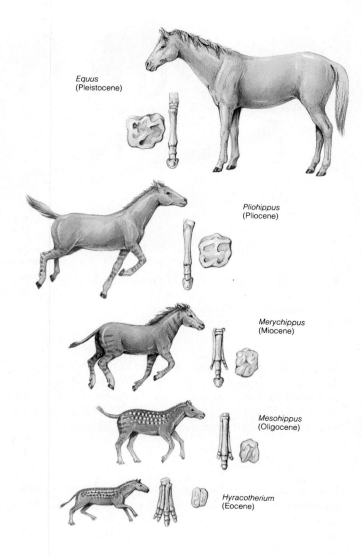

Equus
(Pleistocene)

Pliohippus
(Pliocene)

Merychippus
(Miocene)

Mesohippus
(Oligocene)

Hyracotherium
(Eocene)

◆ FIGURE 24.27 Evolution of horses. *(a)* Summary chart showing the recognized genera of horses and their evolutionary relationships. Note that during the Oligocene, two separate lines emerged, one leading to three-toed browsing horses and the other to one-toed grazers. *(b)* Simplified diagram showing some of the evolutionary trends from *Hyracotherium* to the present-day horse, *Equus.* Important trends shown here include an increase in size, loss of toes, and development of high-crowned teeth with complex chewing surfaces.

sloths, giant camels, and beavers nearly 2 m tall at the shoulder. Kangaroos standing 3 m tall, wombats the size of rhinoceroses, leopard-sized marsupial lions, and large platypuses characterize the Pleistocene fauna of Australia. In Europe and parts of Asia lived cave bears, elephants, and the giant deer commonly called the Irish elk with an antler spread of 3.35 m (◆ Fig. 24.30). The evolutionary trend toward large body size was perhaps an adaptation to the cooler temperatures of the Pleistocene. Large animals have proportionately less surface area compared to their volume and thus retain heat more effectively than do smaller animals.

In addition to mammals, some other Pleistocene vertebrate animals were of impressive proportions. For example, the giant moas of New Zealand and the elephant birds of Madagascar were very large, and Australia had giant birds standing 3 m tall and weighing nearly 500 kg and a lizard 6.4 m long and weighing 585 kg. The tar pits of Rancho La Brea in southern California contain the remains of at least 200 kinds of animals. Many of these are fossils of dire wolves, saber-toothed cats, and other mammals, but some are the remains of birds, especially birds of prey, and a giant vulture with a wingspan of 3.6 m.

◆ FIGURE 24.28 Evolution of the carnivorous mammals. Miacid ancestors gave rise to all present-day placental carnivores. Some of the relationships shown here are well documented by fossils and by studies of present-day animals, but some relationships are yet to be firmly established.

▼ PRIMATE EVOLUTION

Several evolutionary trends in the order **Primates** help define the order and are related to its *arboreal,* or tree-dwelling, ancestry. These include changes in the skeleton and mode of locomotion, an increase in brain size, a shift toward smaller, fewer, and less specialized teeth, and the evolution of stereoscopic vision and a grasping hand with opposable thumb. Not all of these trends occurred in every primate group, nor did they evolve at the same rate in each group. In fact, some primates have retained certain primitive features, whereas others show all or most of these trends.

The primate order is divided into two suborders (● Table 24.1). The *prosimians,* or lower primates, include lemurs, lorises, and tarsiers, while the *anthropoids,* or higher primates, include the monkeys, apes, and humans. Primitive primates may have evolved by the Late Cretaceous, but they were undoubtedly present by the Early Paleocene. By Eocene time large primates had appeared, and fairly modern-looking lemurs and tarsiers are known from Asia and North America. By Oligocene time, primitive New and Old World monkeys had evolved in South America and Africa, respectively. The *hominoids,* the group containing apes and humans (Table 24.1), evolved during the Miocene.

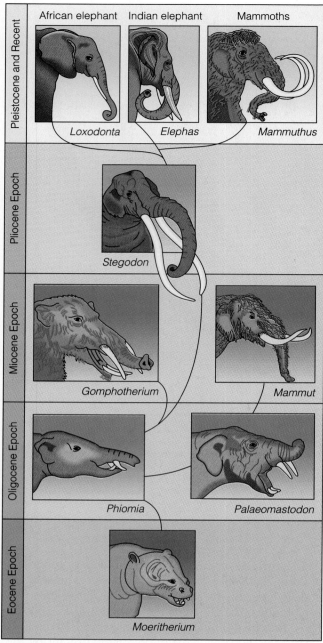

Pleistocene and Recent
African elephant — Loxodonta
Indian elephant — Elephas
Mammoths — Mammuthus

Pliocene Epoch
Stegodon

Miocene Epoch
Gomphotherium
Mammut

Oligocene Epoch
Phiomia
Palaeomastodon

Eocene Epoch
Moeritherium

◆ FIGURE 24.29 Simplified family tree of the elephants. Increase in size, and development of large tusks and a long proboscis were some of the evolutionary trends in elephants.

Hominids

The **hominids** (family Hominidae), the primate family that includes present-day humans and their extinct ancestors (Table 24.1), have a fossil record extending back to only about 4 million years ago. Several features distinguish them from other hominoids. Hominids are bipedal; that is, they have an upright posture which is indicated by several modifications in their skeleton (◆ Fig. 24.31 a and b). In addition, they show a trend toward a large and internally reorganized brain (◆ Fig. 24.31 c–e). Other features include a reduced face and reduced canine teeth, omnivorous feeding, increased manual dexterity, and the use of sophisticated tools.

Many anthropologists think that these hominid features evolved in response to major climatic changes that began during the Miocene and continued into the Pliocene. During this time vast savannas replaced the African tropical rain forests where the lower primates and Old World monkeys had been so abundant. As the savannas and grasslands continued to expand, the hominids made the transition from true forest dwelling to life in an environment of mixed forests and grasslands.

Australopithecines

Australopithecine is a collective term for all members of the genus *Australopithecus.* Currently four species are recognized: *A. afarensis, A. africanus, A. robustus,* and *A. boisei.* Many paleontologists accept the evolutionary scheme in which *A. afarensis,* the oldest known hominid, is ancestral to *A. africanus* and the genus *Homo,* as well as the side branch of australopithecines represented by *A. robustus* and *A. boisei* (◆ Fig. 24.32).

Australopithecus afarensis (◆ Fig. 24.33), which lived from about 4 to 2.75 million years ago, was fully bipedal and had a brain size of 380 to 450 cubic centimeters (cc). Its brain size is greater than the 300 to 400 cc of a chimpanzee, but much smaller than that of present-day humans (1,350 cc average). The skull of *A. afarensis* retained many apelike features, including massive brow ridges and a forward-jutting jaw, but its teeth were intermediate between those of apes and humans.

A. Afarensis was succeeded by *Australopithecus africanus,* which lived from 3 to about 1.6 million years ago. Differences between the two species are minor, although *A. africanus* was slightly larger, had a flatter face, and its brain size of 400 to 600 cc was somewhat larger. Both *A. afarensis* and *A. africanus* differ mark-

◆ FIGURE 24.30 Restoration of the giant deer *Megalocereos giganteus,* commonly called the Irish elk. It lived in Europe and Asia during the Pleistocene. Large males had an antler spread of about 3.35 m. (From the Field Museum, Chicago, Neg. # CK1T.)

● **TABLE 24.1** Classification of the Primates

Order Primates: Lemurs, lorises, tarsiers, monkeys, apes, humans
  Suborder Prosimii: Lemurs, lorises, tarsiers (lower primates)
  Suborder Anthropoidea: monkeys, apes, humans (higher primates)
    Superfamily Hominoidea: apes, humans
      Family Hominidae: humans

edly from the so-called robust species *A. robustus* (2.3 to 1.3 million years ago) and *A. boisei* (2.5 to 1.2 million years ago), neither of which had any evolutionary descendants.

Although there is no doubt that the australopithecines are hominids, there is disagreement on their relationship to the genus *Homo.* According to one evolutionary scheme they are ancestral to *Homo* (Fig. 24.32), but a recent discovery has prompted another

scheme (◆ Fig. 24.34, page 706). In this scheme the two australopithecine lineages and the human lineage evolved independently from ancestors whose fossils have not yet been discovered.

The Genus Homo

The earliest member of our own genus **Homo** is *Homo habilis,* which existed from 2 to 1.4 million years ago. Its brain size was about 700 cc, but in many other respects it resembled *A. africanus.* The evolutionary transition from *H. habilis* to *H. erectus* appears to have occurred in a very short time, between 1.8 and 1.6 million years ago.

In contrast to the australopithecines and *H. habilis,* which are unknown outside Africa, *Homo erectus* was a widely distributed species, having migrated from Africa during the Pleistocene (◆ Fig. 24.35, page 706). The brain size of 800 to 1,300 cc, though much larger than that of *H. habilis,* was still less than the average for *Homo sapiens* (1,350 cc). *H. erectus'* skull was thick-walled, its face was massive, it had prominant brow ridges, and its teeth were slightly larger than those of present-day humans, but otherwise it closely resembled *H. sapiens.*

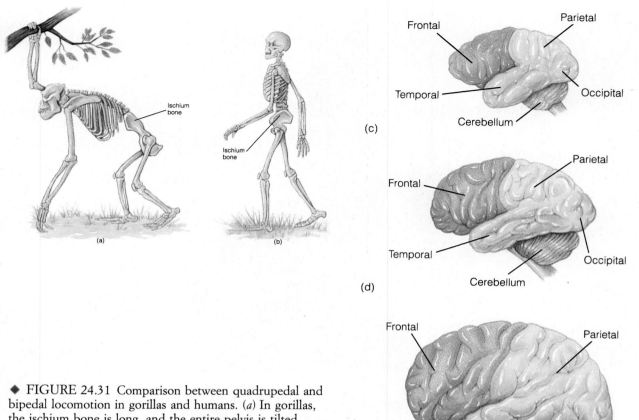

◆ FIGURE 24.31 Comparison between quadrupedal and bipedal locomotion in gorillas and humans. (*a*) In gorillas, the ischium bone is long, and the entire pelvis is tilted toward the horizontal. (*b*) In humans, the ischium bone is much shorter, and the pelvis is vertical. (*c* through *e*) An increase in brain size and organization is apparent in comparing the brains of (*c*) a New World monkey, (*d*) a great ape, and (*e*) a present-day human.

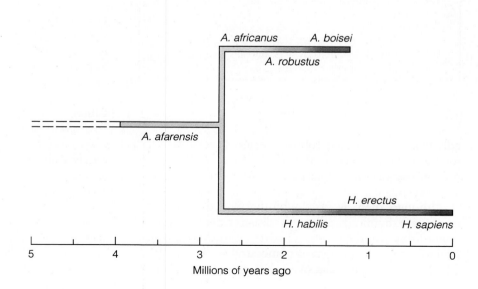

◆ FIGURE 24.32 The evolutionary scheme in which humans and the later australopithecines split from a common ancestor less than 3 million years ago.

◆ FIGURE 24.33 Recreation of a Pliocene landscape showing members of *Australopithecus afarensis* gathering and eating various fruits and seeds.

The archaeological record indicates that *H. erectus* made tools, and probably used fire and lived in caves.

Sometime between 300,000 and 200,000 years ago, *H. sapiens* evolved from *H. erectus*. The transition seems to have been gradual, with a variety of types existing at the same time. The early or archaic members of *H. sapiens* had rounder and higher skulls, a more delicately structured face, and relatively small teeth and jaws. The brain size soon averaged about 1,350 cc, and these archaic forms evolved into present-day humans.

The most famous of all fossil humans are the **Neanderthals** who inhabited Europe and the Near East from about 150,000 to 32,000 years ago (◆ Fig. 24.36). They may be a variety or subspecies of our own species

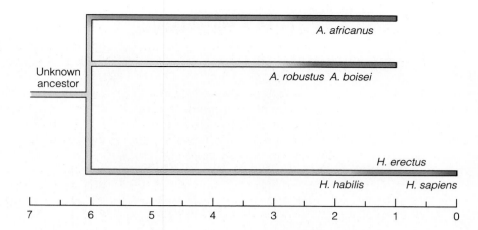

◆ FIGURE 24.34 The evolutionary scheme in which humans and the two main australopithecine groups evolve from creatures that have not yet been discovered.

◆ FIGURE 24.35 Recreation of a Pleistocene setting in Europe in which members of *Homo erectus* are using fire and stone tools. (From the Field Museum, Chicago, Neg. # A76851c.)

◆ FIGURE 24.36 Archaeological evidence indicates that Neanderthals lived in caves and participated in ritual burials as depicted in this painting of a burial ceremony that occurred approximately 60,000 years ago at Shanidar Cave, Iraq.

(*H. sapiens neanderthalensis*) or a separate species (*H. neanderthalensis*). The most notable difference between Neanderthals and present-day humans occurs in the skull. Neanderthal skulls were long and low with heavy brow ridges, a projecting mouth, and a weak, receding chin. Their brain was slightly larger than our own and somewhat differently shaped, and the body was more massive and heavily muscled than ours. Based on specimens from more than 100 sites, we now know that Neanderthals were not much different from us, only more robust.

About 35,000 years ago, humans closely resembling present-day Europeans moved into the region inhabited by Neanderthals and completely replaced them. **Cro-Magnons,** the name given to the successors of Neanderthals in France, lived from about 35,000 to 10,000 years ago; during this time the development of art and technology far exceeded anything the world had seen before. Using paints made from manganese and iron oxides, Cro-Magnon people painted hundreds of scenes on the ceilings and walls of caves in France and Spain.

⩔ PLEISTOCENE EXTINCTIONS

Near the end of the Pleistocene many terrestrial mammals became extinct. Although this extinction was modest compared to previous ones, it was unusual in that it affected mostly mammals weighing more than 40 kg. Furthermore, its effects were much greater in North and South America and Australia than in the Old World, where comparatively few extinctions occurred.

Two hypotheses for this extinction are currently being debated. The first holds that the large mammals became extinct because they could not adapt to the rapid climatic changes at the end of the Pleistocene. According to the second hypothesis, the mammals were killed off by human hunters, a hypothesis known as *prehistoric overkill.*

Researchers favoring a climatic cause point to the rapid changes in climate and vegetation that occurred during the Late Pleistocene. While rapid changes in climate with their accompanying effects on vegetation can certainly result in changes in animal populations, the hypothesis encounters several problems. First, why didn't the large mammals migrate to more suitable habitats as the climate and vegetation changed? After all, many other animal species did. The second argument against this hypothesis is that previous changes in climate during the Pleistocene were not marked by episodes of mass extinctions.

Paul Martin of the University of Arizona at Tucson, the leading proponent of the prehistoric overkill hypothesis, argues that the mass extinctions in North and South America and Australia coincided closely with the arrival of humans in each area. According to Martin, hunters had a tremendous impact on the faunas of North and South America about 11,000 years ago because the animals had no previous experience with humans. The same thing happened much earlier in Australia soon after people arrived about 40,000 years ago. No large-scale extinctions occurred in Africa and most of Europe because animals in those regions had long been familiar with humans.

Several arguments can also be made against the prehistoric over-kill hypothesis. One problem is that it is hard to imagine how a few hunters could have decimated so many species of large mammals, even if they hunted mostly females and young animals. A second problem is that present-day hunters concentrate on smaller, abundant, and less dangerous animals. And finally, few human artifacts are found among the remains of extinct animals in North and South America, and there is usually little evidence that the animals were hunted.

The reason for the extinctions of large mammals at the end of the Pleistocene is still unresolved and probably will be for some time. It may turn out that the extinction resulted from a combination of many different circumstances.

>◇◇◇◇◇◇◇◇◇◇◇◇◇◇◇◇◇◇◇◇◇◇◇◇◇>◇◇◇◇◇◇◇◇◇◇◇◇◇◇◇◇◇◇◇◇◇◇◇

Chapter Summary

1. Cenozoic orogenic activity occurred mostly in two major belts—the Alpine-Himalayan orogenic belt and the circum-Pacific orogenic belt. Each belt is composed of smaller units called orogens.
2. The North American Cordillera is a complex mountainous region extending from Alaska into Mexico. Its

Cenozoic evolution included deformation during the Laramide orogeny, extensional tectonics that formed the basin-and-range structures, intrusive and extrusive volcanism, and uplift and erosion.
3. Subduction of the Farallon plate beneath North America resulted in the vertical uplifts of the Laramide orogeny.

The Laramide orogen is centered in the middle and southern Rockies, but Laramide deformation occurred from Alaska to Mexico.

4. The westward drift of North America resulted in its collision with the Pacific-Farallon ridge. Subduction ceased, and the continental margin became bounded by major transform faults, except where the Juan de Fuca plate continues to collide with North America.

5. Sediments eroded from Laramide uplifts were deposited in intermontane basins, on the Great Plains, and in a remnant of the Cretaceous epeiric sea in North Dakota. A seaward-thickening wedge of sediments pierced by salt domes on the Gulf Coastal Plain contains large quantities of oil and natural gas.

6. Cenozoic uplift and erosion were responsible for the present topography of the Appalachian Mountains. Much of the sediment eroded from the Appalachians was deposited on the Atlantic Coastal Plain.

7. During the Pleistocene Epoch, glaciers covered about 30% of the land surface. About 20 warm-cold Pleistocene climatic cycles are recognized from paleontologic and oxygen isotope data derived from deep-sea cores.

8. Several intervals of widespread glaciation, separated by interglacial periods, occurred in North America. The other Northern Hemisphere continents were also affected by widespread Pleistocene glaciation.

9. Marine invertebrate groups that survived the Mesozoic extinctions continued to expand and diversify during the Tertiary.

10. The Paleocene mammalian fauna was composed of Mesozoic holdovers and a number of new orders. This was a time of diversification among mammals, and several orders soon became extinct. Most living mammalian orders were present by the Eocene.

11. Placental mammals owe much of their success to their method of reproduction. Shrewlike placental mammals that appeared during the Cretaceous were the ancestral stock for the placental adaptive radiation of the Cenozoic.

12. The perissodactyls and artiodactyls evolved during the Eocene. Ungulate adaptations include modifications of the teeth for grinding up vegetation and limb modifications for speed.

13. The evolutionary history of horses is particularly well documented by fossils. The earliest horse, *Hyracotherium,* and the present-day horse, *Equus,* differ considerably, but a continuous series of intermediate fossils shows that they are related.

14. The primates probably evolved during the Late Cretaceous. Several trends help characterize primates and differentiate them from other mammalian orders. These include a change in overall skeletal structure and mode of locomotion, an increase in brain size, stereoscopic vision, and evolution of a grasping hand with opposable thumb.

15. The earliest known hominids are the australopithecines, a fully bipedal group that evolved in Africa nearly 4 million years ago. Currently, four australopithecine species are known: *Australopithecus afarensis, A. africanus, A. robustus,* and *A. boisei.*

16. The human lineage began about 2 million years ago in Africa with the evolution of *Homo habilis,* which survived as a species until about 1.4 million years ago. It evolved into *Homo erectus.*

17. *Homo erectus* evolved from *Homo habilis* about 1.8 million years ago and was the first hominid to migrate out of Africa. By 1 million years ago, *H. erectus* had spread to Europe, India, China, and Indonesia. *H. erectus* used fire, made tools, and lived in caves.

18. Sometime between 300,000 and 200,000 years ago, *Homo sapiens* evolved from *Homo erectus.* These early or archaic humans may be ancestors of Neanderthals.

Important Terms

artiodactyl
australopithecine
Basin and Range Province
browser
Cro-Magnon
Farallon plate
glacial stage
grazer

hominid
Homo
interglacial stage
Laramide orogeny
marsupial
monotreme
Neanderthal
North American Cordillera

perissodactyl
placental
primate
Quaternary Period
Tejas epeiric sea
Tertiary Period
ungulate

Review Questions

1. An area of North America currently being deformed by tensional forces is the:
   a. ___ Colorado Plateau; b. ___ Basin and Range Province; c. ___ Columbia River basalts; d. ___ Interior Lowlands; e. ___ Atlantic Coastal Plain.

2. Which of the following areas within the North American Cordillera was little deformed during the Laramide orogeny?
   a. ___ Colorado Plateau; b. ___ Coast Ranges; c. ___ Laramide orogen; d. ___ Sierra Nevada; e. ___ San Andreas fault.

3. The Tertiary history of the Appalachians involved mostly:
   a. ___ uplift and erosion; b. ___ subduction and island arc collision; c. ___ tension and block-faulting; d. ___ compression and folding; e. ___ carbonate deposition and origin of salt domes.

4. During the Tertiary Period, the _____ Sea was the last of the epeiric seas to invade North America.
   a. ___ Sauk; b. ___ Alleghenian; c. ___ Tejas; d. ___ Antler; e. ___ Absaroka.

5. Which of the following statements is correct?
   a. ___ the Laramide orogen is farther inland than is typical of most orogens; b. ___ volcanism occurs when a subducted plate moves beneath another at a low angle; c. ___ the Tejas epeiric sea covered most of North America during the Early Tertiary; d. ___ the Great Bahama Bank has been an area of Tertiary detrital sediment deposition; e. ___ the Gulf Coastal Plain Tertiary rocks are mostly carbonates.

6. How many years ago did the Pleistocene Epoch begin?
   a. ___ 6.6 million; b. ___ 2 million; c. ___ 1.6 million; d. ___ 1 million; e. ___ 10,000.

7. Which of the following is not one of the North American glacial stages?
   a. ___ Wisconsin; b. ___ Nebraskan; c. ___ Kansan; d. ___ Iowan; e. ___ Illinoian.

8. Lakes that formed far from glaciers during times of glaciation in what are now dry areas are known as:
   a. ___ proglacial; b. ___ pluvial; c. ___ playa; d. ___ peneplain; e. ___ none of these.

9. An animal with teeth adapted for eating grass is a:
   a. ___ grazer; b. ___ primate; c. ___ proboscidean; d. ___ browser; e. ___ marsupial.

10. The oldest known horse is:
    a. ___ *Ceratotherium*; b. ___ *Pliohippus*; c. ___ *Paraceratherium*; d. ___ *Hyracotherium*; e. ___ *Moeritherium*.

11. The large body size of Pleistocene mammals may have been an adaptation to:
    a. ___ increased predation; b. ___ more seasonal climates; c. ___ cooler temperatures; d. ___ higher elevations; e. ___ longer summers.

12. Increased volcanism and more dissolved silica in seawater probably account for the Miocene abundance of:
    a. ___ foraminifera; b. ___ diatoms; c. ___ angiosperms; d. ___ scleractinian corals; e. ___ bryozoans.

13. Which of the following is a hypothesis for Pleistocene extinctions?
    a. ___ meteorite impact; b. ___ prehistoric overkill; c. ___ reduced area of continental shelves; d. ___ freezing; e. ___ extensive volcanism.

14. Which of the following evolutionary trends characterize primates?
    a. ___ grasping hand with opposable thumb; b. ___ stereoscopic vision; c. ___ increase in brain size; d. ___ change in overall skeletal structure; e. ___ all of these.

15. The oldest hominids belong to which genus?
    a. ___ *Australopithecus*; b. ___ *Homo*; c. ___ *Ramapithecus*; d. ___ *Dryopithecus*; e. ___ *Pithacanthropus*.

16. Which were the first hominids to migrate out of Africa?
    a. ___ *Australopithecus robustus*; b. ___ *Australopithecus boisei*; c. ___ *Homo habilis*; d. ___ *Homo erectus*; e. ___ *Homo sapiens*.

17. How does the Laramide orogen differ from typical arc orogens?

18. What sequence of events was responsible for the origin of the Queen Charlotte and San Andreas transform faults?

19. Describe the sedimentary facies of the Gulf Coastal Plain. What event or process controlled facies patterns?

20. Briefly outline the Cenozoic history of the Appalachian Mountains.

21. What type of continental margin is represented by the Atlantic continental margin of North America, and how did it form?

22. What direct evidence is there for Pleistocene glaciation?

23. How did glaciers indirectly affect areas far removed from the ice sheets?

24. Explain how pluvial and proglacial lakes differ.

25. Briefly summarize the important Paleocene and Eocene evolutionary events among mammals.

26. How do placental mammals differ from marsupials and monotremes?

27. What are ungulates? How have their limbs been modified for speed?

28. Explain how the fossil record demonstrates that animals as different as horses and rhinoceroses share a common ancestor.

29. Discuss the evidence for the two hypotheses for Pleistocene extinctions.

30. Discuss the two hypotheses concerning the evolutionary history of the hominids.

Additional Readings

Armentrout, J. M., M. R. Cole, and H. Terbest, Jr. 1979. Cenozoic paleogeography of the western United States. *Pacific Coast Paleogeography Symposium 3.* Los Angeles: Society of Economic Paleontologists and Mineralogists.

Blumenschine, R. J., and J. A. Cavallo. 1992. Scavenging and human evolution. *Scientific American* 267, no. 4:90-97.

Brace, C. L. 1991. *The stages of human evolution.* 4th ed. Englewood Cliffs, N.J.: Prentice-Hall.

Brown, M. H. 1990. *The search for Eve.* New York: Harper and Row.

Carroll, R. L. 1988. *Vertebrate paleontology and evolution.* New York: W. H. Freeman.

Colbert, E. H., and M. Morales. 1991. *Evolution of the vertebrates.* 4th ed. New York: John Wiley & Sons.

Flores, R. M., and S. S. Kaplan, eds. 1985. Cenozoic paleogeography of the west-central United States. *Rocky Mountain Paleogeography Symposium 3.* Denver, Colo.: Society of Economic Paleontologists and Mineralogists.

Fulton, R. J., ed. 1989. *Quaternary geology of Canada and Greenland.* The geology of North America, vol. K-1. Ottawa, Canada: Geological Survey of Canada.

Hooper, P. R. 1982. The Columbia River basalts. *Science* 215:1463-68.

Kurten, B. 1988. *Before the Indians.* New York: Columbia University Press.

Molnar, P. 1986. The geologic history and structure of the Himalaya. *American Scientist* 74, no. 2:144-54.

Morrison, R. B., ed. 1991. *Quaternary nonglacial geology: Conterminous U.S.* The geology of North America, vol. K-2. Boulder, Colo.: Geological Society of America.

Nilsson, T. 1983. *The Pleistocene: Geology and life in the Quaternary ice age.* Holland: D. Reidel Publishing Co.

Park, A. 1990. Giants once ruled Australia, fossil discoveries reveal. *Smithsonian* 10, no. 10:133-43.

Savage, R. J. G. 1986. *Mammal evolution: An illustrated guide.* New York: Facts on File Publications.

Storch, G. 1992. The mammals of island Europe. *Scientific American* 266, no. 2:64-69.

Sutcliffe, A. J. 1985. *On the trace of Ice Age mammals.* Cambridge, Mass.: Harvard University Press.

Thomson, K. S. 1992. The challenge of human origins. *American Scientist* 80, no. 6:519-22.

Thorne, A. G., and M. Wolpoff. 1992. The multiregional evolution of humans. *Scientific American* 266, no. 4:76-83.

Wilson, A. C., and R. L. Cann. 1992. The recent African genesis of humans. *Scientific American* 266, no. 4:68-73.

Wolpoff, M., and A. Thorne. 1991. The case against Eve. *New Scientist* 130, no. 1774:37-41.

CHAPTER
25

Chapter Outline

Petroleum is one of the important natural resources the world depends on. This
petroleum refinery is in Torrence, California.

Geology, the Environment, and Natural Resources

Prologue

The Aral Sea in the Central Asian desert region of the former Soviet Union (♦ Fig. 25.1) is a continuing environmental and human disaster. In 1960 it was the world's fourth largest lake. Since then it has steadily shrunk so that today it is only the sixth largest. During this time the Aral Sea decreased in size from 67,000 km² to about 34,000 km², and its volume has fallen from 1,090 km³ to 300 km³. Presently, it is divided into two separate basins, a small northern basin and a larger southern basin. At its present rate of reduction, the Aral Sea could disappear within the next 30 years.

What could have caused such a disaster? For thousands of years the Aral Sea was fed by two large rivers, the Amu Dar'ya and Syr Dar'ya. The headwaters of both rivers begin in the mountains more than 2,000 km to the southeast, and flow northward through the Kyzyl Kum and Kara Kum deserts into the Aral Sea. Until the early part of this century, a balance existed between the water supplied by the Amu Dar'ya and the Syr Dar'ya rivers and the rate of evaporation of the Aral Sea, which has no outlet. In 1918, however, it was decreed that waters from the two rivers supplying the Aral Sea would be diverted to irrigate millions of acres of cotton so the Soviet Union could become self-sufficient in cotton production. As a result of this decision, the Aral Sea has become an ecological and environmental nightmare affecting some 35 million people.

To gain an appreciation of the magnitude of this disaster, consider that as late as the 1960s the Aral Sea was home to 24 native species of fish and supported a major fishing industry. The town of Muynak was the major fishing port and produced 3% of the Soviet Union's annual catch. The fishing industry provided about 60,000 jobs, with approximately 10,000 fishermen working out of Muynak. Today, Muynak is more than 20 km from the shoreline of the Aral Sea, and there are no native fish species left in the Aral.

As the Aral Sea shrinks, vast areas of the former sea bottom are exposed. This new sediment, far from being rich and productive, contains large amounts of sodium chloride and sodium sulfate. The concentration of salts is so high that only one plant species grows in it. It is too salty for anything else.

As the winds blow across the near-barren land, salt and dust are picked up and carried throughout the Aral region. These salts cause great damage to the cotton crops and other vegetation of the Aral basin and also exact a heavy toll on humans. The dry, salty dust has caused respiratory and eye diseases to increase dramatically during the past 30 years, as well as the reported number of cases of throat cancer. In addition, the drinking water supply has become so polluted that many people suffer from intestinal disorders.

As a result of the diversion of water from the Amu Dar'ya and Syr Dar'ya rivers and the resulting reduction in the Aral Sea, desertification has become a major problem in the Aral basin. This has caused a change in the weather patterns of the region, so that it is now colder in the winter and hotter and drier in the summer.

In spite of the damage done to the region by the dying of the Aral Sea, irrigation for cotton production

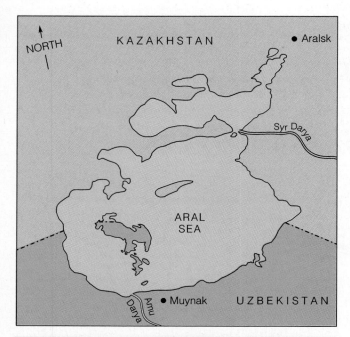

◆ FIGURE 25.1 Location and 1985 space shuttle view of the Aral Sea.

continues. However, the soil in many of the fields has become increasingly salty due to lower stream and groundwater levels, requiring more irrigation to maintain the same production levels. This means that even more water is diverted from the Aral Sea, causing it to shrink still further.

Can the Aral Sea be saved? Only by a concerted and cooperative effort by the countries that comprise the Aral basin. Realizing this, the now independent states of Kazakhstan, Uzbekistan, Kyrgyzstan, Tadzhikistan, and Turkmenistan took the first steps in 1992 by signing a formal agreement on sharing the waters of the Amu Dar'ya and Syr Dar'ya. In addition, these states entered into discussions to create a council to oversee and coordinate the management of the basin's resources. While achieving full restoration of the Aral Sea is probably not possible, maintaining its current surface level and even raising the surface level of the small northern Aral Sea is feasible. If this could be accomplished salinity levels would be reduced, making it possible to reintroduce native fish species and revive commercial fishing.

⩔ INTRODUCTION

One theme of this book is that the Earth is a complex, dynamic planet that has changed continually since its origin some 4.6 billion years ago. These changes are the result of internal and external processes that interact and affect each other, leading to the present-day features we observe. One cause of these changes is plate tectonics. As we have discussed, the movement of plates has had a profound effect on the formation of landscapes, the distribution of mineral resources, and atmospheric and oceanic circulation patterns, as well as the evolution and diversification of life.

Whereas an obvious interrelationship exists between the environment and the Earth's biota, some scientists think that an even more fundamental relationship is present, one in which the biosphere itself is so completely self-regulating as to make continued life possible (see Perspective 25.1). Although this view, known as the Gaia hypothesis, has not been completely accepted by most scientists, there can be little doubt that the Earth's various components do not act in isolation, but are interconnected such that when one part of a system changes, it has an effect on the other parts of the system.

Accordingly, we must understand that changes we make in the global ecosystem can have wide-ranging effects that we might not be aware of. For this reason, an understanding of geology, and science in general, is of paramount importance so that minimal disruption to the ecosystem occurs. On the other hand, we must also remember that humans are part of the ecosystem and, like all other life forms, our presence alone affects the ecosystem. We must therefore act in a responsible

THE GAIA HYPOTHESIS

············ ············

In 1785, James Hutton said, "I think the Earth is a superorganism." More recently, in 1972, James Lovelock, a British scientist-inventor, proposed his controversial Gaia hypothesis, which he updated in 1988. According to Lovelock and other proponents of this hypothesis, "the dynamic forces of life so dominate our planet that life has a controlling influence on the oceans and atmosphere." In other words, life has shaped the environment in order to keep it within a comfortable range rather than adapting to an otherwise benign environment.

There is little doubt that interrelationships exist between life-forms and the environment. No one would deny that microorganisms aid in such processes as soil formation and deposition of some rocks, or that green plants obtain carbon dioxide from the atmosphere and release oxygen as a waste product. The destruction of tropical rain forests is much in the news, because such destruction results in changes in the relative proportions of gases in the atmosphere—that is, more carbon dioxide and less oxygen. These changes could produce dramatic climatic effects such as global warming.

Although most scientists accept such relationships as those just mentioned, the Gaia hypothesis, named for the Greek Earth goddess, proposes an even more fundamental relationship between organisms and the environment. According to Lovelock, rather than adapting to an evolving environment determined solely by physical and chemical processes, organisms have the capacity to control the environment, especially the atmosphere and oceans, in such ways as to make continued life possible. As one might guess, it is this idea that the biosphere as a whole is self-regulating that many scientists find hard to accept.

Life exists within a narrow range of physical and chemical conditions, and Lovelock proposes that feedback mechanisms exist that control the environment to suit the needs of organisms. The Earth's temperature, for example, has stayed within the narrow limits suitable for the existence of life for at least the last 3.5 billion years, even though the Sun now produces about two-thirds more heat and is brighter than it was when the first organisms existed on Earth.

To demonstrate how organisms maintain environmental parameters within narrow limits, Lovelock proposed a mathematical model that he called Daisyworld, in which an imaginary planet is populated only by white and black daisies. If the temperature on Daisyworld rises, the black daisies absorb too much heat and die, thus leaving mostly white daisies that reflect more heat and cool the planet down. When Daisyworld cools sufficiently, black daisies thrive again and absorb more heat. In short, there is a feedback mechanism for temperature control.

Unquestionably, biologic processes are important, but those who accept the Gaia hypothesis also claim that the proportions of various gases in the atmosphere are kept in balance by feedback mechanisms. They point out that the present-day atmosphere is dominated by nitrogen and oxygen, both reactive gases that should have long ago combined with other elements to form nitrates. Furthermore, they claim that without life, carbon dioxide would have become the dominant atmospheric gas, as it is on Venus. Such feedback mechanisms, according to proponents of the Gaia hypothesis, indicate that the Earth is a giant self-regulating body in which there is an intimate connection between the evolution of the living and nonliving components of the planet.

As some critics point out, however, the composition of the atmosphere has changed through geologic time. And many of these changes were detrimental to life, as indicated by the various mass extinctions seen in the fossil record. If, according to the Gaia hypothesis, the biosphere was able to regulate the environment to suit itself, why have there been periods of biological instability?

As one would expect, there are strong objections to the Gaia hypothesis. Many biologists dismiss it because it is teleological; that is, it appeals to design or purpose in nature and thus cannot be tested. Some geologists point out that plate tectonics alone can control the Earth's temperature through the recycling of carbon dioxide (see Perspective 19.2).

While the Gaia hypothesis is, to say the least, controversial, it remains to be seen whether it will eventually become an acceptable theory. As in any scientific endeavor, new and radical ideas must demonstrate their worth in the competitive field of hypothesis, evidence testing, and prediction. Perhaps Gaia will be supported as scientists investigate its theoretical postulates, or it may be rejected or modified, depending on future discoveries. In any case, Gaia has forced scientists to critically evaluate the relationship between life and the global environment.

manner, based on sound scientific knowledge, so future generations will inherit a habitable environment.

The concept of *sustainable development* has received increasing attention, particularly since the United Nations Conference on Environment and Development met in Rio de Janeiro, Brazil during the summer of 1992. This important concept links satisfying basic human needs with safeguarding our environment to ensure continued economic development.

If we are to have a world in which poverty is not widespread, then we must develop policies that ensure continuing economic development along with management of our natural resources. Meeting the needs of an increasing global population will result in increased demand for food, water, and natural resources, particularly nonrenewable mineral and energy resources. As the demand for these resources increases, geologists will play an increasingly important role in locating them, as well as ensuring protection of the environment for the benefit of future generations.

❧ SCIENCE AND THE CITIZEN

We live in an age of ever increasing complexity, in which scientific and technological innovations are emerging at an astonishingly rapid rate. New discoveries in medicine, chemistry, and electronics are announced almost daily. Advances in computer technology have revolutionized the way we live and work. Almost every aspect of modern society involves the application of science and technology.

As jobs become more technologically oriented, it is imperative that everyone know more science and how it affects our lives. Unfortunately, according to a 1985 U.S. National Science Board report, the last time most high school students ever take a math or science course is the tenth grade. Furthermore, students in the United States spend only one-half to one-third as much time learning science as do students in Germany, the former Soviet Union, China, and Japan. If our nation is to compete in the global marketplace, we must have a scientifically literate work force.

It is becoming a cliché to point out that the American public knows and understands very little science. However, in 1988, only one American in five knew what DNA was, yet we are debating whether and under what conditions the genetic code of organisms should be purposely altered, and spending millions of dollars to map the human genome. Furthermore, about 50 percent of American adults said they did not understand the concept of radiation. And yet we are asked to vote on measures to build or close down nuclear power plants, debate the merits of building a radioactive waste dump in Nevada's Yucca Mountain (see Perspective 13.2), and decide if we should have our homes checked for high concentrations of radon (see Perspective 17.1).

Based on a study conducted in 1985, John Miller, director of the Public Opinion Laboratory at Northern Illinois University, estimated that only 5% of the U.S. public was scientifically literate. According to Dr. Miller, scientific literacy means understanding the scientific method, knowing the common vocabulary of science, and appreciating the social impact of scientific advances.

The essential point of Dr. Miller's survey was that only 1 in every 20 Americans understands science and the way science works. This does not mean that we must all become scientists. It does point out, however, that as scientifically illiterate consumers and citizens, we run the risk of falling victim to charlatans. How can we make informed decisions about the many critical environmental issues affecting us if we cannot separate fact from fiction and logically follow debates about issues involving science and technology?

It is equally important that the scientific community do a better job of informing and explaining the benefits of their research to society. Science does not exist in a vacuum. It proceeds on the basis of the scientific method (see Chapter 1), and everyone needs at least a rudimentary knowledge of this method and the way science works to make intelligent and informed decisions. Without such a basic understanding of science and its implications for society, we risk losing the many potential benefits science and technology can provide.

One of the benefits of studying geology is that one gains an appreciation of the geological processes that ultimately affect us all. Most readers of this book will not go on to become professional geologists, but may become involved in geological decisions in a peripheral way, as a member of a planning board, for instance, or as a property owner with mineral rights. In such cases, a knowledge of geology is imperative if one is to make informed decisions. Furthermore, many professionals must deal with geological issues as part of their jobs. For example, lawyers are becoming more involved in issues ranging from ownership of natural resources to environmental impact reports. As government takes a greater role in environmental issues and regulations, there is an even greater need for people to become knowledgeable in the geological sciences.

▼ GEOLOGY AND THE ENVIRONMENT

Geology is playing an increasingly important role in environmental issues and regulation. While geologists have traditionally been involved in the exploration for mineral and energy resources, they are also now being asked to apply their expertise to many of the environmental problems facing the world today.

Probably the greatest challenge to the environment is population growth. It is projected that the world's population will grow by 1.7 billion people during the next two decades, bringing the Earth's human population to 7 billion. While this may not seem to be a geological problem, we must remember that these people must be fed, housed, and clothed, and all with a minimal impact on the environment. Some of this population growth will be in areas that are already at risk from such geological hazards as earthquakes, volcanic eruptions, and mass wasting. Safe and adequate water supplies must be found and kept from being polluted. More oil, gas, coal, and alternative energy resources must be discovered and utilized to provide the energy to fuel the economies of nations with ever-increasing populations. New mineral resources must be found. In addition, ways to reduce usage and reuse materials must be found so as to decrease dependence on new sources of these materials.

Thus there is a great need for well trained and competent geologists, not only to find new sources of natural resources, but also to ensure that the environment is properly managed. There will also be a great need in the future for people with a strong background in the geological sciences to advise policy makers in drafting laws concerning the environment.

When such environmental issues as acid rain, the greenhouse effect, and the depletion of the ozone layer are discussed and debated, it is important to remember that they are not isolated topics, but part of a larger system that involves the entire Earth. Another important point to keep in mind is that the Earth goes through cycles of a much longer duration than the human perspective of time. While it may have disastrous effects on the human species, global warming and cooling is part of a larger cycle that has resulted in numerous glacial advances and retreats during the past 1.6 million years. In fact, geologists can make important contributions to the debate on global warming because of their geological perspective. Long-term trends can be studied by analyzing deep-sea sediments, changes in sea level during the geologic past, and the distribution of plants and animals through time.

Having examined the various aspects of geology and the evolution of the Earth and its biota, we will now address several of the important environmental issues concerning our planet that were only peripherally addressed earlier. While the subjects examined are by no means the only important ones, they help illustrate many of the geological processes and interrelationships we have discussed throughout this book.

Overpopulation

Overpopulation is the greatest environmental problem facing the world today because the human population is exceeding the Earth's ability to support it (◆ Fig. 25.2). An ecosystem's carrying capacity, or ability to support a given number of organisms, is limited by at least three factors: its food supply, its resource supply,

◆ FIGURE 25.2 Overpopulation is the greatest environmental problem facing the world today. Until the world's increasing population is brought under control, scenes like this one will become more common.

and its ability to assimilate the pollution produced by the organisms.

As the global human population increases, nations are finding it increasingly difficult to maintain adequate food supplies. Crop yields can only be increased so much, and as the population continues to grow, productive farm lands are being taken over by villages, towns, and cities, which only increases the pressure on the remaining land. The farmlands that remain are relying on the ever increasing use of fertilizers and pesticides to increase food production, which leads to pollution of water supplies by runoff and depletion of natural minerals in the soil.

In addition to crops, raising livestock as a food source is a common economic activity in many parts of the world even though it is a much more energy- and resource-intensive activity than crop production. Unfortunately, in some of the drier areas the number of animals has been increasing to the point that they now exceed the land's capacity to support them. As a result, the vegetation cover that protects the soil has been depleted, causing accelerated soil erosion and the loss of valuable grazing land (see Fig. 15.2).

Some of the world's greatest population densities occur in countries of the Pacific Rim, or Ring of Fire, so named because of the intense volcanic and seismic activity that encircles the Pacific Ocean (see Fig. 9.1). In these countries, productive farmland is frequently destroyed by volcanic eruptions. Many of the populous island nations in the Ring of Fire, however, are located in tropical climates where weathering occurs rapidly, replenishing the nutrients and forming new productive soils.

In addition to food and water, human populations require a great many other resources, such as energy and building materials. Many of these resources, including oil, gas, coal, and minerals, are finite and nonrenewable (see Chapter 2). Exploration for natural resources has been geologists' main activity in the past, and will continue to be an important aspect of the science in the future. As natural resources dwindle and become more difficult to find and extract, geologists will need increasingly sophisticated techniques to ensure the world has an adequate supply. Because many of the natural resources essential to society today are finite, we must find ways to economically recycle these resources and use a larger percentage of renewable resources.

The last factor in determining an ecosystem's carrying capacity is its ability to assimilate pollution. We are all aware of air pollution, solid waste disposal, contami-nated groundwater, and acid rain, to name only a few of the problems of an increasingly industrialized planet. In natural ecosystems, wastes are usually diluted to harmless levels and reused and recycled through nutrient cycles, so that a natural balance exists between populations and waste production.

Humans produce tremendous amounts of wastes, however, many of which do not quickly or easily degrade. Furthermore, many of these pollutants are lethal and contaminate the soil, air, and water. According to a study by the Conservation Foundation, every citizen of the United States produces 22,500 kg of domestic, agricultural, and industrial waste and pollution per year. This represent 25% of the world's pollution, and yet the U.S. makes up only 6% of the world's population. The problem is certainly not limited to the United States. The industrialized countries of the world make up a minority of the world's total population, yet are responsible for much of the pollution. Some of the worst environmental disasters, such as Chernobyl, have occurred in the industrialized countries.

The problem of overpopulation and its effects on the global ecosystem are varied. For many of the poor and nonindustrialized countries, the problem is too many people and not enough food. For the more developed and industrialized countries, it is too many people rapidly depleting both the nonrenewable and renewable natural resource base. And in the most industrialized countries, it is people producing more pollutants than the environment can safely recycle on a human time scale. In all cases, it is an environmental imbalance created by a human population exceeding the Earth's carrying capacity.

Global Warming

Carbon dioxide is produced as a by-product of respiration and the burning of organic material. As such it is a component of the global ecosystem and is constantly being recycled as part of the carbon cycle (see Fig. 17.24). The concern in recent years over the increase in atmospheric carbon dioxide has to do with its role in the greenhouse effect. Recall from Perspective 19.2 that the recycling of carbon dioxide between the crust and atmosphere is an important climatic regulator because carbon dioxide, as well as other gases such as methane, nitrous oxide, chlorofluorocarbons, and water vapor, allow sunlight to pass through them but trap the heat reflected back from the Earth's surface. Heat is thus retained, causing the temperature of the Earth's surface and, more importantly, the atmosphere, to increase, producing the greenhouse effect.

Until the Industrial Revolution began during the mid-eighteenth century, the contribution by humans to the global temperature pattern was negligible. With industrialization and its accompanying burning of tremendous amounts of fossil fuels, carbon dioxide levels in the atmosphere have been steadily increasing since about 1850. In fact, atmospheric levels of carbon dioxide are presently 25% higher than a century ago, and at their current yearly rate of increase will double from their present concentrations within the next 60 to 70 years. This increase in atmospheric carbon dioxide is largely attributed to the burning of coal, oil, and natural gas, which releases carbon dioxide as a by-product of combustion.

Recent research also indicates that deforestation of large areas, particularly in the tropics, is another cause of increased levels of carbon dioxide. This is because plants use carbon dioxide in photosynthesis and thus remove it from the atmosphere. With a decrease in the global vegetation cover, less carbon dioxide is removed from the atmosphere.

Carbon dioxide is not, however, the only gas responsible for an increased greenhouse effect. Methane and chlorofluorocarbons are also major contributors to global warming patterns. Methane is a natural gas that is 20 times as effective as carbon dioxide in trapping heat in the atmosphere. Current atmospheric levels of methane are double those of the pre-industrialized period. Methane is a by-product of manure and the decomposition of crops such as rice under oxygen-poor conditions, and is also produced by the digestive processes of livestock and termites. Termites are particularly abundant in deforested areas of the world, resulting in the combined effect of increased methane production and reduced carbon dioxide in the atmosphere when a forest is cut down.

Chlorofluorocarbons are well known for their effect on the degradation of the ozone layer, but also are thought to be responsible for about 20% of the global warming potential. Chlorofluorocarbons are released from air conditioners, refrigerators, and many spray cans, and are at least 10,000 times as effective as carbon dioxide in trapping atmospheric heat. As most countries agree to reduce or eliminate chlorofluorocarbons, the contribution of these gases to global warming should decrease.

Because of the increase in human-produced greenhouse gases during the last two hundred years, many scientists are concerned that a global warming trend has already begun, with the consequence of such an increase being severe global climatic shifts. A graph of global mean temperatures since the late 1800s tends to substantiate this claim. It shows a rise in temperature of 0.6°C during this time, with the six warmest years during the 1900s occurring during the 1980s. Other scientists, however, point out that the warming observed before 1940 may simply be a recovery and adjustment from the effects of the "Little Ice Age" that occurred from about 1500 to the mid- to late-1800s (see the Prologue to Chapter 14).

Most computer models based on the current rate of increase in greenhouse gases show the Earth warming as a whole by 2°C to 5°C during the next 75 years. Such a temperature change will be uneven, however, with the greatest warming occurring in the higher latitudes. As a consequence of this warming, rainfall patterns will shift dramatically (♦ Fig. 25.3). This will have a major effect on the largest grain producing areas of the world, such as the American Midwest. Drier and hotter conditions will intensify the severity and frequency of droughts, leading to more crop failures and increased food prices. With such shifts in climate, the Earth may experience an increase in desertification (see the Prologue to Chapter 15).

With a rise in global temperature, glaciers would melt, contributing to rising sea levels and greatly affecting coastal areas where many of the world's large population centers are located. For example, approximately 17% of Bangladesh would be flooded, as would many of the extensive lowland rice-producing regions in other Asian countries. Many low-lying cities such as Houston and New Orleans would have to be ringed with dikes and levees or abandoned.

Can anything be done to reverse this apparent warming trend? The answer is yes, but it will require major shifts in the way people live and changes in the economic structure of the industrialized countries (♦ Table 25.1). Among the ways to reduce the buildup of greenhouse gases are to reduce the wholesale logging of the tropical rainforests and begin reforesting areas that are now denuded of vegetation. Sharp reductions in the consumption of fossil fuels through conservation measures and a switch to alternative energy sources must also occur. While it will take time to reduce the trend of global warming, immediate changes now will have long-term benefits.

We cannot leave the subject of global warming without pointing out that many scientists are not con-

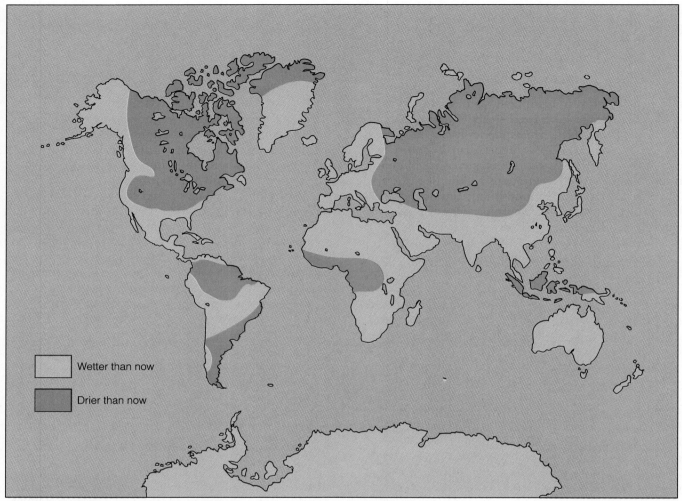

◆ FIGURE 25.3 Changes in the world's rainfall patterns resulting from global warming.

Wetter than now

Drier than now

vinced that the global warming trend is the direct result of increased and industrialization-related human activity. They point out that while there has been an increase in greenhouse gases, there is still uncertainty about their rate of generation and rate of removal, and about whether the 0.6°C rise in global temperature during the past century is the result of normal climatic variations through time or the result of human activity. Furthermore, they point out that even if there is a general global warming during the next 75 years, it is not certain that the dire predictions made by proponents of global warming will come true. The Earth, as we know, is a remarkably complex system, with many feedback mechanisms and interconnections throughout its vari-

ous subsystems. It is very difficult to predict just what all of the consequences to atmospheric and oceanic circulation patterns would be as a result of global warming.

Acid Rain

Atmospheric pollution is one of the consequences of industrialization. Several of the most industrialized nations, including the United States, Canada, and the former Soviet Union, have reduced their emissions into the atmosphere, but many developing nations continue to increase theirs. Some of the consequences of atmospheric pollution include smog, possible disruption of the ozone layer, global warming, and acid rain.

Reduce the population growth rate.
Switch from coal-fired and oil-fired power plants to natural gas.
Implement technologies that burn coal more efficiently.
Expand cogeneration—processes that trap waste heat and put it to good use.
Boost automobile efficiency.
Expand mass transit.
Develop alternative liquid fuels for the transportation sector.
Improve the efficiency of industry.
Make new and existing homes more energy-efficient.
Build many new homes that use solar energy for space heating.
Reduce global deforestation.
Begin a massive reforestation effort.
Phase out all chlorofluorocarbons as soon as possible.
Reduce consumption of unnecessary items.
Expand recycling efforts.

Recall that water and carbon dioxide in the atmosphere react to form carbonic acid that dissociates and yields hydrogen ions and bicarbonate ions (see Chapter 5). The net effect of this reaction is that all rainfall is slightly acidic. Thus, acid rain is the direct consequence of the self-cleansing nature of the atmosphere; that is, many suspended particles of gases in the atmosphere are soluble in water and are removed from the atmosphere during precipitation.

Several natural processes, including volcanism and the metabolism of soil bacteria, introduce gases into the atmosphere that cause acid rain. Human activities, however, produce added atmospheric stress. For example, the burning of fossil fuels has added excess carbon dioxide to the atmosphere. Nitrogen oxide from internal combustion engines and nitrogen dioxide, which is formed in the atmosphere from nitrogen oxide, react to form nitric acid. Although carbon dioxide and nitrogen gases contribute to acid rain, the greatest culprit is sulfur dioxide, which is primarily released by burning coal that contains sulfur. Once in the atmosphere, sulfur dioxide reacts with oxygen to form sulfuric acid, the main component of acid rain.

The phenomenon of acid rain was first recognized in England by Robert Angus Smith in 1872, about a century after the beginning of the Industrial Revolution. It was not until 1961, however, that acid rain became a public environmental concern. At that time it was realized that acid rain is corrosive and irritating, kills vegetation, and has a detrimental effect on surface waters. Since then the effects of acid rain have been recognized in Europe (especially in Eastern Europe where so much coal is burned), the eastern United States, and eastern Canada. During the past three decades acid rain has been getting worse, particularly in Europe and eastern North America (♦ Fig. 25.4). During the last decade, however, the developed countries have made efforts to reduce the impact of acid rain; in the United States the Clean Air Act of 1990 outlined specific steps to reduce the emissions of pollutants that cause acid rain.

The areas most affected by acid rain invariably lie downwind from coal-burning power plants or other industries that emit sulfur gases. Chemical plants and smelters (plants where metal ores are refined) discharge large quantities of sulfur gases and other substances such as heavy metals. The effect of acid rain in these areas may be modified by the local geology (♦ Fig. 25.5). For example, if an area is underlain by limestone or alkaline soils, the acid rain tends to be neutralized by the limestone or soil. Areas underlain by granite, on the other hand, are acidic to begin with and have little or no effect on the rain.

The effects of acid rain vary. Small lakes become more acid as they lose the ability to neutralize the acid rainfall. As the lakes increase in acidity, various types of organisms disappear, and, in some cases, all life-forms eventually die. Acid rain also causes increased weathering of limestone and marble (recall that both are soluble in weak acids) and, to a lesser degree, sandstone. Such effects are particularly visible on buildings, monuments, and tombstones; a notable example is Gettysburg National Military Park in Pennsylvania, which lies in an area that receives some of the most acidic rain in the country.

Whereas the effects on vegetation in the immediate areas of sulfur-gas-emitting industries are apparent, some people have questioned whether acid rain has much effect on forests and crops distant from such sources. But many forests in the eastern United States show signs of stress that cannot be attributed to other causes. In Germany's Black Forest, the needles of firs, spruce, and pines are turning yellow and falling off.

Currently, about 20 million tons of sulfur dioxide are released yearly into the atmosphere in the United

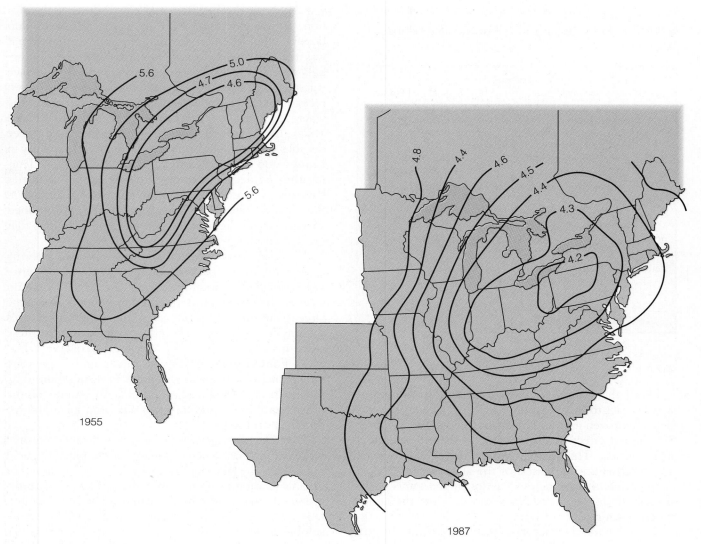

◆ FIGURE 25.4 Distribution and strength of acid rain in the midwestern and eastern United States in 1955 and 1987.

States, mostly from coal-burning power plants. Power plants built before 1975 have no emission controls but the problems they pose must be addressed if emissions are to be reduced to an acceptable level. The most effective way to reduce emissions from these older plants is with flue-gas desulfurization, a process that removes up to 90% of the sulfur dioxide from exhaust gases. There are, however, drawbacks to flue-gas desulfurization. One is that some plants are simply too old to be profitably upgraded; the 85-year-old Phelps

Dodge copper smelter in Douglas, Arizona, closed in 1987 for this reason. Other problems with flue-gas desulfurization include disposal of sulfur wastes, the lack of control on nitrogen gas emissions, and reduced efficiency of the power plant, which must burn more coal to make up the difference.

Other ways to control emissions include the conservation of electricity; the less that is used, the lower the emissions of pollutants. Natural gas contains practically no sulfur, but converting to this alternate energy source

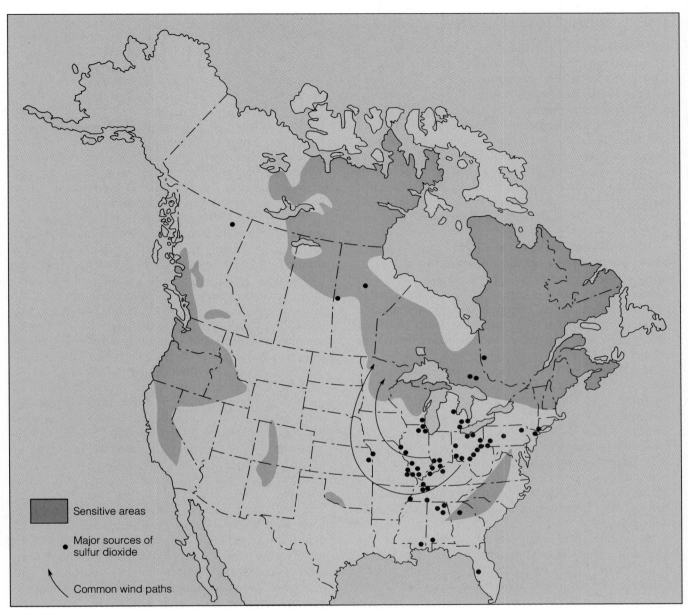

◆ FIGURE 25.5 Acid-sensitive areas in North America and major acid rain-producing sources.

Sensitive areas

• Major sources of sulfur dioxide

Common wind paths

would require the installation of expensive new furnaces in existing plants.

Acid rain, like global warming, is a global problem that knows no national boundaries. Wind currents may blow pollutants from the source in one country to another where the effects are felt. Developed nations have the economic resources to reduce emissions, but many underdeveloped nations cannot afford to do so. Furthermore, many nations have access to only high-sulfur coal and cannot afford to install flue-gas desulfurization devices. Nevertheless, acid rain can be controlled only by the cooperation of all nations contributing to the problem.

Nuclear Waste Disposal

Most environmentalists have long opposed the use of nuclear power, primarily because of the potential for catastrophic accidents occurring at the plants, and problems with radioactive waste disposal. However, with the realization that such problems as global warming and acid rain are partially caused by the burning of fossil fuels, nuclear power may soon be experiencing a resurgence, particularly if acceptable safety features can be incorporated into the plants and the problem of waste disposal can be satisfactorily resolved. Presently, uranium inventories remain high around the world, and little change in demand is expected during the next few years.

A discussion of the safety features of nuclear power plants is beyond the scope of this book, but an examination of the environmental problems associated with nuclear waste disposal is certainly appropriate. One of the problems of nuclear energy is finding sites suitable for storing radioactive waste products, most of which must be isolated from the environment for thousands of years.

Near the end of 1987, Congress passed the Nuclear Waste Policy Amendments Act, authorizing the U.S. Department of Energy's Civilian Radioactive Waste Management Program to evaluate the potential of Nevada's Yucca Mountain to serve as the nation's first high-level radioactive waste dump (see Perspective 13.2). Since then, scientists from many disciplines have been analyzing the site to determine if it can isolate high-level radioactive waste from the environment for 10,000 years, the minimum time such waste will remain dangerous.

In addition to concerns about the geology of the Yucca Mountain area, including possible volcanic activity, the depth of the water table, and the likelihood of a change in climate in the area during the next 10,000 years, an additional factor must be considered, and that is its economic potential. The question is whether the proposed site has a natural resource potential greater than the surrounding areas, or whether it presently has economically extractable resources. One of the reasons for concern about potentially valuable natural resources is the possibility of disturbing the nuclear waste and causing environmental damage due to direct or indirect exploration activities. Studies done to date indicate a low probability of such natural resources as oil and gas, and a low potential for precious and base metals.

One of the reasons for concern about the proper disposal of nuclear waste is the environmental catastrophies that are now being discovered to have occurred in uranium mining areas in eastern Europe. For example, after years of supplying the former Soviet Union with more than 220,000 tons of uranium ore, the environmental, human, and monetary costs of East Germany's uranium industry are finally coming due.

For years, East Germany was the world's third largest producer of uranium ore (following the United States and Canada), shipping most of it to the Soviet Union and helping it become a nuclear superpower. With the collapse of the Soviet Union, East Germany's uranium industry also collapsed as there was no longer a market for its low-grade uranium ore.

Within East Germany's uranium mining area vast amounts of low-grade uranium ore were mined and processed with virtually no regard for the environmental devastation resulting from such operations. Huge open pits as deep as 165 m were dug during mining operations, with the uranium tailings piled up into hills nearby, many as high as 150 m. Winds would carry the radioactive dust from these tailings and deposit it on the surrounding towns and villages. In addition to the radioactive dust, many of the mines generated waste highly concentrated in heavy metals, pyrite, and arsenic.

As a result of the mining and processing activities, the soil and groundwater in many areas of East Germany is highly contaminated. The German government has allocated about 1 billion marks (U.S. $682 million) per year during the next 15 years to deal with the massive clean-up of the area. Unfortunately, many officials think that this expenditure will still not be enough.

In addition to the environmental clean-up costs, the costs in terms of human suffering are still to be determined. Although detailed health records of the miners were kept, the overall effect on the health of the local residents is not known. To date, more than 6,000 confirmed cases of lung cancer have been reported among mine workers who worked in the uranium mines during the 1950s. Some of the mining areas have high rates of certain cancers, but as yet no effort has been made to see if these are statistically significant and if they can be related to the mining and processing activities.

Another recent example of unregulated nuclear waste concerns a huge waste pile of 1,200 tons of uranium and 4 million tons of uranium mining residue dumped in a pond at a secret Soviet military plant 187 km east of Tallinn, Estonia. The pond covers 82 acres and has radiation levels 500 times greater than accepted limits. There are no safety measures at this location and there is great concern that the pond could overflow during stormy weather, sending a stream of radioactive water into the Gulf of Finland. During its operation from 1940 to 1991, more than 4 million tons of uranium, most of it imported from Hungary and Czechoslovakia, were processed at the plant.

The four examples just discussed point out the need for qualified geologists and people trained in the geological sciences to help solve many of our major environmental problems.

▼ GEOLOGY AND THE SEARCH FOR NATURAL RESOURCES

The search for natural resources has been the main area of employment for geologists in the past and will continue to be in the future. The industrialized economies will always need energy and minerals to sustain themselves, and finding those resources is the job of geologists. No country is self-sufficient in all of the mineral resources it needs and uses, and therefore must import some of them. For example, the United States is rich in a variety of mineral resources (● Table 25.2) but still must import much of its oil and many of its metals. Such a dependence on other nations for certain of its basic economic needs translates into political policy in protecting those areas or regions. The 1991 Persian Gulf war was fought, in part, to make sure Kuwait's oil was available to the West.

Today, geologists use increasingly sophisticated techniques in their worldwide exploration for natural resources. Remote sensing and computer technology are just two of the ways geologists are applying modern technology in their work. The application of plate tectonic theory to the formation and distribution of natural resources is helping geologists to focus their attention on areas that have a high potential for economic success. In this section we will focus on coal, oil and natural gas, and metals.

● **TABLE 25.2** Value of Selected Mineral Resources Produced in the United States

Mineral Resources	Production Value (millions of dollars)	Principle Producing States
Energy Resources		
Coal	20,987.0	WY, KY, WV, PA, IL
Natural gas	30,096.0	TX, LA, OK, NM
Petroleum	37,447.4	AK, TX, CA, LA, OK
Uranium	336.8	WY, NM, TX
Industrial Minerals		
Bromine	144.0	AR, MI
Cement, Portland	3,575.9	CA, TX, PA, MI
Cement, masonry	243.9	FL, IN, PA, AL
Clays	1,400.8	GA, OH, NC, TX
Gypsum	109.2	OK, IA, MI, TX
Phosphate rock	887.8	FL, NC, ID, UT
Salt	680.2	LA, TX, NY, OH
Sulfur	430.8	TX, LA
Metals		
Copper	3,771.6	AZ, NM, UT, MT
Gold	2,831.3	NV, CA, SD, UT
Iron ore	1,716.7	MN, MI, MO, UT
Lead	315.2	MO, ID, CO, MT
Molybdenum	266.9	AZ, CO, MT, VT
Silver	349.3	NV, ID, MT, AZ
Zinc	324.2	TN, NY, MO, MT

SOURCE: Statistical Abstract of the United States, 1991.

Coal

Because of dwindling supplies of oil in the United States and increasing reliance on imported oil, coal has been championed as a major source of energy, and one that is free of foreign control. In fact, coal supplies about 24% of the nation's current energy needs, compared to 18% in 1977. It is estimated that the United States has coal reserves (that part of the resource base that can be economically extracted) of 243 billion metric tons, and at the current rate of production of 1 billion metric tons per year, a roughly 243-year supply of coal.

In recent years, however, this estimate of coal reserves has been challenged, and it is possible that the seemingly inexhaustible supply is not as large as it once appeared. In part the problem has been that the Department of Energy relies on data bases compiled by the United States Geological Survey and state geological surveys, which are not uniform in what they consider to be mineable coal or in the classification standards used. The Department of Energy, however, has interpreted the information provided and applied it as best it could to the national standard to produce an annual summary of the nation's mineable coal reserves. It is this figure that is usually quoted when referring to the nation's coal reserves.

One of the problems with this figure is that coals that have neither current or future economic mining potential are included in it. In addition to these coals, there are those that are economically mineable, but may be unavailable to mining because of land use restrictions or because they do not meet current emission standards. In a joint study conducted by the United States Geological Survey and the Kentucky, Virginia, and West Virginia state geological surveys, it was found that about half of the original coal reserves in the initial 12 study areas are not available for mining. This study has been expanded to other states, and initial results show similarly disturbing trends in the actual amount of coal available for mining. What once seemed like a secure and almost inexhaustible source of energy is no longer the only answer to our energy needs.

Oil and Natural Gas

During the 1850s, the demand for petroleum was increasing in the United States as people sought a cheap alternative to other sources to be used for lighting, as a lubricant for machinery, and as an ingre-dient in liniments. In 1859, Edwin L. Drake drilled an oil well 21 m deep at Titusville, Pennsylvania, and began pumping 10 to 35 barrels of oil per day (one barrel is equal to 42 gallons). The United States quickly became the world's leading producer, a position it maintained until 1965. In 1992 it was in third place, behind the Commonwealth of Independent States and Saudi Arabia (◆ Fig. 25.6).

In little more than one hundred years after Drake drilled his well, the United States had become a net petroleum importer. In 1992, the United States imported more than 6 million barrels of oil a day, which represented 47% of its daily consumption. The cost of importing foreign oil was $50 billion and accounted for almost 50% of the United States' trade deficit. Even with increased conservation and improved energy saving devices, it won't be long before the United States imports more than half of its oil from overseas, with much of it coming from the Persian Gulf region.

Even though petroleum was discovered as early as 1908 in Iran, the Persian Gulf did not become a significant petroleum-producing area until the economic recovery after World War II. Following the war, Western Europe and Japan in particular became dependent on Persian Gulf oil, and still rely heavily on this region for most of their supply. The United States is also dependent on imports from the Gulf, but receives significant quantities of petroleum from other sources such as Mexico and Venezuela. Currently, fully 40% of all petroleum imports in the world come from the Persian Gulf countries.

Although large concentrations of petroleum occur in many areas of the world, more than 60% of all proven reserves are in the Persian Gulf region (◆ Fig. 25.7).

◆ FIGURE 25.6 The top 10 oil-producing countries for 1992. Numbers indicate barrels of oil produced daily.

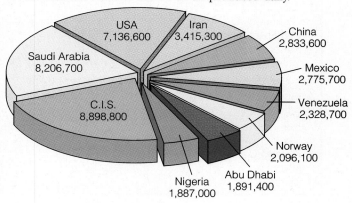

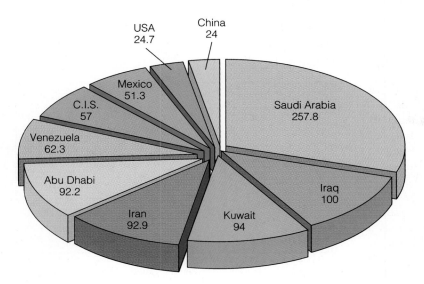

◆ FIGURE 25.7 The top 10 countries in proven oil reserves in 1992. Numbers indicate billions of barrels of oil.

Furthermore, some of the oil fields are gigantic; at least 20 are expected to yield more than 5 billion barrels of oil each, and seven had surpassed this figure by 1983.

Many nations including the United States are heavily dependent on imports of Persian Gulf oil, a dependence that will increase in the future. It is projected that the world's demand for petroleum will increase another 15% by the end of the century. With the decline in production in the United States and the Commonwealth of Independent States, the world will be even more dependent on oil from the Persian Gulf, a fact that will be important in the global political arena.

The third major energy source besides oil and coal is natural gas. One of the major problems faced in promoting natural gas as an energy source has been establishing the infrastructure to get it from the gas fields to market. The use of natural gas in the United States has been steadily growing during the past decade, in part because of large reserves in North America (● Table 25.3), and is expected to continue to increase during the 1990s.

According to the International Energy Agency, the demand for natural gas in Western Europe is projected to increase by one-third, or an additional 100 billion m³, by the end of this century. Natural gas has already surpassed oil as the main fuel used by power generating plants in Europe, and its use as a source for heating is also increasing. The European Bank for Reconstruction and Development estimates that natural gas usage will double in Western Europe in the next 20 years.

● TABLE 25.3 Natural Gas Reserves in North America

Country	Billions of Cubic Feet	Percentage
Mexico	70,900	21.2
United States	167,062	50.1
Canada	95,734	28.7

SOURCE: Oil and Gas Journal, v. 90, p. 44–45 (Dec. 28, 1992).

Waiting to serve that market are Russia's largely untapped natural gas reserves in Siberia. These reserves are the largest in the world and are estimated to represent 40% of the world's total natural gas reserves. Western Europe presently relies upon Russia for about 25% of its natural gas needs, and could become even more dependent should Russia be able to develop Siberia's natural gas potential.

The problem lies in getting the gas from Siberia to its markets in Europe. It is estimated that it will cost about U.S. $15 billion to construct a pipeline from western Siberia to the German-Czechoslovakian border where it can be fed into Europe's natural gas pipeline grid. Such an undertaking presents enormous economic and political risks. The two major ones are Russia's ability to raise the money to build a pipeline, and whether it would be profitable at the low prices natural gas presently commands.

Even if the pipeline is built, there is no guarantee there will be enough new customers for the gas. Several countries expect to increase production and build pipelines of their own to serve Europe in the near future. For example, Algeria plans to double its natural gas exports to Europe by the end of this century and is expanding its present pipeline capacity to Italy and building a new pipeline to Spain. Iran has 13% of the world's natural gas reserves, second only to Russia, and is planning a pipeline through Turkey to serve Europe.

Currently, oil is still the major source of energy in the world, but as pipeline grids are constructed and natural gas fields are brought on-line, natural gas looms as a major threat to oil's dominance in the world energy market. While there will always be a need for oil, clean burning natural gas is an attractive alternative to oil as a source for power and heating needs.

Metals

Metals are an essential part of the world's economy. Recall that the distribution and concentration of various metal ore bodies are controlled by plate tectonics (see Chapter 9), and geologists are increasingly using plate tectonic theory in their search for metals. Recall also that different periods in Earth history are characterized by different types of metal deposits (see Chapters 20, 21, 23, and 24). For example, banded iron formations and much of the world's gold occurs in Precambrian rocks (see Chapter 20). Because of the importance of plate tectonics to the formation of various metal ores, we will briefly examine its role in relation to Canada's mineral wealth (● Table 25.4), specifically in the Canadian Shield (see Chapter 20).

Canada is a leading producer of iron ore, nickel, zinc, lead, gold, and silver, many of which are found in the Canadian Shield. In fact, all of Canada's iron and nickel, nearly 90% of its gold, 50% of its copper and silver, and about 30% of its zinc comes from the Canadian Shield.

As the history of the Canadian Shield becomes better known, it appears that it formed as a result of various continental collisions and breakups along plate boundaries. Continental formation is marked by orogenic activity involving deformation (see Chapter 10), magma generation (see Chapter 3), and regional metamorphism (see Chapter 6). Breakup was accompanied by rifting and the emplacement of mafic dikes (see Chapter 9). During these accretionary and rifting tectonic phases, the various metal ores formed.

● **TABLE 25.4** 1990 Production of Leading Metals in Canada

Metal	Millions of Dollars
Copper	2,500
Zinc	2,480
Gold	2,380
Nickel	2,050
Iron	1,350
Uranium	970
Lead	260
Silver	256
Platinum	230
Molybdenum	140

SOURCE: Canada Year Book, 1992.

The vast Precambrian iron ore deposits of the Canadian Shield formed as a result of the precipitation of dissolved iron and silica in shallow marine waters during the quiescent intervals between major orogenies, mainly during the Proterozoic Eon. All of Canada's iron ore is mined in the Canadian Shield. More than 90% of Canada's gold comes from Archean and Early Proterozoic terranes of the Canadian Shield. These terranes formed during the various accretionary orogenic events occurring along convergent plate boundaries. Metal-rich magmas and fluids were produced during convergence which upon cooling concentrated the gold and other metals now mined.

In addition to the metals formed as a result of convergence, the Canadian Shield also has massive volcanic-associated sulfide deposits containing copper and zinc. These deposits are believed to be the result of major volcanic events occurring along mid-oceanic ridges or in back-arc basins corresponding to the beginning of orogenic episodes during the Archean and Proterozoic eons.

Current knowledge about the history of the Canadian Shield indicates that most of its metal resources were formed during three major orogenic events and their intervening quiescent periods. These three orogenic events occurred during the Late Archean, Early Proterozoic, and Middle Proterozoic eons.

▼ THE EVOLUTION OF THE EARTH

Having examined the role and impact of geology in relation to the environment and the search for natural resources, we can now focus our attention on what lessons we can learn from the study of the Earth's geologic and biologic history. One of the fundamental concepts to be learned from the study of historical geology is that the Earth continuously changes or evolves. One cannot study the history of the Earth and its biota without appreciating the role of time and how imperceptible changes can have dramatic effects when viewed from the perspective of geologic time. In some ways, a geologic history of the Earth can be summed up by saying that the only thing certain is change.

As we discussed in Chapter 1, one of the cornerstones of geology is the principle of uniformitarianism. This principle is based on the premise that present-day processes have operated throughout geologic time, albeit at sometimes different rates than at present. What makes uniformitarianism such a powerful principle is that it allows geologists to use present-day processes as the basis for interpreting the past and for predicting future events. We can use our understanding of the underlying principles and processes of geology to help us predict volcanic eruptions or earthquakes, or to avoid areas of potential disasters like landslides or floods. We have also learned that the Earth is composed of various interrelated systems, and that when we make changes to one part of a system we affect the other parts of the system, sometimes with undesirable results.

Another aspect of uniformitarianism is the cyclical nature of Earth history. In terms of plate tectonics, many geologists now view the movement of plates in terms of a supercontinent cycle—the formation and breakup of continental masses through time (see Perspective 9.1). It also appears that the Earth has undergone cyclic occurrences of glacial and interglacial episodes and we must keep this perspective of climatic change in mind when debating whether the rise in mean global temperatures during the past century is due to human activity or part of a longer glacial-interglacial cycle. Lastly, the history of the cratons can be studied from the perspective of transgressive and regressive cycles of epeiric seas throughout geologic time. Again, the unstated but underlying principle of uniformitarianism plays a pivotal role in deciphering the Earth's history.

When we examine the history of life, we again see the theme of change. The history of life is the result of natural selection operating on random mutations ever since life first evolved sometime about 3.5 billion years ago. Thus the history of life represents a series of unique events. As Stephen Jay Gould so eloquently stated in *Wonderful Life,* "Wind back the tape of life to the early days of the Burgess Shale; let it play again from an identical starting point, and the chance becomes vanishingly small that anything like human intelligence would grace the replay."* What this means is that evolution is irreversible and, like history, does not exactly repeat itself.

Another important implication derived from studying the history of life is the interaction between the living and nonliving components of the global ecosystem and how each part affects the other. Just as plate tectonics plays an important role in the formation and distribution of natural resources, it plays a pivotal part in the evolution of life. The distribution of plants and animals is controlled to a large extent by the distribution of continents and oceans. Atmospheric and oceanic circulation patterns are also controlled by plate movement. We have cited numerous examples of the role of plate tectonics and its effect on the evolution of life. For example, recall that the Pennsylvanian Period was a time when seedless vascular plants were particularly diverse and abundant in vast coal-forming swamps (see Chapter 22). These swamps, which were the perfect environment for the life cycle of seedless vascular plants, formed as a result of transgressions and regressions of epeiric seas over low-relief continents. The transgressions and regressions were caused by glacial and interglacial events as Gondwana moved over the South Pole.

The study of the cause of the mass extinctions at the end of the Cretaceous Period led researchers to realize that a nuclear war could lead to what has popularly been called nuclear winter. The somewhat serendipitous discovery of a 2.5-cm-thick layer of iridium at the Cretaceous-Tertiary boundary in Italy led to the hypothesis that a collision with a meteorite 66 million years ago was responsible for the mass extinctions observed in the fossil record at that time (see Chapter 23). It is thought that the tremendous amount of dust released into the atmosphere from the impact of the meteorite, combined with the soot produced by massive forest fires, resulted in a decrease in light reaching the Earth's surface. This caused a reduction in photosynthesis and a lowering of global temperatures that disrupted the global ecosystem, leading to mass extinctions.

Applying the same reasoning to the amount of dust and soot that would be thrown into the atmosphere as the result of a nuclear war, scientists concluded that the

. . . .
*S. J. Gould, *Wonderful Life* (New York: Norton, 1989), p. 14.

same thing could happen again, resulting in major crop failures which would lead to chaos for today's ecosystem. We have already examined the effects of volcanic eruptions on global climates (see Perspective 4.1) and how volcanic ash and gases can affect global temperatures. Thus an event that happened 66 million years ago may have important implications for the continued survival of the human race.

The study of geology is more than learning numerous facts about the Earth. Geology is an integral part of our lives. As individuals and societies, the standard of living that we enjoy is directly dependent on the consumption of natural resources and interaction with the environment. An appreciation of geology and its relationship to the environment is critical if we, as a species, are to continue to exist on this planet.

Chapter Summary

1. The Earth is a complex, dynamic planet that has been continually changing since it formed 4.6 billion years ago.
2. Geology is part of the human experience and a basic understanding of it is important for dealing with the many environmental problems and issues facing society today.
3. In addition to their traditional role in finding new deposits of natural resources, geologists are also playing an increasingly important role in environmental issues and regulations.
4. The greatest environmental problem facing the world today is overpopulation. The Earth's carrying capacity is being overwhelmed by an increasingly greater number of humans. The result of this is misery and suffering for a large part of the human population.
5. The Earth's carrying capacity is limited by three factors: food supply, resource supply, and the ability to assimilate pollution produced by organisms. All three factors can be addressed from a geologic perspective.
6. The contribution by humans to global warming is currently being debated. While the global mean temperature has risen 0.6°C since the late 1800s, it is not at all clear whether this is due to human activities, especially the burning of fossil fuels, or part of the natural global temperature cycle. The consequences of a continued rise in global temperatures, however, may be catastrophic.
7. Acid rain is another problem brought on by industrialization. The areas most affected by acid rain usually lie downwind from coal-burning plants or other sulfur-emitting industries. The most effective way to minimize the effects of acid rain is to reduce sulfur emissions.

8. Another geologically-oriented problem affecting society is nuclear waste disposal. As a result of mining and processing activities, vast areas of Eastern Europe are highly contaminated and polluted. The geologic community in the United States is presently studying the feasibility of Nevada's Yucca Mountain as a high-level radioactive waste repository.
9. The search for natural resources has been the main area of employment for geologists in the past and will continue to be in the future.
10. The three main energy sources are coal, oil, and natural gas. Coal has long been touted as a major source of energy in the world and particularly in the United States because of its abundance.
11. Oil continues to be the energy source of choice for most nations of the world because of its abundance and relatively low cost. However, most of the world's oil reserves occur in the Persian Gulf region, a fact that will become increasingly important in the global political arena.
12. As oil becomes scarcer, natural gas will make up a larger portion of the energy picture. The major problem in the utilization of natural gas has been in establishing an infrastructure to get it from the gas fields to market.
13. Metals are an essential part of the world's economy. Their distribution and concentration are controlled by plate tectonics.
14. One of the fundamental concepts to be learned from the study of historical geology is that the Earth and life are continually evolving.

Review Questions

1. The greatest environmental problem facing the world today is:
   a. ___ decreasing food supply; b. ___ increasing pollution; c. ___ overpopulation; d. ___ depletion of natural resources; e. ___ none of these.

2. Which of the following factors limits the carrying capacity of an ecosystem?
   a. ___ food supply; b. ___ resource supply; c. ___ its ability to assimilate the pollution produced by organisms; d. ___ all of these; e. ___ none of these.

3. Some of the world's greatest population densities occur in countries in:
   a. ___ North America; b. ___ the Pacific Rim;
   c. ___ South America; d. ___ Europe; e. ___ Africa.
4. One of the by-products of respiration and the burning of organic material and an important cause of global warming is:
   a. ___ carbon dioxide; b. ___ sulfur dioxide; c. ___ carbon monoxide; d. ___ hydrogen; e. ___ nitrogen.
5. The three major contributors to global warming are:
   a. ___ nitrogen, hydrogen, and oxygen; b. ___ carbon dioxide, methane, and chlorofluorocarbons; c. ___ carbon monoxide, carbon dioxide, and methane; d. ___ methane, ethane, and carbon monoxide; e. ___ nitrogen, chlorofluorocarbons, and hydrogen.
6. The mean global temperature has increased by how many degrees since the late 1800s?
   a. ___ 0.6°C; b. ___ 1.0°C; c. ___ 1.6°C; d. ___ 2.0°C; e. ___ 6°C.
7. The major contributor to acid rain is:
   a. ___ carbon dioxide; b. ___ carbon monoxide; c. ___ nitrogen; d. ___ sulfur dioxide; e. ___ none of these.
8. Which of the following fossil fuels does the United States have enough reserves of to last for at least another century?
   a. ___ geothermal energy; b. ___ natural gas; c. ___ oil; d. ___ nuclear energy; e. ___ coal.
9. What region of the world contains more than 60% of all proven oil reserves?
   a. ___ North America; b. ___ Commonwealth of Independent States; c. ___ South America; d. ___ Persian Gulf; e. ___ Australia.
10. Roughly 40% of the world's natural gas reserves occur in what country?
    a. ___ United States; b. ___ Canada; c. ___ Russia; d. ___ Iran; e. ___ Iraq.
11. Banded iron formations and much of the world's gold occur in what age deposits?
    a. ___ Precambrian; b. ___ Paleozoic; c. ___ Mesozoic; d. ___ Cenozoic; e. ___ Recent.
12. Discuss why it is important that people have a better understanding of science and how it works.
13. What are some of the benefits of studying geology?
14. Discuss the criteria that must be met before Nevada's Yucca Mountain can be used as a high-level nuclear waste disposal site.
15. Discuss why the principle of uniformitarianism is one of the cornerstones of geology.
16. Discuss the advantages and disadvantages of each of the following as an energy resource: oil, natural gas, coal, and nuclear energy.
17. Discuss the role of plate tectonics in the formation and distribution of metallic ores.
18. Discuss the importance of geology in solving many of our current environmental problems.
19. Discuss how geology can help determine whether global warming is part of a long-term temperature trend, or primarily the result of human activity during the past 200 years.
20. Discuss the causes of and possible remedies for the acid rain problem.

>=<>=<>=<>=<>=<>=<>=<>=<>=<>=<>=<>=<>=<>=<>=<>=<>=<>=<>=<>=<>=<>=<>=<>=<>=<

Additional Readings

Brookins, D. G. 1990. *Mineral and energy resources: Occurrence, exploitation, and environmental impact.* Columbus, Ohio: Merrill.

Davis, G. R. 1990. Energy for planet Earth. *Scientific American* 263, no. 3: 55–74.

Edmonds, J., and J. M. Reilly. 1985. *Global energy: Assessing the future.* New York: Oxford Univ. Press.

Ellis, W. S. 1990. A Soviet sea lies dying. *National Geographic* 177, no. 2: 73–93.

Holloway, M. 1993. Sustaining the Amazon. *Scientific American* 269, no. 1: 90–100.

Howell, D. G., K. J. Bird, and D. L. Gautier. 1993. Oil: Are we running out? *Earth* 2, no. 2: 26–33.

Howell, D., F. Cole, M. Fanelli, and K. Wiese. 1993. Natural gas. *Earth* 2, no. 5: 52–59.

Jones, P. D., and T. M. L. Wigley. 1990. Global warming trends. *Scientific American* 263, no. 2: 84–91.

Lovelock, J. E. 1988. *The ages of Gaia: A biography of our living Earth.* New York: Norton.

Micklin, P. P. 1993. The shrinking Aral Sea. *Geotimes* 38, no. 4: 14–18.

Mohnen, V. A. 1988. The challenge of acid rain. *Scientific American* 259, no. 2: 30–38.

Rowland, F. S. 1989. Chlorofluorocarbons and the depletion of stratospheric ozone. *American Scientist* 77, no. 1: 36–45.

Sawkins, F. J. 1984. *Metal deposits in relation to plate tectonics.* New York: Springer-Verlag.

Skinner, B. J. 1986. *Earth resources.* Englewood Cliffs, N.J.: Prentice-Hall.

World Commission on Environment and Development. 1987. *Our common future.* New York: Oxford University Press.

English-Metric Conversion Chart

English Unit	Conversion Factor	Metric Unit	Conversion Factor	English Unit
Length				
Inches (in)	2.54	Centimeters (cm)	0.39	Inches (in)
Feet (ft)	0.305	Meters (m)	3.28	Feet (ft)
Miles (mi)	1.61	Kilometers (km)	0.62	Miles (mi)
Area				
Square inches (in^2)	6.45	Square centimeters (cm^2)	0.16	Square inches (in^2)
Square feet (ft^2)	0.093	Square meters (m^2)	10.8	Square feet (ft^2)
Square miles (mi^2)	2.59	Square kilometers (km^2)	0.39	Square miles (mi^2)
Volume				
Cubic inches (in^3)	16.4	Cubic centimeters (cm^3)	0.061	Cubic inches (in^3)
Cubic feet (ft^3)	0.028	Cubic meters (m^3)	35.3	Cubic feet (ft^3)
Cubic miles (mi^3)	4.17	Cubic kilometers (km^3)	0.24	Cubic miles (mi^3)
Weight				
Ounces (oz)	28.3	Grams (g)	0.035	Ounces (oz)
Pounds (lb)	0.45	Kilograms (kg)	2.20	Pounds (lb)
Short tons (st)	0.91	Metric tons (t)	1.10	Short tons (st)
Temperature				
Degrees Fahrenheit (°F)	−32° × 0.56	Degrees Celsius (Centigrade) (°C)	× 1.80 + 32°	Degrees Fahrenheit (°F)

Examples: 10 inches = 25.4 centimeters; 10 centimeters = 3.9 inches
100 square feet = 9.3 square meters; 100 square meters = 1080 square feet
50°F = 10.08°C; 50°C = 122°F

Classification of Organisms

Any classification is an attempt to make order out of disorder and to group similar items into the same categories. All classifications are schemes that attempt to relate items to each other based on current knowledge and therefore are progress reports on the current state of knowledge for the items classified. Because classifications are to some extent subjective, classification of organisms may vary among different texts.

The classification that follows is based on the five-kingdom system of classification of Margulis and Schwartz.* We have not attempted to include all known life forms, but rather major categories of both living and fossil groups.

Kingdom Monera

Prokaryotes
Phylum Anaerobic Photosynthetic Bacteria—(Archean-Recent)
Phylum Cyanobacteria—Blue-green algae or blue-green bacteria (Archean-Recent)

Kingdom Protoctista

Solitary or colonial unicellular eukaryotes
Phylum Acritarcha—Organic-walled unicellular algae of unknown affinity (Proterozoic-Recent)
Phylum Bacillariophyta—Diatoms(Jurassic-Recent)
Phylum Charophyta—Stoneworts (Silurian-Recent)
Phylum Chlorophyta—Green algae (Proterozoic-Recent)

. . . .

*Margulis, L., and K. V. S. Schwartz, 1982. *Five Kingdoms.* New York: W. H. Freeman and Co.

Phylum Chrysophyta—Golden-brown algae, silicoflagellates and coccolithophorids (Jurassic-Recent)
Phylum Euglenophyta—Euglenids (Cretaceous-Recent)
Phylum Myxomycophyta—Slime molds (Proterozoic-Recent)
Phylum Phaeophyta—Brown algae, multicellular, kelp, seaweed (Proterozoic-Recent)
Phylum Protozoa—Unicellular heterotrophs (Cambrian-Recent)
Class Sarcodina—Forms with pseudopodia for locomotion (Cambrian-Recent)
Order Foraminifera—Benthonic and planktonic sarcodinids most commonly with calcareous tests (Cambrian-Recent)
Order Radiolaria—Planktonic sarcodinids with siliceous tests (Cambrian-Recent)
Phylum Pyrrophyta—Dinoflagellates (Silurian?, Permian-Recent)
Phylum Rhodophyta—Red algae (Proterozoic-Recent)
Phylum Xanthophyta—Yellow-green algae (Miocene-Recent)

Kingdom Fungi

Phylum Zygomycota—Fungi that lack cross walls (Proterozoic-Recent)
Phylum Basidiomycota—Mushrooms (Pennsylvanian-Recent)
Phylum Ascomycota—Yeasts, bread molds, morels (Mississippian-Recent)

Kingdom Plantae

Photosynthetic eukaryotes

Division* Bryophyta—Liverworts, mosses, hornworts (Devonian-Recent)

Division Psilophyta—Small, primitive vascular plants with no true roots or leaves (Silurian-Recent)

Division Lycopodophyta—Club mosses, simple vascular systems, true roots and small leaves, including scale trees of Paleozoic Era (lycopsids) (Devonian-Recent)

Division Sphenophyta—Horsetails, scouring rushes, and sphenopsids such as the Carboniferous *Calamites* (Devonian-Recent)

Division Pteridophyta—Ferns (Devonian-Recent)

Division Pteridospermophyta—Seed ferns (Devonian-Jurassic)

Division Coniferophyta—Conifers or cone-bearing gymnosperms (Carbonaceous-Recent)

Division Cycadophyta—Cycads (Triassic-Recent)

Division Ginkgophyta—Maidenhair tree (Triassic-Recent)

Division Angiospermophyta—Flowering plants and trees (Cretaceous-Recent)

Kingdom Animalia

Nonphotosynthethic multicellular eukaryotes (Proterozoic-Recent)

Phylum Porifera—Sponges (Cambrian-Recent)

Order Stromatoporoida—Extinct group of reef-building organisms (Cambrian-Oligocene)

Phylum Archaeocyatha—Extinct spongelike organisms (Cambrian)

Phylum Cnidaria—Hydrozoans, jellyfish, sea anemones, corals (Cambrian-Recent)

Class Hydrozoa—Hydrozoans (Cambrian-Recent)

Class Scyphozoa—Jellyfish (Proterozoic-Recent)

Class Anthozoa—Sea anemones and corals (Cambrian-Recent)

Order Tabulata—Exclusively colonial corals with reduced to nonexistant septa (Ordovician-Permian)

Order Rugosa—Solitary and colonial corals with fourfold symmetry (Ordovician-Permian)

Order Scleractinia—Solitary and colonial corals with sixfold symmetry. Most colonial forms have symbionic dinoflagellates in their tissue. Important reef builders today (Triassic-Recent)

Phylum Bryozoa—Exclusively colonial suspension feeding marine animals that are useful for correlation and ecological interpretations (Ordovician-Recent)

Phylum Brachiopoda—Marine suspension feeding animals with two unequal sized valves. Each valve is bilaterally symmetrical (Cambrian-Recent)

Class Inarticulata—Primitive chitino-phosphatic or calcareous brachiopods that lack a hinging structure. They open and close their valves by means of complex muscles (Cambrian-Recent)

Class Articulata—Advanced brachiopods with calcareous valves that are hinged (Cambrian-Recent)

Phylum Mollusca—A highly diverse group of invertebrates (Cambrian-Recent)

Class Monoplacophora—Segmented, bilaterally symmetrical crawling animals with cap-shaped shells (Cambrian-Recent)

Class Amphineura—Chitons. Marine crawling forms, typically with 8 separate calcareous plates (Cambrian-Recent)

Class Scaphopoda—Curved, tusk-shaped shells that are open at both ends (Ordovician-Recent)

Class Gastropoda—Single shelled generally coiled crawling forms. Found in marine, brackish, and fresh water as well as terrestrial environments (Cambrian-Recent)

Class Bivalvia—Mollusks with two valves that are mirror images of each other. Typically known as clams or oysters (Cambrian-Recent)

Class Cephalopoda—Highly evolved swimming animals. Includes shelled sutured forms as well as non-shelled types such as octopus and squid (Cambrian-Recent)

Order Nautiloidea—Forms in which the chamber partitions are connected to the wall along simple, slightly curved lines (Cambrian-Recent)

Order Ammonoidea—Forms in which the chamber partitions are connected to the wall along wavy lines (Devonian-Cretaceous)

. . . .

*In botany, division is the equivalent to phylum.

Order Coleoidea—forms in which the shell is reduced or lacking. Includes octopus, squid, and the extinct belemnoids (Mississippian-Recent)

Phylum Annelida—Segmented worms. Responsible for many of the Phanerozoic burrow and trail trace fossils (Proterozoic-Recent)

Phylum Arthropoda—The largest invertebrate group comprising about 80% of all known animals. Characterized by a segmented body and jointed appendages (Cambrian-Recent)

Class Trilobita—Earliest appearing arthropod class. Trilobites had a head, body, and tail and were bilaterally symmetrical (Cambrian-Permian)

Class Crustacea—Diverse class characterized by a fused head and body and an abdomen. Included are barnacles, copepods, crabs, ostracodes, and shrimp (Cambrian-Recent)

Class Insecta—Most diverse and common of all living invertebrates, but rare as fossils (Silurian-Recent)

Class Merostomata—Characterized by four pairs of appendages and a more flexible exoskeleton than crustaceans. Includes the extinct eurypterids, horseshoe crabs, scorpions, and spiders (Cambrian-Recent)

Phylum Echinodermata—Exclusively marine animals with fivefold radial symmetry and a unique water vascular system (Cambrian-Recent)

Subphylum Crinozoa—Forms attached by a calcareous jointed stem (Cambrian-Recent)

Class Crinoidea—Most important class of Paleozoic echinoderms. Suspension feeding forms that are either free-living or attach to sea floor by a stem (Cambrian-Recent)

Class Blastoidea—Small class of Paleozoic suspension feeding sessile forms with short stems (Ordovician-Permian)

Class Cystoidea—Globular to pear-shaped suspension feeding benthonic sessile forms with quite short stems (Ordovician-Devonian)

Subphylum Homalozoa—A small group with flattened, asymmetrical bodies with no stems. Also called carpoids (Cambrian-Devonian)

Subphylum Echinozoa—Globose, predominantly benthonic mobile echinoderms (Ordovician-Recent)

Class Helioplacophora—Benthonic, mobile forms, shaped like a top with plates arranged in a helical spiral (Early Cambrian)

Class Edrioasteroidea—Benthonic, sessile or mobile, discoidal, globular or cylindrical shaped forms with five straight or curved feeding areas shaped like a starfish (Cambrian-Pennsylvanian)

Class Holothuroidea—Sea cucumbers. Sediment feeders having calcareous spicules embedded in a tough skin (Ordovician-Recent)

Class Echinoidea—Largest group of echinoderms. Globe or disk shaped with movable spines. Predominantly grazers or sediment feeders. Epifaunal and infaunal (Ordovician-Recent)

Subphylum Asterozoa—Stemless, benthonic mobile forms (Ordovician-Recent)

Class Asteroidea—Starfish. Arms merge into body (Ordovician-Recent)

Class Ophiuroidea—Brittle star. Distinct central body (Ordovician-Recent)

Phylum Hemichordata—Characterized by a notochord sometime during the life history. Modern acorn worms and extinct graptolites (Cambrian-Recent)

Class Graptolithina—Colonial marine hemichordates having a chitinous exoskeleton. Predominantly planktonic (Cambrian-Mississippian)

Phylum Chordata—Animals with notochord, hollow dorsal nerve cord, and gill slits during at least part of their lifecycle (Cambrian-Recent)

Subphylum Urochordata—Sea squirts, tunicates. Larval forms have notochord in tail region.

Subphylum Cephalochordata—Small marine animals with notochords and small fish-like bodies (Cambrian-Recent)

Subphylum Vertebrata—Animals with a backbone of vertebrae (Cambrian-Recent)

Class Agnatha—Jawless fish. Includes the living lampreys and hagfish as well as the extinct armored ostracoderms (Cambrian-Recent)

Class Acanthodii—Primitive jawed fish with numerous spiny fins (Silurian-Permian)

Class Placodermii—Primitive armored jawed fish (Silurian-Permian)

Class Chondrichthyes—Cartilaginous fish such as sharks and rays (Devonian-Recent)

Class Osteichthyes—Bony fish (Devonian-Recent)

Subclass Actinopterygii—Ray-finned fish (Devonian-Recent)

Subclass Sarcopterygii—Lobe-finned, air-breathing fish (Devonian-Recent)

Order Crossoptergii—Lobe-finned fish that were ancestral to amphibians (Devonian-Recent)

Order Dipnoi—Lungfish (Devonian-Recent)

Class Amphibia—Amphibians. The first terrestrial vertebrates (Devonian-Recent)

Subclass Labyrinthodontia—Earliest amphibians. Solid skulls and complex tooth pattern (Devonian-Triassic)

Subclass Salientia—Frogs, toads, and their relatives (Triassic-Recent)

Subclass Condata—Salamanders and their relatives (Triassic-Recent)

Class Reptilia—Reptiles. A large and varied vertebrate group characterized by having scales and laying an amniote egg (Pennsylvanian-Recent)

Subclass Anapsida—Reptiles whose skull has a solid roof with no openings (Pennsylvanian-Recent)

Order Cotylosauria—Captorhinomorphs or earliest reptiles (Pennsylvanian-Triassic)

Order Chelonia—Turtles (Triassic-Recent)

Subclass Euryapsida—Reptiles with one opening high on the side of the skull behind the eye. Mostly marine (Permian-Cretaceous)

Order Protorosauria—Land living ancestral euryapsids (Permian-Cretaceous)

Order Placodontia—Placodonts. Bulky, paddle-limbed marine reptiles with rounded teeth for crushing mollusks (Triassic)

Order Ichthyosauria—Ichthyosaurs. Dolphin-shaped swimming reptiles (Triassic-Cretaceous)

Subclass Diapsida—Most diverse reptile class, characterized by two openings in the skull behind the eye. Includes lizards, snakes, crocodiles, thecodonts, dinosaurs, and pterosaurs (Permian-Recent)

Infraclass Lepidosauria—Primitive diapsids including snakes, lizards, and the large Cretaceous marine reptile group called mosasaurs (Permian-Recent)

Order Mosasauria—Mosasaurs (Cretaceous)

Order Plesiosauria—Plesiosaurs (Jurassic-Cretaceous)

Order Squamata—Lizards and snakes (Triassic-Recent)

Order Rhynchocephalia—The living tuatara *Sphenodon* and its extinct relatives (Jurassic-Recent)

Infraclass Archosauria—Advanced diapsids (Triassic-Recent)

Order Thecodontia—Thecodontians were a diverse group that was ancestral to the crocodilians, pterosaurs, and dinosaurs (Permian-Triassic)

Order Crocodilia—Crocodiles, alligators, and gavials (Triassic-Recent)

Order Pterosauria—Flying and gliding reptiles called pterosaurs (Triassic-Cretaceous)

Infraclass Dinosauria—Dinosaurs (Triassic-Cretaceous)

Order Saurischia—Lizard-hipped dinosaurs (Triassic-Cretaceous)

Suborder Theropoda—Bipedal carnivores (Triassic-Cretaceous)

Suborder Sauropoda—Quadrupedal herbivores, including the largest known land animals (Jurassic-Cretaceous)

Order Ornithischia—Bird-hipped dinosaurs (Triassic-Cretaceous)

Suborder Ornithopoda—Bipedal herbivores, including the duck-billed dinosaurs (Triassic-Cretaceous)

Suborder Stegosauria—Quadrupedal herbivores with bony spikes on their tails and bony plates on their backs (Jurassic-Cretaceous)

Suborder Pachycephalosauria—Bipedal herbivores with thickened bones of the skull roof (Cretaceous)

Suborder Ceratopsia—Quadrupedal herbivores typically with horns or a bony frill over the top of the neck (Cretaceous)

Suborder Ankylosauria—Heavily armored quadrupedal herbivores (Cretaceous)

Subclass Synapsida—Mammal-like reptiles with one opening low on the side of the skull behind the eye (Pennsylvanian-Triassic)

Order Pelycosauria—Early mammal-like reptiles including those forms in which the vertebral spines were extended to support a "sail" (Pennsylvanian-Permian)

Order Therapsida—Advanced mammal-like reptiles with legs positioned beneath the body and the lower jaw formed largely of a single bone. Many therapsids may have been endothermic (Permian-Triassic)

Class Aves—Birds. Endothermic and feathered (Jurassic-Recent)

Class Mammalia—Mammals. Endothermic animals with hair (Triassic-Recent)

Subclass Prototheria—Egg-laying mammals (Triassic-Recent)

Order Docodonta—Small, primitive mammals (Triassic)

Order Triconodonta—Small, primitive mammals with specialized teeth (Triassic-Cretaceous)

Order Monotremata—Duck-billed platypus, spiny anteater (Cretaceous-Recent)

Subclass Allotheria—Small extinct early mammals with complex teeth for grinding food (Jurassic-Eocene)

Order Multituberculata—The first mammalian herbivores and the most diverse of Mesozoic mammals (Jurassic-Eocene)

Subclass Theria—Mammals that give birth to live young (Jurassic-Recent)

Order Symmetrodonta—Small, primitive Mesozoic therian mammals (Jurassic-Cretaceous)

Order Upantotheria—Trituberculates (Jurassic-Cretaceous)

Order Creodonta—Extinct ancient carnivores (Cretaceous-Paleocene)

Order Condylartha—Extinct ancestral hoofed placentals (ungulates) (Cretaceous-Oligocene)

Order Marsupialia—Pouched mammals. Opossum, kangaroo, koala (Cretaceous-Recent)

Order Insectivora—Primitive insect-eating mammals. Shrew, mole, hedgehog (Cretaceous-Recent)

Order Xenungulata—Large South American mammals that broadly resemble pantodonts and uintatheres (Paleocene)

Order Taeniodonta—Includes some of the most highly specialized terrestrial placentals of the Late Paleocene and Early Eocene (Paleocene-Eocene)

Order Tillodontia—Large, massive placentals with clawed, five-toed feet (Paleocene-Eocene)

Order Dinocerata—Uintatheres. Large herbivores with bony protuberances on the skull and greatly elongated canine teeth (Paleocene-Eocene)

Order Pantodonta—North American forms are large sheep to rhinoceros-sized. Asian forms are as small as a rat (Paleocene-Eocene)

Order Astropotheria—Large placental mammals with slender rear legs and stout forelimbs and elongate canine teeth (Paleocene-Miocene)

Order Notoungulata—Largest assemblage of South American ungulates with a wide range of body forms (Paleocene-Pleistocene)

Order Liptoterna—Extinct South American hoofed-mammals (Paleocene-Pleistocene)

Order Rodentia—Rodents. Squirrel, mouse, rat, beaver, porcupine, gopher (Paleocene-Recent)

Order Lagomorpha—Hare, rabbit, pika (Paleocene-Recent)

Order Primates—Lemur, tarsier, loris, monkey, human (Paleocene-Recent)

Order Edentata—Anteater, sloth, armadillo, glyptodont (Paleocene-Recent)

Order Carnivora—Modern carnivorous placentals. Dog, cat, bear, skunk, seal, weasel, hyena, raccoon, panda, sea lion, walrus (Paleocene-Recent)

Order Pyrotheria—Large mammals with long bodies and short columnar limbs (Eocene-Oligocene)

Order Chiroptera—Bats (Eocene-Recent)

Order Dermoptera—Flying lemur (Eocene-Recent)

Order Cetacea—Whale, dolphin, porpoise (Eocene-Recent)

Order Tubulidentata—Aardvark (Eocene-Recent)

Order Perissodactyla—Odd-toed ungulates (hoofed placentals). Horse, rhinoceros, tapir, titanothere, chalicothere (Eocene-Recent)

Order Artiodactyla—Even-toed ungulates. Pig, hippo, camel, deer, elk, bison, cattle, sheep, antelope, entelodont, oredont (Eocene-Recent)

Order Proboscidea—Elephant, mammoth, mastodon (Eocene-Recent)

Order Sirenia—Sea cow, manatee, dugong (Eocene-Recent)

Order Embrithopoda—Known primarily from a single locality in Egypt. Large mammals with two gigantic bony processes arising from the nose area (Oligocene)

Order Desmostyla—Amphibious or seal-like in habit. Front and hind limbs well developed, but hands and feet somewhat specialized as paddles (Oligocene-Miocene)

Order Hyracoidea—Hyrax (Oligocene-Recent)

Order Pholidota—Scaly anteater (Oligocene-Recent)

Glossary

A

aa A lava flow with a surface of rough, jagged angular blocks and fragments.

abrasion The process whereby exposed rock is worn and scraped by the impact of solid particles.

Absaroka sequence A widespread sequence of Upper Mississippian to Lower Jurassic sedimentary rocks bounded above and below by unconformities; deposited during a transgressive-regressive cycle of the Absaroka Sea.

absolute dating The process of assigning an actual age to geologic events. Various radioactive decay dating techniques yield absolute ages. (*See relative dating*)

abyssal plain The flat surface of the sea floor, covering vast areas beyond the continental rises of passive continental margins.

Acadian orogeny A Devonian orogeny in the northern Appalachian mobile belt resulting from a collision of Baltica with Laurentia.

acanthodian Any of the fish first having a jaw or jawlike mechanism; a class of fishes (class Acanthodii) appearing during the Early Silurian and becoming extinct during the Permian.

acondrite A type of stony meteorite lacking condrules; composition similar to that of terrestrial basalt.

active continental margin A continental margin that develops at the leading edge of a continental plate where oceanic lithosphere is subducted.

adaptive radiation The adaptation of species of related ancestry to various aspects of the environment; the branching out of organisms of related ancestry into new habitats.

aftershock An earthquake caused by adjustments along a fault following a larger earthquake. Major earthquakes are usually followed by numerous aftershocks.

Alleghenian orogeny Pennsylvanian to Permian orogenic event during which the Appalachian mobile belt from New York to Alabama was deformed; occurred in the area of the present-day Appalachian Mountains.

allele Alternative form of a gene controlling the same trait.

allopatric speciation Model for the origin of a new species from a small population that has become geographically isolated from its parent population.

alluvial fan A cone-shaped deposit of alluvium; generally deposited where a stream flows from mountains onto an adjacent lowland.

alluvium A general term for detrital material deposited by streams.

alpha decay A type of radioactive decay involving the emission of a particle consisting of two protons and two neutrons from the nucleus of an atom; emission of an alpha particle decreases the atomic number by two and the atomic mass number by four.

altered remains Fossil remains that have been changed from their original composition or structure or both. (*See unaltered remains*)

amniote egg An egg in which the embryo develops in a liquid-filled cavity called the amnion. The embryo is also supplied with a yolk sac and a waste sac. The amniote egg is shelled in reptiles, birds, and egg-laying mammals and is retained but modified in all other mammals.

anaerobic A term referring to organisms that are not dependent on oxygen for respiration.

analogous organ Body part, such as wings of insects and birds, that serve the same function, but differ in structure and development. (*See homologous organ*)

Ancestral Rockies Late Paleozoic uplift in the southwestern part of the North American craton.

angiosperm Vascular plants having flowers and seeds; the flowering plants.

angular unconformity An unconformity below which older strata dip at a different angle (usually steeper) than the overlying younger strata. (*See disconformity and nonconformity*)

anticline An up-arched fold characterized by an axial plane that divides it in half.

Antler orogeny A Late Devonian to Mississippian orogeny that affected the Cordilleran mobile belt; deformation extended from Nevada to Alberta, Canada.

aphanitic A fine-grained texture in igneous rocks in which the individual mineral grains are too small to be seen without magnification. An aphanitic texture results from rapid cooling of magma.

Appalachian mobile belt A mobile belt along the eastern margin of the North American craton; extends from Newfoundland to Georgia; probably continuous to the southwest with the Ouachita mobile belt.

aquiclude Any material that prevents the movement of groundwater.

aquifer A permeable layer that allows the movement of groundwater.

archaeocyathid A benthonic sessile suspension feeder that lived during the Cambrian and constructed reeflike structures.

Archean Eon A part of Precambrian time beginning 3.8 billion years ago, corresponding to the age of the oldest known rocks on Earth, and ending 2.5 billion years ago. (*See Proterozoic Eon*)

arête A narrow, serrated ridge separating two glacial valleys or adjacent cirques.

artesian system A system in which groundwater is confined and builds up high hydrostatic (fluid) pressure.

artificial selection The practice of selective breeding of plants and animals for desirable traits.

artiodactyl Any member of the order Artiodactyla, the even-toed hoofed mammals. Living artiodactyls include swine, sheep, goats, camels, deer, bison, and musk oxen.

aseismic ridge A long, linear ridge or broad plateaulike feature rising as much as 2 to 3 km above the surrounding sea floor and lacking seismic activity.

ash Uncemented pyroclastic material measuring less than 2 mm that is erupted by a volcano.

assemblage range zone A type of biozone established by plotting the overlapping ranges of fossils that have different geologic ranges; the first and last occurrences of fossils are used to establish assemblage range zone boundaries.

assimilation A process in which a magma reacts with preexisting rock with which it comes in contact.

asthenosphere The part of the mantle that lies below the lithosphere; behaves plastically and flows.

atom The smallest unit of matter that retains the characteristics of an element.

atomic mass number The total number of protons and neutrons in the nucleus of an atom.

atomic number The number of protons in the nucleus of an atom.

aureole A zone surrounding an igneous intrusion in which contact metamorphism has taken place.

australopithecine A term referring to several extinct species of the genus *Australopithecus* that existed in South and East Africa during the Pliocene and Pleistocene epochs.

autotrophic Describes organisms that synthesize their organic nutrients from inorganic raw materials; photosynthesizing bacteria and plants are autotrophs. (*See heterotrophic*)

B

back-arc basin A basin formed on the continent side of a volcanic island arc; thought to form by back-arc spreading; the site of a marginal sea, e.g., the Sea of Japan.

backshore The area of a beach that is usually dry, being covered by water only by storm waves or exceptionally high tides.

bajada A broad alluvial apron formed at the base of a mountain range by coalescing alluvial fans.

Baltica One of six major Paleozoic continents; composed of Russia west of the Ural Mountains, Scandinavia, Poland, and northern Germany.

banded iron formation (BIF) Sedimentary rocks consisting of alternating thin layers of silica (chert) and iron minerals (mostly the iron oxides hematite and magnetite).

barchan dune A crescent-shaped dune whose tips point downwind; found in areas with generally flat dry surfaces with little vegetation, limited supply of sand, and nearly constant wind direction.

barchanoid dune A dune intermediate between transverse and barchan dunes; typically forms along the edges of a dune field.

barrier island An elongate sand body oriented parallel to a shoreline, but separated from the shoreline by a lagoon.

basal slip A type of glacial movement that occurs when a glacier slides over the underlying surface.

basalt plateau A large plateau built up by numerous lava flows from fissure eruptions.

base level The lowest limit to which a stream can erode.

basin The circular equivalent of a syncline. All of the strata in a basin dip toward a central point.

Basin and Range Province An area centered on Nevada but extending into adjacent states and northern Mexico; characterized by Cenozoic block-faulting.

batholith The largest of intrusive bodies, having at least 100 km² of surface area. Most batholiths are discordant and are composed chiefly of granitic rocks.

baymouth bar A spit that has grown until it completely cuts off a bay from the open sea.

beach A deposit of unconsolidated sediment extending landward from low tide to a change in topography or where permanent vegetation begins.

beach face The sloping area below the berm that is exposed to wave swash.

bedding The layering in sedimentary rocks. Layers less than 1 cm thick are laminae, whereas beds are thicker.

bed load The coarser part of a stream's sediment load; consists of sand and gravel.

benthos Any organism that lives on the bottom of seas or lakes; may live upon the bottom or within bottom sediments.

berm The backshore area of a beach consisting of a platform composed of sediment deposited by waves; berms are nearly horizontal or slope gently in a landward direction.

beta decay A type of radioactive decay during which a fast-moving electron is emitted from a neutron and thus is converted to a proton; results in an increase of one atomic number, but does not change atomic mass number.

Big Bang A model for the evolution of the universe in which a dense, hot state is followed by expansion, cooling, and a less dense state.

biostratigraphic unit A unit of sedimentary rock defined by its fossil content.

biozone The fundamental biostratigraphic unit (e.g., range zone and concurrent range zone).

bipedal Walking on two legs as a means of locomotion.

body fossil The actual remains of any prehistoric organism; includes shells, teeth, bones, and, rarely, the soft parts of organisms. (*See trace fossil*)

bonding The process whereby atoms are joined to other atoms.

bony fish A class of fishes (class Osteichthyes) that evolved during the Devonian; the most common fishes; characterized by an internal skeleton of bone; divided into two subgroups, the ray-finned fishes and lobe-finned fishes.

Bowen's reaction series A mechanism that accounts for the derivation of intermediate and felsic magmas from a mafic magma. It consists of a discontinuous branch of ferromagnesian minerals that change from one mineral to another over specific temperature ranges and a continuous branch of plagioclase feldspars whose composition changes as the temperature decreases.

brachiopod Any member of a group of bivalved, suspension-feeding, marine, invertebrate animals.

braided stream A stream possessing an intricate network of dividing and rejoining channels. Braiding occurs when sediment transported by the stream is deposited within channels as sand and gravel bars.

breaker A wave that oversteepens as it enters shallow water until the crest plunges forward.

browser An animal that eats tender shoots, twigs, and leaves. Compare with *grazer.*

butte An isolated, steep-sided, pinnacle-like erosional structure found in arid and semiarid regions; formed by the breaching of a resistant cap rock, which allows rapid erosion of the less resistant underlying rocks.

C

caldera A large, steep-sided, circular or oval volcanic depression usually formed by summit collapse resulting from the underlying magma chamber being partly drained.

Caledonian orogeny A Silurian-Devonian orogeny that occurred along the northwestern margin of Baltica resulting from the collision of Baltica with Laurentia.

Canadian shield The Precambrian shield of North America; exposed mostly in Canada, but outcrops occur in Minnesota, Wisconsin, Michigan, and New York.

capillary fringe The area extending irregularly upward a few centimeters to several meters from the water table.

captorhinomorph The oldest known reptiles; evolved during the Early Pennsylvanian; ancestors of all other reptiles, thus commonly called the stem reptiles.

carbon 14 dating technique An absolute dating method that relies upon determining the ratio of C^{14} to C^{12} in a sample; useful back to about 70,000 years ago; can be applied only to organic substances.

carbonaceous chondrite A type of stony meteorite; same as ordinary chondrites except they contain about 5% organic compounds including inorganically produced amino acids.

carbonate mineral A mineral that contains the negatively charged carbonate ion $(CO_3)^{-2}$ (e.g., calcite [$CaCO_3$] and dolomite [$CaMg(CO_3)_2$]).

carbonate rock A rock containing predominately carbonate minerals (e.g., limestone and dolostone).

carnivore-scavenger Any animal that depends on other animals, living or dead, as a source of nutrients.

cartilaginous fish Fishes such as living sharks, rays, and skates, and their extinct relatives that have a skeleton composed of cartilage.

cast A replica of an object such as a shell or bone formed when a mold of that object is filled by sediment or mineral matter. (*See mold*)

Catskill Delta The Devonian clastic wedge deposited adjacent to the highlands that formed during the Acadian orogeny.

cave A naturally formed subsurface opening that is generally connected to the surface and is large enough for a person to enter.

chemical sedimentary rock Rock formed of minerals derived from the ions taken into solution during weathering.

chemical weathering The process whereby rock materials are decomposed by chemical alteration of the parent material.

China One of six major Paleozoic continents; composed of all of southeast Asia, including China, Indochina, part of Thailand, and the Malay Peninsula.

chondrule A small, round mineral body formed by rapid cooling; found in chondritic meteorites.

chordate All members of the phylum Chordata; characterized by a notochord, a dorsal, hollow nerve cord, and gill slits at some time during the animal's life cycle.

chromosome Complex, double-stranded, helical molecule of deoxyribonucleic acid (DNA); specific segments of chromosomes are genes.

cinder cone A small steep-sided volcano that forms from the accumulation of pyroclastic material around a vent.

circum-Pacific belt A zone of seismic and volcanic activity that nearly encircles the margins of the Pacific Ocean basin; the majority of the world's earthquakes and volcanic eruptions occur within this belt.

cirque A steep-walled, bowl-shaped depression formed by erosion by a valley glacier.

clastic wedge An extensive accumulation of mostly clastic sediments eroded from and deposited adjacent to an uplifted area; clastic wedges are coarse-grained and thick near the uplift and become finer-grained and thinner away from the uplift, e.g., the Queenston Delta.

cleavage The breaking or splitting of mineral crystals along smooth planes of weakness. Cleavage is determined by the strength of the bonds within minerals.

column A cave deposit formed when stalagmites and stalactites join.

columnar joint A type of jointing that forms columns in some igneous rocks. The joints commonly form a polygonal (usually hexagonal) pattern.

complex movement A combination of different types of mass movements in which one type is not dominant; most complex movements involve sliding and flowing.

composite volcano A volcano composed of pyroclastic layers, lava flows typically of intermediate composition, and mudflows. Composite volcanoes, also called stratovolcanoes, are steep-sided near their summits (up to 30°), but decrease in slope toward their base where they are generally less than 5°.

compound A substance resulting from the bonding of two or more different elements.

compression Stress resulting when rocks are squeezed by external forces directed toward one another.

concordant Refers to plutons whose boundaries are parallel to the layering in the country rock.

cone of depression The lowering of the water table around a well in the shape of a cone; results when water is removed from an aquifer faster than it can be replenished.

contact metamorphism Metamorphism in which a magma body alters the surrounding country rock.

continental-continental plate boundary A type of convergent plate boundary along which two continental lithospheric plates collide (e.g., the collision of India with Asia).

continental crust The continental rocks overlying the upper mantle and consisting of a wide variety of igneous, sedimentary, and metamorphic rocks. It has an overall granitic composition and an overall density of about 2.70 g/cm³.

continental drift The theory that the continents were once joined into a single landmass that broke apart with the various fragments (continents) moving with respect to one another; proposed by Alfred Wegener in 1912.

continental glacier A large glacier covering a vast area (at least 50,000 km²) and unconfined by topography. Also called an ice sheet.

continental margin The area separating the part of a continent above sea level from the deep-sea floor.

continental rise The gently sloping area beyond the base of the continental slope.

continental shelf The area between the shoreline and continental slope where the sea floor slopes very gently in a seaward direction.

continental slope The relatively steep area between the shelf-slope break (at an average depth of 135 m) and the more gently sloping continental rise or oceanic trench.

convergent evolution The development of similarities in two or more distantly related organisms as a consequence of adapting to a similar lift-style, e.g., ichthyosaurs and porpoises. (*See parallel evolution*)

convergent plate boundary The boundary between two plates that are moving toward one another; three types of convergent plate boundaries are recognized. (*See continental-continental plate boundary, oceanic-continental plate boundary, and oceanic-oceanic plate boundary*)

coprolite Fossilized feces.

Cordilleran orogeny A protracted episode of deformation affecting the western margin of North America from Jurassic to Early Cenozoic time; typically divided into three separate phases called the Nevadan, Sevier, and Laramide orogenies.

Cordilleran mobile belt A mobile belt in western North America bounded on the west by the Pacific Ocean and on the east by the Great Plains; extends north-south from Alaska into central Mexico. (*See North American Cordillera*)

core The interior part of the Earth which begins at a depth of about 2,900 km; probably composed mostly of iron and nickel; divided into an outer liquid core and an inner solid core.

Coriolis effect The deflection of winds to the right of their direction of motion (clockwise) in the Northern Hemisphere and to the left of their direction of motion (counterclockwise) in the Southern Hemisphere due to the Earth's rotation.

correlation Demonstration of the physical continuity of rock units or biostratigraphic units in different areas, or demonstration of time equivalence as in time-stratigraphic correlation.

covalent bond A bond formed by the sharing of electrons between atoms.

crater A circular depression at the summit of a volcano resulting from the extrusion of gases, pyroclastic materials, and lava; connected by a conduit to a magma chamber below the Earth's surface.

craton The name applied to the relatively stable part of a continent; consists of a Precambrian shield and a platform, a buried extension of a shield; the ancient nucleus of a continent.

cratonic sequence A widespread sequence of sedimentary rocks bounded above and below by unconformities; deposited during a transgressive-regressive cycle of an epeiric sea, e.g., the Sauk sequence.

creep The imperceptible downslope movement of soil or rock; it is the slowest type of flow.

crest The highest part of a wave.

Cretaceous Interior Seaway An interior seaway that existed during the Late Cretaceous; formed when northward-transgressing waters from the Gulf of Mexico joined with southward-transgressing water from the Arctic;

effectively divided North America into two large landmasses.

Cro-Magnon A race of *Homo sapiens* that lived mostly in Europe from 35,000 to 10,000 years ago.

cross-bedding Sedimentary rocks containing beds that were deposited at an angle to the surface upon which they were accumulating, as in desert dunes, are cross-bedded.

crossopterygian A specific type of lobe-finned fish; possessed lungs; ancestral to amphibians.

crust The outermost layer of the Earth; the upper part of the lithosphere, which is separated from the mantle by the Moho; divided into continental and oceanic crust.

crystal settling The physical separation of minerals from a magma by crystallization and gravitational settling.

crystalline solid A solid in which the constituent atoms are arranged in a regular, three-dimensional framework.

Curie point The temperature at which iron-bearing minerals in a cooling magma attain their magnetism.

cyclothem A vertical sequence of cyclically repeated sedimentary rocks resulting from alternating periods of marine and nonmarine deposition; commonly contain a coal bed.

cynodont A type of therapsid (advanced mammal-like reptile); the ancestors of mammals.

D

daughter element An element formed by the radioactive decay of another element, e.g., argon 40 is the daughter element of potassium 40.

debris avalanche A complex movement that often occurs in very steep mountain ranges; typically starts out as a rockfall.

debris flow A type of mass movement containing larger-sized particles and less water than a mudflow.

deflation The removal of loose surface sediment by the wind.

deflation hollow A shallow depression of variable dimensions that results

from the differential erosion of surface materials by wind.

delta An alluvial deposit formed at the mouth of a stream.

depositional environment Any area in which sediment is deposited; a depositional site that differs in physical aspects, chemistry, and biology from adjacent environments.

desert Any area that receives less than 25 cm of rain per year. Typically, a desert has poorly developed soil and is mostly or completely devoid of vegetation.

desert pavement A surface mosaic of close-fitting pebbles, cobbles, and boulders found in many dry regions; formed by the removal of sand-sized and smaller particles by wind.

desertification The expansion of deserts into formerly productive lands.

detrital sedimentary rock Rock consisting of detritus, the solid particles of preexisting rocks.

differential pressure Pressure that is not applied equally to all sides of a rock body; results in distortion of the body.

dike A tabular or sheetlike discordant pluton.

dip A measure of the maximum angular deviation of an inclined plane from horizontal; measured perpendicular to the strike direction.

dip-slip fault A fault on which all movement is parallel with the dip of the fault plane. (*See normal fault and reverse fault*)

discharge The total volume of water in a stream moving past a particular point in a given period of time.

disconformity An unconformity above and below which the strata are parallel. (*See angular unconformity and nonconformity*)

discontinuity A marked change in the velocity of seismic waves indicating a significant change in Earth materials or their properties.

discordant Refers to plutons whose boundaries cut across the layering of country rock.

dissolved load That part of a stream's load that consists of ions taken into solution by chemical weathering.

divergent evolution The diversification of a species into two or more descendant species. (*See adaptive radiation*)

divergent plate boundary The boundary between two plates that are moving apart; new oceanic lithosphere forms at the boundary; characterized by volcanism and seismicity.

divide A topographically high area that separates adjacent drainage basins.

DNA (deoxyribonucleic acid) The substance of which chromosomes are composed; the genetic material of all organisms except bacteria.

dome A circular equivalent of an anticline. All strata in a dome dip away from a central point.

drainage basin The surface area drained by a stream and its tributaries.

drainage pattern The regional arrangement of channels in a drainage system.

dripstone Various cave deposits resulting from the deposition of calcite.

drumlin An elongated hill of till measuring as much as 50 m high and 1 km long; formed by the movement of a continental glacier.

dry climate A climate that occurs in the low and middle latitudes where the potential loss of water by evaporation exceeds the yearly precipitation; covers 30% of the Earth's land surface and is divided into semiarid and arid regions.

dune A mound or ridge of wind-deposited sand.

dynamic metamorphism Metamorphism associated with fault zones where rocks are subjected to high differential pressures.

E

earthflow A flow that moves from the upper part of a hillside, leaving a scarp, and flows slowly downslope as a thick, viscous, tongue-shaped mass of wet regolith.

earthquake The vibration of the Earth caused by the sudden release of energy, usually as a result of the displacement of rocks along faults.

Ediacaran faunas A collective name for all Late Proterozoic faunas containing animal fossils similar to those of the Ediacara fauna of Australia.

elastic rebound theory A theory that explains how energy is released during earthquakes. When rocks are deformed, they store energy and bend. When the inherent strength of the rocks is exceeded, they rupture, releasing the energy in the form of an earthquake.

elastic strain Strain in which the material returns to its original shape when stress is relaxed.

electron A negatively charged particle of very little mass that encircles the nucleus of an atom.

electron capture A type of radioactive decay in which an electron is captured by a proton, and converted to a neutron; results in a loss of one atomic number, but no change in atomic mass number.

electron shell Electrons orbit rapidly around the nuclei of atoms at specific distances known as electron shells. Each shell can only accommodate a certain number of electrons; when a shell is filled, electrons move to the next shell farther from the nucleus.

element A substance composed of all the same atoms; it cannot be changed into another element by ordinary chemical means.

emergent coast A coast where the land has risen with respect to sea level.

end moraine A pile of rubble deposited at the terminus of a glacier. (See *recessional moraine and terminal moraine*)

epeiric sea A broad shallow sea that covers part of a continent; six epeiric seas covered parts of North America during the Phanerozoic Eon, e.g., the Sauk Sea.

epicenter The point on the Earth's surface vertically above the focus of an earthquake.

erosion The removal of weathered material.

esker A long sinuous ridge of stratified drift formed by deposition by running water in tunnels beneath stagnant ice or in meltwater channels on the surface of a glacier.

eukaryotic cell A type of cell with a membrane-bounded nucleus containing chromosomes; also contains such organelles as plastids and mitochondria that are absent in prokaryotic cells. (See *prokaryotic cell*)

eupantothere Any member of a group of mammals that included the ancestors of both marsupial and placental mammals.

evaporite A sedimentary rock that forms by inorganic chemical precipitation of minerals from solution.

Exclusive Economic Zone An area extending 371 km seaward from the coast of the United States and its territories in which the United States claims all sovereign rights.

exfoliation dome A large rounded dome of rock resulting from the process of exfoliation.

F

Farallon plate A Late Mesozoic-Cenozoic oceanic plate that was largely subducted beneath North America; remnants of the Farallon plate are the Juan de Fuca and Cocos plates.

fault A fracture along which movement has occurred parallel to the fracture surface.

felsic magma A type of magma containing more than 65% silica and considerable sodium, potassium, and aluminum, but little calcium, iron, and magnesium. (See *mafic and intermediate magma*)

ferromagnesian silicate A silicate mineral containing iron and magnesium or both. Such minerals are commonly dark colored and denser than nonferromagnesian silicates.

fetch The distance the wind blows over a continuous water surface.

fiord A long, narrow glacial valley below sea level.

firn A granular type of snow formed by the melting and refreezing of snow.

firn limit The elevation to which snow recedes during a wastage season.

fission track dating The process of dating samples by counting the number of small linear tracks (fission tracks) that result when a mineral crystal is damaged by rapidly moving alpha particles generated by radioactive decay of uranium.

fissure eruption An eruption in which lava or pyroclastic material is emitted along a long, narrow fissure or group of fissures.

floodplain A low-lying, relatively flat area adjacent to a stream that is covered with water when the stream overflows its banks.

fluid acitivity An agent of metamorphism in which water and carbon dioxide promote metamorphism by increasing the rate of chemical reactions.

focus The place within the Earth where an earthquake originates and energy is released.

foliated texture A texture of metamorphic rocks in which the platy and elongate minerals are arranged in a parallel fashion.

footwall block The block that lies beneath a fault plane.

fore-arc basin A basin between a volcanic island arc and the subduction complex of an oceanic trench; typically contains a diverse assortment of generally flat-lying detrital sediments derived from the weathering and erosion of island arc volcanoes.

foreshore The area of a beach covered by water during high tide but exposed during low tide.

formation The basic lithostratigraphic unit; a unit of strata that is mappable and that has distinctive upper and lower boundaries.

fossil The remains or traces of prehistoric life preserved in rocks of the Earth's crust. (See *body fossil and trace fossil*)

fracture A break in a rock resulting from intense applied pressure.

Franklin mobile belt The most northerly mobile belt in North America; extends from northwestern Greenland westward across the Canadian Arctic islands.

frost action The disaggregation of rocks by repeated freezing and thawing of water in cracks and crevices in rocks.

G

gene The basic unit of inheritance; a specific segment of a chromosome. (*See allele*)

gene pool The total of all alleles of all genes available to an interbreeding population.

geologic time scale A vertical geologic chart arranged such that the designation for the earliest part of geologic time appears at the bottom, and progressively younger designations appear in their proper chronologic sequence.

geology The science concerned with the study of the Earth; includes studies of Earth materials (minerals and rocks), surface and internal processes, and Earth history.

geothermal energy Energy that comes from the steam and hot water trapped within the Earth's crust.

geothermal gradient A temperature increase with depth. It is about 25°C/km near the Earth's surface, but varies from area to area.

geyser A hot spring that intermittently ejects hot water and steam.

glacial budget The expansion and contraction of a glacier in response to accumulation and wastage.

glacial drift A collective term for all sediment deposited by glacial activity, including material deposited directly by glacial ice (till) and material deposited in streams derived from the melting of ice (outwash).

glacial erratic A boulder transported by a glacier from its original source.

glacial groove A deep straight scratch on a rock surface formed by the movement of sediment-laden glaciers over bedrock. Glacial grooves are much deeper than glacial striations.

glacial ice Ice that has formed from firn.

glacial polish A smooth glistening bedrock surface formed by the movement of a sediment-laden glacier over it.

glacial stage A time of extensive glaciation. At least four glacial stages are recognized in North America, and six or seven are recognized in Europe.

glacial striation A straight scratch on a rock surface caused by the movement of sediment-laden glaciers. Glacial striations are rarely more than a few millimeters deep.

glacier A mass of ice on land that moves by plastic flow and basal slip.

***Glossopteris* flora** A Late Paleozoic flora found only on the Southern Hemisphere continents and India; named after its best known genus, *Glossopteris.*

Gondwana One of six major Paleozoic continents; composed of the present-day continents of South America, Africa, Antarctica, Australia, and India, and parts of other continents such as southern Europe, Arabia, and Florida; began fragmenting during the Triassic Period.

graded bedding A type of sedimentary bedding in which an individual bed is characterized by a decrease in grain size from bottom to top.

graded stream A stream possessing an equilibrium profile in which a delicate balance exists between gradient, discharge, flow velocity, channel characteristics, and sediment load such that neither significant erosion nor deposition occurs within the channel.

gradient The slope over which a stream flows. Gradient generally varies from steep to gentle along the course of a stream, being steep in the upper reaches and gentle in the lower reaches.

granite-gneiss complex One of the two main types of rock bodies characteristic of areas underlain by Archean rocks.

grazer Any animal that crops low-growing vegetation, especially grasses.

greenstone belt A linear or podlike association of volcanic and sedimentary rocks particularly common in Archean terranes; typically synclinal and consists of lower and middle volcanic units and an upper sedimentary rock unit.

Grenville orogeny An area in the eastern United States and Canada that was accreted to Laurentia during the Late Proterozoic.

ground moraine The sediment liberated from melting ice as a glacier's terminus retreats.

groundwater The water stored in the open spaces within underground rocks and unconsolidated material.

guide fossil Any easily identifiable fossil that has a wide geographic distribution and a short geologic range; used to determine the geologic ages of strata and to correlate strata of the same age.

guyot A flat-topped seamount of volcanic origin rising more than 1 km above the sea floor.

gymnosperm The flowerless, seed-bearing land plants.

H

half-life The time required for one-half of the original number of atoms of a radioactive element to decay to a stable daughter product, e.g., the half-life of potassium 40 is 1.3 billion years.

hanging valley A tributary glacial valley whose floor is at a higher level than that of the main glacial valley.

hanging wall block The block that overlies a fault plane.

headland The seaward-projecting part of a shoreline that is eroded on both sides due to wave refraction.

heat An agent of metamorphism; heat comes from increasing depth in the crust, magma, and applied pressure.

herbivore An animal that is dependent on plants as a source of nutrients.

Hercynian orogeny Pennsylvanian to Permian orogeny in the Hercynian mobile belt of southern Europe.

heterotrophic Describes organisms such as animals that depend on preformed organic molecules from the environment as a source of nutrients. (*See autotrophic*)

hiatus The interval of geologic time not represented by strata in a sequence of rock layers containing an unconformity.

hominid Abbreviated form of Hominidae; the family to which humans belong. Such bipedal primates as *Australopithecus* and *Homo* are hominids.

Homo The genus to which humans belong; includes *Homo erectus* and *Homo sapiens.*

homologous organs Body parts in different organisms that have a similar structure, similar relationships to other organs, and similar development, but do not necessarily serve the same function, e.g., the wing of a bird and the forelimbs of whales and dogs. (*See analogous organ*)

horn A steep-walled, pyramidal peak formed by the headward erosion of cirques.

hot spot Localized zone of melting below the lithosphere; detected by volcanism at the surface.

hot spring A spring in which the water temperature is warmer than the temperature of the human body (37°C).

hydraulic action The power of moving water.

hydrologic cycle The continuous recycling of water from the oceans, through the atmosphere, to the continents, and back to the oceans.

hydrolysis The chemical reaction between the hydrogen (H^+) ions and hydroxyl (OH^-) ions of water and a mineral's ions.

hypothesis A provisional explanation for observations. Subject to continual testing and modification. If well supported by evidence, hypotheses are then generally called theories.

I

Iapetus Ocean A Paleozoic ocean basin that separated North America from Europe; the Iapetus Ocean began closing when North America and Europe began moving toward one another, and it was eliminated when the continents collided during the Late Paleozoic.

igneous rock Any rock formed by cooling and crystallization of magma, or by the accumulation and consolidation of volcanic ejecta such as ash.

incised meander A deep, meandering canyon cut into solid bedrock by a stream.

index mineral A mineral that forms only within specific temperature and pressure ranges. Index minerals allow geologists to recognize low-, intermediate-, and high-grade metamorphic zones.

infiltration capacity The maximum rate at which a soil or sediment can absorb water.

inselberg An isolated steep-sided erosional remnant that rises above a surrounding desert plain.

intensity The subjective measure of the kind of damage done by an earthquake as well as people's reaction to it.

interglacial stage A time between glacial stages when glaciers cover much less area and global temperatures are warmer than during a glacial stage.

intermediate magma A magma having a silica content between 53 and 65% and an overall composition intermediate between felsic and mafic magmas.

internal drainage A type of drainage found in semiarid and arid regions in which a stream drains into a central low area without exiting.

ion An electrically charged atom produced by adding or removing electrons from the outermost electron shell.

ionic bond A bond that results from the attraction of positively and negatively charged ions.

irons A group of meteorites composed primarily of iron and nickel alloys and accounting for about 6% of all meteorites.

isograd A line on a map connecting the first appearances of a particular index mineral and thus indicating equal metamorphic intensity.

isostasy *See principle of isostasy.*

isostatic rebound The phenomenon in which unloading of the Earth's crust causes it to rise upward until equilibrium is again attained.

isotope All atoms of a chemical element have the same number of protons in the nucleus, but may have variable numbers of neutrons; those with different numbers of neutrons are isotopes of that element, e.g., carbon 12 and carbon 14.

J

joint A fracture along which no movement has occurred, or where movement has been perpendicular to the fracture surface.

Jovian planet Any of the four planets (Jupiter, Saturn, Uranus, and Neptune) that resemble Jupiter. They are all large and have low mean densities, indicating that they are composed mostly of lightweight gases, such as hydrogen and helium, and frozen compounds, such as ammonia and methane.

K

karst topography A topography developed largely by groundwater erosion and characterized by numerous caves, springs, sinkholes, solution valleys, and disappearing streams.

Kaskaskia sequence A widespread sequence of Devonian and Mississippian sedimentary rocks bounded above and below by unconformities; deposited during a transgressive-regressive cycle of the Kaskaskia Sea.

Kazakhstania One of six major Paleozoic continents; a triangular-shaped continent centered on Kazakhstan (part of Asia).

key bed A rock unit that is sufficiently distinctive to allow identification of the same rock unit in different areas.

L

labyrinthodont Any of the amphibians, from the Devonian to the Triassic, characterized by labyrinthine wrinkling and folding of their teeth.

laccolith A concordant pluton with a mushroomlike geometry.

lahar A volcanic mudflow.

Laramide orogeny The Late Cretaceous to Early Cenozoic phase of the Cordilleran orogeny; responsible for many of the structural features of the present-day Rocky Mountains.

lateral moraine The sediment deposited as a long ridge of till along the margin of a glacier.

laterite A soil formed in the tropics where chemical weathering is intense and leaching of soluble minerals is complete.

Laurasia A Late Paleozoic, Northern Hemisphere continent composed of the present-day continents of North America, Greenland, Europe, and Asia.

Laurentia The name given to a Proterozoic continent that was composed mostly of North America and Greenland, parts of northwestern Scotland, and perhaps parts of the Baltic shield of Scandinavia.

lava Magma at the Earth's surface.

lava dome A bulbous, steep-sided structure formed by very viscous magma moving upward through a volcanic conduit.

lava flow A stream of magma flowing over the Earth's surface.

lithification The process by which sediment is transformed into sedimentary rock.

lithosphere The outer, rigid part of the Earth consisting of the upper mantle, oceanic crust, and continental crust; lies above the asthenosphere.

lithostatic pressure Pressure resulting from the weight of the overlying rock; it is applied equally in all directions.

lithostratigraphic unit A unit of sedimentary rock, such as a formation, defined by its lithologic characteristics rather than its biologic content or time of origin.

lobe-finned fish A type of fish in which the fin contains a series of articulating bones and is attached to the body by a fleshy shaft; one of the two major subgroups of bony fishes.

loess Windblown silt and clay deposits; derived from three main sources—deserts, Pleistocene glacial outwash deposits, and floodplains of streams in semiarid regions.

longitudinal dune A long, parallel ridge of sand aligned generally parallel to the direction of the prevailing wind; forms where the sand supply is somewhat limited.

longshore current A current between the breaker zone and the beach that flows parallel to the shoreline and is produced by wave refraction.

longshore drift The movement of sediment along a shoreline by longshore currents.

Love wave (L-wave) A surface wave in which the individual particles of the material only move back and forth in a horizontal plane perpendicular to the direction of wave travel.

low-velocity zone The zone within the mantle between the depths of 100 and 250 km where the velocity of both P- and S-waves decreases markedly; it corresponds closely to the asthenosphere.

M

mafic magma A silica-poor magma containing between 45 and 52% silica and proportionately more calcium, iron, and magnesium than an intermediate or felsic magma.

magma Molten rock material generated within the Earth.

magma mixing The process whereby magmas of different composition are mixed together producing a modified version of the parent magmas.

magnetic anomaly Any change, such as a change in average strength, of the Earth's magnetic field.

magnetic reversal The phenomenon in which the north and south magnetic poles are completely reversed.

magnitude The total amount of energy released by an earthquake at its source.

mantle The zone surrounding the core and comprising about 83% of the Earth's volume; it is less dense than the core and is thought to be composed largely of peridotite.

mantle plume A stationary column of magma that originates deep within the mantle and slowly rises to the Earth's surface to form volcanoes or flood basalts.

marine regression The withdrawal of the sea from a continent or coastal area resulting in the emergence of land as sea level falls or the land rises with respect to sea level.

marine terrace An old wave-cut platform now elevated above sea level.

marine transgression The invasion of coastal areas or much of a continent by the sea resulting from a rise in sea level or subsidence of the land.

marsupial mammal Any of the pouched mammals such as opossums, kangaroos, and wombats. At present, marsupials are common only in Australia.

mass wasting The downslope movement of material under the influence of gravity.

meandering stream A stream possessing a single, sinuous channel with broadly looping curves.

mechanical weathering The breaking of rock materials by physical forces into smaller pieces that retain the chemical composition of the parent material.

medial moraine A moraine formed where two lateral moraines merge.

Mediterranean belt A zone of seismic and volcanic activity that extends westerly from Indonesia through the Himalayas, across Iran and Turkey, and through the Mediterranean region of Europe; about 20% of all active volcanoes and 15% of all earthquakes occur in this belt.

mesa A broad, flat-topped erosional remnant bounded on all sides by steep slopes; forms when the resistant cap rock is breached, allowing rapid erosion of the less resistant underlying sedimentary rock.

metamorphic facies A group of metamorphic rocks characterized by particular mineral assemblages formed under the same broad temperature-pressure conditions.

metamorphic grade The rocks within a metamorphic zone, all of which are the same grade, i.e., low, medium, or high grade.

metamorphic rock Any rock type altered by high temperature and pressure and the chemical activities of fluids is said to have been metamorphosed, e.g., slate, gneiss, marble.

metamorphic zone The region between isograds.

meteorite A mass of matter of extraterrestrial origin that has fallen to the Earth.

Midcontinent rift A Late Proterozoic intracontinental rift within Laurentia; contains thick accumulations of extrusive igneous rocks and detrital sedimentary rocks.

Milankovitch theory A theory that explains cyclic variations in climate as a consequence of irregularities in the Earth's rotation and orbit.

mineral A naturally occurring, inorganic, crystalline solid having characteristic physical properties and a narrowly defined chemical composition.

mobile belt Elongated area of deformation as indicated by folds and faults; generally located adjacent to a craton, e.g., the Appalachian mobile belt.

modern synthesis A synthesis of the ideas of geneticists, paleontologists, population biologists, and others to yield a neo-Darwinian view of evolution; includes chromosome theory of inheritance, mutation as a source of variation, and gradualism.

Modified Mercalli Intensity Scale A scale having values ranging from I to XII that is used to measure earthquake intensity based on damage.

Mohorovičić discontinuity The boundary between the crust and the mantle. Also called the Moho.

mold A cavity or impression of an organism or part thereof in sediment or sedimentary rock, e.g., a mold of a clam shell. (*See cast*)

monocline A simple bend or flexure in otherwise horizontal or uniformly dipping rock layers.

monomer A comparatively simple organic molecule, such as an amino acid, that is capable of linking with other monomers to form polymers. (*See polymer*)

monotreme The egg-laying mammals; only two types of monotremes now exist, the platypus and spiny anteater of the Australian region.

mud crack A sedimentary structure found in clay-rich sediment that has dried out. When such sediment dries, it shrinks and forms intersecting fractures.

mudflow A flow consisting of mostly clay- and silt-sized particles and more than 30% water; most common in semiarid and arid environments.

multicellular organism Any organism consisting of many cells as opposed to a single cell; possesses cells specialized to perform specific functions such as reproduction and respiration.

mutation Any change in the genetic determinants or genes of organisms; some of the inheritable variation in populations upon which natural selection acts arises from mutations in sex cells.

N

native element A mineral composed of a single element.

natural levee A ridge of sandy alluvium deposited along the margins of a stream channel during floods.

natural selection A mechanism proposed by Charles Darwin and Alfred Russell Wallace to account for evolution; as a result of natural selection, organisms best adapted to their environment are more likely to survive and reproduce.

Neanderthal A type of human that inhabited Europe and the Near East from 150,000 to 32,000 years ago; considered by some to be a variety or subspecies (*Homo sapiens neanderthalensis*) of *Homo sapiens* and by some as a separate species (*Homo neanderthalensis*).

nekton Actively swimming organisms, e.g., fishes, whales, and squids. (*See plankton*)

neutron An electrically neutral particle found in the nucleus of an atom.

Nevadan orogeny Late Jurassic to Cretaceous phase of the Cordilleran orogeny; most strongly affected the western part of the Cordilleran mobile belt.

nonconformity An unconformity in which stratified sedimentary rocks above an erosion surface overlie igneous or metamorphic rocks. (*See angular unconformity and disconformity*)

nonferromagnesian silicate A silicate mineral that does not contain iron or magnesium. Nonferromagnesian silicate minerals are generally light colored and less dense than ferromagnesian silicate minerals.

nonfoliated texture A metamorphic texture in which there is no discernible preferred orientation of mineral grains.

nonvascular Plants lacking specialized tissues for conducting fluids.

normal fault A dip-slip fault resulting from tensional forces in which the hanging wall block has moved downward relative to the footwall block. (*See reverse fault*)

North American Cordillera A complex mountainous region in western North America extending from Alaska into central Mexico. (*See Cordilleran mobile belt*)

nucleus The central part of an atom consisting of one or more protons and neutrons.

nuée ardente A mobile dense cloud of hot pyroclastic materials and gases ejected more or less horizontally from a volcano. Because a nuée ardente is denser than air, it rushes down the slope of a volcano engulfing everything in its path.

O

oblique-slip fault A fault having both dip-slip and strike-slip movement.

oceanic-continental plate boundary A type of convergent plate boundary along which oceanic lithosphere and continental lithosphere collide; characterized by subduction of an oceanic plate beneath a continental plate and by volcanism and seismicity.

oceanic crust The crust underlying the ocean basins. It ranges from 5 to 10 km thick, is composed of gabbro and basalt, and has an average density of 3 g/cm^3.

oceanic-oceanic plate boundary A type of convergent plate boundary along which two oceanic lithospheric plates collide and one is subducted beneath the other.

oceanic ridge A submarine mountain system found in all of the oceans; it is composed of volcanic rock (mostly basalt) and displays features produced by tensional forces.

oceanic trench A long, narrow feature restricted to active continental margins and along which subduction occurs.

ooze Deep-sea pelagic sediment composed mostly of shells of marine animals and plants.

ophiolite A sequence of igneous rocks thought to represent a fragment of oceanic lithosphere; composed of peridotite overlain successively by gabbro, sheeted basalt dikes, and pillow basalts.

ordinary chondrite The most abundant type of stony meteorite; composed of high-temperature ferromagnesian minerals such as olivine and some pyroxenes.

organic evolution *See theory of evolution.*

organic reef A wave-resistant limestone structure with a structural framework of animal skeletons, e.g., stromatoporoid reef or coral reef.

ornithischian Any dinosaur belonging to the order Onithischia; characterized by a birdlike pelvis. (*See saurischian*)

orogen A linear part of the Earth's crust that was deformed during an orogeny. (*See orogeny*)

orogeny The process of forming mountains, especially by folding and thrust faulting; an episode of mountain building.

ostracoderm The "bony-skinned" fish; first appeared during the Late Cambrian and thus are the oldest known vertebrates; characterized by a lack of jaws and teeth and presence of bony armor.

Ouachita mobile belt A mobile belt located along the southern margin of the North American craton; probably continuous with the Appalachian mobile belt.

Ouachita orogeny An orogeny that deformed the Ouachita mobile belt during the Pennsylvanian Period.

outgassing The process whereby gases derived from the Earth's interior are released into the atmosphere by volcanic activity.

outwash plain The sediment deposited by the meltwater discharging from the terminus of a continental glacier.

oxbow lake A cutoff meander filled with water. Oxbow lakes form when meanders become so sinuous that the thin neck of land separating adjacent meanders is cut off during a flood, leaving a cutoff meander.

oxidation The reaction of oxygen with other atoms to form oxides or, if water is present, hydroxides.

P

pahoehoe A type of lava flow with a smooth, ropy surface.

paleomagnetism The remanent magnetism in ancient rocks that records the direction and strength of the Earth's magnetic field at the time of their formation.

Pangaea The name proposed by Alfred Wegener for a supercontinent that existed at the end of the Paleozoic Era and consisted of all the Earth's landmasses.

Panthalassa Ocean The Late Paleozoic worldwide ocean that surrounded the supercontinent Pangaea.

parabolic dune A crescent-shaped dune in which the tips point upwind; forms where the vegetation cover is broken and deflation produces a blowout.

parallel evolution The development of similarities in two or more closely related but separate lines of descent as a consequence of similar adaptations. (*See convergent evolution*)

parent element An unstable element that by radioactive decay is changed into a stable daughter element. (*See daughter element*)

parent material The material that is being chemically and mechanically weathered to yield sediment and soil.

passive continental margin The trailing edge of a continental plate consisting of a broad continental shelf and a continental slope and rise. A vast, flat abyssal plain is commonly present adjacent to the rise. Passive continental margins lack intense seismic and volcanic activity.

pedalfer A soil that develops in humid regions and has an organic-rich A horizon and aluminum-rich clays and iron oxides in horizon B.

pediment An erosional bedrock surface of low relief gently sloping away from a mountain base; most pediments are covered by a thin layer of debris or by alluvial fans or bajadas.

pedocal A soil characteristic of arid and semiarid regions with a thin A horizon and a calcium carbonate-rich B horizon.

pegmatite A very coarse-grained igneous rock usually of granitic composition commonly associated with plutons.

pelagic clay Generally brown or reddish deep-sea sediment composed of clay-sized particles derived from the continents and oceanic islands.

pelycosaur Pennsylvanian to Permian "finback reptiles"; possessed some mammalian characteristics.

perched water table A water table that forms where a local aquiclude occurs within a larger aquifer; water migrating through the zone of aeration is stopped by the local aquiclude, and a localized zone of saturation "perched" above the main water table is created.

period The fundamental unit in the hierarchy of time units; a part of geologic time during which a particular sequence of rocks designated as a system was deposited.

perissodactyl Any member of the order Perissodactyla, the odd-toed hoofed mammals; living perissodactyls include horses, rhinoceroses, and tapirs. (*See artiodactyl*)

permafrost Ground that remains permanently frozen; covers nearly 20% of the world's land surface.

permeability A material's capacity for transmitting fluids.

phaneritic A coarse-grained texture in igneous rocks in which the mineral grains are easily visible without magnification. A phaneritic texture results from the slow cooling of a magma.

phenocryst The larger mineral grains in a porphyritic texture.

photochemical dissociation A process whereby water molecules in the upper atmosphere are disrupted by ultraviolet radiation, yielding oxygen (O_2) and hydrogen (H).

photosynthesis The metabolic process of synthesizing organic molecules from water and carbon dioxide, using the radiant energy of sunlight captured by chlorophyll-containing cells.

phyletic gradualism An evolutionary concept holding that a species evolves gradually and continuously through time to give rise to new species. (*See punctuated equilibrium*)

pillow lava Bulbous masses of basalt resembling pillows. It forms when lava is rapidly chilled beneath water and is characteristic of much of the igneous rock in the upper part of the oceanic crust.

placental mammal Any of the mammals that have a placenta to nourish the embryo; fusion of the amnion of the amniote egg with the walls of the uterus forms the placenta; most mammals, living and fossil, are placentals.

placoderm The "plate-skinned" fishes; Late Silurian through Permian; characterized by jaws and bony armor especially in the head-shoulder region.

planetesimal An asteroid-sized body that along with other planetesimals aggregated to form protoplanets.

plankton Animals and plants that float passively, e.g., phytoplankton and zooplankton. (*See nekton*)

plastic flow The flow that occurs in response to pressure and causes permanent deformation.

plastic strain The result of stress in which a material cannot recover its original shape and retains the configuration produced by the stress such as by folding of rocks.

plate An individual piece of lithosphere that moves over the asthenosphere.

plate tectonic theory The theory that large segments of the outer part of the Earth (lithospheric plates) move relative to one another; lithospheric plates are rigid and move over the asthenosphere, which behaves much like a very viscous fluid.

playa A dry lake bed found in deserts and characterized by mudcracks and chemically precipitated rocks such as

rock gypsum; formed by the evaporation of water in a playa lake.

playa lake Broad, shallow, temporary lake that forms in an arid region and quickly evaporates to dryness.

plunging fold A fold with an inclined axis.

pluton An intrusive igneous body that forms when magma cools and crystallizes within the Earth's crust.

plutonic (intrusive igneous) rock Igneous rock that crystallizes from magma intruded into or formed in place within the Earth's crust.

point bar The sediment body deposited on the gently sloping side of a meander loop.

polymer A comparatively complex organic molecule, such as a nucleic acid and protein, formed by monomers linking together. (*See monomer*)

porosity The percentage of a material's total volume that is pore space.

porphyritic An igneous texture with mineral grains of markedly different sizes that results from a two-stage cooling history. The larger grains are phenocrysts, and the smaller ones are referred to as groundmass.

Precambrian A widely used informal term referring to all rocks stratigraphically beneath rocks of the Cambrian System and to all geologic time preceding the Cambrian Period.

Precambrian shield A vast area of exposed Precambrian rocks on a continent; areas of relative stability for long periods of time, e.g., the Canadian shield.

pressure release A mechanical weathering process in which rocks that formed deep within the Earth, due to a release of pressure, expand upon being exposed at the surface.

pressure ridge A buckled area on the surface of a lava flow that forms because of pressure on the partly solid crust of a moving flow.

primary producer Those organisms in a food chain, such as green plants and bacteria, upon which all other members of the food chain depend directly or indirectly; those organisms not dependent on an external source of nutrients. (*See autotrophic*)

Primates The order of mammals that includes prosimians (lemurs and tarsiers), monkeys, apes, and humans; characteristics include large brain, stereoscopic vision, and grasping hand.

principle of cross-cutting relationships A principle used to determine the relative ages of events; holds that an igneous intrusion or fault must be younger than the rocks that it intrudes or cuts.

principle of fossil succession The principle based on the work of William Smith that holds that fossils, and especially assemblages of fossils, succeed one another through time in a regular and determinable order.

principle of inclusions A principle that holds that inclusions, or fragments, in a rock unit are older than the rock unit itself, e.g., granite fragments in a sandstone are older than the sandstone rock unit.

principle of isostasy The theoretical concept of the Earth's crust "floating" on a dense underlying layer.

principle of lateral continuity A principle that holds that sediment layers extend outward in all directions until they terminate.

principle of original horizontality A principle developed by Nicolas Steno that holds that sediment layers are deposited horizontally or very nearly so.

principle of superposition A principle that holds that younger rocks are deposited on top of older layers.

principle of uniformitarianism A principle that holds that we can interpret past events by understanding present-day processes; based on the assumption that natural laws have not changed through time.

prokaryotic cell A type of cell having no nucleus and lacking such organelles as plastids and mitochondria; cells of bacteria and cyanobacteria (blue-green algae) are prokaryotic. (*See eukaryotic cell*)

Proterozoic Eon That part of Precambrian time beginning 2.5 billion years ago and ending 570 million years ago. (*See Archean Eon*)

proton A positively charged particle found in the nucleus of an atom.

punctuated equilibrium An evolutionary concept that holds that a new species evolves rapidly, perhaps in a few thousands of years, then remains much the same during its several millions of years of existence.

P-wave A compressional, or push-pull wave; the fastest seismic wave and one that can travel through solids, liquids, and gases; also known as a primary wave.

P-wave shadow zone The area between 103° and 143° from an earthquake focus where little P-wave energy is recorded by seismographs. The P-wave shadow zone results from the fact that the Earth has a solid inner core.

pyroclastic (fragmental) texture A fragmental texture characteristic of igneous rocks formed by explosive volcanic activity.

pyroclastic material Fragmental material such as ash explosively ejected from a volcano.

Q

quadrupedal Referring to locomotion on all four legs. (*See bipedal*)

quartzite-carbonate-shale assemblage A suite of sedimentary rocks characteristic of passive continental margins, but also known from intracratonic basins and back-arc basins.

Quaternary Period A term for a geologic period or system comprising all geologic time or rocks from the end of the Tertiary to the present; consists of two epochs or series, the Pleistocene and the Recent (Holocene).

Queenston Delta The clastic wedge resulting from the erosion of the highlands formed during the Taconic orogeny; deposited on the west side of the Taconic Highlands.

quick clay A clay that spontaneously liquefies and flows like water when disturbed.

R

radioactive decay The spontaneous decay of an atom to an atom of a different element by emission of a particle from its nucleus (alpha and beta decay) or by electron capture.

rainshadow desert A desert found on the lee side of a mountain range; forms because moist marine air moving inland forms clouds and produces precipitation on the windward side of the mountain range such that the air descending on the leeward side is much warmer and drier.

rapid mass movement A type of mass movement involving a visible movement of material; usually occurs quite suddenly and the material moves very quickly downslope.

ray-finned fish A subclass (Actinopterygii) of the bony fish (class Osteichthyes). A term describing the way the fins are supported by thin bones that project from the body.

Rayleigh wave A surface wave in which the individual particles of material move in an elliptical path within a vertical plane oriented in the direction of wave movement.

recessional moraine A moraine formed by a retreating glacier; it marks the location where the terminus of a glacier has stabilized and till was deposited. (*See end moraine and terminal moraine*)

recharge The addition of water to the zone of saturation.

red beds Sedimentary rocks, mostly sandstone and shale, with red coloration due to the presence of ferric oxides.

reef A moundlike, wave-resistant structure composed of the skeletons of organisms.

reflection The return of some of a seismic wave's energy when it encounters a boundary separating materials of different density or elasticity within the Earth.

refraction The change in direction and velocity of a seismic wave when it travels from one material into another of different density and elasticity.

refractory element Any element, such as iron, magnesium, silicon, or aluminum, that condenses easily at high temperature.

regional metamorphism Metamorphism that occurs over a large area and is usually the result of tremendous temperatures, pressures, and deformation within the deeper parts of the Earth's crust.

regolith The layer of unconsolidated rock and mineral fragments that covers almost all the Earth's surface.

relative dating The process of determining the age of an event relative to other events; involves placing geologic events in their correct chronologic order, but involves no consideration of when the events occurred in terms of numbers of years ago. (*See absolute dating*)

reserve That part of the resource base that can be extracted economically.

resource A concentration of naturally occurring solid, liquid, or gaseous material in or on the Earth's crust in such form and amount that economic extraction of a commodity from the concentration is currently or potentially feasible.

reverse fault A dip-slip fault resulting from compressional forces in which the hanging wall block has moved upward relative to the footwall block. (*See normal fault*)

Richter Magnitude Scale An open-ended scale that measures the amount of energy released during an earthquake.

rip current A narrow surface current that flows out to sea through the breaker zone.

ripple mark Wavelike (undulating) structure produced in granular sediment such as sand; formed by wind, unidirectional water currents, or wave currents.

rock A consolidated aggregate of minerals or particles of other rocks; although they are exceptions to this definition, coal and natural glass are also considered rocks.

rock cycle A sequence of processes through which Earth materials may pass as they are transformed from one rock type to another.

rockfall A common type of extremely rapid mass movement in which rocks of any size fall through the air.

rock-forming mineral A common mineral that comprises a significant portion of a rock.

rock glide A type of rapid mass movement in which rocks move downslope along a more or less planar surface.

rock varnish A thin, red, brown, or black shiny coating on the surface of many desert rocks; composed of iron and manganese oxides.

rounding The process by which the sharp corners and edges of sedimentary particles are abraded during transport and become rounded.

runoff The surface flow of streams.

S

saltwater incursion The displacement of fresh water by salt water as a result of excessive pumping of groundwater in coastal areas.

Sauk sequence A widespread sequence of sedimentary rocks bounded above and below by unconformities; deposited during a latest Proterozoic to Early Ordovician transgressive-regressive cycle of the Sauk Sea.

saurischian Any dinosaur belonging to the order Saurischia; characterized by a lizardlike pelvis. (*See ornithischian*)

scientific method A logical, orderly approach that involves data gathering, formulating and testing of hypotheses, and proposing theories.

sea-floor spreading The theory that the sea floor moves away from spreading ridges and is eventually subducted and consumed at convergent plate margins.

seamount A structure of volcanic origin rising more than 1 km above the sea floor.

sediment Weathered material derived from preexisting rocks.

sediment-deposit feeder Any animal that ingests sediment and extracts the nutrients from it.

sedimentary facies Any aspect of a sedimentary rock unit that makes it recognizably different from adjacent sedimentary rocks of the same, or approximately same, age, e.g., a sandstone facies.

sedimentary rock Any rock composed of sediment. The sediment may be particles of various sizes such as gravel or sand, the remains of animals or plants as in coal and some limestones, or derived from chemicals in solution by organic or inorganic processes.

sedimentary structure Any structure in sedimentary rock such as crossbedding, mud cracks, and animal burrows.

seedless vascular plant A type of land plant with vascular tissues for transport of fluids and nutrients throughout the plant; reproduces by spores rather than seeds, e.g., ferns and horsetail rushes.

seismogram The record of earthquake waves made by a seismograph.

seismograph An instrument that detects, records, and measures the various vibrations produced by an earthquake.

seismology The study of earthquakes.

sequence stratigraphy The study of rock relationships within a time-stratigraphic framework of related facies bounded by widespread unconformities.

Sevier orogeny The Cretaceous phase of the Cordilleran orogeny that affected the continental shelf and slope areas of the Cordilleran mobile belt.

shear strength The resisting forces helping to maintain slope stability; includes the slope material's strength and cohesion, the amount of internal friction between grains, and any external support of the slope.

shear stress The result of forces acting parallel to one another but in opposite directions; results in deformation by displacement of adjacent layers along closely spaced planes.

shield volcano The largest type of volcano; has a low rounded profile and is composed mostly of basalt flows.

shoreline The line of intersection between the sea or a lake and the land.

Siberia One of six major Paleozoic continents; composed of Russia east of the Ural Mountains, and Asia north of Kazakhstan and south of Mongolia.

silica A compound of silicon and oxygen atoms.

silica tetrahedron The basic building block of all silicate minerals. It consists of one silicon atom and four oxygen atoms.

silicate A mineral containing silica.

sill A tabular or sheetlike concordant pluton.

sinkhole A depression in the ground that forms in karst regions by the solution of the underlying carbonate rocks or the collapse of a cave roof.

slide A type of mass movement involving movement of material along one or more surfaces of failure.

slow mass movement Mass movement that advances at an imperceptible rate and is usually only detectable by the effects of its movement.

slump The downslope movement of material along a curved surface of rupture; characterized by the backward rotation of the slump block.

soil Regolith consisting of weathered material, water, air, and organic matter that can support plants.

soil horizon A distinct soil layer that differs from other soil layers in texture, structure, composition, and color.

solifluction The slow downslope movement of water-saturated surface sediment; most common in areas of permafrost.

solar nebular theory A theory for the evolution of the solar system from a rotating cloud of gas.

solution A reaction in which the ions of a substance become dissociated from one another in a liquid, and the solid substance dissolves.

Sonoma orogeny A Permian-Triassic orogeny caused by the collision of an island arc with the southwestern margin of North America.

sorting A term referring to the degree to which all particles are about the same size in sediment or sedimentary rock, e.g., well-sorted, poorly sorted.

spatter cone A small, steep-sided cone that forms when gases escaping from a lava flow hurl globs of molten lava into the air that fall back to the surface and adhere to one another.

species A population of similar individuals that in nature can reproduce and produce fertile offspring.

spheroidal weathering A manifestation of chemical weathering in which rock, even if rectangular to begin with, weathers to form a spheroidal shape.

spit A continuation of a beach forming a point that projects into a body of water, commonly a bay.

spreading ridge A location where plates are separating and new oceanic lithosphere is forming.

spring A place where groundwater flows or seeps out of the ground.

stalactite An icicle-shaped carbonate structure hanging from a cave ceiling; forms as a result of precipitation from carbonate-saturated dripping water.

stalagmite A carbonate projection that rises from a cave floor; forms from carbonate-saturated water dripping from a cave ceiling.

stock A discordant pluton with a surface area less than 100 km². Many stocks are simply the exposed parts of much larger plutons.

stones A group of meteorites composed of iron and magnesium silicate minerals and comprising about 93% of all meteorites.

stony-irons A group of meteorites composed of nearly equal amounts of iron and nickel and silicate minerals and comprising about 1% of all meteorites.

stoping A process in which rising magma detaches and engulfs pieces of the surrounding country rock.

strain Deformation caused by stress.

stratified drift Drift displaying both sorting and stratification.

stream Runoff that is confined to channels regardless of size.

stream terrace An erosional remnant of a floodplain that formed when a stream was flowing at a higher level.

stress The force per unit area applied to a material such as rock within the Earth's crust.

strike The direction of a line formed by the intersection of a horizontal plane with an inclined plane, such as a rock layer.

strike-slip fault A fault involving horizontal movement in which blocks on opposite sides of a fault plane slide sideways past one another. (*See dip-slip fault*)

stromatolite A structure in sedimentary rocks, especially limestones, produced by entrapment of sediment grains on sticky mats of photosynthesizing bacteria; a biogenic sedimentary structure.

subduction zone A long, narrow zone at a convergent plate boundary where an oceanic plate descends relative to another plate, e.g., the subduction of the Nazca plate beneath the South American plate.

submarine canyon A steep-sided canyon cut into the continental shelf and slope.

submarine fan A cone-shaped sedimentary deposit that accumulates on the continental slope and rise.

submergent coast A coast along which sea level rises with respect to the land or the land subsides.

Sundance Sea A wide seaway that existed in western North America during the Middle Jurassic Period.

superposed stream A stream that once flowed on a higher surface and eroded downward into resistant rocks, while still maintaining its course.

suspended load The smallest particles carried by a stream, such as silt and clay, which are kept suspended by fluid turbulence.

suspension feeder An animal that consumes microscopic plants, animals, or dissolved nutrients from water.

S-wave A shear wave that moves material perpendicular to the direction of travel, thereby producing shear stresses in the material it moves through; also known as a secondary wave, an S-wave only travels through solids.

S-wave shadow zone Those areas more than 103° from an earthquake focus where no S-waves are recorded. The S-wave shadow zone indicates that the outer core must be liquid because S-waves cannot travel through liquid.

syncline A down-arched fold characterized by an axial plane that divides it in half.

system The fundamental unit in the time-stratigraphic hierarchy of units; the Devonian System refers to rocks deposited during a specific interval of geologic time, the Devonian Period.

T

Taconic orogeny An Ordovician orogeny that resulted in deformation of the Appalachian mobile belt.

talus Weathered material that accumulates at the bases of slopes.

Tejas epeiric sea A Cenozoic epeiric sea that was largely restricted to the Atlantic and Gulf Coastal plains and parts of coastal California, but did extend into the continental interior in the Mississippi Valley.

tension Forces acting in opposite directions along the same line.

terminal moraine The outermost end moraine, marking the greatest extent of a glacier. (*See end moraine and recessional moraine*)

terrestrial planet Any of the four innermost planets (Mercury, Venus, Earth, and Mars). They are all small and have high mean densities, indicating that they are composed of rock and metallic elements.

Tertiary Period A term for a geologic period or system comprising all geologic time or rocks from the end of the Cretaceous to the beginning of the Quaternary. The Tertiary consists of five epochs or series: Paleocene, Eocene, Oligocene, Miocene, and Pliocene.

thecodontian An informal term for a variety of Permian and Triassic reptiles that had teeth set in individual sockets. Small, bipedal thecodontians are the probable ancestors of dinosaurs.

theory An explanation for some natural phenomenon that has a large body of supporting evidence; to be considered scientific, a theory must be testable (e.g., plate tectonic theory).

theory of evolution The theory that all living things are related and that they descended with modification from organisms that lived during the past.

therapsid Permian to Triassic reptiles that possessed mammalian characteristics and thus are called mammal-like reptiles; one group of therapsids, the cynodonts, gave rise to mammals.

thermal convection cell In plate tectonics, a type of circulation of material in the asthenosphere during which hot material rises, moves laterally, cools and sinks, and is reheated and reenters the cycle.

thermal expansion and contraction A type of mechanical weathering in which the volume of rocks changes in response to heating and cooling.

thrust fault A type of reverse fault in which the fault plane dips less than 45°.

tide The regular fluctuation in the sea's surface in response to the gravitational attraction of the Moon and Sun.

till All sediment deposited directly by glacial ice.

time-distance graph A graph showing the average travel times for P- and S-waves for any specific distance from an earthquake's focus.

time-stratigraphic unit A unit of strata that was deposited during a specific interval of geologic time, e.g., the Devonian System, a time-stratigraphic unit, was deposited during that part of geologic time designated as the Devonian Period.

time unit Any of the units such as eon, era, period, epoch, and age used to refer to specific intervals of geologic time.

Tippecanoe sequence A widespread sequence of sedimentary rocks bounded above and below by unconformities; deposited during a Middle Ordovician to Early Devonian transgressive-regressive cycle of the Tippecanoe Sea.

tombolo A type of spit that extends out into the sea and connects an island to the mainland.

trace fossil Any indication of prehistoric organic activity such as tracks, trails, burrows, borings, or nests. (*See body fossil*)

Transcontinental Arch An area consisting of several large islands extending from New Mexico to Minnesota that was above sea level during the Cambrian transgression of the Sauk Sea.

transform fault A type of fault that changes one type of motion between plates into another type of motion.

transform plate boundary Plate boundary along which plates slide past one another, and crust is neither produced nor destroyed; a type of strike-slip fault along which oceanic ridges are offset.

transport The mechanism by which weathered material is moved from one place to another, commonly by running water, wind, or glaciers.

transverse dune A long ridge of sand perpendicular to the prevailing wind direction; forms in areas where abundant sand is available and little or no vegetation exists.

tree-ring dating The process of determining the age of a tree or wood in structures by counting the number of annual growth rings.

trilobite A group of benthonic, detritus-feeding, extinct marine invertebrate animals (phylum Arthropoda), having skeletons of an organic compound called chitin.

trophic level The complex interrelationships among producers, consumers, and decomposers in a community of organisms. There are several trophic levels of production and consumption through which energy flows in a feeding hierarchy.

trough The lowest point between wave crests.

tsunami A destructive sea wave that is usually produced by an earthquake but can also be caused by submarine landslides or volcanic eruptions.

turbidity current A sediment-water mixture denser than normal seawater that flows downslope to the deep-sea floor.

U

unaltered remains Fossil remains that retain their original composition and structure. (*See altered remains*)

unconformity An erosion surface that separates younger strata from older rocks. (*See angular unconformity, disconformity, and nonconformity*)

ungulate An informal term referring to the hoofed mammals, especially the orders Artiodactyla and Perissodactyla.

U-shaped glacial trough A valley with very steep or vertical walls and a broad, rather flat floor. Formed by the movement of a glacier through a stream valley.

V

valley glacier A glacier confined to a mountain valley or perhaps to an interconnected system of mountain valleys.

valley train A long, narrow deposit of stratified drift confined within a glacial valley.

vascular A term referring to land plants possessing specialized tissues for transporting fluids.

velocity A measure of the downstream distance water travels per unit of time. Velocity varies considerably among streams and even within the same stream.

ventifact A stone whose surface has been polished, pitted, grooved, or faceted by the wind; a common product of wind abrasion.

vertebrate Any animal having a segmented vertebral column; members of the subphylum Vertebrata; includes fishes, amphibians, reptiles, mammals, and birds.

vesicle A small hole or cavity formed by gas trapped in a cooling lava.

vestigial structure A body part that serves little or no function, e.g., the dewclaws of dogs; a vestige of a structure or organ that was well developed and functional in some ancestor.

viscosity A fluid's resistance to flow.

volatile element Element such as hydrogen, helium, ammonia, and methane that condenses at very low temperatures.

volcanic island arc A curved chain of volcanic islands parallel to a deep-sea trench where oceanic lithosphere is subducted causing volcanism and the origin of volcanic islands.

volcanic neck An erosional remnant of a volcanic pipe after a volcano has eroded away.

volcanic pipe The conduit connecting the crater of a volcano with an underlying magma chamber.

volcanic (extrusive igneous) rock An igneous rock that forms when magma is extruded onto the Earth's surface and cools and crystallizes or when pyroclastic materials become consolidated.

volcanism The process whereby magma and its associated gases rise through the Earth's crust and are extruded onto the surface or into the atmosphere.

volcano A conical mountain formed around a vent as a result of the eruption of lava and pyroclastic materials.

W

water table The surface separating the zone of aeration from the underlying zone of saturation; the configuration of the water table is generally a subdued replica of the overlying land surface.

water well A well made by digging or drilling into the zone of saturation.

wave base A depth of about one-half wave length, where the orbital motion of water particles is essentially zero.

wave-cut platform A beveled surface that slopes gently in a seaward direction; formed by the retreat of a sea cliff.

wave height The vertical distance from wave trough to wave crest.

wave length The distance between successive wave crests or troughs.

wave period The time required for two successive wave crests (or troughs) to pass a given point.

wave refraction The bending of a wave so that it more nearly parallels the shoreline.

weathering The physical breakdown and chemical alteration of rocks and minerals at or near the Earth's surface.

Y

yardang An elongated and streamlined ridge that looks like an overturned ship's hull; formed by wind erosion and typically found grouped in clusters aligned parallel to the prevailing wind direction.

Z

zone of accumulation Part of a glacier where additions exceed losses and the glacier's surface is perennially covered by snow.

zone of aeration The zone above the water table that contains both water and air within the pore spaces of the rock or soil.

zone of saturation The zone below the zone of aeration in which all the pore spaces are filled with groundwater.

zone of wastage The part of a glacier where losses from melting, sublimation, and calving of icebergs exceed the rate of accumulations.

Zuni sequence An Early Jurassic to Early Paleocene sequence of sedimentary rocks bounded above and below by unconformities; deposited during a transgressive-regressive cycle of the Zuni Sea.

Answers to Multiple-Choice and Fill-in-the-Blank Review Questions

CHAPTER 1
1. c; 2. a; 3. b; 4. c; 5. d; 6. e; 7. a; 8. d; 9. c; 10. b;
11. a; 12. c; 13. d; 14. a; 15. a; 16. e; 17. b; 18. a.

CHAPTER 2
1. b; 2. e; 3. c; 4. d; 5. b; 6. c; 7. b; 8. a; 9. c; 10. b;
11. a; 12. a; 13. b; 14. e; 15. c.

CHAPTER 3
1. b; 2. a; 3. d; 4. a; 5. c; 6. d; 7. d; 8. e; 9. b; 10. d;
11. a; 12. a; 13. d.

CHAPTER 4
1. a; 2. c; 3. a; 4. e; 5. b; 6. b; 7. e; 8. b; 9. a; 10. c;
11. a; 12. c; 13. d; 14. a; 15. d; 16. e; 17. d.

CHAPTER 5
1. e; 2. c; 3. d; 4. a; 5. d; 6. b; 7. c; 8. a; 9. e; 10. b;
11. a; 12. e; 13. c; 14. b.

CHAPTER 6
1. c; 2. e; 3. a; 4. c; 5. a; 6. c; 7. d; 8. c; 9. d; 10. b;
11. e; 12. b; 13. d; 14. b; 15. a; 16. e; 17. b; 18. d.

CHAPTER 7
1. c; 2. a; 3. d; 4. a; 5. b; 6. d; 7. e; 8. b; 9. b; 10. c;
11. c; 12. a; 13. e; 14. c; 15. b.

CHAPTER 8
1. b; 2. d; 3. a; 4. e; 5. c; 6. d; 7. b; 8. b; 9 a; 10. c;
11. d; 12. e; 13. b.

CHAPTER 9
1. d; 2. a; 3. e; 4. c; 5. e; 6. b; 7. c; 8. d; 9. b; 10. c;
11. a; 12. b; 13. c; 14. b; 15. divergent;
16. oceanic-oceanic convergent; 17. transform;
18. oceanic-continental convergent.

CHAPTER 10
1. b; 2. c; 3. d; 4. a; 5. b; 6. c; 7. c; 8. d; 9. c; 10. a;
11. a; 12. d; 13. a; 14. c; 15. b; 16. b.

CHAPTER 11
1. e; 2. e; 3. b; 4. d; 5. c; 6. a; 7. e; 8. c; 9. a; 10. e.

CHAPTER 12
1. d; 2. a; 3. c; 4. e; 5. b; 6. c; 7. a; 8. c; 9. b; 10. d;
11. a; 12. c; 13. d; 14. b; 15. c; 16. c; 17. e; 18. a;
19. c; 20. d.

CHAPTER 13
1. a; 2. c; 3. b; 4. d; 5. e; 6. d; 7. e; 8. b; 9. d; 10. e;
11. e; 12. b.

CHAPTER 14
1. c; 2. a; 3. b; 4. c; 5. e; 6. b; 7. c; 8. b; 9. e; 10. c;
11. b; 12. a; 13. c; 14. b.

CHAPTER 15

1. d; 2. b; 3. a; 4. c; 5. e; 6. d; 7. c; 8. a; 9. b; 10. e; 11. c; 12. b; 13. d; 14. a; 15. e.

CHAPTER 16

1. e; 2. a; 3. d; 4. b; 5. d; 6. b; 7. a; 8. c; 9. a; 10. c; 11. d; 12. e; 13. c; 14. c; 15. d; 16. a.

CHAPTER 17

1. c; 2. c; 3. a; 4. e; 5. d; 6. a; 7. c; 8. e; 9. d; 10. b; 11. c; 12. e; 13. b.

CHAPTER 18

1. b; 2. d; 3. b; 4. c; 5. d; 6. b; 7. b; 8. c; 9. a; 10. e; 11. c; 12. e; 13. e; 14. b; 15. d.

CHAPTER 19

1. a; 2. d; 3. e; 4. c; 5. b; 6. a; 7. c; 8. e; 9. d; 10. a; 11. c; 12. e; 13. e; 14. a; 15. a; 16. e; 17. d; 18. b; 19. c; 20.b

CHAPTER 20

1. b; 2. e; 3. e; 4. c; 5. d; 6. c; 7. a; 8. b; 9. c; 10. b; 11. b; 12. e; 13. c; 14. b; 15. d.

CHAPTER 21

1. e; 2. d; 3. a; 4. b; 5. e; 6. d; 7. a; 8. c; 9. b; 10. b; 11. a; 12. b; 13. a; 14. c; 15. a; 16. d; 17. d; 18. c; 19. c; 20. e.

CHAPTER 22

1. b; 2. d; 3. e; 4. e; 5. c; 6. a; 7. c; 8. d; 9. c; 10. d; 11. a; 12. d; 13. b; 14. d; 15. c; 16. c; 17. b; 18. e; 19. c; 20. b.

CHAPTER 23

1. e; 2. c; 3. c; 4. c; 5. d; 6. b; 7. b; 8. c; 9. b; 10. a; 11. c; 12. c; 13. b; 14. b; 15. b; 16. d.

CHAPTER 24

1. b. 2. a; 3. a; 4. c; 5. a; 6. c; 7. d; 8. b; 9. a; 10. d; 11. c; 12. b; 13. b; 14. e; 15. a; 16. d.

CHAPTER 25

1. c; 2. d; 3. b; 4. a; 5. b; 6. a; 7. d; 8. e; 9. d; 10. c; 11. a.

Index

O

Oblique-slip fault, 252
Obsidian, 39, 51, 58–59
Ocean Drilling Program, 190
Oceanic-continental convergent plate
  boundary,
    mountain building and, 254, 256
    western North America, 647
Oceanic-continental plate boundary,
  226–227
Oceanic crust, 8, 259–260
    age and distance from oceanic ridges,
      218–219, 222
    composition, volume, density,
      259–260
Oceanic lithosphere, 222–223
Oceanic-oceanic plate boundary,
  225–226
    mountain building, 254–255
Oceanic ridges, 197–199. *See also*
  Mid-Atlantic Ridge;Spreading ridge
    oceanic crust and, 218–219, 222
Oceanic trenches,
    earthquakes and, 195–196
    volcanism and, 196
Oceanographic research, 190–191
Oceanography, 4
Oceans,
    Archean, 537–539
    circulation patterns and breakup of
      Pangaea, 640–642
    volume of Earth's water, 300
Octopuses, 692
Oil, 123, 201. *See also* Petroleum
Oil shale, 95–96
Old Faithful geyser, Yellowstone
  National Park, Wyoming, 90, 348, 349
Oldham, R. D., 177
Old Red Sandstone, 588–589
Oligocene epoch, 677
Olivine, 28,-32, 40, 52, 100
    metamorphosed into serpentine,
      135–136
Olympus Mons, 508
*On the Origin of Species by Means of
  Natural Selection* (Darwin), 13
Ooids, 113
Oolitic limeonite, 654
Oolitic limestones, 113, 115
Ooze, 200
Ophiolites, 227, 260
    Archean, 527
    Proterozoic, 533, 535
Optical pyrometer, 49
Ordinary chondrite, 499–500

Ordovician, 569–571, 588, 605, 609
    continents, 562–563
    extinction, 611, 612
    land plants, 622
    marine community, 609–611
    North America, 569, 570
    reefs, 570
    sandstone, 572
Ore deposits, contact metamorphism
    and, 146–148
Organic evolution, 3, 13, 15, 464. *See
    also* Evolution
Organic reefs, 569, 573
Organisms,
    classification of, 479–481
    earliest, 543–544
    at hydrothermal vents, 189–190
    soil formation, 105
    weathering and, 97, 98
Original horizontality, principle of, 434
Origin of Continents and Oceans, The
    (Wegener), 210
Origin of Species, The (Darwin), 466, 481
Ornithischia, 657, 658
Ornithopods, 657–659
Orogenesis, 253–259. *See also* Mountain
    building
Orogens, 220, 528, 530
Orogeny, Cenozoic, 678
Orthoclase, 26, 29, 38
    hydrolysis of, 100–101
Ostracoderms, 616–617
Ouachita mobile belt, 560, 561, 563,
    589–591
Ouachita mountains, 563
Ouachita orogeny, 565, 590
Our Wandering Continents (du Toit), 210
Outflow channels, 299
Outgassing, 501, 504, 538–539
Outlet glaciers, 361
Outwash plains, 369, 372
Overloading, mass wasting and, 273
Overpopulation, 717–718
Overturned fold, 244, 245
Oxbow lakes, 306–307
Oxidation, 100
Oxides, 30
Oxygen, in atmosphere, 538–540
Ozone, 539

P

Pachycephalosaurs, 657–660
Pacific plate, 230
Pahoehoe lava flow, 76
Paleocene Epoch, 684, 694–695

Paleocurrents, 119
Paleogene, 678
Paleogeography, 4
    Cambrian (North America), 569, 570
    Carboniferous, 563, 566–567
    Devonian (North America), 563, 566,
      576–577
    maps, constructing, 562
    Mesozoic (North America), 642–644,
      665–666
    Mississippian (North America), 580
    Ordovician (North America),
      562–563, 569, 570
    Paleozoic, 560–565
    Pennsylvanian (North America), 582
    Permian (North America), 563–564,
      567, 585
    Sauk sequence, 569–570
    Silurian (North America), 562–563,
      574
Paleomagnetism, 216–217, 562
Paleozoic,
    Absaroka sequence, 580–586
    amphibians, 619–621, 623
    black shale, 576–577
    cratons and mobile belts, 560–561
    evolution of North America, 565–568
    invertebrate marine life, 606–614
    Kaskaskia sequence, 572–579
    microplate and formation of pangaea,
      590–592
    mineral resources, 592–593
    mobile belts history, 586–590
    paleogeography, 560–565
    plant evolution, 625–630
    reptile evolution of, 621–625
    summary of evolutionary/geologic
      events, 632–633
    summary of geologic events, 596–599
    Tippecanoe sequence, 569–572
    vertebrate evolution, 615–625
Palisades, Hudson River, New York, 61
Pangaea, 9, 210, 212, 220
    breakup of, 638–642
    effects of breakup, 640–642
    formation of, 533, 563, 590, 592
Panthalassa, 563
Parabolic dunes, 389, 392
Parallel evolution, 473–475
Parent element, 447
Parent material, 96
Partial melting, 88–89
Passive continental margins, 195–196
Peat, 115, 691
Pedalfers, 104–105
Pediments, 401–402

Photo Credits

CHAPTER 1

Opener: Photo courtesy of NASA. **1.1a**: ©Patricia K. Armstrong/ Visuals Unlimited. **1.1b**: American Association of Petroleum Geologists/IBM. **1.2**: Collection of the New York Public Library: Astor, Lenox, and Tilden Foundations. **1.4**: Victor Royer. **1.5, 1.7, 1.10**: Carlyn Iverson. **1.6**: Rolin Graphics. **1.8**: Precision Graphics. **1.11**: Precision Graphics. From A. R. Palmer, "The Decade of North American Geology, 1983 Geologic Time Scale." *Geology* (Boulder, Colo.: Geological Society of America, 1983): 504. Reprinted by permission of the Geological Society of America. **Table 1.2**: Modified from R. V. Dietrich and R. Wicander, *Minerals, Rocks, and Fossils* (New York: John Wiley & Sons, Inc., 1983): 160, Table iv-2.

CHAPTER 2

2.2, 2.3: Layout by Georg Klatt; final inking by Elizabeth Morales-Denney. **2.4, 2.5, 2.6, 2.7, 2.9, 2.10, 2.16, 2.17, Review Question 17**: Precision Graphics. **2.8, 2.20**: Precision Graphics. From R. V. Dietrich and Brian J. Skinner, *Gems, Granites, and Gravels: Knowing and Using Rocks and Minerals* (New York: Cambridge University Press, 1990): 39, Figure 3.4; and 97, Figure 6.1. Reprinted with permission of Cambridge University Press. **2.15b**: Photo courtesy of Ward's Natural Science Establishment, Inc. **2.21**: Publication Services. Reprinted with permission from B. J. Skinner, "Mineral Resources of North America," *Geology of North America*, v. A, 1989, p. 577 (Fig. 2).

CHAPTER 3

3.2: Photo of painting by Herbert Collins, courtesy Devil's Tower National Monument. **3.3, 3.4, 3.7, 3.8, 3.9, 3.11, 3.18, 3.22, 3.23, 3.24**: Precision Graphics. **3.12**: Precision Graphics. Modified from R. V. Dietrich, *Geology and Michigan: Forty-nine Questions and Answers.* 1979. **Perspective 3.1, Figure 1**: Rolin Graphics. From W. R. Dickinson and W. C. Luths, "A Model for Plate Tectonic Evolution of Mantle Layers." *Science* 174 (22 October 1971): 402, Figure 1. Copyright ©1971 by the AAAS. Reprinted by permission of the AAAS and W. R. Dickinson. **3.19**: Martin G. Miller/Visuals Unlimited. **Perspective 3.2, Figure 1**: Photo ©1985 by Wendell E. Wilson. **3.26, 3.27**: Carlyn Iverson.

CHAPTER 4

4.1: Rolin Graphics. From R. I. Tilling, U.S. Geological Survey. **4.2**: D. R. Crandall, U.S. Geological Survey. **4.3**: Keith Ronnholm. **4.4, 4.5, 4.6b**: U.S. Geological Survey. **4.6a**: T. J. Takahashi, U.S. Geological Survey. **4.7a**: D. W. Peterson, U.S. Geological Survey. **4.7b**: J. B. Stokes, U.S. Geological Survey. **4.9**: Copyright by Marie Tharp, 1977. Reproduced by permission of Marie Tharp, 1 Washington Ave., South Nyack, NY 10960. **4.10**: Reuters/Bettmann Archive. **4.11**: J. P. Lockwood, U.S. Geological Survey. **Perspective 4.2, Figure 1**: Precision Graphics. From R. I. Tilling, U.S. Geological Survey. **4.12a-d**: Precision Graphics. From Howel Williams, *Crater Lake: The Story of Its Origin* (Berkeley: University of California Press): Illustrations from p. 84. Copyright ©1941 Regents of the University of California, ©renewed 1969 Howel Williams. **4.13, 4.14a, 4.15a, 4.16, 4.19, 4.25**: Precision Graphics. **4.14b**: Solarfilma/ GeoScience Features. **4.15b**: Lawrence R. Solkoski, consulting geologist, Vancouver, B.C., Canada. **4.17**: I. C. Russell, U.S. Geological Survey. **4.18**: Photo courtesy of Ward's Natural Science Establishment, Inc. **4.21**: Rolin Graphics. Modified from R. I. Tilling, C. Heliker, and T. L. Wright, *Eruptions of Hawaiian Volcanoes: Past, Present, and Future.* 1987. U.S. Geological Survey. **4.22**: Precision Graphics. W. G. Ernst, *Earth Materials*, ©1969, p. 107. Reprinted by permission of Prentice-Hall, Inc., Englewood Cliffs, NJ 07632. **4.23**: Rolin Graphics. From "Hot Spots on the Earth's Surface," copyright ©1976, by Scientific American, Inc., George V. Kelvin, all rights reserved. **4.24**: Carlyn Iverson.

CHAPTER 5

5.2: R. L. Elderkin, U.S. Geological Survey. **5.3, 5.9, 5.10, 5.11, 5.12, 5.14, 5.20, 5.25, 5.26, 5.28, 5.32, 5.35, 5.36**: Precision Graphics. **5.5**: Precision Graphics. From A. Cox and R. R. Doell, "Review of Paleomagnetism." *GSA Bulletin* 71 (1960): 758, Figure 33. **5.6**: National Geophysical Data Center photo archives. **Perspective 5.1, Figure 2**: N. K. Huber, U.S. Geological Survey. **5.15**: Walt Anderson/Visuals Unlimited. **Perspective 5.2, Figure 1**: Rolin Graphics. Modified from Donald Worster, *Dust Bowl* (New York: Oxford University Press, 1979): 30. **Perspective 5.2, Figure 2**: Kansas State Historical Society. **5.17**: Science VU/Visuals Unlimited. **5.19, 5.33**: Rolin Graphics. **5.22c**: Rex Elliot. **5.34**: Alan L. Mayo, GeoPhoto Publishing Company.

CHAPTER 6

6.1a, 6.3, 6.7, 6.11a: Precision Graphics. **6.1b**: Precision Graphics. From G. Rapp, Jr., and J. A. Gifford, eds., *Archaeological Geography* (New Haven, Conn: Yale University Press, 1985): 338, Figure 13.3. Reprinted by permission of Yale University Press and Norman Herz. **6.2**: Courtesy of the Arthur M. Sackler Museum, Harvard University Art Museums, Cambridge, Massachusetts. Fund in memory of John Randolph Coleman III, Harvard class of 1964. **6.4, 6.22 (top)**: Rolin Graphics. **Perspective 6.1, Figure 1**: Smithsonian Institution. **6.5a, 6.10**: Precision Graphics. From C. Gillen, *Metamorphic Geology* (London: Chapman & Hall, 1982): 24 and 73, Figures 2.3 and 4.4. Reprinted by permission of Chapman & Hall and C. Gillen. **6.19**: Rolin Graphics. From H. L. James, *GSA Bulletin* 66 (1955): 1454, Plate 1. Reprinted by permission of the Geological Society of America. **6.20**: Precision Graphics. From *AGI Data Sheets*, 3d ed., compiled by J. T. Dutro, Jr., R. V. Dietrich, and R. M. Foose. Copyright 1989; Data Sheet 35.4. Courtesy of the American Geological Institute Alexandria, Virginia. **6.21**: Carlyn Iverson. **6.22 (bottom)**: Carlyn Iverson. From "Effects of Late Jurassic–Early Tertiary Subduction in California." *Late Mesozoic and Cenozoic Sedimentation and Tectonics in California,* San Joaquin Geological Society Short Course (1977): 66, Figure 5-9. **Table 6.1**: From C. Gillen, *Metamorphic Geology* (London: Chapman & Hall, 1982): 70, Table 4–1. Reprinted by permission of Chapman & Hall and C. Gillen.

CHAPTER 7

Opener: J. P. Stacy, U.S. Geological Survey. **7.1a**: Martin E. Klimek, Marin *Independent Journal.* **7.1b**: E. V. Leyendecker, U.S. Geological Survey, National Geophysical Data Center, NOAA, Boulder, Colorado. **7.2a; 7.4b, c; 7.5; 7.10; Perspective 7.1, Figure 1; 7.15; 7.22; 7.24; 7.25; 7.26; 7.27; Perspective 7.2, Figure 1**: Precision Graphics. **7.2b**: U.S. Geological Survey. **7.3**: Reproduced by permission of the Trustees of the Science Museum, London. **7.4a**: *Earthquake Information Bulletin 181*, U.S. Geological Survey. **7.6**: Rolin Graphics. Data from National Oceanic and Atmospheric Administration. **7.7**: Carlyn Iverson. **7.8**: J. K. Hillers, U.S. Geological Survey. **7.9**: Precision Graphics. From *Nuclear Explosions and Earthquakes: The Parted Veil* by Bruce A. Bolt. Copyright ©1976 by W. H. Freeman and Co. Reprinted by permission. **7.11**: Precision Graphics. Data from C. F. Richter, *Elementary Seismology.* Copyright ©1958 by W. H. Freeman and Co. Reprinted with permission. **7.12**: Rolin Graphics. **7.13, 7.19**: Rolin Graphics. From M. L. Blair and W. W. Spangle, *U.S. Geological Survey Professional Paper 941-B.* 1979. **7.14 and Perspective 7.1, Figure 2**: Precision Graphics. From *Earthquakes* by Bruce A. Bolt. Copyright ©1988 by W. H. Freeman and Co. Reprinted by permission. **Perspective 7.1, Figure 3**: M. Celebi, U.S. Geological Survey. **7.16, 7.18**: National Geophysical Data Center, NOAA, Boulder, Colorado. **7.17**: Hebei Provincial Seismological Bureau, U.S. Geological Survey. **7.20**: Precision Graphics. Reprinted with permission from *Geotimes* 10 (1966): 17. **7.21**: Victor Royer. **7.23, 7.28**: Precision Graphics. From G. C. Brown and A. E. Musset, *The Inaccessible Earth* (London: Chapman & Hall, 1981): 17 and 124, Figures 12.7a and 7.11. Reprinted by permission of Chapman & Hall. **Perspective 7.2, Figure 2**: Precision Graphics. From Keith G. Cox, "Kimberlite Pipes." Original illustration by Adolph E. Brotman. Copyright © April 1978 by Scientific American, Inc. All rights reserved. **7.29**: Precision Graphics. Andrew Christie/Copyright ©1987 Discover Publications.

CHAPTER 8

Opener: Woods Hole Oceanographic Institution. **8.1, 8.3, 8.4, 8.5, 8.7, 8.14**: Precision Graphics. **Perspective 8.1, Figure 2. 8.8; 8.10**: Rolin Graphics. **8.2**: Precision Graphics. From U.S. Geological Survey. **Perspective 8.1, Figure 1**: Rolin Graphics. From Phyllis Young Forsyth, *Atlantis: The Making of a Myth* (Montreal: McGill-Queen's University Press): 13, Figure 2. **8.6**: Precision Graphics. Modified from Bruce C. Heezen and Charles D. Hollister, *The Face of the Deep* (New York: Oxford University Press, 1971): 297, Figure 8.15. Used with permission. **8.9**: Precision Graphics. From B. C. Heezen, M. Tharp, and M. Ewing, "The Floors of the Oceans, Part 1, The North Atlantic." *Geological Society of America Special Paper 65.* 1959. **8.11**: Carlyn Iverson. **8.12**: Dr. Bruce Heezen, Lamont-Doherty Geologic Observatory, Columbia University, courtesy Scripps Institution of Oceanography, University of California, San Diego. **8.13**: ©Bruce Berg/Visuals Unlimited. **8.15**: ©Douglas Faulkner, Science Source/Photo Researchers. **8.16**: Rolin Graphics. From U.S. Geological Survey.

CHAPTER 9

Opener: NASA. **9.1 (left), 9.10, 9.12a, 9.19, 9.23, 9.28, Review Question 15**: Precision Graphics. **9.1 (right); Perspective 9.1, Figure 1; 9.18; 9.20; 9.21; 9.22; 9.24; 9.27 (left); 9.29**: Carlyn Iverson. **9.2**: U.S. Geological Survey. **9.4**: Bildarchiv Preussischer Kulturbesitz. **9.5**: Rolin Graphics. From E. Bullard, J. E. Everett, and A. G. Smith, "The Fit of the Continents Around the Atlantic." *Philosophical Transactions of the Royal Society of London* 258 (1965). Reproduced with permission of the Royal Society, J. E. Everett, and A. G. Smith. **9.6**: Rolin Graphics. Reprinted with permission of Macmillan College Publishing Company from *General Geology*, 5/e, by Robert J. Foster. Copyright ©1988 by Macmillan College Publishing Company, Inc. **9.7, 9.8, 9.16, 9.27 (right), 9.30**: Rolin Graphics. **9.9**: Rolin Graphics. Modified from E. H. Colbert, *Wandering Lands and Animals* (1973): 72, Figure 31. **9.11**: Carto-Graphics. Reprinted with permission from A. Cox and R. R. Doell, "Review of Paleomagnetism," *GSA*, v. 71, 1960, p. 758 (Fig. 33), Geological Society of America. **9.12b, 9.14**: Publication Services. Reprinted with permission from A. Cox, "Geomagnetic Reversals." *Science*, v. 163, p. 240 (Fig. 4), Copyright ©1969 by the AAAS. **9.13**: Photo courtesy of ALCOA. **9.15**: Rolin Graphics. From *The Bedrock Geology of the World* by Larson and Pitman. Copyright ©1985 by W. H. Freeman and Company. Reprinted with permission. **9.17**: Woods Hole Oceanographic Institution. **9.25 (art)**: Carto-Graphics. **9.26**: Rolin Graphics. Data from J. B. Minster and T. H. Jordan, "Present-day Plate Motions." *Journal of Geophysical Research* 83 (1978): 5331—51. Copyright by the American Geophysical Union. Used with permission.

CHAPTER 10

10.1: Precision Graphics. From *Structural Geology of North America* by A. J. Eardley. Copyright ©1951 by Harper & Row, Publishers, Inc. Copyright ©1951 by A. J. Eardley. Reprinted by permission of Harper Collins Publishers. **10.2**: U.S. Geological Survey. **10.3; 10.5; 10.6; 10.7; 10.9; 10.10; 10.11; 10.12a, b; 10.13; 10.15; Perspective 10.1, Figure 3; 10.20; 10.23; 10.24; 10.28; 10.29**: Precision Graphics. **10.4, 10.12c, 10.19**: John S. Shelton. **10.14**: S. W. Lohman, U.S.

Geological Survey. **10.16, 10.17, 10.21, 10.27**: Rolin Graphics. **10.18**: B. Bradley, University of Colorado Geology Department. National Geophysical Data Center photo archives. **10.22**: Michael Thomas Associates. Reprinted by permission of the Geological Society and A. M. Spencer from A. M. Spencer, ed., *Mesozoic-Cenozoic Orogenic Belts* (Bath: Geological Society Publishing House, 1974). **10.25**: Rolin Graphics. From Peter Molnar, "The Geologic History and Structure of the Himalaya." *American Scientist* 74: 144–154, Fig. 4, pp. 148–149. Reprinted by permission of *American Scientist*, journal of Sigma Xi, the Scientific Research Society. **10.26**: Rolin Graphics. From Zvi Ben-Avraham, "The Movement of Continents." *American Scientist* 69: 291–299, Fig. 9, p. 298. Reprinted by permission of *American Scientist*, journal of Sigma Xi, the Scientific Research Society. **10.30**: Rolin Graphics. From R. F. Flint, *Glacial and Quaternary Geology* (New York: John Wiley & Sons, Inc., 1971): 363, Figure 13–13. Reprinted by permission of John Wiley & Sons, Inc.

CHAPTER 11

Opener: Courtesy Hong Kong Government Information Services. **11.1 left; Perspective 11.1, Figure 1 (left); 11.12 top; 11.14 art; 11.15 top**: Rolin Graphics. **11.1 right**: George Plafker, U.S. Geological Survey. **11.2**: U.S. Geological Survey. **11.3; 11.4a, b; 11.5a-c; 11.7; 11.8; 11.11; 11.13; 11.15a; 11.17; 11.18a; 11.19 top, a; 11.22a; 11.23; 11.24a; 11.25; 11.26a; 11.27a**: Precision Graphics. **11.6**: Boris Yaro, *Los Angeles Times*. **Perspective 11.1, Figure 1 (right)**: T. Spencer/Colorific. **11.10**: W. R. Hansen, U.S. Geological Survey. **11.12 bottom**: John S. Shelton. **11.14b**: Steven R. Lower, GeoPhoto Publishing Company. **11.15b, 11.20b, 11.22b**: B. Bradley and the University of Colorado's Geology Department. National Geophysical Data Center photo archives. **11.19b**: Department of the Army, U.S. Army Engineer District, Alaska Corps of Engineers. **11.20a**: Rolin Graphics. From O. J. Ferrians, Jr., R. Kachadoorian, and G. W. Greene, *U.S. Geological Survey Professional Paper 678*. 1969. **11.21**: O. J. Ferrians, Jr., U.S. Geological Survey. **Perspective 11.2, Figure 1**: Rolin Graphics. Reprinted with permission of ASCE. From G. A. Kiersch, "Vaiont Reservoir Disaster." *Civil Engineering* 34 (1964). **Perspective 11.2, Figure 2**: UPI/Bettmann. **Perspective 11.2, Figure 3**: Precision Graphics. Reprinted with permission of ASCE. From G. A. Kiersch, "Vaiont Reservoir Disaster." *Civil Engineering* 34 (1964). **11.26b**: John D. Cunningham/Visuals Unlimited. **11.27b**: Dell R. Foutz/Visuals Unlimited.

CHAPTER 12

12.1: JPL/NASA. **12.2**: Martin G. Miller/Visuals Unlimited. **12.3; 12.19a, b**: Rolin Graphics. **12.4; 12.5; 12.6; 12.7; 12.8; 12.11; 12.15; Perspective 12.1, Figures 1 and 2; 12.16; 12.17; 12.18a; 12.22; 12.23; 12.24; 12.25; 12.27; 12.28; 12.29; 12.30**: Precision Graphics. **12.13, 12.31**: John S. Shelton. **12.19c**: Rolin Graphics. From W. L. Fisher et al., *Delta Systems in the Exploration for Oil and Gas — A Research Colloquium* (1969). **12.20**: Alan L. Mayo, GeoPhoto Publishing Company. **12.26**: Petley Studios. **Perspective 12.2, Figure 2**: Precision Graphics. From *Natural Bridges*. National Park Service.

CHAPTER 13

Opener: Sarah Stone/Tony Stone Worldwide. **13.1a, c**: Rolin Graphics. From *Trapped* by Robert K. Murray and Roger W. Brucker.

Copyright ©1979 by Murray and Brucker, copyright renewed. Used by permission. **13.1b**: Brown Brothers. **13.2, 13.3, 13.4, 13.6, 13.8, 13.9, 13.11, 13.14, 13.16, 13.19, 13.22, 13.26**: Precision Graphics. **13.5**: G. E. Seaburn, U.S. Geological Survey. **13.7**: Linda D. Mayo, GeoPhoto Publishing Company. **13.10**: J. R. Stacy, U.S. Geological Survey. **13.12**: Alice Thiede, Carto-Graphics. **13.13**: Frank Kujawa, University of Central Florida, GeoPhoto Publishing Company. **13.15b**: John S. Shelton. **Perspective 13.1, Figure 1**: Ed Cooper. **Perspective 13.1, Figure 2**: W. L. McCoy. **13.17**: Daniel W. Gotshall/Visuals Unlimited. **13.18**: Rolin Graphics. From J. B. Weeks et al., *U.S. Geological Survey Professional Paper 1400-A*. 1988. **13.20 top**: Rolin Graphics. **13.20 bottom**: U.S. Geological Survey. **13.21**: Long Beach Department of Oil Properties, Long Beach, California. **Perspective 13.2, Figure 1**: Precision Graphics. Modified from *U.S. News and World Report* (18 March 1991): 72–73. **13.24**: British Tourist Authority.

CHAPTER 14

Opener: David Hiser, Photographers/Aspen. **14.1a**: Offentliche Kunstammlung, Kupferstichkabinett Basel. **14.1b**: Reproduced courtesy of the Board of Directors of the Budapest Museum of Fine Arts. **14.2, 14.7, 14.8, 14.13, 14.14, 14.25, 14.28**: Precision Graphics. **14.3, 14.21**: Rolin Graphics. **14.4, 14.22**: Engineering Mechanics, Virginia Polytechnic Institute and State University. **14.5**: Frank Awbrey/Visuals Unlimited. **14.6; Perspective 14.1, Figure 1**: Alice Thiede, Carto-Graphics. **14.9**: National Park Service photograph by Ruth and Louis Kirk. **14.18**: Bob and Ira Spring, Kirkendall-Spring Photographers. **14.19**: Swiss National Tourist Office. **14.20**: Alan Kesselheim/Mary Pat Ziter, ©JLM Visuals. **14.24, 14.26a**: John S. Shelton. **Perspective 14.1, Figures 2 and 3**: P. Weis, U.S. Geological Survey.

CHAPTER 15

Opener; 15.12b; Perspective 15.2, Figure 2; 15.24; 15.29a: John S. Shelton. **15.1, 15.18**: Alice Thiede, Carto-Graphics. **15.2**: Steve McCurry/Magnum Photos. **15.3; 15.5a; 15.8a, b; 15.9; 15.10; 15.12a; 15.13a; 15.14a; 15.16a; 15.21; 15.27a**: Precision Graphics. **15.4, 15.7, 15.25**: Martin G. Miller/Visuals Unlimited. **Perspective 15.1, Figure 1**: Image processed by Mary A. Dale-Bannister, Washington University, St. Louis. **Perspective 15.1, Figures 2 and 3**: NASA. **15.13b, 15.14b, 15.26**: Alan L. and Linda D. Mayo, GeoPhoto Publishing Company. **15.15**: Willard Clay/Tony Stone Worldwide. **15.17**: Steve McCutcheon/Visuals Unlimited. **15.19**: Alex Teshin Associates. Based on F. K. Lutgens and E. J. Tarbuck, *The Atmosphere: An Introduction to Meteorology* (Englewood Cliffs, New Jersey: Prentice-Hall, 1979): 150, Figure 7.3. Reprinted by permission of Prentice-Hall, Inc., Englewood Cliffs, NJ. **15.20**: Rolin Graphics. **Perspective 15.2, Figure 1**: Precision Graphics. From C. B. Hunt and D. R. Mabey, *U.S. Geological Survey Professional Paper 494A* (1966): A5, Figure 2. **Perspective 15.2, Figure 3**: John D. Cunningham/Visuals Unlimited. **Perspective 15.2, Figure 4**: U.S. Borax.

CHAPTER 16

16.1, 16.2: Courtesy of the Rosenberg Library, Galveston, Texas. **16.3; 16.4; Perspective 16.1, Figure 2; 16.23**: Rolin Graphics. **16.5, 16.9, 16.10a, 16.11, 16.12a, 16.13a, 16.16, 16.17, 16.18a**: Precision

...aphics. **16.6, 16.7, 16.10b, 16.15, 16.18c**: John S. Shelton. **16.8b**: Michael Slear. **16.14**: NASA. **Perspective 16.1, Figure 1**: Rolin Graphics. From *U.S. Geological Survey Circular 1075*. **Perspective 16.1, Figure 3**: U.S. Army Corps of Engineers. **Perspective 16.1, Figure 4**: Steve Starr/SABA.**16.18b**: Nick Harvey. **16.19**: GEOPIC, Earth Satellite Corporation. **16.22**: Karl Kuhn.

CHAPTER 17

17.1: Darwen and Vally Hennings. Modified from *Geologic Time*. 1981. U.S. Geological Survey. **17.2**: Precision Graphics. From A. R. Palmer, "The Decade of North American Geology, 1983 Geologic Time Scale." *Geology* (Boulder, Colo.: Geological Society of America, 1983): 504. Reprinted by permission of the Geological Society of America. **17.5; 17.6; 17.9a; 17.10a; 17.11a; 17.12; 17.13; 17.16; 17.18; Perspective 17.1, Figure 1; 17.20b; 17.21; 17.22; 17.24; Review Question 30**: Precision Graphics. **17.7**: Precision Graphics. From *The Story of the Great Geologists* by Carroll Lane Fenton and Mildred Adams Fenton. Copyright ©1952 by Carroll Lane Fenton and Mildred Adams Fenton. Used by permission of Doubleday, a division of Bantam Doubleday Dell Publishing Group, Inc. **17.8, 17.17, 17.26, 17.28**: Publication Services. **17.14**: Publication Services. From *History of the Earth: An Introduction to Historical Geology*, second edition, by Bernhard Kummel. Copyright ©1970 by W. H. Freeman and Company. Reprinted by permission. **17.15**: Precision Graphics. From *Geologic Time*. 1981. U.S. Geological Survey. Photos by Reed Wicander. **Perspective 17.1, Figure 2**: Rolin Graphics. Data from Environmental Protection Agency. **17.19**: Precision Graphics. Data from S. M. Richardson and H. Y. McSween, Jr., *Geochemistry — Pathways and Processes* (Englewood Cliffs, New Jersey: Prentice-Hall, 1989). **17.20a**: Precision Graphics. From Don L. Eicher, *Geologic Time*, 2d ed., ©1976, p. 120. Reprinted by permission of Prentice-Hall, Inc., Englewood Cliffs, New Jersey, 07632. **17.23**: Photo courtesy of Charles W. Nueser, U.S. Geological Survey. **17.25**: Precision Graphics. Reprinted with permission from Stokes and Smiley, *An Introduction to Tree Ring Dating*, ©The University of Chicago Press.

CHAPTER 18

Opener: The Granger Collection, New York. **18.1, 18.4, 18.5, 18.7, 18.8, 18.11, 18.19**: Precision Graphics. **18.2; 18.3; 18.12; 18.13; Perspective 18.1, Figure 3b; 18.16; 18.17**: Carlyn Iverson. **18.6; 18.10; Perspective 18.1, Figure 2; 18.26; 18.27**: Publication Services. **18.9**: Publication Services. From *Principles of Paleontology*, 2d ed., by D. M. Raup and S. M. Stanley. Copyright ©1978 by W. H. Freeman and Company; reprinted by permission. **Perspective 18.1, Figure 1**: Publication Services. Reprinted with permission from Starr and Taggart, *Biology: The Unity and Diversity of Life*, 1989, p. 556 (Fig. 37.2a). ©Wadsworth Publishing Company. **Perspective 18.1, Figure 3a**: Publication Services. Reprinted with permission from Weishampel, Dodson, et al., *Dinosauria: Paleobiology of the Terrible Lizards*, ©1990 The Regents of the University of California. **18.14**: Precision Graphics. Reprinted with permission from L. S. Dillon, *Evolution: Concepts and Consequences*, 2d ed., 1978, p. 250 (Fig. 12–6). **18.15**: Precision Graphics. Reproduced by permission of the Society of Evolution from *Evolution*, v. 28, p. 448. **18.18**: John and Judy Waller. **18.21**: Photo ©M. W. Tweedie/Photo Researchers.

18.22c: American Museum of Natural History. **18.23a**: Photo ©J. Kolvula/Science Source, Photo Researchers, Inc. **18.23b**: Courtesy of the George C. Page Museum. **18.24a**: American Museum of Natural History. **18.24b**: Photo ©Sovfoto/V. Khristoforov.

CHAPTER 19

Opener: Palomar Observatory, California Institute of Technology. **19.1, 19.11b**: JPL/NASA. **19.2; 19.9a; 19.10b; 19.12a; 19.13a, c-e; 19.14a, b**: NASA. **19.3a**: John and Judy Waller. **19.3b, 19.6a, 19.17**: Precision Graphics. **19.4, 19.20**: Publication Services. **19.5**: Precision Graphics. From Eicher/McAlester, *History of the Earth*, ©1980, p. 11. Reprinted by permission of Prentice-Hall, Inc., Englewood Cliffs, NJ 07632. **19.6b, d**: Ken Nichols, University of New Mexico, Institute of Meteoritics. **19.6c, e, f**: Brian Mason, Smithsonian Institution. **19.7**: D. J. Roddy, U.S. Geological Survey. **19.8**: Paul Dimare. **Perspective 19.1, Figure 1**: Carto-Graphics. **Perspective 19.1, Figure 2**: TASS from Sovfoto. **Perspective 19.2, Figure 1**: Carlyn Iverson. **19.9b, 19.10d, 19.11a, 19.12b, 19.13b, 19.14c, 19.15, 19.16, 19.18**: Victor Royer. **19.9c**: Image from NASA's Mariner 10 Spacecraft provided by National Space Science Data Center. **19.10a**: Finley Holiday Film. **19.10c**: AP/Wide World Photos. **19.19** Copyright UC Regents; UCO/Lick Observatory image. **19.21**: W. Benz and W. Slattery, Los Alamos National Laboratories.

CHAPTER 20

20.1: Herb Orth, Life Magazine. ©Time Warner Inc.; painting by Chesley Bonestell, ©The Estate of Chesley Bonestell. **Table 20.1**: Publication Services. Reprinted with permission from H. L. James, "Stratigraphic Commission: Note 40—Subdivision of Precambrian: An Interim Scheme to Be Used by USGS," *AAPG*, v. 56, no. 6, 1972, p. 1130 (Table 1), and from Harrison and Peterman, "North American Commission on Stratigraphic Nomenclature: Report 9 — Adoption of Geochronometric Units for Divisions of Precambrian Time," *AAPG*, v. 66, no. 6, 1982, p. 802 (Fig. 1), American Association of Petroleum Geologists. **20.2**: Carto-Graphics. Reprinted with permission from A. M. Goodwin, "The Most Ancient Continental Margins," in *The Geology of Continental Margins*, Burk and Drake (eds.), 1974, p. 768 (Fig. 1), ©Springer-Verlag. **20.4**: Carto-Graphics. Reprinted with permission of the Geological Survey of Canada. **20.5a; 20.19; Perspective 20.2, Figure 2; 20.32f**: Precision Graphics. **Table 20.2, Table 20.3, 20.20, 20.24a-d**: Publication Services. **20.6**: Publication Services. Reprinted from K. C. Condie, *Plate Tectonics and Crustal Evolution*, 2d ed., ©1982, p. 87, with permission from Pergamon Press, Ltd., Headington Hill Hall, Oxford OX3 OBW, U.K. **20.7**: Carto-Graphics. Reproduced with permission from P. F. Hoffman, "United Plates of America, The Birth of a Craton: Early Proterozoic Assembly and Growth of Laurentia." *Annual Review of Earth and Planetary Sciences*, v. 16, p. 544, ©1988 by Annual Reviews, Inc. **20.8**: Carto-Graphics. Reprinted with permission from Medaris, Byers, Mikelson, and Shanks (eds.), "Proterozoic anorogenic granite plutonism of North America," *Proterozoic Geology: Selected Papers* from an International Symposium, p. 135 (Fig. 1). **20.9**: Publication Services. Reprinted with permission from J. C. Green, *Tectonophysics*, v. 94, p. 414 (Fig. 1), ©1983, Elsevier Science Publishers. **20.10a**: Publication Services. Reprinted with permission from Daniels, *Geology*, GSA Memoir

156, 1982, the Geological Society of America. **20.11**: Publication Services. Reprinted with permission from R. P. E. Poorter, "Precambrian Paleomagnetism of Europe and the Position of the Balto-Russian Plate Relative to Laurentia," in *Precambrian Plate Tectonics*, A. Kroner (ed.), p. 603 (Fig. 24–2), ©1981 Elsevier Science Publishers. **Perspective 20.1, Figure 1a**: Publication Services. From Kontinen, *Precambrian Research*, ©1987 Elsevier Science Publishers. Reprinted with permission. **Perspective 20.1, Figures 1b, c**: Asko Kontinen, Geological Survey of Finland. **20.12**: Carto-Graphics. Reprinted with permission from Ian W. D. Dalziel, "Pacific margins of Laurentia and East-Antarctica-Australia as a conjugate rift pair: Evidence and implications for an Eocambrian supercontinent," *Geology*, v. 19, June 1991, p. 600 (Fig. 3), Geological Society of America. **20.13**: Carto-Graphics. Reprinted with permission from G. M. Young, "Tectono-Sedimentary History of Early Proterozoic Rocks of the Northern Great Lakes Region," in *Early Proterozoic Geology of the Great Lakes Region*, Medaris (ed.), GSA Memoir 160, 1983, p. 16 (Fig. 1), Geological Society of America. **20.15**: Carto-Graphics. **20.17, 20.18**: Publication Services. Reprinted with permission from L. A. Frankes, *Climates Throughout Geologic Time*, pp. 39, 88, ©1979 Elsevier Science Publishers. **20.22**: Publication Services. Reprinted with permission from S. L. Miller, "The Formation of Organic Compounds on the Primitive Earth," in *Modern Ideas of Spontaneous Generation*, Nigrelli (ed.), Annals of the New York Academy of Sciences, v. 69, Art. 2, Aug. 30, 1957, p. 261 (Fig. 1). **20.27**: Precision Graphics. From Lynn Margulis, "Symbiosis and Evolution," copyright ©1971 by Scientific American, Inc. All rights reserved. **Perspective 20.2, Figure 1**: Precision Graphics. From "Symbiosis and Evolution," by Lynn Margulis. Copyright ©1971 by Scientific American, Inc. All rights reserved. **Table 20.4**: Publication Services. Reprinted with permission from Kontinen, *Precambrian Research*, ©1987 Elsevier Science Publishers. **20.30**: Publication Services. Reprinted with permission from Kontinen, *Precambrian Research*, ©1987 Elsevier Science Publishers.

CHAPTER 21

Opener: National Park Service. **21.1, 21.6, 21.7, 21.10, 21.13, 21.14, 21.16, 21.18, 21.20a, 21.21, 21.22**: Carto-Graphics. **21.2, 21.3, 21.4**: Carto-Graphics. Topography reprinted with permission from Bambach, Scotese, and Ziegler, "Before Pangaea: The Geographies of the Paleozoic World," *American Scientist*, v. 68, no. 1, 1930. **21.5**: Publication Services. Reprinted with permission from L. L. Sloss, "Sequences in the cratonic interior of North America," *GSA Bulletin*, v. 74, 1963, p. 110 (Fig. 6), Geological Society of America. **21.8a; 21.9b; 21.12; 21.19a, c; 21.20b; 21.25; 21.26; 21.27; 21.29**: Precision Graphics. **21.8b**: Wisconsin Department of Natural Resources. **21.11a**: Precision Graphics. Reprinted with permission from Mesolella, Robinson, McCormick, and Ormiston, "Cyclical Deposition of Silurian Carbonates and Evaporites in Michigan Basin," *AAPG Bulletin*, v. 58, no. 1, 1974, p. 40, American Association of Petroleum Geologists. **21.15a, Table 21.1, Table 21.2**: Publication Services. **21.23**: Rubin's Studio of Photography, the Petroleum Museum, Midland, Texas. **21.28**: Precision Graphics. From Briggs and Roeder, *A Guidebook to the Sedimentology of Paleozoic Flysch and Associated Deposits, Ouchita Mountains and Arkoma Basin, Oklahoma, c. 1975* (Fig. 6). Reprinted by permission of the Dallas Geological Society, Inc. **21.30**: Rolin Graphics. From the U.S. Geological Survey.

CHAPTER 22

Opener: Smithsonian Institution. Transparency No. 86–13471A. **22.1, 22.7, 22.20, 22.22, 22.24, 22.27, 22.28, 22.32**: Precision Graphics. **22.3, 22.16, 22.18, Table 22.3, Table 22.4**: Publication Services. **22.4, 22.19, 22.21, 22.23, 22.25, 22.26, 22.29, 22.30, 22.33, 22.35**: Carlyn Iverson. **22.5**: Publication Services. Reprinted with permission from H. Tappan, "Microplankton, Ecological Succession and Evolution, Part H," in *Proceedings of the North American Paleontological Convention*, E. Yochelson (ed.), 1970–71, p. 1064 (Fig. 2). **22.6**: Carnegie Museum of Natural History. **22.8**: Smithsonian Institution. **22.12**: ©American Museum of Natural History, Trans. #K10257. **22.13**: ©American Museum of Natural History, Trans. #K10269. **22.14**: Publication Services. Reprinted with permission from Raup and Sepkoski, "Mass Extinctions in the Marine Fossil Record," *Science*, v. 215, 1982, p. 1502, ©1982 by the AAAS. **22.15**: Precision Graphics. Adapted with permission from drawing by Ralph Buchsbaum in V. Pearse, J. Pearse, M. Buchsbaum, and R. Buchsbaum, *Living Invertebrates*, 1987, Boxwood Press; reprinted with permission. **22.17**: Precision Graphics. Reprinted with permission from Romer, *The Vertebrate Story*, ©The University of Chicago Press.

CHAPTER 23

23.2: Carto-Graphics. Reprinted with permission from Dietz and Holden, "Reconstruction of Pangaea: Breakup and Dispersion of Continents, Permian to Present," *Journal of Geophysical Research*, 1970, v. 75, no. 26, pp. 4939–4956, ©1970 by the American Geophysical Union. **23.3**: Carto-Graphics. Reprinted with permission from K. Burke, "Atlantic Evaporites formed by evaporation of water spilled from Pacific, Tethyan, and Southern oceans," *Geology*, v. 3, 1975, p. 614 (Fig. 1), the Geological Society of America. **23.4, 23.5, 23.6, 23.7, 23.16a**: Carto-Graphics. **23.8, 23.11, 23.12, Table 23.2, Table 23.3**: Publication Services. **23.9, 23.13, 23.14, 23.22, 23.23**: Precision Graphics. **23.10**: Publication Services. Reprinted with permission from Schweickert and Cowan, "Early Mesozoic Tectonic Evolution of the Western Sierra Nevada, California," *GSA Bulletin*, v. 86, 1975, p. 1334 (Fig. 3), ©1975 the Geological Society of America. **Perspective 23.1, Figure 1**: ©Stephen J. Kraseman/Photo Researchers. **23.17**: Publication Services. Reprinted with permission from R. L. Armstrong, "Sevier Orogenic Belt in Nevada and Utah," *GSA Bulletin*, v. 79, 1968, p. 446 (Fig. 5), ©1968 the Geological Society of America. **23.21**: Smithsonian Institution, Photo no. 106677. **23.24, 23.25, 23.26, 23.30**: Carlyn Iverson. **23.27**: Precision Graphics. Reprinted with permission from R. Monastersky, "Chinese Bird Fossil: Mix of Old and New," *Science News*, v. 138, 1990, p. 247. **23.28, 23.29**: Precision Graphics. Reprinted with permission from J. A. Hopson, "The Mammal-Like Reptiles: A Study of Transitional Fossils," *The American Biology Teacher*, v. 49, no. 1, 1987, pp. 18, 22 (National Association of Biology Teachers: Reston, VA).

CHAPTER 24

24.1, 24.25, 24.32, 24.34: Publication Services. **24.2**: Carto-Graphics. Compiled from several sources. **24.3**: Publication Services. Reprinted with permission from Dickinson and Snyder, "Plate Tectonics of the Laramide Orogeny," in *Laramide Folding Associated with Block Faulting in the Western United States, GSA Memoir 151*, 1978,

p. 359, the Geological Society of America. **24.4**: Carto-Graphics. Reprinted with permission from W. R. Dickinson, "Cenozoic Plate Tectonic Setting of the Cordilleran Region in the U.S.," in *Cenozoic Paleogeography of the Western U.S., Pacific Coast*, Symposium 3, 1979, pp. 2, 4. **24.6b**: Ward's Natural Science Establishment, Inc. **24.6c, 24.7, 24.18, 24.26, 24.28**: Precision Graphics. **24.8**: Precision Graphics. Reprinted with permission from J. H. Stewart, *GSA Memoir 152*, 1978, p. 24 (Fig. 1–16), the Geological Society of America. **24.10**: Publication Services. Reprinted with permission from W. R. Dickinson, "Cenozoic Plate Tectonic Setting of the Cordilleran Region in the U.S.," in *Cenozoic Paleogeography of the Western United States, Pacific Coast*, Symposium 3, 1979, p. 2 (Fig. 1). **24.12**: Publication Services. Reprinted with permission from S. W. Lowman, "Sedimentary Facies in the Gulf Coast," *AAPG*, v. 33, no. 12, 1949, p. 1972 (Fig. 23), American Association of Petroleum Geologists. **24.13**: Publication Services. Reprinted with permission from Friedman and Sanders, *Principles of Sedimentology*, 1978, p. 371. Copyright ©1978 John Wiley and Sons, Inc. **24.14**: Carlyn Iverson. From D. Johnson, *Stream Sculpture on the Atlantic Slope*, 1931, Columbia University Press. **24.15**: Precision Graphics. Reprinted with permission from Crow and Sheridan, "U.S. Atlantic Continental Margin: A Typical Atlantic Type or Passive Continental Margin," *The Geology of North America*, v. 1–2, pp. 2, 4: U.S. Geological Survey. **24.17**: Rolin Graphics. **24.19**: Publication Services. From Wright, H. E., Jr., and Frey D. G. (eds.), *Quaternary of the United States*, p. 266 (Fig. 1). Copyright ©1965 by Princeton University Press. Reprinted with permission of Princeton University Press. **Perspective 24.1, Figure 1**: Rolin Graphics. From V. K. Prest,

Geology and Economic Minerals of Canada: Department of Energy, Mines, and Resources Economic Geology Report 1, 5/e (1970): 90–91, Figure 7–6. Reproduced with the permission of the Minister of Supply and Services Canada, 1991. **24.22, 24.31**: Carlyn Iverson. **24.24a**: Precision Graphics. From *The Vertebrate Body*, 3d ed., Shorter version by Alfred S. Romer, Fig. 224, copyright ©1962 by Saunders College Publishing, a division of Holt, Rinehart, and Winston, Inc. and renewed 1990 by Alfred S. Romer; reprinted by permission of the publisher. **24.24b**: Precision Graphics. Reprinted with permission from D. A. Peterson, "Osteology of *Oxydactulas*," *Annals of Carnegie Museum,* v. 2, no. 3, 1904, ©Carnegie Museum of Natural History, Pittsburgh, PA. **Perspective 24.2, Figure 1**: University of Nebraska State Museum. **24.27a**: Publication Services. Reprinted with permission from B. J. MacFadden, "Patterns of Phylogeny and Rates of Evolution in Fossil Horses," *Paleobiology*, v. 11, no. 3, 1985, p. 247 (Fig. 1), ©The Paleontological Society. **24.17b**: Carlyn Iverson. **24.29**: Precision Graphics. Reprinted with permission from L. S. Dillon, *Concepts and Consequences*, 1978, p. 256 (Fig. 12–13). **24.33**: Darwen and Vally Hennings. **24.36**: Painting by Ronald Bowen; photo courtesy of Robert Harding Picture Library.

CHAPTER 25

Opener: Ed Degginger. **25.1a, 25.3, 25.4, 25.5**: Precision Graphics. **25.1b**: NASA. **25.2**: AP/Wide World Photos. **25.6, 25.7**: Precision Graphics. Data from *Oil and Gas Journal*, v. 90, pp. 44–45 (Dec. 28, 1992).